Pharmaceutical Powder Compaction Technology

DRUGS AND THE PHARMACEUTICAL SCIENCES

A Series of Textbooks and Monographs

edited by

James Swarbrick
AAI, Inc.
Wilmington, North Carolina

ADDITIONAL VOLUMES IN PREPARATION

Pharmaceutical Powder Compaction Technology

edited by
Göran Alderborn
Christer Nyström

Uppsala University
Uppsala, Sweden

CRC Press
Taylor & Francis Group
Boca Raton London New York

CRC Press is an imprint of the
Taylor & Francis Group, an **informa** business

CRC Press
Taylor & Francis Group
6000 Broken Sound Parkway NW, Suite 300
Boca Raton, FL 33487-2742

First issued in paperback 2019

© 2008 by Taylor & Francis Group, LLC
CRC Press is an imprint of Taylor & Francis Group, an Informa business

No claim to original U.S. Government works

ISBN-13: 978-0-8247-9376-0 (hbk)
ISBN-13: 978-0-367-40157-3 (pbk)

Library of Congress Cataloging-in-Publication Data

Pharmaceutical powder compaction technology.
    edited by Goran Alderborn and Christer Nystrom.
    p. ; cm. -- (Drugs and the pharmaceutical sciences ; v. 71)
    Includes bibliographical references and index.
    ISBN-13: 978-0-8247-9376-0 (alk. paper); ISBN-10: 0-8247-9376-5 (alk. paper)
    1. Tablets (Medicine). I. Alderborn, Goran.
  II. Nystrom, Christer.
  III. Series.

    RS201.T2P465 1995
    615'.19--dc20
                                                96-24571

Visit the Taylor & Francis Web site at
http://www.taylorandfrancis.com

and the CRC Press Web site at
http://www.crcpress.com

Preface

Oral administration is the dominant method of delivering drugs to the human systemic blood circulation because of its safety and simplicity. Thus, great interest has been focused within pharmaceutical science on the design of oral dosage forms with optimal therapeutic properties. The prevailing oral dosage form today is the tablet due to its elegance. Tablets of various types and biopharmaceutical properties—from conventional, disintegrating tablets, to advanced modified release systems—exist, but their common denominator is the way in which they are formed, i.e., powder compaction. Physical and technological aspects of this process, from a pharmaceutical point of view, are the theme of this book.

The complexity of the compaction process—what at first sight seems to be a simple mechanical operation—was recognized early. Problems still exist in large-scale production of tablets, such as low tablet strength, capping, limited use of direct compression, and sensitivity to batch variability of starting materials. Moreover, the use of basic physical data in formulation work in order to predict tableting behavior of particles such as compressibility (ability to reduce in volume) and compactibility (ability to cohere into compacts) is limited. Thus, tablet formulation must still be based to a large extent on empirical knowledge rather than on scientific theory.

An improved theoretical understanding of the compaction process will enable a more rational approach to the formulation of tablets. However, the investments in research on the physics of the compaction process have, in relative terms, been limited in universities and the pharmaceutical industry. In spite of this, a large number of publications on the theme of the formation of tablets by compaction exist today in the pharmaceutical literature. This literature can be broadly classified into three categories: (1) reports on specific formulations and their compactibility and on formulation solutions to compaction-related problems, (2) studies on mechanisms of and theories for the compression and the compaction of pharmaceutical powders (such studies also include articles dealing with the development and evaluation of methods for theoretical studies), and (3) evaluation, with recognized methods and theories, of the compression and compaction behavior of pharmaceutical tableting excipients.

In the older literature, publications were focused mainly on the practical aspects of the preparation of tablets. However, since the late 1940s, articles focused on the theoretical aspects of the compaction process have been presented in the pharmaceutical scientific literature. As a consequence of the growing interest in directly compactable formulations, new excipients with improved tableting performance have been developed and the compaction characteristics of these have been the object of scientific studies. Despite this growing literature on the physics and technology of powder compaction, the interest in bringing together the accumulated knowledge in the form of comprehensive reference works has hitherto been limited. It is thus a great pleasure for the editors of this volume to present a book on theoretical and practical aspects of the process of forming compacts by powder compression. This is, to our knowledge, the first book devoted entirely to this theme. It has been made possible by the contribution of chapters from researchers throughout Europe and North America. To achieve the high level needed, only recognized scientists, representing academia or the pharmaceutical industry, have been involved, and each contributor has been encouraged to focus on his or her field of expertise. The role of the editors has been to primarily select topics and authors for the contributions and to find a suitable structure for the book. The consequence of this is that different concepts and beliefs in the field of powder compaction are presented and discussed in the book, and we have not attempted to hide this diversity. This diversity reflects the complexity of studying and establishing theories for the handling and processing of "real" materials. Moreover, there are also different traditions with respect to the nomenclature used in the discussion on powder compaction, and this inconsistency among researchers in this respect is also reflected in this book. The editors allowed each author to use terms in accordance with his or her tradition. However, to improve the stringency in the use of the nomenclature for the future, a short list of definitions follows this preface.

During the preparation of this book, some topics within the area of pharmaceutical powder compaction have not been dealt with as separate chapters, as they are not covered extensively in the literature. Examples of such topics are energy aspects of the formation of tablets, physical instability in compacts during storage, and mathematical expressions for the tensile strength of compacts. However, these topics are discussed and references are given in some of the chapters of this book.

Although great progress in the theoretical understanding of the compaction process has been made since the late 1940s, the need for further research is obvious. It is our hope that this volume can contribute to and stimulate such intellectually challenging research.

We are very grateful to Marcel Dekker, Inc., for taking the initiative to prepare a book on pharmaceutical powder compaction technology. We express our sincere appreciation especially to Sandra Beberman and Ted Allen, for pleasant cooperation during the preparation of this book, for their qualified contributions, and for their support and patience with us in our role as editors.

We are also very grateful to all contributors to this volume, for their positive attitude to share their expertise in the field of powder compaction and for the time and effort taken to write articles of high quality. Without their collaboration and contributions, the writing of this book would never have been accomplished.

Finally, we would like to thank Mrs. Eva Nises-Ahlgren for qualified administrative work in connection with the preparation of this book.

Göran Alderborn
Christer Nyström

Nomenclature

Below are the definitions of some terms commonly used in relation to powder compaction. It should be noted that within this book the terms are not used strictly in accordance with these definitions, as some authors have used the terms differently. The list is narrow in that it presents only the terms that seem to cause the most confusion in discussions of powder compaction. Less ambiguous terms, describing mechanical properties of and bonding mechanisms between particles, are defined within the individual chapters and in other books on material science and chemical engineering.

Compactibility The ability of a powder bed to cohere into or to form a compact. Usually described in terms of tablet strength as a function of applied compaction stress.

Compaction The transformation of a powder into a coherent specimen of defined shape by powder compression.

Compressibility The ability of a powder bed to be compressed (be reduced in volume) due to the application of a given stress.

Compression The reduction in volume of a powder bed due to the application of a stress, e.g., loading or vibration.

Consolidation Mostly used synonymously with compaction. The term has also been used to describe compression of powders.

Elastic deformation of particles Time-independent, recoverable deformation of a particle. Deformation occurs parallel to a contraction of the particle.

Hardness The resistance of a specimen against penetration into the surface of the specimen.

Particle deformation The change in shape of a particle during compression. Can be quantified with some shape factor for the particle as a function of applied stress during compression.

Plastic deformation of particles Time-independent, permanent deformation of a particle. Degree of deformation is thus controlled by the applied stress and independent of the time of loading. Deformation occurs without a change in particle volume.

Particle fragmentation The fracturing of a particle into a number of smaller, discrete fragments during compression. Can be quantified as the change in particle size or particle surface area with applied stress during compression.

Time-dependent deformation of particles Degree of deformation of a particle is controlled by the applied stress and the time of loading.

Viscoelastic deformation of particles Time-dependent recoverable deformation of a particle.

Viscous deformation of particles Time-dependent permanent deformation of a particle.

Contents

Contents

Contributors

Göran Alderborn, Ph.D. Division of Pharmaceutics, Uppsala University, Uppsala, Sweden

N. Anthony Armstrong, Ph.D. Welsh School of Pharmacy, University of Wales, Cardiff, United Kingdom

Gerad K. Bolhuis, Ph.D. Department of Pharmaceutical Technology and Biopharmacy, University of Groningen, Groningen, The Netherlands

Jean-Daniel Bonny, Ph.D. School of Pharmacy, University of Basel, Basel, Switzerland

Zak T. Chowhan, Ph.D. Department of Formulation Development, Syntex Research Institute of Pharmaceutical Sciences, Palo Alto, California

Peter N. Davies, Ph.D.* School of Pharmacy, University of London, London, United Kingdom

---

*Current affiliation: Roche Products, Welwyn Garden City, Hertfordshire, United Kingdom

Wendy C. Duncan-Hewitt, Ph.D. Faculty of Pharmacy, University of Toronto, Toronto, Ontario, Canada

John T. Fell, Ph.D. Department of Pharmacy, University of Manchester, Manchester, United Kingdom

Claus Führer, Ph.D. Institut für Pharmazeutische Technologie der Technischen Universität Braunschweig, Braunschweig, Germany

Everett N. Hiestand, Ph.D. * Upjohn Company, Kalamazoo, Michigan

Arne W. Hölzer, Ph.D. Astra Hässle AB, Mölndal, Sweden

Jukka Ilkka, M.Sc. Department of Pharmaceutical Technology, University of Kuopio, Kuopio, Finland

Per-Gunnar Karehill, Ph.D. Astra Hässle AB, Mölndal, Sweden

Ruth Leu, Ph.D. School of Pharmacy, University of Basel, Basel, Switzerland

Hans Leuenberger, Ph.D. School of Pharmacy, University of Basel, Basel, Switzerland

Fritz Müller, Ph.D. Pharmazeutisches Institut der Universität Bonn, Bonn, Germany

J. Michael Newton, Ph.D. School of Pharmacy, University of London, London, United Kingdom

Christer Nyström, Ph.D. Division of Pharmaceutics, Uppsala University, Uppsala, Sweden

Petteri Paronen, Ph.D. Department of Pharmaceutical Technology, University of Kuopio, Kuopio, Finland

Fridrun Podczeck, Ph.D. School of Pharmacy, University of London, London, United Kingdom

---

*Retired. Residing in Galesburg, Michigan

Gert Ragnarsson, Ph.D.*   Pharmacia Biopharmaceuticals, Stockholm, Sweden

Ron J. Roberts, Ph.D.   Zeneca Pharmaceuticals, Macclesfield, Cheshire, United Kingdom

Ray C. Rowe, Ph.D.   Zeneca Pharmaceuticals, Macclesfield, Cheshire, United Kingdom

Martin Wikberg, Ph.D.   Kabi Pharmacia Therapeutics Uppsala, Uppsala, Sweden

----

*Current affiliation: Astra Draco AB, Lund, Sweden

1

Interparticulate Attraction Mechanisms

Claus Führer

Institut für Pharmazeutische Technologie der Technischen Universität Braunschweig, Braunschweig, Germany

I. INTRODUCTION

A deciding factor in the technology of all dispersed systems is the influence of interparticulate attractions (see also [15,16]). Examples are the influence of the interparticulate attraction on the stability of emulsions and suspensions or on the flowability and compactibility of powders. The interparticular forces producing these attractions are a function of the nature of the particles and the interparticular distances. Normally the deciding properties are located on the particle surface or at particle interfaces.

In the following these considerations are applied to various phenomena in powder technology. A powder can be described as a special case of a dispersed system solid in gas, where the solid particles remain in contact. Because of this the powder may also be regarded to understand as a more or less bicontinuous system. The permanent interparticular contact is caused by external forces such as gravity and by interparticular interactions. Interparticular attractions between faces of the same material are called cohesion forces, independent of the nature of the interactions, whereas the attractions between faces of different materials are called adhesion forces [1,2]. Both depend on the same principles, and the distinction between them is thus sometimes very arbitrary. The adhesivity (resp. cohesivity) of a powder bed can be measured indirectly with a shearing cell (see also [3]).

Shearing forces may overcome the forces of interaction. Since the forces of interaction normally have their origin on the particle surfaces, whereas the shearing forces mostly depend on the mass of the particles, the phenomena become more significant with decrease of the particle diameter. Depending on the intensity of the mechanical treatment and on the special properties of the powder bed, the resulting irreversible deformations can be classified into three groups:

1. The contact areas change during the deformation from particle to particle, preserving the coherency of the solid phase in its totality. In this case one observes the typical flow properties of a powder bed.
2. A second possibility is the fracture of the powder bed into two or more parts. This corresponds to a fracture of a solid body. In this case the powder does not flow, it breaks.
3. If the solid phase is completely dispersed in the surrounding gas with no or only few interparticular contacts, the system has lost the typical properties of a powder and represents an aerosol.

Compression of a powder bed with partial elimination of the gas phase enhances the enlargement of the interparticular contact areas and thus the strength of the powder bed, which may finally be transformed in this way to a solid body (e.g., a tablet).

II. DIFFERENT TYPES OF INTERPARTICULATE ATTRACTIONS

The interparticular forces are roughly classified according to their origin as electrostatic or molecular interactions. For further classifications see also [4]. This classification is also arbitrary, and the names of the two types are not quite correct. Corresponding to this classification one understands electrostatic interactions as interactions of relatively highly charged particles or highly charged faces, being active over a relatively long distance. Molecular interactions normally also have the nature of electrostatic phenomena, but with small charges or polarizations on molecules or especially only on some atomes of the molecules, which are responsible for the valencies of secondary bonds. Because of the limited charge they have only a very short distance of activity (see also [5]).

A. Electrostatic Interactions

Two small single particles 1 and 2 with a distance r between them much larger than their diameters and with the electrostatic charges q_1 and q_2 interact with force F_{12}:

$$\frac{F_{12} = q_1 q_2}{4 \pi \epsilon_0 r^2}$$

where ϵ_0 is the permitivity constant. The force is negative and attractive if the two charges have opposite signs. If both charges are positive or if both are negative, the force is repulsive. In the case of an assembly of particles with different charges, the equation holds for every pair of particles (Fig. 1).

Contrary to chemical linkages, the electrostatic interactions cannot be saturated. The total influence of the assembly on particle 1 can be expressed by the vector equation (see also Fig. 1)

$$F_{1i} = F_{12} + F_{1a} + F_{14t} + \ \cdots$$

Such equations can be used in an aerosol where the distances between particles are relatively large. In a powder bed the particles touch each other, and the electrostatic force influencing one particle is thus affected by interactions with the close surrounding particles. Because of the small distances or even direct contacts, the resulting forces can be relatively high. If the attractions are of the order of the weight of the particles or even higher, the powder cannot flow.

1. Triboelectrostatic Charge

A turbulent movement of a powder, especially in mixers, causes very frequent collisions of the particles mutually and with the surfaces of the vessel and the tools. With each collision normally an interfacial charge transfer of electrons occurs (Fig. 2a). Because of the intensive movement of the material causing the charge transfer, it seems sensible to nominate this event as a

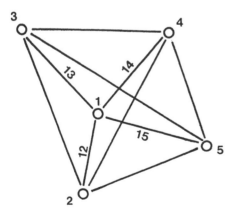

Fig. 1 Electrostatic interactions in an assembly of particles.

triboelectrostatic charge transfer. The direction of this charge transfer is given by the difference of the electron affinity of the two touching materials in the contact area at the moment of collision. The respective electron affinity depends not only on the chemical state of the materials but also on the crystallographic state, including the type of the lattice and distribution of lattice defects in the surface. But it depends also on the special state of the material under the influence of the high-energy transfer concentrated in the touching regions at the moment of collision.

The charge remains on the points of the transition if the respective contact area breaks under the influence of the further movement of the powder bed (Fig. 2b) and the particles separate again. Subsequent collisions of the charged particles have extremely low probabilities to find just the same points on the surface for the new interparticular contacts. Consequently, it is assumed that every or nearly every collision causes a new interparticular charge transfer (Fig. 2c). With each collision a new charge will be transfered. If the charges have no mobility within the surface, the process continues. Consequently, the density of the charged spots on the surface rises. Also the probability rises that an already charged spot collides with the surface of a particle. Thus the charge transfer will be hindered more and more.

Particles with highly polar surfaces have, under normal conditions, a water sorption layer which is able to distribute the transfered charge over the surface. The thicker this layer, the higher is its conductivity and the more easily is this charge distribution effected on the surface. The distribution becomes relatively homogeneous on the surface of an amorphous

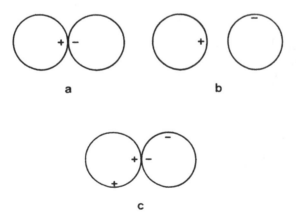

Fig. 2 Formation of the triboelectrostatic charge (schematically). a, First collision; b, separation; c, second collision.

spherical particle (Fig. 3a). But in the case of crystalline nonspherical particles, especially of milled crystalls with pointed edges, extremely high values of charge will be concentrated in the tops of the edges (Fig. 3b).

A relatively homogeneous charge distribution over the surface of the whole particle hinders more charge transfers. The process may reach an equilibrium. But with nonhomogeneous charge distributions, as schematicaly demonstrated in Fig. 3b, the total charge of the particle, especially in the region of edges, may rise to extreme values.

Mostly the surrounding gas is not indifferent [6]. In a humid climate the polar substances normally have a multimolecular water sorption layer and have a relatively frequent exchange of the water molecules of the particle surface with those of the atmosphere. Consequently, the charge is not merely distributed over the surface, but it can also leave the particle. Therefore, the total charge of the particles will be relatively low compared with materials without polar groups in the surface and no water sorption layer. In this case the total charge of each particle may rise to very high values depending on the intensity of the movement.

This shows that mixing two different powdered materials results in a disproportion of the charge within the powder bed. The two materials

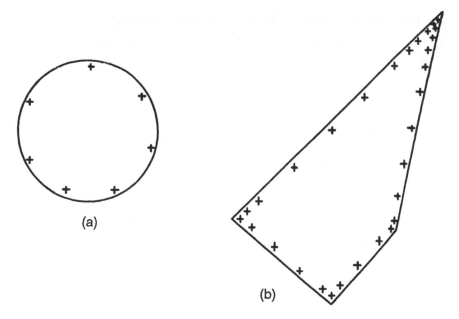

Fig. 3   Charge distribution under the influence of conductivity in the surface. a) Amorph spherical particle, b) milled crystal.

attract each other because of their opposite polarities. Depending on the composition of the mixture and the intensity of the movement, one can observe an increase in the bulk density and a decrease of the flowability.

In case of uniform particles of the same material a powder bed charges in a uniform sense with intensive movements, whereas the machine takes over the opposite charge. A similar effect can be observed if large amounts of air pass the powder bed, as for instance in a fluid bed. The powder bed and the air have opposite polarities. The machine usually has a certain conductivity, and the charged air leaves the system. Because of this the charge on the powder, compared with the ground potential, may rise to some 10,000 volts. In extreme dangerous cases, especially with nonpolar fine powders of large specific surface area, violent dust explosions may occur.

Powders charged with a uniform polarity are characterized by low bulk densities because of their elevated interparticular repulsion forces.

The charge distribution in a powder bed of a mixture allows ordered mixtures to be obtained and in this way to overcome the statistical optimal homogeneity. A uniformly charged powder allows, because of the elevated interparticular repulsions, sticking to be avoided in flow procedures. But charged powders do have the great disadvantage of bad charge reproducibility and bad stability.

2.   *Permanent Electrostatic Charge on the Particle Surface*

The permanent electrostatic charge is not induced by the mechanical treatment of the powder or the interaction of the particles but is a consequence of the surface structure of the powder. This charge is immobile and can only be suppressed by oppositely charged materials.

Unlike amorphous materials, the surface of a crystalline material is not uniform. Each face cuts the crystallographic lattice of the particle and represents a special case of a crystallographic plan of the lattice. The face, consequently, has its own structure, depending to its orientation against the lattice. It contains only some special atoms or atomic groups of the material in a state of order. Hence, each face has its own energy level, its own surface or interface tension, and, with the presence of polar functional groups, its own polarity.

a.   Ionized Surfaces

Due to the special chemical and crystallographic state of the surface of solid particles, surfaces can be charged, even if each particle in total is neutral. This is possible especially if the material itself is a salt or is coated by a ionized tensid.

With crystalline salts the state of order gives theoretically the possibility of the formation of crystallographic faces with high positive or high negative charge. Depending on the conditions of crystallization, it is to be expected that one can find neutral faces on the surface of the same particle along with positive or negative charged faces. With sodium chloride the octaeder faces (111) are positively or negatively charged, depending on the last layer of the ions forming the crystal surface (sodium or chloride). Such faces have a very high energy level and may exist only under certain conditions.

The surfaces of precipitates of salts are occupied initially by those ions which have been in stoichiometric surplus. Thus, a precipitation of silver chloride, for instance, yields particles with a surface occupied by chloride ions, if after the precipitation the mother solvent still has a certain cloride concentration. In contrast to a surplus of silver ions the precipitate has an outer layer of silver ions. In the first case the surface of the particle is negatively charged, whereas in the second case the surface is positively charged.

Amorphous materials also may have uniform charged ions in their surface which are responsible for electrostatic attractions or repulsions. A significant example is the attraction of the particles in a mixture of ion-exchange resins of opposite type. In a regenerated mixed-bed ion-exchange column, one observes a relatively highly ordered mixture, where the anion-active and the cation-active particles are in alternating places. This demonstrates the possibility of higher homogeneities in charged mixtures than the statistical optimum.

The coat of ionized tensid molecules around a particle of relatively low polarity has an orientation where the polar groups of the tensid molecules arc located in the surface and the opposite nonpolar part covers the outer layer of the particle. If the particles are completely surrounded by such surface-active molecules, neighboring particles will come under strong interparticular repulsion forces. Thus the tensid has a disagglomerating effect on the powder.

b. Presence of Polar Functional Groups in the Surface

A crystalline material with highly ordered functional polar groups can also be the reason for a permanent charge on the surface. But the field intensity surrounding these groups is too low to cause electrostatic attraction or repulsion forces to influence decisively the mechanical properties of a powder bed.

B. Molecular Interactions

Highly charged particles and faces give interactions over long distances. The molecular interactions are also of an electrostatic nature, but they have only a very limited range. They are secondary bonds or van der Waals attractions that can be classified as ion-dipole, dipole-dipole, or van der Waals–London interactions.

Dry particles with no water sorption layer allow only very few molecular interparticular interactions. To form a molecular interaction, a short interatomic distance and a certain steric arrangement between the respective functional groups are required. Because of the roughness of the particle surfaces, the probability of such an interparticular arrangement is extremely low. When two bodies are placed in contact, there are, if any, very few functional groups which are by accident in that suitable opposite position to allow interparticular linkages (Fig. 4a). These few interactions are much too weak to resist the usual mechanical stress that makes a powder flow. Only small forces are needed to overcome the few linkages of interaction. Thus, even polar substances with a relatively high concentration of free secondary dipole-dipole valencies in the surface often have very low coherency in the dry state, which means good flow properties. The enlargement of the microscopic contact areas is proportional to the normal force, because the contact regions deform plastically under the great stress that develops in these regions (Fig. 4b).

In a humid atmosphere the surfaces of polar materials adsorb water. The thickness of the sorption layer depends on the polarity of the surface

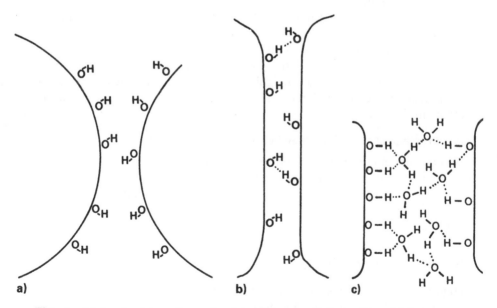

Fig. 4 Molecular interactions of polar substances: a) dry surface, b) dry surface after plastic deformation, c) with water sorption layer.

and the humidity of the atmosphere. The water molecules are linked to the surface and among themselves by hydrogen bonds. But the surface layer does not have the properties of a liquid. Information on the structure of the sorption layers is given by the arrangement of the polar groups in the surface of the particles. Thus even with a multimolecular layer the water molecules are not able to form the normal cluster arrangement. With the sorption layer the particle gets a new highly polar outer surface. The polar groups in this face have, contrary to the groups of the dry particles, a high flexibility, rising with the thickness of the sorption layer (Fig. 4c).

If two such particles come together, the water sorption layers might possibly interact over a high concentration of hydrogen bonds (Fig. 4c). The interparticular layer may be deformed, and finally the two particles have a joint water sorption layer. The result is a strong interparticular attraction. In consequence of this consideration one can understand the strong dependence of the coherency of polar materials on the humidity of the surrounding atmosphere.

C. Capillary Forces

A wetted powder contains, in addition to the sorption water, water as a separate phase, with the typical properties of a liquid. If the angle of contact between the water and the surface is small, a limited amount of water takes a preferred place in the interparticular contact regions with the formation of a meniscus corresponding to Fig. 5. A simple explanation of

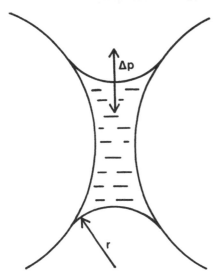

Fig. 5   Capillary forces between two particles.

the capillary forces may be given by the following: The concave deformation of the liquid surface requires a pressure difference between the atmosphere and the liquid, corresponding to

$$\Delta p = \frac{2\sigma}{r}$$

where Δp is the pressure difference, σ is the surface tension of the liquid phase, and r is the radius of the concave deformation. The pressure in the liquid phase must be less than the atmospheric level. Consequently, one can observe an interparticular attraction. The high surface tension of pure water gives strong attractions. On the other hand, on adding a surfactant the attractions become weak.

The attraction is influenced by the enlargement of the interparticular contact area and the radius of the meniscus as well as the surface tension. The radius of the meniscus depends not only on the angle of contact between the water and the solid surface but also, because of the more or less round form of the particle, on the amount of water between the particles. Very small amounts of water produce a surface with a concavity of small radius but of small interparticular contact area. Large amounts have large contact areas but also large radii of the meniscus. Since the functions of the enlargement of the contact area and the radius of meniscus are different, the attraction forces reach a maximum with a certain amount of the wetting liquid. This corresponds to the fact that the wet granulation requires a certain amount of water to give optimal mechanical properties in the mass, corresponding obviously to the maximum dilatant rheological behavior.

The importance of the surface tension in this effect demonstrates that all ingredients lower the surface tension of the water.

III. CONSEQUENCES FOR PRACTICAL PHARMACEUTICAL TECHNOLOGY

In powder technology the interparticular interactions are important for the flow properties and bulk density of a powder, the mixing behavior, granulating, and tabletting or capsule filling. Only some of these will be discussed.

A. Flow Properties, Bulk Density

Interparticulate attraction in any case makes the flow properties worse. On the other hand, electrostatic repulsions may facilitate the flow of a powder bed. For the rheology of powders see [7–10].

Despite this, triboelectrostatic charges should be avoided, especially for powders provided for direct compression to tablets or for capsule filling. Both the tablet and capsule-filling machine regulate the dose volumetrically. In order to have tablets or capsules with a sufficiently constant mass, the bulk density of the powder must be as constant as possible during the whole fabrication of the batch. The bulk density of powders is very sensitive to electrostatic charge. The triboelectrostatic charge is not reproducible, does not distribute homogeneously within a batch of a powder, and changes rapidly during storage and the operation of tablet compression. Because of this the triboelectrostatic charge influences in a negative way the uniformity of the bulk density and with this the uniformity of dosage.

It is in many cases not possible to exclude absolutely any charge transfer by special measures during the handling of a powder (e.g., in milling, mixing, or drying operations), but there are various ways to diminish the triboelectrostatic charge transfer or to discharge the material.

1. The powder should have a sufficiently polar surface and a water sorption layer; i.e., it should have been handled or stored under the influence of air of sufficient humidity (e.g., 70% rh). If the material contains polar groups due to its chemical state, these groups should be present in the surface. With crystalline materials the crystallization conditions often allow variation of the polarity of the surface. A certain aid is given by the rough rule that the more polar the medium of crystallization, the more polar is the crystal surface. The temperature of crystallization also sometimes has an influence: the lower the temperature, the higher the polar character of the medium and the higher the polarity of the crystallizate.

2. The polarity of the surface can be enhanced by small concentrations of excipients (e.g., Aerosil), but also in the extreme case of nonpolar surfaces, with amphiphilic substances of the O/W type. They distribute on the surface coating the material with a more or less statistical orientation. In this way a certain concentration of polar groups in the surface may obtained. The typical O/W orientation of, e.g., emulsions are not to expected because for powders the air as gas does not attract the polar groups of the amphiphilic molecules.

3. Triboelectrostatic charged materials can be discharged by treatment with ionized air, such as air that has passed an electric field of high voltage. A second possibility is treatment with humid air, but many substances do not allow this. Both methods do not change permanent charges on the surface, and their efficiency is very limited.

4. All charges discussed, which are located in the particle surface-independent of their nature—can be compensated by the addition of amphiphilic excipients. The optimal excipient must be found empirically by experiment. The aim is to find a substance whose polar groups have an intensive affinity for the respective charged groups. In this case they take orientation where the polar groups are located on the original particle

surface and the nonpolar groups form the outer surface. Very often one choses substances with W/O character. These are more universal, whereas the more selective O/W surfactants may have a much better efficiency.

5. The possibility also exists of coating the particles with Aerosil in order to isolate more or less charged regions of the surface, against direct contact with the respective regions of other particles.

These considerations show that the normally used excipients in powder technology (e.g., lubricants) diminish the negative influence of the charges. But they demonstrate also that after the treatment of a powder with these excipients, intensive powder movements must be avoided. For the influence of interparticular attraction on the fluidization of a powder bed see [11–13].

B. Mixing Behavior

The mixing operation is, in the case of mutually indifferent substances and where the components have the same physical properties, a pure entropical operation. The optimal homogeneity of such mixtures is given by a pure statistical distribution of the components. In case of attractions or repulsions within the material this distribution can be disturbed.

A two-component system where one component has a positive charge and the other a negative occurs because of the interparticular attraction of the two different components, which causes better homogeneities than the statistical one. This effect is supported because of the repulsion between particles of the same component. In the case of a 1:1 mixture the material tends to an alternating state of order. It is sometimes useful to compare these states of order with the state of order of a crystalline substance. To do this the term "particle crystalline state" is proposed. The state of order represented under this definition is a more or less polycrystalline one. In the case where one component has a very high concentration and the other a low, the low-concentration component is separated and surrounded by the high-concentration one. In a statistical distribution the probability always exist that two or even more particles of the low-concentration substance are together. The charged material represents something like a solution. Between the two extremes—the particle crystalline state and the solution—homogeneous transitions exist.

It is evident that in any case the quality of a two-component mixture improves with the charge disproportion. The discussed disadvantages of triboelectrostatic charges as badly reproducible and unstable are not very important for the improvement of homogeneity.

Multicomponent systems are much more complicated. Already, in a pure statistical state, each component in the mixture has its own quality of distribution. Different homogeneities exist for one and the same mixture, depending on the respective questioned components. The influence of charges on the quality of such a mixture depends on the type of charge distribution. Consequently, for some components the effect is less than for others. One can also imagine that a multicomponent mixture of two or more components becomes worse under the influence of a triboelectrostatic charge.

C. Granulation

After having produced a mixture of a suitable quality, the aim of granulation is to agglomerate the particles to produce new particles of enhanced size. Each particle of the granulate should represent the whole composition of the mixture. For this purpose the interparticular attraction needed for the agglomeration must be uniform throughout the whole mass, independent of the character of the components.

The type of interparticular attraction used for this purpose must be relatively nonspecific with regard to the properties of the surfaces of the different materials. This is possible with interactions within the water sorption layers or with capillary effects. The water sorption layer is primarily used for dry granulations as the compaction, whereas the wet granulation uses the capillary forces. It is evident that both methods requires polar surfaces of all components. The distribution of the different components in terms of the particle sizes in a granulate is often a mirror of the polarity of the materials. Excipients used for granulating and tabletting are normally highly polar in their surfaces, whereas the active drug substances very often are much less polar. Consequently they have a much smaller tendency to agglomerate. Because of this one often finds preferred the active drug substances in the smallest particles of the granulate.

A good wettability (i.e., a small angle of contact) is a precondition for agglomeration with capillary forces. To enhance the wettability of the active drug substances, one sometimes adds surface-active substances. Since the agglomeration by wet granulation also depends on the surface tension of the granulation liquid, it may happen that the particle size distribution of the granulate shifts toward smaller particle sizes because of a poorer agglomeration tendency in the whole mixture. But often it is possible to find a suitable compromise between the wettability differences of all components and the tendency of the whole mass to agglomerate.

With dry granulation, one presses the powder (e.g., in a compactor), thus increasing the interparticular contact areas. It is in any case to be assumed that to some extent plastic deformations will occur. The interparticular linkages are qualitatively the same as those one observes in tablet compression. Since for most tablets adhesion and cohesion forces are mostly required to ensure short disintegration times, one expects with dry granulation methods that these forces are responsible for the strength of the granulate.

Consequently one has to require that the surface of all components be polar with a multimolecular water sorption layer. Nonpolar-active drug substances can be included in the compacted mass but do not actively participate in the solidification. Because of the water sorption layer required, the mass should be in equilibrium with a humid atmosphere of about 60–70% rh.

IV. SOLID BRIDGES

With mechanical treatment—tabletting or dry granulation—and with wet granulation, the interparticular interfaces of the contact areas within a powder bed may disappear by sintering or by recrystallization [14]. Thus, the particles grow together and lose their individuality. The material of these areas passes by diffusion over the contact interface, resulting in a more or less weak transition of the composition and structure between the primary particles. The strength of the solid bridges corresponds to the internal strength of the particles. It is not possible by mechanical or other treatments to divide these aggregates into their original particles. To have a separation, a fracture of the solid bridge must occur. Thus, the resulting particles are not identical to the original. This is the significant difference of the solid bridges to the above-mentioned interparticular interactions. Agglomerates formed by the latter may be separated into their primary particles.

Solid bridges are always to be expected if the material has, over a limited time period, a certain mobility in the contact areas. This mobility may be caused by a fused state under the influence of heat or a dissolved state under the influence of a multimolecular solvent sorption layer, preferably water.

With mechanical treatment, elevated energy levels in the interparticulate contact areas occur, which may overcome the molecular interactions. This occurs at first in the preferred crystallographic gliding or fracture planes but with higher levels also in the stronger planes. Because of the broad energy distribution in the stressed substance, regions with energy levels sufficient to destroy the crystallographic state exist, i.e., to transfer the material into an amorphous liquid state. Extremes of those energy levels are the hot spots, i.e., very small regions where the energy of the material corresponds to extreme temperatures being able even to decompose the molecules.

The consequence of the broad energy distribution and the steep gradients of energy within the material is that the energy level may change very fast. The half-life of the hot spots is very short. Passing the maximum energy level, the material solidifies and may recrystallize during the following decrease of energy. Naturally, such solid bridges have a high concentration of lattice defects or remain even in the amorphous state.

A special case of solid bridges is the interparticular interactions with strong binders. One assumes that the material of the particles and the binder diffuse across the interfaces. Thus, the interface becomes a more or less broad and diffuse region. This consideration demonstrate a relationship of this type of interaction to the normal interparticular interaction without binder.

REFERENCES

1. A. Martin, J. Swarbrick, and A. Cammarata, *Physical Pharmacy*, Lea & Febiger, Philadelphia, PA, 1983.
2. O. Molerus, *Particle Technol.* (Conference), Amsterdam NL, June 3–5, 1980, p. 932.
3. W. Schütz and U. H. Schubert, *Chem. Eng. Tech. 48*: 567 (1976).
4. H. Rumpf, *Chem. Eng. Tech. 30*: 144 (1958).
5. E. N. Hiestand, *J. Pharm. Sci. 55*: 1325 (1966).
6. K. Rietema, J. Boonstra, G. Schenk, and A. H. M. Verkooyen, *Particle Technol.* (Conference), Amsterdam NL, 1980, p. 981.
7. T. Y. Chem, W. P. Walawender, and L. T. Fan, *Powder Technol. 22*: 89 (1979).
8. J. S. M. Botterill and B. B. Abdal-Halim, *Powder Technol. 23*: 67 (1979).
9. S. Stemerding and G. W. J. Wes, *Chem. Weekblad, 75*: 8 (1979).
10. O. Molerus, *Powder Technol. 20*: 161 (1978).
11. S. M. P. Musters and K. Rietema, *Powder Technol. 18*: 239 (1977).
12. S. M. P. Musters and K. Rietema, *Powder Technol. 18*: 249 (1977).
13. K. Schügerl, *Fluidisation*, (J.F. Davidson and D. Harrison, eds.), Academic Press, London, 1971, Chap. 6.
14. C. Führer, *Labo-Pharma-Probl. Tech.* (1977), 25(269), 759–62.
15. 2nd World Congress Particle Technology, Kyoto, Japan 1990 Society of Powder Technology, Japan; VDI, Germany; Fine Particle Soc., USA.
16. K. Iinoya, J. K. Beddow, and G. Jimbo, *Powder Technology*, Hemisphere, Washington, DC, 1981.

2

The Importance of Intermolecular Bonding Forces and the Concept of Bonding Surface Area

Christer Nyström

Uppsala University, Uppsala, Sweden

Per-Gunnar Karehill

Astra Hässle AB, Mölndal, Sweden

I. INTRODUCTION

A. Stages in the Compaction Process

When a force is applied on a powder bed consisting of more or less nonporous particles, a number of mechanisms become involved in the transformation of the powder into a porous, coherent compact with a well-defined shape. A compaction process is normally described by a number of sequential phases [1–3]. Initially, the particles in the die are rearranged resulting in a closer packing structure. At a certain load, the packing characteristics of the particles or a high interparticulate friction between particles will prevent any further interparticulate movement. The subsequent reduction of compact volume is therefore accompanied by elastic and plastic deformation of the initial particles [4–7]. Elastic deformation is a reversible, while the plastic is an irreversible deformation of the whole or a part of a particle. For many materials these particles are then fragmented. Fragmentation can be defined as a dividing up of a particle into a number of smaller, discrete parts [8,9]. The particle fragment will then normally find new positions, which will further decrease the compact volume. When the applied pressure is further increased, the smaller particles formed could again undergo deformation [3]. Thus, a single particle may pass through one or several of these processes several times during a compression. As a consequence of the compression of the powder, particle surfaces are brought into close proximity to each other and interparticulate attraction or bonds will be formed.

The volume reduction processes consume energy (endothermal processes) and will normally increase the amount of particle surface area capable of forming interparticulate attraction forces. Bond formation is however, an exothermal process, thereby releasing energy [10]. During ejection, when the load is reduced, many materials produce laminated (capped) compacts or results in compacts of pronounced low strength. These observations indicate the importance of the elastic component of tableting materials.

To summarize, the following processes are involved in the compaction of a powder:

1. Particle rearrangement
2. Elastic deformation of particles
3. Plastic deformation of particles
4. Fragmentation of particles
5. Formation of interparticulate bonds

Examples of materials consolidating by plastic deformation are sodium chloride, starch, and microcrystalline cellulose [8,11,12]. Fragmenting materials are, for example, crystalline lactose, sucrose, and Emcompress [8,13–16]. However, all materials possess both an elastic and a plastic component. The volume reduction mechanism which will dominate for a specific material is dependent on factors such as temperature and compaction rate. Lower temperatures and faster loading [17,18] during compression will generally facilitate consolidation by fragmentation. Pharmaceutical materials normally consolidate by more than one of these mechanisms [3], which emphasizes the need for adequate characterization techniques.

B. Physical Description of a Tablet

The axial compaction of pharmaceutical powders results in anisotrope and inhomogeneous compacts or tablets; i.e. a tablet shows varying values of some characteristics, porosity, density (e.g. [19]), bonding, mechanical strength in different directions and parts (Fig. 1).

For normal compaction pressures, not exceeding 300–500 MPa, the final compact porosity is 1% to 25%, depending on the powder compressibility. Two extreme models could be used to describe the distribution of

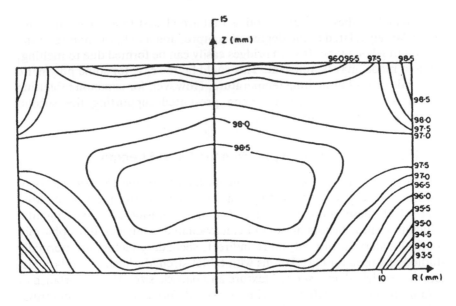

Fig. 1 Density variation in the cross section of a die compact. (From Ref. 19).

A Swiss Cheese   A particulate dispersion in air

Fig. 2 Models for describing the physical structure of a pharmaceutical tablet.

this gas phase (Fig. 2). First, the air could be regarded as a disperse phase of individual units incorporated in a solid continuous phase (like a Swiss cheese). Then the pores are to be considered as intra particulate pores in a large particle (the tablet). Second, the air could be regarded as a continuous phase, in which solid particulate units are dispersed. In this case, the individual solid particles are separated by some distance and the tablet contains continous pores like a loose powder plug. Then the tablet can be penetrated by a flowing medium and characterized on, e.g., permeability properties [20] and permeametry surface area.

Which of these extreme models that are closest to a correct description is largely related to the degree of compression and to the nature of the dominating bond type. If solid bridges easily can be formed due to melting, the first model may be relevant. This could be the case for some polymeric materials with a low melting temperature. However, for common tabletting materials strong evidence has been presented supporting the second model, as will be discussed below.

C. Primary and Secondary Factors for Tablet Strength

Two factors could be regarded as primary factors for the compactability of powders [21–27]: the dominating bond mechanism and the surface area over which these bonds are active. Owing to considerable experimental difficulties, these factors have not been evaluated in any detail for pharmaceutical materials. Instead, more indirect, secondary factors are normally studied and used for correlations with tablet strength. Such secondary factors are particle shape, surface texture and particle size. The importance of volume reduction mechanisms (i.e., elastic deformation, plastic deformation and particle fragmentation) have also been studied in detail.

D. Bonding Surface Area

The term *bonding surface area* is often defined as the effective surface area taking part in the interparticulate attraction. In the case of solid bridges, the term corresponds to the true interparticulate contact area, while for intermolecular forces the term is more difficult to define. It can seldom be estimated from direct measurements of the surface area of the starting material. This is especially obvious for strongly fragmenting materials [28]. Furthermore, in practice, many powders possess, in addition to their external visible surface area, an internal surface area. This internal surface area is small for dense crystalline solids such as sodium chloride, but in porous bodies such as microcrystalline cellulose and Emcompress the internal surface area may be considerably greater than the external surface area [8].

Thus the bonding surface area is a function of several secondary factors [29]. Apart from the complex origin of a bonding surface area, this property is also difficult to define exactly. Consequently, experimental determinations are rare in the literature. Instead of a direct measure of the bonding surface area, the secondary factors listed in Table 1 have been measured and used for the correlation to tablet strength [29].

Of special importance for the final bonding surface is probably the particle elasticity of the materials. This property is normally not measured, but the axial elastic recovery of the tablet is determined. Extensive

Table 1 Factors Influencing the Surface Area of Tablet Particles and the Bonding Surface Area in Tablets

Tablet particle surface area		Bonding surface area	
Before compaction	After compaction	During compaction	After Compaction
Particle size	Particle size	Particle size	Particle size
Particle shape	Particle shape	Particle shape	Particle shape
	Fragmentation	Fragmentation	Fragmentation
		Plastic deformation	Plastic deformation
		Elastic deformation	Elastic deformation
			Elastic recovery
			Friction properties
			Bond strength

particle elasticity could cause a drastic decrease in tablet strength, due to the breakage of interparticulate bonds, thereby reducing the bonding surface area [3].

E. Bonding Mechanisms

The general bonding mechanisms co- or adhering particles have been classified by Rumpf to be of mainly five types [30]:

1. Solid bridges (sintering, melting, crystallization, chemical reactions, and hardened binders)
2. Bonding due to movable liquids (capillary and surface tension forces)
3. Non-freely-movable binder bridges (viscous binders and adsorption layers)
4. Attractions between solid particles (molecular and electrostatic forces)
5. Shape-related bonding (mechanical interlocking)

This general classification has been widely accepted in the literature. In the case of compaction of dry, crystalline powders, it has been suggested that the mechanisms of importance could be restricted to classes 1 and 4 [15] and perhaps class 5 [31]. However, it cannot be excluded that the presence of liquids in a compact might be of significance for the tablet strength [27,32]. It can be discussed, though, if the change in tablet strength is due to an effect of liquid on the compressibility of the powder or on the nature of the particle-particle interactions.

However, the dominating bond types adhering particles together in compression of dry powders could for simplicity be limited to three types [31] (Table 2):

1. Solid bridges (due to, e.g., melting)
2. Distance attraction forces (intermolecular forces)
3. Mechanical interlocking (between irregularly shaped particles)

The first type corresponds to strong bonds, where a true contact is established between adjacent particles. The second group could roughly be described as weaker bonds acting over distances.

The term *intermolecular forces* is used in this chapter as a collective term for all bonding forces that act between surfaces separated by some distance. Thus, the term includes van der Waals forces, electrostatic forces, and hydrogen bonding [33]. The dominant interaction force between solid surfaces in the van der Waals force of attraction [34–36]. This force operates in vacuum, gas, and liquid environments up to a distance of approximately 100–1000 Å. Hydrogen bonding is predominantly an electrostatic interaction and may occur either intramolecularly or intermolecularly [37]. These bonds are of special importance for many direct compressible binders such as Avicel, Sta-Rx 1500, and lactose. Electrostatic forces arise during mixing and compaction due to triboelectric charging. These electrostatic forces are neutralized with time by electrostatic discharging. For compacts stored at ambient relative humidity or in liquids, this is a relatively fast process owing to the high diffusivity of the charges in the liquid or the adsorbed liquid layers.

Table 2 Some Specifications of Bonding Mechanisms in Compacted Dry Powders

Type (−)	Dissociation energy (kcal/mol)	Separation distance at equilibrium (Å)	Maximum attraction distance (Å)
Solid bridges			
Covalent homopolar	50–150	<2 ⎫	
Covalent heteropolar	100–200	⎬	<10
Ionic	100–200	<3 ⎭	
Intermolecular forces			
Hydrogen	2–7 ⎫		
van der Waals	1–10 ⎬	3–4	
Electrostatic	—	—	100–1000
Mechanical interlocking	—	—	—

Solid bridges that contribute to the overall compact strength can be defined as areas of real contact, i.e., contact at an atomic level between adjacent surfaces in the compact. Different types of solid bridges have been proposed in the literature, such as solid bridges due to melting, self-diffusion of atoms between surfaces, and recrystallization of soluble materials in the compacts [38–41]. Solid bridges can be detected by electrical resistivity measurements [42,43]. Electric conductance is found in compacts manufactured from metal powders or polycrystalline materials. Most of the electric conductance arises from valency vacancies and impurities in the crystal [43]. The amount of electricity that travels between the different crystals in solid and liquid bridges is in the ideal case proportional to the area of real contact between the surfaces. Calculations of the contact area between metallic surfaces have indicated that the area of real contact is relatively small compared with the available geometrical surface area [42].

The term *mechanical interlocking* is used to describe the hooking and twisting together of the packed material. It has been claimed [31] that materials bonding predominantly by this mechanism require high compression forces and have low compact strength and an extremely long disintegration time. However, a more limited description of this bonding mechanism is that it is dependent on the shape and surface structure of the particles; i.e., long needle-formed fibers and unregular particles have a higher tendency to hook and twist together during compaction compared with smooth spherical ones.

II. REPORTED MEANS OF ESTIMATING THE DOMINATING BOND MECHANISMS

A. Surface Specific Tablet Strength

Since solid bridges seem to be relatively strong bonds, while intermolecular forces are weaker, though acting over distances, a ratio between the compact strength and the surface area of the starting compound could perhaps be used to distinguish the two bonding mechanisms [23]. The surface specific strength would then give a high value for a material bonding predominantly with solid bridges and a low value for a material bonding with long-range forces as the dominating bond type. In Table 3 the compact strength is expressed in relation to the surface area of the respective materials.

The coarse sodium chloride qualities gave high values, and iron, Sta-Rx 1500, sodium bicarbonate, Avicel PH 101, and sodium chloride <63 μm gave lower values. It is of special interest that the surface specific strength of Avicel values is of the same order as the other materials in this group. This indicates that the high absolute tablet strength of Avicel probably is caused by a high surface area taking part in the bonding.

Table 3   Surface Specific Radial Tensile Strength of Test Materials Compacted at 150 MPa

Material (−)	Radial tensile strength δ_x (kPa)	Powder surface area s (cm^2)	Surface specific tensile strength, δ_x/s (kPa/cm^2)
Sodium chloride 425–500 μm	1150	44[a]	26.1
Sodium chloride 250–355 μm	793	68[a]	11.7
Sodium chloride <63 μm	2990	612[b]	4.9
Iron	1880	1490[b]	1.3
Avicel PH 101	7330	1610[b]	4.6
Sodium bicarbonate 90–150 μm	571	503[b]	1.1
Sta-Rx 1500 90–150 μm	514	358[b]	1.4

[a] Surface area of the amount of material corresponding to a tablet, as measured by gas adsorption.
[b] Surface area as measured by permeametry.

The results indicate that all of the investigated materials are predominantly bonding with long-range forces; however, two of the materials (i.e., sodium chloride 425–500 μm and sodium chloride 250–355 μm) are also bonding with a significant contribution from solid bridges.

B.   The Use of Lubricant Films

The effect of mixing sodium chloride, iron, Avicel PH 101, sodium bicarbonate, or Sta-Rx 1500 with small amounts of magnesium stearate on compact strength was investigated [24]. The effect of removing magnesium stearate by soaking the compacts in an organic solvent was also studied in an attempt to regain the initial compact strength [24].

The main effect of the magnesium stearate film on compact strength is to reduce bonding with intermolecular forces. If intermolecular forces were the only bond type present and no fragmentation of the particles or rupture of the lubricant film occurred, a compact strength close to zero would be expected. For materials bonding with solid bridges, relatively high local stresses may be formed within the compact, as suggested for coarse sodium chloride [23]. Such materials ought then to be capable of penetrating the lubricant film prior to the formation of solid bridges. The compact strength should level off at higher concentrations of magnesium stearate, thereby ideally reflecting the relative contribution of solid bridges to the total compact strength.

The proportion of lubricant added to each test material corresponded to values above and below the theoretical amount required to form a monomolecular coat on each material [58].

The plot (Fig. 3) of the remaining strength (%) against the weight of added magnesium stearate in relation to the compound surface area shows that the radial tensile strength decreases for all materials with increasing amounts of magnesium stearate. This is in agreement with earlier reports [15,44], where this effect was explained in terms of the formation of a molecular lubricant film around the compound particles.

At concentrations between 0.2 and 1 $\mu g/cm^2$, the compact strength

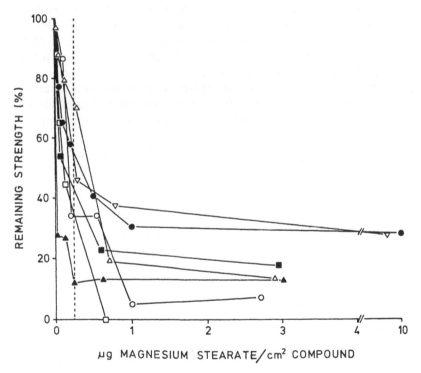

Fig. 3 The effect of surface specific lubricant concentration on remaining strength. ($\triangle$) Avicel PH 101; (■) iron; (○) sodium chloride <63 μm; (●) sodium chloride 250–355 μm; ($\triangledown$) sodium chloride 425–500 μm; (▲) sodium bicarbonate; (□) Sta-Rx 1500. (From Ref. 23).

leveled off and reached a constant value for all materials except Sta-Rx 1500. This is in agreement with the lubricant concentration predicted to give a molecular film around the particles on the assumption that magnesium stearate corresponds to a molecular surface area of 400 m²/g [44]. This is equivalent to 0.25 μg magnesium stearate/cm compound and is denoted by a dotted line in Fig. 3.

The results indicate that a shift in main bond type has taken place; i.e., intermolecular forces have been filtered out by the magnesium stearate film and that the dominating bond type involves solid bridges when the plateau has been reached. If this assumption is correct the relative contribution by solid bridges could be estimated.

The coarser fractions of sodium chloride appear to bond with a relatively high proportion of solid bridges, while for the other materials the contribution is less. However, for Avicel PH 101 an incomplete surface coverage by the lubricant and a limited particle fragmentation can be expected which will influence the interpretation of the results. This assumption was supported by the fact that the constant level found for Avicel PH 101 at 150 MPa was absent when the material was compressed at 35 MPa. Then zero compact strength was obtained. According to Sixsmith [12] there is limited fragmentation tendency below 50 MPa. The level found for Avicel PH 101 at 150 MPa is thus best explained by an incomplete surface coverage with magnesium stearate after compaction due to fragmentation of the Avicel PH 101 particles rather than solid bonds penetrating the surface film.

All lubricated tablets were then tested on soaking by immersing them in a mixture of isoamylalcohol and chloroform (1/1) for seven days. For all materials, soaking the unlubricated compacts resulted in little change in the mechanical strength, indicating that the solvent did not cause any specific interaction, such as dissolution followed by recrystallization. For all lubricated compacts except sodium bicarbonate the soaking procedure resulted in a reduction in the amount of magnesium stearate present in the compacts and with the exception of sodium bicarbonate, the lubricated compacts increased significantly in strength after the soaking (Figs. 4 and 5). Theoretically, if magnesium stearate functions as a filter for intermolecular attraction and thereby reduces the long-range bonding forces, the compact strengths ought to be fully restored after soaking, provided all lubricant is removed and that the proportion and degree of solid bridges have not been altered due to the admixing of the lubricant. However, the soaked compacts never attained the same strength as the unlubricated compacts. The most probable explanation for this is incomplete soaking of the specimens tested. For sodium chloride <63 μm (Fig. 4), the treatment resulted in an almost complete recovery of compact strength.

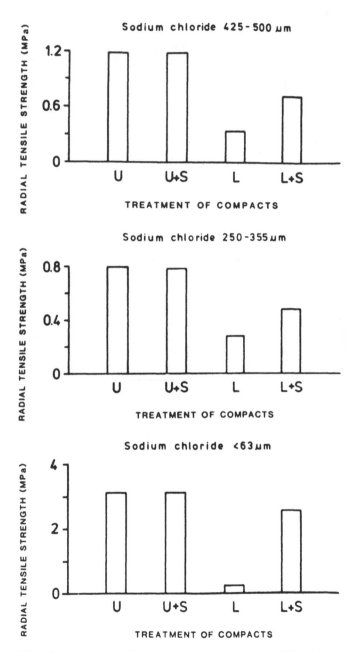

Fig. 4 Radial tensile strength for unlubricated (U), unlubricated plus soaked (U+S), lubricated (L) and lubricated plus soaked (L+S) compacts of sodium chloride. (From Ref. 23).

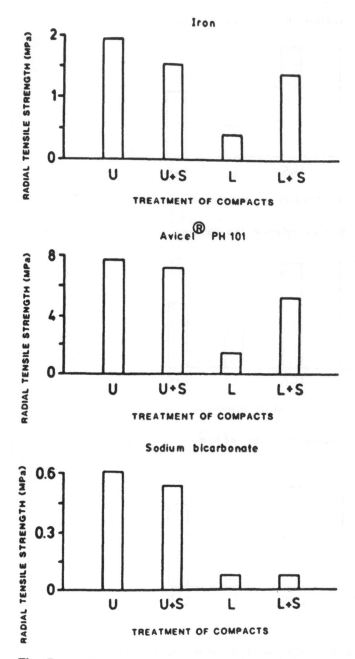

Fig. 5 Radial tensile strength for unlubricated (U), unlubricated plus soaked (U+S), lubricated (L) and lubricated plus soaked (L+S) compacts of iron, Avicel PH 101 and sodium bicarbonate. (From Ref. 23).

The results from the soaking experiments indicate that intermolecular forces, acting over distances, contribute substantially to the strength of the tested compacts. These forces could be filtered out by the addition of a film-forming lubricant. However, this effect seems to be reversible in the sense that removal of the lubricant increases the strength again. Intermolecular forces thus constitute the dominating or only bonding mechanism for the investigated materials. Only for sodium chloride is there substantial evidence for the existence of solid bridges.

C. Powder Compaction in Liquids with Different Dielectric Constants

If a surface is covered with a liquid, the liquid will act as a barrier for the adhesion forces and hence reduce the interaction between surfaces in close proximity. However, if more liquid condenses on the surface, liquid bridges may be formed between surfaces in close contact. These bridges are strong, and could thus increase the compact strength. However, it should be noted that only the interfacial forces at the liquid-gas interface contribute to the bonding force between the grains. As soon as the liquid completely envelops the compact, all capillary bonding force vanishes [38].

Shirvinskii reported that the adhesion forces for a number of different particulate materials (silicon, aluminum, and calcite) adhering to surfaces of silicon, quartz, Teflon, and steel decreased with increasing dielectric constant for liquid binary media or along a curve with a minimum plateau value [45]. The basic idea of this theory is that the forces interacting between surfaces in close contact are considered to be due to a fluctuating electromagnetic field. Because of the quantum mechanical fluctuations, this field is always present in the interior of a material medium, and it also extends beyond its boundaries. This theory is applicable to any material independent of its molecular nature [46,47].

From the Hamaker and Lifshits theories some general conclusions related to the forces interacting in a medium between particles can be drawn:

a. The London–van der Waals forces between two particles of the same material dispersed in a fluid are always attractive, provided there is no marked orientation of the fluid molecules. If the particles are of different composition, the resultant force may be repulsive in its nature [46,48].

b. The London–van der Waals forces between any two bodies in vacuum are always attractive [46,48].

c. A decrease in London–van der Waals forces with an increase in the dielectric constant of the medium could be expected [45,46].

It seems, therefore, reasonable to determine the tablet strength for a number of pharmaceutical materials compressed in liquids with different dielectric constants, and to compare these strength values with the compact strength values obtained in both ambient air and in vacuum. The importance of solid bridges or mechanical interlocking in relation to bonding with intermolecular forces could then be obtained since these latter forces are at a minimum for liquids with dielectric constants of 10 to 20 [45].

1. Apparatus for Compaction in Liquids

Tablets were compressed in liquids in a specially designed apparatus manufactured of steel (Fig. 6) equipped with flat-faced punches of diameter 1.13 cm [24].

The apparatus was filled with liquid to the top of the die. All liquids used in this study were preconditioned with the solid material prior to compaction to obtain a saturated solution. Excess liquid was then removed from the die with a pipette, making it possible to add the powder in dry form. An amount of powder sufficient to give a compact height of approxi-

Fig. 6 Tablet compression apparatus: (1) upper punch, (2) load cell, (3) compression chamber, (4) lower punch, and (5) die. (From Ref. 24).

mately 0.3 cm at 150 MPa pressure was poured into the die. Liquid was then added to the apparatus to completely wet the powder mass. After 1 min the upper punch was inserted in the die and the compression chamber was mounted in the hydraulic press. The upper punch pressure was then raised to 150 MPa over 10 s. This procedure gave two effects. First, the liquid was able to leave the tablets when the load increased without disturbing the consolidation of the compact, and, second, it increased the time for plastic flow in the tablet. Ejection of the tablet from the die was performed in the liquid by turning the die and applying pressure with the hydraulic press on the lower punch. The compact was not exposed to air during the compression and ejection phase since the compression chamber was completely filled with liquid. After ejection the tablets were removed and stored in the saturated liquid for 24 h before the tensile strength was measured. The compaction procedure described was carried out in a ventilated safety box.

Two plastically deforming materials (sodium chloride, Avicel PH101) and one material undergoing extensive fragmentation (lactose) were compressed both in air at ambient conditions and in liquids with different dielectric constants [24].

The use of liquids to reduce bonding with intermolecular forces has one important limitation for compounds easily soluble in water (e.g., sodium chloride and lactose). The solubility of these materials increases drastically with increasing dielectric constant of the test liquids [49]. A high solubility increases the risk that solid bridges and surface properties of the compact can change due to dissolution. This risk is difficult to grade and evaluate. However, for liquids with a dielectric constant below 10 to 20 the solubility is limited and subsequently does not significantly affect the bonding properties of the tested materials [24].

2. Apparatus for Characterization of Radial Tensile Strength in Vacuum

A vacuum tensile strength tester was constructed [24] (Fig. 7). The tester was manufactured from a tube of stainless steel with two movable vacuum tight end plates. On each side of the tube two bellows were attached with vacuum tight welding. Inside the bellows the tablet-crushing device moved freely with a minimum of friction to transmit the force to the tablets in the tester. The force was manually transmitted to the crushing device by a traction wheel and measured with a piezoelectric crystal attached to the bellows. The tester could hold five compacts, stored in a sliding magazine. Prior to each measurement, the tablets were degassed for 24 h or longer at a pressure of less than 10^{-4} mm Hg. For each material, 10 tablets were

Fig. 7   Vacuum tensile strength tester: (1) load cell, (2) traction wheel, (3) end
plate, (4) magazine, (5) bellow, and (6) tube. (From Ref. 24).

measured in vacuum and 10 in air in the apparatus. The radial tensile
strength was then calculated.

The results (Table 4) showed that all investigated materials increased
in radial tensile strength in vacuum. This was explained by the removal of
condensed material, primarily water from the particle surfaces in the compact [50]. Similar results have been reported for water absorption on
degassed silica compacts [51]. They found that the compact strength decreased proportionally to the surface coverage of water. This suggests that
the strength increase in vacuum is primarily caused by an increased surface
interaction due to removal of adsorbed water vapor and surface contamination, which act as a filter to reduce bonding with intermolecular forces [24].
A completely clean surface correspond to a dielectric constant of unity
(i.e., maximal interaction between surfaces in close contact).

In Fig. 8 the compact strength in liquids is compared with the radial
tensile strength in vacuum presented, and in Fig. 9 the corresponding remaining strength values (%) are given. Since the tensile strength of compacts
prepared in air can be regarded as a starting point before transfer takes place
to a vacuum or a liquid environment, the compact strength measured under
ambient conditions is denoted by a different symbol (X) in the figure.

Table 4 Radial Tensile Strength Measured under Ambient Conditions and in Vacuum of Compacts Prepared at 150 MPa

Material (−)	Tensile strength (MPa)[a]		Increase in tensile strength (MPa)
	Ambient	Vacuum	
Avicel PH101	6.37 (0.81)	8.08 (0.73)	1.71
Sodium chloride, coarse	1.01 (0.13)	1.28 (0.20)	0.27
Sodium chloride, fine	4.13 (0.56)	5.00 (0.56)	0.87
Lactose, coarse	0.73 (0.08)	0.80 (0.10)	0.07
Lactose, fine	1.03 (0.22)	1.40 (0.21)	0.37

[a] Standard deviations are given in parentheses.

All materials decreased with increasing dielectric constant in the interval tested. This is similar to the results obtained above with small amounts of magnesium stearate [23]. The formation of a plateau was suggested to be typical for a material bonding with at least two different bond types, say solid bridges and intermolecular forces [21,23]. When a stable plateau is obtained, the compact strength will in the ideal case be determined by solid bridges or mechanical interlocking alone.

It seems therefore that the results [24] for both size fractions of sodium chloride support the existence of solid bridges for this material. A similar result was obtained for coarse lactose compacts which also formed a plateau. Although lactose is described as a brittle material, the participation of plastic flow in the densification process has been reported [3]. A limited contribution of bonding with solid bridges was therefore regarded as a possible explanation for the plateau for coarse lactose. For the fine fraction of lactose, no coherent compacts were formed when the liquid dielectric constant exceeded approximately seven, indicating that intermolecular forces is the only bonding mechanism for this size fraction.

Avicel PH 101 showed a continuous decrease in tablet strength down to zero with increasing dielectric constant. Thus bonding with intermolecular forces seems to be the dominating bonding mechanism for this material.

In a recent study [52] also the effect of compaction pressure on the relative contribution of solid bridges was tested for sodium chloride (Fig. 10). In these experiments only one liquid (butanol) was used with a dielectric constant of 17.8. The results demonstrate that the development of solid bridges is increased with compaction pressure.

Also earlier some studies have been reported [53,54], where the concept of using liquids for filtering out, predominantly the intermolecular

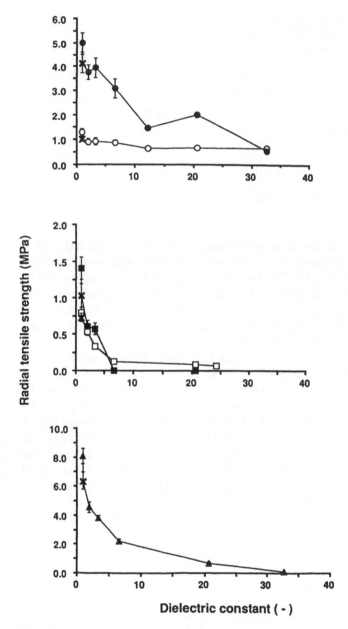

Fig. 8 The effect of the dielectric constant of the test liquid on the compact strength. (○) sodium chloride coarse; (●) sodium chloride fine; (□) lactose coarse; (■) lactose fine; (▲) Avicel PH101, compressed at 150 MPa. Error bars represent confidence intervals of the mean for 95% probability. For some of the result, precision is better than can be denoted in figure. (From Ref. 24).

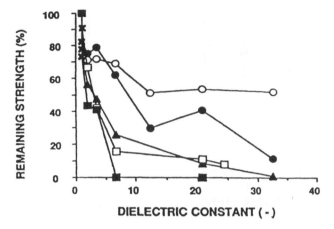

Fig. 9 The remaining compact strength in liquids compared with the radial tensile strength in vacuum. Compact strength in ambient surrounding is marked with X. Symbols as in Fig. 8. (From Ref. 24).

forces, was applied. However, in those reports, the compacts were prepared in conventional manner, i.e., in a dry state, and then subsequently soaked in different organic liquids. Although, this technique might be applicable in cases where relatively porous tablets, without any distinct peripheral high-density "seal," are at hand, it has been established in our laboratory to be an unsuitable technique for most pharmaceutical materials. The main problem is the limited possibility for liquids to penetrate the compact and to reach the interior of the tablet.

III. REPORTED MEANS OF ESTIMATING BONDING SURFACE AREA

A. Relations between Tablet Surface Area and Bonding Surface Area

For many materials it has been possible to relate an improved tablet strength to, e.g., a decrease in particle size or a change to more irregularly shaped particles. It has also been claimed that materials which deform plastically bind better than materials undergoing elastic deformation or extensive particle fragmentation [55,56]. Plastic deformation is thus believed to be of special importance for the formation of a large bonding surface area (e.g., [31]).

Some fundamental studies in the field of metallurgy [42] have indicated that the surface area taking part in the attraction between compact

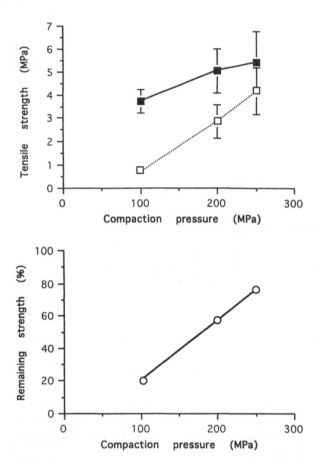

Fig. 10 Upper graph: Effect of the compaction pressure on the compact strength of sodium chloride 20–40 μm. Compaction under ambient conditions ε = 1 (■), compaction in butanol ε = 17.8 (□). Lower graph: The remaining compact strength of sodium chloride 20–40 μm in butanol compared with the compact strength under ambient conditions.

particles is relatively small, being only a minor fraction of the geometrical surface area available. Studies on pharmaceutical materials, using gas adsorption techniques, have, however, suggested that larger surface areas are involved [11,28,57,58]. In these studies, a decrease in tablet surface area with compaction load has been regarded as a reflection of the surface area utilized for the bonding between particles.

With few exceptions, the relations between compaction load and measured tablet surface are a presented in the literature are relatively complex

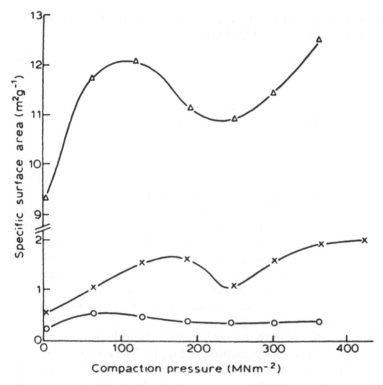

Fig. 11 Specific surface area—compaction pressure profiles of sodium chloride ($\bigcirc$), phenacetin ($\times$) and magnesium carbonate ($\triangle$). (From Ref. 57).

(e.g., [57]) (Fig. 11). Normally, an initial increase in tablet surface area with compaction load is recorded. This has been explained by a fragmentation of the compressed particles, resulting in the formation of new surfaces. With increasing compaction load, more dense compacts will be formed, bringing particle surface areas into closer proximity. The surface area available for the gas molecules will then be dependent upon the penetration capacity of the technique, i.e., mainly the size of the gas molecules used [21]. For gases like nitrogen and krypton, that are normally used in these kind of studies, the molecules will only reach a fraction of the total tablet surface area. Since the distance between solid surfaces needed for the development of bonds could be substantially smaller than the size of the gas molecules, only a part of the "nonavailable" surface area may be utilized for bonding [11]. Additionally, the use of porous materials and the possibility for pores and cracks to be closed during compression may further complicate the evaluation of a bonding surface area [8,20].

Using *gas adsorption* data the following relation has been applied to estimate the surface area utilized for bonding:

$$S_T = S_P + S_F - S_B \tag{1}$$

where

    S_T = tablet surface area as measured by gas adsorption
    S_P = surface area of uncompressed particles
    S_F = new surface area created by fragmentation during compression
    S_B = surface area consumed for bonding

From the discussion above it is, however, obvious that such calculations involve too many uncontrolled factors, especially if porous, granulated materials are tested. These difficulties thus make the conclusion that larger surface areas may be involved in pharmaceutical tablets highly uncertain.

If, however, solid nonporous materials undergoing volume reduction mainly by plastic deformation without any tendency toward fragmentation are used, some kind of proportional relation between compact surface area and bonding surface area ought to be obtained. For iron powder and coarse particulate sodium chloride, two materials belonging to this group, this has been investigated at our laboratory [21].

The characteristics of both iron and sodium chloride compacts, compressed at increasing loads, are presented in Fig. 12. The compact strength increased approximately linearly for both materials tested. The minimum load at which coherent compacts could be obtained was about 100 and 50 MPa for iron and sodium chloride respectively. The slope of the linear part of the strength-pressure profile was approximately three times higher for iron than for sodium chloride, indicating that the iron powder was bonding with a stronger bond type or that the surface area utilized for bonding was higher. Assuming that the major bond types involved are unchanged with increasing pressure, it seems reasonable that an increase in strength is accompanied by a proportional increase in bonding surface area. If the permeametry and gas adsorption techniques are capable of detecting such an increase in bonding surface area, results would be expected showing a decrease in compact surface area after 100 and 50 MPa for the two materials respectively. However, for the iron compacts, the surface areas obtained by both permeametry and gas adsorption (Fig. 12) seem to be fairly constant with increasing pressure. For permeametry, even a slight increase was observed. It could, however, be questioned whether this is an artefact or not. In an earlier study [20], it was shown that measurements on powders compressed at relatively high loads could result in an overestimation of compact surface area. The surface areas measured by gas adsorption showed an appreciable variation, especially at high loads, making it diffi-

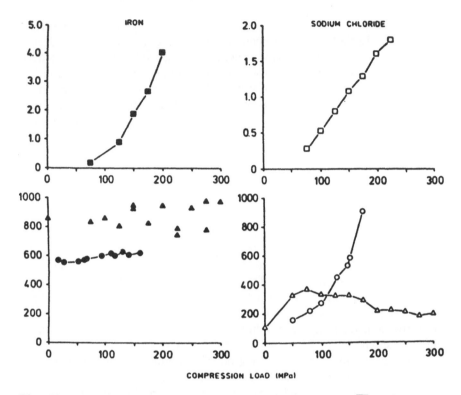

Fig. 12 The effect of compression load on compact strength (☐) and compact surface area as measured by gas adsorption (△) and permeametry (○), for iron (closed symbols) and sodium chloride (open symbols). (From Ref. 21).

cult to draw firm conclusions. The result could not be interpreted as giving a significant decrease in compact surface area with compression load, after 100 MPa. The relatively small difference in surface area obtained by the two techniques investigated supports the idea that the iron particles were essentially nonporous.

The lack of increase in permeametry surface area with increased compression load indicates that the iron particles were not significantly fragmenting during compression. The data obtained for the iron powder therefore indicate that the surface area utilized for interparticulate attraction is equal to or less than the precision of the surface area methods tested. Since it is reasonable to assume that these methods will tend to overestimate the bonding surface area [11] as discussed in the introduction, the results imply that the bonding surface area in the iron compacts is very small [42].

The surface area profiles for sodium chloride (Fig. 12) show a different pattern. Initially, the permeametry surface area shows a moderate increase, whereafter an extensive increase is observed. Although the data obtained at higher pressures presumably correspond to an overestimation of the surface area [20], the results indicate that volume reduction of sodium chloride is accompanied by some tendency toward particle fragmentation. The surface area as measured by gas adsorption shows initially a higher increase with compression load than the corresponding data obtained by permeametry. This probably reflects the fact that cracks and pores were formed during compression. After approximately 75 to 100 MPa, the surface area decreases with an increase in compression load. This decrease in gas adsorption surface area could then be a reflection of an increase in bonding surface area. This means that sodium chloride particles bond with a significantly weaker bond type than the iron particles, resulting in weaker compacts but utilizing a larger surface area for interparticulate attraction, which subsequently is reflected in the decrease of gas adsorption surface area. This decrease could alternatively be interpreted as a sealing of cracks created. Considering the profiles obtained for the iron powder, the latter explanation seems more probable. The results therefore indicate that the bonding surface area for sodium chloride is relatively small.

In an attempt to vary the degree of bonding without changing the compression load and subsequently not affecting the fragmentation tendency or formation of cracks, mixtures of sodium chloride and varying amounts of magnesium stearate were compressed at 150 MPa. The effect of increasing additions of the lubricant on both compact strength and surface area for sodium chloride is presented in Fig. 13.

Using minute additions, the compact strength decreased strongly with increasing magnesium stearate concentrations. This is in agreement with results reported earlier [8,15,23,59], where this effect was explained in terms of the formation of a lubricant film around the compound particles. However, at additions in excess of approximately 0.005 wt.%, the compact strength leveled off and even the use of a relatively high concentration (0.05 wt.%) gave no further reduction in strength. Similar profiles have been obtained for silica compacts [51], where only a small amount of adsorbed water decreases the compact strength considerably, whereafter the strength remained fairly constant when the adsorption of water was further increased. The interpretation of the data obtained is as discussed that two different bond types are involved in the interparticulate attraction [23,24].

One bonding mechanism (e.g., molecular forces of van der Waals type) is very sensitive for surface changes, while the other mechanism (e.g., solid bridges due to ionic bonding) could be established even in the presence of a lubricant film by, due to penetration by point irregularities. When a total surface coverage of sodium chloride particles by magnesium stearate is obtained, the latter bond type alone would determine the compact strength.

As is evident from Fig. 13, the admixture of the different lubricant amounts did not affect the powder surface area as characterized by the gas adsorption technique. However, the gas adsorption surface area of the corresponding compacts showed a decrease with increased quantities of lubricant. This effect was not expected, considering the obtained decrease in compact strength, which was supposed to expose the surface area used for bonding [28]. The result could probably be explained by a reduction in particle fragmentation and crack formation in the systems containing lubricant additions. Similar observations have been reported earlier [57]. The results therefore indicate that the surface area used for bonding is relatively small, and masked by the surface area changes due to fragmentation and crack formation.

The results obtained demonstrate the problems involved when investigating the compaction behavior of pharmaceutical materials. Although sodium chloride in pharmaceutical studies (e.g., [8,15,57]) normally represents a nearly ideal model substance (nonporous and consolidating by plastic deformation), it is evident from the results that sodium chloride must be classified as a complex substance with respect to its compaction behavior. In contrast to the iron powder, sodium chloride particles undergo such changes during compression (fragmentation and formation of cracks and pores) that a simple monitoring of changes in surface area with pressure (Fig. 12) cannot unambiguously be used for the evaluation of bonding surface area. Considering the even more complex nature of most pharmaceutical materials used in some reported studies [8,11,28,57,58], it is questionable whether the interpretations suggested, purporting relatively high bonding surface areas [11,28,57,58], are justified solely on the basis of previously published experimental data.

The results in this study indicate that the surface area taking part in the bonding between particles for sodium chloride, as well as for iron powder, is small in relation to the geometrical surface area available. As discussed, it seems reasonable that the major bonding is facilitated by long-range forces (e.g., molecular forces of van der Waals type), thereby explaining why the two surface area techniques are not capable of detecting any substantial change in bonding surface area with changes in the different parameters tested.

Thus, a tablet which can be described as a dispersion of solid particles in gas phase where the individual particles are separated by some distance

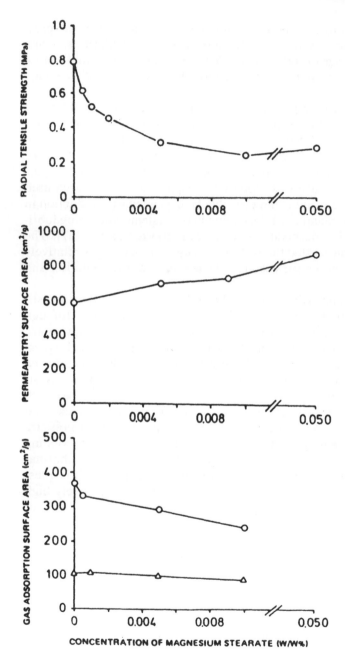

Fig. 13 The effect of lubricant concentration on compact strength and surface area for sodium chloride, compressed at 150 MPa. (From Ref. 21).

cannot probably be assayed on bonding surface area in any direct sense by surface area methods. The possibility of obtaining a proportional relation between measured tablet surface area and the fraction of surface area participating in bonding is then probably limited to relatively simple test systems. So far, such relations have only been reported to a limited extent, e.g., for lactoses undergoing fragmentation during compression [59]. In these studies, a unique relation was claimed to exist between compact strength and surface area as measured by mercury porosimetry (Fig. 14). It was also suggested that the obtained results indicated that the same bonding mechanism must be active in all types and size fractions of the lactoses tested. However, it could be discussed whether mercury porosimetry is the best method to reflect the surface area of tablet particles that potentially could

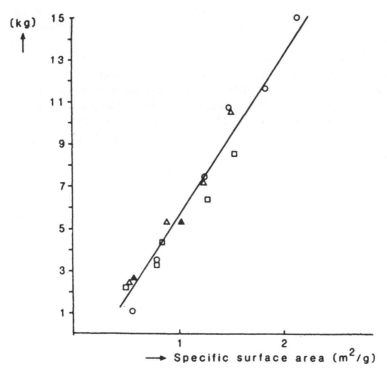

Fig. 14 Crushing strength versus specific surface area for tablets compressed from different types of crystalline lactose; α-lactose monohydrate ($\square$), anhydrous α-lactose ($\bigcirc$), roller-dried β-lactose ($\triangle$), crystalline β-lactose ($\blacktriangle$). (From Ref. 59).

participate in bonding since mercury porosimetry also monitors the surface area of intraparticulate pores. Another approach is to use a permeametric technique. Several studies have demonstrated the usefulness of this technique for the characterization of external surface areas in tablets (e.g., [9,20,60,61]).

B. The Importance of Volume Reduction Mechanisms for Bonding Surface Area

The effective amount of surface area available for bonding is dependent on several material properties [29]. Both the particle characteristics of the starting material and the changes caused by the volume reduction will be determining factors. The problems of measuring and defining a bonding surface area have resulted in a great interest for more indirect, secondary parameters (Table 1). In this section, the influence of volume reduction mechanisms on bonding surface and compact strength will be discussed.

To study the influence of volume reduction mechanisms on flattening of the surface roughness and the creation of bonding surface areas, a model system was developed [26] consisting of two layers that are compressed to form a single tablet (Fig. 15). The first layer was precompressed at a compaction pressure of 25–200 MPa. Powder material was then added on top of the first layer, and the lower punch was adjusted to give a compaction pressure of 200 MPa for the double-layer tablet. The tablet strength was characterized by measuring the axial tensile strength according to Nyström et al. [62].

The axial tensile strength for double layer tablets of all materials are presented in Figs. 16 and 17. For all materials tested, an increase in pressure on the first layer of the double-layer tablet resulted in a decrease in the axial strength. All the plastically deforming materials (sodium chloride, Sta-Rx

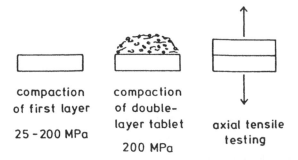

compaction   compaction
of first layer   of double-
                   layer tablet   axial tensile
25 - 200 MPa   testing
                   200 MPa

Fig. 15 Double-layer tablet technique for studying the influence of volume reduction behavior on bonding surface area and tablet strength.

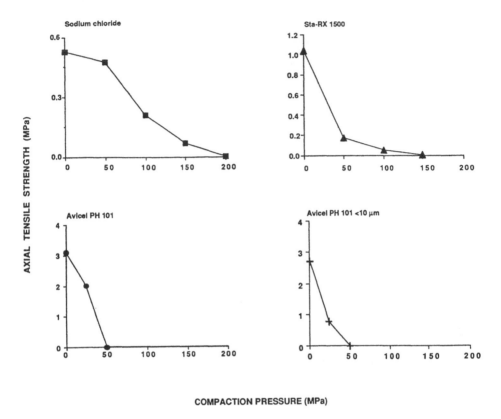

Fig. 16 The effect of compaction pressure of the first portion on the axial tensile strength of the double-layer compact compressed at 200 MPa for plastically deforming materials. (From Ref. 26).

1500, Avicel PH 101, and Avicel PH 101<10 μm) failed in the contact zone between the first and the second layer of the double-layer tablet, indicating that the bonding strength in the contact zone generally was lower than in the intervidual layers. The materials consolidating mainly by fragmentation failed generally in the first layer of the double-layer tablet, indicating that the bonding strength between the two layers was higher than that of the individual layers.

In Fig. 18, the percentage decrease in strength is plotted against the compaction pressure of the first layer of the double-layer tablet for all materials tested. Avicel PH 101 and Sta-Rx1500 showed the greatest sensitivity followed by sodium chloride. The materials undergoing extensive fragmentation (i.e., Emcompress, lactose, and sucrose) are relatively insen-

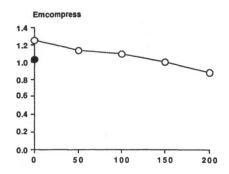

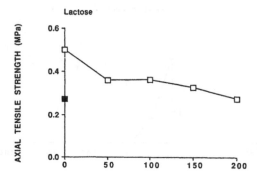

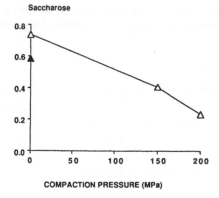

Fig. 17   The effect of compaction pressure of the first portion on the axial tensile strength of the double-layer compact compressed at 200 MPa for materials undergoing volume reduction mainly by fragmentation. The strength of a single tablet compressed two times at 200 MPa is denoted by a filled symbol on the *y*-axis. (From Ref. 26).

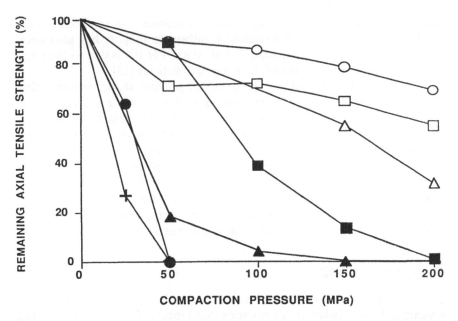

Fig. 18 The remaining axial tensile strength with the increase in compaction pressure of the first portion of the double-layer tablet for all materials tested. Symbols as in Figs. 16–17. (From Ref. 26).

sitive to an increase in compaction pressure compared to the plastically deforming materials. Thus volume reduction by fragmentation seems to be a more efficient means of producing larger surface areas that would promote bonding between particles in the compacts [26,59]. This is especially valid for materials bonding with intermolecular attraction forces—probably the majority of pharmaceutical compounds and excipients.

The so-called plastically deforming materials used in this study were all very sensitive to a decrease in the surface roughness of the first layer of the double-layer tablet; i.e., the plastic deformation of these materials was inadequate for the development of large zones of intimate contact between the layers. It is debatable whether the term *plastic deformation* should be applied to all less fragmenting materials possessing varying degrees of plasticity. Many of the commonly used amorphous tablet binders with pronounced plastic deformation may provide an effective means of creating large interparticulate attraction surface areas. For most of the plastically deforming materials, possessing a moderate plasticity and bonding with intermolecular forces, it seems that a high external specific surface area is a prerequisite for high compactibility. This is best achieved by using fine particle sizes or qualities with high surface roughness, as reported earlier for Sta-Rx 1500 [14] and for sodium bicarbonate [63]. It has also been suggested that the high compactability of Avicel PH 101 is due to the large external specific surface area of the irregular particles [23].

A large surface area and an irregular particle shape will probably promote all bonding mechanisms discussed. Fragmenting materials normally bonding by intermolecular forces [29] do not seem to be severely affected by an increase in compression load since the particles in the second layer can develop large bonding surface areas not only by particle rearrangement together with a limited plastic deformation but also by extensive fragmentation. For the plastically deforming materials used in this study, the initially high surface roughness is reduced after compression, with a subsequent reduction in intermolecular forces, the development of solid bridges (in, e.g., sodium chloride), and mechanical interlocking.

IV. CONCLUSIONS

In this chapter, several pharmaceutical materials have been reported on volume reduction behavior, dominating bond mechanism, and, to a limited extent, bonding surface area. The following general conclusions have been suggested.

1. Intermolecular forces constitute the dominating bond mechanism for pharmaceutical materials.
2. A proportional relation between compact surface area and bonding surface area could in some cases be established.
3. The following material properties will in principle favor a high compact strength:
   - limited elastic deformation
   - high compact surface area
       fine particulate starting materials
       highly fragmenting starting materials
       starting materials possessing high surface roughness
   - extreme plastic deformation (e.g., amorphous binders)

A. Classification of Tabletting Materials

It is believed that for materials undergoing extensive fragmentation, a large number of interparticulate contact points are created. The compaction load per unit area of such contact points will thereby be low. This indicates that mainly relatively weak attraction forces, acting over distances, can be formed. However, due to the large number of bonds or attraction zones, a relatively strong compact could be formed. For less fragmenting materials, a smaller number of contact points are formed, which then would result in strong compacts only if relatively strong attraction forces, such as solid bridges, could be developed. This is probably the situation for compacts of coarse crystalline sodium chloride. If the material exhibits extensive plastic deformability, such as many amorphous binder materials, the number of weak distance forces would probably be much higher and thereby contribute significantly to the compact strength. Also materials having a rough surface texture ought to be capable of forming a relatively large number of weak distance forces, in spite of the fact that such a material does not fragment extensively. If the powder being compressed consists of particles with both a rough surface texture and a pronounced plastic deformability, compacts of extremely high mechanical strength ought to be obtained, due to the large number of weak distance attractive forces developed. Microcrystalline cellulose (e.g., Avicel) could be an example of such a material. A suggestion for how materials can be classified according to their compactability and its dependence on volume reduction and bonding properties is presented in Table 5.

An important parameter that will influence the final compact strength is the proportion of elastic deformation and, consequently, elastic recovery that will take place after compaction. By definition, so-called plastically deforming materials show little elastic deformation. Thus for fragmenting materials it seems important to distinguish between those where the smaller particles formed undergo mainly plastic deformation or elastic deformation. Most of the materials intended for direct compression have in common a minute elastic behavior.

High tablet strength is thus primarily produced by materials possessing a low elastic component during consolidation and having a high bonding surface area that could develop intermolecular forces. To this group of materials belong fine particulate materials, milled materials, and strongly fragmenting materials such as granulations. Excipients belonging to this group may produce a large bonding surface area in various ways; e.g., the material can be highly fragmenting, very plastically deforming, or exhibit a pronounced surface roughness.

B. Advantages of Using Fragmenting Materials

Some possible advantages for the use of materials with a high fragmentation tendency are presented in Table 6.

Several studies have shown that a pronounced material fragmentation will result in a reduced dependence on initial particle size, surface shape and texture, additions and load rate for the compact strength. In this

Table 5 Primary and Secondary Factors Affecting Compact Strength for Pharmaceutical Materials

	Low tablet strength	High tablet strength	
Dominating volume reduction mechanism	Small number of weak attractions (low bonding surface area and bonding predominantly with intermolecular distance forces)	Large number of weak attractions (high bonding surface area and bonding predominantly with intermolecular distance forces)	Small number of strong attractions (small bonding surface area and bonding both with intermolecular distance forces and solid bridges)
Plastic deformation	Not capable of forming solid bridges e.g., sodium bicarbonate	Very fine particulate qualities e.g., fine particulate sodium chloride Pronounced surface roughness e.g., Avicel, Sta-Rx 1500 and milled qualities of materials Very plastically deformable e.g., amorphous binders	Capable of forming solid bridges e.g., coarse particulate sodium chloride
Fragmentation and plastic deformation	Nonexisting	With a small elastic component e.g., lactose Emcompress	
Fragmentation and elastic deformation	With a large elastic component e.g., phenacetin paracetamol		

Table 6 Possible Advantages with Fragmenting Materials

Less sensitive for variations in particle size and shape of starting materials
Less sensitive for admixture of lubricants
Less sensitive for load rate
Less prone to undergo postcompaction strength changes by stress relaxation
High compactability, provided the elastic component of smaller particles (fragments) formed is minute

---

chapter it has also been suggested that fragmenting materials in general will result in a high bonding surface area and compact strength, provided the elastic component of the material is limited.

REFERENCES

1. D. Train, *Inst. Chem. Eng. 35*: 258 (1957).
2. J. T. Carstensen, *Solid Pharmaceutics: Mechanical Properties and Rate Phenomena*, Academic Press, New York, 1980.
3. M. Duberg and C. Nyström, *Powder Technol. 46*: 67 (1986).
4. I. Krycer and D. G. Pope, *Int. J. Pharm. Technol. Prod. Manuf. 3*: 93 (1982).
5. C. Führer and J. Ghadially, *Acta Pharm. Suec. 3*: 201 (1966).
6. C. Führer, E. Nickel, and F. Thiel, *Acta Pharm. Technol. 21*: 149 (1975).
7. C. Führer, *Acta Pharm. Technol. 6*: 129 (1978).
8. M. Duberg and C. Nyström, *Acta Pharm. Suec. 19*: 421 (1982).
9. G. Alderborn, K. Pasanen, and C. Nyström, *Int. J. Pharm. 23*: 79 (1985).
10. D. P. Coffin-Beach and R. G. Hollenbeck, *Int. J. Pharm. 17*: 313 (1983).
11. J. S. Hardman and B. A. Lilley, *Nature 228*: 353 (1970).
12. D. Sixsmith, *J. Pharm. Pharmacol. 34*: 345 (1982).
13. E. T. Cole, J. E. Rees, and J. A. Hersey, *Pharm. Acta, Helv. 50*: 28 (1975).
14. G. Alderborn and C. Nyström, *Acta Pharm. Suec. 19*: 381 (1982).
15. A. H. de Boer, G. K. Bolhuis, and C. F. Lerk, *Powder Technol. 25*: 75 (1978).
16. A. McKenna and D. F. McCafferty, *J. Pharm. Pharmacol. 34*: 347 (1982).
17. R. J. Roberts and R. C. Rowe, *J. Pharm. Pharmacol. 37*: 377 (1985).
18. R. J. Roberts and R. C. Rowe, *J. Pharm. Pharmacol. 38*: 567 (1986).
19. A. Kandeil, M. C. De Malherbe, S. Critchley, and M. Dokainish, *Powder Technol. 17*: 253 (1977).
20. G. Alderborn, M. Duberg, and C. Nyström, *Powder Technol. 41*: 49 (1985).
21. C. Nyström and P. G. Karehill, *Powder Technol. 47*: 201 (1986).
22. Z. T. Chowhan and Y. P. Chow, *Int. J. Pharm. Technol. Prod. Manuf. 2*: 29 (1981).
23. P. G. Karehill, E. Börjesson, M. Glazer, G. Alderborn, and C. Nyström, *Drug. Dev. Ind. Pharm. 19*: 2143 (1993).
24. P. G. Karehill and C. Nyström, *Int. J. Pharm. 61*: 251 (1990).
25. P. G. Karehill and C. Nyström, *Int. J. Pharm. 64*: 27 (1990).

26. P. G. Karehill, M. Glazer, and C. Nyström, *Int. J. Pharm. 64*: 35 (1990).
27. J. J. Benbow, in *Enlargement and Compaction of Particulate Solids* (N. G. Stanley-Wood, ed.), Butterworths, London, 1983, p. 171.
28. N. G. Stanley-Wood and M. S. Shubair, *Powder Technol. 25*: 57 (1980).
29. M. Duberg and C. Nyström, *Int. J. Pharm. Technol. Prod. Manuf. 6*: 17 (1985).
30. H. Rumpf, *Chem. Eng. Tech. 30*: 144 (1958).
31. C. Führer, *Lab. Pharma. Probl. Technol. 269*: 759 (1977).
32. C. Ahlneck and G. Alderborn, *Int. J. Pharm. 54*: 131 (1989).
33. J. N. Israelachvili, *Intermolecular and Surface Forces*, Academic Press, London, 1985, p. 21.
34. B. V. Derjaguin, *Sci. Am. 203*: 47 (1960).
35. B. V. Derjaguin, I. I. Abrikosova, and E. M. Lifshitz, *Quart. Rev. Chem. Soc. 10*: 295 (1956).
36. J. N. Israelachvili and D. Tabor, *Prog. Surf. Membr. Sci. 7*: 1 (1973).
37. J. N. Israelachvili, *Intermolecular and Surface Forces,* Academic Press, London, 1985, p. 98.
38. H. Rumpf, in *Agglomeration* (W. A. Knepper, eds.), Interscience, New York, 1962, p. 379.
39. G. R. B. Down and J. N. McMullen, *Powder. Technol. 42*: 169 (1985).
40. A. G. Mitchell and G. R. B. Down, *Int. J. Pharm. 22*: 337 (1984).
41. C. Ahlneck and G. Alderborn, *Int. J. Pharm. 56*: 143 (1989).
42. F. P. Bowden and D. Tabor, *The Friction and Lubrication of Solids*, Oxford University Press, New York, 1950, p. 25.
43. R. P. Bhatia and N. G. Lordi, *J. Pharm. Sci. 68*: 222 (1979).
44. G. K. Bolhuis, C. F. Lerk, H. T. Zijlstra, and A. H. de Boer, *Pharm. Weekblad 110*: 317 (1975).
45. A. E. Shirvinskii, V. A. Malov, and I. S. Lavrov, *Colloid J. U.S.S.R. 46*: 867 (1984).
46. I. E. Dzyaloshinskii, E. M. Lifshitz, and L. P. Pitaevskii, *Adv. Phys. 10*: 165 (1961).
47. J. N. Israelachvili, *Intermolecular and Surface Forces*, Academic Press, London, 1985, p. 115.
48. H. C. Hamaker, *Physica 4*: 1058 (1937).
49. J. N. Israelachvili, *Intermolecular and Surface Forces*, Academic Press, London, 1985, p. 30.
50. H. K. Sartor, *Proc. Int. Fachtagung Komprimate* 2, Solingen, 1978.
51. D. Dollimore and G. R. Heal, *J. Appl. Chem, 11*: 459 (1961).
52. Å. Svensson, H. Olsson, and C. Nyström, in preparation.
53. M. Luangtana-anan and J. T. Fell, *Int. J. Pharm. 14*: 197 (1990).
54. D. R. Fraser, Powder Advisory Center, London, 149, 1973.
55. G. Milosovich, *Drug Cosmet. Ind. 92*: 557 (1963).
56. J. E. Rees, *Acta Pharm. Suec. 18*: 68 (1981).
57. N. A. Armstrong and R. F. Haines-Nutt, *Powder Technol. 9*: 287 (1974).
58. T. Higuchi, N. A. Rao, L. W. Busse, and J. V. Swintosky, *J. Am. Pharm. Assoc. Sci. Ed. 42*: 194 (1953).

59. H. Vromans, A. H. de Boer, G. K. Bolhuis, C. F. Lerk, and K. D. Kussendrager, *Pharm. Weekblad Sci. Ed. 7*: 186 (1985).
60. G. Alderborn and C. Nyström, *Powder Technol. 44*: 37 (1985).
61. C. Nyström and M. Glazer, *Int. J. Pharm. 23*: 255 (1985).
62. C. Nyström, W. Alex, and K. Malmqvist, *Acta Pharm. Suec. 14*: 317 (1977).
63. G. Alderborn, E. Börjesson, M. Glazer, and C. Nyström, *Acta Pharm. Suec. 64*: 31 (1988).

3

Porosity–Pressure Functions

Petteri Paronen and Jukka Ilkka

University of Kuopio, Kuopio, Finland

I. INTRODUCTION

The powder column is a heterogeneous system consisting of solid particulate material and air. Air can exist both between particles (interparticulate voids) and inside particles (intraparticulate voids). The physical nature of a powder column is different from that of a solid body, as powder can flow and have rheological properties typical of liquids. On the other hand, permanent deformation (plastic flow), reversible deformation (elasticity), and brittle fracturing of particles typical phenomena for solid bodies occur in powders. Thus, the behavior of powders in pharmaceutical processes, e.g., during compression, is often very complicated. In die compaction of powders, materials are subjected to compressive forces which lead to a volume reduction of the powder column. A volume is reduced by decreases in the inter- and intraparticulate pore space. The process of volume reduction is generally divided into different stages: die filling, rearrangement of particles, deformation by elastic changes, permanent deformation by plastic flow, or particle failure by brittle fracturing. The measuring of porosity changes as a function of the compression pressure is a method widely used in describing the compaction processes of powders.

Porosity is a function of the voids in a powder column, and in general all pore space is considered, including both inter- and intraparticulate voids. For porosity measurements, the dimensions and weight of a powder column (i.e., apparent density) and the particle density (referred to often as true density) of the solid material should be known. The porosity, ϵ, can be expressed by the equation

$$\epsilon = 1 - \frac{\rho_A}{\rho_T} \tag{1}$$

where ρ_A is the apparent density of a powder column and ρ_T is the particle density of the compressed material. The value of ρ_A/ρ_T, also referred as D, is regarded as the relative density or the packing fraction, which describes the solid fraction of a porous powder column. A value for the applied pressure, P, while loading a powder column under pressure is a function of the compressional force, F, and the punch tip area, A:

$$P = \frac{F}{A} \tag{2}$$

Instrumentation of tablet presses with force and displacement transducers enables a data acquisition for porosity-pressure analysis. Several equations have been proposed for describing the relationship between the porosity of a powder column and the applied pressure [1–3].

II. DETERMINATION OF POROSITY AND PRESSURE

Measurements for the relationship between porosity and pressure have usually been made during one-sided, uniaxial compressions. In such studies, instrumented eccentric tablet presses were utilized, as well as hydraulic and electrical universal testing machines, and high-speed rotary presses with double-ended compression were seldom used.

 Compressive force is currently measured by strain gauges or piezoelectric transducers mounted directly on the compressing parts of the press. After proper calibration reliable force measurements can be obtained. Acceptable measurements of accuracy and precision of force are possible with modern strain gauges and amplifiers.

 For an accurate determination of porosity, the height and diameter of a powder column must be reliably measured. Any looseness or elastic deformation of the compression devices' parts should be evaluated to allow an accurate measurement of the true displacement. The effects of machine deformation may be diminished by mounting the displacement transducers as close as possible to the compressing punch tip. Furthermore, the connection of a transducer to both the upper and lower punches makes the measurement even more accurate, e.g., by preventing a summation of the nonlinearity of discrete transducers. Several studies have been published on the mounting, calibration, and validation of measuring systems [4–7].

Besides the measuring transducers and their mounting also the amplifiers and analogue digital converters also affect the accuracy and precision of measurements [4,5]. The resolution of an A/D converter could have a remarkable effect on the accuracy of measuring systems. Typically a 12-bit conversion is used, which is enough for obtaining measurements for fully describing a compression process [8]. Smoothing of the original measurement data (e.g., using polynomial fitting) is not needed if the resolution of a measuring system is high enough and the electrical noise is removed by filtering.

The weight and the particle density of the compressed material, as well as the bulk density of loose powder, should also be carefully measured. Preweighed powder samples may be used to improve the repeatability of the measurements. The apparent density of a compact is obtained by dividing the weight by the apparent volume of powder compact. Compaction is often considered to begin at the loose packing state described by the bulk density of powder. The bulk density as measured by pouring a known quantity into a graduated vessel may differ considerably from that obtained by a die-filling process. In addition, the apparent density of a powder column, measured at the moment of the first detected force, has been used for calculations instead of separately measured bulk density [9].

Pycnometric methods, especially helium pycnometry, is preferably used for measurements of the particle density. Adsorbed volatile impurities and even absorbed water may cause variations. In this respect the particle density can not always be regarded as a material constant. In addition, even the particle density may change during compression. Changes are probably rare, and difficult to demonstrate. However, polymorphic transitions and changes in crystallinity are typical phenomena related to changes in density. An elastic deformation of a crystal structure could theoretically lead to reversible changes in density. Obviously, unpredictable and temporary changes in density can cause oddities in porosity-pressure analyses [10].

The volume of a powder column, and its corresponding porosity, can be measured either during compression or after ejection of the compact from a die. Instrumentation of a tablet press enables continuous monitoring of the powder column height during the volume reduction process. This method is referred as "in die" or "at pressure." In the "ejected tablet" or "at zero pressure" method, the dimensions of a compact are measured after ejection. The maximum compression pressure observed is usually taken to correspond to an achieved volume reduction or a certain porosity. A single compression provides only one measurement as a function of compression pressure. A disadvantage of this method is the need to perform several compressions at different pressure levels. Also an "at pressure" method is possible, where only the maximum compression pressure and the corresponding porosity of a powder column in the die are taken for further analysis. Consequently, compressions at different pressure levels are also needed in this method.

The porosity-pressure relationship is greatly influenced by the method in which the dimensions of a powder column or a compact are measured. When pharmaceutical powders are subjected to compression, part of volume reduction in a powder column may occur by reversible, elastic densification which then cause measurable changes in compact dimensions, at the decompression phase and even after ejection. Elastic recovery of a compact has even been used to indicate the elastic properties and bonding of a compressed material. However, the mechanical properties of a solid material may differ from those of a powder column consisting of particulate material and air.

III. POROSITY-PRESSURE EQUATIONS

Numerous mathematical models describing the change of relative density in a powder column as a function of the applied pressure have been derived and adopted from other fields of industry for research in pharmaceutical compression processes [1–3]. Three equations have been widely applied to pharmaceutical purposes, namely Heckel (also called as Athy-Heckel), Kawakita, and Cooper-Eaton. These equations will be considered in this chapter.

A. Heckel Equation

Heckel [11,12] introduced an equation for the densification phenomenon following the first-order kinetics. The equation is

$$\ln \frac{1}{1 - D} = kP + A \tag{3}$$

where k and A are constants obtained from the slope and intercept of the plot $\ln(1/(1 - D))$ versus P, respectively, D is the relative density of a powder column at the pressure P (Fig. 1a).

A is an intercept which is extrapolated from the linear part of the Heckel plot. As the plots were curved at low pressures, Heckel related the constant A to processes of volume reduction which have taken place by (1) die filling and (2) particle rearrangement before deformation and bonding of the discrete particles [11,12]. Densification of a powder by die filling can be expressed as

$$\ln \frac{1}{1 - D_0} \tag{4}$$

where D_0 is the relative density of a powder column at resting pressure, and usually derived from the bulk density. The combined effect of die filling and particle rearrangement at low pressures can be described by the equation

$$A = \ln \frac{1}{1 - D_0} + B \tag{5}$$

where B describes a volume reduction purely by particle rearrangement. Relative densities corresponding the processes above are D_A, which includes both die filling and particle rearrangement, and D_B, which describes only the extent of particle rearrangement.

The relative densities can be related by the equation

$$D_A = D_0 + D_B \tag{6}$$

and D_A may be calculated from

$$A = \ln \frac{1}{1 - D_A} \tag{7}$$

In his original work Heckel studied the densification of metal powder [11]. The slope, k, of the Heckel plot was intended to give a measure of the plasticity of a compressed material. Consequently, greater slopes indicated a greater degree of plasticity of material. The slope was also related to the yield strength, Y, of the material by the equation

$$k = \frac{1}{3} Y \tag{8}$$

Hersey and Rees [13] later defined the reciprocal of k to be the mean yield pressure, P_Y, in order to study whether the fragmentation of particles was the predominant compaction mechanism of powders.

The constants of the Heckel equation are commonly determined by the linear regression analysis by using the least-squares method. However, two difficulties are associated with this procedure. First, correct selection of a linear region of the function, and, second, the different weight values of measurement points on the logarithmic scale. In using the first and second derivatives of the function, the linear region can be correctly se-

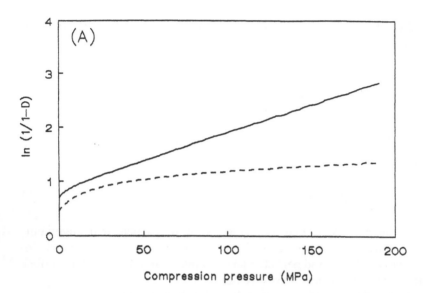

Fig. 1 Graphical presentation of the Heckel (A), Kawakita (B), and Cooper-Eaton (C) equations. The same measurement data obtained by the "at pressure" method for pregelatinized starch (Starch 1500) [——] and dicalcium phosphate dihydrate (Emcompress) [– – –] are plotted in each figure.

lected [9,14]. In the strictly linear regions of the plot the first derivative is constant and the second derivative is zero. The other possibility is to select a range of measurement points where the linear regression coefficient is higher than that decided for the threshold value. These methods are better than straight-line calculations over a constant pressure range without ensuring the actual linearity of the plot. However, if the method of constant range is used the regression coefficient should be reported, and in some cases deductions from the behavior of a compressed material can even be done with a variation in linearity.

The linearity calculations performed by using the least-squares method are based on the assumption that the function's values have an equal weight on the scale used. This is true only on an arithmetic scale but not on logarithmic scales. The problem of different weights could be avoided by using iterative techniques and nonlinear fitting for the minimizing function values. The most common methods used for iterative minimization of nonlinear equations is the simplex technique, which is often based on the Gaussian algorithm. By using commercial software, the proper mathematical procedures for calculations can easily be determined and performed.

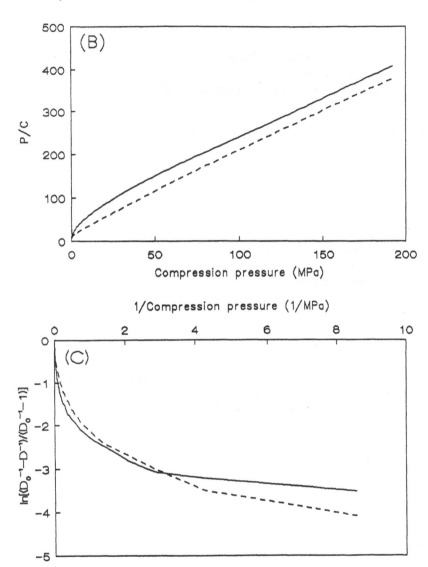

B. Kawakita Equation

Kawakita [15] introduced an equation describing the relationship between the volume reduction of a powder column and the applied pressure. The Kawakita equation is written as

$$C = \frac{V_0 - V}{V_0} = \frac{abP}{1 + bP} \tag{9}$$

where

C = degree of volume reduction
V_0 = initial volume
V = volume of a powder column under the applied pressure P
a, b = constants characteristic to powder being compressed

In addition, C can be described by means of the density and porosity as

$$C = 1 - \frac{\rho_0}{\rho_P} = \frac{\epsilon_0 - \epsilon_P}{1 - \epsilon_P} \tag{10}$$

where

ρ_0 = bulk density
ρ_P = apparent density at pressure P
ϵ_0 = porosity at the bulk state
ϵ_P = porosity at pressure P

Equation (9) can also be rearranged in linear form as

$$\frac{P}{C} = \frac{P}{a} + \frac{1}{ab} \tag{11}$$

From the graphical presentation of P/C versus P, the constants may be evaluated (Fig. 1b). The constant a is given as a reciprocal of the slope from the linear part of the plot and is equivalent to the value of C at infinitely high pressures:

$$C_\infty = \frac{V_0 - V_\infty}{V_0} = a \tag{12}$$

An intercept at the P/C-axis, extrapolated from the linear region of the plot, gives a value for $1/ab$.

The constant a gives an indication of the maximum volume reduction available and is considered to describe the compressibility of a powder, while b is considered to describe an inclination toward volume reduction. However, the actual physical meaning of the constants a and b have been in question [1,16].

Kawakita and co-workers [1,16] have also applied the equation in describing the volume reduction on tapping and vibrating processes. For the former application, the pressure, P, is replaced by the tapping number, N, and for latter by the vibration time, T. For these applications the meaning of the constants a and b become physically more significant than in Eq. [11], since the experimental conditions in tapping and vibrating processes are strictly independent of the powder tested. In die compaction a material may influence the development of the applied pressure.

C. Cooper-Eaton Equation

The Cooper-Eaton equation is [17]

$$\frac{1/D_0 - 1/D}{1/D_0 - 1} = a_1 \exp\left(-\frac{k_1}{P}\right) + a_2 \exp\left(-\frac{k_2}{P}\right) \tag{13}$$

where

D_0 = relative density at zero pressure or the bulk density divided by the particle density

D = relative density at pressure P, or the apparent density of a powder column divided by the particle density

a_1, a_2 = dimensionless constants that indicate the fraction of the theoretical maximal densification which could be achieved by filling voids of the same size (a_1) and of a smaller size (a_2) than the actual particles

The most probable pressures at which the respective densification process would occur are described by k_1 and k_2.

Cooper and Eaton [17] considered the compaction of powders to take place in two stages. First by the filling of voids of the same or larger size than the particles, where the original particles are moved and rearranged during this stage. At the second stage, the voids smaller than the original particle size are filled due to particle deformation. This phase may proceed by elastic deformation, plastic flow, or fragmentation of the compressed particles.

According to Cooper and Easton [17] densification can be described by a biexponential equation. If a nonporous powder column is produced under infinite pressure, then the sum of a_1 and a_2 is unity. If the sum is greater than unity, then a nonporous compact can be achieved at lower pressures. If the sum is less than unity, then other processes operating before complete volume reduction is achieved are indicated.

The Cooper-Eaton plots are typically biphasic linear plots, at least for hard and monodisperse metal and ceramic powders [17]. The measurements are presented graphically by the equation (Fig. 1c)

$$1n\left(\frac{1/D_0 - 1/D}{1/D_0 - 1}\right) \text{ versus } \frac{1}{P} \tag{14}$$

where a and $a_1 + a_2$ can be determined from the ordinate intercepts of the first and the second linear regions, respectively, while k_1 and k_2 can be determined from the slopes of these two linear regions.

Cooper and Eaton [17] have studied the applicability of their equation with metal and ceramic powders. With relatively hard materials and monodisperse powders their equation can clearly distinguish the two putative densification stages, where $a_1 + a_2$ is close to unity with these kinds of materials. The studies with relatively soft materials and polydisperse pharmaceutical powders have pointed out that the densification stages are not always so clearly distinguishable [18–22]. Also the deviation of $a_1 + a_2$ from unity has been more extensive, and values both over and under unity have been reported. Thus it seems that the Cooper-Eaton equation is not as suitable for soft polydisperse powders as it is for hard monodisperse powders. This may result from the densification of the powder column by several simultaneous mechanisms. Thus, it might be impossible to notice the totally separated rearrangement and deformation stages. This is often supported by the poor linear regressions of Cooper-Eaton plots with pharmaceutical materials. On the other hand, the application of linear regression analysis may be questionable, again due to the logarithmic scale of function values and reciprocals of the compression pressure.

The main advantage of the Cooper-Eaton equation is, however, the possibility to accurately study the initial stages of volume reduction, i.e., measurement points at large $1/P$ values. Thus, information can be obtained from the effects of particle surface properties, shape, and size of the densification of powder columns.

So far, there is a lack of a universal compaction equation which would be capable of describing the whole volume reduction process. Most equations seem to be limited to certain conditions in which they are applicable, commonly within a certain range of pressure or porosity. Also, the susceptibility of the equations to experimental variables differ. In the Kawakita equation the constant C is highly dependent on the initial packing state. For example, it is obvious that varying die filling methods or dies with different diameter provide a different values for constants of the Kawakita equation [23]. Also, for the Cooper-Eaton equation, the initial packing state must be determined. Another difficulty with the Cooper-Eaton equation is in measuring the constant a_1, which is highly dependent on the pressure range chosen for the determination [22,23]. In the Heckel equation, the constant k (or $1/k$) is not dependent on the initial packing of the powder column, since the constant measured after the particle rearrangement phase [23]. Moreover the Kawakita equation is valid at low pressure and large to intermediate porosity, whereas the Heckel equation gives the best fit at intermediate to high pressure and low porosity [24]. However, different opinions on the applicability of the equations have been presented, and often the extremes on both pressure and porosity have been found to give a poor fit [2,3,22–24].

IV. APPLICATIONS OF POROSITY-PRESSURE FUNCTIONS

In pharmaceutical tablet compression the ability of the powdered material to form coherent compacts by inter- and intraparticulate bonding is of prime importance. The known compression properties of materials to be tabletted give a basic indication of the tablettability and may allow prediction of the properties of the formulation. With the porosity-pressure functions, a relationship may be created between tablettability or compact formation and the volume reduction properties of powders. In this chapter the applications of the porosity-pressure functions in pharmaceutics are reviewed. The published reports deal mainly with the utilization of the Heckel equation. Thus only a few reports are available for the other equations.

A. Shapes of Heckel Plots

Three types of volume reduction mechanisms of pharmaceutical powders have been distinguished by using the Heckel equation [13,18]. The types are referred as A, B, and C (Fig. 2). Materials were categorized by compressing different particle size fractions of various powders. In type A, size fractions had different initial packing fractions and the plots remained parallel as the compression pressure was increased. In type B, the plots were slightly curved at the initial stages of compaction and later became coincidental. In type C, the plots had an initial steep linear part after which they became coincidental with only trivial volume reduction. Generally, type A behavior was related to the densification by plastic flow, preceded by particle rearrangement. In type B, powder densification occurs by fragmentation of the particles. Differences in initial packing have no effect on further densification as the initial structure of a powder column is completely destroyed by fragmentation of the particles. Type C densification occurs by plastic flow but no initial particle rearrangement is observed.

The effects of experimental variables on the Heckel plots have been studied quite intensively. Rue and Rees [25] and York [26] have published critical notes about the limitations of Heckel plots used for predicting the compaction mechanisms. Rue and Rees [25] pointed out that caution is necessary when materials are classified according to changes in the Heckel plots with different particle sizes. The predominant compaction mechanism may change with the particle size. Transitions from brittle to ductile behavior, and even vice versa, have been described [27–29]. The effect of compression time was also described by Rue and Rees [25]. Using a microfine cellulose powder, an increased volume reduction was observed with increased compression time. This was shown to indicate deformation by plastic flow. In contrast, no increase in volume reduction as a function of

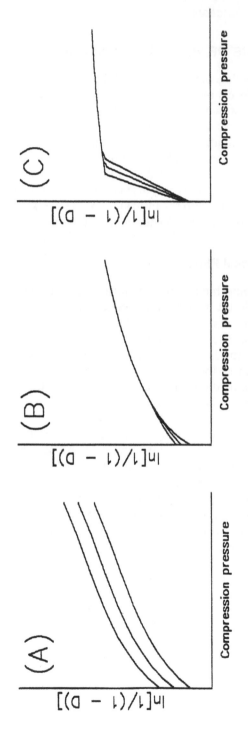

Fig. 2 Different types of compression behavior distinguished by the Heckel equation.

contact time was observed for dicalcium phosphate, this being characteristic of brittle materials. Rue and Rees [25] also proposed measuring the area under the Heckel plot (AUC) to quantify the amount of plastic deformation during compaction.

York [26] reviewed several studies where the densification behavior of crystalline lactose was evaluated and he pointed out that the general form of the Heckel plots was similar in all cases though the numerical values describing the compression process were dependent on experimental variables. York [26] also demonstrated a dramatic effect of die size on the densification behavior with microfine cellulose powder. York [26] has listed several variables such as the state and type of lubrication, rate of compaction, mode of die filling, contact time, dimensions of tools, and techniques used to measure compact dimensions, all which are necessary to take into account in tabletting studies.

Ragnarsson and Sjögren [30,31] found that the die wall friction between powder particles and wall surfaces had a clear effect on the Heckel plots of pharmaceutical powders. The upper punch pressure P in Heckel equation was noticed to be friction dependent and the mean of upper and lower punch pressures may be used instead [30]. In general, the state of lubrication, particle interactions and friction had only minor effects on Heckel plots and calculated parameters such as yield pressure, assuming the mean punch pressure was used instead of the upper punch pressure [31].

Duberg and Nyström [32] have used the initial curvature of Heckel plots as an indication of particle fragmentation. Also the correlation coefficient describing the linearity of the Heckel plots have been used for the same purpose. Thus, nonlinear plots may indicate fragmentation, and linear plots the deformation by plastic flow. Also poor packing of powder due to cohesiveness, small particle size and irregular particle shape may lead to initial curvature in a Heckel plot. In most cases, however, the rearrangement of the relatively regular particles occurs already at low pressures. Thus measurements of the nonlinearity could be a useful tool for categorizing pharmaceutical materials. This is especially true if fragmentation of multicrystalline particles or aggregates of primary particles are concerned, since they are intensively fragmented to smaller particles already at low compression pressures and this results in considerable volume reduction. At higher pressures the volume reduction becomes more difficult, and is expressed by nonlinear Heckel plots. On the other hand, the densification due to plastic flow and elastic deformation of particles, and the volume reduction of a powder column proceed both steadily, since according to Heckel function the porosity reduction occurs exponentially.

The compaction of granulated materials might be considered even more complicated than the compaction of discrete, solid primary particles. Deformation of aggregates or agglomerates is proposed to give an additional mechanism in the compaction of powder columns [33]. In literature this process has gained limited attention. Obviously, a distinction between void volumes associated with a whole powder column, single granule, and primary particle is complicated. Thus Heckel plots, for example, have been found to be difficult to interpret in terms of the physical meaning of yield pressure for granulated materials [34].

A novel treatment of porosity-pressure measurements based on the Heckel equation was proposed by Carstensen and Hou [35] and Carstensen et al. [36]. At infinite pressure, a density of a powder column was considered to approach a value different from the particle density. This was due to the assumption that only pores and voids having a diameter larger than a threshold value, as measured by mercury porosimetry, were considered to take part in volume reduction. As a consequence, a totally nonporous state at infinite pressure is not achieved.

B. Constants of Heckel Equation

In several reports it has been pointed out that the mean yield pressure, determined by the Heckel equation, cannot be taken as the actual yield point of a compressed material [19,22]. The practical value of the mean yield pressure is that it gives a general impression from the deformation tendency of a powder column. In addition, the mean yield pressure values determined by using "at pressure" and "at zero pressure" methods are often very different. This is due to the elastic deformation of a powder column considered only in the "at pressure" method.

Particle shape affects the deformation properties of a powder column. Regular particles tend to move more easily at the initial stages of compression [37]. In the case of ductile materials, irregular particles have been noticed to deform more easily than regular particles [38], and the mean yield pressure values were smaller for irregular particles.

The irregular shape and surface roughness of the particles may support the plastic flow, as deformation begins in surface asperities due to mechanical shear forces. On the other hand, the irregular shape of primary particles may reflect the existence of crystal defects which facilitate the plastic flow. The particle shape of mainly fragmenting materials did not have a remarkable effect on the yield pressure values [38]. However, for irregular particles greater fragmentation propensity measured by air permeability was observed.

Forbes et al. [39] noticed that elongated particles from a series of *para*-aminosalicylate salts had smaller yield pressures than spherical ones.

They concluded that the orientation of elongated particles and therefore the orientation of crystal planes in a powder column was more uniform and preferable for deformation. Thus the applied pressure affects more effectively on the crystal planes of the nonspherical particles.

Roberts and Rowe [9] have studied the effect of punch velocity on compaction properties of pharmaceutical powders by using a compaction simulator. The velocity was found to have a considerable effect on the Heckel plots, and on the constants derived from the equation. However, changes in the plots were greatly dependent on the material being compressed. For materials which were known to deform by plastic flow, an increase was found in the mean yield pressure as the punch velocity increased. This was concluded to result from either a change from ductile to brittle behavior, or a reduction in the amount of plastic deformation due to the time-dependent nature of plastic flow. For microcrystalline cellulose and maize starch the latter mechanism was considered to be more likely, whereas for mannitol and sodium chloride an increase in brittle behavior was suspected. No major change in the mean yield pressure was found as punch velocity was increased for materials deforming mainly by fragmentation.

Roberts and Rowe [9] introduced a strain-rate sensitivity (SRS) index which was calculated by using the mean yield pressure values from compressions with high and low velocities. The equation for the SRS index is

$$SRS = \frac{P_{Y2} - P_{Y1}}{P_{Y2}} \times 100 \tag{15}$$

where P_{Y2} and P_{Y1} are the main yield pressures at the velocities of 300 mm/s and 0.033 mm/s, respectively. According to Roberts and Rowe [9], it is possible to use the SRS index in ranking materials. Materials which are strain-rate sensitive (high SRS index) tend to deform by plastic flow. In further studies Roberts and Rowe [27,28] showed the importance of the relationship between particle size and compression behavior of paticulate materials. For example, a transition for lactose from a brittle to a ductile material was determined to occur at a median particle size of about 20 microns. In another example of a drug material a transition in the deformation mechanism of a phthalazine derivative, from a ductile to a brittle fracture, was observed as the particle size was reduced.

Using the areas under the Heckel plots (AUC), as suggested by Rue and Rees [25], McKenna and McCafferty [40] evaluated the effects of particle size and contact time on the densification of pharmaceutical powders. They concluded that a greater amount of plastic flow (larger AUC) was associated with the smaller particle size of modified starch. A compaction mechanism of spray-dried lactose was found to be brittle fracturing and independent of particle size fraction. These results from lactose contradict those of Roberts and Rowe [27]. A comparison between these two studies is, however, difficult due to the different methods used.

Duberg and Nyström [41] and Paronen [42] have attempted to evaluate the elastic behavior of powder compacts from the decompression phase of the Heckel plots (Fig. 3). The reciprocal of the slope, calculated from the downward portion of the plot, can be utilized as the yield pressure of fast elastic deformation [22,42]. This parameter gives an indication of the fast elastic behavior of a compact. In comparing the Heckel plots, obtained by using both the "at pressure" and the "ejected tablet" methods, conclusions from the total elastic recovery can be drawn. By subtracting the slope

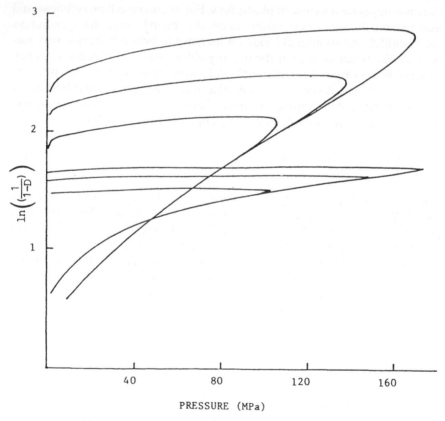

Fig. 3 Heckel plots for microcrystalline cellulose (upper curves) and dicalcium phosphate dihydrate (lower curves) obtained by "at pressure" method using three different compression pressures. (From Ref. 14.)

obtained from the "ejected tablet" method from that obtained from the "at pressure" method, and taking the reciprocal of that value, it is possible to get a parameter that describes the tendency of a compact to recover elastically [22].

C. Other Applications

Most of the reports on porosity-pressure functions are derived from single-component powders. Reports concerning the compressional behavior of multicomponent powder mixtures are, however, more rare [19,43–50]. The determination of the porosity-pressure relationships, and their respective constants, have been widely applied to the industrial R&D of tablets. It would be very useful to be able to predict the compressional behavior of multicomponent powders from the measured behavior of their individual components, though contradicting results for this possibility have been published. Humbert-Droz et al. [44] have noticed a linear relationship between the mean yield pressure and the proportions of the mixture's components. However, Ilkka and Paronen [49] noticed that although the packing fraction values determined from Heckel plots were in a linear relationship with the mixture composition (Fig. 4), there were systematic deviation from the linearity in the mean yield pressures, both positive and negative (Fig. 5). They assumed that the compression behavior of a powder mixture was often dominated by one of the mixture's components.

The applicability of the porosity-pressure functions to powder compaction can be evaluated under extreme conditions, e.g., compressing metal powders or polymeric materials under exceptionally high pressures with fast and slow loading. Page and Warpenius [51] noticed that the equations are differentially sensitive to the densification close to the particle density or infinite density of the material. Due to the logarithmic scale, the Heckel equation is much more inaccurate under these conditions than the Kawakita equation. The Kawakita and Cooper-Eaton equations can be generally found to fit better to whole compression data than the Heckel equation. The reason for this is the scaling and the biphasic nature of Kawakita and Cooper-Eaton equations, respectively.

V. CONCLUSIONS

Porosity-pressure functions have been successfully applied to obtain basic understanding of the compression behavior of particulate materials. The volume reduction process of a powder column is considered as a continuum in which the proportion of air between and inside the solid particles is decreased according to theoretical or empirical equations. Thus, a powder

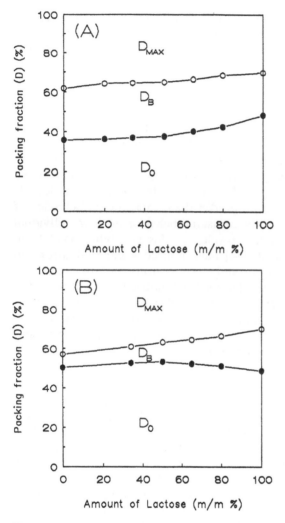

Fig. 4 The packing fractions (%) for the binary mixtures of (A) crystalline lactose and dicalcium phosphate dihydrate, and (B) crystalline lactose and modified starch. Densification due to the die filling D_0 (●) and particle rearrangement D_B (○), respectively. The sections noted with D_0 and D_B correspond the relative part of the densification of powder mixtures achieved by the respective mechanisms. At D_{MAX} a densification of powder column is obtained mainly by deformation mechanisms. Packing fraction of 100% corresponds a nonporous state, and it is not usually achieved during compression. (Data adapted from Ref. 49).

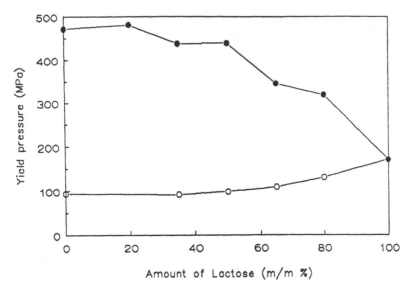

Fig. 5 The mean yield pressures for the binary mixtures of (●) crystalline lactose and dicalcium phosphate dihydrate, and (○) crystalline lactose and modified starch, obtained by "at pressure" method. (Data adapted from Ref. 49).

column is primarily seen as a body having macroscopic dimensions. On the other hand, microscopic factors, e.g., various solid-state, surface, and interface properties, are not directly included into these functions. Currently, especially the Heckel equation has been used to relate the behavior of a powder column to the fundamental properties of the compressed material. However, it is interesting to note that in many of the distinguished compaction studies for pharmaceutical powders, deviations from Heckel's theory have been utilized in obtaining additional information. So, most often the Heckel equation is strictly speaking invalid on most stages of compaction of pharmaceutical powders. It might be advantageous to utilize other porosity-pressure functions and methods of material science contemporaneously with the Heckel equation.

REFERENCES

1. K. Kawakita, and K. H. Lüdde, *Powder Technol. 4*: 61 (1970/71).
2. P. J. James, *Powder Metall. Int. 4*: 82, 145, 193 (1972).
3. V. A. Belousov, V. P. Fedin, *Pharm. Chem. J. 12*: 263 (1978).
4. R. F. Lammens, J. Polderman, and C. J. DeBlaey, *Int. J. Pharm. Tech. Prod. Mfr. 1*: 26 (1979).

5. R. F. Lammens, J. Polderman, C. J. DeBlaey, and N. A. Armstrong, *Int. J. Pharm. Tech. Prod. Mfr. 1*: 26 (1980).
6. M. J. Juslin and P. Paronen, *J. Pharm. Pharmacol. 32*: 796 (1980).
7. A. Y. K. Ho, J. F. Barker, J. Spence, and T. M. Jones, *J. Pharm. Pharmacol. 31*: 471 (1979).
8. P. Ridgway Watt, *Tablet Machine Instrumentation: Principles and Practice*, Ellis Horwood, Chichester, 1988, p. 203.
9. R. J. Roberts and R. C. Rowe, *J. Pharm. Pharmacol. 37*: 377 (1985).
10. H. K. Chan and E. Doelker, *Drug Dev. Ind. Pharm. 11*: 315 (1985).
11. R. W. Heckel, *Trans. Metal. Soc. AIME. 221*: 671 (1961).
12. R. W. Heckel, *Trans. Metal. Soc. AIME. 221*: 1001 (1961).
13. J. A. Hersey and J. Rees, *Nature PS. 230*: 96 (1971).
14. P. Paronen, *Using the Heckel Equation in the Compression Studies of Pharmaceuticals*, Proc. 4th Int. Conf. Pharmaceutical Technology, Paris, pp. 301–307 (1986).
15. K. Kawakita and Y. Tsutsumi, *Bull. Chem. Soc. Japan 39*: 1364 (1966).
16. M. Yamashiro, Y. Yuasa, and K. Kawakita, *Powder Technol. 34*: 225 (1983).
17. A. R. Cooper and L. E. Eaton, *J. Am. Ceramic Soc. 45*: 97 (1966).
18. P. York and N. Pilpel, *J. Pharm. Pharmacol. 25*: 1 (1973).
19. T. R. R. Kurup and N. Pilpel, *Powder Technol. 19*: 147 (1978).
20. N. A. Armstrong and F. S. S. Morton, *Pharm. Weekbl. Sci Ed. 1*: 234 (1979).
21. Z. T. Chowhan and Y. P. Chow, *Int. J. Pharm. 5*: 139 (1980).
22. P. Paronen and M. J. Juslin, *J. Pharm. Pharmacol. 35*: 627 (1983).
23. M. Sheikh-Salem and J. T. Fell, *J. Pharm. Pharmacol. 33*: 491 (1981).
24. R. Ramberger and A. Burger, *Powder Technol. 43*: 1 (1985).
25. J. Rue and J. E. Rees, *J. Pharm. Pharmacol. 30*: 642 (1978).
26. P. York, *J. Pharm. Pharmacol. 31*: 244 (1979).
27. R. J. Roberts and R. C. Rowe, *J. Pharm. Pharmacol. 38*: 567 (1986).
28. R. J. Roberts and R. C. Rowe, *Int. J. Pharm. 36*: 205 (1987).
29. R. J. Roberts, R. C. Rowe, and K. Kendall, *Chem. Eng. Sci. 44*: 1647 (1989).
30. G. Ragnarsson and J. Sjögren, *Acta Pharm. Suec. 21*: 141 (1984).
31. G. Ragnarsson and J. Sjögren, *J. Pharm. Pharmacol. 37*: 145 (1985).
32. M. Duberg and C. Nyström, *Acta Pharm. Suec. 19*: 421 (1982).
33. J. Van der Zwan and C. A. M. Siskens, *Powder Technol. 33*: 43 (1982).
34. M. Wikberg and G. Alderborn, *Int. J. Pharm. 62*: 229 (1990).
35. J. T. Carstensen and X-P. Hou, *Powder Technol. 42*: 153 (1985).
36. J. T. Carstensen, J. M. Geoffroy, and C. Dellamonica, *Powder Technol. 62*: 119 (1990).
37. P. York, *J. Pharm. Pharmacol. 30*: 6 (1978).
38. L. W. Wong and N. Pilpel, *Int. J. Pharm. 59*: 145 (1990).
39. R. T. Forbes, P. York, and R. Davidson, *Compaction Behavior within a Salt Series: Salts of p-Aminosalicylic Acid*, Proc. 10th Pharmaceutical Technology Conf., Vol. 1., Bologna, pp. 181–197, 1991.
40. A. McKenna and D. F. McCafferty, *J. Pharm. Pharmacol. 34*: 347 (1982).
41. M. Duberg and C. Nyström, *Powder Technol. 46*: 67 (1986).

42. P. Paronen, in *Pharmaceutical Technology*: *Tabletting Technology*, Vol. 1. (M. H. Rubinstein, ed.), Ellis Horwood, Chichester, 1987, p. 139.
43. M. Sheikh-Salem and J. T. Fell, *Int. J. Pharm. Tech. Prod. Mfr. 2*: 19 (1981).
44. P. Humbert-Droz, D. Mordier, and E. Doelker, *Acta Pharm. Technol. 29*: 69 (1983).
45. M. Duberg and C. Nyström, *Int. J. Pharm. Tech. Prod. Mfr. 6*: 17 (1985).
46. H. Vromans and C. F. Lerk, *Int. J. Pharm. 46*: 183 (1988).
47. H. C. M. Yu, M. H. Rubinstein, I. M. Jackson, and H. M. Elsabbagh, *Drug Dev. Ind. Pharm. 15*: 801 (1989).
48. J. S. M. Garr and M. H. Rubinstein, *Int. J. Pharm. 73*: 75 (1991).
49. J. Ilkka and P. Paronen, *Predictability of Compressional Behavior of Powder Mixtures Using Heckel Treatment*, Proc. 10th Pharmaceutical Technology Conf., Vol. 1., Bologna, pp. 225–238, 1991.
50. J. S. M. Garr and M. H. Rubinstein, *Int. J. Pharm. 82*: 71. (1992).
51. N. W. Page and M. K. Warpenius, *Powder Technol. 61*: 87 (1990).

4

Force-Displacement and Network Measurements

Gert Ragnarsson*

Pharmacia Biopharmaceuticals, Stockholm, Sweden

I. FORCE-DISPLACEMENT MEASUREMENTS IN COMPRESSION STUDIES

Force-displacement measurement has been one of the most popular methods for studying the compression process during tabletting. Force-displacement curves are obtained from measurements of punch force and displacement. Energy is needed for the compression of materials and formation of strong compacts. It seems logical to correlate the properties of the compact with the energy input rather than the compression pressure. The force-displacement profiles as such may also be useful as a material characteristic in preformulation work or for detecting batch-to-batch variations in the compression properties of materials.

The major reason for the interest is, however, the assumption that it should be possible to correlate the energy input, or work of compression, with the deformation and tablet-forming properties of materials. In many of the studies, the work of compression has been calculated as the total area under the upper punch force versus upper punch displacement curve [1,2], although it was realized in very early studies [3] that this value will include the work to overcome die wall friction.

A number of different methods have been suggested for evaluating and

---

Current affiliation: Astra Draco AB, Lund, Sweden

interpreting force-displacement measurement. The possibility for obtaining data from a variety of materials by carrying out large numbers of accurate studies have improved considerably during the last 10 to 15 years due to the development and improvement of the equipment and sampling technique.

This chapter presents a general description of force-displacement measurements plus discusses its possibilities in the evaluation of the compaction process and material properties. No attempt is made to present a full review of the literature.

II. INSTRUMENTATION

Force-displacement measurements requires accurate recording of upper punch force and displacement as well as lower punch force. The latter merits special discussion and will be treated later. It is possible, theoretically, to gather adequate data by using any of the different types of instrumented machines discussed in the literature (i.e., hydraulic presses, excenter machines, rotary presses, and compaction simulators). The compaction simulator meets the requirements of an ideal testing device as it allows measurements under dynamic conditions simulating normal tabletting and is suitable for small amounts of testing materials. Although studies have been carried out with compaction simulators [4,5] as well as rotary presses [6], most studies and theoretical discussions are based on experiments using single-punch excenter machines.

The instrumentation of tablet machines in general will not be discussed in this chapter. There are, however, several aspects of special importance in force-displacement measurements. The sampling procedure has to be sufficiently sensitive and accurate to register the whole compression cycle and the data sampling system should be able to handle a considerable amount of data. This is achieved, for example, by using high-quality force transducers, amplifiers, inductive displacement transducers and a AD converter connected to a suitable computer [7].

Attempts have been made to eliminate some registration error by using complex calibration methods [8,9]. The need for such complicated and tedious calibrations has been reduced by improved measuring devices and sampling equipment (amplifiers, AD converters, etc.). Nonetheless, the true accuracy of the system should be determined and special attention should be paid to error sources such as electrical noise generators.

Figure 1 illustrates the improvement in the data registration quality by a careful search for and elimination of electrical noise generators, primarily different earth connections [7]. The figure may also serve as a typical example of a force-displacement registration in a single-punch excenter press.

To make accurate force-displacement measurements, it is necessary

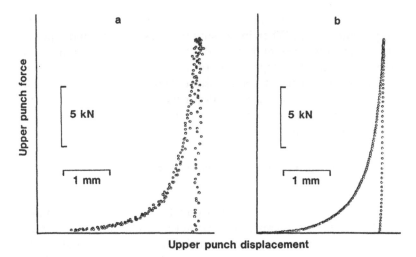

Fig. 1 Typical upper punch force versus upper punch displacement plots illustrating the improvement (b) in the force-displacement registration by elimination of electrical noise generators. (From Ref. 7).

to consider and carefully compensate for the deformation of machine parts such as punches and punch holders. Preferably, the deformation at normal machine rate should be measured. This can be obtained by compressing a flat steel disc between the punches, e.g., a 4–5-mm-thick disc, with a diameter slightly smaller than the die diameter and consisting of a steel quality that is insignificantly deformed within the pressure range of the tablet machine. The deformation during the compression can be represented by a mathematical function. If the deformation is linear, the deformation of the machine parts will be directly proportional to the compaction force, facilitating recalculation of the apparent displacement readings.

It is also advisable to study the displacement registration during the decompression phase. A hysteresis between the displacement during compression and decompression indicates that machine parts undergo time-dependent deformation at normal machine rate. Hysteresis may also indicate time dependent deformation of force transducers (such as piezo-electric force transducers). As will be discussed in more detail below, the area under the force-displacement curve during the decompression phase, i.e., the expansion work, is often used as a measure of the elastic deformation of the tested material. Failure to compensate for the deformation of machine parts may have a considerable impact on the accuracy of the calculated expansion work.

Furthermore, a time-dependent deformation means that the calibration should be carried out at the machine rate actually used in the study and at different pressure levels. When increasing the pressure, the upper punch velocity at impact with the steel disc, or the test material, should increase [10]. This may influence the slope of the force versus displacement line as well as the hysteresis during decompression. When running a real test, compensation should be made for the deformation of the measurement equipment during both compression and decompression, including the hysteresis, at the actual pressure level.

III. QUANTITATIVE ANALYSIS OF FORCE-DISPLACEMENT PROFILES

Figure 2 shows a schematical plot of upper punch force versus upper punch displacement. The work of compression, sometimes also called gross work input or upper punch work, is represented by the total area ABC. (The origin represents the point where the upper punch gets in contact with the material in the die. The areas E_1–E_3 will be used in the discussion to follow.)

Some of this work, or energy, is recovered during decompression as

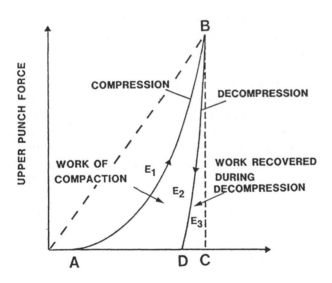

Fig. 2 Schematical plot of upper punch force versus upper punch displacement during compression and decompression including different areas (E_1–E_3) used in the characterization of the compression process.

work performed on the upper punch, which is represented by the area *DBC*. The area *ABD* should represent the apparent net work used in the formation of the compact and the work needed to overcome die wall friction. ("Apparent net work" is a more appropriate term than "net work" since the expansion is not complete.) As mentioned in the introduction, these areas have been related to the deformation properties of tested materials as well as their binding properties. A simplified description of the compression process identifies three components; elastic deformation, plastic deformation and fragmentation, as illustrated in Fig. 3.

The work needed to deform an *elastic* material will be completely recovered during the decompression phase and there will be no net work used in the formation of a compact. On the other hand, *plastic* deformation, with or without *fragmentation*, yields a net work input. Materials that are irreversibly deformed to a large extent should be expected to give a large work input.

The absolute values of the areas in Fig. 2 will of course be dependent on the force level. In a number of papers [11–14] it has been shown that the upper punch force versus displacement generally follows a hyperbolic form and can be described by hyperbolic constants. The constants should be independent of the pressure level and attempts have been made to correlate them with the properties of different mixtures [14].

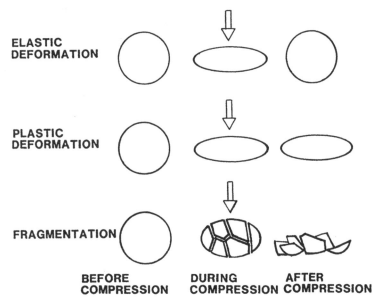

ELASTIC DEFORMATION

PLASTIC DEFORMATION

FRAGMENTATION

| BEFORE COMPRESSION | DURING COMPRESSION | AFTER COMPRESSION |

Fig. 3 Schematical illustration of processes that take place during compression.

A more common approach has been to use ratios between different areas in the force displacement plot. Dürr et al. [15] used the areas E_1, E_2, and E_3 (see Fig. 2) and the sum of these areas ($E_{max} = E_1 + E_2 + E_3$) to characterize the compressibility of powders and compacts. According to the authors' experience, E_1 should be as small as possible and the ratios ($E_2 + E_3$)/E_1 and E_2/E_3 should be as large as possible. An energy ratio (Energieverhältnis) was defined as $E_{2\%}/E_{3\%}$ $E_{1\%}$. To make the value E_2 pressure-independent they used the equation $E_2 = E_2\infty P/(b + P)$, where P is the force, b a constant and $E_2\infty$ the energy transferred to the tablet at a infinitely large pressure. They suggested that a small b (indicating a fast increase in E_2 at increasing pressure) and a large $E_2\infty$ are favorable. In an attempt to reduce the influence of die wall friction, they used the lower punch force in the calculations.

Stamm and Mathis [16] used similar methods to calculate plasticity constants. Plasticity (P1) was calculated as $100E_2/E_2 + E_3$; i.e., the ratio of the work used to form the compact to the total work input. A high P1-value indicates that a large part of the energy input is utilized in irreversible deformation of the material. As the measure is not pressure independent, they characterized a variety of materials by compressing a series of tablets with different tablet hardness. Like Dürr et al. they used lower punch values. They ranked a large number of excipients. The "plasticity" values ranged for example from 58.2 (mannitol) to 94.1 (Avicel PH 101).

Doelker et al. [17,18] used the same plasticity coefficient as Stamm and Mathis but called it PL1 and defined it as the ratio between apparent net work input and the "lower punch work." The apparent net input is equal to the upper punch work minus the expansion work and is thus equivalent to E_2 in the equations above. The "lower punch work" is the area under the lower punch versus upper punch displacement curve. The term "lower punch work" is unfortunate since the lower punch cannot produce any physical work as long as it is stationary. The main reason to use lower punch data in this and the previous studies is, however, that these data have been suggested as being independent of die wall friction [19]. The validity of this approach will be discussed later on.

A second plasticity parameter, PL2, was calculated as the net input-to-upper-punch-work ratio. To calculate the net input they used lower punch values and a double-compression technique suggested by de Blaey and Polderman [19,20]. De Blaey and Polderman assumed that only elastic deformation will take place during a second compression of the tablet. According to this assumption, the compression work during the second compression should represent true expansion work and thus make possible improved calculation of the actual net input used for permanent deformation and particle bonding.

The authors stated that incorporation of a binder gave higher values for each kind of work, i.e., increased the resistance to deformation, and gave higher plasticity coefficients. They suggested that a comparatively greater proportion of energy is consumed by tablets made from granulations. They also concluded that tablet strength was related to the net work input. Furthermore rank orders and not the magnitude of mechanical strength, can be predicted because "the work taken up by the tablet is not a direct indication of the work used for the formation of bonds and of the bonding capacity of the components."

In a large number of studies, force-displacement measurements have been used to characterize the compression properties and to show correlations between net work values, or alternatively different work ratios, and tablet strength [2,5,21–38]. Most studies concluded that improved tablet strength was associated with a large degree of plasticity.

There are sound reasons to consider the accuracy and relevance of the findings reported in many force-displacement studies. As discussed in the instrumentation part of this chapter, it is difficult to measure expansion work. Small errors in the compensation for the deformation of machine parts will have a large impact. The influence of such errors will be very large if ratios between the expansion work and measures of the compaction work such as E_2/E_3 are used. E_3 usually assumes a value which is only a fraction of E_2. Thus dividing a large value by a small one, itself measured with considerable uncertainty, runs the risk of considerable error. Accurate expansion work measurement is made more difficult still by time-dependent, incomplete tablet expansion within the die and die wall friction.

An in depth review of the literature reveals that different techniques and assumptions have been used in the calculation of net work input as well as in the interpretation of the results. It is thus justified to discuss the net work concept in more detail.

IV. CALCULATION OF NET WORK FROM FORCE-DISPLACEMENT PROFILES

De Blaey, Polderman, and co-workers have proposed a definition of the different energy-consuming steps during compression and introduced methods to calculate the net work input. According to their definition [19], work is consumed during compression

    a. For arriving at the closest possible proximity of the particles of the granulate

    b. By friction between the particles

    c. By friction with the die wall

    d. By plastic deformation
    e. By elastic deformation

(Similar lists have been presented by others, such as Führer and co-workers.) It may also be argued that energy used for fragmentation and energy released due to reduced surface energy during bonding (see Sect. VI) should be included when discussing the overall energy balance. This list above, however, been of fundamental importance in the attempts to define net work. The authors further assumed that the steps a and b, which can be regarded as particle rearrangement and interparticulate friction, can be neglected. The work used to overcome die wall friction (c) and the work recovered during decompression (e) should, on the other hand, be subtracted from the gross work input to form a net work (NETW) input. They consequently assumed that the calculated NETW represented the energy used for plastic deformation and bond formation. The NETW calculation should thus be very informative in characterising a material provided that these assumptions are correct.

    As discussed above, we first have to subtract the work needed to overcome friction (FW) and the expansion work (EXPW) from the total work consumption (see Fig. 4).

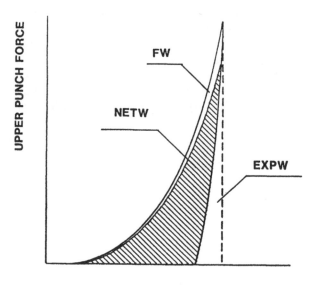

UPPER PUNCH DISPLACEMENT

Fig. 4   Force-displacement plot illustrating the net work (NETW), work of friction (FW) and work recovered during expansion (expansion work, EXPW).

De Blaey and Polderman [19] have suggested that the work of die wall friction may be calculated according to Eq. (I) in Fig. 5. This method was criticized by Järvinen and Juslin [39], who tried to derive a work of friction based on the movement of the particles in contact with the die wall and the force acting on the particles rather than the force and displacement of the upper punch. They derived Eq. (II) in Fig. 5.

Equation (I) in Fig. 5 implies that the FW should be equal to the area limited by the curves D_SA and D_SB in Fig. 5, i.e., the area D_S-A-B. The lower punch reading should consequently be independent of die wall friction and useful as force measure in NETW calculations.

Invoking the second equation results in values approximately half of those obtained by de Blaey and Polderman (at low or moderate friction levels).

The assumption that the frictional force should not coincide with the movement of the upper punch appears reasonable. Only particles in the upper layer of the compact are capable of moving the same distance as the upper punch. Particles at the lower punch surface will be stationary.

The method by de Blaey and Polderman has been the more widely used as discussed previously. The validity of the two methods has been the

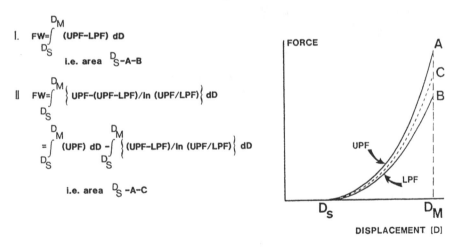

$$\text{I.} \quad FW = \int_{D_S}^{D_M} (UPF - LPF)\ dD$$

i.e. area D_S-A-B

$$\text{II} \quad FW = \int_{D_S}^{D_M} \left\{ UPF - (UPF - LPF)/\ln(UPF/LPF) \right\} dD$$

$$= \int_{D_S}^{D_M} (UPF)\ dD - \int_{D_S}^{D_M} \left\{ (UPF - LPF)/\ln(UPF/LPF) \right\} dD$$

i.e. area D_S -A-C

Fig. 5 Calculation of work needed to overcome die wall friction (FW) according to I (de Blaey and Polderman [19]), II (Järvinen and Juslin [39]), illustrated with a schematical force-displacement plot. The following abbreviations have been used: D, the displacement of the upper punch, measured relative to the lower punch; D_S, the point at which the force rises from zero; D_M, the maximum displacement of the upper punch; UPF, the upper punch force; LPF, the lower punch force.

subject of some discussion, however, [40, 41] since neither method had been confirmed empirically.

A recent series of experiments was carried out to resolve this controversy [34] by holding constant all experimental factors except the die wall friction. Under these conditions true net work should be constant even though the friction varies, provided that FW is calculated correctly.

The experiments were carried out using a die prelubricated by compressing a number of tablets containing a large excess (approx 25%) of magnesium stearate. It is known that magnesium stearate forms a lubricant layer that can resist a number of repeated compressions [42]. When a series of tablets of an unlubricated material are compressed in a prelubricated die, the lubricant layer will gradually be worn off, resulting in a gradual increase in die wall friction. This is illustrated in Fig. 6.

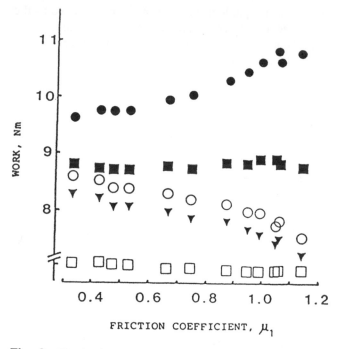

Fig. 6 Work of compression versus friction coefficients of consecutive anhydrous lactose tablets compressed in a prelubricated die. ●, upper punch work; ○, "lower punch work"; □, expansion work; ▼ net work (NETW) when work of friction (FW) is calculated according to de Blaey and Polderman [19]; ■, NETW when FW is calculated according to Järvinen and Juslin [39]. (Adapted from Ref. 34.)

As can be expected, friction increased as the lubricant wore off and increased the total work input. The Järvinen and Juslin method to calculate FW resulted in an approximately constant NETW (filled squares in Fig. 6). The de Blaey and Polderman method produced a decreasing NETW value. Thus the latter method appears to overestimate the influence of die wall friction, which seems theoretically sound.

As predicted by Järvinen and Juslin, the FW values were close to half of those calculated according to Eq. (I) in Fig. 5. This was found to be true not only at low friction levels but over the whole friction range, even at friction levels where further compressing was impossible. As a practical consequence, the mean of the upper punch force (UPF) and the lower punch force (LPF); i.e., the mean punch force (MPF) can be regarded as a compression force measure that is independent of die wall friction. The NETW may thus be calculated by a simple integration of the MPF versus upper punch displacement plot and subtraction of the expansion work (EXPW). The usefulness of MPF as a general friction independent measure of compaction force has been demonstrated in other studies [43,44] and is gaining wider acceptance [5].

The axial expansion of a tablet within a die is imcomplete during the decompression phase which makes the measurement of the EXPW difficult. For example, comparison of the tablet height after ejection and in the end of the decompression phase, have shown that only about 65% of the total axial expansion of anhydrous lactose tablets took place in the die [7]. This value was obtained with a well-lubricated die and decreased to about 40% at high die wall friction.

De Blaey and Polderman suggested [19] that the EXPW should be measured by using a double-compression technique. The tablet will continue its expansion in the die when the first compression is completed and the area under the decompression curve (EXPW$_1$) should thus underestimate the true expansion work. They suggested that the upper punch work during a second compression should represent the true EXPW. A problem with the double-compression technique is that a second compression may not only include elastic deformation but also some plastic deformation [45]. In addition, the properties of the compact may change at repeated compression [46].

When the friction is high, neither the single- nor the double-compression technique appears to give the complete EXPW due to incomplete expansion in axial direction. At low friction, obtained by prelubrication of the die or lubrication of the test material, only small differences between the two methods have been obtained [19].

When all these aspects are considered there appears to be little justification in using the more complicated double-compression technique. I recommend that force-displacement measurements are carried out using the single-compression technique in prelubricated dies giving low die wall friction without interference with the bonding properties of the test material.

The calculated EXPW will underestimate the true expansion work, not only due to incomplete expansion in axial direction. The tablets will also expand in radial direction after ejection. It appears, nevertheless, reasonable to subtract the calculated EXPW in NETW calculation. The error in the NETW calculation due to an underestimated EXPW should in general be small. Empirically, the EXPW tends to be only a few percentage of the total compression work for pharmaceutical materials (e.g., 0.5–10% of the NETW values for a variety of materials in [7]).

The importance of the criticism put forward in this chapter regarding established methods to calculate FW and EXPW is dependent on the aim of the studies, experimental conditions and the general accuracy of the measurement system. It is, for example, reasonable to believe that data registration limitations have affected results and conclusions more than the choice of calculation methods in some of the early studies.

Of more fundamental interest are the basic assumptions about the different processes taking part in powder compaction and their contribution to the NETW.

V. INFLUENCE OF INTERPARTICULATE FRICTION AND BONDING ON CALCULATED NET WORK

As discussed previously, the introduction of the NETW calculation (i.e., [20,21,24]) was an attempt to calculate the work used for plastic deformation and for bond formation. While it was considered necessary to subtract FW and EXPW from the total work input, the work needed for particle rearrangement and interparticulate friction were assumed to be negligible. Further deformation, requiring a net work input, should reflect the inherent deformation properties of the test material.

In a crystal, plastic deformation or fragmentation takes place when the intermolecular forces are exceeded. Schematically, the plastic deformation can be described as taking place along slip planes inside the crystals as, e.g., reviewed by Moldenhauer et al. [47].

A compact is deformed plastically if it undergoes irreversible deformation by plastic flow or fragmentation of the particles (crystals) in the compact and shows little elastic recovery during decompression. It may thus be reasonable to distinguish between plastic deformation of a crystal and of a compact consisting of a large number of particles, unless it can clearly be demonstrated that interactions between the particles play an insignificant role in the compression event.

Attempts have been made to vary the interaction of interparticulate friction and bonding among the particles in a compact with minimal effect on the composition by varying the particle size, the lubrication, and the moisture content [35]. The influence on both NETW and Heckel plots were studied. The Heckel equation, which will be discussed in more detail elsewhere in this book, is one of the experimental powder compression equations that relates applied pressure to the volume change during compression.

For example by adding a small amount of a lubricant, 0.5% magnesium stearate to dicalcium phosphate, Emcompress, a small decrease in the tablet strength was obtained in addition to reducing the interparticulate friction. The result was a small reduction in the NETW (Fig. 7). The reduction in the NETW was not due to changes in the EXPW, which was supported by measurements of the total axial expansion of the tab-

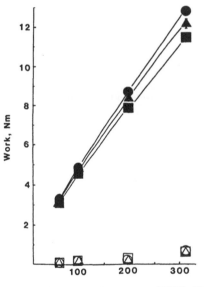

AXIAL ELASTIC RECOVERY, %			
MPP	EC	EC1	EC30
55	2.2	2.1	2.1
96	3.1	3.0	3.0
198	2.4	2.3	2.4
313	3.3	3.4	3.5

Mean punch pressure (MPP), MPa

	Emcompress (EC)	Emcompress + 0.5% Magnesium stearate Mix. time 1 min. (EC1) Mix. time 30 min. (EC)30	
Net work	●	▲	■
Expansion work	○	△	□

Fig. 7 Net work and expansion work when compressing dibasic calcium phosphate dihydrate, Emcompress, with or without small amounts of magnesium stearate (0.5%) admixed for 1 and 30 minutes. (Adapted from Ref. 35.)

lets. As a precaution die wall friction was kept low by using prelubricated dies. Mean punch pressure was used as a measure of the compaction pressure.

The bonding properties were drastically reduced by adding magnesium stearate to starch, Sta-Rx, known to be very sensitive to lubricants. Total elimination occurred after mixing for 30 min (Fig. 8). As can be seen in Fig. 8 this reduced particle to particle interaction resulted in a significant reduction in the NETW. It seems very unlikely that a small amount of a lubricant and its admixing time should affect the ability of the individual units of the starting material to deform plastically. In contrast to earlier suggestions, interparticulate effects during compaction appears to significantly affect the NETW. Tests on other materials verified that decreased particle to particle interaction, obtained by changing the surface properties

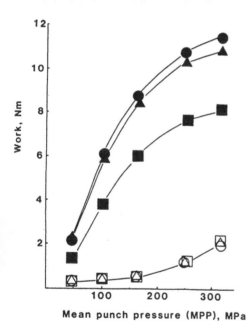

AXIAL ELASTIC RECOVERY, %			
MPP	SX	SX1	SX30
48	17.0	16.9	–
103	16.1	16.4	–
164	15.9	16.0	–
251	17.8	17.8	–
316	19.3	19.6	–

	STA-RX (SX)	STA-RX + 0.5% Magnesium stearate	
		Mix. time 1 min. (SX1)	Mix. time 30 min. (SX30)
Net work	●	▲	■
Expansion work	○	△	□

Fig. 8 Net work and expansion work when compressing corn starch, Sta-RX 1500, with or without small amounts of magnesium stearate (0.5%) admixed for 1 and 30 min. (Adapted from Ref. 35.)

of particles in the compact will reduce the resistance to consolidation and thereby lower the NETW.

There is no simple correlation between the NETW and the deformation properties of a material since the NETW is also substantially affected by the particle to particle interaction. The experiment with lubricated Sta-RX suggests that about 30% of the NETW may be due to such interactions!

Is it possible that particle-to-particle interaction can be this large? Our standard model assumes that the force needed to overcome particle-to-particle interactions is small compared with the force needed to deform particles in the compact. Yet the results of the lubricated Sta-RX experiment appear to violate this assumption.

Let us have another look at the Sta-RX and magnesium stearate experiment, now presented as force displacement curves (Fig. 9). We see how the resistance to compression is actually reduced, especially during the initial part of the compression. The reduction in *force* needed to compress

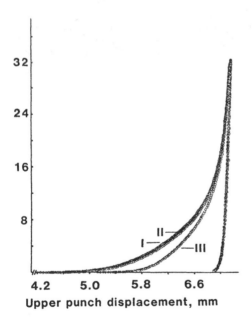

Upper punch force, kN

32

24

16

8

II
I —III

4.2   5.0   5.8   6.6

Upper punch displacement, mm

Fig. 9 Force-displacement curves of Sta-RX compressed with or without small amounts of magnesium stearate. I, pure Sta-RX; II, 0.5% magnesium stearate admixed for 1 min; III, 0.5% magnesium stearate admixed for 30 min. (Adapted from Ref. 35.)

the lubricated material is indeed small, but takes place over a long distance. The result is a substantial reduction in NETW.

The effect of interparticulate friction and bonding may perhaps be explained with the aid of a simplified figure (Fig. 10). When the particles within the die are compressed, irregular rearrangement takes place. In the beginning of the process there will be both particles in mutual contact and interparticulate voidage.

Further particle rearrangement can reduce the porosity of the compact, registered as a punch displacement, without particle fragmentation or plastic deformation. Such rearrangement will, however, probably be counteracted by high bonding or friction forces in the contact points between particles. Increased friction and bonding may thereby increase the work needed to compress the material independent of any changes in the deformation mechanism. Any change in the surface properties that promotes bonding should consequently give an increase in the NETW due to higher total resistance to deformation of the large number of particles that forms a

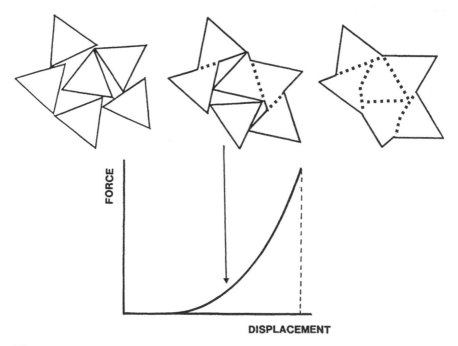

Fig. 10 Schematical figure illustrating the influence of interparticulate friction and bonding on force-displacement plots. Bonding or friction forces in contract points (dots) are suggested to have a significant effect on the work needed for further compression (see text).

compact. If this theory is correct, it is obvious that the importance of the deformation mechanism has been overestimated in many of the earlier NETW measurements while the effect of surface properties, bonding mechanism and bonding strength has been underestimated.

VI. ENERGY BALANCE DURING COMPACTION

Apart from the attempts to correlate force-displacement data with the tablet-forming properties of materials, such curves may be used to study the energy balance in tabletting. All energy used to compress a material will be released as heat if no changes in the energy content of the material takes place. The work of compaction, Wc, will then be equal to the heat released during compaction, Qc. i.e.,

$Ec = Wc - Qc$
Ec = energy change during compression
Wc = work done on the powder (work of compression)
Qc = heat released by the system
(Adapted from Coffin-Beach and Hollenbeck [48])

Führer and Parmentier [13] estimated that about 90% of the work of compression was released as heat.

Coffin-Beach and Hollenbeck [48] have more recently made very interesting studies of the energy balance with the aid of a highly sensitive compression calorimeter. By measuring the work of compression during compaction at a constant pressure, they found that the energy released as heat was *larger* than the energy of compression for all materials tested. They measured only the effect during the compression phase and not during decompression but efforts were made to compensate for energy changes associated with the deformation of machine parts.

The extent to which the heat released *exceeded* the work of compression was termed the *energy of formation* as it was assumed that this energy was equal to the reduction in surface energy due to bonding. For example, microcrystalline cellulose, Avicel, gave a high energy of formation while dicalcium phosphate, Di-Tab, known to fragment to a large extent during compression, gave considerably lower values. It was further suggested that fracture and bonding balanced each other at forces below approximately 10,000 N for dicalcium phosphate, while particle recombine and bond at higher pressures resulting in increased *energy of formation*. The *energy of formation* correlated with the tensile strength for each material but it appears not to be a simple general correlation.

As suggested by the authors, these type of measurements may not only give quantitative evaluation of the energetics of the interparticulate interactions due to bonding but possibly also some information about the true particle area involved in bonding.

Simultaneous measurements of compression work and heat released during compression should be rather tricky, however. During compression the punches will deform elastically. They will thereby act as springs, as very roughly illustrated in Fig. 11. The upper punch is drawn as a relaxed spring at the beginning of compression and as a compressed spring at maximum pressure (the total deformation of machine parts has for example been shown to be over 200 μm at a pressure of 250 MPa using 11.3-mm punches in an excenter machine [7]).

Let us assume, for simplicity, that all elastic deformation of the equipment is caused by the upper punch. The true deformation of the test material, D_I, can then easily be obtained by subtracting the known elastic deformation of the punch making it possible to calculate the compression work Wc_I. Unfortunately, it take some time to measure the heat released and it appears necessary to stop the compaction process at a suitable pressure level. During the heat measurement the material tends toward further deformation, as the deformation process in general is a time-dependent process and the material is under pressure of the elastically deformed punch. The deformation of the material generates additional work of com-

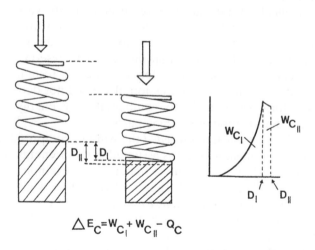

$$\triangle E_C = W_{C_I} + W_{C_{II}} - Q_C$$

Fig. 11 Schematic illustrating the problems of measuring energy change during compression (ΔE_C) under static conditions. The punch, which is elastically deformed, and illustrated as a spring in the figure, will keep the material under pressure and may cause further plastic deformation ($D_{II} - D_I$). This will cause additional compaction work (Wc_{II}) which should be included in the energy balance where Qc is equal to heat released by the system.

pression which should be included in the energy balance equation above. It may be difficult to measure this work as the displacement transducer is mounted some distance from the punch tip.

Figure 11 shows punch deformation. The punch may expand the distance $D_{II} - D_I$ without giving any visible increase in displacement. The deformation should, however, give a decrease in the force reading. By using equipment capable of keeping the punch force exactly constant by a constant slow compression, governed by the deformation rate of the test material, it might be possible to measure the total compression work, i.e., including Wc_{II}. If this extra compression work is not considered, the energy of formation of materials such as cellulose with a very-time-dependent deformation [49] should be overestimated when compared with a material with no or little time-dependent deformation such as dicalcium phosphate [49].

Coffin-Beach and Hollenbeck used the static method in a tablet press with a slow rate of compression. Sufficient stress relaxation may have occurred during the compression to give an acceptable calculation of the total work input (personal communication with the authors).

VII. SUMMARY AND CONCLUSION

Force-displacement measurements have received much attention, especially during the 1970s and early 1980s. The basis for this interest has mainly been the proposed possibility to characterize the compression and deformation properties of materials and to obtain predictive data. Modern data registration and computer systems have given us increasingly accurate raw data. However, there remain pitfalls and problems in the measurement techniques, as well as with the evaluation of obtained data, which keep force-displacement and NETW registration from being uncomplicated tools.

Based on published data it appears reasonable to conclude that the NETW may be useful for detecting batch-to-batch variations in the compaction properties of materials due to its high sensitivity to *both* inter- and intraparticulate properties, good reproducibility and low dependence upon die wall conditions.

Results obtained during the last century indicate that the predominant assumption that the influence of particulate interaction during compression can be neglected is incorrect. Additional methods will be needed [35] to determine whether interlot variations in the NETW consumption during compaction are due to altered surface properties or intraparticulate changes (e.g., polymorphism and lattice defects). This should be kept in mind when considering the predictive value of NETW measurements.

A natural area where NETW calculations should be useful is in the evaluation of crystallographic changes during compaction. In these cases it appears obvious that the work, or energy, used during compression is of far more interest than the maximum force or pressure used in the process. This possibility has been mainly overlooked except for a few studies [50].

Simultaneous measurement of work of compression and heat released during compression is an interesting approach that may benefit from new and very sensitive measurement devices. The technique will not only increase our knowledge about thermal effects in compaction but may, theoretically, yield information about surface energy reduction due to bonding and thus indicate the true bonding surface in tablet formation. It would be of particular interest to see a comparison between the results of such studies and the results from some of the new approaches to measuring the true bonding area which are discussed elsewhere in this book.

REFERENCES

1. H. Heins, W. Ott, and C. Führer, *Pharm. Ind. 31*: 155 (1969).
2. N. A. Armstrong and F. S. S. Morton, *J. Powder Bulk Solids Technol. 1*: 32 (1977).
3. E. Nelson, L. W. Busse, and T. Higuchi, *J. Am. Pharm. Assoc. Sci. Ed. 44*: 223 (1955).
4. S. D. Bateman, The effect of compression speed on the properties of compacts, Ph.D. thesis, School of Pharmacy, Liverpool Polytechnic, 1988.
5. J. S. M. Garr, and M. H. Rubinstein. *Int. J. Pharm. 64*: 223 (1990).
6. R. J. Oates, and A. G. Mitchell, *J. Pharm. Pharmacol. 41*: 517 (1989).
7. G. Ragnarsson, Evaluation of tabletting properties in preformulation and early formulation work, Ph.D. thesis, Uppsala University, 1985.
8. R. F. Lammens, J. Poldermen, and C. J. de Blaey, *Int. J. Pharm. Tech. Prod. Mfr. 1*: 26 (1979).
9. R. F. Lammens, J. Polderman, C. J. de Blaey and N. A. Armstrong, *Int. J. Pharm. Tech. Prod. Mfr. 1*: 26 (1980).
10. P. Colombo, U. Conte, C. Caramella, A. La Manna, J. C. Guyot, A. Delacourte, B. Devise, M. Traisnel, and M. Boniface, *Il Farmaco - Ed.Pr. 33*: 531 (1978).
11. C. Führer, (1965). *Dtsch. Apoth. Ztg. 105*: 1150 (1965).
12. W. Parmentier, Untersuchungen zur Gesetzmässigkeit des Kraftferlaufs bei der Tablettierung, Diss., der Technischen Universität Carolo-Wilhelmina zu Braunschweig, 1974.
13. C. Führer, and W. Parmentier, *Acta Pharm. Technol. 23*: 205 (1977).
14. C. Führer, G. Bayraktar-Alpmen, and M. Schmidt, *Acta Pharm. Technol. 23*: 215 (1977).
15. M. Dürr, D. Hansen, and H. Harwalik, *Pharma. Ind. 34*: 905 (1972).
16. A. Stamm and C. Mathis, *Acta Pharm. Technol. Suppl. 1*: 7 (1976).

17. E. Doelker, *Pharm. Acta Helv. 53(6)*: 182 (1978).
18. E. Doelker, R. Gurny, and D. Mordier, *Acta Pharm. Technol. 26*: 155 (1980).
19. C. J. de Blaey, and J. Polderman, *Pharm. Weekbl. 105*: 241 (1970).
20. C. J. de Blaey, and J. Polderman, *Pharm. Weekbl. 106*: 57 (1971).
21. C. J. de Blaey, M. C. B. van Oudtshoorn, and J. Polderman, *Pharm Weekbl. 106*: 589 (1971).
22. C. J. de Blaey, A. B. Weekers-Andersen, and J. Polderman, *Pharm. Weekbl. 106*: 893 (1971).
23. J. T. Fell, and J. M. Newton, *J. Pharm. Sci. 60*: 1428 (1971).
24. J. Polderman, and C. J. de Blaey, *Farm. Aikak. 80*: 111 (1971).
25. J. Gillard and M. Roland, *Pharm. Acta Helv. 52*: 154 (1971).
26. M. Dürr, *Acta Pharm. Technol. 22*: 185 (1976).
27. Erhardt, and E. Schindler, *Pharm. Ind. 41*: 1213 (1979).
28. Erhardt, and E. Schindler, *Pharm Ind. 42*: 1213 (1980).
29. G. Ragnarsson, and J. Sjögren, *Int. J. Pharm. 12*: 163 (1982).
30. I. Krycer, D. G. Pope, and J. A. Hersey, *Int. J. Pharm. 12*: 113 (1982).
31. I. Krycer, D. G. Pope, and J. A. Hersey, *J. Pharm. Pharmacol. 34*: 802 (1982).
32. I. Krycer, and D. G. Pope, *Powder Technol. 34*: 39 (1983).
33. I. Krycer, and D. G. Pope, *Powder Technol. 34*: 53 (1983).
34. G. Ragnarsson and J. Sjögren, *J. Pharm. Pharmacol. 35*: 201 (1983).
35. G. Ragnarsson, and J. Sjögren, *J. Pharm. Pharmacol. 37*: 145 (1985).
36. N. Kaneniwa, K. Imagawa, and M. Otsuka, *Chem. Pharm. Bull. 32*: 4986 (1984).
37. T. Cutt, J. T. Fell, J. R. Rue, and M. S. Spring, *Int. J. Pharm. 49*: 157 (1989).
38. M. Otsuka, T. Matsumoto, and N. Kaneniwa, *J. Pharm. Pharmacol. 41*: 665 (1989).
39. M. J. Järvinen, and M. J. Juslin, *Farm. Aikak. 83*: 1 (1974).
40. R. F. Lammens, T. B. Liem, J. Polderman, and C. J. de Blaey, *Powder Technol. 26*: 169 (1980).
41. M. J. Järvinen and M. J. Justin, *Powder Technol. 28*: 115 (1981).
42. A. W. Hölzer, and J. Sjögren, *Int. J. Pharm. 3*: 221 (1981).
43. G. Ragnarsson, and J. Sjögren, *Int. J. Pharm. 16*: 349 (1983).
44. G. Ragnarsson, and J. Sjögren, (1984). *Acta Pharm. Suec. 21*: 141 (1984).
45. I. Krycer, D. G. Pope, and J. A. Hersey, *Drug Dev. Ind. Pharm. 8*: 307 (1982).
46. N. A. Armstrong, N. M. A. H. Abourida, and L. Krijgsman, *J. Pharm. Pharmacol. 34*: 9 (1982).
47. H. Moldenhauer, H. Kala, G. Zessin, and M. Dittzen, *Pharmazie 35*: 714 (1980).
48. D. P. Coffin-Beach, and R. G. Hollenbeck, (1983). *Int. J. Pharm. 17*: 313 (1983).
49. J. E. Rees and P. J. Rue, *J. Pharm. Pharmac. 30*: 601 (1978).
50. M. Otsuk, T. Matsumoto, and N. Kaneniwa, *J. Pharm. Pharmacol. 41*: 665 (1989).

5

Viscoelastic Models

Fritz Müller

Pharmazeutisches Institut der Universität Bonn, Bonn, Germany

I. INTRODUCTION

A comparison of load (force, stress) versus time (σ/t) and displacement (strain) versus time curves (ϵ/t) often shows that their maxima do not occur simultaneously (Fig. 1). The load reaches its maximum first and decreases to a significantly lower value when the upper and lower punches are in the closest position (Schierstedt [1], Schierstedt and Müller [2], Caspar [3], Müller and Caspar [4], Ho and Jones [5]). This effect can also be noticed in the time independent load versus displacement-curves (σ/ϵ) where the coordinates of the maximal load and maximal displacement are different [1, 2, 6, Fig. 2].

Experiments involving various agents and excipients show that the described time shift is typical for the behavior of compressed materials (Müller and Caspar [4]). Macromolecular substances, among them important tabletting materials such as starches and celluloses, comply with this empirical experience (Cole, Rees, and Hersey [7]). First observations in this field have been reported by David and Augsburger [8], Wiederkehr [9], and Rees and Rue [10]. These experiments were performed in a static manner, for example by stopping the machine at its lowest point, the dead center position of the upper punch and observing the time-dependent change of the load. These phenomena can only be explained by the influence of the upper punch proceeding into the already compacted powder

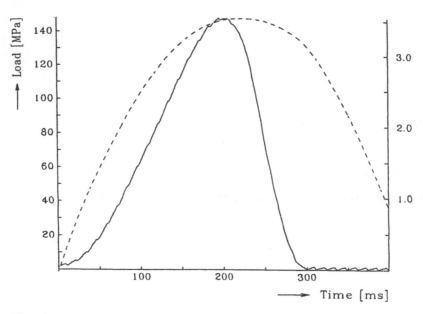

Fig. 1 Load vs. time (solid line) and displacement vs. time (dashed), raw data for Starch.

bed where the resistance diminishes, giving way due to the characteristics of the substance, alternatively there could be entrapped air which will escape as a function of the gas pressure in the die, as supposed by Casahoursat et al. [11,12]. Continuum mechanics assigns this type of phenomena to elasticity, viscosity, viscoelasticity, or viscoplasticity.

Furthermore, the curves indicate, that the upper punch has not yet reached the original position with no load, when the load reaches zero again at the point of punch lift-off. The difference between the initial and final displacement can serve as a measure of plastic deformation.

Elastic deformation is displayed by the symmetry of the load versus time curve or by the declining section of the load versus displacement curves. Using appropriate parameters corresponding to the shape of the load versus time curves facilitates the evaluation of tablet mass deformation. Emschermann [13] and Emschermann and Müller [14] investigated various tablet excipients using a single-punch machine. The authors subdivided the area under the load versus time curve into two sections: one from the origin to the maximum of compression force and the other one from there to the end of the process. The ratio of the two areas represents a measure of curve symmetry. It allows us to discern substances, that

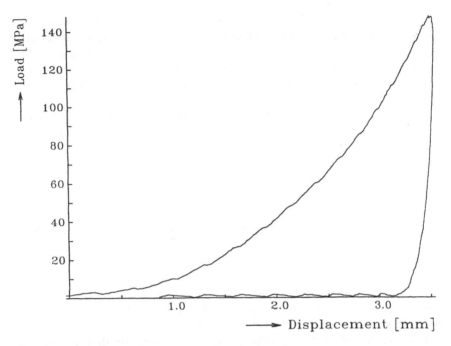

Fig. 2 Load vs. displacement, raw data for Starch.

undergo plastic rather than elastic deformation, from substances which behave vice versa. An ideally elastic material should produce a symmetrical curve and an area ratio of unity. Tenter [15] and Schmidt et al. [16] applied this area evaluation principle to load versus time curves of a rotary press in some papers. Since pinpointing the compaction maximum is difficult in this case due to the trapezoidal formed curve, they calculated the point of subdivision to the time coordinate of the center of gravity under the curve. Subsequently, depending on the mass deformation, the area ratio now assumed values below and above unity. More recently the authors divided the area into four sections in order to get more information [17]. Dietrich and Mielck [18] introduced sensitive quantitative parameters to give an interpretation of the shape of a modified Weibull function to get time-independent results. All attempts to obtain the influence of time on the slope of a Heckel-plot [19] as strain rate sensivity (Roberts and Rowe [20]) are efforts to characterize the viscoelastic properties in this indirect manner.

One can expect that all properties are superimposed upon each other, so that components that arise dynamically through the tabletting process should be isolated part by part. Therefore, a thoroughly detailed investigation of force relaxation phenomena, creep characteristics along with the development of mechanical models and calculation methods is an important field of theory in tabletting to predict the load versus displacement curve shape, or vice versa.

For example, the Brinell value is suitable for judging the plastification characteristics of a tablet. It involves measuring the imprint caused by a round object that is pressed against the upper or lower surface of a tablet with a given force (Jetzer, Leuenberger, and Sucker [21]).

The load versus displacement curves show that tablet materials are neither ideally elastic, nor ideally viscous (or plastic). The behavior of pharmaceutically used substances lies between these features. The physical description of these properties is based on the theory of continuum mechanics. David and Augsburger [8] first applied this kind of theoretical models on tablets. For this purpose they loaded tablets in a static way (i.e., using constant stress or strain).

Once pharmaceutical powder is compressed to test tablets with certain dimensions, as is typical in the field of technical material testing, various methods can be applied. Roberts and Rowe [20] offered a relationship between indentation hardness and yield pressure that provides a method for calculating the elasticity of compacted substances. To ensure compatibility, the authors suggested extrapolating elasticity moduli for porous samples to a state of zero porosity.

An important subject within the theory of tabletting is the exact investigation of the phenomena of stress relaxation and creep compliance, especially the development of mechanical models and mathematical methods to predict load versus displacement curves. In this way it should be possible to get information about the physics of flow properties during the compaction of materials.

Müller and Caspar [4, 22] undertook high resolution load versus displacement measurements using a single-punch machine. The authors showed a specific time shift of load and displacement maxima, due to the properties of the substances and drew attention to the fact that relaxation phenomena are also detectable during the dynamic process of compression. They adapted the section up to the load maximum by (static) one-dimensional simple models and found all material features, the height of the compressed tablets and the maximum of the load to be important. Only the particle size was found not to be relevant. The parameters were poorly reproducible and sometimes negative.

Sutanto [23] compared various mathematical methods to estimate different viscoelastic parameters using load versus time and displacement versus time curves on single-punch machines. He used an explicit solution of the constitutive equation, the linear creep compliance, to characterize the materials and described the stress relaxation under a dynamic load. These functions were found by deconvolution of the load versus time curves or the displacement versus time curves respectively. The obtained creep compliances showed irregularities such as two maxima which could be a result of two different, but superimposed events like consolidation and creep compliance and made good physical sense only in the first phase of compressing. Subsequently the deconvolution degenerated. Fischer [24] tested the applicability of the deconvolution method and found it not to be satisfactory, even with error-free test data.

Danielson et al. [25] and Rippie and Danielson [26] studied the decompression and the postcompression phase of substances with known differing compression characteristics using a rotary machine. The so-called postcompressing phase is the phase after compression when the tablet remains for some minutes in the die. Using load versus time curves deduced from radial and axial load measurements and with displacement versus time functions deduced mathematically from the rotary machine's geometry, they adjusted three-dimensional viscoelastic models where the reexpansion volume was related to an elastic process and the deformation to viscoelastic behavior. The arithmetic evaluation was resolved with a constitutive equation, i.e., the stress-strain-relation, correlating with the respective model. The parameters determined were found to relate to the maximum compression load and some turned out to be negative (eluding a physical explanation). Mathematical methods for three-dimensional models have been discussed by Morehead [27], who reduced these three dimensions to the cylindrical dimensions of tablets. He introduced in this paper viscoelastic properties in relation to work of compaction.

This short review shows the difficulties to obtain reasonable figures to characterize viscoelastic effects during tabletting. Hence it seems to be important to show the theoretical background of this behavior. Because there are many effects, one depending on another, this chapter may have a more philosophical touch then a technical report should have. Nevertheless it is the only way to understand the information hidden in the displacement versus time and the load versus time curves of a tabletting process.

A. Simple Models Explaining Viscosity and Elasticity

A mathematical procedure is given by adjusting mechanical models to the load versus time and displacement versus time curves in order to quantify material characteristics. Mathematical models incorporating viscous, elastic, and viscoelastic substances make physical sense and are easily understandable. Real existing materials react in a rather complex way. Therefore an attempt to cover all observed reactions with a simple constitutive equation, i.e., a relation of internal and external forces to its properties, is hardly possible. Constitutive equations are suited rather to define "ideal substances," such as the "ideal elastic material," the "ideal viscous fluids," or the "viscoelastic materials." The behavior of real substances can be attributed to these material models after initial and boundary conditions as well as applicable regions have been determined. Under these circumstances one has to consider that in powders the flow properties are bound to unknown borders as given by plasticity and viscosity (yield values).

Cases in which a complete description of a material is unnecessary or inaccessible, other effects, for example thermal can be uncoupled in order to simplify the theory. This would, however, not imply a neglect of the influence of temperature upon mechanical processes. It is expected that the temperature and/or temperature distribution is known or can be investigated separately. As the constitutive equations now hold only static and kinematic variables they are called stress-strain-relationships.

Stress (σ) describes surface forces which act perpendicularly on small surface elements. Strain (ϵ) as simple shear, tension or compression during tabletting describes local deformation related to original dimensions.

1. Elasticity (Spring)

Linear elastic behavior is represented by a spring. When a massless linear spring is pulled apart or compressed, the necessary force (load) is proportional to the relative expansion or compression. The relative change of length ($\Delta l/l_0$) is a consequence of the action of force without time delay and independent of the present state of load. Once a load is removed the spring returns immediately to its original state (l_0). Then Hooke's law is applicable:

$$\sigma = E\epsilon \tag{1}$$

where

σ = stress
ϵ = relative strain $\Delta l/l_0$
E = elasticity modulus

For the practical use of Hooke's law in the field of tabletting one has to consider with the effect of Poisson's law, because in a die the material cannot increase wider and wider in diameter like a free handled body will do (Bauer [28]).

2. *Viscosity (Dashpot)*

The dashpot exemplifies viscosity clearly. Force is necessary to move a piston through a cylinder filled with viscous fluid. The velocity of motion is proportional to this force. This linear behavior is described by Newton's law:

$$\sigma = F\dot{\epsilon} \tag{2}$$

where

σ = stress
$\dot{\epsilon}$ = $d\epsilon/dt$
F = viscosity modulus

The change in length depends on the time interval. A constant stress σ_0 in an interval Δt results in a strain of ϵ:

$$\epsilon = \frac{\sigma_0}{F} \Delta t \tag{3}$$

Since there is no restoring force, the dashpot remains elongated when the load is removed.

B. Composite Models Explaining Viscoelasticity

Viscoelastics are substances, the behavior of which lies betweeen elastic solids and viscous fluids. Elastic (Hookean) solids and viscous (Newtonian) fluids show great differences in their deformation behavior. Deformed elastic materials return to their original shape, once the load is removed. On the other hand, fluids remain in their state. Furthermore elastic deformation (ϵ) and stress (σ) are directly proportional to one another, whereas a fluid's internal stress (σ) depends on its rate of deformation (ϵ).

1. *Maxwell Model*

The serial connection of the two basic elements, spring and dashpot, is marked in a way that the entire elongation of such a model is equal to the sum of elongations of its components. As all components are accessible through the same amount of stress (σ) for the serial connection of a spring and a dashpot, out of the so-called Maxwell model, the following elongation results:

$$\epsilon = \epsilon_D + \epsilon_S \tag{4}$$

From $\sigma = E\epsilon_D$ for the spring and $\sigma = F\dot{\epsilon}_S$ for the dashpot follows for a Maxwell model a relationship between force and elongation (stress and strain) with the derivative of ϵ with respect to time:

$$\dot{\epsilon} = \dot{\epsilon}_D + \dot{\epsilon}_S = \frac{\dot{\sigma}}{E} + \frac{\sigma}{F} \tag{5}$$

The standardized constitutive equation as the mathematical stress-strain relationship is as follows:

$$\sigma + p_1\dot{\sigma} = q_1\dot{\epsilon} \tag{6}$$

The microconstants E for the elasticity (Young's) modulus and F for viscosity, which can be easily interpreted physically, are related to the hybrid parameters p and q in the constitutive equation, whose significance is recognizable only in the most simple model, for example in this Maxwell model:

$$p_1 = \frac{F}{E} \quad \text{and} \quad q_1 = F. \tag{7}$$

2.   Kelvin (Voigt) Model

In the Kelvin (Voigt) model the spring and the dashpot are connected in parallel, so the elongation (ϵ) of both parts is equal at any time in the entire model. In such an arrangement the stress σ is distributed in different shares on the spring and the dashpot. Shares of the spring (σ_S) and shares of the dashpot (σ_D) are

$$\sigma_S = E\epsilon; \; \sigma_D = F\dot{\epsilon} \tag{8}$$

and add up to the total stress σ:

$$\sigma = \sigma_S + \sigma_D \tag{9}$$
$$\sigma = E\epsilon + F\dot{\epsilon} \tag{10}$$

With $q_0 = E$ and $q_1 = F$ the constitutive equation of this model becomes

$$\sigma = q_0\epsilon + q_1\dot{\epsilon} \tag{11}$$

The stress (σ) in the Kelvin element is proportional both to the elongation of the spring and to the rate of deformation of the dashpot.

3.   Generalized Models

All higher-order models can be attributed to the two basic combinations: the "simple Kelvin model" and the "simple Maxwell model." The "generalized Kelvin model" results from the serial connection of n Kelvin models, from which one or, at maximum, two may be reduced to a dashpot and/or to a spring. The total extension of such a model is the sum of the elongations of the *n*-Kelvin models. Writing the constitutive equation of a single Kelvin model,

$$\sigma = E\epsilon + F\dot{\epsilon} \tag{12}$$

in operator form:

$$\sigma = \{E + F\delta_t\}\epsilon \tag{13}$$

(with $\{\delta_t\} \equiv \delta/\delta t$ the linear differential time operator) adding up the ϵ_i of the *n* elements, when solving for ϵ the result for a solid is as follows:

$$\epsilon = \frac{\sigma}{E_1 + 1/F_1\delta_t} + \frac{\sigma}{E_2 + 1/F_2\delta_t} + \ldots + \frac{\sigma}{E_n + 1/F_n\delta_t} \tag{14}$$

The "generalized Maxwell model" results from parallel connection of *n* Maxwell models, one or two of which can be reduced to a dashpot and/or to a spring. In this model the *n* strains σ_i of the single elements add up to the total strain σ:

$$\sigma = \frac{\dot{\epsilon}}{\delta_t/E_1 + 1/F_1} + \frac{\dot{\epsilon}}{\delta_t/E_2 + 1/F_2} + \ldots + \frac{\dot{\epsilon}}{\delta_t/E_n + 1/F_n} \tag{15}$$

A practical problem are the indices of the reduced elements. It would be convenient to use the same index for both reduced elements (Table 1).

Newton's law dictates that a sudden extension of a dashpot at the time $t = 0$ requires an infinite load. The peak of the load that induces such a theoretic expansion is described mathematically with a single step function $\Delta(t)$ and its derivative, the Dirac delta function $\delta(t)$.

The Heaviside unit step function $\Delta(t)$ is divided into two sections, one with $t < 0$ where $\Delta(t) = 0$, and a second with $t > 0$, where $\Delta(t) = 1$. The derivative $d[\Delta(t)/dt] = \Delta'(t) = \delta(t)$ of the single step function, the so-called Dirac delta function, must be therefore

$$\delta(t) = 0 \text{ for } t \neq 0 \tag{16}$$
$$\delta(t) = \infty \quad \text{ for } t = 0 \tag{17}$$

and so the integral becomes unity:

$$\int_{-\infty}^{+\infty} \delta(t)\, dt = \int_{0-}^{0+} \delta(t)\, dt = 1 \tag{18}$$

Table 1 Identification of the First 12 Simple Viscoelastic Models. The indices for the degenerated elements are always 1.

Fluids	Solids

11: $1p$ model (Newton)

F_1

12: $1p$ model (Hooke)

E_1

21: $2p$ model (Maxwell)

F_1

E_1

22: $2p$ model (Kelvin)

F_2

E_2

31: $3p$ model

F_2 F_1

E_2

32: $3p$ model

F_2

E_2 E_1

41: $4p$ model

F_2 F_1

E_1 E_2

42: $4p$ model

F_2 F_3

E_2 E_3

51: $5p$ model

F_2 F_3 F_1

E_2 E_3

52: $5p$ model

F_2 F_3

E_2 E_3 E_1

61: $6p$ model

F_2 F_3 F_1

E_1 E_2 E_3

62: $6p$ model

F_2 F_3 F_4

E_2 E_3 E_4

Physical experiments can only emulate extension leaps within finite time intervals. In principle it is acceptable, that these intervals are small compared to the total observation time, but the integral should be equal to unity. With a continuous function $f(t)$ and $t > 0$

$$\int_{-\infty}^{t} f(t')[\delta(t')] \, dt' = f(t)[\Delta(t)] \tag{19}$$

with t' as the variable of integration results the stress relaxation for example of a Maxwell model

$$\sigma(t) = E\varepsilon_0/e^{-Et/F} [\Delta(t)] = E\epsilon_0 e^{-\tau t} [\Delta(t)] \tag{20}$$

with $E/F = \tau$ as the relaxation time or $1/\tau$ as the retardation time and for a Kelvin model:

$$\sigma(t) = E\epsilon_0[\Delta(t)] + F\epsilon_0[\delta(t)] \tag{21}$$

The solution for the strain is analogous (Mase [29]).

II. MATHEMATICAL METHODS

A. Construction of Constitutive Equations

The equations for the simple models show how to formulate the unknown stress-strain behavior of a new model. For models from the series of the "generalized Kelvin models" the following statements apply:

1. The elongation ϵ of the Kelvin models adds up to its specific increase for each element.
2. The total strain σ is applied to each element.
3. The following elements can be serially connected:
   a. n Kelvin elements and possibly
   b. one spring as a reduced Kelvin element and/or
   c. one dashpot as a reduced Kelvin element.

1. Microconstants and Hybrid Parameters

For calculations of parameters during tabletting experiments and for the physical understanding of the process, the coherence between the hybrid parameters (p_i and q_i) in the constitutive equations and the microconstants (E_i and F_i) in the models are of great interest. Basically it is always easier to calculate the hybrid parameters than the microconstants than vice versa (Table 2). In the latter case for a $4p$ model up to four solutions for a microconstant can be obtained and their physical validity has to be tested to find a real solution (Sutanto [23], Stahn [30], Sirithunyalug [31]). Due to this calculation and the analog distribution of the elements within the model a simplified method results for all models, which belong to a group with just n elements, where a spring, a dashpot, or a Kelvin model represents an element.

The following terms are an example for the Laplace-transformed figures (23) of Eq. (22):

$$\epsilon = \frac{\sigma}{E_1 + 1/F_1\delta_t} + \frac{\sigma}{E_2 + 1/F_2\delta_t} + \frac{\sigma}{E_3 + 1/F_3\delta_t} \tag{22}$$

Table 2 Hybrid Parameters, Calculated with the Microconstants for Elasticity (E_i) and Viscosity (F_i)

Model	Hybrid parameters

11 $\quad q_1 = F_1$

12 $\quad q_0 = E_1$

21 $\quad p_1 = \dfrac{F_1}{E_1}, q_1 = F_1$

22 $\quad q_0 = E_1, q_1 = F_1$

31 $\quad p_1 = \dfrac{F_1 + F_2}{E_2}, q_1 = F_1, q_2 = \dfrac{F_1 F_2}{E_2}$

32 $\quad p_1 = \dfrac{F_2}{E_1 + E_2}, q_0 = \dfrac{E_1 E_2}{E_1 + E_2}, q_1 = \dfrac{E_1 F_2}{E_1 + E_2}$

41 $\quad p_1 = \dfrac{E_1 F_1 + E_1 F_2 + E_2 F_1}{E_1 E_2}, p_2 = \dfrac{F_1 F_2}{E_1 E_2}, q_1 = F_1, q_2 = \dfrac{F_1 F_2}{E_2}$

42 $\quad p_1 = \dfrac{F_2 + F_3}{E_2 + E_3}, q_0 = \dfrac{E_2 E_3}{E_2 + E_3}, q_1 = \dfrac{E_2 F_3 + E_3 F_2}{E_2 + E_3}, q_2 = \dfrac{F_2 F_3}{E_2 + E_3}$

51 $\quad p_1 = \dfrac{E_2 F_1 + E_2 F_3 + E_3 F_2 + E_3 F_1}{E_2 E_3}, p_2 = \dfrac{F_1 F_2 + F_1 F_3 + F_2 F_3}{E_2 E_3}$

$\quad\quad q_1 = F_1, q_2 = \dfrac{E_2 F_1 F_3 + E_3 F_1 F_2}{E_2 E_3}, q_3 = \dfrac{F_1 F_2 F_3}{E_2 E_3}$

52 $\quad p_1 = \dfrac{E_1 F_2 + E_1 F_3 + E_2 F_3 + E_3 F_2}{E_1 E_2 + E_1 E_3 + E_2 E_3}, p_2 = \dfrac{F_2 F_3}{E_1 E_2 + E_1 E_3 + E_2 E_3}$

$\quad\quad q_0 = \dfrac{E_1 E_2 E_3}{E_1 E_2 + E_1 E_3 + E_2 E_3}, q_1 = \dfrac{E_1 E_2 F_3 + E_1 E_3 F_2}{E_1 E_2 + E_1 E_3 + E_2 E_3}$

$\quad\quad q_2 = \dfrac{E_1 F_2 F_3}{E_1 E_2 + E_1 E_3 + E_2 E_3}$

61 $\quad p_1 = \dfrac{E_1 E_3 F_1 + E_1 E_3 F_2 + E_1 E_2 F_3 + E_1 E_2 F_1 + E_2 E_3 F_1}{E_1 E_2 E_3}$

$\quad\quad p_2 = \dfrac{E_1 F_1 F_3 + E_1 F_1 F_2 + E_1 F_2 F_3 + E_2 F_1 F_3 + E_3 F_1 F_2}{E_1 E_2 E_3}$

$\quad\quad p_3 = \dfrac{F_1 F_2 F_3}{E_1 E_2 E_3}, q_1 = F_1, q_2 = \dfrac{E_2 F_1 F_3 + E_3 F_1 F_2}{E_2 E_3}, q_3 = \dfrac{F_1 F_2 F_3}{E_2 E_3}$

62 $\quad p_1 = \dfrac{E_1 F_2 + E_1 F_3 + E_2 F_1 + E_2 F_3 + E_3 F_1 + E_3 F_2}{E_1 E_2 + E_1 E_3 + E_2 E_3}, p_2 = \dfrac{F_1 F_2 + F_1 F_3 + F_2 F_3}{E_1 E_2 + E_1 E_3 + E_2 E_3}$

$\quad\quad q_0 = \dfrac{E_1 E_2 E_3}{E_1 E_2 + E_1 E_3 + E_2 E_3}, q_1 = \dfrac{E_1 E_3 F_2 + E_1 E_2 F_3 + E_2 E_3 F_1}{E_1 E_2 + E_1 E_3 + E_2 E_3}$

$\quad\quad q_2 = \dfrac{E_1 F_2 F_3 + E_2 F_1 F_3 + E_3 F_1 F_2}{E_1 E_2 + E_1 E_3 + E_2 E_3}, q_3 = \dfrac{F_1 F_2 F_3}{E_1 E_2 + E_1 E_3 + E_2 E_3}$

for a model with $n = 3$ elements (6p solid).

$$\bar{\epsilon} = \bar{\sigma} \left[\frac{1}{E_1 + sF_1} + \frac{1}{E_2 + sF_2} + \frac{1}{E_3 + sF_3} \right] \tag{23}$$

Simple calculations show the transformed strain-stress relation expressed as a sequence of sums:

$$\bar{\epsilon}(E_1 + F_1)(E_2 + sF_2)(E_3 + sF_3) = \bar{\sigma}[(E_2 + sF_2)(E_3 + sF_3)$$
$$+ (E_1 + sF_1)(E_3 + sF_3) + (E_1 + sF_1)(E_2 + sF_2)] \tag{24}$$

Multiplication and rearrangements of terms in the latter equation lead to a polynomial equation (Sutanto [23], Stahn [30]):

$$\bar{\epsilon}[q_0's^0 + q_1's^1 + q_2's^2 + q_3's^3] = \bar{\sigma}[p_0's^0 + p_1's^1 + p_2's^2] \tag{25}$$

It is obvious that by the rules of transformation and by comparison of the coefficients, a parameter p_k', which is a factor of s^k in the transformed equation and the parameter p_i' in the inverse transformation corresponds to the factor of σ_i. The parameters q_i' are also related as shown in the following generalized equations:

$$\bar{\epsilon} \sum_{k=0}^{m} p_k' s^k = \bar{\sigma} \sum_{k=0}^{n} q_k' s^k \tag{26}$$

$$\sum_{i=0}^{m} p_i' \frac{\partial^i \sigma}{\partial t^i} = \sum_{i=0}^{n} q_i' \frac{\partial^i \epsilon}{\partial t^i} \tag{27}$$

Described in standard form, the number of microconstants of a model (the number of all springs and dashpots) is equal to the number of parameters belonging to the constitutive equation. For "solids," $q_0 \neq 0$, and for "fluids," $q_0 = 0$; for "solids" and "fluids" with spontaneous elastic answer, $q_n = 0$ for the nth element.

Therefore, for each three-element model the constitutive equation can easily be found without mathematical effort, for instance:

$$\bar{\epsilon} \sum_{k=0}^{3} p_k' s^k = \bar{\sigma} \sum_{k=0}^{2} q_k' s^k \tag{28}$$

and in the inverse general transformation:

$$\sum_{i=0}^{2} p_i' \frac{\partial^i \sigma}{\partial t^i} = \sum_{i=0}^{3} q_i' \frac{\partial^i \epsilon}{\partial t^i} \tag{29}$$

For example, in reduced models the parameters p_i' and q_i', which can be deduced from the 6p solid, can be determined by setting spring and/or dashpot constants to zero.

For physical reasons (no negative parameters) there are relations in form of inequalities between the microconstants E_i and F_i. The relations which are a result of the classification save the use of complicated calculations to determine the inequalities (Table 3). If the equalities of a non-reduced Kelvin model with n elements are known, the inequalities of the derived models appear by setting the parameter q_0 and/or q_n to zero according to the scheme.

2. Laplace Transformation

For each element (string, dashpot, Kelvin) the stress-strain relation can be performed separately. The mathematical problem to connect the individual differential equations to the stress-strain equation of the total model can be solved by means of a Laplace transformation.

The solutions for all models are differential equations of the general form:

$$p_0' \, \sigma + p_1' \, \dot{\sigma} + p_2' \, \ddot{\sigma} + \ldots = q_0' \, \epsilon + q_0' \, \epsilon + q_1' \, \dot{\epsilon} + q_2' \, \ddot{\epsilon} + \ldots \quad (30)$$

Table 3 Inequalities of the First 12 Models

Model	Inequalities
11	$q_1 > 0$
12	$q_0 > 0$
21	$p_1, q_1 > 0$
22	$q_0, q_1 > 0$
31	$p_1 q_1 > q_2$
32	$q_1 > p_1 q_0$
41	(1) $p_1^2 > 4p_2$, (2) $p_1 q_1 q_2 > p_2 q_1^2 + q_2^2$
42	(1) $q_1^2 > 4q_0 q_2$, (2) $p_1 q_1 > p_1^2 q_0 + q_2$
51	(1) $q_2^2 > 4q_1 q_3$, (2) $p_1^2 > 4p_2$
52	(1) $q_1^2 > 4q_0 q_2$, (2) $p_1^2 > 4p_2$
61	(1) $q_2^2 > 4q_1 q_3$, (2) $\dfrac{3p_1 p_3 - p_2^2}{3p_3^2} < 0,$ (3) $\left(\dfrac{1}{27} \dfrac{p_2^3}{p_3^3} - \dfrac{1}{6} \dfrac{p_1 p_2}{p_3^2} + \dfrac{1}{2p_3} \right)^2 + \left(\dfrac{3p_1 p_3 - p_2^2}{9p_3^2} \right)^3 < 0$
62	(1) $4p_2 < p_1^2$, (2) $\dfrac{3q_0 q_2 - q_1^2}{3p_0^2} < 0,$ (3) $\left(\dfrac{1}{27} \dfrac{q_1^3}{q_0^3} + \dfrac{1}{6} \dfrac{q_1 q_2}{q_0^2} - \dfrac{q_3}{2q_0} \right)^2 + \left(\dfrac{3q_0 q_2 - q_1^2}{9q_0^2} \right)^3 < 0$

If both sides of the differential equation are divided by p_0', the standard form of the constitutive equation is as follows with $p_0 = 1$ as unity:

$$\sigma + p_1\dot{\sigma} + p_2\ddot{\sigma} + \ldots = q_0\epsilon + q_1\dot{e} + q_2\ddot{e} + \ldots \tag{31}$$

Terms of stress and strain are combined in the most generalized form:

$$\sum_{i=0}^{m} p_i \frac{\partial^i \sigma}{\partial t^i} = \sum_{i=0}^{n} q_i \frac{\partial^i \epsilon}{\partial t^i} \tag{32}$$

Flügge [32] stated that the number m of the stress terms has to be smaller or equal to the number n of strain terms; otherwise the differential equations show a behavior of the substance different from linear viscoelastic materials.

B. Complex Modulus and Complex Compliance (Dynamic Modulus and Dynamic Compliance)

Besides the work with viscoelastic models, the calculation of their parameters and physical interpretation of their components, for example in the standard experiments, there is another more common method for experimental exploration with a different kind of mathematical description of the models. This method was introduced to pharmaceutics primarily for the characterization of gels or ointments.

The experimental approach is to apply an oscillating (for example, a sinusoidal) stress to the sample and to observe the strain as a response (or vice versa) which may be altered in phase and amplitude. With a loading stress, which is very similar to the tabletting load function of an excenter:

$$\sigma(t) = \sigma_{\max} \sin(\omega t) \tag{33}$$

there will be a resulting strain response

$$\epsilon(t) = \epsilon_{\max} \sin(\omega t + \delta) \tag{34}$$

The angular velocities ω of stress and strain are the same, but within the strain response, there is a lagtime like the maximum differences in load and displacement curves, defined by the loss angle δ as a phase shift. A large field for experimental alterations is opened, because it is easy to apply an alternating displacement to a sample and obtain a load as a response. For example, in the field of tabletting, experiments can be made with the first compression or repeated compressions within the same die. Using the whole tablet in a compaction simulator as often as possible studying the resulting strain could be another way. But the common way, for instance, for a plastic material is to apply an oscillatory strain to one end of a rod and to measure the torque on the other end, which means the rod is exposed to alternating positive and negative stress. In the case of compacting a tablet in a die, only the first quarter of the first oscillation is relevant to describe the whole oscillation. For the experimental estimation of the parameters on the other hand an equilibrium between elastic and viscous movements is necessary. Therefore, for a correct experiment one needs several full oscillations:

$$E^* = \frac{\sigma_{max}}{\epsilon_{max}} \quad \text{absolute dynamic modulus} \tag{35}$$

$$G' = \frac{\sigma_{max} \cos \delta}{\epsilon_{max}} \quad \text{storage modulus (elasticity)} \tag{36}$$

$$G'' = \frac{\partial_{max} \sin \delta}{\epsilon_{max}} \quad \text{loss modulus (viscosity)} \tag{37}$$

$$J' = \frac{\epsilon_{max} \cos \delta}{\sigma_{max}} \quad \text{storage compliance (elasticity)} \tag{38}$$

$$J'' = \frac{\epsilon_{max} \sin \delta}{\sigma_{max}} \quad \text{loss compliance (viscosity)} \tag{39}$$

$$\eta' = \frac{G''}{\omega} \quad \text{real part of viscosity} \tag{40}$$

$$\eta'' = \frac{G'}{\omega} \quad \text{complex part of viscosity} \tag{41}$$

This kind of description is a consequence of the generalized formulation in complex form of stress and strain which always leads to two solutions in real and imaginary parts.

The nomenclature concerning storage and loss moduli indicates the close vicinity to problems of energy. A short theoretical introduction is given by Pipkin [33]. A first attempt to characterize the work of tabletting using viscoelastic parameters was done by Morehead [27].

Very important for discussing tabletting problems is the change of storage and loss moduli with the change of the frequency of the compression device, that is the change with respect to the speed of the punch within the die. When the frequency increases, the loss modulus G'' may be lowered and, vice versa, the storage modulus G' increases. Computing the numbers of a simple model like a Kelvin or Maxwell model, one can see the different influence of a free spring, a free dashpot, or a Kelvin element (Table 4).

Table 4 Interaction of Viscous and Elastic Features in Model, Depending on the Speed of Deformation

	Spring	Dashpot	Maxwell	Kelvin	$3p$ fluid	$3p$ Solid
G'	$E \to$ const	$0 \to$ const	$0 \to E$	$E \to$ const	$0 \to$ max	min $\to E_1$
G''	$0 \to$ const	$0 \to$ max	$0 <<$ max $>> 0$	$0 \to$ max	min $\to$ max	min $<$ max $>$
δ	$0 \to$ const	$\pi/2 \to$ const	$\pi/2 \to 0$	$0 \to \pi/2$	$\pi/2 >$ min $< \pi/2$	$0 < \pi/2 > 0$
η'	$0 \to$ const	$F \to$ const	$F \to 0$	$F \to$ const	max $\to$ min	max $\to$ min
σ_{max}	const	$<<$ max	$<$ max	$<<$ max	min $\to$ max	min $\to$ max

In a spring model the stress is always directly proportional to the strain, so the displacement versus time and strain versus time curves respectively do not change with altering frequencies. This means that the loss angle δ is zero and therefore the loss modulus G'' and the loss compliance J'' are zero too, because sin δ is zero. On the other side, cos δ is unity, so the storage modulus G' and the storage compliance J' are important to characterize the elastic properties of a model.

The behavior of a dashpot is also easy to understand. In this case the resulting stress is always proportional to the rate of deformation, so at any frequency the loss angle is $\pi/2$. That means sin δ is at its maximum and cos δ is zero. The characteristics are the loss modulus G'' or the loss compliance J''. Because in a dashpot the resulting stress is proportional to a given frequency of deformation the maximum load will increase with ascending radian frequency.

Both elementary models are the base for composite models like Kelvin, Maxwell, or higher-order models. To obtain in these models the relation of strain and stress, a series of different frequencies must be used to estimate the loss and storage moduli. With a plot of G', G'', J', J'' or η' versus frequency the viscoelastic properties become evident. A reasonable approach is to use log/log plots to get simple relations if they exist.

In a Maxwell model (Fig. 3) the storage modulus G' will approach Young's modulus E for the spring in the model with higher speed (higher frequencies) of tabletting. With lower speed the viscosity η' approaches the viscosity modulus, which is the same as the microconstant F of the dashpot in the model. The interaction between elasticity and viscosity in a Maxwell model becomes evident in the loss modulus G'', which has a maximum at a medium value of the frequency. On the other hand, the loss angle δ decreases from $\pi/2$ at lowest frequency to zero at high frequencies and the maximum of the load will rise.

In the Kelvin model (Fig. 4), the dashpot and the spring are connected in parallel. So they must always be in the same position, but the

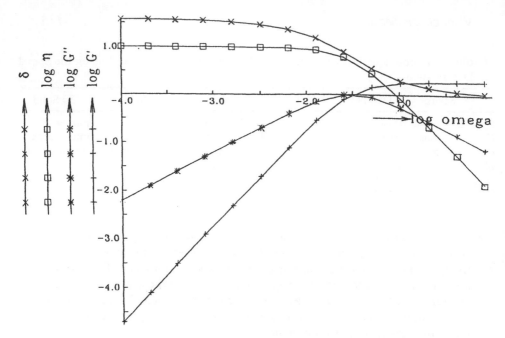

Fig. 3 Change of storage modulus G', loss modulus G'', viscosity η', and loss angle δ of a Maxwell model with time (theoretic data with $E = 2$ and $F = 10$).

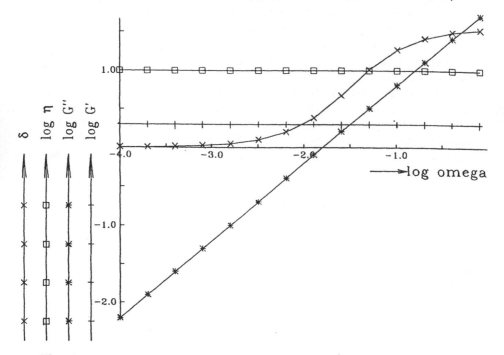

Fig. 4 Change of storage modulus G', loss modulus G'', viscosity η', and loss angle δ of a Kelvin model with time (theoretic data with $E = 2$, $F = 10$).

stress in each element is different and the storage modulus G' representing elasticity and η' representing viscosity do not vary with altering frequencies. But the load needed in this example to obtain the same elongation as at a low speed is in this example 25 times higher, and the loss modulus G'' will rise also. The loss angle δ changes from zero at low speed to $\pi/2$ at the highest frequency, and the maximum load rises to a significantly higher value.

In 3p models the loss angle δ shows a minimum or a maximum in contrast to the simpler models which have been discussed. G'' rises in the 3p fluid and in the 3p solid we see a maximum.

All these relations show on the one hand the possibilities to obtain information and on the other hand the difficulties to apply the theory to the process of powder compaction. The main problem is the nonequilibrium in the first quarter of the oscillation in a tabletting process. Correct values could only be estimated in the following full oscillations when elastic and viscous properties are in equilibrium and the strain is alternating with the positive and negative parts of the theoretical oscillations as shown in Figs. 5 and 6 for a Maxwell model. Here it seems to be very important to look at the first quarter of the oscillation which is highly influenced by the initial conditions.

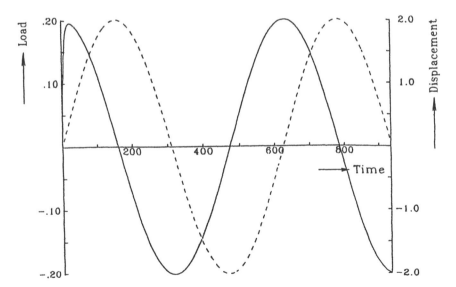

Fig. 5 Load vs. time (solid line) and displacement vs. time (dashed line) curves for a Maxwell model (theoretic data with $E = 2$, $F = 10$) at slow speed.

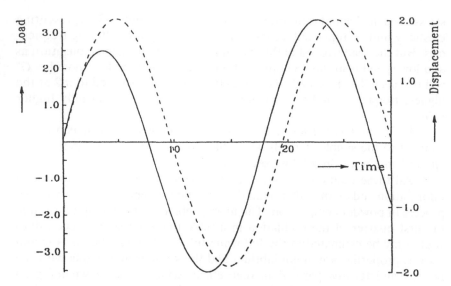

Fig. 6 Load vs. time (solid line) and displacement vs. time (dashed line) curves for a Maxwell model (theoretic data with $E = 2$ and $F = 10$) at high speed.

C. Explicit Solutions

1. *Linear Creep Compliance*

The strain-stress behavior of different materials is completely described by their constitutive equations. But the physical features of materials can better be understood by application and physical interpretation of a standard test: The explicit solutions, which were derived from the constitutive equations, characterize these materials as a function of time in a creep compliance with constant stress or a stress relaxation test with constant strain. In the standard experiment one of the variables is constant, so the derivatives vanish and the explicit solutions are simple.

As we only consider the linear viscoelastic materials, the change in the relative extension ϵ is proportional to the constant stress σ_0 and can be described as follows:

$$\epsilon(t) = \sigma_0 J(t) \tag{42}$$

The function $J(t)$, the linear creep compliance,

$$J(t) = \frac{\epsilon(t)}{\sigma_0} \tag{43}$$

describes the elongation or compression of a material under unitary (constant) load. The diagram of individual behavior of material, the creep compliance function, is equivalent to the mathematical description with constitutive equations with a constant stress (Table 5). To get a specific answer to an analysis of viscoelastic materials the creep compliance must be estimated as a response function and should be fitted by a physically obvious model.

Table 5 Creep Compliance, Calculated with Microconstants (E_i and F_i)

Model	Creep compliance
11	$\dfrac{t}{F_1}$
12	$\dfrac{1}{E_1}$
21	$\dfrac{1}{E_1} + \dfrac{t}{F_1}$
22	$\dfrac{1}{E_2}(1 - e^{-(E_2/F_2)t})$
31	$\dfrac{1}{E_2}(1 - e^{-(E_2/F_2)t}) + \dfrac{t}{F_1}$
32	$\dfrac{1}{E_1} + \dfrac{1}{E_2}(1 - e^{-(E_2/F_2)t})$
41	$\dfrac{1}{E_1} + \dfrac{1}{E_2}(1 - e^{-(E_2/F_2)t}) + \dfrac{t}{F_1}$
42	$\dfrac{1}{E_2}(1 - e^{-(E_2/F_2)t}) + \dfrac{1}{E_3}(1 - e^{-(E_3/F_3)t})$
51	$\dfrac{1}{E_2}(1 - e^{-(E_2/F_2)t}) + \dfrac{1}{E_3}(1 - e^{-(E_3/F_3)t}) + \dfrac{t}{F_1}$
52	$\dfrac{1}{E_1} + \dfrac{1}{E_2}(1 - e^{-(E_2/F_2)t}) + \dfrac{1}{E_3}(1 - e^{-(E_3/F_3)t})$
61	$\dfrac{1}{E_1} + \dfrac{1}{E_2}(1 - e^{-(E_2/F_2)t}) + \dfrac{1}{E_3}(1 - e^{-(E_3/F_3)t}) + \dfrac{t}{F_1}$
62	$\dfrac{1}{E_2}(1 - e^{-(E_2/F_2)t}) + \dfrac{1}{E_3}(1 - e^{-(E_3/F_3)t}) + \dfrac{1}{E_4}(1 - e^{-(E_4/F_4)t})$

Barry [34] estimated viscosity, elasticity, and several serial Kelvin elements from the creep compliance by numerical or graphical methods using the standardized experiment with constant stress. The creep compliance shows a very common solution for the generalized Kelvin model with two reduced elements:

$$J(t) = \frac{1}{E_1} + \frac{1}{F_1} t + \frac{1}{E_i} \sum_{i=2}^{n} \left[1 + e^{-(F_i/E_i)t} \right] \tag{44}$$

Under a tensile stress real materials will expand and under pressure they will be compressed. This behavior corresponds with the monotonously increasing course of the creep compliance. Therefore the microconstants E_i and F_i are always positive. The rules of conversion show that also the hybrid parameters can only assume positive values. These consequences do not become evident if one only deals with differential equations. Among the microconstants there are relationships existing in form of inequations (Table 3), which must hold for physically meaningful models. Under the given conditions of the standard experiment the creep compliance is the explicit solution of the constitutive equation. The solution of the differential equations can be obtained in most cases by means of Laplace transformation.

2.   Linear Relaxation Function

The second explicit solution of the differential equation describes the behavior during a relaxation experiment. The stress relaxation function $G(t)$ gives the response of a stressed (compressed or expanded) material as a function of time, related to a given constant strain. The solution of the constitutive equation is determined in the same way as that of the creep compliance, but it is more complicated to calculate its parameters (Table 6). This applies also for the transformed function $\overline{G}(s)$ according to Laplace. Physically, in the time-dependent domain, no simple relation between the creep compliance and the relaxation function can be found. But after a Laplace transformation the functions have been converted to the very simple relation (Flügge [32]):

$$\overline{J}(s)\overline{G}(s) = s^{-2} \tag{45}$$

In the stress relaxation functions of different models no simple and clear relations can be found as in the functions of the creep compliance. This is because the relaxation of force in more complex models, parts of elongation ϵ_i of the springs, dashpots, and Kelvin elements equalize differently in time and direction, until a forceless condition is reached. Using the hybrid parameters, an arrangement of terms of the stress relaxation function can be found, which allows one to formulate the stress relaxation functions of the different models with one, two, or three elements (Stahn [30]). But

Table 6 Stress Relaxation Function, Calculated with Hybrid Parameters (p_i and q_i)

Model	Stress relaxation function
11	$q_1 \delta(t)$
12	q_0
21	$\dfrac{q_1}{p_1} e^{-t/p_1}$
22	$q_0 + q_1 \delta(t)$
31	$\dfrac{q_2}{p_1} \delta(t) + \dfrac{1}{p_1}\left(q_1 - \dfrac{q_2}{p_1}\right) e^{-t/p_1}$
32	$\dfrac{q_1}{p_1} e - t/p_1 + q_0(1 - e^{-t/p_1})$
41	$\dfrac{1}{\sqrt{p_1^2 - 4p_2}}\left((q_1 + q_2\alpha)e^{\alpha t} - (q_1 + q_2\beta)e^{\beta t}\right)$
42	$q_0 + \dfrac{q_2}{p_1} \delta(t) - \left(q_0 - \dfrac{q_1}{p_1} + \dfrac{q_2}{p_1^2}\right) e^{-t/p_1}$
51	$\dfrac{q_3}{p_2} d(t) + \dfrac{1}{\sqrt{p_1^2 - 4p_2}}\left((q_1 + q_2\alpha + q_3\alpha^2)e^{\alpha t} - \left(q_1 + q_2\beta + q_3\beta^2\right)e^{\beta t}\right)$
52	$q_0 + \dfrac{1}{\sqrt{p_1^2 - 4p_2}}\left(\left(\dfrac{q_0}{\alpha} + q_1 + q_2\alpha\right)e^{\alpha t} - \left(\dfrac{q_0}{\beta} + q_1 + q_2\beta\right)e^{\beta t}\right)$
61	no solution
62	$q_0 + \dfrac{q_3}{p_2} \delta(t) + \dfrac{1}{\sqrt{p_1^2 - 4p_2}}\left(\left(\dfrac{q_0}{\alpha} + q_1 + q_2\alpha + q_3\alpha^2\right)e\alpha t - \left(\dfrac{q_0}{\beta} + q_1 + q_2\beta + q_3\beta^2\right)e^{\beta t}\right)$
	with $\alpha, \beta = \dfrac{-p_1 \pm \sqrt{p_1^2 - 4p_2}}{2p_2}$

both relations, creep compliance as well as stress relaxation function, hold at the same point and the same time.

D. Estimation of Parameters from Compaction Dynamics

1. Constitutive Equation

Differential equations like the constitutive equation are usually solved explicitly resulting in the creep compliance or the stress relaxation function.

In this case it is possible to provide a physical interpretation for the relation between the variables. If one only wants the numbers of the parameters, for example to compare different methods of tablet compaction, it is not necessary to know the complex solution. Therefore the constitutive equation was reassembled to get 100 to 300 numerical equations related from all the 100–300 data of a load versus time and displacement versus time curves:

$$\sigma_i = -p_1\dot{\sigma}_i - p_2\ddot{\sigma}_i - \ldots + q_0\epsilon_i + q_1\dot{\epsilon}_i + q_2\ddot{\epsilon}_i + \ldots \qquad (46)$$

The displacement ϵ_i, the speed $\dot{\epsilon}_i$, the acceleration $\ddot{\epsilon}_i$, the change of load $\dot{\sigma}_i$ and the second $\ddot{\sigma}_i$ or higher derivatives with respect to time of the ith data point must be known at every time. If all these values are known at each of the 100 to 300 data points and an estimate of the experimental load σ_i^* at this point, the sum of squares becomes

$$\sum_{i=1}^{n} (\sigma_i - \sigma_i^*)^2 \equiv \min \qquad (47)$$

which can be minimized by the method of maximum likelihood or similar methods.

The *Numerical Recipes* [35] contain very helpful source modules in Fortran or Pascal to find the numerical solution of the constitutive equation. The main problem is to find the smoothing function for load and displacement and their first to third derivatives with sufficient accuracy. The numerical value of the calculated load will be added from up to five terms in the constitutive equation and must be subtracted from the estimated load. In order to find adequately fine calculated data points, we used polynomial smoothing algorithms, which can easily be derivated in double precision (Sutanto [23]).

The advantage of this procedure is that it is not necessary to find accurate boundary and initial conditions as for the explicit solutions (i.e., Dirac delta function). The disadvantage is the high precision required by the polynomials. So we used polynomials of 9th to 15th degree. Once programmed, they allowed to get the significance of the polynomial in a first run and to get the best model out of every combination up to four-(sometimes even six-) parameter models in a second run with, e.g., by multiple linear or nonlinear regression. Other functions for load or strain like sums of sine and cosine or exponential functions from a Fourier transformation may be feasible for smoothing too. In all cases it should be easy to obtain the derivatives.

To find satisfying solutions for the $4p$ to $6p$ models, the calculated parameters must be tested for their significance: If the parameters are greater than 1000, the reciprocal will be so small that the calculated effect of the element will vanish (Sirithunyalug [31]). A small residual amount may remain and will add to another (reduced) element, so that another uncertainty will arise.

2. *Explicit Solutions of the Constitutive Equation*

a. **Superposition Principle**

For the mathematical analysis of the displacement versus time or stress versus time behavior it is important that the explicit solutions, i.e., the creep compliance and/or the stress relaxation function, which describe the material in a static situation, must be applicable to characterize the dynamic of the process of powder compaction for example in a single-punch machine. Therefore the Boltzmann superposition principle is used. The linear behavior of an elastic or viscoelastic material is determined by the total effect of a sum of causes, which equals the sum of the single effects of a single cause, but at consecutive times.

For example, a compaction curve from $t = t_0$ to $t = t_n$ is a response to a sum of n changing-of-stress events in single steps and can be described as follows:

$$\epsilon_n = \sigma_0 J(t_n - t_0) + \sigma_1 J(t_n - t_1) + \sigma_2 J(t_n - t_2) + \ldots + \sigma_{n-1} J(t_n - t_{n-1}) \quad (48)$$

In general,

$$\epsilon_n = \sum_{i=1}^{n} \sigma_{i-1} J(t_n - t_{i-1}) \quad (49)$$

If the complete stress history of a material is known from minus infinity to the time t_0, the creep compliance or the stress relaxation curves can be calculated from the load versus time and displacement versus time curves. But it is much easier to get a solution if the initial conditions start at zero stress and at zero strain. Figure 7 shows, how a load curve is built up from single steps of a discrete creep compliance ($4p$ fluid) corresponding to the displacement curve. In the range of decreasing load, steps of negative load also are superimposed.

In earlier publications of Rippie [26] and Danielson [25], Neuhaus [36] and Sutanto [23] the sections of increasing and decreasing load were treated separately. A simple example shows that the direction of the load is of no consequence for the mathematical execution of superposition (Stahn [30]).

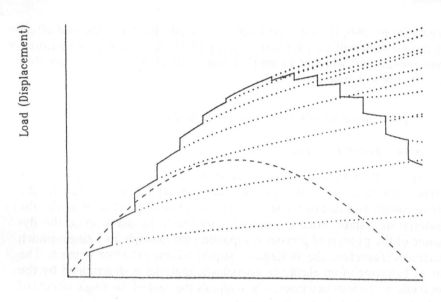

Time

Fig. 7   Schematic construction of displacement vs. time curve for a 4p fluid with a load vs. time curve (dashed line) and 15 consecutive creep compliances (dotted lines). Envelope curve as a first approximation (solid line).

b.   Hereditary Integral

When the entire process is divided into many small steps, this leads finally to an integral, which describes the history of the powder compacting process. The hereditary integral is defined by the equation

$$\epsilon(t) = \sigma(0)J(t) + \int_0^t J(t - t') \frac{d\sigma'}{dt'} \, dt' \tag{50}$$

The explicit solution for tabletting problems becomes possible with a substitution of $J(t - t')$ and $d\sigma'$ and then following the rules of integration of products.

J(0), the initial condition of the creep compliance (for example, free spring with a spontaneous answer), and for $J(t)$ the terms of the creep compliance will be used. The equation for the 4p fluid (Rieger [37]) with the derivative of the creep compliance for a 4p fluid (in brackets) is as follows:

$$\epsilon(t) = 1/E_1\sigma(t) + \int_0^t \sigma(t') \left[\frac{1}{F_1} + \frac{F_2}{E_2^2} e - (F_2/E_2)(t - t') \right] dt' \tag{51}$$

The calculation of the load curve $\sigma(t)$ is performed by an analog transformation. The integration can be done for example according to Romberg's method.

c. Convolution Processes

Principally in

$$\epsilon(t) = \sigma_0 J(t) + \int_0^t \sigma(t)' J(t - t')\, dt' \tag{52}$$

the integral is a convolution operation (53), whereby the load appears as an input function, the creep compliance as a response function, and the displacement curve as the output function. According to Stepanek [38] in

$$\sigma(t) * J(t) = \epsilon(t) \tag{53}$$

where the asterisk stands for the convolution operator. A similar notation is used to find the inverse function

$$J(t) = \epsilon(t) \overset{*}{*} \sigma(t) \tag{54}$$

where $\overset{*}{*}$ represents the deconvolution operator. Therefore (53) is the displacement curve and (54) is the creep compliance.

Deconvolution. There are different ways to find the solution of $\epsilon(t)$ by deconvolution: with help of the Laplace transformation, by development of time series (FFT), or by a numerical method. Here we use the numerical step-by-step method to solve the deconvolution problem. The cumulative displacement curve is the output function obtained from the response function as the creep compliance and the load curve as the input function, split to fractional intervals. As the equation for the displacement is calculated by means of repeated creep compliances, the deconvolution is calculated vice versa. The creep compliance itself always keeps the same shape after each step of deconvolution of load. Therefore,

$$t_0 \qquad J(t) = 0 \qquad \text{resp. } \frac{1}{E} \text{ with an elastic response}$$

$$t_1 \qquad J(t) = \epsilon_1/\sigma_0$$

$$t_2 \qquad J(t) = (\epsilon_2 - J_1\, d\sigma_1)/\sigma_0$$

$$t_3 \qquad J(t) = (\epsilon_3 - (J_1\, d\sigma_2 + J_2\, d\sigma_1))/\sigma_0$$

$$\vdots \qquad \vdots$$

$$t_n \qquad J(t) = (\epsilon_n - \sum_{i=1}^{n-1} J_i\, d\sigma_{n-i})/\sigma_0 \tag{55}$$

This formula is not difficult to program, but load and displacement have to be continuous functions and the time intervals must be small and constant. Therefore experimental data have to be smoothed and divided to equal intervals. As long as the data of displacement and load are only interpolated, a fifth- to ninth-degree polynomial is used to obtain the graphical construction of the creep compliance $J(t)$. The shape of the resulting curve with the height and the position of the maximum is dependent on the substance and the maximum load. Very often the curves cannot be interpreted because they are distorted by oscillations and degenerate. Reasons for this include limited accuracy of the numerical transaction and the uncertainty of the initial conditions. In case of tabletting the load versus time and the displacement versus time curves are not ideal, because the consolidation, reducing the volume, adds to the viscoelastic features. So the deconvolution can never produce an ideal creep compliance. The parameters of the exposed creep compliance are found by a Simplex process or by regression using the method of the steepest slope. The advantage of the method of deconvolution is the "once only" single exposure of the creep compliance, which is then subject to the regression to get the parameters of the chosen model.

Convolution. The method of convolution is more reliable. Here a first guess of a parameter sequence for the creep compliance of the given model is estimated. The entire displacement function is calculated with the parameters of an assumed creep-compliance function. By iteration with the Simplex algorithm improved parameters are searched until the result (sum of squares) is satisfactory. Before a new cycle with a new set of parameters for the creep compliance starts, it has to be checked, whether the new parameters will meet the inequalities. The calculation will be stopped, if the variation of the parameters does not improve the fit of the curve.

If the calculation converges—this should be the case after 300–800 steps of variation—the iteration is stopped. Another advantage of convolution is the possibility to stay within the physically feasible region of the parameter space with its equalities and inequalities discussed above. If the load maximum is exceeded, the steps of the load become negative. The alterations of the displacement curves change their directions and are taken from those parts of the displacement which emerged from earlier positive steps of load and which increase even further (Fig. 7). The best possible model can be found by repetition of the calculations with different candidates for the model. It is very helpful to follow the overlays visually on the screen. The smaller the steps of load and the greater the number of overlays, the smoother the envelope curve will be, and the better it will also fit the Dirac delta function. The number of steps is chosen according to the number of datapoints (200–300) except for rough calculations. The first result shows that the calculated load versus time and displacement versus time curves reproduce the real initial situation in a qualitatively and quantitatively correct way: the load maximum is located in front of the displacement curve maximum, the extent of the time shift results from the choice of the model and the parameters of the substance. In the same way the load curve is adapted, according to the displacement curve. The corresponding displacement curves are calculated from the load curves by a shifted addition of the estimated stress relaxation function curves. To check the correct program function, alternate calculations of load and displacement curves should be performed with different initial parameters. The advantage compared to the deconvolution is the invariance of the given creep compliance, because in this case the stable function will be used to find the "not ideal" load curve as similar as possible.

III. REMARKS CONCERNING PRACTICAL APPLICATIONS

The numerical methods are difficult, extremely expensive in programming and different for each element. The main problem however is the numerical stability of the regression methods. Therefore it is a good practice to prove the numeric stability with a set of well-defined, known theoretical data (Sirithunyalug [31]).

For the use of theoretical numbers there must be a given function to simulate the strain or the stress respectively. This function should be easy to transform into the Laplace domain. This could be for example a quadratic polynomial, a sum of two or three exponential terms, a simple or composite sine function, just to get a similar behavior as a displacement (resp. load) curve. The function must however have enough derivatives to meet the chosen model. The most natural approximation is the calculated function with the geometry of the press. The next step is to choose a model to balance either the stresses or the strains of the single elements and then to transform the equations to Laplace. Now it is possible to calculate the theoretical stress (resp. strain) function with the given set of parameters (the microconstants of the different elements). It is convenient to use the data to prove the smoothing function and to find the best approximation to the parameters, which are now well known. In this point every inequation must be satisfied and all parameters must be positive! The next criterion is the variance of the calculated parameters. The results must be in a reasonable order of magnitude. But the main problem lies in a personal decision: The resolution of the numerical calculation of the constitutive equation is based on the degree of the polynomial. So it is recommended to use polynomials of every power of the polynomial from 3 to 15 to find the best approximation of the well-known parameters and a reasonable sum of squares (there is a minimum), besides the above-mentioned properties of the inequalities. With this feeling for the accuracy the same procedure should be chosen in the main experiment.

With the criterion for the regression being the sum of squares one should be aware that, e.g., the constitutive equation is composed of a sum of two to six single terms, each of which is a different product. Hence, there exist billions of variations of terms, but only one meets the true value. That means the shape of the curve should be characteristic for the model which could be achieved by a high number of data points, a typical segment of the curve and in the case of convolution as the most effective method by simultaneous calculation of the added sums of squares from the creep compliance and the stress relaxation function, because both functions are "true" at all points and times.

Practical investigations for estimation of viscoelastic properties should be validated with materials that display the characteristics expected in the experiments. Using convolution or integral methods, one needs the initial condition, which is the most difficult prerequisite for a solution of a model. In practice the viscoelastic behavior is overlapped with the consolidation phase of the materials in the die. Both events are power consuming. We always failed to find this point in tabletting experiments. The best approach is a precompression to a distinct load and restart the machine to get the final compaction and decompression phase. On rotary machines we used the precompression station [39]. Another approximation is to substitute the displacement with the corresponding analogon such as the relative porosity like in the experiments by Heckel. In this case the dimension of the displacement will be lost and the parameters become relative (Heikamp [39]).

Stahn [30] validated the experimental methods using a sintered polyethylene powder. In this way he obtained well-defined initial conditions and compressed without a die, avoiding the irregularities produced according to Poisson's law.

In Figs. 5 and 6 we see how the type of a model can change its behavior with the frequency, that is with the punch velocity (Müller [40]). So the Maxwell model converts from a single spring to a real Maxwell model and at the highest speed to a simple dashpot (Table 4).

It seems important to find the best model at a given speed which shows viscoelastic properties out of all possible combinations. If the chosen model is too complex, then it will degenerate. A single dashpot or a single spring assume such big numbers that in a calculation of, for instance, the creep compliance, there will be no effect of these elements because the reciprocals of the parameters are nearly zero and eliminate these terms. If there are too many Kelvin elements, this has two consequences: First, a huge parameter of the Kelvin spring brings the whole element to zero, or a huge parameter of the Kelvin dashpot eliminates the exponential term in the creep compliance. In this case only the effect of the spring is important. If both parameters are large and of the same magnitude, a free dashpot will occur (Sirithunyalug [31]).

The equation of Hooke is theoretically unlimited for a massless linear spring and is valid only in one dimension. Real elastic materials show a constant ratio in the change in length and diameter during their pull and push strain in their elastic section. In contrast, during compaction the expansion in diameter will be confined by the die. For these reasons the pure linear models of the theory of viscoelasticity cannot be perfect, as long the model is not three-dimensional (Bauer [28]).

On the other hand, the region of validity for the ideal viscoelastic behavior will probably be very small, because there are only some microns in the compaction bed where such effects are evident. Therefore, one needs an excellent resolution of the electronic displacement registration chain.

Analyses of the creep compliance, the relaxation functions or the constitutive equations of the tabletting processes require an exact synchronization of measurement for displacement and load. Thus the dependence from speed and acceleration becomes evident. The correction of the elasticity of the punch and of other parts of the machine is estimated by a punch-to-punch compression [1,2], which should be used for correction of every point of displacement in relation to the stress. The real displacement with respect to the time can be calculated from the corrected displacement and load measurements in the punch-to-punch compression. Therefore the assembled electronic modules and amplifiers to get and transmit the signals need special care. A simple test for good synchronization is a sudden stroke with a hammer to the upper side of the excenter and the time delay in the registration device between the first occurrence of stress and strain (Bauer [28]).

The investigations must discriminate elastic, viscous, and viscoelastic components during the phase of stress and release. In the present state of knowledge no one can expect to get precise results from the experiments. But it is easy to do "forward calculations" to realize the puzzle of stress and strain.

Theoretical investigations with oscillating load show new difficulties depending on the tabletting rate (Müller [40]): While elastic properties always remain proportional to the load, in contrast viscous needs a maximal load to hold the displacement constant (as in a displacement versus time curve) and viscoelastic (Maxwell or Kelvin model) materials give different pictures depending on the speed of tabletting as shown in Figs. 3 to 6.

In real experiments one will never see the theoretical punch upset to calculate l_0 in the sense of viscoelasticity. It is only possible to compare results within the same powder mass, for example at different speeds or of different maximal loads.

The only way to avoid all these difficulties seems to be the use of a well-equipped compaction simulator with precompression and an oscillating load to characterize the materials in the equilibrium state of viscous and elastic properties at different frequencies to determine the material constants. But even in this manner there remains the nightmare of more or less entrapped air. Casahoursat and co-workers [11,12] stopped the press and observed the stress while the displacement was varied by a crank of the lower punch. The authors defined the exponential decay of stress as the effect of escaping air.

So far as there is no better chance, we recommend the most primitive method, suggested by [1,4,22,40]: To use the time delay between load and displacement as a characteristic and try to find a function out of (35) to (41) to extrapolate the history prior to the load maximum, just to overcome the consolidation phase.

NOTATION

δ	loss angle
$\delta(t)$	Dirac delta function
$\{\delta^{(t)}\}$	linear differential time operator
$\Delta(t)$	Heaviside step function
ϵ	strain (displacement) ($\Delta l/l_0$) in the time domain
ϵ_0	initial strain (displacement)
ϵ_{max}	maximal strain (displacement) in oscillation experiments
ϵ_S	strain of a spring ($\Delta l/l_0$)
ϵ_D	strain of a dashpot ($\Delta l/l_0$)
$\overline{\epsilon}$	strain in the Laplace domain
$\dot{\epsilon}$	derivative of strain with respect to time
$\ddot{\epsilon}$	second derivative of strain with respect to time
E^*	absolute (complex) dynamic modulus
E_i	$i = 1$ to n, elasticity (Young's) modulus in the 1st to nth element; in reduced elements, $i = 1$ always
F_i	$i = 1$ to n, viscosity modulus in the 1st to nth element; in reduced element, $i = 1$ always
$G(t)$	stress relaxation function in the time domain
$\overline{G}(s)$	stress relaxation function in the Laplace domain
G'	storage modulus (real component)
G''	loss modulus (imaginary component)
$J(t)$	creep compliance in the time domain
$\overline{J}(s)$	Creep compliance in the Laplace domain
J'	storage compliance (real component)

J''	loss compliance (imaginary component)
η'	real component of viscosity
η''	complex component of viscosity
p_i'	absolute hybrid constants related to stress
p_i	relative hybrid constants related to stress
q_i'	absolute hybrid constants related to strain
q_i	relative hybrid constants related to strain
s	Laplace variable
σ	stress (load) in the time domain
σ_0	initial stress (load)
σ_{max}	maximal stress (load) in oscillation experiments
$\bar{\sigma}$	stress in the Laplace domain
σ_S	stress (load) of the spring
σ_F	stress (load) of the dashpot
$\dot{\sigma}$	derivative of stress with respect to time
$\ddot{\sigma}$	second derivative of stress with respect to time
t	time
t'	variable for integration
τ	relaxation time
$1/\tau$	retardation time
ω	radiant frequency

REFERENCES

1. D. Schierstedt, Rückdehnung der Tabletten während der Kompression, Diss. rer. nat., Bonn, 1982.
2. D. Schierstedt and F. Müller, *Pharm. Int. 44*: 932 (1982).
3. U. Caspar, Viskoelastische Phänomene während der Tablettierung, Diss. rer. nat., Bonn, 1983.
4. F. Müller and U. Caspar, *Pharm. Ind. 46*: 1049 (1984).
5. A. Y. K. Ho and T. M. Jones, *J. Pharm. Pharmacol. Suppl. 40*: 75 (1988).
6. N. M. A. H. Abourida, Ph.D. thesis, University of Wales, 1980.
7. E. T. Cole, J. E. Rees, and J. A. Hersey, *Pharm. Acta Helv. 50*: 28 (1975).
8. S. T. David and L. L. Augsburger, *J. Pharm. Sci. 66*: 155 (1977).
9. C. Wiederkehr von Vincenz, Instrumentierung und Einsatz einer Rundlauf-Tablettenpresse zur Beurteilung von pharmazeutischen Pressmaterialien, Diss., ETH Zürich, 1979.
10. J. E. Rees and P. J. Rue, *J. Pharm. Pharmacol. 30*: 601 (1978).
11. L. Casahoursat, G. Lemagnen, and D. Larrouture, *Drug Dev. Ind. Pharm. 14(15–17)*: 2179 (1988).
12. L. Casahoursat, G. Lemagnen, D. Larrouture, and A. Etienne. *Study of the Visco-elastic Behaviour of Some Pharmaceutical Powders*, II Congresso Int. diciencas farmaceuticas, Barcelona, 1987.

13. B. Emschermann, Korrelation von Presskraft und Tabletteneigenschaften, Diss. rer. nat., Bonn, 1978.
14. B. Emschermann and F. Müller, *Pharm. Ind. 43*: 191 (1981).
15. U. Tenter, Preßkraft- und Weg-Zeit-Charactreristik von Rundlaufpressen. Diss. rer. nat., Marburg, 1986.
16. P. C. Schmidt, S. Ebel, H. Koch, T. Profitlich, and U. Tenter. *Pharm. Ind. 50*: 1409 (1988).
17. P. C. Schmidt and H. Koch, *Pharm. Ind. 53*: 508 (1991).
18. R. Dietrich and J. B. Mielck, *Pharm. Ind. 46*: 863 (1984)
19. R. W. Heckel, *Trans. Metall. Soc. 221*: 1001 (1961).
20. R. J. Roberts and R. C. Rowe, *Int. J. Pharm. 37*: 15 (1987).
21. W. Jetzer, H. Leuenberger, and H. Sucker, *Pharm. Technol. 7(4)*: 33 (1983).
22. U. Caspar and F. Müller, *Pharm. Acta Helv. 59*: 329 (1984).
23. L. Sutanto, Berechnung von Parametern zur Beschreibung der Tablettierung, Diss. rer. nat., Bonn, 1986.
24. W. Fischer, Berechnung von Parametern zur Viskoelastizität bei der Tablettierung, Diss. rer. nat., Bonn, 1989.
25. D. W. Danielson, W. T. Morehead, and E. G. Rippie, *J. Pharm. Sci. 72*: 342 (1983).
26. E. G. Rippie and D. W. Danielson, *J. Pharm. Sci. 70*: 476 (1981).
27. W. T. Morehead, *Drug Dev. Ind. Pharm. 18(6,7)*: 659 (1992).
28. A. Bauer, Untersuchungen zur Prozeßdatengewinnung, Viskoelastizität und Struktur von Tabletten, Diss. rer. nat., Bonn, 1990.
29. G. E. Mase, *Theory and Problems of Continuum Mechanics*, McGraw-Hill, 1970.
30. P. Stahn, Modellrechnungen zur Parametrisierung von Tablettierkurven an Polyethylenproben, Diss. rer. nat., Bonn, 1990.
31. J. Sirithunyalug, Validierung der Berechnungsoperationen zur Bestimmung viskoelastischer Eigenschaften von Tabletten. Diss. rer. nat., Bonn, 1992.
32. W. Flügge, *Viscoelasticity*, 2nd ed., Springer-Verlag, Berlin, Heidelberg, New York, 1975.
33. A. C. Pipkin, *Lectures on Viscoelasticity Theory*, Springer-Verlag, New York, Heidelberg, Berlin, 1972.
34. B. W. Barry, in *Advances in Pharmaceutical Sciences*, Bean e.a., Vol. 4, Academic Press, London, 1974.
35. W. Press, B. P. Flannery, S. A. Teukolsky, and W. T. Vetterling, *Numerical Recipes*, 2nd ed., Cambridge University Press, 1986.
36. J. Neuhaus, Viskoelastische Phänomene beim Tablettieren auf Rundläuferpressen, Diss. rer. nat., Bonn, 1985.
37. H. Rieger, personal communication. RHRZ Universität Bonn.
38. E. Stepanek, *Praktische Analyse linearer Systeme durch Faltungsoperationen*, Akad. Verlagsges. Geest & Portig K.G., Leipzig, 1976.
39. H. P. Heikamp, Bestimmung viskoelastischer Eigenschaften von Pulvermischungen auf einer Rundläufertablettenpresse, Diss. rer. nat., Bonn, 1990.
40. A. Müller, Probleme bei der Auswertung von Weg-Spauuwigs-Daten an Hanol viskoelastischer Modelle bei der Tablettierung. Diss. rer. nat., Bonn, 1994.

6

Application of Percolation Theory and Fractal Geometry to Tablet Compaction

Hans Leuenberger, Ruth Leu, and Jean-Daniel Bonny

School of Pharmacy, University of Basel, Basel, Switzerland

I. SHORT INTRODUCTION TO PERCOLATION THEORY AND FRACTAL GEOMETRY

Percolation theory [1] and fractal geometry [2] represent novel powerful concepts which cover a wide range of applications in pharmaceutical technology [3]. Both concepts provide new insights into the physics of tablet compaction and the properties of compacts [4–12].

A. Percolation Theory

Different types of percolation can be distinguished: random-site, random-bond, random-site-bond, correlate chain, etc. Generally, percolation theory deals with the number and properties of clusters [1]. A percolation system is considered to consist of sites in an infinitely large real or virtual lattice. Applying the principles of random-site percolation to a particulate system, a cluster may be considered as a single particle or a group of similar particles which occupy bordering sites in the particulate system (see Figs. 1a,b). In the case of bond percolation, a group of particles is considered to belong to the same cluster only when bonds are formed between neighboring particles.

In random-bond percolation, the bond probability and bond strength between different components can play an important role. The bond proba-

P = 0,50

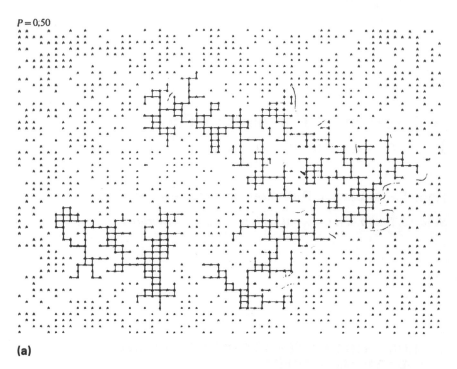

(a)

Fig. 1 (a) Example for percolation on a square lattice for $p = 0.50$ [1]. Occupied sites are shown with asterisk; empty sites are ignored. Two clusters are marked by lines. (b) Example for percolation on a square lattice for $p = 0.6$ [1]. The "infinite cluster" is marked by lines.

bility p_b can assume values between 0 and 1. When $p_b = 1$, all possible bonds are formed and the tablet strength is at its maximum; i.e., a tablet should show maximal strength at zero porosity when all bonds are formed. In order to form a stable compact it is necessary that the bonds percolate to form an "infinite" cluster within the ensemble of powder particles filled in a die and put under compressional stress. Tablet formation can be imagined as a combination of site and bond percolation phenomena. It is evident that for a bond percolation process the existence of an infinite cluster of occupied sites in a lattice is a prerequisite. Figure 2 shows the phase diagram of a site-bond percolation phenomenon [13].

Site percolation is an important model of a binary mixture consisting of two different materials. In the three-dimensional case, two percolation thresholds, p_c, can be defined: a lower threshold, p_{c1}, where one of the components just begins to percolate, and a second, upper percolation

$P = 0.60$

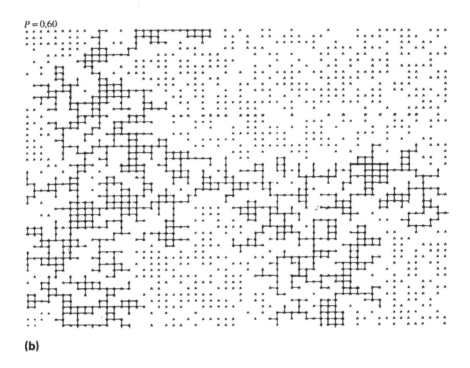

(b)

threshold, p_{c2}, where the other component ceases to have an infinite cluster. Between the two thresholds the two components form two interpenetrating percolating networks. Below the lower or above the upper percolation threshold, the clusters of the corresponding components are finite and isolated. Thus, in site percolation of a binary powder mixture, p_c corresponds to a critical concentration ratio of the two components. From emulsion systems these concentrations are well known where oil-in-water or water-in-oil emulsions can be prepared exclusively.

Table 1 shows critical volume-to-volume ratios for well-defined geometrical packing of monosized spherical particles. The critical volume-to-volume ratios depend on the type of percolation and the type of lattice. In the case of real powder systems the geometrical packing is a function of the particle size, the particle size distribution, and the shape of the particles.

As different types of packing of monosized spherical particles show different porosities, a powder system which has porosity ϵ can be represented in an idealized manner as an ensemble of monosized spheres having hypothetical mean diameter x and a mean coordination number k corresponding to hypothetical (idealized) geometrical packing. Table 2 shows

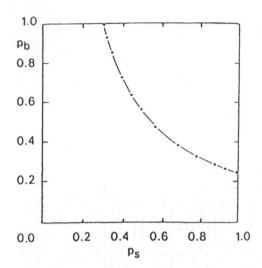

Fig. 2 Phase diagram of random site-bond percolation in the simple cubic lattice; p_s = site probability, p_b = bond probability (Monte Carlo simulation [13]). When $p_s = 1$, $p_b = 0.249$ = bond percolation threshold; when $p_b = 1$, $p_s = 0.312$ = site percolation threshold.

Table 1 Selected Percolation Thresholds for Three-Dimensional Lattices

Lattice type	Site	Bond
Diamond	0.428	0.388
Simple cubic	0.312	0.249
Body-centered cubic	0.245	0.179
Face-centered cubic	0.198	0.119

Source: From Ref. 1.

Table 2 Coordination Numbers of Isometric Spherical Particles for Different Packing Structures

Lattice type	Coordination number	Porosity
Diamond	4	0.66
Simple cubic	6	0.48
Body-centered cubic	8	0.32
Face-centered cubic	12	0.26

Source: From Ref. 19.

the coordination number of isometric spherical particles of different packing structures.

Using the simplified model of powder systems mentioned, the following equation was developed [14]:

$$k = \frac{\pi}{\epsilon} \tag{1}$$

for porosities in the range $0.25 < \epsilon < 0.5$. This equation is a rough estimate and does not hold for compacts, where usually $\epsilon < 0.25$.

At a percolation threshold some property of a system may change abruptly or may suddenly become evident. Such an effect starts to occur close to p_c and is usually called a critical phenomenon. As an example, the electrical conductivity of a tablet consisting of copper powder mixed with Al_2O_3 powder may be cited. The tablet conducts electricity only if the copper particles form an "infinite" cluster within the tablet, spanning the tablet in all three dimensions.

In case of a pharmaceutical tablet consisting of an active drug substance and excipients, the principle of function is not the electric conductivity and the tablet usually does not consist of a binary powder system compressed. However, often also in case of a complex tablet composition the system can be reduced to a type of binary powder system dividing the drug and excipients, etc., involved in two classes of function, such as material which is swelling or is easily dissolved in water. Thus, in case of a mixture KCl-StaRX 1500 cornstarch the two percolation thresholds expected are well recognized as a function of the disintegration time of the tablet [3] (see Fig. 3).

B. Fractal Geometry

Fractal geometry is related to the principle of self-similarity; i.e., the geometrical shape is kept identical independent of the scale, magnification, or power of resolution [2]. In practice the range of self-similarity may, however, sometimes be limited to only a few orders of magnitude.

A typical case of fractal geometry is the so-called Coastline of Britain Problem [2]: the length of the coastline is continuously increasing with increasing power of resolution, i.e., with a smaller yardstick to measure the length. Thus a log-log plot of the length of the coastline as a function of the length of the yardstick to perform a polygon approximation yields a straight line with slope $1-D_1$, where D_1 is equal to the fractal dimension of the coastline. Coastlines with perfect self-similarity can also be constructed mathematically (see Fig. 4). The fractal dimension of a coastline is thus in between the Euclidean dimension 1 for a straight line and 2 for a surface.

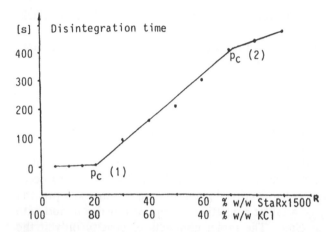

Fig. 3 Percolation thresholds for the compacted binary mixture KC1-StaRX 1'500; tablet property: disintegration time [3].

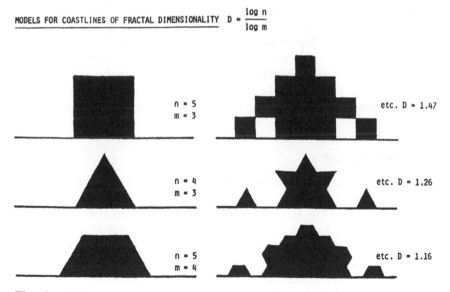

Fig. 4 Different self-similar structures as coastline models; m = theoretical number of equal parts of the unit length projected on a straight line, n = theoretical number of equal parts of unit length describing the coastline structure [5].

It is also fruitful to imagine a fractal surface dimension D_s describing the roughness of a surface. Such a description again includes the prerequisite of a self-similar shape independent of the scale. In this respect the introduction of a surface fractal is very advantageous in powder technology: the result of a measurement of the specific surface of a powder is, as is well known, dependent on the power of resolution of the apparatus (e.g., Blaine, mercury intrusion porosimetry, nitrogen gas adsorption BET method, etc.). Thus the result of a measurement with a chosen method is not able to describe adequately the roughness of the surface. However, if this roughness shows at least within a certain range an approximative self-similarity, it is possible to describe the surface by indicating a value for the specific surface and a value for the surface fractal D_s. Consequently, it is possible to know the specific surface data for different powers of resolution applying a log-log plot of the specific surface as a function of the yardstick length, describing the power of resolution, where the slope of the resulting straight line is equal to 2-D_s. As the slope is negative—i.e., the surface is larger for a smaller yardstick length—the value for D_s is between 2 and 3.

In case of porous material it is also possible to define a volume fractal. This concept is based on the fact that as a function of the power of resolution to detect a pore volume or pore size the void volume is increased. It is evident that in a practical case the porosity of a material attains a limiting value; i.e., the self-similarity principle is only valid within a limited range. Based on a mathematical self-similar model of pores—i.e., a Menger sponge (see Fig. 5)—the relationship between the accessible void space (sum of pore volumes) and the power of resolution of the pore size was established [9]. For this purpose the mercury intrusion porosimetry is the method of choice as the pore volume; i.e., the void space of a tablet is filled with mercury as a function of the mercury intrusion pressure, which is related to the accessible pore size. It is, however, necessary to keep in mind that as a consequence of percolation theory not all of the pores are accessible in the same way and some of them are not accessible at all; i.e., there may be closed pores present and pores of increased size may be hidden behind pores of smaller size, a fact that is responsible for the hysteresis loop between filling up and draining off the mercury from the void space within the tablet. These reservations have to be taken into account when the volume fractal dimension D_v of a porous network is determined.

In case of a porous network the solid fraction—i.e., relative density $\rho_r = 1 - \epsilon$ of the tablet (determined according to the pore volume fraction ϵ (d) filled up by mercury)—as a function of the pore diameter d is related to the volume fractal D_v as follows:

$$\log \rho_r = (3 - D_v)\log d + c \qquad (2)$$

with c = constant.

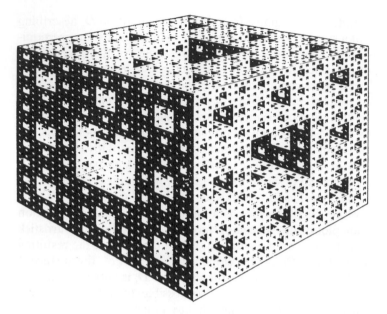

Fig. 5   Menger sponge with fractal dimension of 2.727 (idealized three-dimensional network of a pore system) [2].

It is a unique property of the Menger sponge that its surface and volume fractals D_s and D_v are identical and equal to 2.727 = log 20/log 3.

On the other hand, in case of an agglomerate or aggregate of the size L consisting of identical primary particles of diameter δ, the following relationship holds:

$$\log \rho_r = (3 - D_v)\log\left(\frac{\delta}{L}\right) + c \qquad (3)$$

with c = constant.

A volume fraction ϕ can be attributed to such an aggregate. The above equation plays an important role in the gelification of, e.g., silica particles, which is by nature a percolation process. It has to be taken into account, however, that secondary aggregates of volume fraction ϕ (fractal blobs) and size L and not the primary individual silica particles of diameter δ are the percolating units. Such frail structures may have fractal dimensions below D_v = 2. In case of Aerosil 200 aggregates a fractal dimension D_v = 1.77 was determined [15]. Thus, depending on the structure of an aggregate, the range of D_v overlaps the range of linear and surface fractals, introduced in a first step to trigger the imagination. This is not a contradiction, as the definitions of linear, surface, and volume fractals are arbitrarily related to the Euclidean dimensions to which we are better accustomed. In fact, the electron micrograph of an Aerosil aggregate shows a chainlike structure leading to a fractal dimension of 1.77 as mentioned. Other types may have a fractal dimension close to 2 or even 3. Well known by the work of Mandelbrot [2], the concept of fractal geometry has numerous applications in other fields. In this chapter the concept of fractal geometry is treated only in respect to the physics of tablet compaction and the resulting tablet properties.

II. THE FORMATION OF A TABLET [10]

Filling of the die: For simplicity, it is imagined that the volume of the die is spanned by a three-dimensional lattice. The lattice spacing is assumed to be of the order of a molecular diameter. Thus granules represent clusters of primary particles and primary particles are considered as clusters of molecules. After pouring the particles/granules to be compacted into the die, the lattice sites are either empty forming pores or occupied by molecules forming clusters with site occupation probability p_s. This site occupation probability is equal to the relative density $\rho_r = 1 - \epsilon = $ porosity.

A. Loose Powder Compacts

To fill hard gelatine capsules, one of the principles consists in performing a loose powder compact as a unit dosage form. As only relatively low compressional force is applied, no brittle fracture or plastic flow is expected. However, at the relative density ρ_r in the range of $\rho_p \leq \rho_r \leq \rho_t$ ($\rho_p = $ poured, $\rho_t = $ tapped relative density) bonds are already formed at the contact points throughout the powder bed. This process can be considered as a bond percolation problem. For a sufficient strength of this powder bed only a weak compressional stress σ_c is needed.

In this simplified model it is assumed that the particle size of the powdered substances is sufficiently small to form a cohesive compact under a weak compressional stress. In practice the residual moisture content and the individual substance specific capacity to form weaker or stronger bonds have to be taken into account. Thus a bond percolation threshold ρ_0 is expected in the range $\rho_p \leq \rho_0 \leq \rho_t$.

B. Dense Powder Compacts

1. Pharmaceutical Tablets

Tablets represent the majority of solid dosage forms on the pharmaceutical market. This position is due to its elegance and convenience in application.

Thus, a tablet usually has, among other properties, smooth surfaces, low friability, and sufficient strength (e.g., tensile strength or deformation hardness). For the production of such tablets the compressional stress σ_c needs to be important enough to induce plastic flow and/or brittle fracture of the primary granules or particles, i.e., to produce simultaneously new surfaces and bonds in this dense powder compact.

2. Process of Uniaxial Compression

The usual tableting machines work according to the principle of uniaxial compression. Thus the upper and lower surface areas of the tablet remain constant during the compression process; the thickness of the tablet is reduced with application of the compressional stress σ_z in the z-direction. Because of the initial high porosity of the powder bed, the radial transmission σ_r of the main stress cannot be calculated easily.

In the following simplified model of uniaxial compression, the radial stress σ_r need not be specified explicitly. However, a lateral displacement or rearrangement of particles occupying former pore sites is allowed. The compression process is now studied starting from a loose powder compact with a relative density ρ_p and a moving upper punch. Again a three-dimensional lattice with lattice spacing of molecular diameter spanning the die volume is imagined. During the compression the number of sites to be occupied is constantly reduced and the material (particles, granules, i.e., cluster of molecules) is available to occupy remaining sites. According to the principle of uniaxial compression the mean particle-particle separation distance is more reduced in the z-direction than in the lateral directions. Thus it can be assumed that in the beginning a one-dimensional bond percolation (i.e., a chain of molecules) is responsible for the stress transmission from the upper to the lower punch due to the repulsion forces of the electron shell (Born repulsion forces). After the rearrangement of the particles/granules at a relative density $\rho_r = \rho_r^*$ an important buildup of stress occurs as particles/granules can no longer be displaced easily. At this relative density ρ_r^* the compact can be considered as the "first" dense tablet.

On a molecular scale the molecules react in a first approximation as a hard-core spheres model and span as an infinite cluster the die. This situation is typical for a site percolation process. Above the percolation threshold $\rho_c = \rho_r^*$ still-empty lattice sites can be occupied due to brittle fracture and/or plastic flow of particles. Thus at higher compressional stress one can imagine that new bonds are formed with a certain bond formation probability p_b and sites are occupied with a site occupation probability p_s, typical for a site-bond percolation phenomenon. Figure 2 represents the phase diagram of a site-bond percolation process. Thus, due to the complex situation during the formation of a tablet, no sharp percolation threshold is expected.

3. Stress Transmission in the Die

It is well known that the compressional stress is transmitted from the upper punch to the lower punch by means of particle-particle contact in the powder bed. Thus, one may expect that the stress is conducted similar to electric current. As a consequence it is of interest to measure at the same time the stress and the electric current transmitted. This experimental work was realized by Ehrburger et al. [15], using as a conducting material different types of carbon black and for comparison the electrical insulating silica particles Aerosil 200. The physical characterization of the material tested is compiled in Table 3.

Silica was chosen because this material is often used to study the gelation process which can be adequately described by percolation theory. From nonlinear regression analysis the parameters of the following power laws were determined:

$$C = C_0(\rho_r - \rho_c)^t \tag{4}$$

$$\sigma_c = \sigma_0(\rho_r - \rho_0)^\tau \tag{5}$$

Where

C = conductivity [(ohm cm)$^{-1}$]
C_0 = scaling factor
σ_c = compressional stress transmitted
σ_0 = scaling factor
ρ_c, ρ_0 = percolation thresholds
t, τ = experimentally determined scaling exponents, expected to equal the electrical conductivity coefficient $\mu = 2$.

Table 3 Physical Characterization of Aerosil 200 and Different Types of Carbon Black [15]

Material	BET surface (m²/g)	Density (g/cm³)			Fractal dimension	
		True	Poured	Tapped		
Aerosil 200	202	2.2	0.016	0.022	1.77	±0.05
Types of carbon black						
Noir d'acétylène	80	1.87	0.027	0.041	1.99	±0.10
TB #4500	57	1.83	0.043	0.066	1.76	±0.12
TB #5500	206	1.84	0.025	0.043	1.83	±0.13
Sterling FT	15	1.85	0.27	0.40	3.0	

Table 4 Stress Transmission and Conductivity of Aerosil 200 and Different
Types of Carbon Black [15]

Material	Stress transmission			Conductivity		
	ρ_0	τ	ρ_{max}	ρ_c	t	ρ_{max}
Aerosil 200	0.025	1.5±0.1	0.06	—	—	—
Types of carbon black						
Noir d'acétylène	0.032	2.9±0.1	>0.24	0.024	1.9±0.1	≥0.24
TB #4500	0.050	2.2±0.1	0.15	0.040	1.8±0.1	≥0.27
TB #5500	0.033	2.1±0.1	0.10	0.019	1.8±0.1	>0.2
Sterling FT	0.27	3.9±0.2	>0.56	0.27	3.4±0.2	>0.56

The results of these investigations are compiled in Table 4. The authors [15] concluded that both the stress transmission and the conductivity
follow the power laws of percolation. The value of ρ_{max} should indicate the
range of relative densities ρ_r: $\rho_0 < \rho_r < \rho_{max}$ and $\rho_c < \rho_r < \rho_{max}$ where this
power law is still valid. Table 4 shows rather low values for the percolation
thresholds ρ_0 and ρ_c. This fact can be explained that in the process of
percolation secondary agglomerates (aggregates of size L) consisting of
primary carbon black or silica particles (of size δ) are responsible for the
stress transmission. Taking into account the fractal geometry of these aggregates, i.e., the volume fractal D_v, Ehrburger et al. [15] obtained a good
estimate for the ratio L/δ, using the following equation based on the percolation threshold ρ_0:

$$\rho_0 = 0.17\left(\frac{L}{\delta}\right)^{D_v-3} \tag{6}$$

The values calculated for L/δ were in good agreement with estimates
obtained from independent experiments (BET and porosimetry measurements).

III. TABLET PROPERTIES

A number of tablet properties are directly or indirectly related to the
relative density ρ_r of the compact. According to percolation theory the
following relationship holds for the tablet property X close to the percolation threshold p_c:

$$X = S(p - p_c)^q \tag{7}$$

Where

$$X = \text{tablet property}$$
$$p = \text{percolation probability}$$
$$p_c = \text{percolation threshold}$$
$$S = \text{scaling factor}$$
$$q = \text{exponent}$$

In case of a tablet property X the values of S and q are not known a priori. In addition, the meaning of p and p_c has to be identified individually for each property X. In the case of site percolation the percolation probability p is identical to the relative density ρ_r as mentioned earlier. For obvious reasons one can expect that the tensile strength σ_t and the deformation hardness of P are related to the relative density ρ_r in accordance with the percolation law (Eq. (7)).

Unfortunately, in a practical case only the experimental values of σ_t, P, and ρ_r of the tablet are known. In this respect, it is important to take into account the properties of the power law (Eq. (7)) derived from percolation theory: (1) the relationship holds close to the percolation threshold but it is unknown in general how close; (2) the a priori unknown percolation threshold p_c and the critical exponent q are related. Thus there is a flip-flop effect between p_c and q, a low p_c value is related to a high q value, and vice versa. As a consequence, the data evaluation based on nonlinear regression analysis to determine S, p_c, and q may become very tedious or even impossible.

In selected cases such as percolation in a Bethe lattice, the percolation exponent q is equal to unity. A list of selected exponents for 2, 3, and infinite dimensions (i.e., Bethe lattice) known from first principles is compiled in Table 5 and cited from Stauffer [1]. Details should be read there concerning the relevant equation for the property of the system described.

Table 5 Percolation Exponents for Two Dimensions, Three Dimensions, and in the Bethe Lattice and the Corresponding Quantity

Exponent	Dimension			Quantity/property
	2	3	Bethe	
α	$-2/3$	-0.6	-1	Total number of clusters
β	5/36	0.4	1	Strength of infinite network
γ	43/18	1.8	1	Mean size of finite clusters
ν	4/3	0.9	1/2	Correlation length
μ	1.3	2.0	3	Conductivity

In case of the tablet properties such as tensile strength σ_t and deformation hardness P no meaningful results can be obtained without additional expertise. It was a rewarding endeavor [10] to combine the following two equations derived earlier [16] with the well-known Heckel equation [17]:

$$\sigma_t = \sigma_{t\ max}(1 - e^{-\gamma_t \sigma_c \rho_r}) \tag{8}$$

$$P = P_{max}(1 - e^{-\gamma \sigma_c \rho_r}) \tag{9}$$

Where

$$\begin{aligned}
\sigma_{t\ max} &= \text{maximum tensile strength at } \epsilon \rightarrow 0 \\
P_{max} &= \text{maximum deformation hardness at } \epsilon \rightarrow 0 \\
\sigma_c &= \text{compressional stress} \\
\gamma, \gamma_t &= \text{compression susceptibility}
\end{aligned}$$

and

$$\ln \frac{1}{1-\rho_r} = a + b\sigma_c \tag{10}$$

Where a, b = constants specific to the particulate material compressed and σ_c = compressional stress. This derivation does not take into account the exact value of ρ_r (usually between 0.6 and 1) in the exponents of Eqs. (8) and (9). The combination of Eqs. (8) and (10) and Eqs. (9) and (10) yielded the following general relationships:

$$\sigma_t = \frac{\sigma_{t\ max}}{1 - \rho_c}(\rho_r - \rho_c) \tag{11}$$

$$P = \frac{P_{max}}{1 - \rho_c}(\rho_r - \rho_c) \tag{12}$$

with ρ_c = critical relative density (percolation threshold).

It is evident that Eqs. (11) and (12) are formally identical with the fundamental law of percolation theory (Eq. (7)):

$$\sigma_t = S'(\rho_r - \rho_c) \qquad \text{with } S' = \frac{\sigma_{t\ max}}{1 - \rho_c} \tag{13}$$

$$P = S(\rho_r - \rho_c) \qquad \text{with } S = \frac{P_{max}}{1 - \rho_c} \tag{14}$$

and with the exponent $q = 1$ corresponding to a percolation in a Bethe lattice [1].

The general relationships (11), (12) can be specified, on the one hand, for the formation of loose compacts and, on the other hand, for the formation of dense compacts.

For loose compacts, i.e., at a low-pressure range, ρ_c equals the bond percolation threshold ρ_0. P_{max} and $\sigma_{t\ max}$ of the low-pressure range do not correspond to the maximal possible deformation hardness or tensile strength respectively at the relative density $\rho_r = 1$, but describe the strength of the substance specific particle-particle interaction at low relative densities, where the primary particles have not yet lost their identity. The scaling factors S and S' are a measure of the strength of this interaction. In this range the compact can still be separated into the original particles.

For the formation of dense compacts, i.e., at a median pressure range, ρ_c equals the site percolation threshold ρ_r^*. In this case P_{max} and σ_{tmax} correspond to the maximum deformation hardness or tensile strength respectively of the substance at $\rho_r \to 1$.

At higher relative densities of the tablet the pore network may no longer form an infinite cluster. Thus another percolation threshold p_c has to be expected for $\rho_r = \rho_\pi$. It is evident that this threshold is important for, e.g., the disintegration time. Due to the complexity of the tablet formation again no sharp percolation threshold is expected at ρ_π.

The experimental methods and the data evaluation are described in detail in a recent paper [10]. For the physical characterization of the materials used see Table 6. The experimentally determined Heckel plot is approximated by two linear sections: a linear section for low and another for median pressures. The percolation thresholds ρ_0 and ρ_r^* are calculated on the basis of the intercepts of these two linear sections of the Heckel plot. It

Table 6 Physical Characterization of the Substances Used

	Densities					Mean particle size (μm)
	True (g/mL)	Poured (g/mL)	ρ_p	Tapped (g/mL)	ρ_t	
Avicel PH 102 FMC 2843	1.58	0.325	0.206	0.439	0.278	97.0
Caffeine anhydrous Sandoz 88828	1.45	0.323	0.223	0.417	0.288	55.0
Emcompress $CaHPO_4 \cdot 2H_2O$ Ed. Mendell & Co. E27B2	2.77	0.714	0.258	0.870	0.314	106.0
Lactose α-monohydrate DMV 171780	1.54	0.562	0.365	0.735	0.477	53.2
PEG 10'000 Hoechst 605331	1.23	0.568	0.462	0.719	0.585	135.8
Sta-RX 1'500 Sandoz 86823	1.50	0.606	0.404	0.741	0.494	68.1

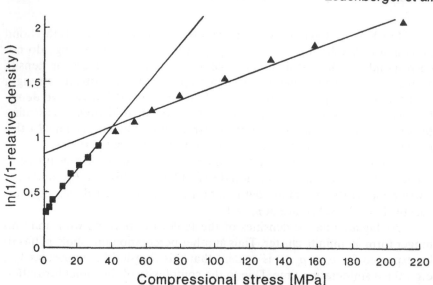

Fig. 6 Heckel plot of microcrystalline cellulose (Avicel) in order to determine the percolation thresholds for loose and dense compacts (ρ_0 and ρ_r^*).

is evident that the Heckel equation is an approximation of the pressure-density profile and the application of two linear sections leads to a better fit of the relationship found in reality (see Figs. 6–8). According to Eqs. (13) and (14) σ_t and P are plotted against the relative density and linearized by two regression lines for the same ranges as in the Heckel plot (see Figs. 9–14). In Table 7 the experimentally determined percolation thresholds from the Heckel plot (Eq. (10)), from the plot of tensile strength against relative density (Eq. (13)), and from the plot of deformation hardness against relative density (Eq. (14)) are summarized. Table 8 shows the values of S, S' and the resulting values for $\sigma_{t\,max}$ and P_{max} at the median pressure range. The values of $\sigma_{t\,max}$, P_{max}, γ, and γ_t according to Eqs. (8) and (9) are compiled in Table 9 and the corresponding plots are shown in Figs. 15 and 16. The comparison of the results and estimated standard deviations indicates that Eqs. (13) and (14) offer in general more reliable estimates for the $\sigma_{t\,max}$ and P_{max} values than Eqs. (8) and (9). However, the squared correlation coefficients of the evaluation according to Eqs. (13) and (14) are lower than according to Eqs. (8) and (9). This is due to the fact that the linear model is not as flexible as the exponential one. That is the reason why it can happen that the exponential model can lead to unreasonable values of the model parameters with large standard errors in spite of the good

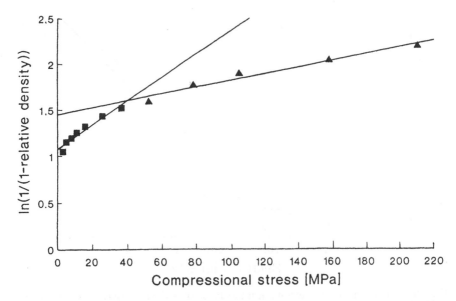

Fig. 7 Heckel plot of lactose in order to determine the percolation thresholds for loose and dense compacts ($\rho_{0\text{ and }}\rho_r^*$).

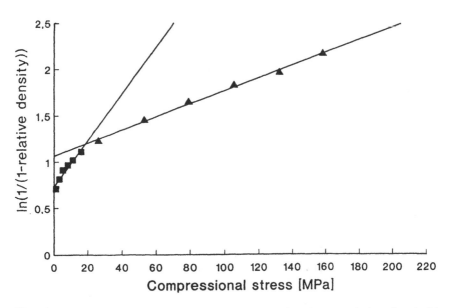

Fig. 8 Heckel plot of caffeine in order to determine the percolation thresholds for loose and dense compacts (ρ_0 and ρ_r^*).

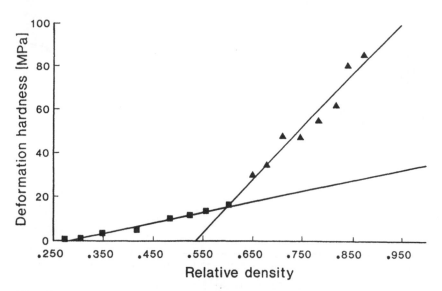

Fig. 9 Tablet property: deformation hardness of Avicel compacts as a function of relative density according to Eq. (14).

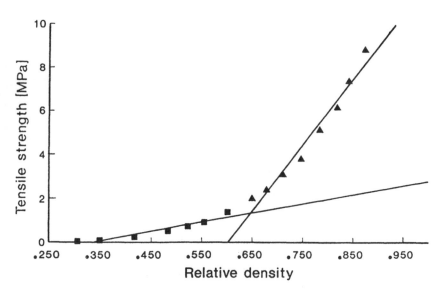

Fig. 10 Tablet property: tensile strength of Avicel compacts as a function of relative density according to Eq. (13).

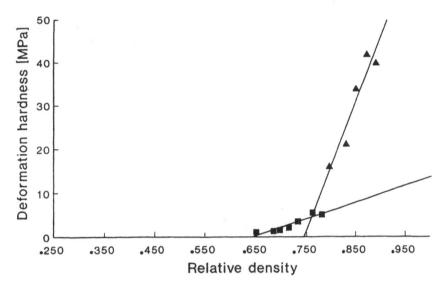

Fig. 11 Tablet property: deformation hardness of lactose compacts as a function of relative density according to Eq. (14).

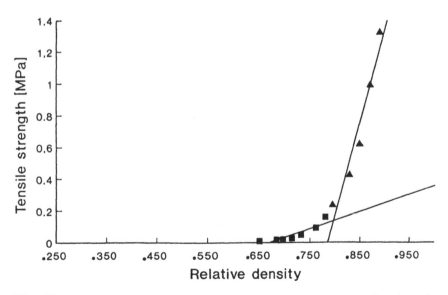

Fig. 12 Tablet property: tensile strength of lactose compacts as a function of relative density according to Eq. (13).

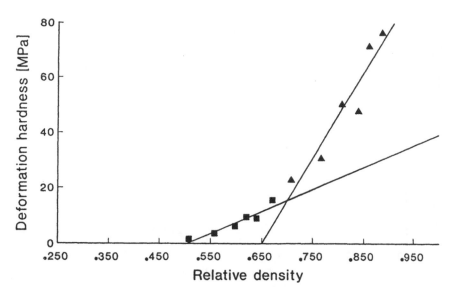

Fig. 13 Tablet property: deformation hardness of caffeine compacts as a function of relative density according to Eq. (14).

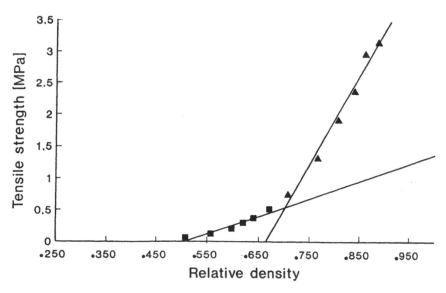

Fig. 14 Tablet property: tensile strength of caffeine compacts as a function of relative density according to Eq. (13).

Table 7 Comparison of Values for ρ_c Experimentally Determined According to Eqs. (10), (13), and (14) for (a) Loose and (b) Dense Compacts

(a) Loose compacts (i.e., low-pressure range $\rho_c = \rho_0$)					
Heckel plot Eq. (10)		Tensile strength Eq. (13)		Deformation hardness Eq. (14)	
$\rho_0 \pm s$	r^2	$\rho_0 \pm s$	r^2	$\rho_0 \pm s$	r^2
Avicel	0.275±0.013	0.922	0.333±0.023	0.911	0.278±0.011	0.980
Caffeine	0.521±0.028	0.939	0.464±0.014	0.946	0.500±0.010	0.952
Emcompress	0.429±0.025	0.924	0.454±0.016	0.854	0.448±0.020	0.801
Lactose	0.660±0.026	0.945	0.659±0.013	0.820	0.646±0.015	0.841
PEG 10'000	0.591±0.028	0.966	0.644±0.016	0.897	0.576±0.008	0.978
Sta-RX 1'500	0.506±0.015	0.976	0.521±0.010	0.917	0.499±0.012	0.927

(b) Dense compacts (i.e., median-pressure range $\rho_c = \rho_r^*$)					
$\rho_r^* \pm s$	r^2	$\rho_r^* \pm s$	r^2	$\rho_r^* \pm s$	r^2
Avicel	0.572±0.039	0.982	0.602±0.013	0.968	0.532±0.025	0.941
Caffeine	0.657±0.023	0.996	0.703±0.019	0.914	0.650±0.028	0.901
Emcompress	0.566±0.006	0.996	0.581±0.008	0.965	0.586±0.018	0.808
Lactose	0.765±0.052	0.966	0.775±0.010	0.930	0.757±0.011	0.934
PEG 10'000	0.807±0.147	0.943	0.813±0.011	0.969	0.823±0.033	0.745
Sta-RX 1'500	0.619±0.067	0.982	0.688±0.014	0.976	0.595±0.070	0.820

squared correlation coefficients. In the linear model the data are fitted by a straight line which results in poorer fits but at the same time more reasonable values for the parameters P_{max} and $\sigma_{t\,max}$.

The description of other tablet properties such as disintegration time [3], dissolution rate, etc., according to percolation theory are limited as long as the theoretical models are not established to allow more to be known about the respective percolation exponents. However, in the special case of a matrix-type slow release system [8,9] it is possible to apply simultaneously the concept of percolation theory and fractal geometry. This is the topic of the next section.

IV. DRUG DISSOLUTION FROM A MATRIX-TYPE CONTROLLED RELEASE SYSTEM

A. Ants in a Labyrinth and Drug Dissolution Kinetics

Molecules of an active substance, which are enclosed in a matrix-type controlled release system, may be called ants in a labyrinth [1] trying to

Table 8 S- and S'-values for Low- and Median-Pressure Ranges and P_{max}- and $\sigma_{t\,max}$-Values Resulting from the S- and S'-Values for Median Pressures According to Eqs. (13) and (14)

| | Tensile strength (Eq. (13)) | | |
| | Low-pressure range | Median-pressure range | |
	$S' \pm s$ (MPa)	$S' \pm s$ (MPa)	$\sigma_{t\,max} \pm s$ (MPa)
Avicel	4.199 ± 0.586	30.232 ± 2.261	12.023 ± 0.564
Caffeine	18.924 ± 2.253	2.443 ± 0.376	0.726 ± 0.075
Emcompress	0.467 ± 0.086	3.451 ± 0.379	1.446 ± 0.135
Lactose	0.705 ± 0.165	10.364 ± 1.421	2.332 ± 0.240
PEG 10'000	1.621 ± 0.316	14.322 ± 1.475	2.688 ± 0.141
Sta-RX 1'500	1.142 ± 0.172	12.167 ± 1.359	3.807 ± 0.282

| | Deformation hardness (Eq. (14)) | | |
| | Low-pressure range | Median-pressure range | |
	$S \pm s$ (MPa)	$S \pm s$ (MPa)	$P_{max} \pm s$ (MPa)
Avicel	48.17 ± 2.77	239.22 ± 24.55	111.85 ± 6.12
Caffeine	99.16 ± 11.19	309.34 ± 51.33	108.39 ± 10.19
Emcompress	21.60 ± 4.82	262.17 ± 73.77	108.63 ± 26.20
Lactose	40.72 ± 8.85	333.41 ± 44.34	81.17 ± 7.50
PEG 10'000	18.45 ± 1.25	139.80 ± 47.18	24.72 ± 4.50
Sta-RX 1'500	32.03 ± 4.49	74.69 ± 24.71	30.25 ± 5.11

escape from an ordered or disordered network of connected pores. For site occupation probabilities p_s far above the percolation threshold p_c, the random walk distance R of such an ant is related to time t as follows: $R^2 = Dt$, where D is the diffusivity. In a more general form this diffusion law can be expressed as $R \propto t^k$ with $k = 0.5$ for p_s above p_c. For p_s far below p_c, R approaches a constant for large times, i.e., $k = 0$. Right at the critical point the value of k ranges between these two extremes and is about 0.2 in three dimensions. This process is called anomalous diffusion.

From these first principles one can conclude that there is at least one percolation threshold, i.e., p_{c1}, where the active drug is completely encapsulated by the water-insoluble matrix substance and where the usual square-root-of-time law for the dissolution kinetics is no longer valid. As in general, in a three-dimensional system two percolation thresholds can be expected, an experiment was set up to elucidate this phenomenon. For

Table 9 Evaluation of the Data According to Eqs. (8) and (9) over the Whole Pressure Range

	Tensile strength (Eq. (8))		
	$\sigma_{t\,max} \pm s$ (MPa)	$\gamma_t \times 10^{-3} \pm s$ (MPa^{-1})	r^2
Avicel	14.44 ± 0.59	5.22 ± 0.30	0.999
Caffeine	4.32 ± 0.34	9.61 ± 1.26	0.995
Emcompress	5.89 ± 12.95	0.57 ± 1.29	0.995
Lactose	494.53 ± 28849	0.01 ± 0.80	0.994
PEG 10'000	3.34 ± 0.39	11.93 ± 2.15	0.992
Sta-RX 1'500	5.26 ± 0.70	3.98 ± 0.79	0.990
	Deformation hardness (Eq. (9))		
	$P_{max} \pm s$ (MPa)	$\gamma \times 10^{-3} \pm s$ (MPa^{-1})	r^2
Avicel	93.22 ± 5.65	12.34 ± 1.44	0.984
Caffeine	94.66 ± 15.64	10.99 ± 3.22	0.968
Emcompress	725.14 ± 10187	0.32 ± 4.60	0.944
Lactose	58.56 ± 12.15	7.45 ± 2.51	0.958
PEG 10'000	37.06 ± 17.73	8.85 ± 5.95	0.924
Sta-RX 1'500	20.38 ± 1.44	19.93 ± 4.20	0.945

this purpose, a highly water-soluble model drug (caffeine anhydrous) and a plastic water-insoluble matrix substance (ethyl cellulose) were chosen. The materials and methods used are described in detail by Bonny and Leuenberger [12]. Here, only the theoretical background and conclusions are summarized. In this experiment the drug content was varied from 10% to 100% (w/w), and the drug dissolution from one flat side of the tablets was studied. In order to measure this intrinsic dissolution rate the tablet was fixed into a paraffin matrix to leave only one side accessible for the dissolution medium (distilled water).

For low drug concentrations, i.e., low porosity of the matrix, most of the drug is encapsulated by the plastic matrix and the release is incomplete. At the lower percolation threshold p_{c1} the drug particles begin to form a connective network within the matrix and according to the theoretical considerations the diffusion should be anomalous. At the upper percolation threshold p_{c2} the particles which should form a matrix start to get isolated within the drug particles and the tablet disintegrates.

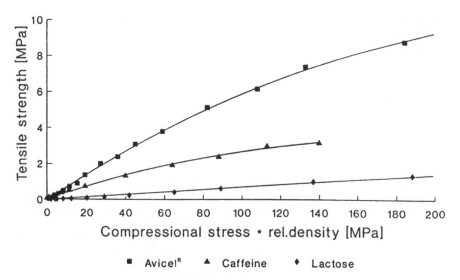

Fig. 15   Plot of the tensile strength σ_t versus the relative density ρ_r according to Eq. (8) for Avicel, caffeine, and lactose.

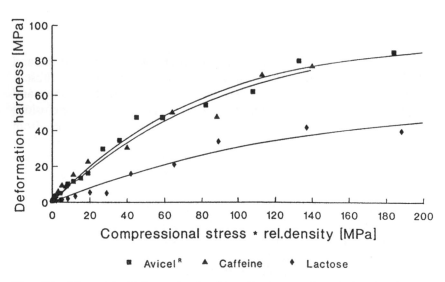

Fig. 16   Plot of the deformation hardness P_{max} versus the relative density ρ_r according to Eq. (9) for Avicel, caffeine, and lactose.

The concentration of the drug particles within the matrix can be expressed as site occupation probability p_s. The amount of drug $Q(t)$ released from one tablet surface after the time t is proportional to t^k and the exponent k depends on the percolation probability p_s:

Case 1: $p_s < p_{c1}$. Only the few particles connected to the tablet surface can be dissolved and $Q(t)$ reaches a constant value.

Case 2: $p_s \approx p_{c1}$. Anomalous diffusion with $k \approx 0.2$ in three dimensions [1].

Case 3: $p_{c1} < p_s < p_{c2}$. Normal matrix-controlled diffusion with $k = 0.5$.

Case 4: $p_{c2} < p_s$. Zero-order kinetics with $k = 1$.

Between the two percolation thresholds p_{c1} and p_{c2} the particles of drug and matrix substance form a bicoherent system; i.e., the drug release matches the well-known square-root-of-time law of Higuchi [18] for porous matrices:

$$Q(t) = \sqrt{\frac{D_0 \epsilon}{\tau} C_s (2A - \epsilon C_s)t} = \sqrt{DC_s(2A - \epsilon C_s)t} \tag{15}$$

where

Q = cumulative amount of drug released per unit exposed area
D_0 = diffusion coefficient of drug in permeating fluid
ϵ = total porosity of empty matrix
τ = tortuosity of matrix
C_s = solubility of drug in permeating fluid
A = concentration of dispersed drug in tablet
D = apparent or observed diffusion coefficient

Close to the percolation threshold the observed diffusion coefficient obeys a scaling law which can be written as [1]

$$D \propto (p_s - p_{c1})^\mu \tag{16}$$

where

p_s = site occupation probability
p_{c1} = critical percolation probability (lower percolation threshold)
μ = conductivity exponent = 2.0 in three dimensions [1]

In the case of a porous matrix p_s can be expressed by the total porosity ϵ of the empty matrix and p_{c1} corresponds to a critical porosity ϵ_c, where the pore network just begins to span the whole matrix. Equation (16) can then be written as

$$D \propto (\epsilon - \epsilon_c)^\mu \quad \text{or} \quad D = \kappa D_0(\epsilon - \epsilon_c)^\mu \tag{17}$$

with κD_0 = scaling factor.

In the cases where the dissolution kinetics are in agreement with Eq. (15) the dissolution data can be linearized by plotting $Q(t)$ versus $\sqrt{t}$ giving a regression line with the slope b,

$$b = \sqrt{DC_s(2A - \epsilon C_s)} \tag{18}$$

which leads to

$$D = \frac{b^2}{C_s(2A - \epsilon C_s)} \tag{19}$$

Combining Eqs. (17) and (19) and assuming $\mu = 2.0$ results in

$$\frac{b^2}{C_s(2A - \epsilon C_s)} = \kappa D_0(\epsilon - \epsilon_c)^{2.0} \tag{20}$$

or

$$\beta = \frac{b}{\sqrt{2A - \epsilon C_s}} = \sqrt{\kappa D_0 C_s}(\epsilon - \epsilon_c) \tag{21}$$

i.e., the tablet property β is determined by a linear relationship of ϵ,

$$\beta = c(\epsilon - \epsilon_c) = -c\epsilon_c + c\epsilon \tag{22}$$

where the constant c equals $\sqrt{\kappa D_0 C_s}$.

By the help of Eq. (22) ϵ_c can easily be calculated by using a nonlinear, or even a linear, regression analysis, giving a slope of c and an intercept of $-c\epsilon_c$.

The results of the intrinsic dissolution test of the tablets with the different caffeine loadings are plotted in Fig. 17. A change in dissolution kinetics can be assumed between 70% and 80% of caffeine loading. In order to test this assumption the release data are evaluated according to the model $Q(t) = a + b\sqrt{t}$ by a simple linear regression to clarify for which loadings the square-root-of-time law is fulfilled. To analyze the diffusion mechanism the data are also evaluated according to $Q(t) = a' + b't^k$ by a nonlinear least square fit. The results are compiled in Table 10.

Comparing the squared correlation coefficients of the $\sqrt{t}$ evaluation, there is a clear decrease in the grade of correlation for caffeine loadings higher than 70%; i.e., the drug loadings from 80% to 100% caffeine are no longer in good agreement with the $\sqrt{t}$ law. For low drug concentrations only a small amount of drug is released and the dissolution curve runs nearly parallel to the abscissa. In these cases the correlation coefficient cannot be used as an indicator for compliance with the model.

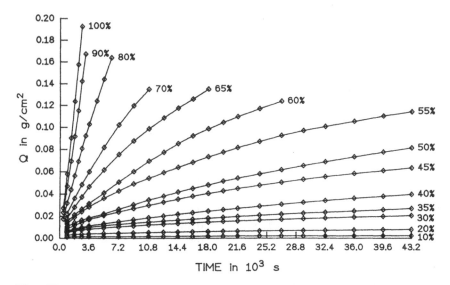

Fig. 17   Cumulative amount $Q(t)$ of caffeine released per unit area as a function of time for tablets with caffeine loadings between 10% (w/w) and 100% (w/w).

The estimation of k according to the model $Q(t) = a' + b't^k$ yields values for k between 0.17 and 1.09. For 35% to 55% of caffeine the exponent k ranges between 0.41 and 0.61, which is in good agreement with the $\sqrt{t}$ kinetics with $k = 0.5$. For higher loadings there is a clear change from $\sqrt{t}$ kinetics to zero-order kinetics with $k = 1$. Both evaluations show that the upper percolation threshold p_{c2} lies between 70% and 80% of caffeine.

For the quantitative determination of the lower percolation threshold p_{c1}, i.e., the critical porosity ϵ_c, Eqs. (21) and (22) are used. The needed data are summarized in Table 11.

The initial porosity ϵ_0 before leaching is calculated from the apparent volume V_{tot} and the true volume V_t of the tablet constituents:

$$\epsilon_0 = \frac{V_{tot} - V_t}{V_{tot}} \tag{23}$$

ϵ_d is the porosity corresponding to the volume occupied by the drug substance in the matrix and is calculated as follows:

$$\epsilon_d = \frac{m_d}{\rho_d V_{tot}} \tag{24}$$

Table 10 Evaluation of Dissolution Data and Percolation
Thresholds

Drug content	$Q(t) = a + b\sqrt{t}$		$Q(t) = a' + b't^k$	
	b	r^2	k	r^2
10	0.006	0.9858	0.17	0.9615
20	0.027	0.9975	0.27	0.9932
Lower percolation threshold expected				
30	0.085	0.9989	0.37	0.9983
35	0.116	0.9994	0.41	0.9973
40	0.189	0.9954	0.52	0.9963
45	0.320	0.9987	0.54	0.9992
50	0.415	0.9929	0.61	0.9980
55	0.593	0.9989	0.56	0.9979
60	0.858	0.9936	0.67	0.9979
65	1.15	0.9983	0.66	0.9991
70	1.56	0.9941	0.74	0.9996
Upper percolation threshold expected				
80	2.55	0.9863	0.84	0.9998
90	4.11	0.9752	1.02	0.9991
100	5.38	0.9784	1.09	>0.9999

Drug content in % (w/w)
b = slope in 10^{-3}g cm^{-2} s$^{-1/2}$
k = dimensionless exponent
r^2 = squared correlation coefficient

with m_d = total amount of drug present in the tablet and ρ_d = true density of drug. ϵ is the total porosity of the empty matrix:

$$\epsilon = \epsilon_0 + \epsilon_d \tag{25}$$

D is calculated according to Eq. (19), and the tablet property β according to Eq. (21).

For estimating ϵ_c with the help of Eq (22), the data for ϵ and β for 35% to 55% of caffeine are used, because in this range there is the best agreement with the normal diffusion law, i.e., $\sqrt{t}$ kinetics with $k = 0.5$ (see Table 10). The nonlinear regression yields $\epsilon_c = 0.35 \pm 0.01$ and $c = (2.30 \pm 0.15) \cdot 10^{-3}$ g$^{1/2}$ cm$^{-1/2}$ s$^{-1/2}$. A linear regression analysis leads to the same result. The critical porosity of 0.35 corresponds to a caffeine content of about 28% (w/w). Figure 18 shows the plot of β versus ϵ where the point of intersection with the abscissa just indicates ϵ_c.

Table 11 Calculation of D and the Tablet Property β

Drug content	ϵ_0	ϵ_d	ϵ	A	b	D	β
10	0.134	0.078	0.212	0.110	0.006	0.0045	0.013
20	0.128	0.158	0.286	0.225	0.027	0.0437	0.041
30	0.121	0.242	0.363	0.344	0.085	0.282	0.104
35	0.118	0.285	0.403	0.405	0.116	0.446	0.130
40	0.116	0.328	0.444	0.466	0.189	1.03	0.198
45	0.109	0.375	0.484	0.532	0.320	2.58	0.313
50	0.110	0.418	0.528	0.594	0.415	3.88	0.384
55	0.106	0.465	0.571	0.660	0.593	7.13	0.520
60	0.103	0.512	0.615	0.727	0.858	13.5	0.717
65	0.098	0.562	0.660	0.798	1.15	22.2	0.918
70	0.099	0.608	0.707	0.863	1.56	37.7	1.197
80	0.092	0.708	0.800	1.006	2.55	86.4	1.811
90	0.088	0.811	0.899	1.151	4.11	196	2.729
100	0.080	0.920	1.000	1.306	5.38	296	3.353

Drug content in % (w/w)
ϵ_0 = initial tablet porosity
ϵ_d = porosity due to drug content
ϵ = total porosity of matrix
A = concentration of dispersed drug in tablet in g cm^{-3}
b = slope in 10^{-3} g cm^{-2} s$^{-1/2}$
D = apparent diffusion coefficient in 10^{-6} cm^2 s^{-1}
β = tablet property in 10^{-3} g$^{1/2}$ cm$^{-1/2}$ s$^{-1/2}$

In Table 1 selected percolation thresholds [1] for three-dimensional lattices (as volume-to-volume ratios) and in Table 2 the corresponding coordination numbers [19] for isometric spherical particles are compiled. Comparing the experimentally determined percolation threshold of 0.35 (volume-to-volume ratio) with the theoretical values shows good agreement with the simple cubic lattice, which has a site percolation threshold of 0.312. In the tablet a brittle and a plastic substance of different grain size are compacted together so that the postulation of a lattice composed of isometric spheres is only a rough estimate, but it points out that the magnitude of ϵ_c is reasonable.

B. Fractal Dimension of the Pore System of a Matrix-Type Slow Release System

If the matrix-type slow release dosage form of the preceeding chapter is removed from the dissolution medium after, e.g., a maximum of 60%

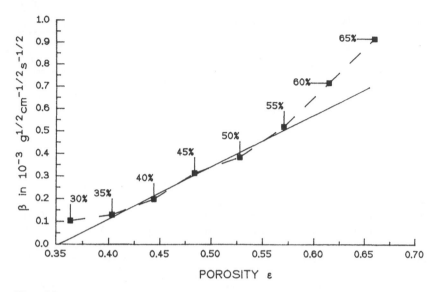

Fig. 18 Tablet property β as a function of porosity ϵ (power law) according to percolation theory.

(w/w) of drug dissolved to guarantee the physical stability of the remaining carcass, the open pore system left can be analyzed by mercury intrusion porosimetry to determine the fractal dimension. It is evident that the pore structure depends on the particle size distribution of the brittle model drug caffeine anhydrous originally embedded in the plastic ethyl cellulose matrix. The volume fractals [9] of leached tablets with an initial caffeine content of 50% (w/w) range between 2.67 and 2.84, depending on the particle size distribution of the water-soluble drug. The system which contained caffeine with a broad particle size distribution (125–355 μm) yielding a fractal volume dimension of 2.734 is closest to the dimension of the Menger sponge ($D_v = 2.727$).

C. Fractal Dimension of the Porous Network of a Fast Disintegrating Tablet

In the thesis of Luy [20] tablets were produced from granules which showed a very high porosity and internal surface. Those granules were obtained with a novel process technology: vacuum fluidized-bed spray granulation [21,22]. Independent of the granule size distribution the resulting tablets released 100% of the soluble drug (solubility in water < 0.01%) within 5 min. The disintegration time was smaller than 1 min. The fractal dimension of the porous network ranged from 2.82 to 2.88.

V. CONCLUSIONS

The concept of percolation theory and fractal geometry allowed new insights into the physics of tablet compaction and the properties of the tablets. The results attained so far are promising and should stimulate further research in this field.

LIST OF SYMBOLS

A	concentration of the dispersed drug in the tablet
b	slope of the regression line in the plot of $Q(t)$ versus $\sqrt{t}$
C_s	solubility of the drug in the permeating fluid
D	diffusivity or apparent diffusion coefficient
D_0	diffusion coefficient of the drug in the permeating fluid
D_1	linear fractal dimension
D_s	surface fractal dimension
D_v	volume fractal dimension
k	mean coordination number
L	size of agglomerate or aggregate
p	percolation probability
p_b	bond formation probability
p_c	percolation threshold
p_{c1}	lower percolation threshold
p_{c2}	upper percolation threshold
P	deformation hardness (Brinell hardness)
$P_{\max}$	maximum deformation hardness at $\epsilon \to 0$
p_s	site occupation probability
q	scaling exponent
Q	cumulative amount of drug released per unit exposed area
S	scaling factor
S'	scaling factor
t	scaling exponent
β	tablet property as defined in Eq. (21)
δ	diameter of particle; size of primary particle
ϵ	total porosity of the empty matrix
ϵ_o	initial porosity
ϵ_c	critical porosity of the matrix
μ	conductivity exponent
ρ_c	percolation threshold
$\rho_{\max}$	maximal relative density where the percolation power law still is valid
ρ_p	poured relative density

ρ_r relative density

ρ_π percolation threshold of the pore network

ρ_r^* site percolation threshold, relative density of the "first" dense tablet

ρ_t tapped relative density

ρ_0 percolation threshold

σ_t tensile strength

$\sigma_{t\,max}$ maximum tensile strength at $\epsilon \to 0$

τ scaling exponent; tortuosity of the matrix

REFERENCES

1. D. Stauffer, *Introduction to Percolation Theory*, Taylor and Francis, London and Philadelphia, 1985.
2. B. B. Mandelbrot, *The Fractal Geometry of Nature*, Freeman, San Francisco, 1982.
3. H. Leuenberger, B. D. Rohera, and C. Haas, *Int. J. Pharm. 38*: 109 (1987).
4. L. E. Holman, and H. Leuenberger, *Int. J. Pharm. 46*: 35 (1988).
5. H. Leuenberger, L. E. Holman, M. Usteri, and S. Winzap, *Pharm. Acta Helv. 64*: 34 (1989).
6. D. Blattner, M. Kolb, and H. Leuenberger, *Pharm. Res. 7*: 113 (1990).
7. L. E. Holman and H. Leuenberger, *Powder Technol. 64*: 233 (1991).
8. H. Leuenberger, J. D. Bonny, and M. Usteri, Proc. Second World Congress Particle Technology, Kyoto, Japan, 1990.
9. M. Usteri, J. D. Bonny, and H. Leuenberger, *Pharm. Acta Helv. 65*: 55 (1990).
10. H. Leuenberger, and R. Leu, *J. Pharm. Sci. 81*: 976 (1992).
11. J. D. Bonny, and H. Leuenberger, Proc. Int. Symp. Control. Rel. Bioact. Mater., 18, Controlled Release Society, Inc., 1991.
12. J. D. Bonny, and H. Leuenberger, *Pharm. Acta Helv. 66*: 160 (1991).
13. D. Stauffer, A. Coniglio, and M. Adam, *Adv. Pol. Sci. 44*: 103 (1982).
14. W. O. Smith, P. D. Foote, and P. F. Busang, *Phys. Rev. 34*: 1272 (1929).
15. F. Ehrburger, S. Misono, and J. Lahaye, *Conducteurs granulaires, théories, caractéristiques et perspectives*, journée d'études Oct. 10, Paris, Textes de communication, pp. 197–204, 1990.
16. H. Leuenberger, *Int. J. Pharm. 12*: 41 (1982).
17. R. W. Heckel, *Trans. Metall. Soc. AIME, 221*: 671 (1961).
18. T. Higuchi, *J. Pharm. Sci. 52*: 1145 (1963).
19. P. J. Sherrington, and R. Oliver, *Granulation*, Heyden, London, 1981, p. 34.
20. B. Luy, *Vakuum-Wirbelschicht*, Ph.D. thesis, University of Basel, Basel, 1991.
21. B. Luy, P. Hirschfeld, and H. Leuenberger, *Drugs Made in Germany, 32*: 68 (1989).
22. H. Leuenberger, B. Luy, and P. Hirschfeld, Proc. Preworld Congress Particle Technology, Gifu, Japan, 1990.

7

Mechanical Strength

Peter N. Davies* and J. Michael Newton

School of Pharmacy, University of London, London, United Kingdom

I. INTRODUCTION

The determination of the mechanical strength of pharmaceutical compacts is carried out for several reasons:

1. As an in-process control to ensure that tables are sufficiently strong to withstand handling yet reman bioavailable
2. To assist in obtaining a fundamental understanding of compaction mechanisms
3. To aid in the characterization of the mechanical properties of the compacted material

For many years the strength of pharmaceutical compacts has been determined in terms of the force required to fracture a specimen across its diameter. The fracture load obtained is usually reported as a "hardness value." This is an unfortunate use of a term which has a specific meaning in material science, associated with indentation. Mechanical strength would be a better term to use. Such a test does not take into account the mode of failure or the dimensions of the tablet; i.e., the result is not a fundamental property of the compact. An ideal test would allow comparison to be made

---

*Current affiliation: Roche Products, Welwyn Garden City, Hertfordshire, United Kingdom.

between samples of different shapes or sizes. An essential feature of such a test is that the geometry of the specimens and the loading must be such that a calculable stress state prevails at the section where fracture occurs so that the fracture stress can be readily calculated from the fracture load.

II. DIRECT TENSILE TESTING

Pharmaceutical powder compacts tend to be brittle, that is fracture is not preceded by significant permanent deformation. For this reason the simple tensile specimen is not ideal for these materials and is rarely used. Tensile tests have been performed on brittle materials, such as gypsum, by preparing dumbbell-shaped specimens. The enlarged ends are held in special grips and an axial tensile load applied. The results of such uniaxial tests are open to question. Photoelastic studies show large stress concentrations in the gripped portions of the specimen which can cause fracture elsewhere than in the central cross section. In tests with concrete conglomerates, the fracture usually occurs at the weakest point between the grips and the central cross section [1]. Strength calculations using this testing procedure neglect these stress concentrations. Also, small misalignments in the grips can add a bending component to the applied tensile stress. Ductile materials would correct this problem by plastic flow, without significantly affecting the results. Brittle materials have little or no capacity to flow plastically and the bending stress can seriously lower the measured strength [2]. In the case of gypsum the material can be introduced into the mold while in a fluid state before it hardens, making the preparation of samples relatively simple. The preparation of compacted powder specimens would be problematic, making the method unsuitable for the testing of pharmaceutical compacts.

An axial tensile strength test has been used to determine the tensile strength of round plane-faced tablets [3]. The tablets were fixed to a pair of platens by a cyanoacrylate adhesive and strained in tension until fracture occurred. No analysis of the stress state induced by this method is reported.

III. FLEXURAL TESTS

In the flexural test or bending test a specimen in the form of a parallel beam is subjected to three- or four-point bending and the maximum tensile stress estimated from the load at fracture. A feature of such a test is that under the correct conditions of loading the specimen will be subjected to a pure longitudinal tensile stress along a line on the opposite surface to that on which the load is applied.

In general for a beam subjected to bending the tensile fracture stress, σ_f, can be calculated from the following expressions [4]:

$$\sigma_f = \frac{My_{max}}{I} \tag{1}$$

where

> M = bending moment at fracture
> y_{max} = transverse coordinate measure in the plane of bending (Fig. 1)
> I = second moment of bending

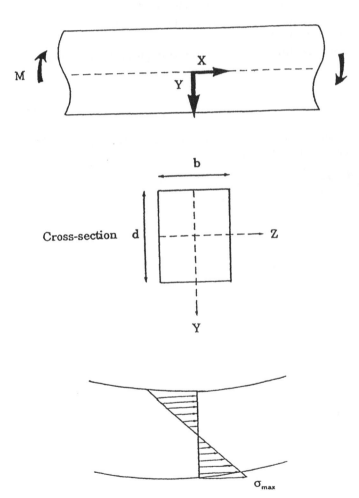

Fig. 1 Diagrammatic representation of the stresses induced by beam bending tests.

For a beam of rectangular cross section, width b, and depth d,

$$I = \frac{bd^3}{12}$$ (2)

and

$$y_{max} = \frac{d}{2}$$ (3)

The maximum bending moment, M, is dependent on the loading configuration used. In practice, one of two loading models is used: three-point loading or four-point loading (Fig. 2).

For the symmetrical three-point loading configuration the maximum bending moment is

$$M = \frac{WL}{4}$$ (4)

where W is the fracture load and L is the distance between supports. Thus at failure the tensile fracture stress is

$$\sigma_f = \frac{WL}{4} \cdot \frac{d}{2} \cdot \frac{12}{bd^3}$$ (5)

leading to

$$\sigma_f = \frac{3WL}{2bd^2}$$ (6)

The maximum bending moment obtained using the four-point loading configuration is

$$M = \frac{Wa}{2}$$ (7)

thus at failure,

$$\sigma f = \frac{Wa}{2} \cdot \frac{d}{2} \cdot \frac{12}{bd^3}$$ (8)

leading to

$$\sigma_f = \frac{3Wa}{bd^2}$$ (9)

Modifications for asymmetric cross sections can be readily introduced.

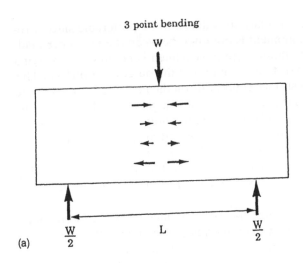

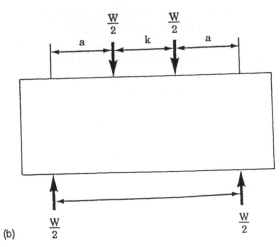

Fig. 2 Loading configurations for beam bending tests: (a) three-point bending, (b) four-point bending.

Four-point loading is usually considered to be preferable since a region of constant bending moment is obtained between the two inner loading points. The stress distribution in the specimen is nonuniform, varying from zero at the neutral axis to a maximum at the outer edge surface. This accentuates the effects of surface conditions on the measured strength, and results obtained using this test are considerably higher than the true tensile strength [5]. Beams of rectangular cross section prepared from pharmaceutical materials have been tested using this method to characterize the material properties of compacted powders [6, 7]. Rectangular beams are not a conventional tablet shape and the density distributions developed during beam formation are unlikely to be the same as those produced in a right circular cylinder [8].

The three-point flexure test has been applied to plane-faced round tablets [9]. This was a manually operated test, which meant the application of stress was variable and operator dependent. While the design of the test produced a bending moment there was also likely to be a large contribution of shear stresses at failure.

David and Augsburger [10] also used the three-point flexure to measure the tensile strength of round tablets. One problem with such an approach is that the dimensions of the tablets are unsuitable, generally being too stubby to allow a valid application of the equations derived from classical theories of bending; shear stresses would be important in such specimens. It is unlikely that the tensile fracture stresses obtained by diametral compression and flexure will be equal [11]. It is, therefore, surprising that David and Augsburger [10] found results that correlated very well. This could be explained by the fact that the mathematical analysis was incomplete. Stanley and Newton [12] stated that some of the mathematical analysis was based on incorrect assumptions.

Capsule-shaped tablets have been tested using a three-point flexure test [13]. The breaking load was used to express the strength of the compacts and no attempt was made to calculate the tensile stresses. Stanley and Newton [12] calculated the theoretical tensile stresses induced by the three-point flexure tests on capsule specimens. It was concluded that a reasonable approximation of the tensile stresses could be obtained using the equation

$$\sigma_x = \frac{3WI}{2d^2} \left[\frac{6 + 2h_c}{6A_c + bd} \right] \tag{10}$$

where A_c is the area of the curved segment of the capsule, h_c is the height of that segment, and other terms are as in Eqs. (2) and (4).

The flexure test has been used to characterise the strength of compacts prepared from pharmaceutical materials. The test does not, however, lend itself to the testing of circular compacts or thin square compacts where the shear stresses significantly affected the results obtained.

IV. DIAMETRAL COMPRESSION TEST

A. Analytical Solution

A method of testing brittle materials that does not suffer from the previous disadvantages is the diametral compression test. This consists of a simple flat-faced disc specimen which is subjected to two diametrically opposed point loads. The diametral compression test was developed independently at the same time by Barcellos and Carneiro [14] in Brazil and by Akazawa [15] in Japan and is referred to as the "Brazilian disc" or indirect tensile test, the indirect referring to the fact that a tensile fracture is obtained from compressive loading.

The test has been used to measure the tensile strength of concrete [11], coal [5], gypsum [16], and pharmaceutical compacts [17]. A complete analytical solution exists for the stress state induced by this loading configuration.

Assuming plane stress and considering a circular disc with concentrated loads on the diameter, three general equations can be used to express the stress conditions at all points within the disc (Fig. 3) [18]:

$$\sigma_x = \frac{-2P}{\pi t} \left(\frac{(R-y)x^2}{r_1^4} + \frac{(R+y)x^2}{r_2^4} - \frac{1}{D} \right) \tag{11}$$

$$\sigma_y = \frac{-2P}{\pi t} \left(\frac{(R-y)^3}{r_1^4} + \frac{(R+y)^3}{r_2^4} - \frac{1}{D} \right) \tag{12}$$

Considering points between OC on the horizontal X axis where $y = 0$,

$$\tau_{xy} = \frac{2P}{\pi t} \left(\frac{(R-y)^2 x}{r_1^4} - \frac{(R+y)^2 x}{r_2^4} \right) \tag{13}$$

$$r_1 = r_2 = \sqrt{x^2 + R^2} \tag{14}$$

Both stresses vanish at the circumference and reach maximum values at the center. The stresses at the center are

$$\sigma_x = \frac{2P}{\pi t D} \tag{15}$$

$$\sigma_y = \frac{-6P}{\pi t D} \tag{16}$$

Therefore, the compressive strength needs to be at least three times the tensile strength to ensure a tensile failure.

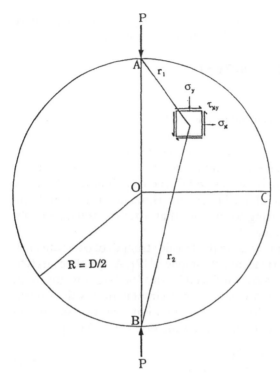

Fig. 3 Diagram of stress distribution across loaded diameter for a cylinder compressed between two line loads.

Along the Y-axis AB where $x = 0$,

$$r_1 = R - y \tag{17}$$

$$r_2 = R + y \tag{18}$$

$$\tau_{xy} = 0 \tag{19}$$

$$\sigma_x = \frac{2P}{\pi t D} \tag{20}$$

$$\sigma_y = \frac{-2P}{\pi t} \left(\frac{2}{D-2y} + \frac{2}{D+2y} - \frac{1}{D} \right) \tag{21}$$

This indicates that tensile failure of the specimen could start at any point along the vertical AB axis due to the even distribution of tensile stresses.

The compressive stress on this axis increases from $\sigma_y = -6P/\pi D t$ at the center to $\sigma y = \infty$ at the loading points. With a concentrated load, the specimen will fail at the loading points due to the compressive stresses and not in the central part of the specimen due to the tensile stress.

B. Effect of Loading Conditions on the Diametral Compression Test

In practice, the theoretical condition of lines of contact between specimen and platens cannot be achieved; instead the load is distributed over areas of contact. Comparison of Fig. 4a, b shows, that provided the area of contact is small in relation to the diameter of the disc, this only affects the stress distribution near the ends of the loaded diameter and the tensile stress is $2P/\pi Dt$ over a large fraction of the loaded diameter [2].

It has been shown that tensile stresses can be held uniform across a reasonable proportion of the loaded diameter if the width of the contact area does not exceed one-tenth of the specimen diameter [19]. The solution for the tensile stresses can only be used to calculate the tensile strength of a specimen provided it fails in tension. This is characterized by failure along the loaded diameter (Fig. 5). The fracture does not always extend right to the ends of the diameter. A second fracture pattern, the triple-cleft failure has also been identified as being failure in tension [2]. Compressive failure occurs at the specimen surface immediately beneath the loads where the compressive stresses are at a maximum and appears as local crushing. If this crushing is not extensive it may only result in an increase in the area over which the load is applied so that ultimate failure may be in shear or tension.

As shown in Fig. 4 the maximum shear stresses occur beneath the surface. The exact location and magnitude of these stresses depend on the distribution of applied loads. Shear failures start at an angle to the loaded diameter and are followed by secondary failures causing an irregular fracture pattern. Thus, the validity of using the diametral compression test under a given set of conditions to determine a tensile strength from a fracture load can be easily assessed by examining the specimen fragments after failure [20]. The mode of failure is affected by the width of the contact area. Some materials such as concrete [1] and ceramics [2] require relatively soft packing pieces to be placed between the platens and specimens to ensure adequate load distribution and therefore failure in tension. Other specimens such as autoclaved plaster deform sufficiently at the points contact to ensure that failure occurs in tension [16]. The failure is also influenced by the type and extent of padding. The effect of padding can be predicted by mathematical study, although experimental evidence indicates that such predictions are only general at best [1].

The material properties of the platen will modify the stress distribution within the disc. Photoelastic studies using epoxy resin discs showed

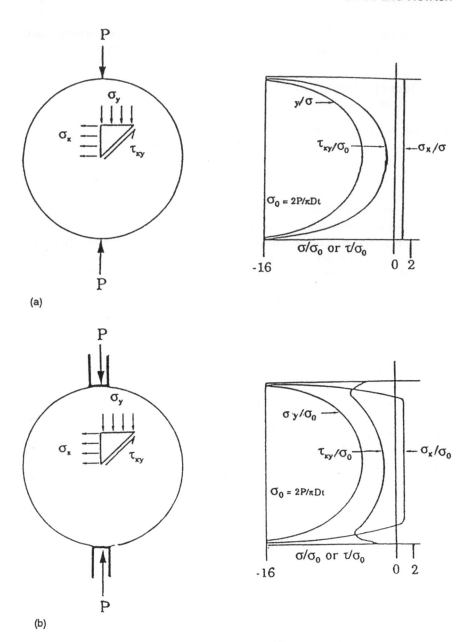

Fig. 4 (a) Diagram of stress distribution across loaded diameter for a cylinder compressed between two line loads (b) Diagram of stress distribution across loaded diameter for a cylinder compressed between two platens with uniform contact pressures.

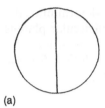

(a)

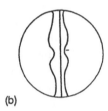

(b)

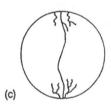

(c)

Fig. 5 Types of failure induced by the diametral compression test: (a) simple tensile failure, (b) triple-cleft failure (tensile failure), (c) failure due to shear at platen edges.

that platens made of different materials such as steel, rubber, and cardboard produced different stress patterns [20]. Experimentally determined fracture strengths indicated that the lower the elastic moduli of the platens or padding the higher was the apparent tensile strength of the disc and the greater the variance of the results. This may be explained by the increased contact area causing a reduction in the volume of material subjected to the maximum tensile stress. Such effects appear to be material and condition dependent as Wright [11] and Fell and Newton [17] reported a decrease of variability with an increase in padding material.

The shape of the platens is also of importance in determining the outcome of the diametral compression test. Concave semicircular platens have been used to induce tensile failure in graphite specimens where formerly only compressive failure occurred. A modified solution for the stresses produced by the platens is required in this case [21].

C. Effect of Loading Rate on Diametral Compression Test

The rate at which the load is applied to specimens can affect the results obtained. Increasing the rate of loading of concrete cylinders resulted in higher observed tensile strengths [1]. Tablets containing lactose and microcrystalline cellulose were tested at loading rates corresponding to cross-head movements of 0.05 to 5 cm min^{-1} [22]. An increase in the loading rate produced a significant increase in the breaking strength, although the standard deviation of replicate values was apparently unaffected. It was also concluded that discrepancies in tensile strength values obtained by different testing instruments may be partially attributed to differences in rates of loading.

The effect of the loading rate on the value of tensile strength obtained for compacts of acetylsalicylic acid and ammonium sulfate has been investigated [23]. Increasing the loading rate produced a decrease in the tensile strength obtained for compacts of both materials. For ammonium sulfate, the change in strength with strain rate was a linear function of the log of the strain rate. The response of a material to loading rate changes appears to be dependent on the mechanism of failure. Variations in strain rate can lead to changes in the failure mode. At low strain rates some materials may fail in a ductile manner. The faster the strain rate the more likely the failure is to be brittle in nature.

D. Tensile Strength Testing of Non-Plane-Faced Circular Compacts

The diametrical compression test has also been used to measure the tensile strength of round convex-faced compacts. The loads required to produce tensile failure of flat-faced and deep convex tablets were compared to produce an empirical equation for tensile strength [24]. The method requires that a linear relation exists between the breaking load and the compaction pressure for the convex tablets and does not allow the isolated determination of the tensile strength of convex-faced tablets.

An equation for the tensile strength of convex-faced tablets was calculated utilizing the central thickness of the tablet [25]. The equation was a correction for the difference in the cross sectional area between a convex-faced and plane-faced specimen which has no theoretical basis in relation to elastic theory.

The stresses induced in convex-faced specimens by the diametral compression test have been examined photoelastically [26]. Formulae were derived that enabled the tensile strength of a convex-faced specimen to be determined from a knowledge of the fracture load and the specimen dimensions. The formulae were verified by comparing the loads required to produce tensile failure of flat-faced and convex-faced gypsum specimens [27]. The tensile strength of a convex specimen may be determined from the fracture load by the equation:

$$\sigma_f = \frac{10P}{\pi D^2} \left(2.84 \frac{t}{D} - 0.126 \frac{t}{C_L} + 3.15 \frac{C_L}{D} + 0.001 \right)^{-1} (22)$$

where

P = fracture load of the convex-faced disc
D = diameter
C_L = cylinder length
t = disc overall thickness

This approach was used to study the influence of the curvature on the mechanical strength of aspirin tablets, when it was established that the "normal" concave punches gave the optimum strength [28].

The diametrical loading of flat-faced discs with a groove across the diameter of one face has been examined photoelastically [29]. The tensile stress distribution was affected by the depth of the groove and also the orientation of the groove with respect to the loaded diameter.

V. LINE LOADING OF RECTANGLES

An analysis of the line loading of rectangles was performed by using both point loading and loading with indentors of different widths [30] (Fig. 6). The analysis showed that in general the average tension σ_x along the loaded line was given by

$$\sigma_x = \frac{2P}{\pi Bt} \tag{23}$$

where B is the length of the line of loading. This expression is independent of the proportions of the rectangle, with the middle portion of the loaded line subjected to approximately uniform normal stress.

The maximum tensile stress is the important quantity in determining failure; for point loading this tensile stress was calculated to rise to a peak

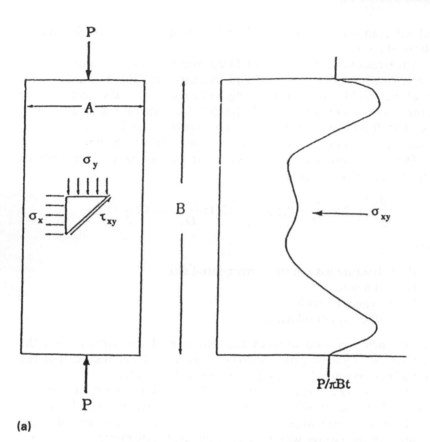

(a)

Fig. 6 (a) Diagram of stress distribution across loaded line for a rectangle ($A/B = 0.4$) compressed between two line loads. (b) Diagram of stress distribution across loaded line for a rectangle ($A/B = 1$) compressed between platens ($a/A = 1$).

value near to the loading platens, before becoming compressive in the immediate vicinity of the indentor (Fig. 6). However, for broad indentors ($a_i/B = 0.1$, where a_i = width of indentor), there was little if any rise in the tensile stress along the loaded line (Fig. 6). The stress remained uniform for a considerable portion of the section before becoming compressive in the loading zone. Hence, for the latter condition, the tensile strength σ_f would be given by

$$\sigma_f = \frac{2P}{\pi Bt} \tag{23'}$$

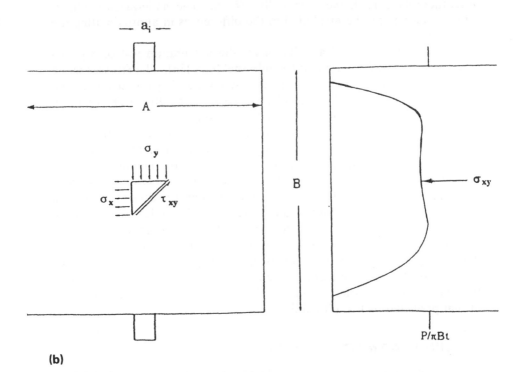

(b)

The tensile strength of concrete cubes was assessed experimentally by using an indentor method [31]. The indentors were used with padding and, under these conditions, good correlation was found between the determined tensile strength of the cubes using Eq. (23') and the determined tensile strength of cylinders using Eq. (20). Subsequently, a finite element method was used [32] to assess the indentor method. Equation (23') was shown to be a good approximation of the tensile strength of the material. The indentor test was examined by photoelasticity for squares [5]. The width of the indentor, a_i, was varied and the resulting stress state calculated. As the indentor width was increased (from $a_i/B = 0.1$ to $a_i/B = 0.75$) with the load kept constant, the tensile stress along the loaded line decreased. However, when the applied stress was kept constant the maximum tensile stress reached its greatest value at $a_i/B = 0.4$.

A theoretical solution for indentor loading was attempted by Jordan and Evans [33]. Four approximate solutions for the elastic stresses in a square were obtained. These solutions were compared with the photo-elastic analysis of Berenbaum and Brodie [5]. Only one solution was found to be in approximately good agreement with the experimental results; the other theoretical results were markedly different. The inadequacies of theoretical solutions can be attributed to the differences in approximating the boundary conditions.

A different theoretical solution for the indentation test on squares was attempted by Sundara and Chandrashekhara [34]. The stresses along the central loaded axis were evaluated by considering the shape of the pressure distribution under the indentor. Again difficulties were found in the theoretical modeling of the stress distribution.

The photoelastic results of Berenbaum and Brodie [5] are used in preference to the theoretical results above when a reliable value for a tensile strength is required [35]. However, the experimental results of Berenbaum and Brodie [5] and the theoretical solution of Goodier [30] give approximately the same tensile strength for indentor widths in the region $a_i/b = 0.1$ to 0.2. Therefore, when these particular indentor widths are used, Eq. (23') can be used to determine the tensile strength.

Cube-shaped specimens have also been tested by loading across a diagonal (i.e., applying the load to opposite edges rather than faces). Analysis of this test configuration by the finite element method indicated that the maximum tensile stress along the loaded axis is approximately 0.77 of that predicted by Eq. (20) [32].

VI. LINE LOADING OF ELLIPSES

A theoretical analysis for the stress distribution in an elliptical disc due to concentrated axial loading showed that the tensile stresses were a function of the width-to-length ratio and were at a maximum for a circular disc [36]. The tensile stress along the loaded axis, σ_x, was calculated to be uniform for the majority of the central portion, but rises to a maximum value at the edge of ellipse at the loading point. The stress σ_x was not calculated to become compressive under the loading zone, this being in contrast to the experimentally determined stress distribution of Frocht [37], indicating a limitation of the analysis of Brisbane [36].

A finite element method was used [38] to predict the stress distribution when ellipses and rectangles were loaded along their central axis. Photoelastic results of Phillips and Mantei [39] and the theoretical results of Brisbane [36] and of Goodier [30] were used to demonstrate the validity of the solutions obtained.

The analyses showed that any elliptical disc with a width-to-length ratio of 0.9 to 1.1 would have the entire Y-axis loaded to 95% or more of the maximum tensile stress. This was because of the nonuniform stresses produced along the Y-axis, particularly once the shape of the ellipse departed significantly from that of a circle.

VII. STATISTICAL TREATMENT OF TENSILE STRENGTH DETERMINATIONS

The stress solutions presented for all the aforementioned tensile strength determinations assume the specimen is isotropic. Such an assumption is unlikely to be true for any pharmaceutical compact as studies have shown that there are distinct density distributions within a compact [40]. Furthermore, within a brittle material there are structural and material defects in the form of interstitial cavities, fractured particles, and interparticular boundaries. Since these defects are randomly distributed throughout the material and are of random severity, there is an inherent variability in the strength of nominally identical brittle specimens. This variabilitly requires that tensile strength test results for brittle materials be treated statistically. The Weibull probability distribution is often used for this purpose [41]. An important attribute of the Weibull hypothesis is that the mean fracture stress of a batch of nominally identical compacts is a characteristic of the specimen and not of the material itself and it is size dependent. The larger the specimen the more likely it is that it will contain a flaw of a given severity, and consequently the smaller will be the mean fracture stress of a batch of such specimens. The predicted relationship between the mean fracture stresses and volumes of two batches A and B is

$$\frac{\sigma_{f_A}}{\sigma_{f_B}} = \left(\frac{V_B}{V_A} \right)^{1/m} \tag{24}$$

where V is the specimen volume and m is the Weibull modulus. This equation assumes that the materials of the different batches are physically identical, which is unlikely to be the case for two batches of tablets. Stanley and Newton [42] prepared six sizes of lactose compacts and compared the experimental mean fracture stresses obtained with those predicted by the Weibull distribution function. For each batch the strength variability was satisfactorily represented by the Weibull distribution, although the effect of size could not be predicted by Eq. (24). A similar conclusion was obtained when comparing two sizes of sodium chloride tablets [43].

VIII. MECHANICAL PROPERTIES RELATED TO STIFFNESS: DETERMINATION OF YOUNG'S MODULUS

Tensile strength has been the most common expression used to describe the strength of compacts, but tensile strength values alone are not sufficient to fully characterize a material's mechanical properties. The stress-strain behavior of a material has been used to measure the toughness and stiffness of materials.

Young's modulus, E, defined as the ratio of stresses, σ, to strain, ϵ, describes the stiffness of a material. Young's modulus of Avicel PH-101 has been determined using the four-point flexure test [44]. In such an arrangement the tensile stress and the associated strain are

$$\sigma = \frac{3Wa}{2bd^2} \tag{25}$$

$$\epsilon = \frac{4\delta d}{k^2} \tag{26}$$

where δ is the vertical displacement of the midpoint of the beam and k is the distance between the loading points on the upper surface of the beam. It was demonstrated that Young's modulus increases as the porosity decreases. Determination of Young's moduli for a range of pharmaceutical excipients demonstrated that there is no relationship between Young's modulus and the tensile strength of a material [6]. To enable comparison to be made of Young's modulus values of porous materials. Sprigg's equation has been used to determine empirically Young's modulus at zero porosity [45]:

$$E_S = E_0 \exp(-cp) \tag{27}$$

where

E_0 = Young's modulus at zero porosity
E_S = specimen's Young's modulus value at porosity p
c = material constant

The ability to calculate Young's modulus from the molecular structure via the cohesion energy density was established by Roberts et al. [46] for a series of organic solids. Attempts to predict Young's modulus of binary powder mixtures, however, were less succesful [47], although the approach proposed by Nielsen [48] appeared to be the most appropriate.

IX. FRACTURE MECHANICS

The four-point bending test has been used to characterize the fracture toughness of Avicel PH-101 [7]. For a crack to grow under static loading the stress must be high enough to initiate fracture and the energy released must be at least as much as that required to form the new surfaces. The stress field near the crack tip will be proportional to the general stress in the material and the square root of the crack length and is called the stress intensity factor (K). This is related to the rate of strain-energy release, G, with crack growth by the elastic modulus of the material:

$$K^2 = EG \qquad (28)$$

The crack will grow when the stress has been raised sufficiently for K and G to reach their critical values K_{IC} and G_{IC}. K_{IC} can be determined by carrying out four-point bending tests on beam-shaped specimens in which a notch has been made in the surface at the point of maximum tension. The value can be calculated from the dimensions of the beam, the maximum load, and the notch depth when this load is reached.

Similar studies have been performed by York et al. [49], who obtained lower values of K_{IC} for Avicel PH-101 than those reported by Mashadi and Newton [7]. The differences were attributed to Avicel PH-101 being a material with a rising crack growth resistance curve. For such materials the method of crack induction and notch geometry become critical. The fracture mechanics approach has been used to characterize the mechanical properties of microcrystalline cellulose [50].

Roberts and Rowe [51] described a method of determining the critical stress intensity factor using circular compacts into which a radial crack was cut. Values for K_{IC} were obtained by two methods: (1) edge opening in which the cracked compacts were gripped in the jaws of a tensometer and pulled apart and (2) a diametral compression test performed with the crack positioned along the line of loading. Of the two techniques the edge opening was preferred since it gave the most stable crack propagation.

X. WORK OF FAILURE

The diametral compression test has also been used to measure the "toughness" of compacts [52]. In this case the change in length of the loaded diameter was monitored throughout testing. Values were recorded for force and the corresponding change in diameter. The product of these values, the area under a force-displacement curve, was called the work of failure, W_f:

$$W_f = \int P \, dx \qquad (29)$$

where dx is the rate of change of distance x.

Tablets with a high work of failure were considered to deform plastically under compressive loading, thus requiring a relatively large platen displacement to produce failure while brittle materials require only a small displacement to produce failure.

Since most specimens fail in tension during the diametral compression test a further expression, the normalized work of failure (NWF), was introduced to convert the applied load, P, to a tensile stress [53]:

$$\text{NWF} = \frac{2}{\pi Dt} \int P \, dx \tag{30}$$

As the specimen fails in tension it might be argued that the increase in the transverse diameter would be a more appropriate measurement to determine the work done. This technique is significant because it demonstrates that materials do behave differently under load, exhibiting different degrees of deformability. This may be due, in part, to the hardness of the material rather than the inherent strength of the compact. Work of failure measurements have been extended to the flexure test [54] and axial tensile strength tests [55].

XI. BRITTLE FRACTURE INDEX

The work-of-failure test has been used to distinguish between ductile and brittle materials. A further method used to measure the brittleness of materials is the brittle fracture index (BFI) [56]. Two sets of square compacts are prepared, one set containing a small circular hole along the axis. The compacts are tested by an indenter test with a platen to compact width ratio of 0.4, and the tensile strengths of the compacts with and without the central hole are compared. Under the test conditions the hole acts as a stress concentrator. Elasticity theory predicts that the stress concentration factor is approximately 3.2 for a hole in an isotropic solid. However, for most pharmaceutical materials the ratio of the tensile strengths of the two types of compact is less than 3. This is believed to be because of the relief of the highly localized stresses by plastic deformation, i.e., the material is not completely brittle. Thus, the ratio of the tensile strength of a compact without a hole, σ_f, to the tensile strength with a hole, σ_{fo}, may indicate its ability to relieve stress by plastic deformation. Based on this the brittle fracture index has been defined:

$$\text{BFI} = 0.5 \left(\frac{\sigma_f}{\sigma_{fo}} - 1 \right) \tag{31}$$

Roberts and Rowe [57] have used the BFI to measure the brittleness of circular compacts using the diametral compression test and have obtained values in good agreement with those of Hiestand and Smith [56].

XII. INDENTATION HARDNESS

While tensile strength measurements describe the global strength of a specimen, indentation hardness describes the "local" plasticity of a material. Hardness may be defined as the resistance of a solid to local permanent deformation, and is usually measured by a nondestructive indentation or scratch test.

The most widely used methods in determining hardness are static indentation methods. These involve the formation of a permanent indentation on the surface of the material to be examined. Normally the diameter of the impression is determined and, from it, the hardness is calculated by means of the formula

$$\text{BHN}(Q) = \frac{2W}{\pi D_i(D_i - \sqrt{D_i^2 - d_i^2})} \tag{32}$$

where

W = applied load
D_i = diameter of the spherical indenter
d_i = diameter of the indentation (Fig. 7)

The Brinell hardness number (BHN) is not a constant for a material but depends on the load and diameter of the indenter [58]. The method has been used to determine the hardness at various points across the diameter of aspirin compacts [59] and direct compression excipients [60]. Studies on the surface hardness distribution over tablet faces with different face curvatures indicated that, as the degree of curvature increased, the hardness of the outer portions of the compact increased relative to the center of the compacts [61].

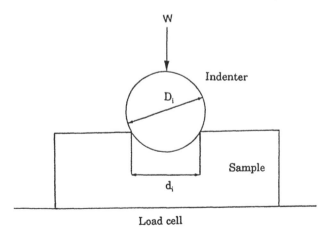

Fig. 7   Diagram of the indentation hardness test.

Hardness tests have been used to measure the elastic recovery. Under load the indenter, radius of curvature r_1, penetrates to a depth h_1 giving an impression of diameter d_i. When the load is removed, elastic recovery occurs and the radius of curvature of the impression increases to r_f, its depth decreasing to h_2. The change $h_1 - h_2 = h$ is a measure of the elasticity. Using this method it is possible to derive a value for the elastic modulus of elasticity [59].

Hardness values of tablets have also been determined using dynamic tests. In dynamic methods, either a pendulum is allowed to strike from a known distance or an indenter is allowed to fall under gravity onto the surface of the test material. The hardness is determined from the rebound height of the pendulum or the volume of the resulting indentation.

The volume of the indentation is directly proportional to the kinetic energy of the indenter. It therefore implies that the material offers an average pressure of resistance to the indenter equal to the ratio

$$\frac{\text{Energy of impact}}{\text{Volume of indentation}}$$

This ratio has the dimensions of pressure and is sometimes referred to as the dynamic hardness number. Hiestand et al. [35] have used the dynamic method to characterize a range of pharmaceutical materials. The indentation hardness, Q, was divided by the tensile strength of the tablet to form a bonding index;

$$\text{Bonding index} = \frac{\text{Indentation hardness}}{\text{Tensile strength}}$$

The objective of the bonding index is to estimate the survival, during decompression of areas of true contact that were established at maximum compression stress.

There are several disadvantages associated with the measurement of the indentation hardness of a specimen. Pharmaceutical compacts consist of particles compacted with voids between them. It is very difficult to decide whether an indentation has been made into a particle, a void or a combination of the two, raising doubts as to what the indentation figure is describing. Furthermore, if the elastic recovery is time dependent, it is difficult to assess when the recovery measurement should be made.

XIII. PRACTICAL STRENGTH TESTING

The choice of method used to determine the mechanical strength of pharmaceutical compacts should be related to the information required. If required for in-process control, it is preferable to provide a system which is reproducible and accurate and which possesses sufficient sensitivity to identify differences between tablets. Several types of instrument are available. Unfortunately, most simply crush the tablet and do not allow the assessment of the mode of failure of the tablet. Such devices will be very insensitive and hence may be apparently reproducible. The variable mode of failure could, however, allow such variations which are not directly related to the properties of the tablet. Simple crushing of tablets of complex shapes can be misleading. If good controlled breaking with a consistent known mode of failure can be induced, the need for complex evaluation of poor-quality systems, as described by Bavitz et al. [62], would be eliminated. There are instruments which can produce this consistent mode of failure and are adaptable for different shapes. When applied to the determination of a fundamental understanding of compaction mechanisms or characterization of the mechanical properties of materials, then only tests which are fundamentally sound should be applied. Even here there are still problems. To undertake tests, the powders have to be compacted. The process of preparation is influenced by the mechanical properties of the material. Hence the final test procedure is an assessment of the mechanical properties of a specimen compounded by its method of formation. For example, the tensile strength of different quantities of lactose at a range of formation pressures was found not to be independent of the sample weight [63]. Subsequent characterization of the tablet formation process by the area under the pressure/time curve rather than the maximum formation value provided a better comparison of tensile strength values for the different quantities of material [64].

There does not appear as yet to be any single method of material characterization which adequately describes the ability of material to form tablets and hence a range of properties, as suggested by York [65], would appear to be more appropriate.

SYMBOLS

a	horizontal distance between upper and lower loading points in four-point bending test.
a_i	indenter width
A	width of rectangle
A_c	area of the curved segment of a capsule-shaped compact
b	width of beam
B	length of the line of loading of rectangle specimen
BFI	brittle fracture index
BHN	Brinell hardness number

c	constant
C_L	cylinder length
d	depth of rectangular beam
d_c	contact width
d_i	indentation diameter
dx	change of distance x
D	diameter of disc specimen
D_i	indenter diameter
E	Young's modulus
E_0	Young's modulus at zero porosity
E_s	Young's modulus of specimen
G	strain energy release
G_{IC}	critical strain energy release
h_c	height of the curved segment of a capsule-shaped specimen
h_1	initial depth of penetration of indenter
h_2	final depth of penetration of indenter
h	$h_1 - h_2$
I	second moment of bending
k	distance between the loading points on upper surface of beam
K	stress intensity factor
K_{IC}	critical stress intensity factor
L	distance between lower supports in the three-point bending test
m	Weibull modulus
M	bending moment
NWF	normalized work of failure
p	porosity
P	applied load
Q	deformation hardness
r_1 r_2	distance from loading points to point within disc specimen
r	initial radius of indenter
r_f	final radius of curvature of indentation
R	radius of disc specimen
t	thickness of disc specimen
V_A	volume of specimen A
V_B	volume of specimen B
W	applied load, tensile test, indentation test
W_f	work of Failure
x	distance along x axis

y_{max}	transverse coordinate measure in the plane of bending
y	distance along y axis
δ	vertical displacement of the midpoint of beam during beam bending tests
ϵ	strain
σ_f	tensile failure stress
σ_{fO}	tensile fracture of specimen with a hole
σ_{fA}	tensile strength of specimen A
σ_{fB}	tensile strength of specimen B
σ_x	tensile strength
σ_y	compressive stress
$\tau_{x,y}$	shear stress
ν	Poisson's ratio

REFERENCES

1. N. B. Mitchell, *Mater. Res. Stand. 1*: 780–788 (1961).
2. A. Rudnick, A. R. Hunter, and F. C. Holden, *Mater. Res. Stand. 3*: 283–289 (1963).
3. C. Nyström, A. Wulf, and K. Malmquist, *Acta Pharm. Suec. 14*: 317–320 (1977).
4. P. Stanley, Post-Graduate School on Production Processes in Tablet Manufacture, School of Pharmacy, University of London.
5. R. Berenbaum, and I. Brodie, *Brit. J. Appl. Phys. 10*: 281–287 (1959).
6. M. S. Church and J. W. Kennerley, *J. Pharm. Pharmacol. 35*: Suppl. 43P (1983).
7. A. B. Mashadi and J. M. Newton, *J. Pharm. Pharmacol. 39*: 961–965 (1987).
8. B. Charlton and J. M. Newton, *Powder Technol. 41*: 123–134 (1985).
9. C. J. Endicott, W. Lowenthal, and H. M. Gross, *J. Pharm. Sci. 50*: 343–346 (1961).
10. S. T. David, and L. L. Augsburger, *J. Pharm. Sci. 63*: 933–936 (1974).
11. P. J. F. Wright, *Mag. Concrete Res. 7*: 87–96 (1955).
12. P. Stanley and J. M. Newton, *J. Pharm. Pharmacol. 32*: 852–854 (1980).
13. G. Gold, R. N. Duvall, and B. T. Palermo, *J. Pharm. Sci. 69*: 384–386 (1980).
14. F. L. L. B. Barcellos and A. Carneiro, *R.I.L.E.M. Bull.* No. 13: 97–123 (1953).
15. T. Akazawa, *R.I.L.E.M. Bull.* No. 16: 11–23 (1953).
16. R. Earnshaw and D. C. Smith, *Austral. Dental J. 11*: 415 (1966).
17. J. T. Fell, and J. M. Newton *J. Pharm. Sci. 59*: 688–691 (1970).
18. J. P. Den Hartog, *Advanced Strength of Materials*, McGraw-Hill, New York, 1952.
19. R. Peltier, *R.I.L.E.M. Bull.* No. 19: 33–74 (1954).
20. E. Addinall and P. Hackett, *Civil Eng. Pub. Works Rev. 59*: 1250–1253 (1964).

21. H. Awaji and S. Sato, *J. Eng. Mater. Tech. 101*: 139–147 (1979).
22. J. E. Rees, J. A. Hersey, and E. T. Cole, *J. Pharm. Pharmacol. 22*: Suppl. 65S–69S (1970).
23. J. M. Newton, S. Ingham, and O. O. Onabajo, *Acta Pharm. Technol. 32*: 61–62 (1986).
24. J. M. Newton, G. Rowley, and J. T. Fell, *J. Pharm. Pharmacol. 24*: 503–504 (1972).
25. S. Esezebo and N. Pilpel, *J. Pharm. Pharmacol. 28*: 8–16 (1976).
26. K. G. Pitt, J. M. Newton, and P. Stanley, *J. Phys. D: Appl. Phys. 22*: 1114–1127 (1989).
27. K. G. Pitt, J. M. Newton, and P. Stanley, *J. Mater. Sci. 28*: 2723–2728 (1988).
28. K. G. Pitt, J. M. Newton, R. Richardson, and P. Stanley, *J. Pharm. Pharmacol. 41*: 289–292 (1989).
29. J. M. Newton, P. Stanley, and C. S. Tan, *J. Pharm. Pharmacol. 29*: Suppl. 40P. (1977).
30. T. N. Goodier, *Trans A.S.M.E. 54*: 173–183 (1932).
31. S. Nillson, *R.I.L.E.M. Bull.* No. 11: 63–65 (1961).
32. J. P. Davies and D. K. Bose, *A.C.I. J. 8*: 662–669 (1968).
33. D. W. Jordan and I. Evans, *B. J. Appl. Phys. 13*: 75–79 (1962).
34. K. T. Sundara and K. Chandrashekhara, *Brit. J. Appl. Phys. 13*: 501–507 (1962).
35. E. N. Hiestand, J. E. Wells, C. B. Peot, and J. F. Ochs, *J. Pharm. Sci. 66*: 510–519 (1977).
36. J. J. Brisbane, *J. Appl. Mech. Trans. A.S.M.E. 85*: 306–309 (1963).
37. M. M. Frocht, *Photoelasticity,* Vol. 2, Wiley, New York, 112–129 (1948).
38. F. J. Appl, *J. Strain Anal. 7*: 178–185 (1972).
39. H. B. Phillips and C. L. Mantei, *Expl. Mech. 9*: 137–139 (1969).
40. D. Train, *J. Pharm. Pharmacol. 8*: 745–761 (1956).
41. W. Weibull, *J. Appl. Mech. 18*: 293–297. (1951).
42. P. Stanley and J. M. Newton, *J. Powder Bulk Solids Technol. 1*: 13–19 (1978).
43. J. W. Kennerley, J. M. Newton, and P. Stanley, *J. Pharm. Pharmacol. 29*: 39P Suppl., (1971).
44. M. S. Church and J. W. Kennerley, *J. Pharm. Pharmacol. 34*: Suppl. 50P (1982).
45. J. C. Kerridge and J. M. Newton, *J. Pharm. Pharmacol. 38*: Suppl. 79P (1986).
46. R. J. Roberts, R. C. Rowe, and P. York, *Powder Technol. 65*, 139–146 (1991).
47. F. Bassam, P. York, R. C. Rowe, and R. J. Roberts, *Powder Technol. 65*: 103–111 (1991).
48. L. E. Nielsen, *J. Appl. Phys. 41*: 4626–4627 (1970).
49. P. York, F. Bassam, R. C. Rowe, and R. J. Roberts, *Int. J. Pharm. 66*: 143–148 (1990).
50. G. F. Podczeck and J. M. Newton, *Die Pharmazie 47*: 462–463 (1992).
51. R. J. Roberts and R. C. Rowe, *Int. J. Pharm. 52*: 213–219 (1989).
52. J. E. Rees and P. J. Rue, *Drug Dev. Ind Pharm. 4*: 131–156 (1978).

53. J. E. Rees, P. J. Rue, and S. C. Richardson, *J. Pharm. Pharmacol.* 2P.: Suppl. 38P (1977).
54. A. E. Moschos and J. E. Rees, *J. Pharm. Pharmacol.* 38: Suppl. 32P (1986).
55. P. J. Jarosz and E. L. Parrott, *J. Pharm. Sci.* 71: 607–614 (1982).
56. E. N. Hiestand and D. P. Smith, *Powder Technol.* 38: 145–149 (1984).
57. R. J. Roberts and R. C. Rowe, *J. Pharm. Pharmacol.* 38: 526–528 (1986).
58. H. Leuenberger and B. D. Rohera, *Pharm. Res.* 3: 12–22 (1986).
59. K. Ridgway, M. E. Aulton, and M. H. Rosser, *J. Pharm. Pharmacol.* 22: Suppl. 70S–78S (1970).
60. M. E. Aulton, H. G. Tebby, and P. J. P. White, *J. Pharm. Pharmacol.* 26: Suppl. 59P–60P (1974).
61. M. E. Aulton and H. G. Tebby, *J. Pharm. Pharmacol.* 27: Suppl. 4P (1975).
62. J. F. Bavitz, N. R. Bohidar, J. I. Karr, F. A. Restaino, *J. Pharm. Sci.* 62: 1520–1524 (1973).
63. J. M. Newton, G. Rowley, J. T. Fell, D. G. Peacock, and K. Ridgway, *J. Pharm. Pharmacol.* 23: 1955–2015 (1971).
64. J. M. Newton and G. Rowley, *J. Pharm. Pharmacol.* 24: 250–257 (1972).
65. P. York, *Drug Dev. Ind. Pharm.* 18: 677–721 (1992).

8

Tablet Surface Area

N. Anthony Armstrong

Welsh School of Pharmacy, University of Wales, Cardiff, United Kingdom

I. INTRODUCTION

The process by which a particulate solid is transformed by the application of pressure to form a coherent compact or tablet can essentially be divided into two stages: consolidation and bond formation.

    1. Consolidation. As pressure is applied, the porosity of the bed of particles is reduced. Initially this is achieved by particle rearrangement, for which only a low pressure is required. Subsequently when rearrangement is effectively complete, further consolidation is achieved by particles undergoing fragmentation or deformation, or most probably both fragmentation and deformation in varying degrees, depending on the solid. Thus, for example, microcrystalline cellulose is primarily a deforming material, dicalcium phosphate dihydrate fragments, and lactose holds an intermediate position.

    2. Bond formation. After particles, by whatever mechanism, have been brought into a sufficiently close proximity with each other, a coherent compact cannot be formed unless some form of bonding occurs between particles. Fuhrer [1] concluded that three types of bond are applicable to tablets: solid bridges, intermolecular forces, and mechanical interlocking.

    Solid bridges have been defined as areas of physical contact between adjacent surfaces. They can occur due to melting followed by resolidification or by dissolution of solid materials followed by recrystallization [2,

3]. Measurements of conductivity of metal powder compacts indicate that only a small proportion of the total surface of particles is in contact in this intimate fashion [4].

If two surfaces are sufficiently close to each other, they will exhibit mutual attraction. It should be noted that unlike solid bridges, the particles need not necessarily be in contact with each other. Intermolecular forces include van der Waal's forces, hydrogen bonding, and electrostatic forces. The first is the most important, and they can be significant at particle separations of up to 10 nm.

The incidence and importance of mechanical interlocking obviously depends on the size and shape of the particles. Smooth spherical particles will have little tendency to interlock, whereas irregularly shaped particles might be expected to do so [5].

II. DETERMINATION OF SURFACE AREA OF COMPACTED POWDER SYSTEMS

The term "surface area" of a particulate solid can be defined in a variety of ways:

1. As the visible or "outer" surface.
2. As the external surface, which is the sum of the surface areas of all the particles, the latter being regarded as a group of nonporous units.
3. As the total area including fine structure and pores within the particles. In a compacted system, there will inevitably be an internal structure.

Surface area is quantified by the calculation of the specific surface area, which is the surface area per gram of solid.

Methods of determination such as those described below differ in the numerical values of surface area which they yield, and hence there is no "correct" figure. Essentially the methods differ in their degree of penetration into porous structure.

A. Methods of Surface Area Determination

A considerable number of methods are available for the determination of surface area of particulate systems. At its simplest, the surface area can be calculated from particle size data, making assumptions of the shape and polydispersity of the powder. However only three methods, namely gas adsorption, mercury porosimetry, and gas permeametry, have been used to any appreciable extent to investigate the surface area of compacted powder systems.

1. Adsorption Methods

These measure the total area of solid which is accessible to the adsorbate, usually a gas such as nitrogen, helium, or krypton. It follows therefore that any internal structure to which the adsorbate has access will be included in the final result. In principle, the method involves determining the quantity of adsorbate which is required to cover all the available surface with a layer one molecule thick. Langmuir [6] postulated that at equilibrium the amount of adsorbate adsorbed was constant at a specific temperature and pressure. He assumed that only a single layer of adsorbate was present and derived the equation

$$\frac{P}{V} = \frac{1}{bV_m} + \frac{P}{V_m} \tag{1}$$

where

V = volume of gas adsorbed at pressure P
V_m = volume required to form a monolayer
b = constant

A plot of P/V against P yields V_m.

However, the adsorbed gas is unlikely to be only one molecule thick. This point was considered by Brunauer, Emmett, and Teller [7], who made three assumptions:

1. Multilayer adsorption occurs, and each layer obeys the Langmuir isotherm.
2. The average heat of adsorption of the second layer is the same as the third and subsequent layers and is equal to the heat of condensation of the liquid.
3. The average heat of adsorption in the first layer is different from that in the second and subsequent layers.

They derived the so-called BET equation:

$$V = \frac{V_m cP}{(P_0 - P)[1 + (c - 1)P/P_0]} \tag{2}$$

where

V = total volume of adsorbed gas on the surface of the adsorbate under conditions of standard temperature and pressure

V_m = volume of adsorbed gas if the entire surface were covered with
a layer one molecule thick
P = partial pressure of adsorbate
P_0 = saturation pressure of adsorbate (thus P/P_0 is relative pressure)
c = constant

Equation 2 can be rearranged to

$$\frac{P}{V(P_0 - P)} = \frac{1}{V_m c}\left(\frac{c-1}{V_m c}\right)\frac{P}{P_0} \tag{3}$$

Thus a graph of V versus P/P_0 gives a straight line where the slope, a, equals $(c - 1)/V_m c$ and the intercept, b, equals $1/V_m c$. Hence $V_m = 1/(a + b)$.

Since the volume which one mole of gas occupies at STP is 22.412 × 10^3 cm^3, the volume of the monolayer can be converted into moles. One gram-mole of gas at STP contains 6.023 × 10^{23} molecules (Avogadro's number), so the number of molecules in the monolayer can be calculated. Then if the cross-sectional area of each molecule is known (for example, the cross-sectional area of nitrogen is 16.2 × 10^{-20} m^2), then the area occupied by the monolayer can be determined.

The underlying theory of the BET approach and alternative treatments has been described by Allen [8]. In general, methods can be divided into two types: static and dynamic. In the former, the adsorbate is stationary, whereas in the latter, adsorption takes place from a moving stream of adsorbate. A number of commercial instruments are available and have been used on tabletted systems, e.g., Sorptometer (Perkin-Elmer) and Quantasorb (Quantachrome). An essential aspect of surface area measurement by this technique is in surface preparation. All solid surfaces are covered with a physically adsorbed film, which must be removed before any quantitative measurements can be made. This is termed "degassing" and is usually achieved by maintaining the solid in vacuo or in a stream of adsorbate, perhaps at elevated temperatures. It naturally follows that the degassing process should not in itself alter the nature of the surface.

2.   Mercury Porosimetry

Mercury intrusion methods have been widely used in powder technology for a number of years.

For a nonwetting liquid to rise up a narrow capillary, pressure must be applied. Assuming a capillary with a circular cross section, then the Washburn equation [9] applies:

$$\Delta P - \frac{2\gamma \cos \theta}{r} \tag{4}$$

where

ΔP = increase in pressure
γ = surface tension of liquid
r = radius of capillary
θ = angle of contact between liquid and capillary wall

Mercury is forced into the pores of a powder bed or tablet, and from the volume of mercury utilized and the pressure needed to force the mercury into the pores a distribution of pore sizes can be established.

Assuming cylindrical pores, the surface area S can be calculated from the equation

$$S = 4 \sum_i \left(\frac{V_i}{d_i} \right) \tag{5}$$

where V_i is the volume of mercury needed to fill pores of diameter d_i.

It follows that if the pores are not cylindrical (i.e., circular in cross section), then results can only be comparative. There are two other potential problems with this technique. The first is the compressibility of the solid. If there are pores which do not connect with the surface, then these pores will be deformed when under pressure without making a contribution to the pore size distribution. Furthermore such deformation will also affect the pore size distribution of those pores which are accessible to the mercury. A further complication is the presence of "ink bottle" pores, which have constricted openings into large void volumes. The presence of pores of this shape in a compacted system is highly likely.

3. Permeametry Methods

Consider air flowing through a powder bed under the influence of a pressure drop. The principal resistance to airflow is at the particle surfaces, and so the permeability of the powder bed is an inverse function of its surface area.

If the powder bed is regarded as a system of parallel capillaries whose diameters are dependent on the average particle size, then the flow rate through a single capillary is given by Poiseuille's equation

$$\frac{V}{t} = \frac{\pi r^4 \Delta P}{8 L \eta} \tag{6}$$

where

V = volume of air that flows in t sec
r = radius of capillary
ΔP = pressure drop across length L
η = viscosity of air

The average pore diameter in a powder depends on the particle diameter (d) and the degree of consolidation, which is related to the porosity (ϵ). Furthermore in a real powder bed the capillaries are not parallel cylinders of uniform radius and length. They are tortuous, and hence their length is greater than that of the powder bed. Neither are they circular in cross section.

These factors are taken into account by the Kozeny-Carman equation (Eq. (7)), which is the basis of most permeability methods [10].

$$S_K^2 = \frac{\Delta PtA}{k_1 LV\eta} \cdot \frac{\epsilon^3}{(1-\epsilon)^2} \tag{7}$$

where

> A = cross-sectional area of bed
> S_K = surface area
> k_1 = constant, which includes shape and tortuosity factors, usually 5

It has been pointed out [11] that for very fine powders, the rate of gas flow through the compact is greater than would be predicted by Eq. (7). This is because of slip flow The contribution of molecular slip (S_M) to the surface area is given by

$$S_M = \frac{\Delta PtA}{LVP}\left(\frac{\epsilon^2}{1-\epsilon}\right)k_2\sqrt{\frac{RT}{M}} \tag{8}$$

where

> R = gas constant
> T = absolute temperature
> M = molecular mass of air

Hence the total surface area S_V can be calculated by combining Eqs. (7) and (8) to give

$$S_V = \frac{S_M}{2} + \sqrt{\frac{S_M^2}{4} + S_K^2} \tag{9}$$

The applicability of Eq. (7) to the determination of surface area has been discussed by Buechi and Soliva [12], who pointed out some of the assumptions which must be made before this equation can be used. The use of the Kozeny-Carman equation in connection with tablets has been criticized by Selkirk [13], but nevertheless this approach has found considerable application.

III. THE RELATIONSHIP BETWEEN SURFACE AREA AND TABLET PROPERTIES

There is a considerable body of published work on the surface area determination of tablets. A list of solids, tablets of which have been the subject of surface area studies, is given in Table 1, together with the method of surface area determination used and the appropriate reference. However there are very few studies which use more than one method of measurement. Therefore it is convenient to subdivide the topic according to the technique of measurement which was used. Work which uses more than one technique is of particular interest and will be discussed more fully later.

A. Surface Area Measurement in Tablets by Gas Adsorption

In 1952, a series of publications began to emerge from workers at the University of Wisconsin under the direction of Professor T. Higuchi. These papers had a profound effect on the subsequent course of tabletting research. In an early paper in this series [14], the relationship between compression force and the specific surface area of the tablet was examined. Using a static nitrogen adsorption method, the specific surface area of tablets made from sulfathiazole granulated with starch was determined.

The results are shown in Fig. 1. The shape of the curve was explained as follows. At lower forces (below 2500 lb, equivalent to 156 MPa), the net surface area of the tablet rose because granules were breaking down to give smaller particles with a higher specific surface area. Above 2500 lb, closer association of the particles became the more dominant process, even though further granule breakdown was undoubtedly occurring. The particles became so closely associated that the surface between them was not available to molecules of nitrogen.

The suggestion that the shape of the compression pressure-surface area curve could be explained by changes in the particle size within the tablet received experimental confirmation from the work of Armstrong and Haines-Nutt [15], who used a Perkin-Elmer Sorptometer. These workers found that when tablets of magnesium carbonate were placed in water, they rapidly disintegrated into their constituent particles. It therefore was possible to prepare tablets from particles of known size, measure the surface area of the tablets, and then disintegrate them so that the particle size after compression could be determined. Figure 2 shows the weight fractions of particles in the 20–30 μm and 90–100 μm fractions before and after compression. The "before-compression" data is represented by the intercepts of the lines on the ordinate (i.e., at zero pressure). Smaller particles

Table 1 Solids, Tablets of Which Have Been the Subject of Surface Area Study

Ascorbic acid	p	25
Bentonite	ga	20
Calcium phosphate dihydrate (Emcompress)	ga	21
Calcium phosphate dihydrate (Emcompress)	mp	32, 35
Calcium phosphate dihydrate (Emcompress)	p	25
Coal	ga	19
Dextrose monohydrate	ga	16
Iron powder	p	29, 30
α-Lactose (amorphous)	mp	35
α-Lactose (anhydrous)	ga	36, 37, 40
α-Lactose (anhydrous)	mp	32, 36
α-Lactose monohydrate	ga	21, 36, 37, 38, 40
α-Lactose monohydrate	mp	33, 34, 35, 36
α-Lactose monohydrate	p	25, 27
β-Lactose	ga	36, 37, 40
Lactose, spray-dried	mp	32, 35
Magnesium carbonate	ga	15, 17, 18
Magnesium oxide	ga	20
Magnesium trisilicate	ga	20
Microcrystalline cellulose (Avicel PH101)	mp	32
Microcrystalline cellulose (Avicel PH101)	p	30
Paracetamol	ga	16
Paracetamol	p	25
Phenacetin	ga	16
Sodium bicarbonate	p	25, 26, 27, 30
Sodium chloride	ga	21, 23
Sodium chloride	p	23, 25, 27, 29, 30
Sodium citrate	ga	21
Sodium citrate	p	25, 26, 27
Starch 1500 (Sta-Rx)	ga	21
Starch 1500 (Sta-Rx)	p	30
Sucrose	ga	19, 21, 23
Sucrose	p	23
Sulfathiazole/starch	ga	14

ga = surface area determined by gas adsorption; p = permeametry; mp = mercury porosimetry.

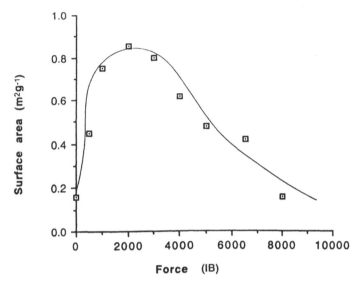

Fig. 1 The effect of compression force on the specific surface area of sulfathiazole tablets. (Reproduced from Reference 14 with permission of the copyright owner, the American Pharmaceutical Association.)

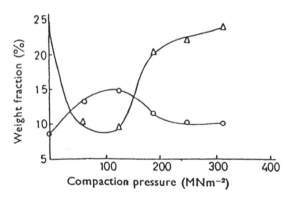

Fig. 2 Percent weight fraction-compression pressure profile of two size fractions of magnesium carbonate; ○, 20–30 μm; △, 90–100 μm. (Reproduced from Reference 15 with permission of the copyright holder.)

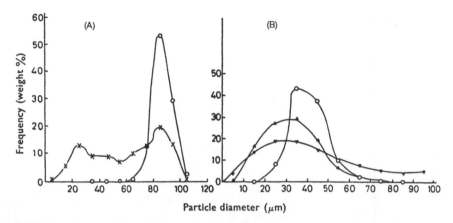

Fig. 3   Particle size distribution of (A) a nominal 73–89 μm fraction; (B) a nominal 33–45 μm fraction of magnesium carbonate; $\bigcirc$, before compression; $\times$, after compression at 62.5 MN m^{-2}; $\blacktriangledown$, after compression at 250 MN m^{-2}. (Reproduced from Reference 15 with permission of the copyright holder.)

show an increase, followed by a decrease, and the reverse is shown by the larger particles. Magnesium carbonate was then fractionated, compressed at either 62.5 MN m^{-2} or 250 MN m^{-2}, and the tablets disintegrated as before. The 75–89 μm fraction showed particle reduction. However at the higher pressure, considerably more "larger" particles were detected after compression than had been present in the original powder (Fig. 3).

Not all particulate materials show the surface area changes described above. For example, Armstrong and Griffiths [16] showed that for phenacetin and dextrose, in addition to the rise and fall in surface area already described, a second increase occurred at pressures greater than 150 kg cm^{-2} (equivalent to 147 MPa) (Fig. 4). It was shown that this was associated with a high value for elastic recovery of the tablets, which in turn resulted in a high degree of lamination or capping [17]. Furthermore it was shown that the magnitude of the surface area changes, and especially the pressure at which maximum surface area was achieved was associated with the presence of lubricating agents, both liquid [16] and solid [18].

Hardman and Lilley [19] used a dynamic method of surface area measurement by gas adsorption, studying sodium chloride (a plastically deforming material), sucrose (a brittle material) and coal powder, which has a considerable pore structure. They measured the surface area of tablets which had been compressed at pressures up to about 200 MPa. For sodium chloride, there was little change in surface area over the whole

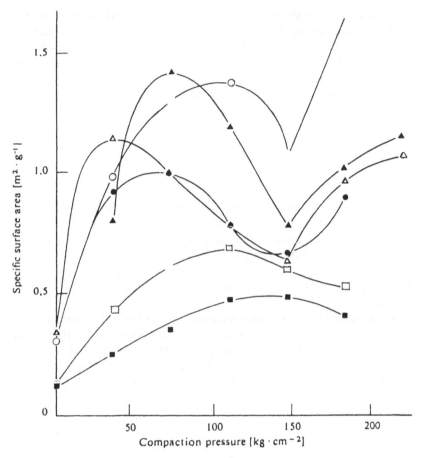

Fig. 4 Surface area-compression pressure profiles of compacts prepared from phenacetin, paracetamol, and dextrose monohydrate: ○, compacts prepared from phenacetin granules, peripheral moisture content 0%; ●, compacts prepared from phenacetin granules, peripheral moisture content 2.5%; □, compacts prepared from paracetamol granules, peripheral moisture content 0%; ■, compacts prepared from paracetamol granules, peripheral moisture content 2.5%; △, compacts prepared from dextrose monohydrate granules, peripheral moisture content 0%; ▲, compacts prepared from dextrose monohydrate granules, peripheral moisture content 6.6%; (Reproduced from Reference 16 with permission of the copyright holder.)

pressure range. Sucrose showed a considerable rise in surface area, attributed to fragmentation, up to about 170 MPa, beyond which the surface area fell. They suggested that this was due to cessation of fracture, accompanied by a small increase in interparticulate contact, some plastic flow also taking place. The surface area of the coal compacts continued to rise over the whole pressure range. Hardman and Lilley attempted to calculate the area of contact between particles by equating this to half the difference in surface area before and after compression. However they pointed out the inability of the nitrogen molecule to penetrate openings of less than 2 nm. Thus surfaces separated by less than this distance would be counted as being in contact.

Further work using nitrogen adsorption was carried out by Stanley-Wood and Johansson using a static method [20]. They attempted to correlate compact density with surface area changes. They found that for magnesium trisilicate and bentonite, surface area fell at pressures up to about 50 MPa and then rose slightly, whereas that for magnesium oxide remained relatively constant over a pressure range from 0 to 500 MPa. They related this difference in behavior to the different consolidation mechanisms of the three solids.

Considerable research has been carried out at Uppsala University on mechanisms by which particles aggregate to form tablets, and surface area measurement has played an important part in this. Duberg and Nystrom [21] evaluated several methods by which particle fragmentation could be assessed. One of these was krypton adsorption, surface area of both powder and tabletted systems being calculated by the BET equation. Tablets were prepared at 165 MPa.

The results they obtained were, at least in part, anomalous. Some substances, such as calcium phosphate dihydrate (Emcompress), which are known to fragment under a compressive load, showed very large increases in surface area after compaction, whereas others, such as starch 1500 (StaRx 1500) and sodium chloride, showed much smaller increases. However, for other substances the change in area was much more difficult to interpret. Duberg and Nyström pointed out that materials with a large number of cracks and pores will give a high value for powder surface area. Furthermore, as particle consolidation increases on compression, that part of the surface area which represents particle contact will be lost. Thus, Duberg and Nyström concluded that gas adsorption was probably an inappropriate method for measuring the degree of fragmentation occurring during compression, and suggested that a technique based on permeametry would be more appropriate. Some of the confusion surrounding their results may stem from the fact that they only measured the surface area of the tablets at one pressure. All previous work had shown that the relationship between compression pressure and surface area is at least parabolic, and the position of the maximum of the parabola is substance dependent. A one-point measurement would not detect this.

Duberg and Nyström concluded that the strength isotropy ratio (the ratio between the axial and radial tensile strengths of the tablet) [22] was a more appropriate means of assessing fragmentation, though they recommended that the scanning electron microscope was essential for a qualitative evaluation of fragmentation.

B. Surface Area Measurement of Tablets by Gas Permeametry

Because of their reservations about measuring tablet surface area by low-temperature gas adsorption, workers at Uppsala University developed a method of measuring surface area by permeametry [23]. Their method was based on the Blaine apparatus [24].

Using sodium chloride and sucrose compressed at pressures between 40 and 300 MPa, they measured surface area by both krypton adsorption and permeametry. The results are shown in Fig. 5. Surface area by

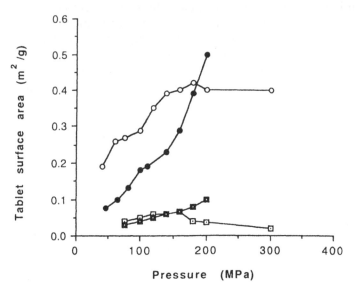

Fig. 5 Weight specific surface area, measured by permeametry (closed symbols) and by the BET method (open symbols) as a function of compaction pressure for sodium chloride (□) and saccharose (○) tablets. (Reproduced from Reference 23 with permission of the copyright holder.)

permeametry increases over the whole of the pressure range for both substances, though the increase for sucrose is much larger. Surface area by gas adsorption shows the parabolic shape reported by earlier workers [14,15] with a maximum at about 150 MPa for both substances. Hence, at the highest pressures studied, there is a considerable difference in magnitude in the results given by the two methods.

Using the permeametry technique described above, Alderborn et al. studied the consolidation behavior of a number of substances [25]. They reasoned that if a solid fragmented under a compressive load, then that should be reflected in an increase in the surface area. They used two materials which deformed plastically (sodium chloride and sodium bicarbonate), four which primarily underwent fragmentation (lactose, sodium citrate, ascorbic acid, and paracetamol) and one (calcium phosphate dihydrate: Emcompress) which fragmented extensively. Tablets were prepared at a pressure range of 40–200 MPa, and the surface areas of the resulting tablets were measured.

Surface area increased for all materials, indicating that some fragmentation occurred during compaction. However the shapes of the surface area–compression pressure curves differed markedly, their slopes being in the same rank order as their propensity to fragment (Fig. 6).

Alderborn [26] then went on to compare the surface areas of intact tablets with those which had been deaggregated by gentle manual abrasion, sodium citrate and sodium bicarbonate being used as solids. He found that, at least at pressures up to 100 MPa, the surface area of the intact tablets differed little from that of the deaggregated tablets. From this finding, he concluded that the actual surface area used for bonding was relatively low.

Alderborn and Nyström [27] investigated if particle size had an effect on the slope of the surface area compression pressure profile, using sodium chloride, sodium bicarbonate, lactose, and sodium citrate as test materials, and five different size fractions of each solid. For all materials and size fractions, the tablet surface area increased with increasing compression pressure. The magnitude of the increase depended on whether the solid's primary means of consolidation was by fragmentation. For all materials, the finer size fractions gave a larger absolute increase in surface area, though the reverse was true if the relative surface area change was considered.

If all the lines were regarded as rectilinear, then the values of the slopes of these lines could be regarded as a quantitative relationship between surface area and compresson pressure, at least up to pressures of about 125 MPa. Alderborn and Nyström then plotted the slopes of these lines against the surface area of the powders (Fig. 7), expressing the slopes

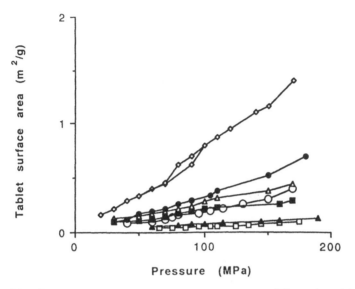

Fig. 6 Specific surface area of sodium chloride (□), sodium bicarbonate (▲), saccharose (○), lactose (●), sodium citrate (■), ascorbic acid (△), paracetamol (◇), and Emcompress (◆), as a function of compaction pressure. (Reproduced from Reference 25 with permission of the copyright holder.)

of these lines as a measure of the rate of increase of tablet surface area with pressure. Thus

$$\frac{ds}{dP} = kS^n \tag{10}$$

where k is a constant describing the fragmentation propensity of the solid. They found a value for n of ⅔ was most appropriate. Thus over the range of pressures 0 to P, with corresponding surface areas of S_1 and S_2, then

$$3(S_2^{1/3} - S_1^{1/3}) = kP \tag{11}$$

They drew attention to the similarity between this equation and that of Bond [28], which relates the energy consumed during size reduction to the particle size of the product.

Attempts have been made using surface area measurement to calculate the extent of the actual area of contact between particles in a tablet [29]. These workers used nonfragmenting particles (iron and sodium chloride) and measured surface area by both permeametry and krypton adsorp-

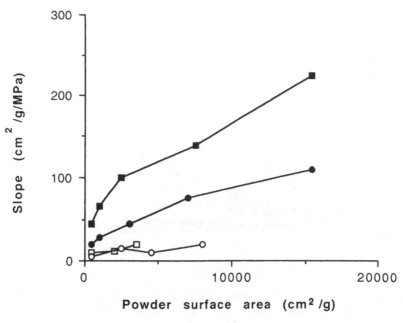

Fig. 7 Effect of powder surface area on the absolute increase in tablet surface area with pressure (i.e., the slope from a plot of tablet surface area as a function of compaction pressure (0–125 MPa)): sodium chloride (○), sodium bicarbonate (□), sodium citrate (●), lactose (■). (Reproduced from Reference 27 with permission of the copyright holder.)

tion. Since no significant change in surface area in the iron particles could be attributed to bonding, it was concluded that the bonding surface of the particles was very small and below the level of detection of the technique.

Sodium chloride showed a different pattern, A fall in surface area measured by gas adsorption was noted at higher pressures, but as the permeability surface area continued to increase, it was concluded that sodium chloride also utilized only a small portion of its surface area for bonding.

Karehill et al. [30] also studied the relationship between the surface area of the solid and the compact strength, using plastically deforming materials—sodium chloride, sodium bicarbonate, starch 1500, microcrystalline cellulose, and iron powder. They concluded that only a very small fraction of the total available surface was used to form solid bridges between adjacent particles, and that tablet strength must be due to relatively long-range forces. The strength of tablets made from some of these substances increased when stored for two days, but no changes in surface area were detected which could account for this change [31].

C. Surface Area Measurement of Tablets by Porosimetry

Another body of work utilizing surface area measurements for research on powders and tablets is that carried out by Professor Lerk and colleagues from the University of Groningen in The Netherlands. This work has been principally concerned with varieties of lactose. Surface area, usually described as pore surface area, was most frequently determined by mercury porosimetry, though gas adsorption was sometimes used.

Vromans et al. found a linear relationship between surface area and compaction load for α-lactose monohydrate and anhydrous α-lactose [32]. Calcium phosphate dihydrate (Emcompress) showed a similar effect. The authors claimed that microcrystalline cellulose did not show an increase in surface area with increased load, but critical examination of their data shows that the latter statement may not be totally valid. Notwithstanding this, their work supports that of other workers [25] in that for fragmenting substances, surface area increased markedly as the load increased, the slope of the line being proportional to the fragmentation propensity of the solid. From this it was concluded that the degree of fragmentation of α-lactose monohydrate was less than that of anhydrous α-lactose.

As tablet strength also increases with compression pressure, it follows that crushing strength must increase with surface area. This is shown in Fig. 8. The significant point about this graph is that all points lie on the same straight line, indicating that tablet strength is related to surface area and hence the degree of fragmentation. Furthermore it would appear that neither the presence of water in the α-lactose monohydrate nor the ratio between α and β lactose has any influence on the tablet strength, suggesting that the same binding mechanism applies to all types of crystalline lactose.

Using a series of sieve fractions of α-lactose monohydrate, de Boer et al. [33] studied the relationship between compaction pressure and compact strength. A nonlinear relationship was obtained, with clearly defined maxima, but stronger tablets were obtained from smaller particles at any given pressure. When the surface area of all the tablets prepared in this study was measured, it was found that a good rectilinear relationship was obtained between surface area and compact strength for all the original size fractions.

In contrast to the behavior of crystalline lactose monohydrate, the pore surface area of amorphous lactose did not increase with compaction pressure. It was therefore inferred that amorphous lactose deformed plastically rather than undergoing consolidation by fragmentation [34]. However,

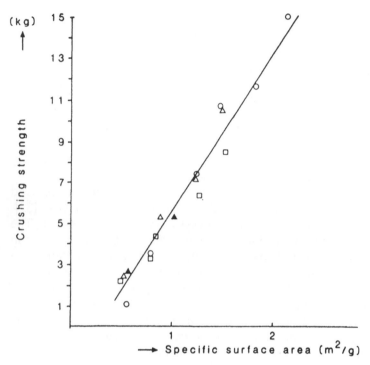

Fig. 8 Crushing strength versus specific surface area for tablets compressed from different types of crystalline lactose: α-lactose monohydrate (□); anhydrous α-lactose (○); roller-dried β-lactose (△); crystalline β-lactose (▲). (Reproduced from Reference 32 with permission of the copyright holder.)

tablet crushing strength showed an increase with compaction pressure. It was thus concluded that for materials which consolidate by deformation, the tablet strength was not related to surface area, as had been found for fragmenting substances. Spray-dried lactose contains about 15% amorphous lactose, which makes a major contribution to tablet strength despite making little contribution to surface area changes. Since little new surface area is generated on compaction, the strength of tablets made from materials such as amorphous lactose are greatly reduced by the addition of magnesium stearate. Compactibility was, however, increased after the solid had been exposed to water vapor. Again this was interpreted as being characteristic of materials which consolidate by deformation rather than fragmentation.

This study was taken further by preparing samples of spray-dried lactose containing differing proportions of amorphous lactose [35]. It was found that the proportion of amorphous lactose present had little effect on the surface are of the tablets, but had a major influence on the tablet strength. It was suggested that the primary role of amorphous lactose was to act as a binder.

Continuing the work with lactose, Vromans and co-workers studied the compactability of a series of lactoses which differed in respect of size, texture, water content, and α/β ratio [36]. They found that the binding capacity was directly related to the surface area of the starting material, surface area being measured by nitrogen adsorption. As shown in Fig. 9, a

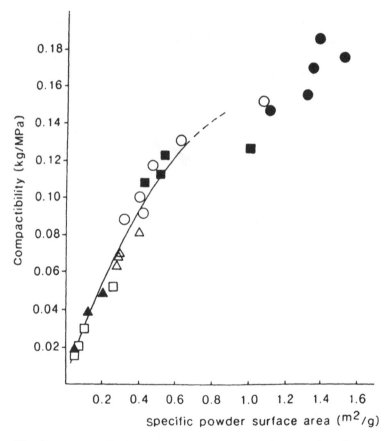

Fig. 9 Compactibility of samples of crystalline lactose plotted versus the powder surface area (S_{N_2}), $\square$, α-lactose monohydrate; $\bigcirc$, anhydrous α-lactose; ●, anhydrous α-lactose; $\triangle$, roller-dried β-lactose; ▲, crystalline β-lactose; ■, compound crystalline lactose. (Reproduced from Reference 36 with permission of the copyright holder.)

rectilinear relationship was obtained between compactibility and surface area, except for highly porous samples of anhydrous lactose, which showed lower compactibility than expected. Compactibility was defined as the slope of the compression pressure-crushing strength curve.

To avoid capillary condensation (see later), they studied tablets prepared at low pressures (37.5 MPa) in which the incidence of small capillaries would be expected to be lower. They found a good rectilinear relationship, except for the porous materials noted above. Furthermore an excellent straight line was obtained by plotting tablet surface area against powder surface area (Fig. 10). It is of interest to note that the slope of this line is very close to unity.

The properties of tablets containing binary mixtures of various types of lactose were studied by Riepma et al. [37]. Using size fractions of the same size for all materials studied, the rectilinear relationship noted earlier between tablet surface area and tablet crushing strength was reported. To

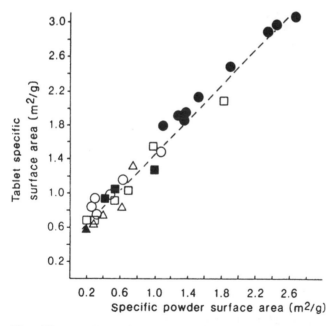

Fig. 10 Specific surface area of tablets compacted at 37.5 MPa versus specific powder surface area of the starting material. Symbols as in Fig. 9. (Reproduced from Reference 36 with permission of the copyright holder.)

avoid capillary condensation, tablets were compressed at a force of only 10 kN (equivalent to about 75 MPa).

Using different size fractions of the same materials, Riepma et al. found again that specific surface and crushing strength were linearly related [38]. The graphs of both surface area against percent fine fraction and tablet strength against fine fraction were very similar, in that they showed a minimum at about 40% fine fraction. The authors explained this by suggesting that the fine fraction possessed less tendency to fragment. A suggestion that smaller particles contain fewer cracks and are therefore less likely to fragment forms part of the Griffith crack theory [39].

As stated earlier, a linear relationship between crushing strength and surface area was found for a wide variety of different types of lactose, and from this it was assumed [32] that the number of bonding sites was proportional to the surface area. Leuenberger et al. [40] derived a theoretical model in which coordination numbers (i.e., the number of isometric spherical particles which form a compact) were calculated. The size of the spheres was obtained from surface area measurements and then by calculating the equivalent diameter of the sphere. The coordination number then depends on the geometrical packing of the particles [41]. Fair agreement was obtained for various types and sizes of lactose, but the authors point out that the calculated diameter of the spheres is very dependent on the method used to measure surface area. Furthermore calculating the equivalent sphere diameter introduces an error, since a comparison of the particle size and surface area data clearly shows that the particles are not spherical.

IV. THE RELATIONSHIP BETWEEN COMPRESSION PRESSURE AND SURFACE AREA IN TABLETTED SYSTEMS

Whether consolidation of a powder bed under pressure takes place by fragmentation or particle deformation, there will be some change in surface area. Then, as the particles assume an even closer relationship, this too will cause surface area changes. It follows that surface area studies may provide important information about the extent and mechanisms of particle consolidation. However as will have been seen from the above discussion, different groups of workers have obtained very different shapes for the graph of surface area against compression pressure, and an explanation of this variation must be sought.

The explanation put forward by the original workers in this area [14] (Fig. 1) is worthy of serious consideration. Three processes which will affect the surface area of the tablet will proceed simultaneously. These are

1. Fragmentation of particles, leading to an increased surface area
2. Increased interparticulate contact, leading to a decrease in surface area
3. Flattening of particle asperities, giving a smoother particle surface and hence a decrease in surface area

The overall shape of the compression pressure-surface area curve is thus determined by the relative importance of these three mechanisms for a given substance at any given pressure. This will be a function of the properties of the substance and in particular its fragmentation propensity. It is to be expected that virtually all particles will undergo some fragmentation either as the result of the imposition of a compressive force or even, at low pressures, by abrasion as the particles slide over one another to give a less porous powder bed, but one which is still composed of discrete particles.

At higher pressures some materials deform significantly, e.g., cellulose derivatives, and a granulating agent, if present, leads to a more deformable product [42].

However, these considerations are complicated by another important factor, namely the method by which the surface area is determined. Each of the three methods of measuring surface area which has been used to examine tablets measures something different. These are the surface to which an adsorbate molecule, e.g., nitrogen, has access, the volume to which pressurized mercury has access, and the surface over which gas flow occurs. Consequently it is not surprising that in the few comparative studies which have been carried out, numerical identity is not achieved. However it is potentially more serious that the overall shape of the curve appears to be dependent on the experimental method used to measure surface area.

Those methods which utilize gas adsorption (see, for example, [14], [16]) usually show a rise in surface area followed by a decrease, whereas the use of permeametry [25] or mercury porosimetry [32] gives a progressive increase in surface area over the whole range of compression pressures studied. Such a divergence in behavior renders more important the few comparative studies which have been carried out, and it is worth examining these in some detail.

Vromans et al. [32] reported a linear relationship between specific surface area as assessed by porosimetry and compression pressure for α-lactose monohydrate and anhydrous lactose, but a curvilinear relationship for β-lactose measured by gas adsorption [36]. Bearing in mind the similarity of behavior of all types of lactose referred to earlier [32], such a difference was considered worthy of further investigation.

Vromans et al. [36] reported that compared to mercury porosimetry results, surface area determined by gas adsorption gave poor reproducibility. The magnitude of the surface area appeared to be related to the degassing and storage conditions the tablets underwent after compression. Tablets examined immediately after compression, i.e., without any pretreatment showed a larger surface area than those which had been subjected to degassing at 120° (Fig. 11). Furthermore the surface area changed after storage for 24 hr, the extent of the change depending on the relative humidity of the storage conditions (Fig. 12). They suggested that this variability was due to capillary condensation, and since the pores in the tablet become progressively smaller as the compression pressure is increased, then the effect of capillary condensation might be expected to be more pronounced in tablets prepared at higher pressures. Hence, doubt must be cast on reported surface area measurements which have used the gas adsorption technique. As Vromans et al. pointed out, it is impossible to prevent some capillary condensation which occurs almost immediately after the tablet is prepared.

These workers also suggested that removing water from the tablet by degassing will have an unavoidable disruptive effect on the tablet structure. Hence for subsequent work using gas adsorption. Vromans and co-workers used tablets compressed at the relatively low pressure of 37.5 MPa, at which capillary condensation would be expected to be low.

The magnitude of the effects of degassing will depend in part on the substance from which the tablet is made. Vromans et al. degassed α-lactose

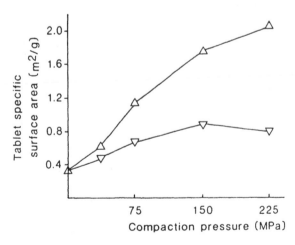

Fig. 11 Specific surface area (S_{N_2}) of tablets compacted from roller-dried β-lactose (100–160 μm) versus the compaction pressure. The tablets were measured immediately after compaction ($\triangle$) or after outgassing at 120°C ($\triangledown$). (Reproduced from Reference 36 with permission of the copyright holder.)

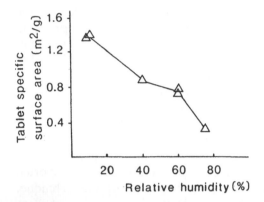

Fig. 12 Specific surface area (S_{N_2}) of tablets compacted from roller-dried β-lactose at 225 MPa plotted in relation to the storage conditions, being the relative humidity at 20°C. (Reproduced from Reference 36 with permission of the copyright holder.)

monohydrate at 120° and an appreciable quantity of lactose might be expected to dissolve and subsequently recrystallize under such circumstances, with obvious effects on the pore structure and surface area of the tablet. Also liberation of the water of crystallization must be considered. However, other workers have used much less soluble solids, e.g., magnesium carbonate [15], and degassed at lower temperatures, so the use of gas adsorption cannot be totally condemned. Nor would the theory of capillary condensation explain the second rise in surface area detected by Armstrong and co-workers [16,18].

Alderborn et al. [23] found similar differences between surface area determinations using gas adsorption and permeametry (see Fig. 5). These authors pointed out that as particles are pressed more closely together, surface area will be lost, due not only to interparticulate contact but also because adsorbate molecules fail to gain access to part of the surface area. Hence, surface area is underestimated. With permeametry, on the other hand, an overestimate may well be obtained, since, as consolidation increases, nonhomogeneity of the compact will also increase. This may well be so, but it is worthy of comment that in all the gas adsorption studies reported, the maximum in the compression pressure surface area curve is always at about 150 MPa, despite the wide range of substances used. Furthermore the two solids used in this work are both freely soluble in water, and so disruption of the particle surfaces might well occur during the degassing process.

REFERENCES

1. C. Fuhrer, *Labo-Pharma Probl. Techn. 25*: 759 (1977).
2. P. York and N. Pilpel, *J. Pharm. Pharmacol. 25*: 1P (1973).
3. W. B. Pietsch, (1969). The strength of agglomerates bound by salt bridges. *Can. J. Chem. Eng. 47*: 403.
4. F. P. Bowden and D. Tabor, *The Friction and Lubrication of Solids*, Oxford University Press, New York, pp. 145–159, 1950.
5. P. J. James, *Powder Metall. 20*: 199 (1977).
6. I. Langmuir, *J. Am. Chem. Soc. 40*: 1361 (1918).
7. S. Brunauer, P. H. Emmett, and E. Teller, *J. Am. Chem. Soc. 60*: 309 (1938).
8. T. Allen, *Particle Size Measurement*, 3rd ed., Chapman and Hall, London, pp. 465–513, 1981.
9. E. W. Washburn, *Phys. Rev. 17*: 374 (1921).
10. B. H. Kaye, *Powder Technol. 1*: 11 (1967).
11. D. T. Wasan, W. Wnek, R. Davies, M. Johnson, and B. H. Kaye, *Powder Technol. 14*: 209 (1976).
12. W. Buechi and M. Soliva, *Powder Technol. 38*: 161 (1983).
13. A. B. Selkirk, *Powder Technol. 43*: 285 (1985).
14. T. Higuchi, A. H. Rao, L. W. Busse, and J. V. Swintosky, *J. Am. Pharm. Assoc. Sci. Ed. 42*: 194 (1953).
15. N. A. Armstrong and R. F. Haines-Nutt, *J. Pharm. Pharmacol. 22*: 8S (1970).
16. N. A. Armstrong and R. V. Griffiths, *Pharm. Acta Helv. 45*: 583 (1970).
17. N. A. Armstrong and R. F. Haines-Nutt, *Powder Technol. 9*: 287 (1974).
18. N. A. Armstrong and R. F. Haines-Nutt, Proc. 1st Int. Conf. on Compaction and Consolidation of Particulate Matter, Brighton, pp. 161–164, 1972.
19. J. S. Hardman and B. A. Lilley, *Proc. R. Soc. London A 333*: 183 (1973).
20. N. G. Stanley-Wood and M. E. Johansson, *Drug Dev. Ind. Pharm. 4*: 69 (1978).
21. N. Duberg and C. Nyström, *Acta Pharm. Suec. 19*: 421 (1982).
22. C. Nyström, W. Alex, and K. Malmquist, *Acta Pharm. Suec. 14*: 317 (1977).
23. G. Alderborn, M. Duberg, and C. Nyström, *Powder Technol. 41*: 49 (1985).
24. R. L. Blaine, *ASTM Bull. 108*: 17 (1943).
25. G. Alderborn, K. Pasanen, and C. Nyström, *Int. J. Pharm. 23*: 79 (1985).
26. G. Alderborn, *Acta Pharm. Suec. 22*: 177 (1985).
27. G. Alderborn and C. Nyström, *Powder Technol. 44*: 37 (1985).
28. F. C. Bond, *Mining Eng. 4*: 484 (1952).
29. C. Nyström and P. G. Karehill, *Powder Technol. 47*: 201 (1986).
30. P. G. Karehill, G. Alderborn, M. Glazer, E. Borjesson, and C. Nyström, Studies on direct compression of tablets. XX. Investigation of bonding mechanisms of some directly compressed materials by addition and removal of magnesium stearate, cited by Karehill, P. G. (1990). Studies in bonding mechanisms and bonding surface area in pharmaceutical compacts. Ph.D. thesis, Uppsala University.
31. P. G. Karehill and C. Nyström, *Int. J. Pharm. 64*: 27 (1990).

32. H. Vromans, A. H. deBoer, G. K. Bolhuis, C. F. Lerk, K. D. Kussendrager, and H. Bosch, *Pharm. Weekbl. Sci. Ed. 7*: 186 (1985).
33. A. H. De Boer, H. Vromans, C. F. Lerk, G. K. Bolhuis, K. D. Kussendrager, and H. Bosch, *Pharm. Weekbl. Sci. Ed. 8*: 145 (1986).
34. H. Vromans, G. K. Bolhuis, C. F. Lerk, K. D. Kussendrager, and H. Bosch, *Acta Pharm. Suec. 23*: 231 (1986).
35. H. Vromans, G. K. Bolhuis, C. F. Lerk, H. van de Biggelaar, and H. Bosch, *Int. J. Pharm. 35*: 29 (1987).
36. H. Vromans, G. K. Bolhuis, C. F. Lerk, and K. D. Kussendrager, *Int. J. Pharm. 39*: 207 (1987).
37. K. A. Riepma, C. F. Lerk, A. H. deBoer, G. K. Bolhuis, and K. D. Kussendrager, *Int. J. Pharm. 66*: 47 (1990).
38. K. A. Riepma, J. Veenstra, A. H. deBoer, G. K. Bolhuis, K. Zuurman, C. F. Lerk, and H. Vromans, *Int. J. Pharm. 76*: 9 (1991).
39. A. A. Griffith, *Philos. Trans. R. Soc. London A221*: 163 (1920).
40. H. Leuenberger, J. D. Bonny, C. F. Lerk, and H. Vromans, *Int. J. Pharm. 52*: 91 (1989).
41. P. J. Sherrington and R. Oliver, *Granulation*, Heyden, London, p. 54, 1981.
42. N. A. Armstrong, N. M. A. H. Abourida, and A. M. Gough, *Pharm. Technol. 6*: 66. (1982).

9

Rationale for and the Measurement of Tableting Indices

Everett N. Hiestand*

Upjohn Company, Kalamazoo, Michigan

I. INTRODUCTION

The impracticability of measuring the properties of individual particles, the real participants of tableting, is a major impediment to obtaining an accurate description of the tableting performance of the powder. The challenge has been to obtain information from the properties of the compact, the "continuum," that will definitively indicate the processing performance of the material. It is believed that the tableting indices[†] developed at the Upjohn Company provide this information.

II. BACKGROUND

A. The Origin of Bond Strength

The forces acting at particle-particle interfaces would be preferred information; but the surface energies and the interaction energies between particles have their origin from the same force. This was demonstrated by using mechanical measurements to estimate solubility parameters [1]. The sur-

---

*Retired. Residing at 11,378 East G Avenue, Galesburg, Michigan 49053
†This is not a review of all indices proposed. Only the system developed by Upjohn is discussed.

face energies are more accessible than the forces per se. Thus, for convenience, surface energies become the subject of discussion. In general, the surface energy of organic materials are in the 25 to 85 ergs/cm² range. Note that the largest is less than four times the smallest value. Clearly, it is impossible to account for the observed variation of bond strength among different materials, variations greater than an order of magnitude (see tensile strengths in Table 1), by considering only the kinds and magnitudes of forces acting. A legendary explanation of the large variation of strength is to assume that the areas of contacts between the particles varies sufficiently to yield these strength differences; however, in Table 1 the tensile strength of compacts at the same solid fraction, i.e., essentially the same interparticle contact area, shows a range greater than 10 to 1. Certainly melting-resolidification could give differences; however, this explanation often collapses when it is recognized that some of the stronger bonding materials decompose before they melt; and melting induced by applied pressure requires the opposite volume change (the Clapeyron-Clausius equation) from that of most organic materials. Furthermore, plastic deformation usually occurs at lower pressures than melting. Clearly, this cannot be a general explanation. While chemical interactions cannot be excluded in all situations, certainly they cannot be invoked to account for the universal adhesion between particles.

Table 1 Indentation Hardness and Tensile Strength Values (kN/cm²) for Various Materials at Selected Dwell Times[a] for H Values and Time Constant of 15 sec for σ_T; Compact Solid Fraction 0.9

Material	H_0	$H_{1.5}$	H_{30}	H_∞	$\sigma_T \times 10$
Avicel[b]	24.1	6.79	6.01	—	8.1
Sorbitol	40.8	3.13	1.39	0.45	1.9
Lactose-Sp.Dr.	33.4	12.6	10.8	6.53	1.2
Sucrose	47.3	—	8.0	—	1.9
Ibuprofen[c]	12.8	—	2.38	—	0.97
Aspirin	1.40	—	0.57	—	0.29
Caffeine	24.9	12.3	8.00	<3.1	3.59
Phenacetin	3.15	2.87	1.99	—	2.78
Acetaminophen	8.8	7.4	7.4	1.4	0.66
$CaSO_4 \cdot 2H_2O$	23.5	—	14.1	—	1.86
Starch (maize)	10.5	—	2.0	—	0.80

[a] Subscripts indicate approximate dwell time in minutes, except ∞ is 24-hr dwell time value.
[b] PH-101, microcrystalline cellulose.
[c] Lot-to-lot variation regularly observed.

   Modern adhesion theories emphasize the role of mechanical properties in determining strength. Co/adhesion strength is the derivative of the maximum incremental work done during the separation of surfaces. The components of the work are (1) the energy needed for the new surfaces formed by the separation, (2) the plastic deformation that occurs as the surfaces separate, and (3) charge separation when the separation is of surfaces containing charged (transfer) sites. Plastic deformation may be either the ductile extension of the entire contact regions or the localized deformation at the interparticle contact perimeters at the receding boundary between two particles. In this process the perimeter is the equivalent of the crack tip region considered in studies of fracture mechanics. Item (3) is the most difficult to estimate since with insulating materials, charge separation is dependent on charge transfer at defect and impurity sites on the crystal surface, which can have dramatic effects even when present at unknown, undetectable levels. However, in most cases, it is believed to be negligible because surface conductivity limits the survival of localized charges.

   Universal attraction (except when close enough for Born repulsion) exists because it is the lowest free-energy state and realized by the particles coming together. The particle rearrangement and plastic deformation produced by compression increase the total number of effective contact sites in a cross section where this low-energy state is attained. All of these processes have been included in models of the processes of tablet bonding [2,3].

B. The Challenge for Tableting Indices

The indices are based on a philosophy that is unique only because it is seldom used in pharmaceutical studies, viz. the indices are determined using measurements on specially prepared test specimens (compacts); only limited data are collected during the making of the compact. A compact, by definition, has sufficient rigidity to be handled without uncontrolled changes occurring as a result of the handling. This makes it a preferred form. However, how does one produce equivalent test compacts from any or all materials that need to be evaluated? This must be done without adding excipients that would "dilute" or modify the property one is trying to measure, and must be done without introducing macroscopic flaws into the compact which would distort the test results. To identify a procedure that works for all materials is a very big challenge; and the making of the test specimen becomes a critical part of the evaluation of tableting properties. It is as important as the choice of measurements made on the test specimen.

III. CONSIDERATIONS FOR STRENGTH MEASUREMENTS

Mechanical properties are the response to an applied stress state. Strength is defined as the stress required to produce a permanent change of shape; both plastic yielding and fracture do this. Thus strength measurements may include response to both compressive and tensile loading. Strength measurements provide the simplest method of characterizing mechanical properties.

A. Fracture Strength

While the tensile strength* is an obvious choice for indicating bond strength, there are numerous complicating factors to be considered. Any macroscopic† defect within the test volume can falsify the results and must be avoided. Also, all compacts are Mohr bodies. Thus, the observed strength is dependent on the hydrostatic stress within the compact, i.e., the stress state. Furthermore, nearly all organic materials exhibit some degree of viscoelastic behavior; therefore, the values of strength obtained depend on the rate of strain during the test. One must use equivalent, not identical, strain rates when the materials being compared exhibit different degrees of viscoelastic properties. Meticulous attention is a requisite to reproduce the exact procedure, including the equivalent strain rate, to obviate these potential problems. Also, to reduce the risk of a macroscopic defect distorting the result, the strength test should subject the minimum volume to the maximum stress that produces the fracture, i.e., minimize the volume from which a crack will originate. Chance defects far removed from this "critical" volume would not significantly distort the results. An additional complication arises from the fact that compacts are not of uniform consolidation‡ throughout, and the magnitudes of the density differences within the compact are dependent on the properties of the material. Clearly, it would be best, when more than one measurement is being used in a comparison, for all measurements to depend on the properties of the same region within the compact. For example, one could measure the properties of the central portion of the compact. Also, strength values increase nearly exponentially with the solid fraction, ρ_r; therefore, one must identify any given tensile strength value, σ_T, with the ρ_r value of the compact measured.

---

*Tensile strength and fracture strength are used herein as synonymous because tensile failure of compacts always is a brittle mechanism.

†For this Chapter macroscopic indicates a size much larger than the particles or pores.

‡Consolidation may be expressed as solid fraction or relative density, which is equal to one minus the porosity of the powder bed or compact.

B. Tensile Strength from Compression

The transverse compression of square compacts provides a tensile test that meets more of the requirements discussed above than any other test procedure known to the author. For the square compacts used with this system, the compact is placed between two platens that cover the central 0.4 of the edge-width of the compact; pads are placed between the platens and the compact to reduce the potential for stress concentrations at the edge of the platen initiating fracture (polymeric platens without pads have been used successfully, also). The maximum tensile stress appears only in the central portion of the test specimen [4,5], the same region used for the indentation hardness described later. It is believed that with this procedure the fracture starts in this volume.

An extreme example will illustrate the need for control of the strain rate. What strain rate should be used to test the fracture strength of a very viscoelastic silicone polymer—specifically, the one sold as (in the United States) "Silly Putty"? Clearly, it must be a very high strain rate; otherwise ductile extension occurs. A convenient strain rate to use in routine testing of most materials would be much too low to cause fracture of the silly putty. However, if a test is to be applicable to all materials, a systematic method to arrive at comparable strain rates must be incorporated into the test procedure when viscoelastic materials are used. With the indices testing, this is done by assuming that the rate of stress increase during the tensile test is approximately exponential with strain. Therefore, one can apply a time constant to the loading-to-fracture process. The time constant is set as a fixed elapse of time between when $1/e$, 0.368, of the fracture force was being applied and the time of fracture. The time interval, the time constant, used in the author's laboratory is 15 sec. (The platen movement rate for a single test is constant but with other materials or other solid fractions the constant rate may be different.) The choice of 15 sec is arbitrary but appears to be suitable for nearly all compacts.* Table 2 shows data for sucrose and sorbitol obtained with different time constants and confirms that the strength varies with strain rate. The setting of the test instrument to produce the desired strain rate will be by trial and error; for this one can use the indentation hardness specimens left over from that test. While the 15-sec rate would not cause silly putty to fracture, one does not expect to make tablets from materials with such unusual properties.

---

*More recent work indicates that a time constant of perhaps 5 sec provides better breaks with some materials than does the 15-sec case. It is important that comparisons be made only using data obtained with the same time constant. (10 sec has been used successfully by Professor McGinity at the University of Texas.)

Table 2   Tensile Strength[a] Variation with Time Constant[b]

Material	Solid fraction	Time constant (sec)	Tensile strength (kN/cm^2)
Sorbitol	0.9055 ±0.0014	43.30 ±1.92	0.1745 ±0.0024
	0.9038 ±0.0008	13.40 ±0.48	0.2049 ±0.0038
	0.9037 ±0.0002	4.70 ±0.19	0.2342 ±0.0040
	0.9044 ±0.0010	1.70 ±0.17	0.2782 ±0.0063
	0.9045 ±0.0006	0.80 ±0.02	0.3021 ±0.0059
Sucrose	0.9071 ±0.0004	26.50 ±0.08	0.2043 ±0.0094
	0.9073 ±0.0002	7.70 ±0.30	0.2274 ±0.0179
	0.9069 ±0.0008	3.32 ±0.08	0.2400 ±0.0085
	0.9068 ±0.0003	1.14 ±0.02	0.2481 ±0.0075
	0.9074 ±0.0005	0.572±0.013	0.2608 ±0.0114

[a] These data were for different lots of powder than used in studies reported in Table 1; some differences observed.
[b] Based on mean of five determinations.

Is tensile strength a sufficient indicator of the tableting properties? The answer is no! Problems of capping and lamination are not solely identified with bond strength. When present, they are fractures at affected "tensile" strength. These result from brittleness and may occur even when the bond strength is excellent. Perhaps a more usable term than tensile strength would be compactibility,* but this too has been used to indicate an absolute strength [6]. To establish a scale for bonding propensity that is applicable to all materials, one might use a dimensionless number based on the tensile strength obtained per unit of compression pressure, σ_c. Unfortunately, the amount of the applied σ_c that is actually contributing to added σ_T varies over a very large range as the ρ_r changes. At very high values of ρ_r, a significant fraction of σ_c in the usual uniaxial die system is counteracting die wall friction or is present as an ineffective hydrostatic stress component within the die. The ineffective portion of the hydrostatic stress is the intraparticle (microregion) hydrostatic stress, which does not produce plastic deformation of the particle† and, therefore, does not produce a change in

---

*It is preferred that compressibility not be confused with compactibility. Gases are highly compressible, but do not readily make compacts; compressibility refers to volume change.

†From strength of materials, one learns that stress is divided into hydrostatic and deviatoric stresses. Only the deviatoric component produces plastic deformation.

bond strength at the contact regions. Only an unknown portion of the total σ_c may be considered to be the bonding active compression stress (BACS).

C. The Indentation Hardness

It is believed that the indentation hardness measurement correlates more consistently with the magnitude of the BACS. It offers several advantages over the total compression pressure. Figures 1 through 5 show data obtained with a few different materials (also see Table 1). Two hardness values are shown: one the impact or, essentially, instantaneous value, H_0, and the other a long indenter dwell time value, H_{30}. Within the solid fraction range readily accessible in most pharmaceutical laboratories, the log H versus ρ_r plots are nearly linear. Note that σ_c usually is between H_0 and H_{30}.

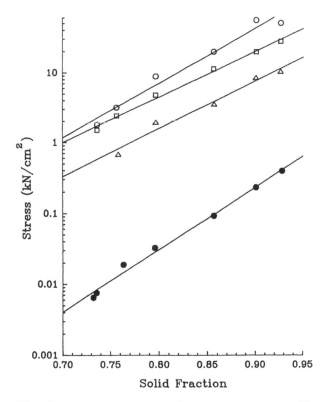

Fig. 1 Strength of sucrose: ○ — impact hardness; □ — compression pressure; △ — hardness, 30-min dwell time; ● — tensile.

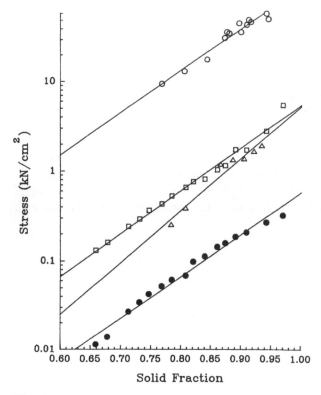

Fig. 2 Strength of sorbitol: ○ — impact hardness; □ — compression pressure; △ — hardness, 30-min dwell time; ● — tensile.

This is assumed to be a result of the difference of dwell time. However, the relative location of the σ_c data and its slope do not fit an understood or predictable pattern. It is assumed that effects from the total confinement of the powder within the die often contribute much to σ_c. While not used in the indices, hardness values for two additional dwell times for the indenter are shown in Table 1.

For use in a bonding index, the indentation hardness test should produce additional consolidation under the indented surface, within the compact, similar to the compression pressure; i.e., the stress must be sufficient to produce the necessary for bonding, plastic deformation of particles. However, the free surface around the indenter may heave, and, thereby, the excessive hydrostatic stress under the indenter is limited. For most of the materials shown in Figs. 1–5 the maximum compression stress

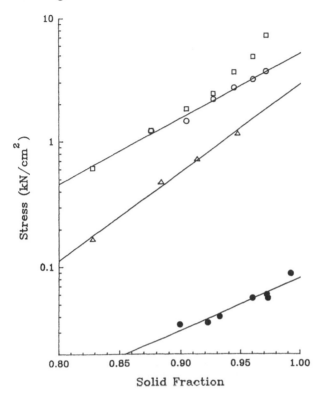

Fig. 3 Strength of aspirin: ○ — impact hardness; □ — compression pressure; △ — hardness, 30-min dwell time; ● — tensile.

applied was not sufficient to introduce extensive departure from a linear log σ_c versus ρ_r relationship. (Earlier experience has shown this departure to occur at high solid fraction for σ_c but not for H_0 [7].) However, with aspirin, log σ_c versus ρ_r clearly is nonlinear even at relatively low solid fractions. The reason for this is not known.

For materials that are difficult to tablet, an extended punch dwell time during the making of the compact may enable one to produce a flawless compact. This freedom to vary the compression pressure in any way needed is important. The use of σ_c in the bonding index would restrict this variation of strain rate used during compaction and could limit the success at making flawless compacts. The indentation hardness measurements do not have these constraints. This is an added reason for using hardnesses instead of σ_c in any bonding index.

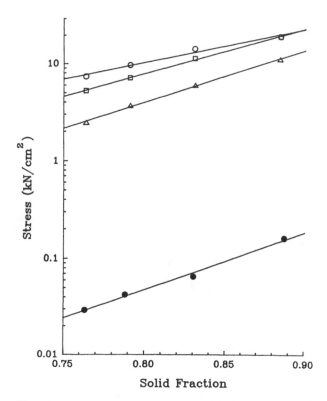

Fig. 4 Strength of calcium sulfate-dihydrate: ○ — impact hardness; □ — compression pressure; △ — hardness, 30-min dwell time; ● — tensile.

IV. THE BONDING INDICES

Based on the above arguments, σ_T/H_0 had been designated [8] as a bonding index; H_0 is the impact hardness. More recently [9], it has been called the worst-case bonding index (BI_w) because $H_0 > H_t$, where t designates a strain rate slower than the impact method of measuring hardness.

A. Hardness and Indenter Dwell Time

H_0 is obtained with a dwell time of the indenter on the compact of less than a millisecond [10, 11]. thus, the viscoelastic decay of the stress during the dwell time is minimal. However, using a different apparatus, the dwell time at maximum penetration for obtaining an indentation hardness can be made any selected length of time. If that dwell time is 30 min, the hardness

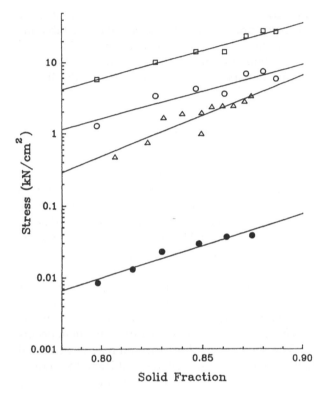

Fig. 5 Strength of acetaminophen: ○ — impact hardness; □ — compression pressure; △ — hardness, 30-min dwell time; ● — tensile.

will be designated as H_{30}. H_{30} would be expected to be less than σ_c. For this case of constant displacement for a selected dwell time, an equivalent strain rate would be hard to estimate, but qualitatively the similarity of long dwell time and a very slow strain rate are obvious.

Since tableting performance of a material is not expected to involve extremely slow strain rates or very long dwell times such as 30 min, a bonding index based on the ratio σ_T/H_{30} would represent a practical [9] best-case bonding index (BI_b). $H_0 > H_{30}$ and $\sigma_T/H_{30} > \sigma_T/H_0$. Real tableting operations probably will correlate closer with the rank order of BI_w than BI_b. However, the theory of tablet bonding indicates that viscoelastic materials produce much better bond than nonviscoelastic materials [2, 3]. Therefore, both the magnitude of BI_w and BI_b should be considered, especially the difference between the two. A large difference is desired, especially when BI_w is on the low side, but it will mean that the tableting performance could be influenced dramatically by the speed of the tableting machine. Table 1 shows data obtained using several dwell times. The choice to use H_{30} for the best case bonding index was arbitrary. Probably an equally useful value could be obtained by comparing shorter dwell time data.

B.   When H_0 Is Inappropriate for Use in the *BI*

The discussion of the BACS concluded that H is a more suitable parameter for the denominator of the bonding index than is σ_c. When making the test specimens, the dwell time of the punch (elapsed time at maximum displacement) is 1 to $1\frac{1}{2}$ minutes. Seldom is $H_0 < \sigma_c$ in the linear portion of the semilog plots; whenever it is, one has a material that has the potential for tableting problems. If $H_0 << \sigma_c$, the material has very undesirable tableting characteristics. Furthermore, the bonding indices based on H values will not be meaningful.

How can H_0 be less than σ_c when the dwell times are so different? A possible, but unproved, explanation can be given. It is believed that when the particles are very hard and the bonding between them is weak, particle movement, sliding over each other, during the formation of the dent occurs at relatively low stresses. The shear strength of the compact is weak and relieves the stress before the applied stresses becomes sufficient to produce plastic deformation of the particles; therefore, $H_0 < \sigma_c$. Fortunately, the observation that $H_0 << \sigma_c$, alerts the experimentalist that he has a special, unusual case, in which the hardness should not be used in the *BI*. When this occurs, the bonding index is estimated for the regions where the log H plot is nearly parallel with the log σ_c plot by using the ratio σ_T/σ_c. An extreme example of this is with acetaminophen shown in Fig. 5; the compression pressure is nearly 4.2 times the impact indentation hardness in the regions of higher solid fraction. If the bonding index were based on the hardness, BI_w would be approximately 0.0075. However, if based on the compression pressure, the estimate for the bonding index becomes approximately 0.0018. The ratio H_0/H_{30} for acetaminophen is only ~1.2, a relatively small value; i.e., it is not very viscoelastic. Nevertheless, one must ask is the 0.0018 value a BI_b or a BI_w bonding index. Since σ_c is based on a $1\frac{1}{2}$-min punch dwell time, it may be neither, but probably it is closer to the best case than to the worst case. If it is taken as the best-case bonding index, the worst case should be $0.0018/(H_0/H_{30}) = 0.0015$.

Acetaminophen has very undesirable tableting characteristics. The fact that one must settle for a substitute for the impact indentation hardness is an indication that the material is a difficult one to tablet. Fortunately, this extreme is a very rare case. With phenacetin and maize starch, σ_c was only slightly larger than H_0; the bonding indices were calculated in the usual way. However, one should recognize that whenever $\sigma_c > H_0$, the material may not tablet satisfactorily. Possibly an index of tableting performance could be defined using this difference; but in the absence of a certain explanation for the relative values of σ_c and H_0, such an index has not been defined. However, it clearly is an indicator of problems that accompany low shear strength.

V. A BRITTLE FRACTURE INDEX

A. Strength When a Macroscopic Defect Is Present

Size is a key consideration for the defect. Microscopically, the perimeter of the true contact regions between particles is a region of stress concentration. However, these are "building units" for the compact and are always present. For the stress at the defect to be treated in a continuum sense, the defect must be large compared to the pores within the compact. The compact fails by brittle fracture even when the interparticle bonding may be a ductile process [2,3]. The macroscopic defect only enhances this brittle behavior. Thus the brittle fracture index is a measure of degree of brittleness. When the brittle fracture index is low, the compact is not sufficiently brittle to be a serious problem in processing.

The Griffith crack theory for the fracture of brittle materials is based on the concept that a crack will be initiated when the incremental change of elastic energy during crack growth provides the incremental gain of surface energy for the new surfaces. Furthermore, it is assumed that the origin of the crack is from a defect site where the elastic stress is concentrated. It is higher than the nominal stress, and therefore it would be the first region to reach the stress level needed for the crack to grow. Since the crack tip may continue to be a stress concentrator, the crack often continues to propagate. Therefore, the material fails at a much lower applied stress than expected from the theoretical bond strength and/or theoretical shear strength. However, if a material is able to relieve by plastic deformation some of the stress in the region of concentrated stress, then the stresses may not build rapidly to the required level for the crack to propagate; i.e., the measured strength will approach closer to the nondefect value. For these cases the Griffith theory is modified to include the absorption of energy by the plastic deformation. Thus, the force required to produce fracture is increased over the Griffith case; i.e., the fracture strength is increased. This is the basis on which a brittle fracture index (BFI) is developed [8, 9, 12]. A stress concentrator, a small (1.1-mm diameter) round hole, is put in the center of the tablet and the measured tensile strength is compared with that of a tablet without a hole. The hole is made during the compression by a retractable, spring-loaded pin extending from the center of one punch. The ratio of the two values depends on the ability to "consume" elastic energy around the hole by plastic deformation, i.e., to increase the work of separation. For this measurement, the platen movement rate is kept the same as for the defect-free compact to which its strength is compared.

Elasticity theory for an isotropic material indicates that for the biaxial stress state induced during the transverse compression tensile test, the stress concentration factor is about 3.2 times.* Since compacts are not isotropic homogeneous bodies, the factor would vary because of these density gradients within the compacts. The specific stress concentration factor for each compact is unknown. For simplicity, the factor 3 was adopted as the stress concentration factor at the edge of the hole in a compact. Thus, it is assumed that in the absence of stress relief by plastic deformation around the hole, a small round hole at the center of the compact would result in an apparent strength of one-third that of a compact without the hole. However, if the material is capable of relieving the stress around the hole by plastic deformation before it fractures, then the stress around the "defect" cannot build up to three times the surrounding stress. Among various materials the stress relief will range from zero to nearly total. Thus the tensile strength of the compact with the hole in it will range from one-third to nearly equal that of the compact without the hole. The brittle fracture index is based on this, but the numbers are approximately normalized by using the equation $BFI = (\sigma_T/\sigma_{T_0} - 1)/2$, where σ_{T_0} is the apparent tensile strength when the hole is present.† Experience teaches that when the BFI is less than 0.20, there will not be a problem with fracture during tableting on a rotary press unless the bonding is very weak. The indices values provide the warning of this problem combination. (It is desirable to have the bonding index greater than 0.005.) Since there is some fracture dependence on the bonding index, both numbers should be considered; i.e., one cannot set absolutes for either number.

VI. A STRAIN INDEX

The magnitude of the elastic recovery following the application of enough stress to produce plastic deformation might be stated as the maximum

---

*The compressive stress at the center of the compact is 3.7 times (and normal to) the induced tensile stress [4].

†Some materials have been observed to have a brittle fracture index greater than one. The scale still is valid. It is a very brittle material and for tableting use will require a diluent that reduces brittleness to an acceptable level, less than ~0.2.

elastic strain for that specific material. For two spheres, if one substitutes $H_0 \pi a^2$ for the force in Hertz's equation for elastic deforming spheres, one obtains the equation $Z_e = 3\pi a H_0/4E'$, where Z_e is the elastic recovery distance of the center of the sphere, a is the chordal radius of the contact of the two surfaces before recovery, and $1/E' = \Sigma(1 - \nu^2)/E$ for the two materials, ν is Poisson's ratio, and E is Young's modulus; or $Z_e/a \propto H_0/E'$. While not a true strain, either Z_e/a or H_0/E' provides a relative value for the elastic strain following plastic deformation. If one views the particles as spherical indenters acting on each other, the possible relationship to the magnitude of the elastic recovery following compression of a compact is apparent; hence, it is a valid strain index. Since the strain index (SI), H_0/E', is obtained from the impact hardness experiment [8] without doing any additional experiments, it is available for use; however, the principal use has been to show that the tendency of compacts to fracture does not correlate with the amount of elastic displacement. Instead, fracture propensity correlates with the brittleness of the material as indicated by the BFI.

VII. EXPERIMENTAL PROCEDURES*

A. The Tablet Press, the Test Specimen

Because compacts made from some materials develop fracture lines, the usual uniaxial compression-decompression cycle is not a satisfactory way to make a test specimen compact. With square compacts, it is possible to split the die across a diagonal so that one can use a triaxial decompression process, i.e., reduce the die wall pressure simultaneously with the punch pressure. This reduces the shear stresses (continuum perspective) within the compact during the decompression. Because compacts are not isotropic homogeneous bodies, the internal shear stresses will not be zero at all points even under the best of conditions. However, if the mean die wall and the mean punch pressures are held equal, the internal shear stresses should be below the strength of the compact, in which case no fracture lines are introduced. Satisfactory test specimens, compacts free from macroscopic defects, can be made from nearly any material by using nearly equal die wall and punch pressure conditions during decompression.

The triaxial decompression press shown in Fig. 6 was built specifically to provide the needed control. This system permits the selection of compact thickness, dwell time of compression, the ratio of die wall and punch

---

*Versions of the apparatuses described and used are covered by U.S. Patents 4,880,373, 4,885,933, and 4,957,003 and owned by The Upjohn Company, Kalamazoo, MI. Applications have been filed in other countries. Licensing to build from Upjohn plans is available at reasonable cost.

Fig. 6   Triaxial decompression tablet press.

pressure, and the duration of decompression. The powder is poured into the die; care is taken to distribute it as evenly as possible. The compression force is applied and the maximum force held constant for a fixed time, perhaps 1 or $1\frac{1}{2}$ min. The triaxial decompression rate is controlled by selecting the time interval over which decompression will occur, usually 1 min is chosen. This operation has been automated by using a computer to control the timing and the pressures during unloading.

B.   Solid Fraction Determination

The apparent density is calculated from the weight and dimensions of the compact, which is divided by the absolute density of the material to give the solid fraction. The tablet dimensions are measured immediately after the indentation hardness test or immediately before the tensile strength test. The average of five thickness values is used. The time for measurement is chosen to minimize the effect of the unavoidable change of dimensions with time that may occur after removal from the die. The solid fraction at the time of the test is the critical value. When the absolute density is unknown, a gas pycnometer is used to determine it.

C. Indentation Hardness Measurements

For the tableting indices, the indentations are produced by a 2.54-cm-diameter steel sphere. The chordal diameter of the dent is kept to less than 4 mm and is made in the center of the square face of the test specimen. Because a large spherical indenter is used to make a small dent, the normal body stresses are oriented within a few degrees of normal to the original surface. (Friction between the compact and the indenter surface would introduce a limited shear stress.) The large indenter is used to avoid the dislodging of particles on the surface. Also it is intended to provide the needed consolidation below the indenter and to diminish the pile up around the indenter.

Figure 7 shows the dynamic indentation hardness apparatus, and Fig. 8 shows the longer dwell time fixture on the drive assembly common to it and the tensile strength test. H_0 is obtained by bouncing the sphere off the surface of the compact. H_{30} is obtained by pushing the spherical indenter into the compact and then holding its position fixed for 30 min while the force decays. Because the average pressure under the indenter is used, the distribution of stress under the indenter, though believed to be very different in the two cases, is not considered. (The stress distribution under the indenter is unknown but will depend on the viscoelastic properties of the material and will be influenced by the dwell time as well as by the porosity.) In practice, a comparison of the magnitudes of H_0 and H_{30} will indicate important differences among materials. An organic material whose properties are $H_0 \geq 5H_{30}$ would be a strongly bonded material. However, if $H_0 \leq 1.2H_{30}$, the bonding will be weak. Since the bonding of a viscoelastic material will be influenced by the dwell time of the press in which the tablet is made, the magnitudes of the differences between the two are needed to indicate the magnitude of this effect; indeed both values are important.

To avoid fracture during the hardness tests the tablet is placed in a split die and some die wall pressure applied before the indentation is made. Especially important with the dynamic test is a backup support, a backup punch is used to push the tablet to the front of the die after some die wall pressure has been applied. Thus, the square tablet is firmly supported on five surfaces during the indentation. An evaluation was done of whether the die wall pressure would influence the hardness value observed. While the answer is yes, for the cases studied the effect was so small that it was insignificant. Of course when the die wall pressure is needed to avoid fracture, the magnitude of the effect cannot be determined easily because the "window" between the attainable (with the current apparatus) die-wall pressure that prevents fracture and that failing to prevent fracture may be small.

Fig. 7   Impact indentation hardness apparatus.

D.   Dent Size, Important Considerations

The relative dent size to compact size must be within certain limits. Arbitrarily, the dent is made to have a chordal diameter between 2 and 4 mm. For this dent size the thickness of the compact should be at least 9.5 mm; otherwise, the supporting punch on the opposite side from the dent will significantly influence the force required to make the dent. The square compacts are 19.05 mm on a side. With this width, the dent must be made in the very center region; otherwise, the die wall could influence the result. Tabor [13] discusses the influence of the elastic deformation of the "hinterland" away from the indenter and indicates that there are effects when Young's modulus for the material being indented is low [13]. With the compacts the discontinuity at the die wall must be considered. The elastic modulus of the steel die wall is much greater than the elastic modulus of the compact. Also, limits on the size of the indentation must be imposed. From experience, the following is considered satisfactory. The distance from the edge of the dent to the specimen edge should not be less than $5a$, the chordal radius, and the thickness should be at least five times the depth of the dent.

Fig. 8 Long dwell time hardness apparatus.

These recommended values are somewhat different from Tabor's recommendations. Possibly because of the porosity changes under the indenter, the requirements are less stringent. Furthermore, reducing the amount of an expensive medicament that is needed for the indices determination is sufficient motivation to accept a slightly smaller than ideal compact for this applied use. This last is important. Formulators need to know the values of the indices for small lots of material used in clinical studies, when the supplies may be very scarce, so these properties can be matched when production scale-up occurs. This avoids clinical studies having to be repeated with a new, different formulation.

E. Dent Size Determination

For both indentation tests the chordal radius of the dent produced must be determined. The simplest procedure for this is to observe the chordal radius, a, using a low-power microscope with oblique angle, surface illumination. Care must be exercised to eliminate differences among operators. The average of five readings, taken by rotating the compact approximately 72° between each reading, is used. If the compact surface is grainy, the microscopic method may not be as accurate as wanted. A more precise value for the chordal radius may be obtained by using a surface roughness analyzer (e.g., Federal Surfanalyzer 150*), an instrument whose probe moves across the dent and yields data to produce a much enlarged profile of the dent (depth is enlarged much more than diameter). Figure 9 shows the arrangement and the probe of the apparatus used with a computer for handling the data. The tip of the probe used with the tablets is a 0.79-mm- (1/32 in.) diameter sphere. (A much smaller probe tip is used for the instrument's original purpose, for surface roughness measurements of metals.) Current practice uses the major portion of the profile to obtain a radius of curvature. A circular segment, least-squares fit is used with the dent depth, measured from the amplified profile, to calculate the chordal radius of the dent. This is done with a menu-driven computer program to handle the data. The design reduces the operator subjectivity of the measurement.

F. Hardness and Strain Index Calculations

H_{30} is the mean pressure under the indenter after a dwell time of one half-hour. Specifically, it is the force observed by the load cell, after the force has decayed for one half-hour, F_{30}, divided by the chordal area of the dent; $H_{30} = F_{30}/\pi a^2$, where a is the chordal radius of the dent.

---

*Federal Products Corporation, 1144 Eddy Street, Providence, RI 02940-9400.

Fig. 9 Apparatus for dent chordal radius measurement.

The equations used to calculate H_0 assume that the work of forming the dent is equal to the hardness, H, times the volume of the dent [10, 11]. Also assumed is Hertz's laws of elasticity during the elastic recovery. Based on experimental evidence [10], Eq. (1) is believed to provide good estimates of the hardness.

$$H_0 = \frac{4mgrh_r}{\pi a^4} \left(\frac{h_i}{h_r} - \frac{3}{8} \right) \tag{1}$$

where

m = indenter mass
r = indenter radius
h_r = indenter rebound height
h_i = indenter initial height

The strain index may be calculated from the same data [8] using Eq. (2):

$$\frac{H_0}{E'} = \frac{5a/6\pi r}{h_i/h_r - \frac{3}{8}} \tag{2}$$

A SI for the relaxed case requires additional data, viz., the radius of curvature of the unloaded surface. Because the radius is changing rapidly immediately after unloading, meaningful data are not easy to obtain. Fortunately H/E' for the relaxed cases is not important for the use of indices.

G. Tensile Strength Measurements

The fixture used for the transverse tensile strength is shown in Fig. 10; it is mounted on the common drive mechanism. For this arrangement, the tensile stress is 0.16 times the mean stress at the platen-compact surface [4]. Since the maximum tensile stress exists at the center of the compact, it is assumed that any crack will start in that region. If stress concentrations at the edge of the platens initiate the fracture, the test has not measured the tensile strength. (The stress at the edge of the platen is not known.) Fortunately the shape of the fragments identify these rare cases. Usually this can be overcome by placing pads, blotter or cardboard, between the compact and the platen. When the failure occurs along a central line normal to the platens, the test has been a measure of the tensile strength. The platen force versus time is recorded and used to obtain the time constant of the tensile fracture test.

VIII. NEEDED PRECISION

For commercial application, it is desirable to reduce the need for many duplicate tests. A realistic demand on the needed precision of the data contributes to this. The criterion is to provide a rugged formulation, one that will manufacture satisfactorily even if some variation of the material properties occurs. The experimental precision of the measurements will vary with the material being evaluated; minimum variation occurs with good tableting materials. Fortunately, there is only limited need for precision for clearly problem materials or for very good ones. The greatest precision may be needed to prove that an acceptable elimination of problems has been attained by the excipients in the formulation. Fortunately, this is the range where little variation of the data occurs.

If one wishes to know how the property varies with solid fraction, semilog plots usually are nearly linear, viz. $\log H$ versus ρ_r and $\log \sigma_T$ versus ρ_r, over the range of interest. These plots can be used to help spot problems with the data, and when nonlinear curve fitting is used, confidence intervals can be estimated. It is useful to establish the minimum target values for the experiments.

A word of caution is needed relative to sampling. There are standard procedures for obtaining representative samples. The use of mixtures, granu-

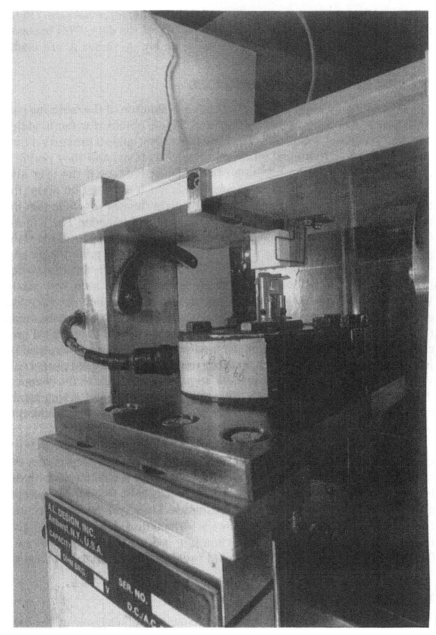

Fig. 10 Apparatus for transverse compression, tensile strength measurement.

lated materials, and direct compaction mixtures requires special care. Often, poor sampling practice introduces much scatter of the data. This becomes obvious when the plots of log H versus ρ_r or log σ_T versus ρ_r are made.

IX. USE OF THE TABLETING INDICES

The tableting indices provide a quantitative evaluation of the tableting performance of a powder or powder mixture. This is obtained without adding "property diluting" excipients. They will detect mechanical property differences among lots that meet all chemical specifications; and they permit a rational approach to the selection of excipients. Perhaps if the user asks himself whether any other system provide as much information about the tableting properties of each individual ingredient, he will conclude that the tableting indices are uniquely useful. Selected data obtained in The Upjohn Company laboratories are shown in Table 3. Hypothetical illustrations of the potential benefits from having a quantitative evaluation of tableting properties have been given [9]. This information could be especially desirable when the dosage is high and the volume of excipient must be kept very small. Without the indices values, it may not be easy to convince the chemical division to produce a product that has better tableting characteristics than the by-chance properties of the most convenient or most economical process. Actually the cost of making the medicament may not be increased by just using carefully controlled operations to produce the needed properties. With the tableting indices, it is possible to provide guidance to the chemical division as they experiment with different crystallization and drying conditions to attain the objective. Also, the indices are useful when evaluating materials available from more than one source.

Values for the BI_w have been observed in the range of 0.001 to 0.04. Currently with only limited studies, BI_b values have been observed in the range 0.002 to 0.14. While tablets can be made from the materials with the lower BI_w values, they may be very soft and excessively friable. The acceptable value is not independent of the brittleness. Usually if the final formulation has a BFI below 0.2, it indicates a formulation that would not give fracture problems. However, if the bonding index value is too low, e.g., 0.001, a slightly lower BFI value would provide a better margin of safety.

X. CONCLUSIONS

The fundamental considerations used to justify the selection of the tableting indices, the strain index, the brittle fracture index, and the bonding indices, have been described. The second bonding index was justified be-

Table 3 Examples of Values Observed for Tableting Indices

Material $\rho_r = 0.9$	$BI_w \times 10^2$	$BI_b \times 10^2$	BFI	$SI \times 10^2$
Avicel[a]	3.4	13.5	0.03	2.3
Sorbitol	0.46	13.7	0.03	0.94
Lactose, Sp. Dr.	0.36	1.1	0.12	1.8
Sucrose	0.40	2.3	0.68	1.5
Ibuprofen[b]	0.76	4.1	0.06	0.6
Aspirin	2.1	5.1	0.19	0.7
Caffeine	1.4	4.5	0.47	1.3
Phenacetin	0.88[c]	1.4	0.43	1.0
Acetaminophen	0.15[d]	0.18[e]	0.03	1.4
$CaSO_4 \cdot 2H_2O$	0.79	1.3	0.08	1.2
Starch (maize)	0.76[c]	0.67	0.80[f]	1.9

[a] PH-101, microcrystalline cellulose.
[b] Lot-to-lot variation regularly observed.
[c] Compression pressure slightly $> H_0$.
[d] $BI_w \times H_{30}/H_0$.
[e] Based on σ_T/σ_c.
[f] At $\rho_r = 0.8$, BFI is 0.44.

cause of the very important role of viscoelasticity in the development of bond strength. Special conditions are described where the indentation hardness should not be used in the bonding index term. Cautious use of the compression pressure is indicated in these cases. The necessary precautions and consideration when using the experimental procedures are described.

REFERENCES

1. H. Leuenberger, *Int. J. Pharm. 27*: 127–138 (1985).
2. E. N. Hiestand, *Int. J. Pharm. 67*: 217–229 (1991).
3. E. N. Hiestand and D. P. Smith, *Int. J. Pharm. 67*: 231–246 (1991).
4. R. Berenbaum and I. Brodie, *Br. J. Appl. Phys. 10*: 281 (1959).
5. E. N. Hiestand and C. B. Peot, *J. Pharm. Sci. 63*: 605–612 (1974).
6. H. Leuenberger, E. Hiestand, and H. Sucker, *Chem.-Ing. Tech. 53*: 45–47 (1981).
7. E. N. Hiestand, Int. Conference on Powder Technology and Pharmacy, Basel, Switzerland, 1978.
8. E. N. Hiestand and D. P. Smith, *Powder Technol. 38*: 145–159 (1984).
9. E. N. Hiestand, *Pharm. Tech. Int. 1*: 22–25 and *Pharm. Tech. 13*(9): 54–66 (1989).

10. E. N. Hiestand, J. M. Bane, Jr., and E. P. Strzelinski, *J. Pharm. Sci. 60*: 758–763 (1971).
11. D. Tabor, *The Hardness of Metals*, Clarendon Press, Oxford, UK, 1951.
12. E. N. Hiestand, J. E. Wells, C. B. Peot, and J. F. Ochs, *J. Pharm. Sci. 66*: 510–519 (1977).
13. D. Tabor, *Rev. Phys. Technol. 1*: 145–179 (1970).

10

Particle Dimensions

Göran Alderborn

Uppsala University, Uppsala, Sweden

I. PHYSICAL STRUCTURE OF PHARMACEUTICAL COMPACTS

A powder can physically be described as a special type of disperse system consisting of discrete, solid particles which are surrounded by or dispersed in air. However, the particles are normally in contact with each other. The interparticulate attractions at the points of contact are relatively weak and the powder is thus characterized by exhibiting a low mechanical strength. The transformation of a powder into a compact is the result of a reduction in the porosity of the powder system, and thus particle surfaces will be brought into close proximity to each other. As a consequence, the number and strength of interparticulate attractions will increase with a subsequent increased coherency of the powder system. The result of the compaction procedure is thus a solid specimen of a certain porosity, normally in the range of 5% to 35% for pharmaceutical compacts.

Although the mechanical strength of a compact is considerably higher than that of a powder, an examination of the literature indicates that it is also adequate to physically describe the compact in terms of or similar to a disperse system, consisting of solid particles surrounded by a gas phase. Several supports for this model can be found in the literature.

1. A comparison between external surface areas of coherent tablets and powders obtained by deaggregation of tablets have indicated a similar

surface area for both these powder systems [1,2]. Hence, although the creation of solid bridges between particles during compression is possible, i.e., the formation of a continuous solid phase between particles, the cross-sectional area of these bridges seems to be small compared to the external surface area of the particles in the compact.

2. The same conclusion has been drawn based on surface area and mechanical strength analysis of compacts after the addition of a lubricant [3]. Furthermore, the removal of magnesium stearate from the compact by dissolution [4] can dramatically increase the compact strength. This indicates that a substantial part of the interparticulate attractions in a compact are formed between solid surfaces which can physically be described as being separated; i.e., separation distances between solid surfaces in the compact are in many cases considerably larger than the separation distances between the molecules and ions which form the solid material.

3. Compaction and the subsequent strength analysis of compacts in liquids [5,6] have also supported that the dominating attraction force between the solid surfaces in the compact is intermolecular attractions forces acting over distances; i.e., the formation of solid bridges between particles is limited.

4. Qualitative inspection of tablets, both upper and fracture surfaces, by electron microscopy has supported [7–10], at least for plastically deforming materials, that the compact consists of discrete particles packed very closely to each other.

5. Finally, direct relationships between the mechanical strength and the pore size characteristics of compacts of nonporous particles [11–14] as well as for compacts of porous, granulated particles [15,16] have been presented. This observation can be explained by the model that the compact consists of small particles cohered to each other and that the size of and separation distance (relative position) between those "compact particles" governs the compact strength.

Thus, a pharmaceutical compact can thereby normally be described as an aggregate of smaller particles which are strongly co- or adhered to each other. The gas phase in the compact can be described as the continuous phase and consists of a three-dimensional network of connected pores. The compact can therefore be physically described by both the characteristics of the interparticulate pore system (e.g., porosity, pore size distribution, and pore surface area) and by the properties of the particles constituting the compact (e.g., surface area, particle size distribution, and the packing characteristics or relative positions of the particles). The distinction between what is normally referred to as a powder or as a compact will in practice be based on the mechanical properties of the assembly of particles or the specimen while the physical structure in a broader sense is similar.

II. THE CONCEPT OF COMPACTIBILITY OF POWDERS

The compactibility of a powder mass has been defined as [17] the ability of the powder to be transformed into a compact of a certain mechanical strength. The compactibility of the powder is thus an essential and fundamental property of a tablet mass and a determining factor for successful tablet production. According to this definition, the term compactibility is related to the ability of the powder to cohere during the actual compression phase (i.e., a process-related parameter). Thus, compactibility is normally assessed by the relationship between compact strength and a process variable, mostly the maximum force or pressure applied on the powder during the compaction. However, other process-related factors, such as the time periods for the different parts of the compaction cycle, can affect the obtained mechanical strength.

The concept of compactibility of a powder is complicated by the fact that the time period which elapses between compaction and strength analysis as well as the storage conditions (primarily the relative humidity of the environment) during this period can dramatically affect the strength of the compact [13,14,18–20]. Thus, the compactibility of a powder can refer to two different powder properties—the ability to cohere during the actual compression phase or the ability to cohere during compaction and subsequent storage of the compact. Therefore, it seems valuable to distinguish those two definitions when the properties of powders are evaluated with respect to their compactibility. Unfortunately, there is only a limited literature on this subject. Thus, in this chapter, the importance of post-compaction strength changes for the relationship between dimensions of particles before compaction and the mechanical strength of the tablets are not considered.

Another complication is that mechanical strength is not a constant property of a compact. A series of reasons for this can be identified, such as the stress direction used during the analysis of compact strength (e.g., tensile strength in radial and axial directions [21,22]) and the type of stress applied to the specimen (e.g., tensile or compressive strength [23]). Whether such aspects of mechanical strength analysis of a specimen are important for the relationship between a particle property and the compact strength is not well known. Thus, it seems practical to treat the concept of the mechanical strength of the compact as a well-defined property of the specimen as long as the experimental procedure used for the evaluation of the compact strength is similar between different studies. In most studies, the diametral compression test has normally been used and presented as either the crushing/breaking strength of the specimen or recalculated to the tensile strength [24]. This seems the most suitable alternative for two reasons: (1) A fracturing of the tablet in a plane parallel to the axis of formation of the tablet will result in a measure which represents the mean strength of the compact [25]. (2) A tensile failure of the compact seems to be the most attractive type of test situation when the strength of a compact is described in terms of the particle-particle interactions in a compact.

III. RELATIONSHIP BETWEEN DIMENSIONS AND VOLUME REDUCTION PROPERTIES OF PARTICLES FOR COMPACT STRENGTH

It has been argued that a suitable physical model of a pharmaceutical compact is an aggregate of smaller particles, which can be described as discrete units strongly cohered to each other. The consequence of this model is that the tensile strength of a given plane within a compact can theoretically be described in terms of the sum of all interparticulate attractions in that plane. When the compact is stressed across that plane, a failure will thus occur when the applied stress exceeds the total attraction strength in the plane, i.e.,

$$\sigma_c = F_b n_b \tag{1}$$

where

σ_c = tensile strength of compact (N/m^2)
F_b = mean bonding force of individual interparticulate bond (N)
n_b = number of interparticulate bonding zones per cross-sectional area (m^{-2})

In practice, the interparticulate bonding forces will vary in strength across the plane and can thus be described as a distribution in interparticulate bonds, i.e., the number of bonds as a function of bonding force.

The validity of Eq. (1) requires probably that the fracture plane be formed instantaneously; i.e., all bonds over the whole cross-sectional area of the fracture plane are separated simultaneously. However, in accordance with fracture mechanics, the failure in practice is often the result of a kinematic process involving the initiation and propagation of the fracture [26]. The existence of a kinematic process during fracturing diametrically has also been demonstrated for pharmaceutical compacts [27]. The practical consequence is that there might be a discrepancy between the theoretical strength of a specimen and the stress needed in practice to fracture a compact, and this discrepancy is probably related to the mechanical properties of the particles constituting the compact. The importance of the kinematic fracturing process for the strength of pharmaceutical compacts in relationship to the particle-particle interactions and the mechanical properties of the particles from which the compact is formed is, however, not yet well understood. In the following discussion, the strength of a compact will for simplicity be described as the sum of the interparticulate attractions in the fracture plane.

The variables F_b and n_b in Eq. (1) have not yet been quantified with respect to pharmaceutical compacts. Fundamental studies on the mechanical strength of compacts have therefore almost entirely been focused on the effect of secondary factors on the compact strength (i.e., material and process factors in relationship to compact strength). The approach here is mainly qualitative, and few attempts to quantitatively relate the properties of the material to the strength of the compact have been presented. However, for loosely packed beds of particles, quantitative expressions which describe the relationship between the strength of the powder bed and the packing characteristics of the particles in the bed have been presented. The theoretical basis of this field has been presented by Rumpf, who developed the following expression [28] for the strength of an aggregate of monodispersed spheres:

$$\sigma = \frac{9(1 - e)kH}{8\pi d^2} \qquad (2)$$

where

σ = tensile strength of aggregate (N/m^2)
e = porosity of aggregate (—)
k = coordination number, i.e., number of contact points for one sphere (—)
H = bonding force at a point of contact (N)
d = diameter of sphere (m)

The factors governing the strength of such powder beds are consequently the size of the particles, their relative positions, and the force of cohesion between particles. Deviations from this simple model make a quantitative description more difficult, but it has been stated that both the distribution in particle size as well as the shape characteristics of the particles affect the mechanical strength of the bed [29].

The physical analogy between beds of powders and compacts discussed above leads to the conclusion that for compacts the dimensions of the particles constituting the compact and their relative positions might also be of direct importance for the mechanical strength of the compact. These characteristics of the compact will probably affect both the bonding force and the number of the interparticulate bonds.

The characteristics of the "compact particles" as well as their relative positions are difficult to assess. However, it is reasonable to assume that the particulate characteristics of the particles before compaction are important for the characteristics of the "compact particles" and thus affects the compactibility of the powder. Thus, an understanding of the effect of a certain particle characteristic for the mechanical strength of compacts can form the basis for a fundamental understanding of the compaction process. Furthermore, this knowledge is also of practical interest in the formulation of direct compactible formulations, e.g., with respect to the importance of batch variations of the raw materials.

The assumption that the particulate characteristics of the particles before compaction are related to the characteristics of the "compact particles" seems to form the basis for attempts to quantify relationships between tablet strength and original particle size. For example, the following expression was used by Shotton and Ganderton [30] to relate compact strength to original particle size:

$$F = Kd^{-a} \tag{3}$$

where

F = force needed to break compact (N)
d = diameter of particle (m)
K, a = constants

The compaction properties of compacts of sodium chloride and hexamin could be described by Eq. (3), but the value of the exponent differed between the materials (Fig. 1). They concluded that this difference was related to the properties of the materials. One such important material property is the volume reduction behavior of the material during the compression.

The compaction of particles into a coherent compact normally requires such high compaction pressures and such a dramatic reduction in volume of the powder bed that the dimensions of the particles can be dramatically changed. The incidence of such changes is often described as the volume reduction mechanisms which a material exhibits while compressed and is probably related to the mechanical properties of the particles (i.e., strength and deformability). A discussion of the effect of particle dimensions on compact strength must thus consider at least the volume reduction properties of the material and be described as a relationship between three variables—original particle dimensions, main volume reduction behavior, and compact strength.

For powders consisting of more-or-less nonporous particles, the main volume reduction mechanisms of the particles during the compaction are particle rearrangement (repositioning), deformation, and fragmentation. If

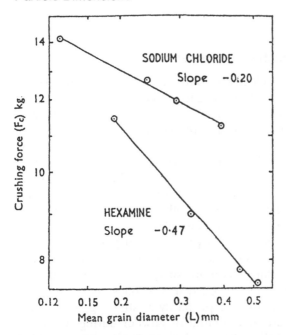

Fig. 1 Log-log relationships between the diametral compression strength of compacts of sodium chloride and hexamin as a function of original particle size. (From Ref. 30 with permission of the copyright holder.)

the first mechanism dominates the volume reduction process (e.g., during the compaction of very fine particulate powders), it is reasonable that the characteristics of the particles before compaction will affect the compact strength in accordance with the general statements which are valid for loosely packed beds of powders. However, a permanent deformation of the whole or parts of a particle as well as particle fragmentation during the compression phase can result in other types of relationships.

The aim of this chapter is to give some examples of experimentally observed relationships between particle dimensions before compaction and the mechanical strength of the resultant compact. It is obvious that the compaction of a powder into a compact and a controlled analysis of compact strength are complex procedures and generalized rules concerning factors of importance for compact strength are therefore difficult to establish. Nevertheless, the aim of this chapter is also to make, based on the model of the physical character of pharmaceutical compacts discussed above, an attempt to present a fundamental explanation for the effect of particle dimensions on compact strength.

IV. PARTICLE SIZE

A. Effect on Volume Reduction Characteristics

It is generally assumed that a change in particle size not only affects the external surface area of particles but also the mechanical properties of the particles. Generally, a decrease in particle size increases its mechanical strength [31]. The underlying reason for the change in particle strength with particle size is generally attributed to a reduced probability of the existence of defects in the crystal structure. These defects can act as sites at which a crack is initiated during a fracture. It has also been reported [32] that when the particle sizes are reduced markedly, a transition from a brittle to a plastic behavior takes place—in other words, a change in the dominating mechanical characteristic of a particle.

The size effects on the mechanical properties of particles indicates that the volume reduction characteristics (i.e., the propensity of the particles to fragment or deform during compression) are also affected by a change in particle size. The effect of particle size on volume reduction behavior of a powder has been assessed primarily by studying the relationship between applied pressure and volume of powder during the actual compression phase by Heckel profiles and force-displacement profiles. Duberg and Nyström [33] found that the linearity of the Heckel profiles (Chapter 3) increased with a decreased particle size during compaction of a series of materials. This can be interpreted as a reduced degree of fragmentation of the particles during compaction.

There are, however, few reports with the object of directly assessing the fragmentation propensity of a material as a function of particle size. By the aid of air permeability measurements of tablets compacted at a series of compaction pressures, Alderborn and Nyström [34] found that the tablet surface area as a function of compaction pressure increased more markedly with a reduction in original particle size (Fig. 2). However, this corresponded to a reduced relative increase in surface area with pressure; i.e., the changes in tablet surface area in relationship to the starting surface area decreased with a decreased particle size. A similar observation was reported by Leuenberger et al. [35] for the compaction of lactose (Fig. 3). It can thus be concluded that the readiness of individual particles to fracture into smaller units is reduced with a reduced particle size, although the absolute change in surface area is larger due to the small dimensions of the involved particles.

As seen in Fig. 2, almost linear relationships between tablet surface area and compaction pressure were obtained, and the slopes of these profiles can thus be used to express the rate of changes in surface area with pressure, dS/dP (S = tablet surface area and P = compaction pressure).

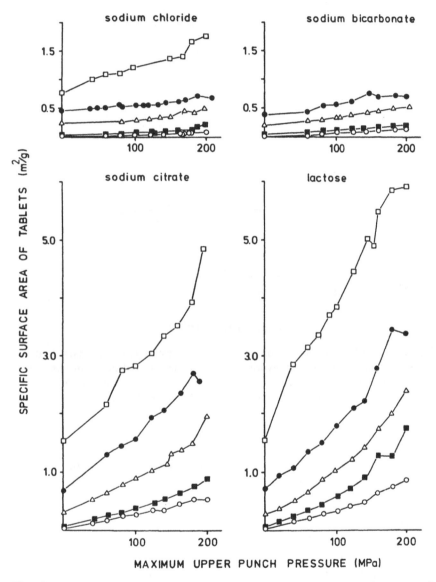

Fig. 2 The effect of compaction pressure on the permeametry surface area of tablets, compacted of different particle size fractions: ○ = 5–10 μm, ■ = 10–20 μm, △ = 20–40 μm, ● = 63–90 μm, □ = 125–180 μm. (From Ref. 34 with permission of the copyright holder.)

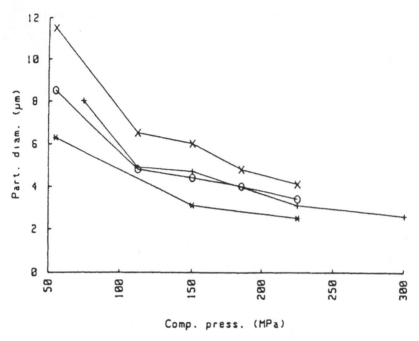

Fig. 3 The effect of compaction pressure on the estimated mean particle diameter in compacts of α-monohydrate lactose, compacted of different particle size fractions: ✻ = 32–45 μm, + = 100–125 μm, ○ = 125–160 μm, ✕ = 315–400 μm. (From Ref. 35 with permission of the copyright holder.)

This rate of change increased with increased original surface area in a nonlinear way, and the relationship between rate of changes and original surface area can be expressed with the differential equation

$$\frac{dS}{dP} = kS^n \tag{4}$$

where

    k = material-related coefficient describing the propensity of the material to fragment

    n = constant

Thus, the compression process is described in terms of a size reduction process, and the pressure required to obtain a given reduction in particle size is related to the starting particle size and the propensity of the material to fragment. In the literature on the comminution of particulate solids, a number of attempts to quantify the relationship between energy consumption and the reduction in particle size during comminution has been presented. The most well-known expressions are the equations given by Kick, Rittinger, and Bond [36]. However, it has been argued [36] that all these expressions are interpretations of a single general equation:

$$\frac{dE}{dD} = cD^n \qquad (5)$$

where

E = energy consumed
D = particle size
c, n = constants

Equations (4) and (5) express the same approach to relate size reduction to the actual particle size at hand and to a process characteristic. In the case of compaction, the choice of pressure or stress applied to the powder seems a reasonable approach to characterize the process since this is the most common parameter in studies on the compression/compaction properties of materials. Furthermore, applied stress is commonly used in relationship to strain in the evaluation of mechanical properties of specimens.

Alderborn and Nyström found that a value for n in Eq. (4) of $\frac{2}{3}$ gave a linear relation between dS/dP and S, and the following expression was developed:

$$S_p^{\frac{1}{3}} - S_0^{\frac{1}{3}} = \frac{kP}{3} \qquad (6)$$

where

S_p = tablet surface area at pressure P (e.g., cm^{-1})
S_0 = original surface area, i.e., powder surface area before compaction (e.g., cm^{-1})
P = compaction pressure (e.g., MPa)

The ratio $\frac{k}{3}$ was denoted the fragmentation propensity coefficient (C_{FP}) and was regarded as a material constant describing the propensity of the material to fragment while compressed independent of original particle size (Fig. 4).

In analogy with the discussion above, it seems reasonable that also the readiness of a particle to deform under stress will be reduced with a reduction in particle size. However, the question regarding the degree of deformation, both plastic and viscoelastic, in relationship to particle size and tablet compaction is not fully understood. However, it has been ob-

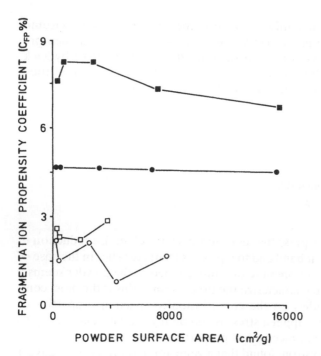

Fig. 4 Relationships between the fragmentation propensity coefficient, calculated according to Eq. (6), and the original powder surface area for sodium chloride (○), sodium bicarbonate (□), sodium citrate (●), and α-monohydrate lactose (■). (From Ref. 34 with permission of the copyright holder.)

served [37] that a reduction in particle size reduces the effect of the rate of compression on the relationship between tablet porosity and applied compression pressure. A reasonable interpretation of the data is a reduced deformability (viscoelastic) of the particles with a reduced particle size.

B. Effect on Tablet Strength

1. Relationships Between Original Particle Size and Compact Strength

It has been claimed that one of the most important variables for the mechanical strength of a compact is the size of the particles before compaction [38], and, as a rule of thumb, it is normally assumed that a decreased original particle size increases the tablet strength. Based on the physical characteristics of a pharmaceutical compact described above, the rule of thumb seems a reasonable statement provided that the compression process only affects the appearance of the particles to a limited extent. The relationship between original particle size and tablet strength has also been extensively studied in the literature. However, an examination of the presented data can give a somewhat confusing picture with respect to this relationship. For some materials (e.g., lactose and sodium chloride [12,30,39–41]) an increased tablet strength with a decreased particle size has been found. However, the reverse, or more complex, relationships have also been reported for compacts of sodium chloride [42,43]. Also for other materials, complex relationships with, for example, a minimum in the compact strength–particle size profile have been presented [42]. Finally, an almost unchanged tablet strength with variations in particle size has been observed for lactose [44], aggregated calcium phosphate [42], and sodium bicarbonate [45].

In the following discussion, some examples of data will be presented in more detail which forms the basis for the mechanistic discussion of the effect of original particle size on compact strength. The procedure used for preparing suitable model materials has normally been the classification of unmilled particles, e.g., by dry sieving.

Alderborn and Nyström [42] studied the effect of particle size on the mechanical strength of tablets of a series of materials consisting of relatively coarse particles (90–1000 μm). The substances were chosen to represent materials with varying volume reduction characteristics during compression, from low- to high-fragmenting materials. The classification of materials with respect to their degree of fragmentation was also later supported by tablet surface area measurements [46].

The most dramatic or complex relationships between original particle size and tablet strength were associated with a change in shape characteristics of the particles with a change in particle size. This was due to the formation of secondary particles with an increased particle size or to a change in the number of primary particles constituting the aggregate. It is obvious that such changes in particle structure can cause a shape-related change in the particle-particle interactions (see below) but also change dramatically the volume reduction behavior of a powder. This change in the appearance of the particles formed the basis for the explanation for the dramatic effect of particle size on the mechanical strength of tablets of dendritic sodium chloride and of Starch 1500.

For the other materials used in the study, the volume reduction properties, as evaluated by the ratio between the tensile strength measured parallel and normal to the compression axis during the tablet formation, did not seem to change dramatically with particle size. Furthermore, the particle shape was similar for all particle sizes used for the respective mate-

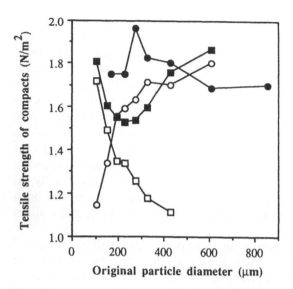

Fig. 5 The effect of original particle size on the tensile strength of compacts of
α-monohydrate lactose (□, compaction pressure of 215 MPa), sodium chloride
(○, compaction pressure of 160 MPa), saccharose (●, compaction pressure of
265 MPa), and Emcompress (■, compaction pressure of 160 MPa). (Drawn
from data given by Alderborn and Nyström [42].)

rial. For these materials, three types of relationships between particle size
and tablet strength were found.

The most common relationship found, e.g., for lactose and sodium
citrate, was an increase in tablet strength with a decrease in particle size, and
the effect was most pronounced at the lower-particle-size range (Fig. 5). A
tablet strength more-or-less independent of particle size was observed for
compacts of aggregated calcium phosphate dihydrate and saccharose (Fig.
5). Finally, for the plastically deforming or low-fragmenting material cubic
sodium chloride, the general tendency was an increased tablet strength with
increased particle size with the most pronounced effect at the lower-particle-
size range (Fig. 5). In a later study [45], a tablet strength more-or-less
independent of particle size was observed for another low-fragmenting material, sodium bicarbonate.

Researchers in Groningen, Holland, have in a series of papers studied
the compaction properties, including the relationship between particle size
and tablet strength, of different types of lactose. Both for α-monohydrate
lactose and α-anhydrous lactose [41], a decreased particle size increased the
compact strength (Fig. 6). The tablet strength for compacts of the anhydrous

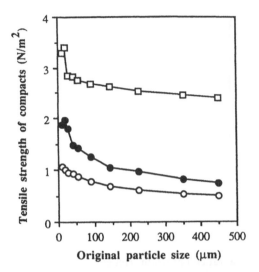

Fig. 6 The effect of original particle size on the tensile strength of compacts of alpha monohydrate lactose (○ = compaction pressure of 151 MPa and, ● = compaction pressure of 226 MPa) and alpha anhydrous lactose (□ = compaction pressure of 151 MPa). (Drawn from data given by Vromans et al., ref. [41].)

lactose was generally higher, which was later attributed [11] to a higher degree of fragmentation during compression. In a later paper [12], the question of the effect of particle size on compact strength was addressed once again during the compaction of α-monohydrate lactose. In this chapter, the surface area of the tablets was measured by mercury intrusion. A more of less linear relationship was found between compact strength and tablet surface area (Fig. 7). Thus, it seems that the compact strength was governed by the surface area of the tablet and the increased tablet strength with decreased original particle size was due to the fact that the finer original particles gave a higher tablet surface area. Thus, a linear relationship between surface area and mechanical strength of compacts has been suggested for compacts of different types of lactoses [11–14]:

$$F_p = kS_p + i \tag{7}$$

where

F_p = fracture or tensile strength of compact at pressure p (N/m^2)
S_p = tablet surface area at pressure p (m^{-1})
k = constant
i = intercept (normally negative for reported values)

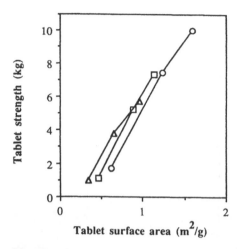

Fig. 7 Relationship between the diametral compression strength and tablet specific surface area, assessed by mercury intrusion, for compacts of alpha monohydrate lactose, compacted of three different particle size fractions: ○ = 32–45 μm, □ = 125–160 μm, △ = 315–400 μm. (Drawn from data given by de Boer et al., ref. [12].)

Equation (7) can be thought of as a relationship between the surface area of the particles constituting the compact and the total area of interparticulate bonds in the compact; i.e., an increased exposed particle surface area increases the bonding area. However, the observed relationship between tablet strength and tablet surface area can alternatively be explained in terms of a relationship between the compact strength and the number of particles exposed in the fracture plane. An increased surface area corresponds to a reduced "compact particle" size and, thus, an increased number of particles (n) exposed within a cross section of the compact, i.e., $n = f(S_p)$. This will, in combination with the porosity of the compact, govern the number of interparticulate bonding zones which must be ruptured when the compact is fractured during strength analysis.

Based on the suggested relationship between tablet strength and tablet surface area (Fig. 7 and Eq. (7)), the effect of type of lactose (Fig. 6) can be explained as follows: an increased degree of fragmentation of the material will stress the relationship between compact strength and original particle size with respect to the absolute difference in strength of compacts of different original particle size.

Results obtained at Uppsala University [47] on the relationship between tablet strength and tablet surface area, as measured by air permeametry, for four materials with varying compression behavior while compacted also generally gave nearly linear relationships. However, the relationship for the respective original particle size did not coincide, as found by Vromans et al. [11–14]. On the contrary, separate relationships dependent on the original particle size were obtained, i.e., an increased original powder surface area gave lines more and more to the right on the abscissa. Such an effect of original powder surface area seems reasonable if one consider the surface area and the strength of the powder as a part of the relationship. The powders represent samples with markedly varying surface areas but, in relative terms, with similar tensile strengths close to zero. The generality of the suggested relationship between tablet strength and tablet surface area, as suggested for the lactose tablets [11–14], can therefore be questioned.

The results discussed show that the relationship between original particle size and tablet strength is complex. However, it seems that the most common effect of a reduced particle size is an increased tablet strength. The results on the compaction of different types of lactose indicate that an increased fragmentation propensity of the material will give a more marked effect of particle size in absolute terms. However, there are reports for materials [42] which are not consistent with these conclusions, e.g., compacts of saccharose, which is a relatively high fragmenting material. In that paper, Alderborn and Nyström suggested that this type of relationship was caused by the high degree of fragmentation of the particles during the compression, which masked differences in particle size before compression. However, alternative explanations for this observation can be found.

Firstly, it is possible that the physical structure, in terms of the size of the compact particles and their relative positions, of the compact is similar, although the original starting particle size varied. A possible explanation for this effect is that the fragmentation propensity was reduced with a decreased particle size, i.e., the concept of fragmentation propensity as a material property cannot be applied to all materials. However, this seems to be a reasonable possibility for a material which consists of aggregates of small primary particles.

Secondly, a more generalized explanation is that the unexpected small influence of original particle size on the compact strength for the fragmenting materials is due to the fact that the original particle size can affect the interparticulate bond structure in a compact in terms of both the number and the bonding force of the interparticulate bonds: i.e., the force needed to rupture the interparticulate bond cannot be treated as a material-specific characteristic but can vary between compacts of different original particle size. This suggestion also forms the basis for the explanation of the observed relationships between tablet strength and original particle size for sodium bicarbonate [45] and sodium chloride [42].

If an increase in original particle size affects the character of the interparticulate bonding zone in such a way that the force needed to rupture the individual interparticulate bond increases, a distribution of bonds in the compact characterized by a reduced number of bonding points but of a higher bonding force will be obtained. The net effect can be a tablet strength independent of particle size (saccharose and sodium bicarbonate) or an increased tablet strength with increased particle size (sodium chloride). In Fig. 8, the effect of particle size on the distribution of bonds in the compact for these two types of idealized situations is schematically illustrated.

Karehill et al. [4] reported ratios between tablet strength and tablet surface area, i.e., a surface-specific tensile strength of tablets, as a measure of the character of the individual interparticulate bonds in compacts of three particle size fractions of sodium chloride. They found (Table 1) an increased ratio with increased particle size, which can be interpreted as support for the suggestion that the force needed to rupture the individual interparticulate bonds in compacts of sodium chloride increases with an increased original particle size. For the other test materials used in the study, the ratios were similar to the observed ratio for the most fine particulate quality of sodium chloride.

An increased particle size will generally reduce the number of contact points in the compact which transmit the applied stress through the powder bed during the compression phase. The result will be an increased stress at the particle-particle contact points which can facilitate local particle defor-

Table 1 Ratio between Tensile Strength and Permeametry Surface Area for Compacts (as a measure of the bonding force of individual bonds between compact particles) of Three Sieve Fractions of Sodium Chloride Compacts at 150 MPa

Sieve fraction (μm)	Ratio (kPa cm^{-2})
<63	0.9
250–355	2.4
425–500	4.2

Source: From Ref. 4

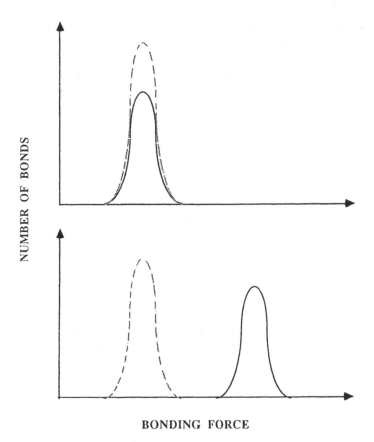

Fig. 8 Hypothetical types of distribution in bonding force illustrating the discussion on the changes in bonding characteristics in tablets compacted of a coarse and a fine particulate fraction of the material. Dashed line = fine particle size, solid line = coarse particle size. Upper panel illustrates a material showing an increased compact strength with decreased original particle size while the lower panel illustrates a material showing a decreased compact strength with decreased original particle size.

mation. The degree of local deformation at the interparticulate junctions can affect the area of bonding for intermolecular forces or the formation of solid bridges between particles, i.e., a bridge constituting of a continuous solid phase of material which bonds particles to each other. The formation of such bridges can be the result of a drastic increase in mobility of the molecules or ions at particle surfaces due to, e.g., increased local temperatures in the compact with a subsequent local surface fusion. It was also suggested above that a change in particle size can affect the deformability of particles. Thus, such a variation in deformability can probably affect the bonding force of the interparticulate bonding zone due to a decreased contact area and thus affect the relationship between original particle size and tablet strength.

In most of the studies discussed above, comparatively coarse particulate materials were used. The preparation of finer particulate materials normally requires a milling procedure before size classification. The bulk of the particles can be affected by inducing dislocations or cracks in the crystal structure, which can affect both the deformability and the fragmentation propensity of the particles. However, more important might be the change in particle surface properties as a result of milling: Firstly, on a molecular level, the surface of the particles will probably show a disordered, amorphous structure [48], and, secondly, on a microscopic level, the surface of the particles will probably show a rough texture with surface protuberances. These two phenomena might lead to a change in the mechanical properties of the surface and markedly increase the surface deformability; i.e., the surface deformability might be higher than the bulk deformability of the particles. Consequently, the bonding force will increase due to an increased area of bonding between particles which is larger than could be predicted from the bulk deformability of the particles. In addition, it has been suggested that a disordering of surfaces, either before or during the compaction, will affect the changes in mechanical strength during storage of the compacts [13,14].

Eriksson and Alderborn [49] studied the effect of particle size on the mechanical strength of sodium chloride tablets compacted of particles of a wide size range. The finest particles were prepared by milling and classification. To avoid the effect of postcompaction strength changes on the interpretation of the data, the compact strength was measured immediately after compaction. An increased tablet strength (Fig. 9) with a decreased original particle size was observed with a more pronounced effect with increased compaction pressure. These results thus contradict the results discussed above for sodium chloride. A possible interpretation is that for this quality of sodium chloride, an increase in the number of interparticulate bonds due to a reduced original particle size governed the compact strength and was more important than a change in the bonding force with a change in original particle size. However, the finest particles were prepared by milling. It could thus be assumed that those particles possess a comparatively rough surface with a disordered solid-state structure. This might lead to a changed local deformability at the particle surface with an improved binding capacity and thus an improved bonding force of the

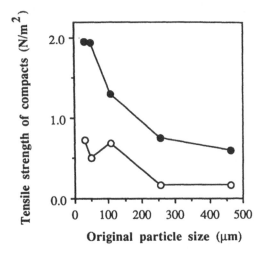

Fig. 9 The effect of original particle size on the tensile strength of compacts of sodium chloride, measured immediately after compaction. ○ = compaction pressure of 75 MPa and ● = compaction pressure of 150 MPa. (Drawn from data given by Eriksson and Alderborn, ref. [49].)

individual bonds. This might thus explain the good compactibility of the fine sodium chloride particles and constitute a general explanation for the commonly accepted statement that smaller particles form stronger tablets.

In the discussion so far, no consideration has been given to elastic deformation of particles. It seems reasonable, however, that if the degree of elastic expansion of particles is limited or does not vary markedly between different particle size fractions, a direct relationship between the interparticulate bonds formed during the actual compression phase and the bonding characteristic in the ejected tablet will be obtained. Thus, the mechanistic discussion on the relationship between volume reduction behavior, original particle size, and compact strength will also be applicable when particles deform elastically to some extent. The results obtained [30] for the mechanical strength of compacts of a capping proned material, Hexamin, support the assumption that in a restricted compaction pressure range the effect of particle size could be predicted also for a material which could be expected to be markedly elastic (Fig. 10). However, after capping, it is not meaningful to relate compact strength to original particle size. It is worth noting, however, that capping occurs at a lower compaction pressure for the finest particulate quality of hexamin.

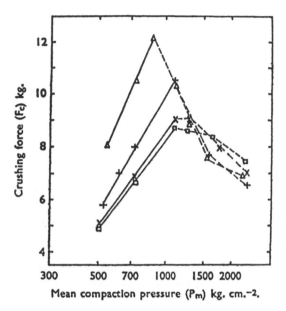

Fig. 10 The effect of compaction pressure on the diametral compression strength of hexamin tablets, compacted of four different particle size fractions: □ = 20–30 mesh, X = 30–40 mesh, + = 40–60 mesh, △ = 60–80 mesh. (From ref. [30] with permission of the copyright-holder.)

2. Effect of Compaction Pressure

In some of the papers on the relationship between compact strength and original particle size, the importance of the compaction pressure has also been addressed. It seems generally that an increased compaction pressure stresses the relationship between original particle size and tablet strength in absolute terms, e.g., for lactose (Fig. 11) [41] and hexamin (Fig. 10) [30]. However, for saccharose and sodium chloride, Alderborn and Nyström [42] observed only a limited effect of compaction pressure on the relationship between compact strength and original particle size (Figs. 12 and 13). A possible interpretation of these results is that the stress applied to the powder during the compression affects both the number of bonds per cross-sectional area of the compact and the bonding force of the individual interparticulate bonds. An increased compaction pressure will probably increase the degree of fragmentation which the particles undergo during compression. Thus, the number of interparticulate bonds per cross-sectional area of compact will increase. However, it seems reasonable that an increased compaction pressure will also increase the degree of particle

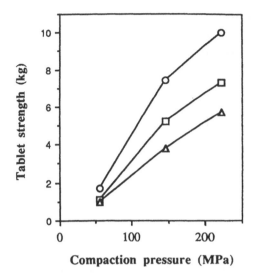

Fig. 11 The effect of compaction pressure on the diametral compression strength of alpha monohydrate lactose tablets, compacted of three different particle size fractions: $\bigcirc$ = 32–45 μm, $\square$ = 125–160 μm, $\triangle$ = 315–400 μm. (Drawn from data given by de Boer et al., ref. [12].)

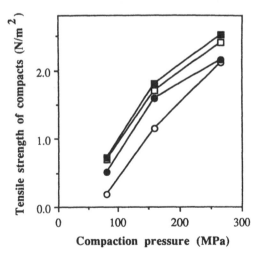

Fig. 12 The effect of compaction pressure on the tensile strength of sodium chloride tablets, compacted of four different particle size fractions: $\bigcirc$ = 90–125 μm, $\bullet$ = 212–250 μm, $\square$ = 355–500 μm, $\blacksquare$ = 500–710 μm. (Drawn from data given by Alderborn and Nyström, ref. [42].)

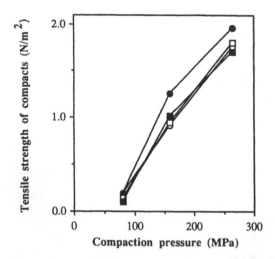

Fig. 13 The effect of compaction pressure on the tensile strength of saccharose tablets, compacted of four different particle size fractions: ○ = 125–212
μm, ● = 250–300 μm, □ = 355–500 μm, ■ = 710–1000 μm. (Drawn from
data given by Alderborn and Nyström, ref. [42].)

deformation with an increased bonding force as the result. Moreover, an
increased difference in bonding force in compacts prepared of particles of
different original size might be obtained. Thus, it is possible that there will
be a relationship between original particle size and compaction pressure
with respect to both the number of bonds and the bonding force.

V. PARTICLE SHAPE

A. Effect of Particle Geometrical Shape on Volume Reduction Properties and Tablet Strength

The shape of a particle is a complex characteristic, and its importance in
relationship to powder properties is therefore difficult to assess. In practice, it is feasible to describe particles as of the same size if they exhibit the
same particle diameter defined in some way, e.g., projected area diameter.
Two particles defined as the same size can, thus, differ with respect to the
relative magnitude of their main dimensions (i.e., length, breadth, and
thickness) as well as with respect to their surface texture. In this chapter, an
attempt is made to distinguish between two main shape characteristics of a
particle. The term "geometrical shape" refers to the main dimensions of a
particle, and the term "surface geometry" to the texture of the surface. It is

then assumed that the dimension of an asperity is small compared to the dimension of the whole particle.

There is a limited number of studies in the pharmaceutical literature which have specifically discussed the relationship between particle shape and compact strength for a specific material. In these studies, two approaches for the preparation of powders which consist of particles with different shapes have been used. In the first case, different qualities of the same material have been produced by crystallization under different conditions or by using different crystallization batches. The other approach is based on the preparation of two shape fractions from the same batch of a material. This has been achieved by either sorting particles by shape or by producing different shapes by a combination of milling and sieving.

Lazarus and Lachman [50] compared the compactibility of different batches of potassium chloride. The compactibility of the different batches varied and it was concluded that the batches consisting of larger and more irregular particles gave tablets of a higher mechanical strength. They also found that the addition of a lubricant to the powders decreased the diametral compression strength of the compacts, indicating that potassium chloride particles fragment to a limited degree during compression.

Shotton and Obiorah [51] found that dendritic sodium chloride gave compacts of a higher mechanical strength compared to the cubic form, which was also shown later [42]. Shotton and Obiorah attributed the observed difference in compactibility to the difference in particle shape; i.e., the more irregular dendritic sodium chloride gave stronger tablets. The compression characteristics of cubic sodium chloride are well studied in the literature, and it has been shown that this material fragments to a limited degree during compression and thus reduces in volume mainly by particle rearrangement and particle plastic deformation. It has, however, been suggested [42] that the dendritic quality of sodium chloride fragments during compression due to its more complex physical character; i.e., dendritic sodium chloride particles are aggregates of small sodium chloride particles, and the aggregates have a complex, porous structure. Hence, the increased compactibility can alternatively be due to a changed compression behavior characterized by a fragmentation of the original particles followed by plastic deformation of the formed sodium chloride particles.

By using a combination of milling-sieving procedures, Alderborn and Nyström [52] and Alderborn et al. [45] prepared two powder qualities for a series of model materials. The two qualities (A and B) for each material had particles of the same sieve size but with different shape characteristics. The model materials were chosen to represent materials with different degrees of particle fragmentation during compression.

The particulate characterization indicated generally an increased irregularity of shape quality B, as evaluated by the Heywood shape coefficient and surface area analysis for powders of sodium chloride, sucrose, and sodium citrate [52]. The same conclusion was drawn from microscopy studies of sodium bicarbonate powders [45].

Compaction of the powders into tablets showed that for the materials which fragmented to a limited degree during compression, the particle shape affected the compact strength; i.e., a more irregular particle improved the compactibility (Figs. 14 and 15). However, for materials which fragmented markedly during compression, the shape of the particles before compaction did not affect compact strength (Fig. 14). Hence, the effect of particle shape on the compact bonding characteristics and, thus, the compact strength is dependent on the volume reduction properties of the material. For a material which fragments to a large extent, the physical structure of the formed compact is to a limited extent affected by variations in particle shape before compaction, provided that the articles can be described as the same particle size.

For the materials which fragmented to a limited extent during compression (sodium chloride and sodium bicarbonate), the increased strength of compacts prepared of the more irregular particles indicates a change in the distribution of interparticulate bonds with a change in original particle shape. It is possible that the bonding characteristics within the compact can be affected by the original shape of the particles with respect to both the number of interparticulate bonds and the force needed to rupture the individual bonds.

It is reasonable that with an increased particle irregularity, the number of possible interparticulate attraction zones in a compact, and consequently the compact strength, will increase, although the packing properties of the particles are similar. It is also possible that during the compression process, the particles can undergo an increased degree of deformation due to an increased particle irregularity, especially at edges and corners of the particles. In this case, the degree of deformation at such sites could be further enhanced due to the existence of lattice defects, primarily dislocations, created during the milling of the material [53]. The consequence of the increased local particle deformation will be an increased bonding force of the attraction zones between compact particles. For sodium chloride, an alternative explanation is that an increased stress or degree of deformation at the interparticulate contact points during the compression process facilitated the formation of solid bridges between particles, which thus increased the bonding force and compact strength.

Both for sodium chloride and sodium bicarbonate, an increased compaction pressure increased the absolute difference in strength of compacts

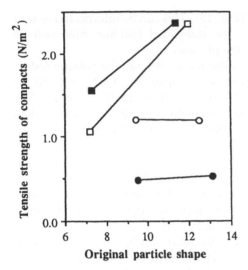

Fig. 14 The effect of original particle shape (expressed as the Heywood surface-volume shape coefficient) on the tensile strength of compacts of sodium chloride I (□), sodium chloride II (■), saccharose (○) and sodium citrate (●). Tablets compacted at 160 MPa. (Drawn from data given by Alderborn and Nyström, ref. [52].)

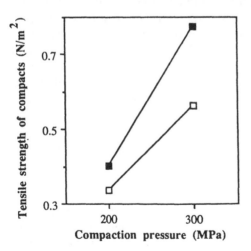

Fig. 15 The effect of compaction pressure on the tensile strength of sodium bicarbonate tablets, compacted of particles of two different shapes: □ = regular particles, ■ = irregular particles. (Drawn from data given by Alderborn and Nyström, ref. [45].)

of different original particle shape (Fig. 15). This can be interpreted as that the difference in the character of the individual bonding zone will be stressed with an increased compaction pressure.

For one of the test materials, sodium bicarbonate, the volume reduction characteristics of the different shape qualities were assessed by porosity-pressure profiles [45], performed as in-die measurements. This evaluation indicated that the volume reduction behavior was more-or-less identical for the two powders. This can be interpreted as support for the explanation that the improved compact strength with increased particle irregularity is related to an effect of particle shape on the number of interparticulate attraction zones in the compact. However, it is possible that a porosity-pressure profile reflects mainly the bulk behavior of the particles; i.e., differences in the degree of local particle deformation can occur without affecting the overall compression profile.

As a result of the milling procedure used during the preparation of the model materials, also the surface characteristics of the particles can change. It has been suggested [48] that the fracturing of a solid, crystalline material can cause a disordering or amorphization of the solid material at the particle surfaces. Such changes in the solid-state structure can cause an improved ability of the material to create interparticulate attractions, e.g., as a result of an increased local deformability or an increased surface energy. Hence, the findings described above might also be explained by a disordering of the material as a result of milling. However, it seems reasonable that during storage of "activated" materials, the disordered regions will crystallize in order to reduce the energy of the particle surfaces. Thus, for the sodium chloride powders, compacts were also made after storage of the powder for one year. The compactibilities of the powders were identical after the storage period. It therefore seems likely the observed effect was related to the changes in particle geometrical shape rather than a disordering of their solid-state structure.

Wong and Pilpel [54] prepared particles with different shape characteristics by the use of a shape sorting table. Also in these studies, materials with different volume reduction characteristics were used. Their observations were consistent with the findings by Alderborn et al; i.e., for materials which fragment markedly during compression, the shape characteristics of the particles did not affect the tablet strength, while the converse applied for materials which showed limited fragmentation (Figs. 16 and 17). In addition to the evaluation of compactibility, the volume reduction characteristics of the test materials were assessed by porosity-pressure profiles and by tablet permeametry surface area–pressure profiles. They observed an increased degree of fragmentation for the fragmenting materials with increased particle irregularity (Table 2). Thus, it seems like more irregular

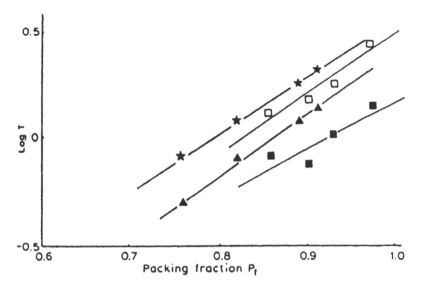

Fig. 16 The effect of the relative compact density on the tensile strength of tablets of sodium chloride (■ = regular shape of original particles, □ = irregular shape of original particles) and Starch 1500 (▲ = regular shape of original particles, ★ = irregular shape of original particles). (From ref. [54] with permission of the copyright-holder.)

particles were slightly more prone to fragment, which might be due to the fact that corners and asperities of the particles correspond to weak parts of the particles which can be fractured or knocked off the rest of the particle relatively easily. However, the observed difference in degree of fragmentation was not marked enough to cause any difference in the mechanical strength of the resultant compacts. The low-fragmenting materials did not show a tendency to an increased degree of fragmentation with a change in particle shape, but the compressibility of the powder increased with increased particle irregularity. They thus concluded that the irregular particles showed a more marked deformation during the compression phase, which especially is due to deformation at particle asperities. This increased deformation can thus produce an increased total bonding strength between particles and increase the tablet strength.

In the papers discussed hitherto, no considerations were taken of postcompaction events in relation to the effect of particle shape on compact strength. As stated, the mechanical strength of compacts can change dramatically during a limited time period after compaction. It is possible that such a phenomenon might be responsible for the observed effects

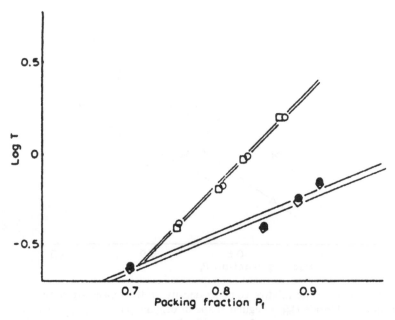

Fig. 17 The effect of the relative compact density on the tensile strength of tablets of lactose ($\Diamond$ = regular shape of original particles, ● = irregular shape of original particles) and Emcompress ($\square$ = regular shape of original particles, $\bigcirc$ = irregular shape of original particles). (From Ref. 54 with permission of the copyright holder.)

Table 2 Fragmentation Propensities, Estimated from Profiles of Tablet Permeametry Surface Area as a Function of Compaction Pressure, and Yield Pressures, Estimated from Profiles of Tablet Porosity (measured in-die and expressed as the Heckel function) as a Function of Compaction Pressure, for Four Materials with Two Shape Characteristics of the Particles Before Compaction

Material	Fragmentation propensity $(m^{-1}\,kN^{-1})$	Yield pressure (MPa)
Sodium chloride, regular	0.303	83.0
Sodium chloride, irregular	0.287	72.3
Starch 1500, regular	0.136	42.0
Starch 1500, irregular	0.128	34.0
Lactose, regular	0.507	154
Lactose, irregular	0.680	150
Emcompress, regular	0.603	264
Emcompress, irregular	0.869	261

Source: From Ref. 54.

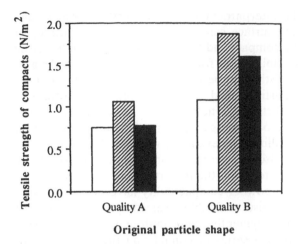

Fig. 18 The effect of original particle shape on the tensile strength of compacts of sodium chloride, compacted at 100 MPa. Open = tablet strength measured immediately after compaction, gray = tablet strength measured after 60-min storage at 0% relative humidity, black = tablet strength measured after 60-min storage at 57% relative humidity. Quality A = regular particles, quality B = irregular particles. (Drawn from data given by Eriksson and Alderborn [49].)

discussed here. Eriksson and Alderborn [49] therefore studied the compactibility of two qualities of sodium chloride powders prepared by the milling-sieving procedure. An increased particle irregularity increased the tablet strength measured directly after compaction (Fig. 18); i.e., the earlier observed differences were not related entirely to postcompaction effects. However, after storage of tablet for one hour at 0% and 57%, the difference in mechanical strength of tablets of the two shape qualities increased further. Hence, the effect of particle shape on tablet strength for sodium chloride seemed to be the combined effect of differences in compaction and postcompaction behavior. It seems also that the relative humidity of the environment can affect the changes in tablet strength which occurs during tablet storage.

B. Effect of Particle Surface Geometry on Tablet Strength

It seems reasonable that also the characteristics of the surface of particles is a factor of importance for the mechanical strength of the compact. The composition and energetics of the surface, affected by, e.g., the presence of contaminants and the solid-state structure of the surface, might affect directly the strength of the interparticulate bonds. Furthermore, as discussed, the geometry of the surface can affect the number of attraction zones between particles in a compact and their volume reduction behavior with respect to their deformability and fragmentation propensity. As also discussed, there is no clear distinction between geometrical shape, related to the main dimensions of particles, and surface geometry with respect to surface roughness. The effect of particle surface geometry on tablet strength is therefore difficult to evaluate due to the difficulty of preparing suitable model materials. Milling, for example, will not change only the shape of the particles, but also the geometry and the energetics of the surface. However, an attempt to assess the importance of the geometry of the surface in relationship to powder compactibility have been presented by Karehill et al. [10] by the use of double-layer compacts. A series of materials, representing both fragmenting and nonfragmenting materials, was compacted into tablets at a series of compaction pressures. An increased compaction pressure resulted in a flatter or less rough upper surface of the compact; i.e., the series of compacts represented material surfaces of different geometry. Without ejecting the tablet from the die, more powder was poured into the die and a double-layer tablet was compacted. This double-layer tablet was ejected and its strength was assessed by axial tensile testing.

For the plastically deforming materials, the double-layer compact failed in the contact zone between the individual tablets while stressed, and the axial tensile strength decreased with increased compaction pressure for the lower tablet (Fig. 19), i.e., a decreased surface roughness. For the fragmenting materials, the double-layer compact failed generally in the lower compact while stressed; i.e., the weakest plane of the whole compact was not in the contact zone between the individual tablets. Nevertheless, an increased compaction pressure reduced the compact strength (Fig. 19), probably due to the creation of weak zones in the first tablet during the second loading.

The results can be interpreted as follows: for materials possessing a limited fragmentation during compression, the possibility of forming interparticulate attractions depends on the geometry of the particle surfaces. An increased roughness of the surface increases the possibility for a particle to find a position at an adjacent surface which promotes the formation of a large number of bonds, i.e., cavities at particle surfaces. If the particle should be able to develop attractions more-or-less independent of geometry of the surface of adjacent particles, a very marked particle deformation is probably necessary. The test materials used by Karehill et al. probably do not show such extensive deformation while loaded. Pharmaceutically rele-

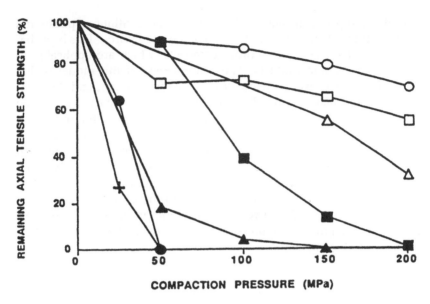

Fig. 19 The axial tensile strength of double-layer compacts, expressed as the relative strength in relationship to the strength of a single-layer compact, as a function of the compaction pressure of the first part of the double-layer compact (○ = Emcompress, □ = Lactose, △ = saccharose, ■ = sodium chloride, ● = Avicel PH 101, + = Avicel PH 101 < 10 μm, ▲ = starch 1500. (From Ref. 10 with permission of the copyright holder.)

vant materials possessing this characteristic might be polymers used as binders in granulated powders.

For fragmenting materials, the results indicate that the fracturing of the particles during compaction creates a situation where the total bonding between particles is more-or-less independent of the geometry of the surface of adjacent particles. Thus, a similar dependence between particle shape and particle surface geometry, on one hand, and the volume reduction characteristic of the material, on the other, in relationship to powder compactibility seems to exist.

VI. CONCLUDING REMARKS

In this chapter, it has been argued that a pharmaceutical compact can be described in terms of a large aggregate of particles which are strongly co- or adhered to each other, i.e., a special case of a disperse system. The gas phase in the compact can thus be described as the continuous phase and constitutes a three-dimensional network of connected pores. This is probably valid at least for relatively porous compacts, which are applicable to "normal" pharmaceutical tablets. The consequence of this model is that bonds within particles will be stronger than bonds between, and the compact will thus fail mainly around, rather than between, particles. At the fracture surface of the compact, a number of closely packed but individual particles will be exposed. The tensile strength of the compact will thus theoretically be governed by the sum of the bonding forces of all individual interparticulate bonds in the failure plane of the compact, i.e., Eq. (1). This distribution in interparticulate bonds in a compact is governed by two main factors: the properties of the material and the characteristics of the compaction process (Fig. 20). Two of the most important material properties for tablet strength are compression behavior of the particles and their dimensions before compaction.

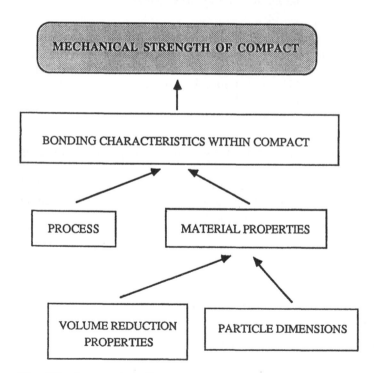

Fig. 20 Proposed qualitative relationships between the mechanical strength of the compact and the dimensions of the particles before compaction.

Concerning the compression behavior of the material it is suggested that both fragmentation and plastic deformation of the particles are bond-forming compression mechanisms. The physical significance of these mechanisms seems to be that fragmentation will affect mainly the number of interparticulate bonds, while deformation will affect mainly the bonding force of the interparticulate bonds. In this chapter it has also been suggested that the dimensions of the particles before compaction can affect both the degree of fragmentation and the degree of deformation which the particles undergo during compression and thus affect the number and the bonding force of the interparticulate bonds.

Concerning the effect of the size of the particles before compaction on the tablet strength, it is normally suggested that the most common type of relationship is that a reduced original particle size will increase the tablet strength, and the magnitude of this effect, in absolute terms, will increase with an increased compaction pressure. Deviations from this pattern are, however, commonly reported both for highly fragmenting and for non-fragmenting materials. This chapter suggests that the relationship between original particle size and tablet strength must be explained in terms of the effect of particle size on the number and bonding force of the interparticulate bonds. A reduced particle size can increase the bonding potential in a cross section of the compact by increasing the number of bonds and by increasing the bonding force of the bonds due to a high surface deformability of rough, disordered particle surfaces (small particles are often prepared by milling). However, a reduced original particle size can also decrease the bonding force due to a decreased local deformation of particles at the interparticulate contact points during the compression. This can be caused be a reduced contact force at the interparticulate contact points during compression and a decreased particle deformability. Examples of mechanistic explanations for relationships between original particle size and compact strength for tablets are given in Fig. 21.

Concerning the effect of the shape of the particles before compaction on the compact strength, it seems that only in the case when the material fragments to a limited degree during the compression phase a change toward more irregular particles increases the compact strength. Furthermore, it is suggested that an increased deformability of the material or an increased propensity to form solid bridges will stress the effect of particle shape on compact strength.

Finally, concerning the geometry of the solid surface, the literature indicates that for fragmenting materials the propensity to establish interparticulate attractions is relatively independent of the geometry of the adjacent surfaces, while the converse applies to nonfragmenting materials. For such materials, an increased irregularity of the adjacent surface im-

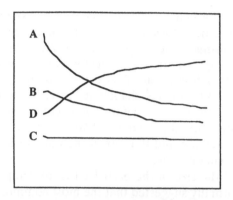

ORIGINAL PARTICLE SIZE

Fig. 21   Examples of explanations for the relationship between original particle size and tablet strength: (A, B) Increased particle surface deformability with
a reduced particle size or marked importance of the number of bonds for tablet
strength, (C) No particle size effect on particle deformability or particle surface
deformability, (D) Increased particle deformability with increased particle size.

Table 3   Proposed Effects on Tablet Strength of Original
Particle Shape and Surface Geometry for Different Types
of Materials[a] with Respect to Volume Reduction
Mechanisms (no capping or lamination tendencies)

Type of material[b]	Effect of increased particle irregularity	Effect of increased surface irregularity
A	No effect	No effect
B	Increase	Marked increase
C	Increase	Limited increase

[a]For materials with a high propensity to form solid bridges while
compacted, it seems that an increased particle irregularity facilitates the formation of solid bridges during the compression process
(e.g., sodium chloride)
[b]A = intermediate or high fragmenting
B = low fragmenting, limited plastic deformability
C = low fragmenting, high plastic deformability

proves the formation of interparticulate attractions. However, an increased
deformability of the material might result in less importance of the surface
geometry. A summary of the proposed effects of original particle shape or
surface geometry on the compact strength for tablets of materials with
different properties is given in Table 3.

REFERENCES

1. G. Alderborn, *Acta Pharm. Suec.* 22:177 (1985).
2. G. Alderborn and M. Glazer, *Acta Pharm. Nord.* 1:11 (1990).
3. C. Nyström and P. G. Karehill, *Powder Technol.* 47:201 (1986).
4. P. G. Karehill, E. Börjesson, M. Glazer, G. Alderborn, and C. Nyström, Studies on direct compression of tablets. XX. Investigation of bonding mechanisms of some directly compressed materials by addition and removal of magnesium stearate, published in thesis by P. G. Karehill, Uppsala University, 1990.
5. M. Luangtana-anan and J. T. Fell, *Int. J. Pharm.* 60:197 (1990).
6. P. G. Karehill and C. Nyström, *Int. J. Pharm.* 61:251 (1990).
7. J. S. Hardman and B. A. Lilley, *Nature* 228:353 (1970).
8. C. Führer, E. Nickel, and F. Thiel, *Acta Pharm. Technol.* 21:149 (1975).
9. G. R. B. Down, *Powder Technol.* 35:167 (1983).
10. P. G. Karehill, M. Glazer, and C. Nyström, *Int. J. Pharm.* 64:35 (1990).
11. H. Vromans, A. H. de Boer, G. K. Bolhuis, C. F. Lerk, K. D. Kussendrager, and H. Bosch, *Pharm. Weekblad Sci. Ed.* 7:186 (1985).
12. A. H. de Boer, H. Vromans, C. F. Lerk, G. K. Bolhuis, K. D. Kussendrager, and H. Bosch, *Pharm. Weekblad Sci. Ed.* 8:145 (1986).
13. C. Ahlneck and G. Alderborn, *Int. J. Pharm.* 56:143 (1989).
14. G. Alderborn and C. Ahlneck, *Int. J. Pharm.* 73:249 (1991).
15. M. Wikberg and G. Alderborn, *Int. J. Pharm.* 63:23 (1990).
16. M. Wikberg and G. Alderborn, *Int. J. Pharm.* 69:239 (1991).
17. H. Leuenberger, *Int. J. Pharm.* 12:41 (1982).
18. J. E. Rees and E. Shotton, *J. Pharm. Pharmacol. Suppl.* 22:17S (1970).
19. P. J. Rue and P. M. R. Barkworth, *Int. J. Pharm. Technol. Prod. Manuf.* 1:2 (1980).
20. P. G. Karehill and C. Nyström, *Int. J. Pharm.* 64:27 (1990).
21. C. Nyström, K. Malmqvist, J. Mazur, W. Alex, and A. W. Hölzer, *Acta Pharm. Suec.* 15:226 (1978).
22. J. M. Newton, G. Alderborn, and C. Nyström, *Powder Technol.* 72:97 (1992).
23. J. M. Newton, G. Alderborn, C. Nyström, and P. Stanley, *Int. J. Pharm.*, to appear.
24. J. T. Fell and J. M. Newton, *J. Pharm. Sci.* 59:688 (1970).
25. G. Alderborn and C. Nyström, *Acta Pharm. Suec.* 21:1 (1984).
26. M. A. Mullier, J. P. K. Seville, and M. J. Adams, *Chem. Eng. Sci.* 42:66 (1987).
27. A. B. Mashadi, Mechanical properties of compacted powders, Ph.D. thesis, University of London, 1988.
28. H. Rumpf, in *Agglomeration* (W. A. Knepper, ed.), Interscience, New York, p. 379, 1962.
29. D. C.-H. Cheng, *Chem. Eng. Sci.* 23:1405 (1968).
30. E. Shotton and D. Ganderton, *J. Pharm. Pharmacol. Suppl.* 13:144T (1961).
31. K. Schönert and H. Rumpf, in *Symposion zerkleinern*, Verlag Chemie, Weinheim, p. 108, 1962.

32. J. J. Benbow, in *Enlargement and Compaction of Particulate Solids* (N.G. Stanley-Wood, ed.), Butterworths, London, p. 161, 1983.
33. M. Duberg and C. Nyström, *Acta Pharm. Suec. 19*:421 (1982).
34. G. Alderborn and C. Nyström, *Powder Technol. 44*:37 (1985).
35. H. Leuenberger, J. D. Bonny, C. F. Lerk, and H. Vromans, *Int. J. Pharm. 52*:91 (1989).
36. G. C. Lowrison, in *Crushing and Grinding*, Butterworths, London, p. 49, 1974.
37. R. J. Roberts and R. C. Rowe, *J. Pharm. Pharmacol. 38*:567 (1986).
38. J. A. Hersey, G. Bayraktar, and E. Shotton, *J. Pharm. Pharmacol. Suppl. 19*:24S (1967).
39. J. T. Fell and J. M. Newton, *J. Pharm. Pharmacol. 20*:657 (1968).
40. A. McKenna and D. F. McCafferty, *J. Pharm. Pharmacol. 34*:347 (1982).
41. H. Vromans, A. H. de Boer, G. K. Bolhuis, C. F. Lerk, and K. D. Kussendrager, *Acta Pharm. Suec. 22*:163 (1986).
42. G. Alderborn and C. Nyström, *Acta Pharm. Suec. 19*:381 (1982).
43. E. Shotton and C. J. Lewis, *J. Pharm. Pharmacol. Suppl. 16*:111T (1964).
44. O. Alpar, J. A. Hersey, and E. Shotton, *J. Pharm. Pharmacol. Suppl. 22*:1S (1970).
45. G. Alderborn, E. Börjesson, M. Glazer, and C. Nyström, *Acta Pharm. Suec. 25*:31 (1988).
46. G. Alderborn, K. Pasanen, and C. Nyström, *Int. J. Pharm. 23*:79 (1985).
47. M. Eriksson and G. Alderborn, *Pharm. Res.*
48. M. J. Kontny, G. P. Grandolfi, and G. Zografi, *Pharm. Res. 4*:104 (1987).
49. M. Eriksson and G. Alderborn, *Int. J. Pharm., 109*: 59 (1994).
50. J. Lazarus and L. Lachman, *J. Pharm. Sci. 55*:1121 (1966).
51. E. Shotton and B. A. Obiorah, *J. Pharm. Pharmacol. Suppl, 25*:37P (1973).
52. G. Alderborn and C. Nyström, *Acta Pharm. Suec. 19*:147 (1982).
53. C. Führer, *Acta Pharm. Technol. Suppl. 6 24*:129 (1970).
54. L. W. Wong and N. Pilpel, *Int. J. Pharm. 59*:145 (1990).

11

Mechanical Properties

Ray C. Rowe and Ron J. Roberts

Zeneca Pharmaceuticals, Macclesfield, Cheshire, United Kingdom

I. INTRODUCTION

The vast majority of drugs when isolated exist as crystalline or amorphous solids. Subsequently they may be admixed with other inactive solids (excipients) and finally compacted to form tablets. The process of compaction involves subjecting the materials to stresses causing them to undergo deformation. The reaction of a material to a deformation stress, σ_d, is dependent on both the mode of deformation and the mechanical properties of the material; e.g.,

(a) for elastic deformation

$$\sigma_d = \epsilon E \qquad (1)$$

where E is Young's modulus of elasticity of the material and ϵ is the deformation strain;

(b) for plastic deformation

$$\sigma_d = \sigma_y \qquad (2)$$

where σ_y is the yield stress of the material, and

(c) for brittle fracture

$$\sigma_d = \frac{AK_{IC}}{\sqrt{d}} \qquad (3)$$

where

K_{IC} = critical stress intensity factor of material (an indication of the stress required to produce catastrophic crack propagation)

d = particle size (diameter)

A = constant depending on geometry and stress application

For compression of rectangular samples with large cracks $A = \sqrt{32/3}$ or 3.27 [1], but for other geometries A varies between 50 and 1 [2].

It is evident from the discussion above that in order to be able to predict the compaction behavior of a material it is essential that methods be derived to measure the following for powdered materials:

(a) Young's modulus of elasticity (E).
(b) Yield stress (σ_y); this is directly related to the indentation hardness, H, since for a plastic material

$$\sigma_y = \frac{H}{3} \tag{4}$$

(c) Critical stress intensity factor (K_{IC}); this is directly related to the fracture toughness, R, since for plane stress

$$K_{IC} = (ER)^{1/2} \tag{5}$$

In this chapter various methods which have been specifically applied to pharmaceutical materials are reviewed.

II. YOUNG'S MODULUS OF ELASTICITY

If an isotropic body is subjected to a simple tensile stress in a specific direction it will elongate in that direction while contracting in the two lateral directions, its relative elongation being directly proportional to the stress. The ratio of the stress to the relative elongation (strain) is termed Young's modulus of elasticity. It is a fundamental property of the material directly related to its interatomic or intermolecular binding energy for inorganic and organic solids respectively and is a measure of its stiffness.

Young's modulus of elasticity of a material can be determined by many techniques several of which have been used in the study of pharmaceutical materials, namely flexure testing using both four- and three-point beam bending, compression testing, and indentation testing on both crystals and compacts.

A. Flexure Testing (Beam Bending)

In flexure testing, a rectangular beam of small thickness and width in comparison to its length is subjected to transverse loads and its central deflection due to bending is measured. The beam may be supported and loaded in one of two ways (Fig. 1). If the beam is supported at two points and is loaded at two points it undergoes what is known as four-point bending, while if it is supported at two points but is loaded at one point it undergoes what is known as three-point bending. Equations for the calculation of Young's modulus from the applied load F and the deflection of the midpoint of the beam ξ can easily be derived:

for four-point bending [3]
$$E = \frac{F}{\xi}\frac{6a}{h^3b}\left(\frac{l^2}{8} + \frac{al}{2} + \frac{a^2}{3}\right) \qquad (6)$$

for three-point beam bending [4]
$$E = \frac{Fl^3}{4\xi h^3 b} \qquad (7)$$

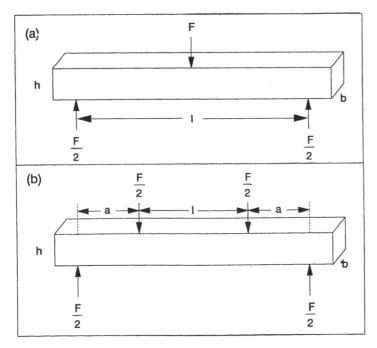

Fig. 1   Geometries for (a) three-point and (b) four-point beam bending: F, applied load; h, beam thickness; b, beam width; l and a, distance between loading points as shown.

where h and b are the height (thickness) and width (breadth), respectively, and l and a are as in Fig. 1. In both cases the value for Young's modulus for the specimen under test can either be calculated from a single point determination for more commonly from the slope of total load (F) versus central deflection (ξ).

The four-point beam bending test was first used for pharmaceutical materials by Church [3] and has since been adapted by Mashadi and Newton [5,6], Bassam et al. [7,8,9], Roberts et al. [10], and Roberts [11]. Generally beams of varying height have been used—100 mm long by 10 mm wide or 60 mm long by 7 mm wide—prepared using specially designed punches and dies. Beams are generally prepared at varying compression pressures to achieve specimens of varying porosity. Although the height of the beam (at constant porosity) is an experimental variable, Bassam et al. [8] have shown that it does not have a significant effect on the measured modulus. Generally low rates of testing are used ($\approx$ 1 mm min^{-1}), but Bassam et al. [8] have shown an independence of loading rate up to rates of 15 mm min^{-1}.

A disadvantage of the four-point beam method is that it invariably requires large specimens and hence large quantities of materials (15–20 g). In addition, high-tonnage presses are needed to prepare the specimens, thus exacerbating problems with cracking and lamination on ejection from the die. It was to overcome these difficulties that Roberts et al. [10] developed a three-point beam testing method that uses beams prepared from 200 mg of material. The beams in this case are 20 mm long by 7 mm wide and are stressed by applying a static load of 0.3 N with an additional dynamic load of ± 0.25 N (at a frequency of 0.17 Hz) using a thermal mechanical analyzer (Mettler Instruments TMA40). In operation a calibration run is first performed to eliminate distortions in the sensor and other parts of the displacement measuring system and then 20 measurements of specimen displacement are undertaken to an accuracy of ± 0.005 μm. Young's modulus of elasticity of the specimen is then calculated from the mean displacement corrected for distortions using Eq. (7) where in this case F is the applied dynamic load. Extensive testing by Roberts et al. [10] and Roberts [11] has shown equivalence between this test and the conventional four-point beam test.

A problem associated with the analysis of data for specimens prepared from particulate solids is in the separation of the material property from that of the specimen property, which by definition includes a contribution due to the porosity of the specimen. All workers have found that for all materials there is a decrease in Young's modulus with increasing porosity (Fig. 2). Numerous equations have been published which describe this relationship [13]. Certain equations are based on theoretical considerations

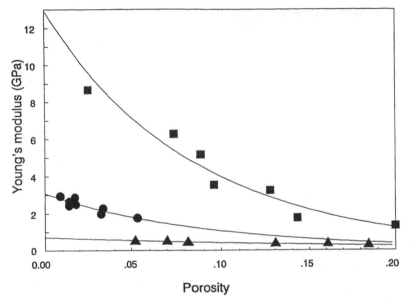

Fig. 2 The effect of porosity on the measured Young's modulus of ▲, PTFE; ●, testosterone propionate; ■, theophylline (anhydrous). (Data generated using three-point beams, taken from Roberts et al. [4].)

[14,15], while others are empirical curve-fitting functions [16,17]. Recently Bassam et al. [8] have reported a comparison of all the equations currently in use for data generated from four-point beam bending on 15 pharmaceutical powders ranging from celluloses and sugars to the inorganic calcium carbonate, and have concluded that the best overall relationship is the modified two-order polynomial [17]

$$E = E_0(1 - f_1P + f_2P^2) \tag{8}$$

where

E_0 = Young's modulus at zero porosity
E = measured modulus of the beams compacted at porosity P
f_1, f_2 = constants

However, such a conclusion may not be universal since during extensive studies on a wider range of materials including drugs, Roberts [11] has concluded that an exponential relationship [16] is the preferred option for data generated using the three-point beam testing method; i.e.

$$E = E_0e^{-bP} \tag{9}$$

Table 1 Young's Modulus at Zero Porosity Measured by Flexure Testing

Material	Particle size (μm)	Young's modulus (GPa)	Reference
Drugs			
Theophylline (anhydrous)	31	12.9	4
Paracetamol DC	120	11.7	11
Caffeine (anhydrous)	38	8.7	4
Sulfadiazine	9	7.7	4
Aspirin	32	7.5	4
Ibuprofen	47	5.0	4
Phenylbutazone	50	3.3	4
Testosterone propionate	85	3.2	4
Sugars			
Sorbitol instant	—	45.0	5
α-lactose monohydrate	20	24.1	4
α-lactose monohydrate	63	3.2	8
Lactose β-anhydrous	149	17.9	8
Lactose β-anhydrous	149	18.5	9
Lactose (spray dried)	—	13.5	12
Lactose (spray dried)	125	11.4	8
Dipac sugar	258	13.4	8
Mannitol	88	12.2	8
Polysaccharides			
Starch 1500	—	6.1	12
Maize starch	16	3.7	8
Celluloses			
Avicel PH-101	50	10.3	6
Avicel PH-101	50	9.7	12
Avicel PH-101	50	9.2	8
Avicel PH-101	50	9.0	7
Avicel PH-101	50	7.8	10
Avicel PH-101	50	7.6	10
Avicel PH-101	50	7.4	10
Avicel PH-102	90	8.7	8
Avicel PH-102	90	8.2	7
Avicel PH-105	20	10.1	7
Avicel PH-105	20	9.4	8
Emcocel	56	9.0	7
Emcocel	56	7.1	8
Emcocel (90M)	90	9.4	7
Emcocel (90M)	90	8.9	8
Unimac (MG100)	38	8.8	7
Unimac (MG100)	38	8.0	8
Unimac (MG200)	103	8.0	7
Unimac (MG200)	103	7.3	8
Elcema (P100)	—	8.6	12

Table 1   Continued

Material	Particle size (μm)	Young's modulus (GPa)	Reference
Inorganics			
Calcium carbonate	8	88.3	8
Calcium phosphate	10	47.8	8
Polymers			
PVC	—	4.4	9
PVC	—	4.1	4
Stearic acid	62	3.8	4
PTFE	—	0.81	10
PTFE	—	0.71	9
PTFE	—	0.71	10

where b is a constant. It is interesting to note that on average the extrapolated values of E_0—Young's modulus at zero porosity—calculated using both equations are only marginally different [7,8].

The analysis of Young's modulus at zero porosity thus provides a means of quantifying and categorizing the elastic properties of powdered materials, Table 1 shows literature data for a variety of pharmaceutical excipients and drugs determined using beam-bending methods. It can be seen that the values vary over two orders of magnitude ranging from hard rigid materials with very high moduli (e.g., the inorganics) to soft elastic materials with low moduli (e.g., the polymeric materials). As a result a rank of increasing rigidity of tabletting excipients can be listed: starch < microcrystalline celluloses < sugars < inorganic fillers with variations in the groups dependent on chemical structure as well as the preparation and pretreatment routes (including particles size).

The effects of particle size are distinguishable within the celluloses and α-lactose monohydrate. In the former there is a small increase with decreasing particle size, while the latter the increase is much greater. Whereas in the former the effect is probably due to an increase in contact area, in the latter the effect is due to specimen defects in that the specimens used by Bassam et al. [8] contained microscopic flaws and cracks. Recent work by Roberts [11] has shown that it is necessary to eliminate all specimens with cracks otherwise the extrapolated modulus values to zero porosity are inconsistent.

For the lactose samples the rank order of increasing rigidity is spray dried < β-anhydrous < α-monohydrate, consistent with the findings of workers describing the compaction properties of the materials using instrumented tabletting machines [18,19]. The variations in the cellulose samples can be attributed to subtle differences in the manufacturing preparative technique. However, it is known that this factor can also affect the equilibrium moisture content of the samples, with Unimac samples attaining a lower equilibrium moisture content than Avicel samples. Since it is known that increasing moisture content can lead to a decrease in Young's modulus for microcrystalline cellulose [8] the differences in modulus between sources listed in Table 1 would be expected to increase if all materials were compared at equivalent moisture contents.

Recently Roberts et al. [4] have investigated the relationship between Young's modulus of a variety of drugs and excipients using three-point beam bending and their molecular structure based on intermolecular interactions using the concept of cohesive energy density (CED). They found a direct relationship of the form

$$E_0 = 0.01699\text{CED} - 2.7465 \tag{10}$$

where CED is expressed in units of MPa and Young's modulus in units of GPa (Fig. 3). The significance of this finding is in the recognition of the validity of both the test method and the data manipulation (i.e., the extrapolation to zero porosity).

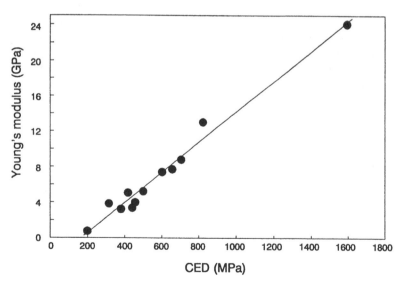

Fig. 3 Young's modulus versus cohesive energy density for various drugs and excipients (Taken from Roberts et al. [4].)

B. Compression Testing

In compression testing a compressive stress is applied to either a crystal or compacted specimen and the corresponding strain measured, the ratio of the stress to strain being a measure of the compressive Young's modulus of elasticity.

The loading system depends on the size of specimen, ranging from a microtensile testing instrument providing a load up to 5 N and a minimum displacement of 5 nm, as used for single crystals by Ridgway et al. [20], to an Instron physical testing instrument providing a load upto 50 kN, as used for flat-faced cylindrical compacts (8 mm diameter) by Kerridge and Newton [21]. Young's modulus in compression can be determined from the equation

$$E = \frac{L}{(X - C)A} \tag{11}$$

where

L = specimen length
X = slope of strain versus stress
C = machine constant (strain versus stress for loading of the machine without a specimen
A = specimen cross-sectional area

While for crystals a single measurement is all that is necessary, for compacts measurements on specimens prepared at different porosities are required. In the latter test Young's modulus at zero porosity can be determined using Eq. (9).

Values for the compressive modulus of a variety of materials in Table 2 are lower than those measured by other techniques. Furthermore, for the

Table 2 Young's Modulus Measured by Compaction Testing

Material	Specimen	Young's modulus (GPa)	Reference
Aspirin	Crystal	0.1	20
Aspirin	Compact	2.5	21
Sodium chloride	Crystal	1.9	20
Potassium chloride	Compact	9.2	21
Avicel PH-102	Compact	4.7	21
Sucrose	Crystal	2.2	20
Salicylamide	Crystal	1.3	20
Hexamine	Crystal	0.9	20

test involving crystals Ridgway et al. [20] indicated that cracks were important and that this factor caused variation in the results. This factor may also account for the low values of modulus when compared with those from flexure testing. In view of this and the fact that the compressive modulus should always be greater than the tensile modulus, the results from these tests must be viewed with some scepticism.

C. Indentation Testing

In indentation testing a hard indenter made of either diamond, sapphire, or steel and machined to a specific geometry (either a square-based pyramid (Vickers indenter) or a spherical ball (Brinell indenter) is pressed under load into the surface of a material either in the form of a crystal or compacted specimen. Although well recognized as a method for measuring hardness (see later), it may also be used to measure Young's modulus, except that in this case it is the recovered depth after removal of the load that is important (Fig. 4). The test may be either static [22,23,24] or dynamic involving a pendulum [25].

For crystals a Vickers indenter is generally used, and in this respect Duncan-Hewitt and Weatherly, [23,24] have used a Leitz-Wetzler Miniload hardness tester on preselected single crystals. Load was applied over 15 sec

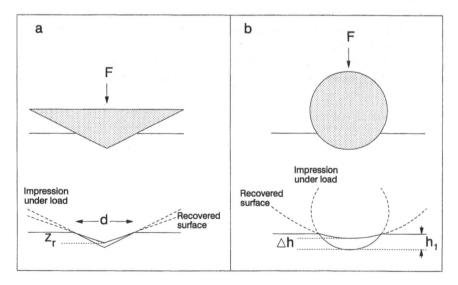

Fig. 4   Loading geometry and recovery of indent for (a) Vickers and (b) Brinell hardness testers used to measure Young's modulus.

and was maintained for 10 sec. In the case of sucrose crystals, hardness anisotropy of the various faces was demonstrated.

The equation used by Duncan-Hewitt and Weatherly [24] involves measuring the recovered depth of indentation and the length of the Vickers diagonal and substituting in the equation of Breval and Macmillan [26]:

$$\left(\frac{2Z_r}{d}\right)^2 = 0.08168 \left[1 - 8.7 \left(\frac{H_v}{E}\right) + 14.03 \left(\frac{H_v}{E}\right)^2 \right] \qquad (12)$$

where

H_v = indentation (see later for equation)
Z_r = recovered depth of the indentation
d = length of Vickers diagonal (Fig. 4).

Data on several materials (Table 3) show values of the same order as those determined by flexure testing. The value of sodium chloride compares favorably with that reported in the literature (37 GPa [27]).

For compacts Ridgway et al. [22] have used a pneumatic microindentation apparatus to measure both the initial depth of penetration (h_1) and recovery (Δh) after application of small loads (4 g) using a sapphire ball indenter 1.5 mm in diameter and substituting in a modified Hertz equation [28]:

$$E(GPa) = 1.034 \frac{F}{\Delta h \sqrt{h_1}} \qquad (13)$$

where h_1 and Δh are in micrometers and F, the indentation load, is in grams. The values quoted for aspirin are very small, two orders of magnitude below the figure given in Table 1 and hence must be viewed with some suspicion.

In the dynamic method used by Hiestand et al. [25] a metre long pendulum with a steel sphere of 25.4 mm diameter strikes the face of a compact of cross-sectional area of 14.52 cm². The modulus of elasticity can be calculated from a knowledge of the indentation hardness, H_b and the

Table 3 Young's Modulus Measured by Indentation on Crystals

Material	Indenter	Young's modulus (GPa)	Reference
Sodium chloride	Vickers	43.0	24
Sucrose	Vickers	32.3	24
Paracetamol	Vickers	8.4	24
Adipic acid	Vickers	4.1	24
Aspirin	Brinell	0.32	22

strain index, ξ_i, a measure of relative strain during elastic recovery that follows plastic deformation using the equation

$$E = \frac{H_b(1 - \nu^2)}{\xi_i} \tag{14}$$

where ν is Poisson's ratio. The strain index ξ_i is calculated using values from the indentation experiment:

$$\xi_i = \frac{5a}{6\pi r} \left(\frac{h_i}{h_r} - \frac{3}{8} \right)^{-1} \tag{15}$$

where

a = chordal radius of the indent
r = radius of spherical indenter
h_i, h_r = initial and rebound heights of indenter

Assuming a Poisson's ratio of 0.3 it is possible, using the calculated values for hardness and strain index [29] to determine Young's modulus for spray-dried lactose and microcrystalline cellulose (Avicel PH-102)—25.5 and 6.2 GPa respectively. These values compare favorably with those obtained from flexure testing (Table 1).

III. INDENTATION HARDNESS AND YIELD STRESS

The hardness and/or yield stress of a material are measures of its resistance to local deformation. For plastic materials the two parameters are directly related, the yield stress being one-third of the hardness (Eq. (4)). While hardness is generally measured by means of indentation, yield stress may be determined by more indirect methods, e.g., data from compaction studies. Both will be discussed in this section.

A. Indentation Hardness Testing

The most common method of measuring the hardness of a material is the indentation method. In this a hard indenter (e.g., diamond, sapphire, quartz, or hardened steel) of specified geometry is pressed into the surface of the material. The hardness is essentially the load divided by the projected area of the indentation to give a measure of the contact pressure. The most commonly used methods for determining the hardness of pharmaceutical materials are the Vickers and Brinell tests. In general the former has been used for measurements on compacts whereas the latter technique has been solely applied to single crystals.

In the standard Brinell test [30] a hard steel ball (usually 2mm in diameter) is pressed normally on to the surface of the material. The load F is applied for a standard period of 30 sec and then removed. The diameter, d, of the indentation is measured and the Brinell hardness, H_b, determined by the relationship

$$H_b = \frac{F}{(\pi D/2)(D - \sqrt{D^2 - d^2})} \tag{16}$$

where D is the diameter of the ball.

In the Vickers test [31] a square-based diamond pyramid is used as an indenter. It is capable of measuring hardnesses over the entire range from the softest to the hardest materials. The Vickers hardness, H_v, is determined from the following equation, where F is applied force and d is the length of the diagonals of the sqare impression:

$$H_v = \frac{2F \sin 68°}{d^2} \tag{17}$$

The specimen thickness should be at least one and a half times the diagonal length with greatest accuracy being obtained with high loads, but loads as small as 0.01 N can be used. However, at lower loads the elastic recovery is of greater importance. For loads lower than 0.5 N the technique is usually described as microhardness testing.

An improved analysis of the Vickers hardness of compacts was performed by Ridgway et al. [32] using a Leitz microhardness tester with loads of between 5 and 2000 g. Accurate diagonal lengths were determined after lightly dusting the indent with graphite powder. Of the materials studied, potassium chloride, hexamine, and urea showed little change in hardness with increasing compaction pressure, while for aspirin and sodium chloride the hardness increased possibly due to work hardening. The maximum hardness values reported were 29, 17, 16, 12, and 8 MPa for sodium chloride, potassium chloride, aspirin, hexamine, urea, respectively. These values are an order of magnitude lower than those generally accepted and hence must be viewed with suspicion.

In many studies on compacts spherical indenters have been used either fitted to commercially available instruments or custom-built equipment. An example of the former is the pneumatic microindentation apparatus described by Ridgway et al. [22]. This instrument can apply loads of between 4 and 8 g using a spherical indenter 1.5 mm in diameter and measure depths of penetration of 1–6 μm. Using this apparatus on aspirin compacts, the authors were able to show that hardness measurements at the center of compacts were higher than those at the periphery, a property for flat-faced compacts confirmed by Aulton and Tebby [33]. However, for compacts prepared using concave punches hardness distribution tends to vary with the degree of curvature [33].

In a modification to the original microindentation apparatus [22], Aulton et al. [34] added a displacement transducer to measure the vertical displacement. They suggested that the elastic quotient index, which is the fraction of the indentation which rebounds elastically, was a measure of the ability of materials to form tablets.

In a further study involving a larger range of materials, Aulton [35] found that hardness measurements were generally higher on the upper face of the compact than on the lower. However, the differences were material dependent and the hardness measurements reported were 62, 54, 51, 36, 19, and 13 MPa for sucrose, Sta-Rx, Emcompress, Avicel PH-101, lactose β-anhydrous, and paracetamol respectively, were low in comparison with other measurements.

Probably the most relevant of all the work carried out on hardness measurement on compacts is that of Hiestand et al. [25], who used a spherical indenter attached to a pendulum, and that of Leuenberger [36] and his co-workers [37,–41], who used a spherical indenter attached to a universal testing instrument. In the dynamic pendulum method of Hiestand et al. [25] a spherical ball of 24.5 mm diameter falls under the influence of gravity and the rebound height and indent dimensions measured. The hardness is calculated from the expression

$$H_b = \frac{4mgrh_r}{\pi a^4} \left(\frac{h_i}{h_r} - \frac{3}{8} \right) \tag{18}$$

where

m = the mass of the indenter
g = gravitational constant
r = radius of the sphere
a = chordal radius of the indent
h_i = initial height of the indenter
h_r = rebound height of the indenter

In the case of the indenter attached to the universal testing instrument [36] loads of 3.92 and 9.81 N were applied to a sphere of 1.76 mm diameter at a velocity of 0.05 cm min^{-1}, indent diameters being determined from scanning photomicrographs.

In both cases indentation hardness was found to be dependent on the compaction pressure and hence relative density (porosity) of the compact in an exponential manner. While Hiestand et al. [25] considered extrapolation to unit relative density to be questionable, Leuenberger [36] has used the relationship between Brinell hardness, H_b, of a compact and its relative density, D, to develop a measure of compactibility and compressibility, namely

$$H_b = H_{b\,max}(1 - e^{-\lambda \sigma_c D})$$ (19)

where

$H_{b\,max}$ = theoretical maximum hardness as the compressive stress σ_c approaches infinity and the relative density of the compact approaches unity

λ = rate at which H_b increases with increasing compressive stress

In the equation Leuenberger [36] has suggested the $H_{b\,max}$ describes the compactibility and λ the compressibility.

Data on indentation hardness using both methods are shown in Table 4. As with the modulus of elasticity values for hardness vary over two orders of magnitude from the very hard materials (e.g., Emcompress) to very soft waxes. The drugs have intermediate hardness values.

It should be noted that the values recorded are also variable due to the

1. Intrinsic variability in the specimens [37]
2. Work hardening as the compaction pressure is increased [40,42]
3. Increase in hardness with increasing indentation load [40]
4. Rate of measurement [39]

The earliest reported study on the hardness of pharmaceutical crystals was carried out by Ridgway et al. [20] using the Leitz microhardness tester with a pyramidal indenter. The crystals were mounted in heat-softened picene wax on a mounting slide to ensure that the surfaces were horizontal. It is interesting to note that the authors observed that aspirin and sucrose showed cracking and regarded this as a problem with the technique. Hardness values from this study are presented in Table 5. In general, softer materials showed the most variation in results due to a decline in the definition and quality of the indent.

The next reported study of crystal hardness using a Vickers hardness tester was carried out by Ichikawa et al. [43]. In this study the majority of the materials (with the exception of sucrose and urea) were recrystallized. Indentation was performed on the crystal face possessing the largest area (i.e., the face that grows the slowest during crystallization) since it was inferred that this face would have the most influence on the compaction properties. Ichikawa et al. [43] considered differences in crystal hardness as a reflection of the mechanism of deformation during compaction. They

Table 4 Indentation Hardness Measured on Compacts

Material	Indentation hardness (MPa)	Reference
Drugs		
Paracetamol DC	265	37
Caffeine (anhydrous)	290	37
Caffeine (granulate)	288	37
Oxprenolol succinate	262	41
Hexamine	232	36
Phenacetin	213	36
Sitosterin	198	36
Metamizol	91	37
Aspirin powder	91	37
Aspirin FC	87	37
Aspirin	60	36
Aspirin	55	36
Ibuprofen (A)	35	36
Ibuprofen (B)	162	36
Sugars		
α-lactose monohydrate	515	36
α-lactose monohydrate	534	37
Lactose β-anhydrous	251	36
Sucrose	1046–1723	36
Sucrose (250–355#)	493	37
Alkali halides		
NaCl	653	36
NaCl	313	37
NaCl (rock salt)	358	37
KCl	99	38
KBr	69	37
Others		
Emcompress	752	37
Avicel PH-102	168	37
Starch 1500	78	41
Sodium stearate	37	40
PEG 4000	36	40
Castor oil (hydrogenated)	32	41
Magnesium stearate	22	40
Sodium lauryl sulphate	10	40

Table 5 Indentation Hardness Measured on Crystals

Material	Indentation hardness (MPa)	Reference
Drugs		
Paracetamol	421	24
Paracetamol	342	43
Sulfaphenazole	289	43
Hexamine	133	20
Hexamine	42	43
Sulfadimethoxine	231	43
Phenacetin	172	43
Salicylamide	151	20
Salicylamide	123	43
Aspirin	87	20
Sugars		
α-lactose monohydrate	523	43
Sucrose	645	24
Sucrose	636	20
Alkali halides		
NaCl	212	20
NaCl	213	24
NaCl	183	43
KCl	177	20
KCl	101	43
Others		
Urea	91	20
Urea	83	43

also showed that the reciprocal of crystal hardness correlated with the slope *K* from the Heckel equation. Data reproduced in Table 5 represent the mean value from three applied loads.

More recently Duncan-Hewitt and Weatherly [23,24] measured the Vickers hardness on single crystals of number of materials using a Leitz-Wetzlar Miniload tester with a load of 147 mN. The surfaces of sucrose were preconditioned, and this may account for their higher hardness values compared with compacts (Table 4), since in this case the materials could be fully work-hardened solids. However, the authors did not indicate whether the other crystals were preconditioned. Furthermore, Duncan-Hewitt and Weatherly [23] showed that the sucrose crystals were anisotropic in that different crystal faces gave different hardness values.

B. Yield Stress from Compaction Studies

Compaction studies, because they mimic the tabletting process, offer an ideal method for assessing the mechanical properties of powders. In powder compaction a specific method used to evaluate the average stress of a material during compression relies on the observations of Heckel [44,45], who found that for materials that plastically deform the relative density of a material, D, could be related to the compaction pressure, P, by the equation

$$\ln\left[\frac{1}{1-D}\right] = KP + A \tag{20}$$

where K and A are constants.

Unfortunately, considerable deviations of the experimental data occur at both low and high pressures due to particle rearrangement and strain hardening, respectively, but at least over the middle pressure range a straight-line relationship exists between $\ln[1/(1-D)]$ and P (Fig. 5).

Equation (20) has been reappraised by Hersey and Rees [46], who suggested that the reciprocal of K can be regarded as numerically equal to the mean yield stress of the powder. However, as pointed out by Roberts et al. [47], this is only a specific case, and the reciprocal of K can be regarded as a mean deformation stress be it a plastic deformation stress (equal to the yield stress) for materials that deform plastically, a fracture stress for mate-

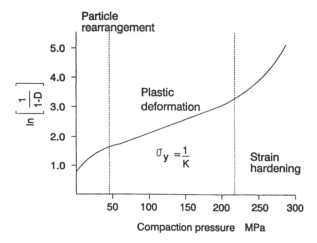

Fig. 5   Schematic diagram of the Heckel plot. (From Refs. 44 and 45.)

rials that undergo fracture, or a combination of the two. This approach implies that provided the experiment is carried out on materials close to or below the brittle-ductile transition (i.e., the material deforms plastically [47]), then the reciprocal of K will be numerically equal to the yield stress of the powder.

Historically two approaches to the analysis of Heckel data have been used and these are generally referred to as "at-pressure" and "zero-pressure" measurements. Pressure/relative density measurements determined during compression are clearly at-pressure measurements, while those from relative density measurements on the compact after ejection are zero-pressure measurements. In the original publication Heckel [44] found that for the metals iron, copper, nickel, and tungsten there was no difference between the two measurements, but for graphite the zero-pressure measurement was higher, attributable to the elastic recovery of the compact causing a lower relative density. Support for this hypothesis can be obtained from studies on the pharmaceutical materials dicalcium phosphate dihydrate (8% increase [49]), lactose (30% increase [18]), microcrystalline cellulose (56% increase [49]), and starch (177% increase [49]) where the magnitude of the percentage increase in yield pressure due to elastic recovery is indirectly related to the modulus of the material, i.e., the larger the increase the lower the modulus.

In the light of the discussion above the findings that other factors such as punch and die dimensions [50,51], state of lubrication [52,53], and speed of compaction [54,55], can have an effect on the measurement, it is not surprising that a great deal of controversy and confusion surrounds the use of data from Heckel plots. However, recent work by Roberts and Rowe [56] has clearly shown that, provided measurements are carried out "at pressure" with lubricated punches and dies on material that is below its brittle/ductile transition and at a very slow speed, the reciprocal of K is identical to the yield stress of the material comparable with that calculated from indentation hardness measurements using Eq. (4).

Although early measurements were generally performed on either instrumented punches and dies in physical testing machines [18] or instrumented single-punch table machines [52] with relatively unsophisticated data capture and analysis, recent measurements have been generally carried out using tablet compression simulators [55,57].

Yield stress data for a number of excipients and drugs are shown in Table 6. The trends mimic those for indentation hardness with the inorganic carbonates and phosphates showing very high values of yield compared to the polymers with low yield values. The variation seen in the drugs may well be due to slight differences in the particle sizes tested.

Table 6 Yield Stresses Measured by Compaction Studies

Material	Experimental details	Yield stress (MPa)	References
Drugs			
Paracetamol DC	Single punch	108	58
Paracetamol DC	Single punch	81	59
Paracetamol DC	Simulator	109	56
Paracetamol	Single punch	79	60
Paracetamol	Single punch	99	61
Paracetamol	Single punch	127	62
Paracetamol	Simulator	102	55
Sulfathiazole	Hydraulic press	109	63
Theophylline (anhydrous)	Single punch	75	61
Aspirin	Single punch	25	60
Aspirin	Single punch	73	62
Tolbutamide	Single punch	24	60
Phenylbutazone	Single punch	24	60
Ibuprofen	Simulator	25	64
Sugars			
α-lactose monohydrate	Hydraulic press	179	65
α-lactose monohydrate	Hydraulic press	183	66
α-lactose monohydrate	Simulator	178	67
Lactose β-anhydrous	Simulator	149	56
Lactose (spray dried)	Simulator	178	57
Lactose (spray dried)	Simulator	147	55
Mannitol	Simulator	90	56
Inorganic			
Calcium phosphate	Simulator	957	56
Calcium carbonate	Simulator	851	56
Calcium carbonate	Hydraulic press	610	68
Magnesium carbonate	Simulator	471	56
Dicalcium phosphate dihydrate	Simulator	431	56
Polymers			
PVC/vinyl acetate	Simulator	70	56
Polyethylene	Simulator	16	56
PTFE	Simulator	12	56
Others			
Sodium chloride	Single punch	89	69
Sodium chloride	Simulator	89	47
Avicel PH-101	Single punch	50	59
Avicel PH-101	Simulator	46	66
Avicel PH-102	Simulator	49	66
Avicel PH-105	Simulator	48	66
Maize starch	Simulator	40	56
Stearic acid	Hydraulic press	4.5	66

Compaction simulators allow the measurement of yield stress over a wide range of punch velocities (Fig. 6). To compare materials, Roberts and Rowe [55] proposed a term—the strain rate sensitivity—describing the percentage decrease in yield stress from a punch velocity of 300 mm sec^{-1} to one of 0.033 mm sec^{-1}. This was later modified [56] to a percentage increase in yield stress over the same punch velocities:

$$\text{SRS} = \frac{\sigma_{y300} - \sigma_{y0.033}}{\sigma_{y0.033}} \tag{21}$$

where σ_{y300} and $\sigma_{y0.033}$ are the yield stresses measured at punch velocities of 300 and 0.033 mm sec^{-1} respectively.

Data on a number of materials are shown in Table 7. Some materials, such as the inorganic carbonates and phosphates, show little rate dependence, while others, such as starch an mannitol, show a large strain rate

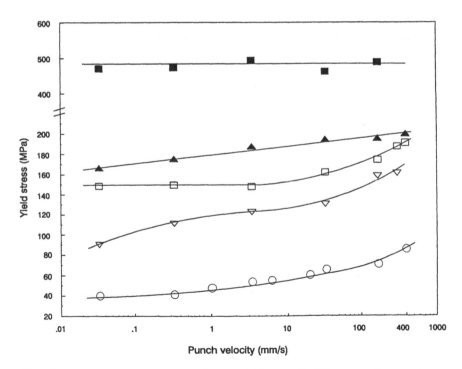

Fig. 6 The effect of punch velocity on yield stress for ■, magnesium carbonate; ▲, α lactose monohydrate; □, lactose β-anhydrous; ▽, mannitol; ○, maize starch.

Table 7 Strain Rate Sensitivities (SRS) of
Some Excipients and Drugs

Material	SRS (%)
Maize starch	97.2
Mannitol	86.5
PVC/vinyl acetate	67.5
Sodium chloride	66.3
Avicel PH-101	63.7
Lactose β-anhydrous	25.5
Lactose (spray dried)	23.8
α-lactose monohydrate (fine grade)	19.4
Paracetamol	11.9
Paracetamol DC	1.8
Heavy magnesium carbonate	0
Calcium carbonate	0
Calcium phosphate	0

dependence consistent with differences in the time-dependent properties of
the material during compaction.

IV. CRITICAL STRESS INTENSITY FACTOR K_{IC}

The critical stress intensity factor, K_{IC}, describes the state of stress around
an unstable crack or flaw in a material and is an indication of the stress
required to produce catastrophic propagation of the crack. It is thus a
measure of the resistance of a material to cracking. Since it is related to the
stress and the square root of crack length it has the dimensions of $MPa\,m^{1/2}$.

All methods used to measure K_{IC} involve specimens containing induced notches and/or cracks and for pharmaceutical materials include
three- or four-point beam bending (commonly known as the single-edged
notched-beam (SENB) test), double torsion, radial-edge cracked tablet or
disc, and Vickers indentation. The choice of the test and its associated
specimen geometry depends on the rate of testing, the ease of formation of
the specimen and the porosity of the specimen.

A. Single-Edge Notched Beam

In this test a prenotched rectangular beam of small thickness and width in
comparison to its length is subjected to transverse loads and the load at
fracture measured. As with the beams used for the determination of

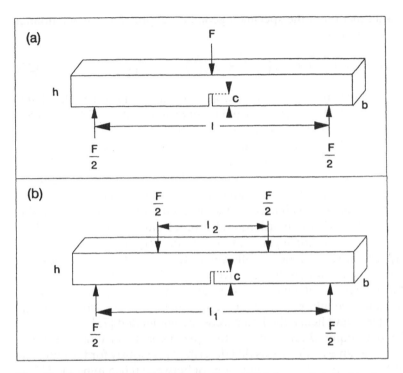

Fig. 7 Geometries for (a) three-point and (b) four-point single-edge notched beam: F, applied load; h, beam thickness; b, beam width; l and a, distances between loading points as shown; c, crack length.

Young's modulus, loading can be by either three- or four-point (Fig. 7). The single-edge notched-beam test has been the subject of much research leading to the specification of standard criteria for test piece geometry dependent on the material to be studied [70,71].

Equations for the calculation of the critical stress intensity factor from the applied load, F, and geometry of the beam can be derived, e.g., for four-point beam bending [5,6]:

$$K_{IC} = \gamma \frac{3Fc^{1/2}(l_1 - l_2)}{2bh^2} \tag{22}$$

and for three-point beam bending [72]:

$$K_{IC} = \gamma \frac{3Fc^{1/2}l}{2bh^2} \tag{23}$$

where γ is a function of the specimen geometry expressed as a polynomial of the parameter c/h:

$$\gamma = A_0 + A_1\left(\frac{c}{h}\right) + A_2\left(\frac{c}{h}\right)^2 + A_3\left(\frac{c}{h}\right)^3 + A_4\left(\frac{c}{h}\right)^4 \qquad (24)$$

where the coefficients A_0, A_1, A_2, A_3, and A_4 have values $+1.99$, -2.47, $+12.97$, -23.17, $+24.8$ for the four-point beam and $+1.93$, -3.07, $+14.53$, -25.11, $+25.8$ for the three-point beam respectively.

The four-point single-edge notched beam was first used for pharmaceutical materials by Mashadi and Newton [5,6] and has since been adopted by York et al. [73]. In all cases large rectangular beams 100 mm long by 10 mm wide of varying height are prepared using the same punches and dies as those used to prepare specimens for the determination of Young's modulus. Notches of varying dimensions and profiles have been introduced by cutting either using a simple glass cutter [5,6] or a cutting tool fitted into a lathe [73]. The latter method has allowed notches of different profiles and dimensions to be accurately cut and investigated. While an arrowhead-type notch did appear to influence the measured value of K_{IC} for beams of microcrystalline cellulose, the effect was much reduced from straight-through notches and hence the latter were recommended [73].

The load required for failure of the specimens under tension is measured using the same testing rig as that described previously for the determination of Young's modulus. Loading rates of between 0.025 mm min^{-1} [5,6] and 100 mm min$^-$ [73] have been used—the latter workers noting a small rise (approx. 10%) in the measured K_{IC} of beams of microcrystalline cellulose for a 100-fold increase in applied loading rate.

As with the measurement of Young's modulus the four-point test requires large beams and consequently large amounts of material. In order to minimize the latter, especially for the measurement of the critical stress intensity factor for drugs under development, Roberts et al. [72] have developed a three-point test using specimen dimensions and testing rig similar to that described earlier for Young's modulus, although in this case a tensometer has been used to stress the specimen. Using beams of dimensions 20 mm long by 7 mm wide of varying height with two types of notches (a V notch induced by a razor blade pressed into the surface and a straight-through notch induced by a small saw blade), the authors were able to show equivalence with data generated by York et al. [73] for the four-point single-edge notched beam.

As can be seen from Fig. 8, specimen porosity has a significant effect on the measured K_{IC}. As porosity decreases, K_{IC} increases, indicating more resistance to crack propagation. It is obvious from the results in Fig. 8 that the relationship between K_{IC} and porosity is not linear as suggested by

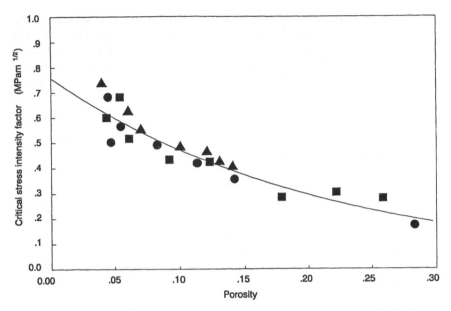

Fig. 8 The effect of porosity on the measured critical stress intensity factor for, ●, V notch; and ■, straight-through notch of Avicel PH-101 under three-point loading. ▲ = data ex York et al. [73] for same material under four-point loading. The line represents Eq. (26) for all points.

Mashadi and Newton [5,6,74]. In this respect York et al. [73] have investigated the application of both the two-term polynomial equation of the type

$$K_{IC} = K_{IC0}(1 - f_1 P + f_2 P^2) \tag{25}$$

and the exponential equation of the type

$$K_{IC} = K_{IC0} e^{-bP} \tag{26}$$

where

$$
\begin{aligned}
K_{IC0} &= \text{critical stress intensity factor at zero porosity} \\
K_{IC} &= \text{measured critical stress intensity factor of the specimen at} \\
&\quad \text{porosity } P \\
b, f_1, f_2 &= \text{constants}
\end{aligned}
$$

These equations are analogous to those used previously for Young's modulus. Both relationships give low standard errors and high correlation coefficients for microcrystalline cellulose from various sources. As with Young's modulus the exponential relationship is the preferred option for pharmaceutical materials [72].

A number of pharmaceutical materials have been measured using both three and four-point beam bending (Table 8). Of all the excipients tested, microcrystalline celluloses exhibited the highest values of K_{IC0}, real differences existing between the materials obtained from different sources. In addition there is also a particle size effect in that, for each of the three

Table 8 Critical Stress Intensity Factors Measured Using a Single-Edge Notched Beam

Material	Critical stress intensity factor ($MPam^{1/2}$)	Particle size (μm)	Reference
Drugs			
Ibuprofen	0.10	47	72
Aspirin	0.16	32	72
Paracetamol DC	0.25	120	72
Paracetamol	0.12	15	72
Sugars			
Lactose β-anhydrous	0.76	149	72
α-lactose monohydrate	0.35	20	72
Sucrose	0.22	74	72
Sorbitol instant	0.47	—	5
Celluloses			
Avicel PH-101	1.21	50	6
Avicel PH-102	0.76	90	73
Avicel PH-101	0.87	50	73
Avicel PH-105	1.33	20	73
Emcocel (90M)	0.80	90	73
Emcocel	0.92	56	73
Unimac (MG200)	0.67	103	73
Unimac (MG100)	0.80	38	73
Avicel PH-102	0.91	90	73
Avicel PH-101	0.99	50	73
Avicel PH-105	1.42	20	73
Emcocel (90M)	0.83	90	73
Emcocel	0.80	56	73
Unimac (MG200)	0.76	103	73
Unimac (MG100)	1.05	38	73
Avicel PH-101	0.76	50	72
Others			
Sodium chloride	0.48	20	72
Adipic acid	0.14	176	72

sources of material, the critical stress intensity factor increased with decreasing particle size. The sugars exhibit intermediate values with relatively low values for the drugs. It is interesting that anhydrous β-lactose has a critical stress intensity factor approximately twice that of α-lactose monohydrate (cf. indentation hardness measurements (Table 4)).

B. Double Torsion Testing

In this test the specimen is a rectangular plate (Fig. 9) with a narrow groove extending its full length supported on four hemispheres. The load is applied by two hemispheres attached to the upper platen. Controlled precracks are introduced in the specimen by preloading until a "pop-in" is observed (a pop-in is a momentary decrease in load and is an indication of crack growth). It should be noted that the groove is necessary to help guide the crack and ensure it remains confined within the groove itself. The critical

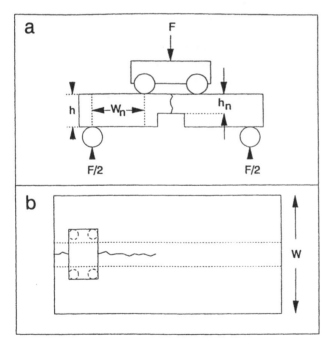

Fig. 9 Geometry for the double-torsion method showing (a) end view and (b) plan view for the measurement of critical stress intensity factor: h, plate thickness; W, plate width; h_n and W_n, distances as shown.

stress intensity factor is then calculated from the load F required to cause catastrophic crack propagation leading to failure by the expression

$$K_{IC} = FW_n \left[\frac{3(1 + \nu)}{Wh^3 h_n} \right]^{1/2} \tag{27}$$

where ν is Poisson's ratio and l, h, h_n, W, and W_n are the dimensions of the specimen given in Fig. 9.

The double-torsion method has only been used for microcrystalline cellulose (Avicel PH-101) and Sorbitol "instant" by Mashadi and Newton [74]. As with the single-edge notched-beam specimens, measurements varied with specimen porosity, and using linear extrapolation Mashadi and Newton [74] calculated values of K_{IC0} of 1.81 and 0.69 MPa·m$^{1/2}$ for the two materials respectively. The higher results obtained for theses two materials as compared to the data obtained for single-edge notched beams (1.21 and 0.47 MPa·m$^{1/2}$ respectively) have been explained in terms of the specific geometries and stress uniformity of the two techniques. However, it is known that values of K_{IC} determined from double-torsion techniques are generally greater than those from notched-beam specimens [75].

A specific practical problem of the double-torsion method for pharmaceutical materials is the preparation of the specimen and the very large compaction pressures needed to produce specimens of low enough porosity. For microcrystalline cellulose, Mashadi and Newton [74] were only able to produce specimens of greater than 25% porosity.

C. Radial Edge Cracked Tablets

All the techniques for the determination of the critical stress intensity factor so far described involve the preparation of compressed rectangular beams or plates requiring special punches and dies and some cases high-tonnage presses. The ideal specimen shape for pharmaceutical materials is the right-angled cylinder or a flat-faced tablet. Such a shape has recently been investigated by Roberts and Rowe [76] using microcrystalline cellulose as a model material. Two test methods were investigated, (a) edge opening (Fig. 10a), (b) diametral compression (Fig. 10b). Both the tests involve the introduction of a precrack into the edge of the disc. In edge opening the critical stress intensity factor is given by [77]

$$K_{IC} = \frac{F}{t} \left[\frac{c}{0.3557 \, (d - c)^{3/2}} + \frac{2}{0.9665(d - c)^{1/2}} \right] \left[\frac{c}{2d} \right]^{-1/2} \tag{28}$$

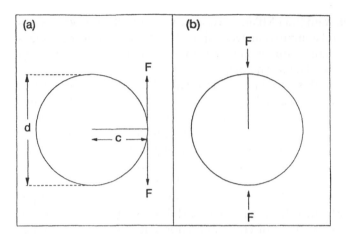

Fig. 10 Geometries for (a) edge opening and (b) diametral compression method for measuring critical stress intensity factors for radially edge cracked tablets: F, applied load; d, tablet diameter; c, crack length.

where F is the peak load for cracking as in Fig. 10a, while for diametral compression the critical stress intensity factor is given by [77]

$$K_{IC} = \frac{F}{dt} \left[\frac{\pi}{2c} \right]^{-1/2} \frac{1.586}{[1 - (c/d)]^{3/2}} \tag{29}$$

where F is the compressive force for cracking (Fig. 10b). In both cases c is the crack length, t is the tablet thickness, and d is the tablet diameter.

In testing microcrystalline cellulose (Avicel PH-101) Roberts and Rowe [76] first prepared tablets of varying porosity using 15mm flat-faced punches on an instrumented tablet press. After precracking using a scalpel blade the cracked tablets were stressed using either edge opening, in which tablets were gripped using the air jaws of a tensometer and pulled apart or diametrally compressed with the crack vertical between the platens of a tensometer. During extensive testing it was concluded that, of the two techniques, edge opening was the preferred option since it gave the most stable crack propagation and the effects of crack length were minimal. However, the diametral compression test was found to have some value provided crack lengths were limited to between c/d values of 0.34–0.6.

Extrapolation of the measured values of critical stress intensity factors of specimens over the porosity range 7–37% using linear, exponential (Eq. (26)), and two-term polynomial (Eq. (25)) gave values for edge opening of 1.91, 2.24, and 2.31 MPam$^{1/2}$, respectively, and for diametral compression 2.11, 2.98, and 2.35 MPam$^{1/2}$, respectively. In all cases values obtained from diametral compression were higher than those from edge opening. However, all values are significantly higher than those determined from measurements on beams and plates.

The reasons for this variation in the results from these different test procedures have been discussed by York et al. [73] specifically for microcrystalline cellulose. Basically the problem lies in the difficulties in the introduction of a two-dimensional sharp crack into a specimen and accurately measuring its length and velocity on the application of load. Ideally K_{IC} should be independent of crack length and for those materials which exhibit flat crack growth resistance curve all methods of measurement should produce equivalent data for K_{IC0}. However, many materials, especially ceramics [78], have been shown to exhibit rising crack resistance curves and hence the method of crack induction and notch geometry become critical. For these measurements of specimens with sawn or machined notch will always produce lower values of K_{IC} than those where the crack is introduced by a controlled flaw. This is the case for the double-torsion method and radially edge-cracked tablet since in these methods the total amount of crack extension at maximum load is always higher. This is direct evidence that microcrystalline cellulose has a rising crack growth resistance curve [76].

D. Vickers Indentation Fracture Test

The most distinctive feature of the indentation of brittle materials by the Vickers or pyramidal indenter is the appearance of cracks emanating from the corners of the indent (Fig. 11), and it is from the measurement of the lengths of these cracks that it is possible to determine the critical stress intensity factor for indentation cracking or K_c.

Before describing the technique it is important to realize the assessment of K_c is strictly semiempirical because, firstly, the analytical solutions of the stress field around indentations have not been solved and only approximate solutions have been derived and secondly, the deformation field is not homogeneous and anisotropy and fracture complicate the problem.

Only two papers have examined the indentation fracture test as a means of determining the critical stress intensity factor of pharmaceuticals [23,24]. In the first of these Duncan-Hewitt and Weatherly [23] evaluated the indentation test using sucrose crystals. Microindentation was performed using a Leitz-Wetzlar Miniload hardness tester (Vickers pyramidal diamond indenter) applying loads of 147 mN, with the indentations and cracks measured using a light microscope (Leitz). Large crystals were prepared (1–4 mm diameter) by slow evaporation of saturated aqueous solution at 23°C

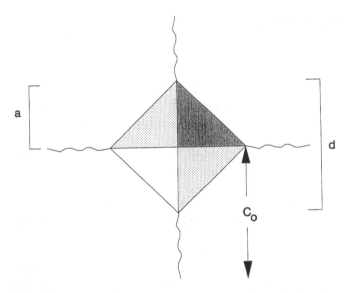

Fig. 11 Schematic diagram showing cracking around a Vickers indent: C_0, crack length; d, length of a diagonal of indent; a, length of half-diagonal indent.

over three to six months. The prismatic crystals were washed in ethanol to remove traces of crystallization solution and were stored under controlled conditions before testing. Furthermore some specific crystal faces [(100), (010), and (001)] were prepared by either abrading with decreasing grades of emery paper or by cleavage. Crystals were mounted in plasticine prior to testing. The authors found that fracture appeared anisotropic and that if crystals were tested immediately after polishing, fracture was either suppressed or significantly decreased. Duncan-Hewitt and Weatherly [23] used two equations to calculate the critical stress intensity factor:

$$K_c = \Phi \left(\frac{E}{H_v} \right)^{1/2} \frac{F}{C_0^{3/2}} \tag{30}$$

as derived by Antis et al. [79] and

$$K_c = \Phi \left(\frac{E}{H_v} \right)^{2/3} \left(\frac{F}{C_0^{3/2}} \right) \left(\frac{\alpha}{l} \right)^{1/2} \tag{31}$$

as derived by Laugier [80], where

  Φ = calibration constant equivalent to 0.016 and 0.0143 in Eqs. (30) and (31) respectively

C_0 = crack length (as in Fig. 11)
l = $C_0 - a$
F = applied load
E = Young's modulus
H_v = Vickers hardness

The authors found that the values of K_c using the two equations gave similar values (the means for the various faces were 0.078 $MPa\dot{m}^{1/2}$ and 0.089 $MPa\dot{m}^{1/2}$ for Eqs. (30) and (31) respectively). Furthermore they reported that Eq. (31) [80] appeared to emphasize the apparent fracture anisotropy. The fracture plane with the lowest value was the (100) in agreement with the easiest to cleave plane (although the (101) has the lowest K_c value) and the plane with the greatest K_c was the (001) plane. The (100) plane has the hardest surface, whereas the (001) plane is the softest [23].

In their most recent paper Duncan-Hewitt and Weatherly [24] published further indentation critical stress intensity factors calculated from the Antis et al. [79] relationship (Eq. (30)). These are shown in Table 9 with the corresponding data from single-edge notched-beam (three-point) beam testing [72]. Although the rank order is the same, the differences in magnitude of the values are large (with the exception of sodium chloride). A possible explanation for these differences is that in the indentation test the theory is not exact and the equations are derived from calibration with ceramics, i.e., materials with considerably different plastoelastic properties than pharmaceutical materials.

Despite its shortcomings the indentation technique has certain advantages over other testing methods since it can be used on small samples, e.g., single crystals; specimen preparation is relatively simple, and the indentation hardness and Young's modulus can be determined simultaneously.

Table 9 Comparison of Critical Stress Intensity Factors Measured Using Indentation (K_c[24]) and Single-Edge Notched Beams (K_{IC0} [72])

Material	K_c(MPa $\dot{m}^{1/2}$)	K_{IC0}(MPa $\dot{m}^{1/2}$)
Sodium chloride	0.50	0.48
Sucrose	0.08	0.22
Paracetamol	0.05	0.12
Adipic acid	0.02	0.14

V. EFFECT OF PARTICLE SIZE—BRITTLE/DUCTILE TRANSITIONS

It has been known for many years that if Heckel plots were constructed for a material of varying particle size the reciprocal of the gradient over the central linear portion of the graph (now defined as the deformation stress, σ_d) varied, either remaining constant or increased as particle size decreased [46,66]. Extensive study [67,48,47] has shown that the effect of particle size is even more complex depending on the material under test (Fig. 12). For a material known to undergo plastic deformation (e.g., microcrystalline cellulose) no effect of particle size could be seen; for a material known to undergo brittle fracture (e.g., dolomite), the deformation stress increased with decreasing particle size, and for materials known to undergo a combi-

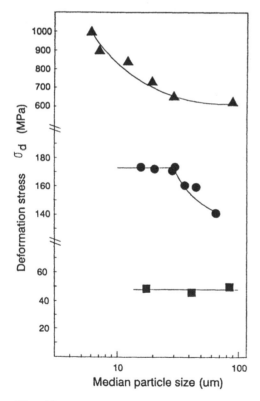

Fig. 12 The effect of particle size on the deformation stress as measured using Heckel plots for ▲, dolomite; ●, α-lactose monohydrate; ■, microcrystalline cellulose. (Adapted from Roberts and Rowe [48].)

nation of brittle fracture and plastic deformation (e.g., α-lactose monohydrate) the deformation stress increased with decreasing particle size to a plateau value.

This is as would be expected from theory (eqs. (2) and (3)) as shown in Fig. 13. The transition from brittle to ductile behavior will occur at a critical size d_{crit} given by the equation [47,48]

$$d_{crit} = \left(\frac{AK_{IC}}{\sigma_y} \right)^2 \tag{32}$$

where $A = 3.27$, for compression of rectangular samples with large cracks [1]. Experiments on the compression of sodium chloride has confirmed the applicability of this equation in tabletting, in that particles above 33 μm in diameter fractured under compression, while particles below this size did not show any fragmentation but deformed plastically under compression [47].

Calculated critical sizes (Eq. (32)) for a variety of excipients and drugs using yield stress data and critical stress intensity factors given in previous tables are shown in Table 10. It can be seen that critical particle sizes can vary over several orders of magnitude. The values appear reasonable in the light of experimental findings for microcrystalline cellulose and

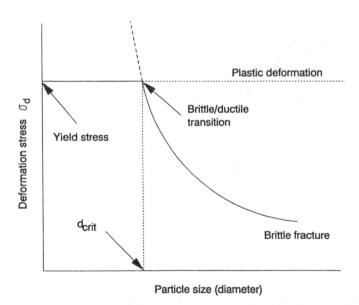

Fig. 13 Schematic diagram showing the effect of particle size on the deformation stress for materials that undergo brittle fracture and/or plastic deformation.

Table 10 Values of Indentation Hardness (H), Critical Stress Intensity Factors (K_{IC0}, from Single-Edge Notched Beams [72]) and Critical Particle Size (d_{crit}) for Some Excipients and Drugs

Material	H (MPa)	K_{IC0} (MPa m$^{1/2}$)	d_{crit} (μm)
Microcrystalline cellulose	168 [37]	0.7569	1949
Lactose β-anhydrous	251 [36]	0.7597	873
Ibuprofen	35 [36]	0.1044	854
Aspirin	87 [20]	0.1561	309
Paracetamol DC	265 [37]	0.2463	83
α-Lactose monohydrate	515 [36]	0.3540	45
Sucrose	645 [24]	0.2239	12
Paracetamol	421 [24]	0.1153	7

α-lactose monohydrate in Fig. 12. The value for microcrystalline cellulose is similar to that determined for other polymeric materials [1]. The differences in the two values for the two lactoses are consistent with the findings of workers describing their compaction properties [19]. Of the three drugs paracetamol has the lowest critical particle size consistent with it being very brittle. However, the addition of a polymeric binder to paracetamol (i.e., Paracetamol DC) increases its critical particle size by an order of magnitude, causing it to become plastic in nature.

VI. EFFECT OF MOISTURE CONTENT

The moisture content of a material can affect its mechanical properties generally by acting as an internal "lubricant" facilitating slippage and plastic flow [81]. Obviously this will lead to a decrease in both yield stress and Young's modulus of elasticity as adequately demonstrated for microcrystalline cellulose [82,8]. In the case of yield stress, increasing the moisture content caused a linear decrease independent of the source and batch of the material under test (Fig. 14). Similar results have been recorded for paracetamol [83].

The effect of moisture content on the critical stress intensity factor is more complex. Bassam et al. [8] have shown that for microcrystalline cellulose there is a decrease with increasing moisture content; i.e., the material became less resistant to crack propagation. Similar results have been found for glass by Wiederhorn [84], who suggested that the effect was caused by water interacting at the crack tip.

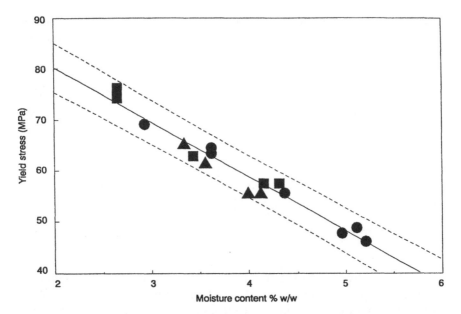

Fig. 14 The effect of moisture content on the yield stress of microcrystalline cellulose for ●, Avicel; ■, Unimac; ▲, Emocel. (Adapted from Roberts and Rowe [82]). The dotted line represents the 95% confidence limits for the fit of the data points.

VII. PREDICTION OF COMPACTION BEHAVIOR

At present a fundamental knowledge and theory describing the compaction of particulate solids, specifically pharmaceuticals, is lacking. While it is generally recognized that the mechanical properties of such materials are critical in directing compaction behavior, a predictive capability is absent. In a recent paper Roberts and Rowe [56] have presented a pragmatic approach to predicting the consolidation mechanism of pharmaceutical materials based on a knowledge of Young's modulus of elasticity, yield stress, hardness, and strain rate sensitivity (Fig. 15). This approach combined with further measurements on critical stress intensity factor enabling the prediction of critical particle sizes, d_{crit} (to account for particle size effects), will enable prediction of the consolidation mechanism of any material of known particle size. This knowledge will allow tablet formulators to take a more scientific approach to formulation. Indeed a combination of a predictive approach based on mechanical property measurements combined with heuristics should enable the development of quality expert systems for tablet formulation.

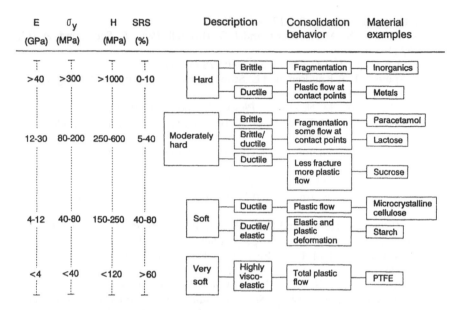

Fig. 15 Relationship between material properties and compaction behavior. (Adapted from Roberts and Rowe [56].)

REFERENCES

1. K. Kendall, *Nature 272*: 710 (1978).
2. K. E. Puttick, *J. Phys. D: Appl. Phys. 12*: 2249 (1980).
3. M. S. Church, Mechanical characterisation of pharmaceutical powder compacts, Ph.D. thesis, Nottingham, 1984.
4. R. J. Roberts, R. C. Rowe, and P. York, *Powder Technol. 65*: 139 (1991).
5. A. B. Mashadi and J. M. Newton, *J. Pharm. Pharmacol. 39*: Suppl., 67P (1987).
6. A. B. Mashadi, and J. M. Newton, *J. Pharm. Pharmacol. 39*: 961 (1987).
7. F. Bassam, P. York, R. C. Rowe, and R. J. Roberts, *J. Pharm. Pharmacol. 40*: Suppl., 68P (1988).
8. F. Bassam, P. York., R. C. Rowe, and R. J. Roberts, *Int. J. Pharm. 64*: 55 (1990).
9. F. Bassam, P. York., R. C. Rowe, and R. J. Roberts, *Powder Technol. 65*: 103 (1991).
10. R. J. Roberts, R. C. Rowe, and P. York, *J. Pharm. Pharmacol. 41*: Suppl., 30P (1989).
11. R. J. Roberts, The elasticity, ductility and fracture toughness of pharmaceutical powders, Ph.D. thesis, Bradford, 1991.
12. R. J. Roberts, and R. C. Rowe, *Int. J. Pharm. 37*: 15 (1987).
13. E. A. Dean, and J. A. Lafez, *J. Amer. Ceram. Soc. 66*: 366 (1983).

14. J. C. Wang, *J. Mater. Sci. 19*: 801 (1984).
15. K. Kendall, M. McN. Alford, and J. D. Birchall, *Proc. R. Soc. Lond. A 412*: 269 (1987).
16. J. M. Spriggs, *J. Am. Ceram. Soc. 44*: 628 (1961).
17. S. Spinner, F. P. Knudsen, and L. Stone, *J. Res. Nat. Bur. Stand. (Eng. Instr.) 67C*: 39 (1963).
18. J. T. Fell, and J. M. Newton, *J. Pharm. Sci. 60*: 1866 (1971).
19. H. Vromans, A. H. De Boer, G. K. Bolhuis, and C. F. Lerk, *Drug Dev. Ind. Pharm. 12* 1715 (1986).
20. K. Ridgway, E. Shotton, and J. Glasby, *J. Pharm. Pharmacol. 21*: Suppl., 19S (1969).
21. J. C. Kerridge, and J. M. Newton, *J. Pharm. Pharmacol. 38:* Suppl., 78P (1986).
22. K. Ridgway, M. E. Aulton, and P. H. Rosser, *J. Pharm. Pharmacol. 22*: 70S (1970).
23. W. C. Duncan-Hewitt, and G. C. Weatherly, *Pharm. Res. 6*: 373 (1989).
24. W. C. Duncan-Hewitt, and G. C. Weatherly, *J. Mat. Sci. Lett. 3*: 1350 (1989).
25. E. N. Hiestand, J. M. Bane, Jnr., and E. P. Strzelinski., *J. Pharm. Sci. 60*: 758 (1971).
26. E. Breval, and N. H. Macmillan, *J. Mater. Sci. Lett. 4*: 741 (1985).
27. G. Simmons, and H. Wang, *Single Crystal Elastic Constants and Calculated Aggregate Properties: A Handbook*. M.I.T., Cambridge, 1971.
28. M. E. Aulton, *Manufact. Chem. Aerosol News 48*: 28 (1977).
29. E. N. Hiestand, *J. Pharm. Sci. 74*: 768 (1985).
30. D. Tabor, *The Hardness of Metals*, Clarendon Press, Oxford, p. 6, 1951.
31. R. Smith, and G. Sandland, *Iron Steel Inst. 1*: 285 (1925).
32. K. Ridgway, J. Glasby, and P. H. Rosser, *J. Pharm. Pharmacol. 21*: 24S (1969b).
33. M. E. Aulton, and H. G. Tebby, *J. Pharm. Pharmacol. 27*: Suppl., 4P (1975).
34. M. E. Aulton, H. G. Tebby, and P. J. P. White, *J. Pharm. Pharmacol. 26*: Suppl., 59P (1974).
35. M. E. Aulton, *Pharm. Acta Hel. 56*: 133 (1981).
36. H. Leuenberger, *Int. J. Pharm. 12*: 41 (1982).
37. W. Jetzer, H. Leuenberger, and H. Sucker, *Pharm. Tech. 7*: April, 33 (1983).
38. W. Jetzer, H. Leuenberger, and H. Sucker, *Pharm. Tech. 7*: Nov., 33 (1983).
39. W. E. Jetzer, W. B. Johnson, E. N. Hiestand, *Int. J. Pharm. 26*: 329 (1985).
40. H. Leuenberger, and B. D. Rohera, *Pharm. Acta Helv. 60*: 279 (1985).
41. B. Galli, and H. Leuenberger, *VDI Ber. 583*: 173 (1986).
42. M. E. Aulton, and I. S. Marok, *Int. J. Pharm. Tech. Prod. Mfr. 2*: 1 (1981).
43. J. Ichikawa, K. Imagawa, and N. Kaneniwa, *Chem. Pharm. Bull. 36*: 2699 (1988).
44. R. W. Heckel, *Trans. Metall. Soc. A.I. M. E. 221*: 671 (1961).
45. R. W. Heckel, *Trans. Metall. Soc. A.I.M.E. 221*: 1001 (1961).
46. J. A. Hersey, and J. E. Rees, *Particle Size Analysis* (M. J. Groves and J. L. Wyatt-Sargent, eds.), The Society for Analytical Chemistry, London, p. 33, 1972.

47. R. J. Roberts, R. C. Rowe, and K. Kendall, *Chem. Eng. Sci. 44*: 1647 (1989).
48. R. J. Roberts, and R. C. Rowe, *Int. J. Pharm. 36*: 205 (1987).
49. P. Paronen, *Pharmaceutical Technology: Tabletting Technology*, Vol. 1 (M. H. Rubinstein, ed.), Ellis Horwood, Chichester, p. 139, 1987.
50. P. York, *J. Pharm. Pharmacol. 31*: 244 (1979).
51. K. Danjo, C. Ertell, and J. T. Cartensen, *Drug Dev. Ind. Pharm. 15*: (1989).
52. H. DeBoer, G. K. Bolhuis, and C. F. Lerk, *Powder Technol. 20*: 75 (1978).
53. G. Ragnarsson, and J. Sjogren, *Acta. Pharm. Suec. 21*: 141 (1984).
54. J. E. Rees, and P. J. Rue, *J. Pharm. Pharmacol. 30*: 601 (1978).
55. R. J. Roberts, and R. C. Rowe, *J. Pharm. Pharmacol. 37*: 377 (1985).
56. R. J. Roberts, and R. C. Rowe, *Chem. Eng. Sci. 42*: 903 (1987).
57. S. D. Bateman, M. H. Rubinstein, R. C. Rowe, R. J. Roberts, P. Drew, and A. Y. K. Ho, *Int. J. Pharm. 49*: 209 (1989).
58. M. S. H. Hussain, P. York, and P. Timmins, *Int. J. Pharm. 70*: 103 (1991).
59. P. Humbert-Droz, D. Mordier, and E. Doelker, *Pharm. Acta Helv. 57*: 136 (1982).
60. P. Humbert-Droz, R. Gurny, D. Mordier, and E. Doekler, *Int. J. Pharm. Tech. Prod. Mfr. 4*: 29 (1983).
61. F. von Podczeck, and U. Wenzel, *Pharm. Ind. 51*: 542 (1989).
62. M. Duberg, and C. Nystrom, *Powder Technol. 46*: 67 (1986).
63. R. Ramberger, and A. Burger, *Powder Technol. 43*: 1 (1985).
64. S. D. Bateman, M. H. Rubinstein, and P. Wright, *J. Pharm. Pharmacol. 39*: Suppl. 66P (1987).
65. H. Vromans, and C. F. Lerk, *Int. J. Pharm. 46*: 183 (1988).
66. P. York, *J. Pharm. Pharmacol. 30*: 6 (1978).
67. R. J. Roberts, and R. C. Rowe, *J. Pharm. Pharmacol. 36*: 567 (1986).
68. O. Ejiofor, S. Esezobo, and N. Pilpel, *J. Pharm. Pharmacol. 38*: 1 (1986).
69. G. Ragnarsson, and J. Sjogren, *J. Pharm. Pharmacol. 37*: 145 (1985).
70. W. F. Brown, and J. E. Srawley, *ASTM Special Tech. Publ.* No. 410 (1966).
71. British Standards Institution, *Plane Strain Fracture Toughness (K_{IC}) of Metallic Materials*, BS 5447, BSI, London, 1977.
72. R. J. Roberts, R. C. Rowe, P. York, *Int. J. Pharm. 91*: 173 (1993).
73. P. York, F. Bassam, R. C. Rowe, and R. J. Roberts, *Int. J. Pharm. 66*: 143 (1990).
74. A. B. Mashadi, and J. M. Newton, *J. Pharm. Pharmacol. 40*: 597 (1988).
75. A. G. Evans, *Fracture Mechanics of Ceramics, Vol. 1: Concepts, Flaws and Fractography* (R. C. Bradt, D. P. H. Hasselman, and F. F. Lange, eds.), Plenum Press, New York, p. 17, 1974.
76. R. J. Roberts, and R. C. Rowe, *Int. J. Pharm. 52*: 213 (1989).
77. K. Kendall, and R. D. Gregory, *J. Mater. Sci. 22*: 4514 (1987).
78. D. Munz, in *Fracture Mechanics of Ceramics, Vol. 6: Measurements, Transformations, and High-Temperature Fracture* (R. C. Bradt, A. G. Evans, D. P. H. Hasselman, and F. F. Lange, eds.), Plenum Press, New York, p. 1, 1983.
79. G. R. Antis, P. Chantikul, B. R. Lawn, and D. B. Marshall, *J. Am. Ceram. Soc. 64*: 533 (1981).

80. M. T. Laugier, *J. Mater. Sci. Lett. 6*: 355 (1987).
81. K. A. Khan, P. Musikabhumma, and J. P. Warr, *Drug Dev. Ind. Pharm. 7*: 525 (1981).
82. R. J. Roberts, and R. C. Rowe, *J. Pharm. Pharmacol. 39:* Suppl., 70P (1987).
83. J. S. M. Garr, and M. H. Rubinstein, *Int. J. Pharm. 81*: 187 (1992).
84. S. M. Wiederhorn, *J. Am. Ceram. Soc. 50: 407* (1967).

12

Granule Properties

Göran Alderborn

Uppsala University, Uppsala, Sweden

Martin Wikberg

Kabi Pharmacia Therapeutics Uppsala, Uppsala, Sweden

I. COMPACTIBILITY OF GRANULATED POWDERS

The traditional way of producing tablets normally involves two size enlargement processes in sequence, i.e., a granulation of the fine particulate drug, often mixed with a filler, followed by the compaction of the granulated powder. The rationale for granulating the powdered drug before tabletting is normally technological. The relationship between bulk density and flowability of a powder mass and the size of the particles constituting the powder is well known and thus, by the aid of a granulation procedure, these characteristics can be improved. However, a granulated powder often exhibits suitable compaction characteristics and powder granulation is thus performed also to ascertain a good compactibility of the tablet mass, i.e., the ability of the powder to be compacted into a compact of a certain mechanical strength [1].

The mechanical strength of a compact is a function of the properties of the material and the treatment of the material during the compaction process, i.e., applied pressure and the time events involved in the tablet formation process. When powders consisting of primary particles are compacted, the mechanical properties of the particles, such as their deformation properties and fragmentation propensity, will be of decisive importance for the mechanical strength of the specimen produced (see Chapters 10 and 11). Also when granules are compacted, the mechanical characteristics of the pri-

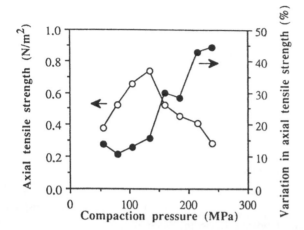

Fig. 1 The axial tensile strength and variability in axial tensile strength of compacts of granules of paracetamol/gelatin (98/2% wt.) as a function of compaction pressure. The reduced strength and increased strength variability of the compacts indicate lamination of the compact (Drawn from data given by Alderborn and Nyström, ref. [2]).

mary particles will probably affect the compactibility of the mass. For example, it is a common experience that capping prone material will show capping tendencies also in the granulated form of the powder (Fig. 1) [2]. However, it has been demonstrated that the compactibility of a granulated powder can be affected by other factors related to the granulation procedure, such as choice of granulation method (Fig. 2) [3,4] and process conditions during the granulation (Fig. 3) [5,6]. For a granulation consisting of a given type of substrate material and a binder, all these factors can be classified into two ways for controlling the compactibility of the granulated powder:

1. Changing the type or amount of binder (normally an amorphous polymer) in the granulation
2. Changing the physical properties of the granules by variations in the process conditions during the granulation procedure or by changing the dimensions of the substrate particles

In practice, formulations are often optimized with respect to their compactibility by the first approach. However, the second approach of optimizing the compaction behaviour is interesting because it allows the design of tablet masses with more limited amounts of excipients and, thus, the production of high-quality tablets at lower cost. Consequently, the pharmaceutical scientist has the possibility to optimize the compaction char-

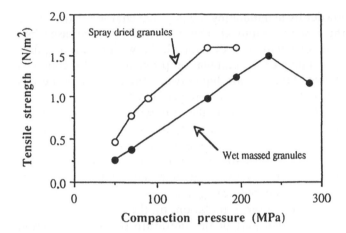

Fig. 2 The tensile strength of compacts of granules of paracetamol and a binder (93/7 % wt.) as a function of compaction pressure. The granules were produced by two different granulation methods. (Drawn from data given by Rue et al., ref. [3].)

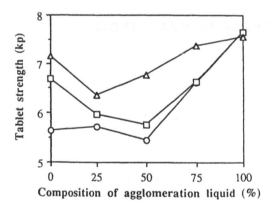

Fig. 3 The effect of the composition of the agglomeration liquid (percentage amount of ethanol in ethanol/water mixtures) on the diametral compression strength of compacts of granules of acetylsalicylic acid/polyvinylpyrrolidone. Compaction pressures: circles = 150 MPa; squares = 225 MPa; triangles = 300 MPa. (Drawn from data given by Wells and Walker, ref. [5].)

acteristics of the powder by manipulating the physical characteristics of the drug particles (in this case the granules). This is normally not the case when a direct compactible formulation is developed. However, this requires firstly, a good understanding of the granulation procedure and secondly, a fundamental knowledge of the relationship between the physical properties of granules and their performance while being compacted.

Although the extensive use of granulated powders in tablet production and, as a consequence of this, the large number of publications on the compactibility of granulated materials, there is still a lack of such fundamental knowledge. The primary reason for this is that there seems to have been few attempts to systematically investigate and understand the relationship between formulation and physical granule properties, on the one hand, and the compactibility of the granulation, on the other.

A survey of a substantial part of the literature on the compaction characteristics of granulated powders has recently been presented by Wikberg [7], and such an overview will not be given here. The aim in this article is to focus on the importance of the physical granule properties for the compactibility of a granulation. This discussion will be based on a selected fraction of papers from the literature.

II. MECHANICAL PROPERTIES OF PHARMACEUTICAL GRANULES

Granules are formed by an aggregation of primary particles into secondary particles. The physical properties of such granules are related to the principle of particle size enlargement and the actual process condition used during the granulation. Today, a broad range of particle size enlargement methods exists and a number of these have been utilized in the production of pharmaceutical granulations. However, the most frequently used granulation methods is based on the formation of secondary particles by the aid of a liquid phase during agitation of particles by convection or by fluidization. The stresses applied to the particles during such agglomeration procedures are probably not high enough to cause deformation or fracturing of the particles. Thus, pharmaceutical granules can be described as a cluster of discrete particles (Fig. 4) and the size and shape characteristics of these primary particles are similar to the original particle properties and therefore definable. An exception is when granules are formed by slugging when marked deformation and fragmentation of the primary particles probably occur during the preparation of the granules.

As a consequence of the technological requirements of a granulation, the most critical physical characteristics of granules for their performance

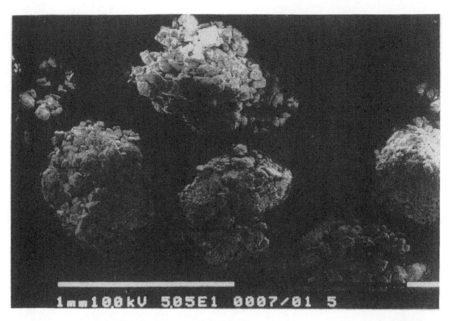

Fig. 4 S.E.M. photomicrograph of a granule of lactose/polyvinylpyrrolidone (95/5 % wt.) produced by wet agglomeration with ethanol as agglomeration liquid. Longer white bar corresponds to a length of 1000 μm.

when processed are their dimensions, porosity, and mechanical strength. These characteristics affect the flowability of the powder, the bulk density, and the ability of the granules to withstand stresses (which can cause fracturing or attrition of the granules) during processing before the actual compaction process (e.g., mixing and transportation of the powders). However, these physical characteristics can also be of importance for the ability of the granules to cohere into compacts.

The porosity of the granules is a reflection of the packing of the primary particles constituting the granules. The porosity is in practice affected by the size of these primary particles as well as the conditions used during the wet agglomeration [8]—i.e., the combined effect of the stress applied to the granules during their formation and the interactions between the primary particles. Generally, a decreased primary particle size tends to increase the granule porosity and an increased degree of liquid saturation and agitation intensity tend to decrease the granule porosity.

The theoretical basis for the prediction of the mechanical strength of an aggregate consisting of defined particles has been presented by Rumpf, who developed the following expression [9] for the tensile strength of an aggregate of monodispersed spheres:

$$\sigma = \frac{9\,(1 - e)kH}{8\pi d^2} \tag{1}$$

where

σ = tensile strength of aggregate (N/m^2)
e = porosity of aggregate (—)
k = coordination number, i.e., number of contact points for one sphere (—)
H = cohesion force at a point of contact (N)
d = diameter of sphere (m)

By assuming that the coordination number is expressed by the ratio π/e [9], the equation is simplified to

$$\sigma \approx \left(\frac{1 - e}{e} \right) \frac{H}{d^2} \tag{2}$$

Although the expression is based on an idealized situation, it can be concluded that the main factors contributing to the strength of an aggregate are the granule porosity and the size of the primary particles. According to the physical character of a pharmaceutical granule described above, it seems reasonable to assume that for a granule which does not include a binder, granule porosity and primary particle size will govern the tensile strength.

According to our experience, few studies in the literature have specifically focused on strength-porosity profiles of pharmaceutical granules irrespective of whether the granules include a binder. In the following discussion, two examples of such a relationship will be discussed.

In Fig. 5, the fracture force-porosity profile for a series of granules of microcrystalline cellulose is shown [10]. Mainly for experimental reasons, the mechanical strength was assessed by diametral compression and the force needed to fracture the granule was used as a measure of the granule strength. The granules did not include a binder and were produced by extrusion-spheronization and generally showed a nearly spherical shape. A reduced porosity increased the fracture force of the granules in an almost linear relationship. Thus, the relationship is generally consistent with Eq. (2), although the physical structure of the granules, with regard to the granule pore structure, varies in a more complex and unpredictable way with the porosity for real granules, compared to idealized granules.

The dominating mechanisms of particle-particle bonds in granules without binders are probably intermolecular attraction forces and solid bridges.

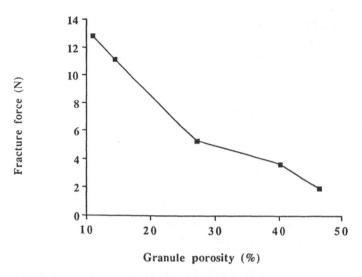

Fig. 5 The effect of granule porosity on the fracture force of spherically shaped microcrystalline cellulose granules produced by extrusion and spheronization. (Drawn from data given by Johansson et al., ref. [10])

However, a binder is normally included in pharmaceutical granules for two reasons: Firstly, in order to secure a sufficient mechanical strength of the granules during processing before the actual compaction procedure, and, secondly, in order to increase the binding capacity of the granules during compaction. The binder is normally mixed with the substrate particles as a relatively viscous liquid, i.e., dissolved in a solvent, although it can also be added as a dry powder. Both the type and the relative amount of binder as well as the procedure used for adding the binder to the substrate particles will affect the mechanical strength of the granules and several reports on these issues exist in the literature ([7],[11]). However, there is still a lack of fundamental understanding of how different binders affect granule strength, although factors such as the interaction between binder and substrate and the mechanical properties of the binder have been addressed.

In an idealized situation, the binder forms a film layer around each substrate particle within the granule. The binder layers can at points be in contact with each other and thus form a binder bridge. However, interactions between binder layers can probably also occur over a certain distance and contribute to the strength of the interparticulate attractions in the granules. When such a granule is stressed, a failure, i.e., a shearing of primary particles or a separation of particles from each other, can be localized to three different places within the granule:

1. At the interface between the substrate and the binder (i.e., an adhesive failure)
2. At the binder-binder interface (i.e., a cohesive failure)
3. Across a binder bridge (i.e., a cohesive failure)

Based on these types of failure, the mechanisms of attraction between particles in a granule can be described as either intermolecular attraction forces (failure 1 and 2) or solid bridges (failure 3). Consequently, the relationship between the strength and the porosity of substrate-binder granules might be dependent on the localization of the failure within the granule and is thus difficult to predict.

The question of the localization of the failure has been the object of examination in the literature. Cutt et al. [12] studied the failure characteristics of granules by SEM examination of fractured granules of glass with different surface properties (hydrophilic and hydrophobic) granulated with various binders. They observed fractures both across binder bridges (primarily for hydrophilic glass beads) as well as at the interface between the binder and the substrate particle (primarily for hydrophobic glass beads). They attributed the localization of the failure to the relative strength of the cohesive and adhesive forces acting within the granules. This in turn is probably related to the dimensions of the binder bridge formed and the relative surface energetics of the binder and the substrate. They also stated that the hydrophilic glass beads gave granules of higher strength, which might be due to the difference in localization of the failure (Table 1). However, the porosity of the granules in relation to strength was not discussed.

Mullier et al. [13] also studied the localization of the failure of granules subjected to stresses. For granules of glass beads and polyvinyl-pyrrolidone, they observed that a failure at the binder-substrate interface predominated when the granules fractured. However, for granules of sand and PVP, the cohesive bridge failure seemed to predominate. They attributed this difference in relationship to the glass particles to the difference in

Table 1 Friability and Fracture Strength of Granules of Glass Beads and Polyvinylpyrrolidon (2.72% w/w)[a]

Glass surface	Granule friability (%)	Granule strength (g)
Hydrophilic	5.18	202
Hydrophobic	13.8	115

[a]The glass beads possessed different surface characteristics.
Source: From Ref. 12.

surface geometry of the substrate particles; i.e., an irregular particle surface promoted the strength of the adhesive binder-substrate interaction.

Wikberg and Alderborn [14] studied the mechanical strength, assessed by diametral compression, of a series of granules consisting of a given combination of a substrate and a binder. Two substrate materials were used in the study (lactose and dipentum), characterized by a difference in substrate particle size and compaction behavior. For both substrate materials used, the granules was considered to respond to the applied force as brittle specimens; i.e., no ideal plastic flow region nor a marked deformation of the granules before fracture was observed. For the lactose granules, a reduced porosity generally increased the granule strength (Fig. 6). The relationship tended to be exponential; i.e., at the upper part of the studied porosity range the effect of porosity was small, while at the lower a marked increase in strength was observed.

For the dipentum granules (Fig. 6), only a slight tendency to an increased granule strength with reduced porosity was observed. A possible explanation for this is that for these granules, the porosity range obtained corresponds to the part of the strength-porosity profile where porosity variations only affect the strength to a limited degree. One can also observe that the dipentum granules generally were of higher strength, probably due

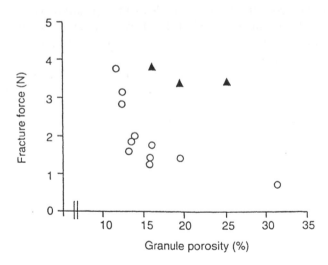

Fig. 6 The effect of granule porosity on the fracture force of granules of lactose/polyvinylpyrrolidone, 95/5 % wt. (cirles) and dipentum/polyvinylpyrrolidone, 94/6 % wt. (triangles). The granules were produced by wet agglomeration. (From ref. [14] with permission of the copyright holder.)

to the lower size of the primary particles of this substrate. This is consistent with the theoretical Eq. (2) as well as with other reports [15] on the strength of lactose granules.

Based on this limited number of publications, the literature also indicates that for binder-substrate granules a reduced porosity and substrate particle size increase the fracture strength of granules. Both these changes in the characteristic of a granule will probably increase the number of bonding points between the primary particles. Thus, a reasonable explanation of this observation is that the granules used in the studies failed by a rupturing of the cohesive binder-binder interactions. The function of the binder will thus be to form a layer around the substrate particles and increase the binding capacity between the substrate particles. However, it is also possible that for substrate-binder granules, failure predominates at the adhesive substrate-binder interface. If this interface constitutes the weakest part of the granule, it seems reasonable that a granule strength more-or-less independent of the granule porosity can be obtained.

As stated above, the compression strength analysis of a series of pharmaceutical granules [14] indicated that these granules could be described as brittle specimens. However, the nature of granules (i.e., a specimen consisting of particles surrounded by a network of pores) indicates that granules may be prone to deform when stressed. It seems reasonable that the primary particles constituing the granules can reposition when subjcted to an external stress which can affect the shape characteristics of the granule (deformation) as well as the granule porosity. In an attempt to assess this propensity of the granules to deform, Wikberg and Alderborn [14] quantified the nearly linear force-displacement curves obtained from the diametral compression measurements of the dipentum granules by calculating the slopes of the profiles (Table 2) (compare also [11]). An increased porosity increased the displacement of the cross-head per applied

Table 2 Deformability, as Measured by the Slope of the Force-Displacement Profile (interquartile range within brackets) During Diametral Compression, of Granules of Different Porosities

Granule porosity (%)	Granule deformability (kN/m)
25.1	69.8 (9.60)
19.4	87.3 (15.2)
15.9	109 (16.1)

Source: From Ref. 14.

force unit which indicates that the porosity of the granules affects their propensity to deform. It seems reasonable that a reduced granule porosity will reduce the void space between the primary particles and, hence, increase the bonding force of the interparticulate attractions. Both the change in interparticulate void space and in the bonding force will probably affect the propensity for densification and deformation of the granule and the stress needed to cause such effects.

The resistance of the granules toward attrition is also of importance during handling and processing of the granules. For the series of lactose granules used in the study in the relationship between granule fracture strength and porosity [14], Wikberg and Alderborn also assessed their resistance to attrition. The friability did not correlate with the porosity or the fracture force of the granules (Fig. 7); i.e., the fracture force and attrition strength of granules are not necessarily related to each other. A probable explanation is that the attrition strength is a mechanical characteristic dependent to a larger extent, compared to the fracture strength, on the shape properties of the granules, primarily the surface geometry. It seems reasonable that particles or small clusters of particles can more easily be abraded from the surface of a granule when the irregularity of the surface increases. It has also earlier been shown [11] that spherical granules possessed a lower friability, compared to more irregular granules produced by traditional wet granulation. Consequently, granules of high porosity can

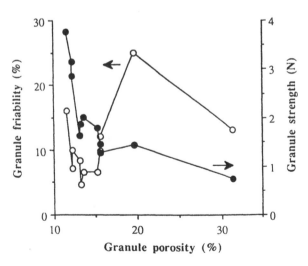

Fig. 7 The effect of granule porosity on the fracture force and the friability of granules of lactose/polyvinylpyrrolidone, 95/5 % wt. (Drawn from data given by Wikberg and Alderborn, ref. [14 and 24].)

probably possess high attritional strength if they are characterized by a regular geometric shape and a smooth surface texture.

The discussion above consequently indicates that the mean separation distance between particles in the granule, and thus the bonding force of the interparticulate attractions, will be of decisive importance for the fracture strength of the granule. However, concerning the attrition strength of granules, the shape characteristics can dominate over the interparticulate bonding force. Some literature reports indicate that a change in the strength of these attractions can also affect the granule friability. For example, it has been reported that an increased amount of binder in the granules can increase the attritional strength of granules [12,16] as well as can a change of agglomeration liquid which increases the solubility of the substrate in the agglomeration liquid [5,17]. Finally, Cutt et al. [12] observed an increased granule friability when the character of the substrate surface was changed from hydrophilic to hydrophobic (Table 1).

III. PHYSICAL STRUCTURE OF COMPACTS OF GRANULES

Pharmaceutical granules are normally relatively porous particles with a porosity in the range of 5–30%. When a bed of such particles is formed by spontaneous packing, the pores within the bed will be distributed between and within the granules (i.e., inter- and intragranular pores). Hence, the pore structure can be described as dualistic. The size distribution in pores is probably in most cases bimodal with a number of large intergranular pores and a number of small intragranular pores. Thus, when the bed of granules is stressed, a failure plane will be obtained between the granules and thus the mechanical properties of the bed of granules are a function of the dimensions and the packing characteristics of the granules.

When granules are compressed by tabletting, the forced reduction in bed porosity will change the characteristics of the pore system of the powder bed. This change will be a reflection of the changes in the physical characteristics and packing of the granules constituting the bed (i.e., the volume reduction behavior of the granules).

It is reasonable that during the initial volume reduction phase, the granules can rearrange in the die and cause a slight reduction in powder bed porosity although the appearance of the granules does not change. However, pharmaceutical granules are normally relatively coarse particles and will probably spontaneously pack to such a low voidage that granule repositioning will contribute only to a limited extent to the total porosity reduction during compression of a bed of granules. As a consequence, irreversible changes in the physical characteristics of the granules will constitute the major part of the volume reduction process.

Van der Zwan and Siskens [18] studied the volume reduction characteristics of granules of a ceramic and a mineral. Their analysis was based mainly on qualitative inspection by SEM of the upper and fracture surfaces of the compacts. For compacts produced at low pressures, individual granules could be clearly distinguished in the compacts but they seemed to be locally deformed at the intergranular contact points. They argued also that the intergranular pore space had already reached a low value at low compaction pressures. Hence, further compression of the mass was associated with a reduction in the porosity of the granules (i.e., granule densification). With increased pressure, it became difficult to distinguish individual granules in the compact.

The authors summarized their findings in the following suggested list of volume reduction mechanisms for granules when compressed:

1. Filling of holes between the granules (i.e., granule rearrangement)
2. Fragmentation and plastic deformation of granules
3. Filling of holes between the primary particles
4. Fragmentation and plastic deformation of primary particles

Mechanism 3 can also be described in terms of a reduction in porosity of the granules—i.e., granule densification. The authors stressed in their discussion that all mechanisms can occur simultaneously in a powder bed during compaction and will not necessarily occur in sequence.

The study by Van der Zwan and Siskens indicated that a compact of granules can be described physically as an aggregate of smaller, cohered granules; i.e., it seems that the granules tend to keep their integrity when compacted. However, with increased compaction pressure, the integrity of the granules will be progressively lost when granules are deformed and fragmented and the intergranular separation distance approaches the distance between the primary particles. In their study, nonpharmaceutical materials were used. However, results have been presented in the literature which support the view that this model of a compact prepared from granules can also be applied to pharmaceutical compacts.

Selkirk and Ganderton [19] compared the pore size distributions, assessed by mercury intrusion, for compacts of a fine powder and the granulated form of the powder. Two materials, sucrose and lactose, were used and the granulations were produced by wet-massing with water. For both materials, similar observations were obtained. The granulation resulted in a considerable widening of the tablet pore structure; i.e., the proportion of coarse pores increased markedly (Fig. 8), compared with compacts formed from the ungranulated particles. Thus, it seems that the granules kept their integrity to some extent when compressed, resulting in intergranular pores which are comparatively coarse. However, the authors

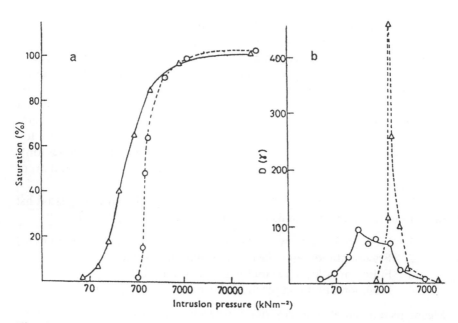

Fig. 8 Cumulative (left graph) and frequency (right graph) pore size distributions for compacts prepared from sucrose granules (circles) and ungranulated sucrose particles (triangles). (From ref. [19] with permission of the copyright holder.)

also pointed out that an increased compaction pressure will probably affect primarily the coarse, intergranular pores and result in a compact of a more even pore size distribution. They also demonstrated, by the aid of tablet air permeability measurements, that the conditions during the granulation procedure, such as amount of agglomeration liquid [20] and slugging pressure [21], affected the pore structure obtained.

Carstensen and Hou [22] studied the pore size distribution of compacts of granules of tricalcium phosphate prepared at a series of compaction pressures. They suggested that the reduction in total tablet porosity, with the subsequent increase in compact strength, was mainly due to the deformation of the granules; i.e., the compact consists of small cohered granules. They argued further that the intragranular pore space seemed to be more-or-less unaffected by the compression procedure.

Wikberg and Alderborn [23] assessed the pore size distribution of compacts of lactose granules by mercury intrusion. The granules were produced by agglomeration with a binder and were of different porosity. The granules of the lower porosity gave a wider pore size distribution with a larger number of coarse pores. The distribution tended to be bimodal

(Fig. 9), which can be interpreted as a reflection of a duality in the pore system of the compacts. For the granules of the higher porosity, a more closed, unimodal distribution was obtained. The observed differences in pore size distributions resulted also in differences in air permeability of the compacts and, thus, the calculated mean pore diameters.

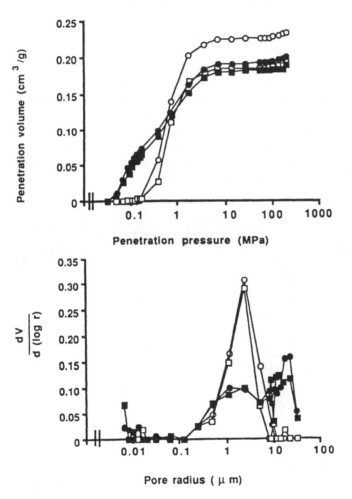

Fig. 9 Cumulative (upper graph) and frequency (lower graph) pore size distributions for compacts prepared from lactose/polyvinylpyrrolidone granules of a porosity of 12.3% (closed symbols) and of 31.3% (open symbols). Compacts were prepared with a compaction pressure of 50 MPa. (From ref. [23] with permission of the copyright holder.)

In another paper [24], Wikberg and Alderborn studied the upper surfaces of compacts of granules of lactose and a high-dosage drug. At relatively low compaction pressures, individual granules could clearly be distinguished at the tablet surface, supporting the model described above. At higher compaction pressures, the authors reported that there was still a tendency that individual granules could be distinguished but the borders between the granules were almost lost. A similar observation was reported by Millili and Schwartz [25] for compacts of pelletized granules of microcrystalline cellulose. This finding was later supported [10] by Johansson et al. by examination of both the upper and the fracture surfaces of compacts of pelletized microcrystalline cellulose granules (Figs. 10 and 11).

In a series of papers [3,26,27], Rue et al. studied the physical properties of a series of granules and of the compacts formed from these granules. The physical structure of the compacts was studied by either a procedure including a removal by dissolution of all compounds in the compact except the binder, followed by an examination of the compact by SEM, or a direct examination of the compact by SEM. They concluded that the compacts of all types of granules used in the study consisted of granules bonded to-

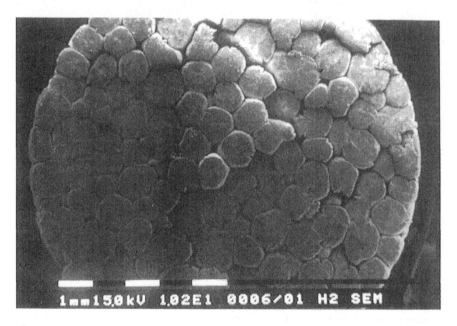

Fig. 10 Upper surface of a compact prepared from microcrystalline cellulose granules of a porosity of 14.4% with a compaction pressure of 100 MPa. White bar corresponds to a length of 1000 μm. (From Johansson et al., ref. [10].)

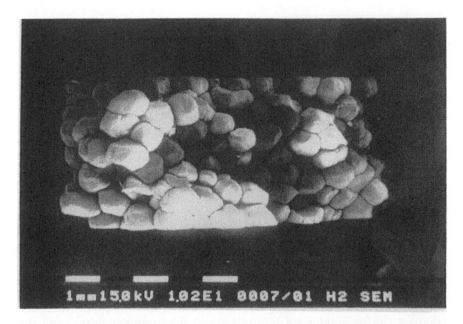

Fig. 11 Fracture surface of a compact prepared from microcrystalline cellulose granules of a porosity of 14.4% with a compaction pressure of 100 MPa. White bar corresponds to a length of 1000 μm. (From Johansson et al., ref. [10].)

gether by intergranular bonds. They also argued that the bonding force of these intergranular bonds will govern the strength of the compact.

Recently, Riepma et al. [28] presented pore size distributions, obtained by mercury intrusion, of tablets of granules formed by slugging. Depending on the applied force during the slugging procedure, uni- and bimodal pore size distributions in the tablets were obtained.

To summarize, an examination of the literature on the physical structure of a compact formed from pharmaceutical granules under normal tabletting conditions indicates that the compact can be described physically as a large aggregate of small, strongly cohered granules. Thus, there is a physical analogy between a powder bed of granules and a compact formed from granules, the main difference being that the mean separation distance between the granules is much smaller and the area of "contact" between two adjacent granules is much larger in the compact. The consequence will be a marked increased coherency of the assembly of granules. The compact can thus be described in terms of the properties of the granules constituting the compact (i.e., size, shape, and porosity) and their relative positions (i.e., intergranular separation distance and number and areas of contact

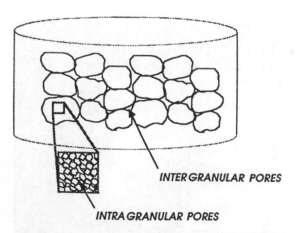

INTER GRANULAR PORES

INTRA GRANULAR PORES

Fig. 12   A simplified schematic illustration of the physical structure from a compact prepared of granules.

zones). This model of the compact implies further that the pores within the compact can be subdivided into inter- and intragranular pores, and the compact will possess a dualistic pore structure. The size distribution of these pores will be bimodal when the intergranular separation distance is large compared to the separation distance between the primary particles constituting the "compact granules." This is probably valid for loosely packed beds of granules and for compacts of granules produced at low compaction pressures. However, for compacts produced at higher compaction pressures, the discrepancy between the two types of pores will probably in many cases be lost, due to the similarity in separation distance between granules and between intragranular particles. A schematic of the suggested physical structure of a compact of granules, illustrating the duality of the pore structure, is given in Fig. 12. A logical consequence of this physical model of a compact of granules is that the dimensions of the granules *before* compaction and the changes in the characteristics of these granules *during* the compression phase, namely their volume reduction behavior, will be of decisive importance for the physical structure of the compact formed.

IV.   ASSESSMENTS OF THE VOLUME REDUCTION PROCESS

Other than inspections by SEM of compacts formed from granules (which is discussed above), there seems to be mainly three methods used in the literature with the aim of characterizing the volume reduction behavior of granules.

A. Pore Structure of Compacts

A common procedure for evaluating the volume reduction behavior of granules was based on the characterization of the pore system of the formed compact, e.g., by mercury intrusion [19,23], liquid penetration [20–21], and air permeability [17,20]. As discussed above, a series of arguments can be found in the literature which support a model of a tablet compacted of granules whereby the compact consists of a number of strongly interacting granules. A reasonable consequence of this model is, at least for compacts produced at relatively low compaction pressures, that the largest pores in the compact are found in the intergranular space. With increased compaction pressure, the size of these pores will approach the size of the intragranular pores and the duality of the pore system will be lost progressively. Thus, the characteristics of the intergranular pore space and how it changes with applied pressure will be a reflection of the changes in the properties of granules and their relative position when compressed. Fast and simple methods for the assessment of the pore size characteristics of a compact are air permeability techniques. This approach was used by Ganderton and Selkirk [20] and later by researchers at Uppsala University for the characterization of the volume reduction behavior of granules.

Wikberg and Alderborn [17] studied the volume reduction process of granules by following the changes in tablet permeability and tablet surface area with compaction pressure over a range of relatively low compaction pressures. The authors used both tablet permeability data and the deduced tablet surface areas, calculated by the Kozeny-Carman equation corrected for slip flow, for the assessment of the volume reduction behavior. The reason for using both these measures was that the calculation of tablet surface area involves an error due to the difficulty of defining the effective porosity of the pore system which take part in the airflow. The background is that, for compacts formed at low pressures, the air will probably flow mainly in the intergranular pores. Thus, the specific surface area of these pores could theoretically be calculated if the porosity of this part of the pore system could be inserted in the permeability equation. However, this effective porosity could not be easily defined. The authors therefore used the apparent particle density of the granules in the calculations but pointed out that this will give an overestimated tablet surface area. They therefore also used the tablet permeability as a complement to the tablet surface area data. The limitation is that a change in air permeability could be governed by either a change in the size characteristics of the pores or a change in the porosity of these pores. It will thus be difficult to relate an observed change in tablet permeability to a specific volume reduction mechanism—i.e., fragmentation or deformation of the granules.

In Fig. 13, the tablet air permeability and the deduced tablet surface area is presented as a function of the compaction pressure for compacts of two granules of different porosity. The tablet permeability is reduced in an exponential way with compaction pressure, while the tablet surface area

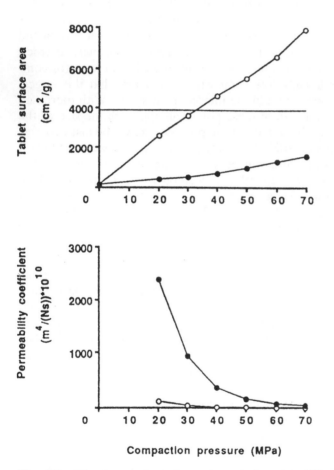

Fig. 13 The changes in surface area and air permeability of compacts of lactose/polyvinylpyrrolidone (95/5 % wt.) granules of a porosity of 12.3% (closed circles) and of 31.3% (open circles) as a function of maximum applied pressure during compaction. The horizontal line in the upper graph represents the surface area of the ungranulated lactose substrate particles. (From ref. [7] with permission of the copyright holder.)

relates almost linearly to the pressure of formation. The reduction in tablet permeability and increase in surface area reflects the formation of a more closed pore structure of the compact. A quantification of these profiles can thus be used as a measure of the volume reduction behavior of the granules. The authors used the area under the curve and the slope of the respective profiles as such measures [17]. The former value can also be seen as a mean tablet permeability over the pressure range studied and can be used to compare different granulations provided that the size and shape characteristics of the original granules are similar; i.e., the original bed of granules will possess a similar air permeability. The profiles in Fig. 13 indicate that the two types of granules respond differently to the applied pressure and it is suggested that the granules which compact into tablets of a more closed pore structure undergo a higher degree of deformation and/ or fragmentation during the volume reduction process.

In the next section the relationship between physical properties of granules, especially size, porosity, and strength of granules, and their volume reduction behavior will be discussed. Most of the interpretations of the degree of fragmentation and deformation which the granules undergo during the compression phase are based on measurements of the air permeability of compacts. Despite the limitation concerning the possibilities to calculate tablet surface areas from permeability results from compacts of granules [17], a permeability method represents a simple technique with high sensitivity for the assessment of the volume reduction characteristics of granules.

B. Tablet Volume—Applied Pressure Relationships

The second approach for assessing the volume reduction behavior of granules is based on the study of the relationship between applied force and punch displacement or tablet volume during the actual compression phase. The techniques used are upper punch force–upper punch displacement profiles [4], upper punch force–die wall force profiles [29], and tablet porosity–applied upper punch pressure profiles [30]. This type of approach can give important information on the ability of the granules to reduce in volume when compressed.

In studies on the compression characteristics of nonporous particles, the tablet porosity–applied pressure relationship according to the Heckel function (see Chapter 3) has been used with the aim of determining the incidence of fragmentation and deformation of particles during the compression process. In this relationship, the pores are considered as reactants as a function of applied pressure. It has been suggested [31,32] that the curvature of the profile over a restricted range of applied pressures can be used as a measure of the degree of fragmentation of the particles and the reciprocal of the slope of the profile as a measure of the degree of deformation of the particles. This way of treating the Heckel profiles has been evaluated for powders of nonporous particles and the whole tablet porosity has been considered to represent the porosity of the reacting pores.

The problem of applying this interpretation also to Heckel profiles from compression of porous particles (i.e., granulated materials) concerns the definition of the reactant pore system. Due to the duality of the pore system, as defined above, a fraction of the porosity represents intragranular pores and the other fraction the intergranular pores. Carstensen and Hou [22] suggested that the pore space of interest in relationship to the Heckel equation is the intergranular pore space and they found that the use of effective particle density (i.e., granule density) values, instead of the apparent particle density, linearized the initial curved part of the Heckel profiles for granules of tricalcium phosphate.

In a study on the compression characteristics of granules of a high-dosage drug, Alderborn et al. [6] observed that quantitative interpretations of Heckel profiles, based on the use of total tablet porosity measurements, gave similar conclusions regarding the volume reduction mechanisms of the granules, as did some other measures. However, in a later paper on the compression characteristics of lactose granules, Wikberg and Alderborn [17] observed the opposite. A possible explanation is that the application of tablet porosity data for the evaluation of volume reduction mechanisms is difficult, due to the complexity of the pore system of the compact. Wikberg and Alderborn also suggested, based on tablet porosity and permeability measurements, that the granules densified markedly in the early phase of the compression process. Thus, it seems that granule densification is an important volume reduction mechanism for granules which is consistent with the observation by Van der Zwan and Siskens [18] (compare also Riepma et al. [28]). The consequence of this is that the effective particle density of the granules, measured before compression, cannot be used to calculate the fraction of the tablet porosity which corresponds to the intergranular pores. It seems therefore reasonable to conclude that the use and interpretations of Heckel data, with respect to the mechanisms involved in volume reduction, should be done with care when granulated materials are studied.

C. Lubricant and Dry Binder Additions

A third approach for studying the volume reduction behavior of particles when compacted is to compare the strength of tablets prepared with and without a second component. Lubricants especially have been used as this second component (e.g., [31–33]) but fine particulate dry binders have been used also for this purpose [34]. Alderborn et al. [6] used both these types of materials to assess the volume reduction characteristics of some granules and in a later study, Wikberg and Alderborn [24] studied the effect of lubricant additions on compact strength. An example is given in Table 3, where the effect of dry binder additions on the relative tablet strength increase for three granulations of a high-dosage drug is summarized. The differences in response to the addition of the excipients indicate that the granules are characterized by differences in their volume reduction behavior, in this case their propensity to expose new surfaces during compression. The same conclusion was drawn from the studies on the effect of lubricant additions [24].

Thus, the approach of assessing the effect of excipients on the compatibility of granules can give valuable information on the incidence of deformation and fragmentation during compression. However, it should be stated that the strength increase and the strength decrease due to the additions of dry binders and lubricants respectively are also related to the distribution of the excipient on the surface of the granules. Vromans et al. [35] pointed out that the degree of strength reduction due to the addition of magnesium stearate was related to the degree of surface coverage of the substrate particles by the lubricant. The surface coverage is a function of the properties of the substrate particles (i.e., their size, shape, and surface roughness [36]) the conditions during the mixing procedure [37] and perhaps also the interaction between substrate and lubricant [38]. The conse-

Table 3 Ratios Between Tensile Strength of Tablets, Compacted from Tablet Masses with and Without Addition of a Dry Binder (denoted a, b, and c) at a Concentration of 8.6% wt[d]

	Tensile strength ratio (—)		
Granulation	a	b	c
I	1.31	1.65	1.49
II	1.27	1.42	1.26
III	1.12	1.28	1.13

a = Avicel PH-101
b = Avicel PH-105
c = Crosslinked polyvinylpyrrolidone
[d]Tablet masses were based on three different granules of a high-dosage drug and a polymeric binder, with the same composition but with different volume reduction behavior.
Source: From Ref. 6.

quence will be that for granules, which might show a large batch variation with respect to the granule surface roughness, the interpretations of strength increase and strength reduction data must be done with care but constitutes an interesting approach for evaluating the volume reduction behavior of granules.

V. PHYSICAL GRANULE PROPERTIES OF IMPORTANCE FOR THE VOLUME REDUCTION PROCESS

A. Granule Porosity and Strength

Ganderton and Selkirk [19–21] studied the pore structure of tablets compacted of granules of lactose and sucrose by air permeability, mercury intrusion, and liquid penetration. Granules with different physical characteristics were prepared by using different methods of granulation (slugging or wet granulation) or by varying the processing conditions during the granulation (slugging pressure or amount of water during agglomeration). The granules consisted generally of one component; i.e., a binder was not included in the granules.

For the granules prepared by wet granulation, an increased amount of agglomeration liquid gave granules which compacted into tablets of a more open and permeable pore structure, i.e., an increased median pore size (Figs. 14 and 15). The authors related this difference in volume reduction behavior of the granules to a decreased porosity and an increased mechanical strength of the granules. However, measures of these granule characteristics were not presented. They suggested that a decreased granule porosity and an increased granule strength reduced both the degree of deformation and fragmentation of the granules when compacted and that these changes in volume reduction behavior of the granules affected the pore structure of the compact formed.

For the granules prepared by slugging [21], an increased slugging pressure gave granules which compacted into more permeable tablets with larger pores. An increased slugging pressure corresponded to a decreased granule porosity and the same explanation of the relationship between porosity and strength of granules, on one hand, and the volume reduction behavior, on the other, was given.

Based on the analysis of liquid penetration into the compacts [20], the authors also concluded that the characteristics of the pores at the surface of the compact were not atypical of the pores in the compact generally. This suggestion supports the validity of the conclusions discussed above concerning the physical structure of compacts of granules based on the observation of the upper surfaces of the compacts.

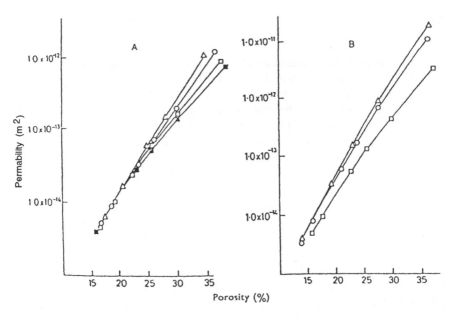

Fig. 14 The air permeability of compacts of lactose granules (left graph) and sucrose granules (right graph) as a function of tablet porosity. The granules were prepared with different amounts of water during the wet agglomeration, i.e. for lactose granules; closed squares = 13%; open squares = 17%; circles = 21%; triangles = 25%, and for sucrose granules: squares = 5%; circles = 7%; triangles = 9%. (From ref. [20] with permission of the copyright holder).

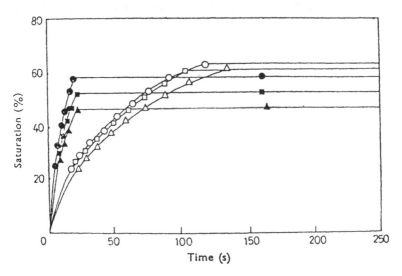

Fig 15 Saturation due to water penetration into compacts of high porosity (closed symbols) and low porosity (open symbols) as a function of penetration time. Compacts were compacted from granules which were prepared with different amounts of water during the wet agglomeration, i.e.: triangles = 5%; squares = 7%; circles = 9%. (From ref. [20] with permission of the copyright holder).

By varying the process conditions during the wet agglomeration of a substrate-binder combination, three granulations of a constant composition were produced by Alderborn et al. [6] and their tabletting properties were subsequently analyzed. The changes in pore structure with compaction pressure was assessed by air permeametry and the surface area–pressure profiles obtained depended on the type of binder solvent used during the agglomeration. The air permeability results correlated well with the magnitude of changes in tablet strength due to the additions of a lubricant and some dry binders (Table 3). Thus, it seems that the changes in pore structure of compacts with applied pressure are associated with the formation of new extragranular surfaces. This formation of new surfaces can be due to either a fragmentation of granules into smaller aggregates or to a marked deformation of the granules.

Alderborn et al. did not address the question of the relationship between the volume reduction behavior of the granules and some other physical characteristics of the granules, such as the porosity. However, Wikberg and Alderborn [17,24] studied the relationship between the porosity of the granules before compaction and their volume reduction behavior, as evaluated by the relationship between air permeability or permeametry surface area and compaction pressure. In these papers, granules of constant proportions of a substrate (lactose) and a binder (polyvinylpyrrolidone) were produced by wet agglomeration in convective mixers. By varying the process conditions during the agglomeration, the porosity of the granules could be varied. The relationship between volume reduction behavior of the granules and the granule porosity was also studied for the dipentum granules used earlier by Alderborn et al. [6].

In both papers, a correlation between granule porosity and the volume reduction behavior of the lactose granules was obtained. In Fig. 16, data from both papers are included in the same graph, and all results fit a unique relationship well. It seems thus that an increased granule porosity before compaction promotes the formation of a pore structure characterized by smaller intergranular pores or by a lower intergranular pore space, i.e., a more closed pore structure. The same type of relationship between granule porosity and tablet pore structure was obtained also for granules of dipentum.

Since a consistent granule size was used in the studies, the observed difference in tablet pore structure due to the variation in granule porosity is caused by variations in the behavior of the granules when compressed. It seems that both differences in degree of fragmentation and deformation between the granules can be responsible for the observed effect. By the use of lubricated granules, Wikberg and Alderborn suggested [24] that the volume reduction process also involved the formation of new extragranular

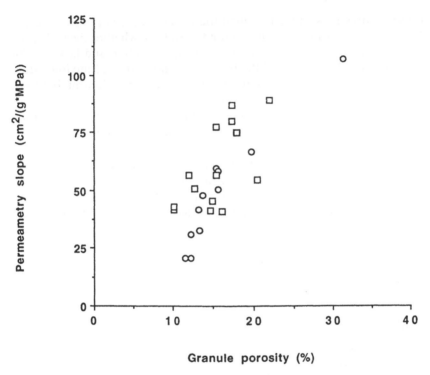

Fig. 16 The slope from table surface area–compaction pressure profiles as a function of the porosity of lactose/polyvinylpyrrolidon (95/5 % wt.) granules produced by wet agglomeration in a planetary mix (squares, from ref. [17]) and in a high shear mixer (circles, from ref. [24]). (Drawn from data given by Wikberg and Alderborn, ref. [17 and 24].)

surfaces. This phenomenon can also be caused by both fragmentation and deformation of the granules during the volume reduction process. In the latter case, a very marked deformation is probably necessary to explain the observed effects, i.e., a deformation which is not comparable with the deformation of nonporous pharmaceutical particles. Marked deformation seems, however, to be a reasonable possibility for granules, since they contain spaces between the substrate particles which allow the primary particles to rearrange, with a subsequent effect on both porosity and shape characteristics of the granules.

Based on the evaluation of the volume reduction behavior of the series of lactose and dipentum granules, Wikberg and Alderborn concluded that both granule fragmentation and deformation occurred during the volume reduction phase. Furthermore, both phenomena probably occurred at low compaction pressures and determined the changes in tablet pore structure with compaction pressure. In an attempt to establish whether the incidence of fragmentation differed between granules of dipentum and lactose, the authors continued to study [14] the mechanical properties of the individual granules, assessed by diametral compression, and the relationship between granule fracture force and tablet permeability.

For the lactose granules, an increased fracture force of the individual granules corresponded to a change in their compression behavior toward a reduced ability to form compacts of a relatively closed pore structure (Fig. 17). The fracture force of a granule can be seen as a measure of the propensity of the granules to fragment into smaller aggregates when loaded. Thus, the good correlation between tablet pore structure and granule strength can be used to support the assumption that the observed

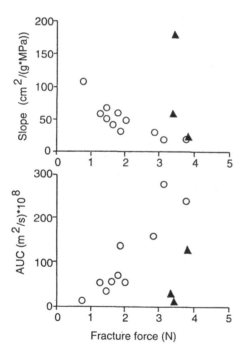

Fig. 17 The slope from tablet surface area–compaction pressure profiles (upper graph) and the area under the curve from tablet permeability-compaction pressure profiles (lower graph) as a function of the fracture force of individual granules of lactose/polyvinylpyrrolidone 95/5 % wt. (circles) and of dipentum/polyvinylpyrrolidone, 94/6 % wt. (triangles). (From ref. [14] with permission of the copyright holder).

differences in tablet pore structure are caused by differences in the degree of fragmentation of the granules during compression; i.e., this mechanism is the significant volume reduction mechanism for the compression process in these granules.

For the dipentum granules, a tendency to a change in tablet pore structure with variation in granule fracture force, was also observed (Fig. 17) but the effect was less pronounced. The results can be interpreted as that the incidence of fragmentation is less pronounced for the dipentum granules compared to the lactose granules and, thus, that densification and deformation of granules is the dominating mechanism, responsible for the changes in pore structure of the compact with applied pressure, for dipentum granules. However, this conclusion concerning the differences in the main volume reduction mechanism was difficult to support by the aid of SEM examinations of the upper tablet surfaces.

In a later paper [10], Johansson et al. studied the volume reduction properties of almost spherical granules of microcrystalline cellulose. Granules of a series of porosities were prepared which, compared to the lactose and dipentum granules, showed a high fracture strength (compare Figs. 5 and 6). The cellulose granules also showed a less distinct failure when compressed diametrically as individual granules, compared to the lactose and dipentum granules; i.e., they could be described as less brittle. This might be related to differences in the mechanical properties of the primary particles within the granules.

Inspection of both upper and fracture surfaces of compacts of the granules indicated that the cellulose granules reduced in volume by deformation (Figs. 10 and 11). This was supported by the fact that the granules also showed a marked sensitivity to addition of a lubricant with respect to compactibility. Also for the cellulose granules, compacts of different pore structure were obtained dependent on the porosity of the granules before compaction (Fig. 18). This finding is thus consistent with the results for lactose and dipentum granules; i.e., an increased granule porosity promotes the formation of a more closed pore structure in the compacts. For the cellulose granules, it is thus suggested that the degree of deformation which the granules undergo during compression to a given applied pressure is governed by the original granule porosity.

Tablets formed at a given pressure tended to be more permeable for compacts prepared from cellulose granules, compared to tablets from lactose and dipentum granules. The differences in size and shape characteristics of the granules before compaction possibly contributes to the difference. However, it might also be related to a difference in the volume reduction behavior of the granules. The granules of microcrystalline cellulose seemed to reduce in volume more or less entirely by densification and

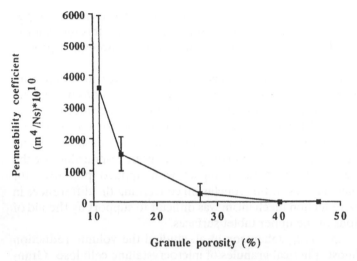

Fig. 18   The permeability coefficient of compacts, prepared at 100 MPa, of microcrystalline cellulose granules of varying porosity. (Drawn from data given by Johansson et al., ref. [10]).

deformation, while for granules of the other materials, fracturing and attrition of granules are probably involved in the volume reduction process to some extent.

The incidence of fragmentation can be affected by both the particulate and the mechanical characteristics of the granules. A change to more irregular granules of a more irregular surface geometry might promote the incidence of fragmentation and attrition during the compression process. A reduction in the granule mechanical strength might also cause a reduction in the propensity of the granules to fracture when compacted. Finally, a change in the mechanical characteristics of the granules (i.e., their brittleness) might also affect their propensity to fail by fracture under load.

The results indicate further that granule porosity will govern the volume reduction phase irrespective of whether the granules show some fragmentation during compression. This might be interpreted that the porosity affects both the degree of granule fragmentation, through a relationship with the granule fracture strength, and the degree of granule deformation. However, it is possible that also for the more brittle granules (lactose and dipentum), the effect of granule porosity on the difference in tablet pore structure could be explained in terms of a relationship between porosity and degree of deformation of granules. However, fragmentation and/or attrition of the granules might still occur to a significant extent during compression, perhaps in the second part of the compression phase. The low sensitivity of the lactose granules of the addition of lubricants can be affected also by difficulties to uniformly distribute the lubricant on the surface of the granules.

B. Intragranular Binder Distribution

It has been suggested [3] that the binder distribution within the individual granules is one physical characteristic of granules which is of decisive importance for their compactibility—i.e., their ability to cohere into compacts. The mechanism behind this effect will be discussed below. However, a localization of a binder at the surface of granules might not only affect the ability of granules to cohere into compacts but also their volume reduction behavior.

Wikberg and Alderborn [39] studied the effect of the intragranular binder distribution on the volume reduction characteristics of granules. Two sets of granules were produced, consisting of the same proportion of substrate and binder (95/5% wt) but with different distribution of the binder within the granules; i.e., in the first set (denoted 5% granules), the binder was relatively homogeneously distributed within the granules, while in the second (denoted 1 + 4% granules), the binder was localized mainly on the granule surface. In addition, a set of granules with a lower amount of binder (1% wt) uniformly distributed in the granules was also studied. These granules were used as "cores" in the preparation of the 1 + 4% granules.

The volume reduction behavior of the granules was evaluated by air permeability measurements. A comparison between the 5% and the 1 + 4% granules indicate that a localization of a substantial amount of binder at the surface of the granules did not affect the assessed changes in pore structure with compaction pressure (Fig. 19). Furthermore, if the 1% and 5% granulations were compared with respect to the relationship between tablet permeability and original granule porosity, the compression characteristics of the granules seemed independent of the amount of binder in the granules. The compressive strength of the granules tended to increase with an increased amount of binder, probably due to an effect of binder content on the bonding force of the interactions between the primary particles within the granules. However, the fracture strength of the individual granules was not markedly affected by the variation in binder content or intragranular binder distribution.

It seems thus that the pore structure of the compacts, prepared from these sets of granules, was governed by the porosity (or the strength) of the granules before compaction, although the physical characteristics of the granules vary with respect to the size of the primary particles, the binder

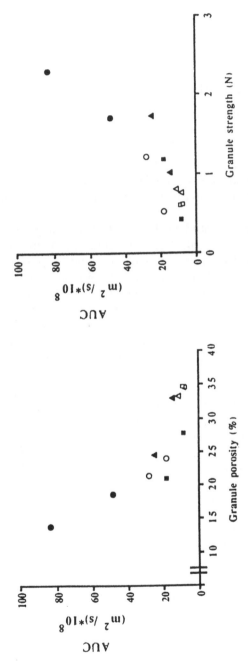

Fig. 19 The area under the curve from tablet permeability–compaction pressure profiles as a function of granule porosity (left graph) and fracture force of individual granules (right graph) for compacts of a series of lactose/polyvinyl-pyrrolidone granules, prepared with different procedures: closed circles = 200 mesh substrate particle size, 5% binder content open circles = 200 mesh substrate particle size, 1+4% binder content closed squares = 200 mesh substrate particle size, 1% binder content closed triangles = 450 mesh substrate particle size, 5% binder content open triangles = 450 mesh substrate particle size, 1+4% binder content open squares = 450 mesh substrate particle size, 1% binder content (Drawn from data given by Wikberg and Alderborn, ref. [39]).

content, and the intragranular binder distribution. The significance of this observation is demonstrated further by comparing the relationship between granule porosity and volume reduction behavior for all lactose granules used by Wikberg and Alderborn [24,39] (Fig. 20). All granules tend to fit a general relationship between the changes in pore structure of compact with applied pressure and granule porosity. However, it must be pointed out that there was a small variation in the original granule size for these granules. The profile (Fig. 20) indicates that the porosity of the granules before compaction is one of the most critical physical characteristics of granules for their volume reduction behavior, and, thus, the pore structure of the compacts formed.

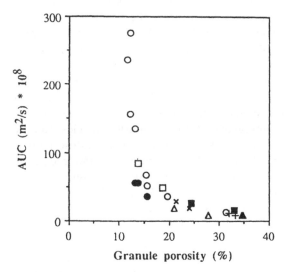

Fig. 20 The area under the curve from tablet permeability–compaction pressure profiles as a function of granule porosity for compacts of a series of lactose/polyvinylpyrrolidone granules, prepared by different procedures:
Open circles = 200 mesh substrate particle size, 5% binder content (agglomeration liquid: ethanol, ref. [24])
Closed circles = 200 mesh substrate particle size, 5% binder content (agglomeration liquid: water, ref. [24])
Open squares = 200 mesh substrate particle size, 5% binder content
Crosses = 200 mesh substrate particle size, 1+4% binder content
Open triangles = 200 mesh substrate particle size, 1% binder content
Closed squares = 450 mesh substrate particle size, 5% binder content
Plusses = 450 mesh substrate particle size, 1+4% binder content
Closed triangles = 450 mesh substrate particle size, 1% binder content
(Drawn from data given by Wikberg and Alderborn, ref. [24 and 39]).

VI. MECHANISMS OF VOLUME REDUCTION OF GRANULES

Based on the discussion above, it is suggested that a series of physical processes is involved in the compaction of pharmaceutical granules, which will affect the physical structure and mechanical strength of the compact formed. A summary of these, classified into two main groups according to the granules and the substrate particles of which the granules consist, is given in Table 4.

During the initial loading phase, some repositioning of granules will probably occur over a range of very low compaction pressures. Thereafter, a further reduction in compact porosity is associated with irreversible changes in the characteristics of the granules. Densification will result from repositioning of the primary particles constituting the granule. A parallel effect to the densification of granules will probably be that the relative dimensions of the granules will change (i.e., a deformation of granules). Thus, these two mechanisms will probably occur simultaneously as a result of the same process: a change in the relative positions of the primary particles within the granules. The loading might also cause a fracturing or fragmentation of granules into smaller aggregates. It is also suggested that primary particles or small aggregates of the primary particles can be abrased from the surface of the granules. This process can be described as an attrition of granules and can be treated as a separate volume reduction mechanism, although it is mechanistically similar to granule fragmentation. Finally, as a consequence of a reduced porosity and size of the granules, the stress exerted on the primary particles will probably increase. The result might be an increased incidence of deformation of the primary particles.

Densification, deformation, and fracturing of granules all seem to be of relevance in the volume reduction process and can occur at low compaction pressures. All these mechanisms require either the formation of a

Table 4 Suggested Physical Mechanisms Involved in the Volume Reduction Process of a Powder Mass Consisting of Porous Granules

A. *Volume reduction mechanisms for the granules*
1. Rearrangement
2. Densification and deformation
3. Fragmentation
4. Attrition
B. *Volume reduction mechanisms for the substrate particles*
1. Deformation (elastic/plastic)
2. Fragmentation

shear plane where primary particles slide against each other, or the formation of a fracture plane causing a more-or-less instantaneous separation of primary particles from each other. However, the dominating mechanism within the compression cycle, including the sequence of the mechanisms involved, has not been established.

The compression induced changes in the physical characteristics of the granules, in terms of their porosity and dimensions, is associated with the following two changes in physical structure of the compact with applied compression stress:

1. A reduction in intergranular separation distance. The most extreme reduction is a situation where the integrity of the granules is lost and the intergranular separation distance is consequently similar to the separation distance between the primary particles; i.e., the duality of the pore structure is lost.

2. A formation of new extragranular surfaces. These surfaces originate from the interior of the granules and will be exposed during the volume reduction process.

These changes in the physical character of the compact will probably affect the formation of intergranular bonds in the compact, with respect to both the bonding force and the number of bonds, and, thus, the compact strength. It is also suggested that the degree of deformation and fragmentation which the granules undergo during compression will be of special importance for the changes in physical structure of the compact which occur during the compression. The degree of deformation and fragmentation seems in many cases to be related to the porosity of the granules before compaction, i.e., the porosity might affect the degree of granule deformation as well as the degree of granule fragmentation, eventually through a relationship with the strength of the granules. However, the question of the importance of the porosity and the strength of granules for their compression behavior is complex and not yet satisfactorily elucidated and at least two factors might affect this relationship. Firstly, it is possible that the strength of the granules can form thresholds below and above which variations in granule porosity and strength do not affect the compression behavior of the granules. At low granule strengths, the granules might collapse entirely when compressed, and below this strength threshold a variation in granule strength and porosity will thus not affect the compression behavior. Furthermore, granules of a very high strength might be very resistant to both deformation and fragmentation, and above this threshold the granules will show similar compression behavior independent of the porosity and strength. Most of the granules discussed in this chapter will consequently belong to a domain between those strength thresholds (which might be the most common situation for granules prepared by wet agglomeration). Secondly, the earlier discussion on the localization of the failure during loading of individual substrate-binder granules is probably applicable also to the process of compaction; i.e., the surface geometry of the substrate particles and the relative surface energies of the binder and the particles in relation to the strength of the binder-binder attractions will determine the localization of the failure plane. It is possible that a localization of the failure at the interface between binder and substrate will have the consequence that the granules compress in a similar way, although the separation distance between the primary particles, and thus the granule porosity, will vary. An increased knowledge of these issues can contribute to an understanding of the apparently conflicting results reported in the literature on the effect of granule strength and granule porosity on the compactibility of granules.

The model of a tablet used in the discussion in this chapter indicates that dimensions of the "compact granules" within the tablet are of relevance for the tablet pore structure. Consequently, not only the compression induced changes in the dimensions of granules but also the original granule dimensions before compaction, primarily the granule size, will affect the pore structure of a compact [28]. Thus, these two factors will in combination govern the changes in physical structure of the compact while being compressed. The discussion on the relationship between volume reduction behavior of granules and the physical structure of the compact formed is summarized qualitatively in Fig. 21.

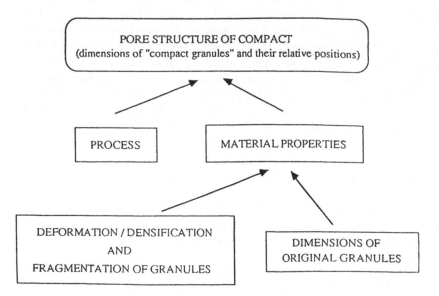

Fig. 21 Proposed main factors contributing to the pore structure of compacts prepared from granules.

VII. PHYSICAL GRANULE PROPERTIES OF IMPORTANCE FOR COMPACT STRENGTH

A. Volume Reduction Behavior

The mechanical strength of a tablet can be described as a function of the bonding force of the bonds between the particles constituting the tablet and the total number of the bonds (see Chapter 10). Both the bonding force and the number of bonding zones will be affected by the dimensions of and the separation distance between the "compact particles."

To achieve small separation distances, it is important that the particles can be brought into close contact without a subsequent elastic recovery; i.e., the volume reduction behavior of the particles will be critical for the relative positions and the area of contact between particles in the compact. A pronounced plasticity of the particles is generally believed to improve the strength of the tablet especially when it occurs at points of interparticle contact [40]. By studying physical characteristics of compacts of different types of lactose [41], Vromans et al. suggested that the degree of fragmentation of particles will be of decisive importance for compact strength.

The degree of deformation or fragmentation which the granules undergo during compression will affect the dimensions of the granules constituting the compact, as well as their relative positions. Consequently, it is reasonable to assume that the volume reduction behavior of the granules will affect the tensile strength of the compact produced.

In Fig. 22, the tensile strength of compacts prepared from nearly spherical granules of microcrystalline cellulose [10], produced by extrusion-spheronization, is shown as a function of the total tablet porosity. The granules showed a range of porosities (Fig. 5) before compaction and the relationship between original granule porosity and volume reduction behavior have been discussed above (Fig. 18). Although the total tablet porosity was similar between the compacts, the tensile strength varied considerably. Thus, it is not possible in this case to explain the differences in compactibility between the granules by differences in the total tablet porosity, i.e., the total degree of bulk volume reduction obtained. The compact strength must be related in this case to the quality of the pore structure.

In Fig. 23, the mechanical strength of tablets produced at 100 MPa from the microcrystalline cellulose granules is shown as a function of the air permeability of the compacts. A direct relationship between the measure of the pore structure and the tensile strength of the compacts was obtained. Similar observations have also been obtained for compacts of a series of lactose-PVP granules [24] of varying porosity (Fig. 24). In this case, the

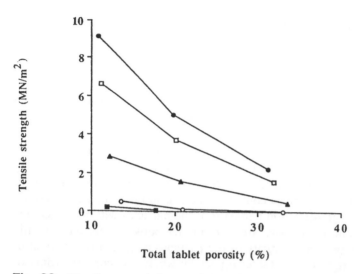

Fig. 22 Tensile strength from compacts prepared of microcrystalline cellulose granules of a range of porosities, as a function of compact porosity. Original granule porosity:
Closed squares = 11.0%; Open circles = 14.4%; Closed triangles = 27.2%;
Open squares = 40.1%; Closed circles = 46.4%
(Drawn from data given by Johansson et al., ref. [10]).

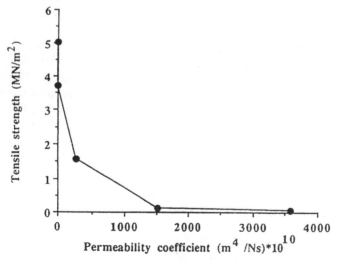

Fig. 23 Tensile strength of compacts, prepared at 100 MPa from microcrystalline cellulose granules of a range of porosities (see Fig. 22), as a function of the air permeability of the compacts. (Drawn from data given by Johansson et al., ref. [10].)

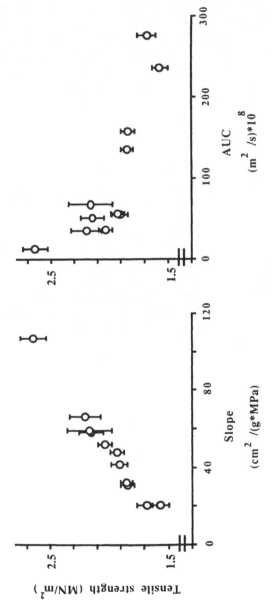

Fig. 24 Tensile strength of compacts, prepared at 150 MPa from a series of lactose-polyvinylpyrrolidone granules of different porosities, as a function of the volume reduction behaviour of the granules (as measured by tablet surface area (slope) or air permeability (AUC) as a function of compaction pressure). Bars represent 95% confidence limits of the mean. (From ref. [24] with permission of the copyright-holder).

pore structure is measured by the changes in tablet permeability with applied pressure. The results indicate that tablets with a pore system characterized in relative terms by a small intergranular separation distance and a large extragranular surface area possess a high mechanical strength.

The results can be explained by assuming that tablets compacted of granules can be described physically in terms of a large aggregate of cohered granules and that the bonding forces of the attractions between the granules is weaker than the bonding forces between the primary particles within the granules. Hence, when the compact is stressed during strength analysis, a fracture plane is created between and around the granules within the tablet and the strength of the tablet will be governed by the sum of the intergranular attractions in the fracture plane. The formation of a closed intergranular pore structure in the compact probably promotes increased areas of contact between granules, and thus increased forces of bonding, as well as an increased number of intergranular bonds. A closed intergranular pore structure will thus correspond to a high tablet strength in a situation where the intergranular bonds constitute the weakest part of the total complex bonding characteristics within a compact.

The formation of a layer of magnesium stearate on the surface of particles/granules is generally expected to decrease the tablet strength by reducing the bonding force of the intragranular bonds; i.e., it acts as a "filter" for intermolecular attractions [42]. The relative decrease in tablet strength for a material has been postulated to be dependent on the degree of fragmentation of the material during compaction, as discussed earlier, provided a similar degree of surface coverage of magnesium stearate on the particles is obtained before compaction. In Fig. 25, the strength of tablets formed from the lactose-PVP granules discussed earlier (Fig. 24) is presented, together with the strength of tablets formed of the same granules which have been lubricated with magnesium stearate, 0.5% wt, before compaction. The importance of the quality of the intergranular pore structure for the tablet strength is stressed by the lubricant addition, in that the compatibility of granules which form tablets with a closed pore system (i.e., a large extragranular surface area and/or a low pore diameter) seems to be reduced less by the addition of magnesium stearate, while the converse applies for granules which form tablets with a more open pore system.

Consequently, the results could be interpreted as supporting the hypothesis that the strength of the tablet will be governed by the sum of the intergranular bonds in the fracture plane. Furthermore, the degree of strength reduction is related to the volume reduction behavior of the granules. It seems reasonable that the formation of new extragranular surfaces during the compression process will compensate for the strength reduction

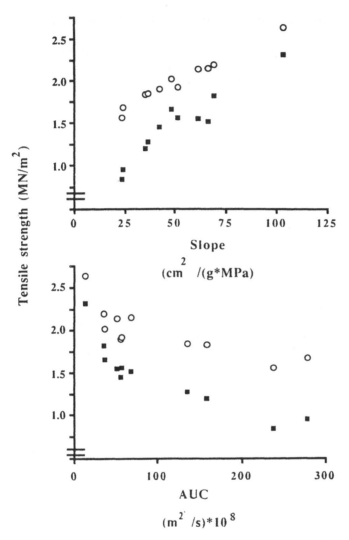

Fig. 25 Tensile strength of compacts, prepared at 150 MPa of a series of lactose-polyvinylpyrrolidone granules of different porosities, as a function of the volume reduction behavior of the granules (as measured by tablet surface area (slope) or air permeability (AUC) as a function of compaction pressure). Open symbols represent unlubricated granules and closed symbols represents granules lubricated with 0.5% wt. magnesium stearate. (Drawn from data given by Wikberg and Alderborn, ref. [24].)

effect of the magnesium stearate addition. Thus, the formation of a closed pore structure is associated with the formation of new extragranular surfaces. A possible interpretation is that the differences in tablet pore structure is caused by differences in the degree of fragmentation of the granules during compression. However, it cannot be excluded that differences in degree of deformation of those granules can be responsible for the observed differences in tablet pore structure and tablet strength reduction caused by the lubricant. To some extent, the degree of surface coverage of the lubricant on the granules might vary with the porosity of the granules and thus affect the degree of tablet strength reduction. Nevertheless, the results clearly indicate that the composition of the extragranular surfaces in the tablet is also of importance for the tablet strength.

B. Granule Size

If the proposed model regarding the physical structure of a tablet made from a granulated material is valid, it is reasonable to assume that the size of the granules before compaction can also affect the strength of the tablet. Although results have been reported on the effect of granule size on tablet strength in the literature, no universal relationship has been found [7].

In Fig. 26, the mechanical strength of tablets prepared from a series of lactose-PVP granules of the same composition (95/5% wt) is shown as a function of the intergranular pore structure of the compacts (as measured by the air permeability of the compacts). The granules were obtained by fractionating three different densified granulations into a series of size fractions between 6 and 1400 μm. In addition, the unfractionated granulation was also compacted. Firstly, it can be seen that the same principal relationship between tablet strength and intergranular pore structure as obtained earlier seems to exist; i.e., tablets with a closed pore system possess a higher mechanical strength. Secondly, no unique relationship between granule size before compaction and the tensile strength of the compact produced was obtained, although there is a tendency that the finer granules formed compacts of a higher tensile strength. The results thus indicate that the combined effect of the original granule size and the degree of granule fragmentation and/or deformation which takes place during the compaction, will govern the pore structure and the tensile strength of the tablet formed. One can also notice that the unfractionated granulation conforms reasonably well to the relationship between tablet strength and tablet permeability.

Riepma et al. [28] found an increased tablet strength with a reduced size of lactose granules, formed by a slugging procedure. The tablet strength correlated positively with the tablet surface area. This surface area

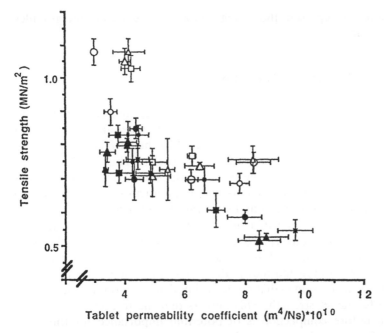

Fig. 26 Tensile strength of compacts, prepared at 60 MPa from a series of lactose-polyvinylpyrrolidone granules of different original size, as a function of the air permeability of the compacts. Three granulations, produced by different processing conditions during the agglomeration (planetary mixer, high speed mixer with low impeller speed and high speed mixer with high impeller speed), were used. Bars represent 95% confidence limits of the mean.
Crosses = unfractionated granulations
Small unfilled circles = 46–63 μm
Large unfilled circles = 63–90 μm
Small unfilled triangles = 90–125 μm
Large unfilled triangles = 125–180 μm
Small unfilled squares = 180–250 μm
Small filled circles = 250–355 μm
Large filled circles = 355–500 μm
Small filled triangles = 500–710 μm
Large filled triangles = 710–1000 μm
Large filled squares = 1000–1400 μm
(From Wikberg and Alderborn, unpublished results).

was considered to represent the extragranular surface area of the granules within the compact.

C. Intragranular Binder Distribution

For compacts prepared of binder-substrate granules, the external surface of the granules which constitute the compact could be equal to the surface of the substrate particles or it could be equal to the surface of a layer of binder which is adherent to the substrate particles. Thus, within tablets compacted of such granules, three types of attractions between the granules can exist: two types of binder-related interactions (i.e., binder to binder and binder to substrate) and substrate-to-substrate interactions. It seems reasonable to assume that intergranular attractions which involve the binder form comparatively strong intergranular bonds, due to the comparatively higher propensity of the binder to deform plastically and thereby increase the contact area and reduce the separation distance, between the granules. Thus, it is assumed that binder-related interactions are of decisive importance for tablet strength. This implies that the intragranular binder distribution before compaction is of potential importance for the compactibility of binder-substrate granules.

In Fig. 27, the mechanical strength of tablets produced from a series of lactose-PVP granules are shown as a function of a measure of their intergranular pore structure. The granules possess different intragranular binder distributions but have similar total binder content [39]. If the mechanical strength of the tablets is compared with respect to the tablet pore characteristics, tablets prepared of granules with a homogeneous intragranular binder distribution were generally of a higher mechanical strength than tablets prepared of granules with a peripheral localization of the binder.

During the compression of granules, new extragranular surfaces can be formed within the compact as a result of fragmentation and deformation of the granules. Thus, the total area of extragranular surfaces in the tablet includes a fraction of new surfaces, created during the compaction process. If the binder is homogenously distributed within the granules before compaction, all extragranular surfaces in the tablet will probably be of a similar composition. Hence, it is suggested that a homogeneous intragranular binder distribution before compaction will give binder interaction zones with a relatively narrow distribution with respect to the bonding force of the individual bonds. For granules with a peripheral localization of binder, a broader distribution in bonding force of the binder interaction zones within the compact will probably result. This distribution is characterized by a

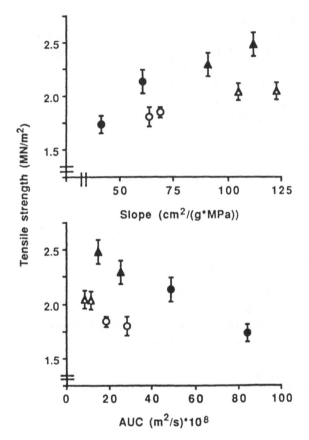

Fig. 27 Tensile strength of compacts, prepared at 150 MPa from a series of granules with different porosity and intragranular binder distributions, as a function of the volume reduction behavior of the granules (as measured by tablet surface area (slope) or air permeability (AUC) as a function of compaction pressure). Open circles = 200 mesh substrate particle size, 5% binder content Closed circles = 200 mesh substrate particle size, 1+4% binder content Open triangles = 450 mesh substrate particle size, 5% binder content Closed triangles = 450 mesh substrate particle size, 1+4% binder content (From ref. [39] with permission of the copyright holder).

fraction of surfaces with a comparatively thick binder layer forming strong binder interaction zones and a fraction of surfaces with a comparatively thin, and perhaps noncontinuous, binder layer forming weaker binder interaction zones. It therefore seems that a homogeneous distribution of binder within the granules promotes compactibility if a significant number of new surfaces, originating from the interior of the granules, are produced during the compression; i.e., the structure of the granules is significantly changed during compression. However, it is also reasonable to assume that if the granules are almost nonfragmenting or nondeformable, a peripheral localization of binder within the granules could be advantageous.

Granules with a more peripheral binder localization were also more sensitive with respect to their compactibility to the admixing of magnesium stearate. The extragranular surfaces in the tablet which correspond to the original granule surface will probably be of vital importance for the compactibility of these granules. If these binder layers are covered with a lubricant before compaction, their bonding potential will be reduced and this could then have a dramatic effect on the tablet strength.

The effect of intragranular binder distribution on the strength of compacts prepared from granules discussed here can be interpreted to be not consistent with the suggestions by Rue et al. (Fig. 2) [3]. They suggested that a peripheral localization of binder in the granules improved the binding properties of the granules and, thus, the compact strength. However, in their study, no attempt was made to relate the compact strength to the pore structure of the compacts and it is possible that low-deformable and low-fragmenting granules were used in the study.

VIII. CONCLUDING REMARKS

This chapter has argued that a tablet compacted of a granulated material can be described physically as a large aggregate consisting of small "compact granules" cohering to each other. The pore system of such a compact is thereby dualistic in nature, although, in practice, both uni- and bimodal pore size distributions will be obtained, when the pore structure of the tablets is analyzed. The structure of the intergranular pores—i.e., the geometry and porosity of the pore system—is a reflection of the physical properties of the granules constituting the compact and their relative positions.

During compaction of the granules, the consequence of the increased proximity between the granules will be the formation of intergranular bonds and a subsequent increased coherency of the assembly of granules. The distribution of intergranular bonds in the compact is probably related to the physical properties and relative positions of the "compact granules." In this chapter, it has been argued that a closed compact pore structure, which is a reflection of a low separation distance between granules and perhaps also of a large extragranular surface area, corresponds to either the formation of large intergranular contact areas, i.e., attraction zones of a high bonding force, or to the formation of a large number of intergranular bonds. Such a pore structure of the compact will consequently correspond to a high total intergranular bonding strength in a given cross section within the compact. However, the relevance of the structure of the intergranular pores for the mechanical strength of the compact will probably be related to the relative bonding force of the inter- and intragranular bonds—i.e., whether a fracture will propagate mainly across or around granules when the compact is stressed.

1. Fracturing of Compact around Granules

A consequence of the duality of the pore structure discussed earlier is that pores between granules are normally larger than pores within the granules. The intergranular bonds formed during compression are, thus, probably often weaker than the bonds within the granules (i.e., between substrate particles). Thus, when a compact is stressed, a fracture plane will be created between and around granules. In this case, the compact strength will probably be governed by the distribution of intergranular bonds in the cross section and the total bonding strength can be described simply as the sum of the bonding forces of the individual intergranular bonds in the failure plane.

The total bonding strength will be related to the number of intergranular bonds per cross-sectional area and the bonding force of the individual bonds. The latter is probably affected by the area of contact between adjacent granules which is developed during the compression process. In addition, the intergranular bonding occurs between surfaces either separated by some distance or other more-or-less fused together. Consequently, the properties of the surface will also affect the bonding force of the bonding zones formed. Two experimental supports for this statement have been discussed in this chapter. Firstly, the addition of a lubricant to the granulation generally reduces the compact strength. Thus, a lubricant which covers the surface of a granule will reduce the strength of the attractions and thereby reduce the tablet strength. However, it is suggested that the ratio between covered and uncovered extragranular surfaces within the compact is related to the degree of fragmentation or deformation of granules during compression and will affect the distribution of bonds in the fracture plane.

Secondly, the distribution of binder within the granules before compaction affected the strength of the compacts; i.e., a homogeneous intragranular binder distribution increased the tablet strength compared to a peripheral localization of binder. It is suggested that the combined effect of the intragranular binder distribution before compaction and the degree of exposure of new surfaces during the compaction phase affects the distribution of intergranular bonds.

In conclusion, in a situation where the fracturing of a compact occurs between the "compact granules," it is suggested that the mechanical strength of the compact is governed by the distribution of intergranular attractions in the fracture plane. This distribution is a reflection of the structure of the intergranular pores and the composition of the exposed extragranular surfaces in the compact. Both these factors can be affected by primary physical characteristics of the granules as discussed earlier in this chapter. A summary of this hypothesis is presented schematically in Fig. 28.

Based on this mechanistic conception, two important criteria for granules with respect to their compactibility can be established (Table 5). One of the challenges for future research in this field is to establish how granules can be manipulated in order to fulfill these criteria (i.e., particle engineer-

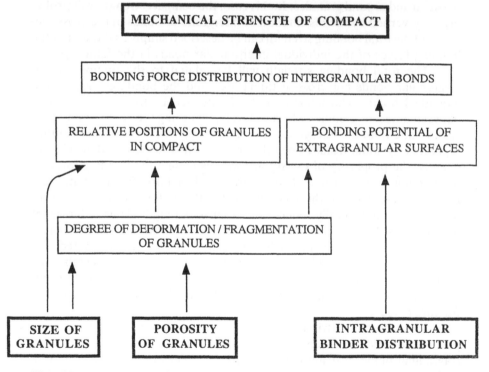

Fig. 28 A schematic description of the qualitative relationship between some granule physical properties, which can be affected by the procedure for preparation of the granules, and the mechanical strength from the compact prepared of the granules.

Table 5 Suggested Criteria for Optimal Substrate-Binder Granules, with Respect to Compactibility

The compression process results in the formation of new extragranular surfaces, low intergranular separation distances, and large intergranular contact areas.
The extragranular surfaces in the compact consist of a film of binder with a high bonding potential.

Table 6 Suggested Important Physical Properties of Substrate-Binder Granules Which Govern Compactibility

A. Physical properties of granule
  1. *Granule porosity and strength*
     Affects the degree of deformation and fragmentation of granules which affects the pore structure of the compact
  2. *Granule size*
     Affects the pore structure of the compact
  3. *Granule shape and surface structure*
     Affects the distribution of lubricant on the granules before compaction (eventually also the pore structure of the compact)
  4. *Intragranular binder distribution*
     Affects the composition of exposed surfaces of "compact granules"
B. Properties of binder
   Deformability of binder
     Affects the bonding force of intergranular attractions
C. Substrate-binder interaction
     Affects probably the volume reduction behavior and the composition of exposed surfaces of "compact granules"

ing). It has been argued in this chapter that the granule porosity is a key factor in this context. The relationship between this granule property and some other physical characteristics, on the one hand, and the compact strength, on the other, has been discussed in qualitative terms in this chapter, and a summary of the effect of granule physical properties on the compact strength is presented in Table 6.

2. *Fracturing of Compact Across Granules*

However, it is possible under certain conditions that the intergranular bonds formed during compression are stronger than the bonds within the granules. The consequence of such a situation is that a fracture will probably propagate across, rather than around, the granules. In that case, the compact strength might probably not be related to the pore structure of the compact, since the intergranular bonds do not represent the weakest part of the compact. Thus, the qualitative discussion summarized in Fig. 28 is not applicable. This type of, probably atypical, situation can be the result of the compaction of granules with a high surface bonding capacity compared to the bonding strength within the granules. This effect can be the consequence of, for example, a localization of the binder more or less entirely on the surface of the granules.

3. Final Comments

The compaction of granules is obviously a complex process and a survey of the literature reveals apparently conflicting results as well as results which lack a mechanistic understanding. In this chapter, an attempt has been made to propose a fundamental understanding of some reported relationships between physical properties of granules and the tensile strength of a tablet formed from those granules. As discussed, the relationship might be restricted to a situation where the compact fails mainly around, rather than across, the granules, when subjected to stress. In addition, the studies have been performed mainly on materials which in themselves compact reasonably well into tablets. The generality of the observations as well as the application of the proposed relationship to materials with complex compaction properties, such as markedly elastic and viscoelastic materials, needs to be resolved. Furthermore, there is also a need to focus on the volume reduction behavior of granules, both with and without binders, in order to understand the relationship between granule porosity and strength and the physical properties of compacts formed from granules.

REFERENCES

1. H. Leuenberger, *Int. J. Pharm. 12*:41 (1982).
2. G. Alderborn, and C. Nyström, *Acta Pharm. Suec. 21*:1 (1984).
3. P. J. Rue, H. Seager, J. Ryder, and I. Burt, *Int. J. Pharm. Technol. Prod. Manuf. 1*:2 (1980).
4. G. Ragnarsson, and J. Sjögren, *Int. J. Pharm., 12*:163 (1982).
5. J. J. Wells, and C. V. Walker, *Int. J. Pharm. 15*:97 (1983).
6. G. Alderborn, P. O. Lång, A. Sågström, and A. Kristensen, *Int. J. Pharm., 37*:155 (1987).
7. M. Wikberg, Studies on relationships between physical granule properties and compaction characteristics of pharmaceutical granulations, Ph.D. thesis, Uppsala University, 1992.
8. H. G. Kristensen, and T. Schaefer, *Drug Dev. Ind. Pharm. 13*:803 (1987).
9. H. Rumpf, in *Agglomeration* (W.A. Knepper, ed.), Interscience, New York, p. 379, 1962.

10. B. Johansson, M. Wikberg, R. Ek. and G. Alderborn, *Int. J. Pharm. 117*:57 (1995).
11. E. Shotton, *Boll. Chim. Farm. 116*:315 (1977).
12. T. Cutt, J. T. Fell, P. J. Rue, and M. S. Spring, *Int. J. Pharm. 33*:81 (1986).
13. M. A. Mullier, J. P. K. Seville, and M. J. Adams, *Chem. Eng. Sci. 42*:66 (1987).
14. M. Wikberg, and G. Alderborn, *STP Pharma Sci 2*:313 (1992).
15. B. M. Hunter, and D. Ganderton, *J. Pharm. Pharmacol. Suppl. 24*:17P (1972).
16. M. H. Rubinstein, and P. A. Musikabhumma, *Pharm. Acta Helv. 53*:125 (1978).
17. M. Wikberg, and G. Alderborn, *Int. J. Pharm. 62*:229 (1990).
18. J. Van der Zwan, and A. M. Siskens, *Powder Technol. 33*:43 (1982).
19. A. B. Selkirk, and D. Ganderton, *J. Pharm. Pharmacol. Suppl. 22*:79S (1970).
20. D. Ganderton, and A. B. Selkirk, *J. Pharm. Pharmacol. 22*:345 (1970).
21. A. B. Selkirk, and D. Ganderton, *J. Pharm. Pharmacol. 22*:86S (1970).
22. J. T. Carstensen, and X.-P. Hou, *Powder Technol. 42*:153 (1985).
23. M. Wikberg, and G. Alderborn, *Int. J. Pharm. 84*:191 (1992).
24. M. Wikberg, and G. Alderborn, *Int. J. Pharm. 69*:239 (1991).
25. G. P. Millili, and J. B. Schwartz, *Drug. Dev. Ind. Pharm. 16*:1411 (1990).
26. H. Seager, I. Burt, J. Ryder, P. J. Rue, S. Murray, N. Beal, and J. K. Warrack, *Int. J. Pharm. Technol. Prod. Manuf. 1*:36 (1979).
27. H. Seager, P. J. Rue, I. Burt, J. Ryder, and J. K. Warrack, *Int. J. Pharm. Technol. Prod. Manuf. 2*:41 (1981).
28. K. A. Riepma, H. Vromans, K. Zuurman, and C. F. Lerk, *Int. J. Pharm. 97*:29 (1993).
29. E. Doelker, and E. Shotton, *J. Pharm. Pharmacol. 29*:193 (1976).
30. N. A. Armstrong, and F. S. S. Morton, *J. Powder Bulk Solids Technol. 1*:32 (1977).
31. A. H. de Boer, G. K. Bolhuis, and C. F. Lerk, *Powder Technol. 20*:75 (1978).
32. M. Duberg, and C. Nyström, *Acta Pharm. Suec. 19*:421 (1982).
33. E. Shotton, and C. J. Lewis, *J. Pharm. Pharmacol. Suppl. 16*:111T (1964).
34. C. Nyström, and M. Glazer, *Int. J. Pharm. 23*: 255 (1985).
35. H. Vromans, G. K. Bolhuis, and C. F. Lerk, *Powder Technol. 54*:39 (1988).
36. L. Roblot-Treupel, and F. Puisieux, *Int. J. Pharm. 31*: 131 (1986).
37. M. S. Hafeez Hussain, P. York, and P. Timmins, *Int. J. Pharm. 42*:89 (1988).
38. R. C. Rowe, *Int. J. Pharm. 41*:223 (1988).
39. M. Wikberg and G. Alderborn, *Pharm. Res. 10*:88 (1993).
40. J. J. Benbow, *Enlargement and Compaction of Particulate Solids* (N. G. Stanley-Wood, ed.), Butterworths, London, p. 161, 1983.
41. H. Vromans, Studies on consolidation and compaction properties of lactose, Ph.D. thesis, University of Groningen, 1987.
42. P. G. Karehill, E. Börjesson, M. Glazer, G. Alderborn, and C. Nyström, Studies on direct compression of tablets. XX. Investigation of bonding mechanisms of some directly compressed materials by addition and removal of magnesium stearate, Published in Thesis by P. G. Karehill, Uppsala University, 1990.

13

Modeling the Compression Behavior of Particle Assemblies from the Mechanical Properties of Individual Particles

Wendy C. Duncan-Hewitt

University of Toronto, Toronto, Ontario, Canada

I. INTRODUCTION

> Gord Meyer, the production supervisor at Pharmapress Canada was sitting over a report and a cup of coffee when the tablet press operator peered around the corner. It was a rainy Friday, typical of the April days in 1989, and the man looked oppressed—more than Gord might have attributed to the wet weather. The press operator spoke quickly and apologetically, "Gord, we're having problems with the new A.S.A. formulation. I know it worked O.K. last time but now it's a mess. Look at these!"

As a pharmaceutics scholar, no doubt you are already formulating a hypothesis about what happened in this problem that I give to my undergraduate students—even without the rest of the information which I provide to them. What conceptual foundation would you use to attack the problem? An effective framework would provide (1) all the parameters required to solve the problem, (2) no unnecessary information, and (3) the relationship between the parameters. The *modeling* process provides one means of establishing a theoretical basis for systematic, scientific problem solving in pharmaceutics. The motivation behind this problem solving—the need to solve and prevent formulation problems such as these—and the inferential reasoning associated with it serves as the foundation for the models of tablet compaction whose *process of development* I will elucidate in this chapter. It is hoped that this discussion will yield the following:

1. A simple explanation of the needs and factor analyses (formal or informal) that must precede any model development process.
2. An appreciation of the need for assumptions and simplifications in the process of model development as well as an early assessment of the *costs* (in terms of predictive power) of these simplifications.
3. A discussion of the experimental methodologies employed.
4. The recognition that models are *not* reality. They are very limited representations of it—they can always be extended, improved, and criticized. You can never prove them to be right—but they can be proven wrong. When presenting a model to the scholarly community, then, it is always best to do so with the spirit of a sandcastle builder—expect the next wave to bring it crashing down.*
5. The realization that modeling is useful *and* aesthetically satisfying.

The elucidation of the models that we have developed to predict tablet densification and stress relaxation will serve as a framework for our pursuit of these aims.

II. NEEDS ANALYSIS

What do we want to achieve? A needs analysis clarifies the goals of the modeling process. For the purposes of the research undertaken in our group, it was the practical need to ensure the quality of compacts that drove our inquiry. The goal of the formulator is to produce a tablet that can resist failure prior to administration yet dissolve reproducibly (and usually quickly and completely) in the gastrointestinal tract. While tablets that meet these specifications often can be fabricated by the relatively empirical methods of granulation or modification of the method and rate of compaction, these solutions can be expensive and they are not always effective. Persistent tablet failure by lamination (splitting horizontally) or by exclussive wear during processes such as coating is confounding.

For us, these problems provided the impetus (1) to *understand* and *predict* the outcome of the compaction process and, (2) to *identify* and

---

* As an aside, it is important to recognize that the paper that is published very seldom reflects the *reality* behind it. As Einstein said: "I think and think for months and years. Ninety-nine times, the conclusion is false. The hundredth time I am right."

control the factors which influence tablet strength. Given the limited amounts of material available from preliminary drug development studies, it would be useful if these objectives could be achieved by *micro*methodology.

This analysis of needs was done on a very informal basis, largely being driven by the academic environment at Toronto in the mid-1980s. As declining research funding and accountability become more prominent issues for the investigator, more formal surveys of the needs of industry and the public would be advisable before significant energies and resources are devoted to any given modeling process.

III. FACTOR ANALYSIS

A. Selection of Material and Process Parameters

In order to model tablet compaction, we must first formulate a hypothesis concerning which factors are necessary and sufficient to serve as parameters in a predictive model. This is where empirical research is critical. If the information is not available in the literature, then it will be necessary to elucidate appropriate correlations experimentally before any modeling process begins. I am indebted to the ample, high-quality tableting research that preceded my work since it made much of this preliminary exploration unnecessary for me.

Empirical evidence (for example, [1–10]) has shown that tablet strength is a function both of the intrinsic properties of the material and of extrinsic factors which operate during compaction, handling, and storage. The intrinsic physical properties which have been implicated as important determinants of tablet strength include:

1. Plastic (irreversible) deformability. High plasticity facilitates the formation of *permanent* particle-particle contact regions during compaction. At a simplistic level the total interparticulate bond strength is given by the product of the contact area and the intrinsic bonding capacity of the materials.
2. Elastic (reversible) deformability. A highly elastic material can store mechanical energy when a stress is applied but will release it again once the stress is removed. This can give rise to residual stresses within the compact during the decompression phase of the compaction cycle which literally can cause the tablet to "tear itself apart."
3. Fracture toughness, which determines the extent to which the particles or the interparticulate contact regions fracture or crush during compaction.

4. Deformation kinetics which determine the relative degree to which elastic, plastic, and brittle behaviors are manifested.
5. Surface free energy, which is an important component of the interparticulate bond strength.

Other material variables which are believed to affect the densification behavior of pharmaceutical materials include particle size, impurities, and crystal habit. I considered them to be less important, or secondary factors for the purpose of creating my models and so neglected any further consideration of them. This assumption was useful, in that it decreased the number of variables that I needed to incorporate and measure. However, there are instances in which these parameters are crucial. Consider, for example, that tablets are known to fail if there are too many fines, or that the difference between steel and iron is the presence of impurities such as carbon.

Process parameters, interparticle and die-wall friction, initial packing of the powder in the die, die size, the range of stresses employed, the compaction rate, the presence or absence of precompaction or other granulation techniques, the concentration of various excipients, and the method and order of mixing [11,12].

The size of this list alone makes it obvious that a truly comprehensive model of tablet compaction would be extremely complex. Therefore, while the ultimate goal is to be able to predict and control the behavior of a specific formulation in an arbitrary compaction procedure, our achievements have been much more modest.

To limit the scope of the modeling process to a practical level, it was decided to remove as many *process variables* as possible by holding them to a constant level.* Therefore the uniaxial compaction of *a single-component* tablet was considered. Furthermore, it was assumed that

1. The load is applied pseudostatically (i.e., slowly enough to appear to be independent of time).
2. The particle size is uniform.
3. The mechanical properties of the particles are isotropic and homogeneous.
4. There is no significant die-wall friction.
5. The density the stress is distributed uniformly through the tablet during compaction.

---

* However, this would not be the optimal approach if one had a compaction simulator at one's disposal—which we did not! If one had precise control over the compaction process, it might make more sense to limit the material parameters (for example, by studying the compaction of lead shot) and model the process variables.

6. No significant elastic strain remains after the compact is ejected from the die so that one can use out-of-die measurements of tablet density and thus make it unnecessary to consider elasticity in the models.

Based on a simple perusal of the tableting literature, the primary material properties that would be expected to play a prominent role in one or more of the simplified compaction situations listed above would be plasticity, elasticity, and fracture toughness.

If one was to consider tablet stress relaxation, one might assume that *compaction was time-independent,* but that relaxation, occurring over tens to hundreds of seconds, was a simple function of the deformation kinetic properties of the material under consideration.

The selection of parameters to be used in the models was informal and intuitive. While this approach was successful in the present case, it is not necessarily the most reliable one, especially if a large number of potentially interrelated factors are being considered. One way to optimize the selection of parameters to be incorporated into a predictive model is the use of factor analysis. This powerful statistical tool reduces the number of necessary variables or dimensions that must be considered without sacrificing any essential information. The method is described in detail by Harman [13].

B. Predictive Costs of Assumptions and Simplifications

Limiting the models to the conditions described above limits, in turn, their applicability and predictive capability. For example, it is not expected that they could be useful in predicting failure due to capping, since this is believed to arise from the operation of factors which we have omitted. Also, the models will not be useful in the selection of excipients or granulation technologies, since we have limited our consideration to single-component tablets. For these reasons, the utility of the models could be questioned by the industrial practitioner. However, I see the models as preliminary—the first step in an iterative approach that eventually will incorporate these and other important compaction parameters. That the models can be shown to have predictive capability in a limited sense provides the justification for continuing the refinement of this approach and verifies that the parameters selected are indeed critical to compaction.

Further assumptions and simplifications were made that are specific to each of the models described below. These will be highlighted, and their significance discussed, as they are introduced.

IV. MATERIALS CHARACTERIZATION

It is necessary to evaluate these material properties *in a manner which is independent of compaction.* Thus, although many scientists have suggested that the inverse slope of the Heckel plot is a measure of the yield or plasticity behavior of a material, this would not be an appropriate measure, since the relationship between yield is itself linked modelistically (and controversially!) to Heckel behavior. However some of the more generally accepted and most easily applied methods of evaluating elasticity (e.g., beam bending, dynamic resonance), plasticity (e.g., uniaxial tension, bending, torsion), and fracture (e.g., 3- or 4-point bending, chevron notch) usually require specimens which are relatively large and which possess a defined geometry. This violates the requirement that the tests be on a *micro* scale to minimize material wastage. Second, single, large, relatively flawless organic crystals are difficult to obtain, and the preparation of suitably oriented test specimens is often difficult or impossible. They are inherently brittle, susceptible to thermal shock, and degrade easily under conditions which are usually employed to produce large single crystals or fully dense polycrystalline specimens of other materials.

Fortunately, the appearance of ceramics that are used in engineering applications has led to the refinement of microindentation as a materials characterization technique. This is because crack propagation, which renders most other mechanical tests of plasticity almost impossible in these materials, is suppressed in this configuration. For ceramics and other brittle materials, the microindentation test is the simplest, and often the only mechanical test available. It is a rapid, convenient, and inexpensive method that has been developed to the extent that it can be used to evaluate all the mechanical parameters above using only a few measurements.

A. Microindentation Test

In the microindentation test, a small, hard (usually steel or diamond) indenter of a defined geometry is pressed into the surface of a test specimen for a specified length of time. The size of the resulting indentation is a function of the load applied to the indenter so that very small indentations, on the order of tens of micrometers in diameter, may be produced under well-controlled conditions.

The variables that must be defined in an indentation test are

a. Specimen size
b. Load
c. Indenter geometry

     d. Environment (composition, temperature, humidity)

     e. Indentation time

Relatively small loads are required to test most pharmaceuticals since organic crystals are generally quite small (<2 mm) and soft. Indentations may be measured using optical microscopy. Criteria which maximize precision have been established [14]. Hardness has been shown to vary as a function of load, even when indenters such as the Vickers or Knoop pyramids (see below) or sharp cones (e.g., Rockwell) are used.* This variability is believed to arise, in part, because the surface of the material may possess properties that are different from the bulk due to adsorption, environmental conditions, etc.

It is important to apply the load slowly and smoothly in the microindentation test, since indentations may be enlarged considerably by any vibrations which occur during the loading procedure. Modern hardness testers are adequate in this regard. However, extraneous vibrations (which cause the same misleading enlargements) must be avoided. It is useful to mount the indentation apparatus in a basement on top of acoustically insulating material.

The specimen must be at least 10 times as thick as the indentation is deep, to avoid the need to consider the role of the method of support when calculating the hardness value. For small specimens, this introduces another variable into the analysis, since the relative roles of bulk and surface deformation must then be considered [16].

The "sharpness" of an indenter plays an important role in governing the mode of deformation of the material during indentation, and also controls the magnitude of the effect of friction. Blunt indenters are preferable because

    1. Their indentation parameters are essentially independent of friction, especially if the indenter is cut from diamond: the coefficient of friction between polished diamond and unlubricated metals ranges from approximately 0.1 to 0.15.

---

*Geometric similarity of indentations at different loads occurs only for indenters whose cross-sections are independent of penetration depth. For these, the flow field should be independent of penetration depth and hardness should be constant. This is not true for spherical indenters. These indenters are often inconvenient if one is comparing a large number of diverse materials, since it is only valid to compare the hardnesses of materials indented to the same depth relative to the radius of the indenter. The benefit of spherical indenters is that they permit a measure of work-hardening [15].

2. The indentations are shallow, therefore the required thickness of the specimen is decreased commensurably.
3. Complications due to fracture behavior occur less frequently.
4. The two indenters that are employed most frequently are blunt (the 136° Vickers pyramid and the asymmetrical Knoop) and their behavior is well documented in the literature. Most advanced theoretical studies and routine applications employ these indenter configurations and instrumentation is readily available.

The specimen must be mounted so that the surface is normal to the direction of indentation. The imperfections of crystals and their small size often make this requirement troublesome because an entire crystal can be wasted if one requires too many "pretest indentations." This problem of precise alignment and movement can be overcome by using a goniometer mount with three axes of rotation that is attached to a two-dimensional micrometer stage. This also simplifies the process of finding an indentation in the future.

B. Hardness and Plastic Flow

The Vickers hardness (H, MPa) is calculated from the mean length of the diagonals of an indentation (Fig. 1) and is equal to the mean stress across the true area of contact:

$$H = \frac{1.854P}{d^2}$$

where P is the load (N) and d is the mean indentation diagonal (mm).

Hardness and the plastic yield stress* (Y) are correlated by a "constraint factor," that is believed to arise from the interaction between the yielding material and the surrounding elastically deforming material. The mechanism of constraint is still not fully understood [17–20]. Empirical correlations between the uniaxial yield stress of metals and their hardness indicate that the constraint factor is constant and approximately equal to 3. According to Marsh [19], the constraint factor for brittle, elastic materials varies with the ratio of the yield stress to elastic modulus of the material:

$$\frac{H}{Y} = 0.28 + 0.6 \ln\left(\frac{0.7E}{Y}\right)$$

Hardness, like plasticity, varies as a function of temperature and indentation time. Therefore one can assess the time-dependence of the

---

* The stress at which significant permanent deformation occurs in a specimen that is tested in a uniaxial tension configuration.

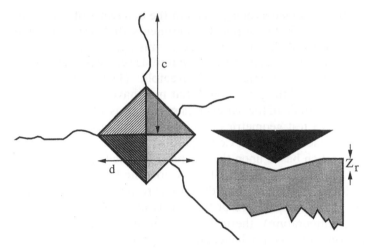

Fig. 1 Geometry of Vicker's indentation with parameters used to calculate material properties: d is the diameter of the indentation, c is the length of a radial crack, and Z_r is the depth of the indentation.

flow behavior of a material by measuring hardness in a controlled environment for a variety of indentation times. The drawbacks to undertaking a deformation kinetic analysis on the data thus obtained are (1) the flow field associated with the contact geometry is extremely complex so that it is difficult to assign a simple strain and strain rate; (2) the extremely long indentation times required makes errors due to extraneous vibrations highly probable. Thus, the measured hardnesses at long times will be extremely variable and often will appear to be smaller than they would be under ideal conditions. Finally, the kinetic analyses of polar substances, especially ones that form hydrogen bonds are complicated by the presence of water which is difficult to control, and which varies as a function of temperature. Flow kinetics will be discussed in more detail in the following section.

C. Kinetics of Flow

The kinetics of deformation are usually characterized by a viscoelastic analysis. A specimen possessing a simple geometry is subjected to simple stressing conditions* and the response is measured as a function of time.

---

*These include constant load (creep), constant strain rate, and constant strain (stress relaxation) configurations, among others.

The behavior is then modeled as being a sum of the behaviors of a series of viscous (dashpots) and elastic (springs) components that are connected both in series and in parallel. The advantage of this approach is that it is simple to perform and well accepted by the materials sciences community. For our purposes, the disadvantages are threefold: (1) the springs and dashpots are essentially "fitting functions" that may have no fundamental meaning; (2) the magnitude of the parameters may vary as a function of the test geometry; (3) the test geometries required are not easily attained in practice in the pharmaceutical field. Consequently, we have opted to adopt an alternative approach based upon the chemical kinetic theory proposed by Eyring [21]. Rapid plastic deformation is governed primarily by the threshold stress for dislocation motion (the critical resolved shear stress for yield), however, if the temperature of the test is increased (>0.4 of the absolute melting temperature), thermal activation usually influences the rate of deformation [20]. The fundamental assumptions of the rate theory of plastic deformation are

1. Energy barriers must be overcome in the flow process which are similar to those limiting the rate of a chemical reaction. Thus flow can be compared with a chemical reaction in which the composition remains constant but the arrangement of the molecules changes.
2. The barrier to deformation is symmetrical at equilibrium.* On average, atoms or molecules in condensed phases occupy equilibrium lattice positions and are oscillating about the minimum of the free-energy wells. When stress is applied work is done on the system and the potential energy of the molecules is increased and stored in a reversible (i.e., elastic) manner. Plastic flow will occur when the atoms move under the combined effect of the applied stress and thermal activation into a new equilibrium valley forming new bonds while breaking the previous ones. Thus when stress is applied the associated elastic work effectively decreases the height of the activation barrier in the forward direction while concurrently increasing the height of the barrier to deformation in the reverse direction (Fig. 2).
3. At high levels of stress, the probability of deformation in the forward direction is so much greater than that in the backward direction that, in principle, a single rate constant can be evaluated.

*The probability of deformation is equal in all directions so that no net deformation is observed.

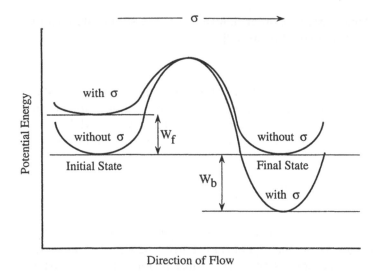

Fig. 2 Activation barrier to flow. When no stress is applied, the barrier is symmetrical and there is no net flow. When a stress is applied, it simultaneously decreases the height of the barrier in the forward direction and increases the height in the backward direction so that there is net flow in the direction of the applied stress.

4. The observed deformation is assumed to be the sum of the contribution of all the mechanisms which are operating concurrently and the observed behavior may be analyzed using nonlinear fitting techniques. However for the sake of simplicity in the discussion which follows, we shall consider a case in which one mechanism operates; a situation which can be realized experimentally by applying a high level of stress and which probably occurs in practice in the microindentation test and during compaction.

According to the equilibrium theory established by Arrhenius, the rate of a reaction is the sum of the rates of molecules going both from reactant to product state in the forward direction and from the product state to the reactant state in the backward direction, therefore the total rate constant K can be expressed as

$$K_f + K_b = A_f \exp\left(\frac{-\Delta E_f}{kT}\right) + A_b \exp\left(\frac{-E_f}{kT}b\right)$$

where A_f and A_b are frequency factors, ΔE_f and ΔE_b are empirical activation energies in the forward and backward directions, k is the Boltzmann constant, and T is the absolute temperature.

For an elementary, first order deformation, the strain rate $\dot{\varepsilon}$ can be derived from Orowan's equation as [20]

$$\dot{\varepsilon} = \alpha \mathbf{b} \rho_m K$$

where α is a geometrical factor relating the active slip system to the shear strain direction, $\mathbf{b}$ is the Burgers vector, ρ_m is the mobile dislocation density. The constant α and $\mathbf{b}$ are known accurately for single crystals, while ρ_m and K are functions of stress, strain, temperature and structure. These concepts are combined as follows:

$$\dot{\varepsilon} = \alpha \mathbf{b} \rho_m \left(A_f \exp \left(\frac{-\Delta E_f}{kT} \right) + A_b \exp \left(\frac{-\Delta E_b}{kT} \right) \right)$$

At equilibrium, the activation barrier is symmetrical so that $\Delta E_f = \Delta E_b = \Delta E_e$, where ΔE_e is the equilibrium barrier height. The net strain rate under these conditions is zero. If a force is applied to the system, the forward and backward barrier heights are changed by the associated work (W):

$$\dot{\varepsilon} = \alpha \mathbf{b} \rho_m \left\{ A_f \exp \left(\frac{-[\Delta E_f - W_f]}{kT} \right) + A_b \exp \left(\frac{-[\Delta E_b + W_b]}{kT} \right) \right\}$$

where W_f is work performed to lower the activation energy barrier in the forward direction, W_b is work performed to increase the activation energy in the backward direction. From the definition of work, stress, and activation volume, the following relationships are obtained:

$$W = Fd$$
$$\sigma = \frac{F}{A}$$
$$W = Ad\sigma$$
$$V_{act} = Ad$$
$$W = \sigma V_{act}$$
$$W_{shear} = \tau V_{act}$$

where F is force, σ is stress, A is area, d is distance, V_{act} is activation volume, and τ is the shear stress.

Combining these equations gives

$$\dot{\varepsilon} = \alpha \mathbf{b} \rho_m \left\{ A_f \exp \left(\frac{-[\Delta E_f - \tau V_{actf}]}{kT} \right) + A_b \exp \left(\frac{-[\Delta E_b + \tau V_{actb}]}{kT} \right) \right\}$$

Assuming that the activation volumes, activation energies and pre-exponential factors are the same in both the forward and backward directions, then

$$\dot{\epsilon} = K \left[\exp \left(\frac{\tau V_{actf}}{kT} \right) - A_b \exp \left(-\frac{\tau V_{actb}}{kT} \right) \right]$$

Noting further that

$$\sinh x = \frac{1}{2(e^x - e^{-x})}$$

then

$$\dot{\epsilon} = \frac{K}{2 \sinh(\tau V_{act}/kT)}$$

At high levels of stress, the deformation is usually controlled primarily by activation over one barrier in the forward direction, so that

$$\dot{\epsilon} = \alpha b \rho_m A_f \exp \left(-\frac{\Delta E_e}{kT} \right) \exp \left(\frac{\tau V_{act}}{kT} \right)$$

The shear stress rate can be substituted if its relationship with the strain rate is known. The Halsey-Eyring three-element model (consisting of a linear spring and a nonlinear dashpot in parallel) often is satisfactory. Deformation kinetic analysis using in this model is performed as follows. In stress relaxation the strain rate of the nonlinear Maxwell element of the three-element model must satisfy the condition that the total strain rate is equal to zero. The following relationship between shear stress rate ($\dot{\tau}$) and strain rate ($\dot{\epsilon}$) then one obtains

$$\epsilon_{tot} = \epsilon_{el} + \epsilon_{visc} = 0$$

$$\epsilon_{el} = -\epsilon_{visc}$$

$$\epsilon_{el} = \frac{\tau}{E}$$

$$-\epsilon_{visc} = \frac{\tau}{E}$$

where E is the apparent elastic modulus which, in practice, is a composite value for the material and testing system in series. Therefore it follows that

$$\dot{\tau} = \alpha b \rho_m A_f E \exp \left(\frac{\Delta E_e}{kT} \right) \exp \left(-\frac{\tau V_{act}}{kT} \right)$$

From the foregoing equation, the experimental activation volume (V_{act}) can be determined from the slope of the line derived by plotting $-\ln$(shear stress rate) versus (average shear stress) at constant temperature. The experimental activation energy (E_{act}) can be calculated from the slope of the line which results from plotting $\ln$(shear stress rate) versus the inverse of the absolute temperature at constant stress [21–23].

The Burgers vector **b** is the unit deformation distance and often equals the dimensions of the unit cell. It is customary to express the activation volume of a deformation mechanism in Burgers vector units. The activation volume can be interpreted physically. For example, if deformation occurs by molecular diffusion, it will be on the order of the molecular size. On the other hand, if flow is limited by the interaction of many dislocations, the activation volume will consist of the entire region containing the interactions and can be of the order of 1000 times the molecular volume.

As discussed above, the microindentation test has been employed to investigate the deformation kinetics of many metals and ceramics, and also fact centered cubic salts such as LiF and NaCl at high levels of stress [24,25]. This fact is particularly fortuitous because the compaction behavior of these materials has been studied exhaustively in the pharmaceutical literature so that their behavior can be used as the basis for the development of the deformation kinetic analysis. While it is often the only test available, it possesses two major drawbacks that must be considered in this instance. Because the deformation field is so complex, consisting approximately of an expanding sphere that presses into the surrounding elastic material, the strain rate is difficult to define, and many different approximations are employed. Verrall et al. [24] used a particularly simple relationship. Noting that the strain $\epsilon \sim 0.08$ for a Vickers indentation, they suggested that the following relationships could hold:

$$\dot{\epsilon} = \frac{0.08}{t}$$

$$\dot{\epsilon}_{shear} = \frac{0.08\sqrt{3}}{t}$$

$$\tau = \frac{H}{3\sqrt{3}}$$

where $\dot{\epsilon}$ and $\dot{\epsilon}_{shear}$ are the strain rate and shear strain rate, respectively, and t is time. The latter equation rises from the relationship between the hardness and the shear strength of a material, assuming a constraint factor of 3 and the von Mises yield criterion [20]. This number should be altered as described above for many brittle materials that possess constraint factors that are substantially smaller than 3. These relationships arise naturally if the change in hardness with time is equivalent to a creep process. The factor of 0.08 is arbitrary. Assuming it is independent of temperature, its magnitude does not affect the results of kinetic analyses.

D. Elastic Behavior During Indentation Testing

While a number of groups have developed procedures to evaluate elastic moduli from microindentation parameters [26], the approach elaborated by Breval and MacMillan [27], which requires the measurement of the recovered depth of an indentation, is one of the most versatile. An indentation in a purely ductile material does not recover elastically. Conversely, the depth of an indentation in a highly elastic material will recover substantially. In the purely elastic limit, no permanent indentation remains (Fig. 1). The final relationship fits the experimental data obtained for polycrystalline specimens tested under uniaxial tension.*

$$\left(\frac{2Z_r}{d} \right)^2 = 0.08168 \left[1 - \frac{8.70H}{E} + 14.03 \left(\frac{H}{E} \right)^2 \right]$$

where Z_r is the recovered depth of indentation (μm), H is the hardness in MPa, and E is the elastic modulus (MPa).

E. Indentation Fracture

Many crystalline materials will fracture in a brittle manner if they are stressed above a critical level.† The value of the fracture stress depends upon the properties of the material, the configuration of the test system, and environmental variables. The science of fracture mechanics was developed to elucidate the fundamental material properties which control fracture behavior, and to discover their relationship with the external variables [28].

The relationship between the stress intensity factor at failure (K_C), often called the fracture toughness, and the fracture stress of a brittle

---

* While the validity of the test when performed on single crystals is unknown, it was used in the present research. Validation of the test is currently being undertaken and preliminary results will be discussed below.

†Extremely ductile materials may elongate almost indefinitely, finally undergoing necking and ductile fracture. These same materials may exhibit brittle fracture behavior, however, if the load is applied quickly enough.

specimen containing a sharp internal crack loaded in simple tension is
expressed as follows:

$$\sigma_f = \left[\frac{K_c^2(1-\nu^2)}{\pi c} \right]^{0.5}$$

where σ_f is the fracture stress, ν is the Poisson ratio, and c is the crack
length.

    While the fracture stress varies with the method of loading and the
size of cracks within a test specimen, K_c is a constant that ideally, depends
only on the mode of crack propagation (e.g., tension, shear, torsion). In
reality, K_c is also a function of nonequilibrium kinetics, deviations of the
crack surface from planarity, plasticity, and other dissipative terms [28].
However, if care is taken during the specification of the conditions under
which the test is performed, K_c may be considered to be a predictive
material property.

    The field of indentation fracture mechanics has been developed re-
cently [29–33]. While exact stress intensity factors have been determined
for many test configurations, only approximate solutions are available for
indentation. Approximate fracture mechanics solutions are *calibrated* using values for K_c determined using other methods [34]. The K_c values used
here were calculated using the equation developed by Anstis et al. [35]
which requires the measurement of radial cracks that form during microindentation (Fig. 1):

$$K_c = 0.016 \left(\frac{E}{H} \right)^{0.5} \frac{P}{c^{1.5}}$$

where P is the load placed on the indenter.

    The Anstis equation has been calibrated using many single crystal
and polycrystalline substances, and has been employed subsequently by
others [36]. The fit is excellent for polycrystalline substances, but not as
good for monocrystals. However, the fit for single crystals is improved if
the calculations are corrected for the angular orientation of the cracks in
the crystals [37].

F. Assumptions in Material Characterization and Possible Consequences

A number of assumptions and simplifications have been made in this section. A brief discussion of each, and the possible consequences with respect
to the predictive capabilities of associated models follows.

1. Use of Hardness to Characterize Deformation

The first, and most critical, assumption is that we can use mechanical parameters derived from crystal hardness measurements to predict the mechanical behavior of compacts at all. The contact configuration is extremely complex, and most of the correlations with mechanical properties derived from simpler experimental configurations are semiempirical, and not immune to dispute. Luckily, the prevailing conditions during tablet compaction and subsequent processing always entail interparticulate contact processes. Therefore, the hardness, apparent elastic modulus, and indentation fracture resistance alone provide valuable information, and analytic solutions are not essential.

2. Vickers Indentation Parameters Are Isotropic and Mimic the Interparticulate Contact Configuration

The deformation and fracture behavior of crystals is anisotropic; for example, slip and fracture occur on defined crystallographic planes, in a limited number of crystallographic directions. An indenter of high symmetry such as the Vickers tends to "average" over almost all possible deformation directions so that for many materials, the Vickers hardness does not vary much as a function of indentation direction. It is not wise to make this assumption, however. It is best to measure crystal hardness in several orientations and on several crystallographic faces to ensure that this assumption is valid. If hardness varies substantially from face to face, the hardness of one face will not predict the deformation behavior of a compact, since the particles contact each other in every direction.

The assumption that parameters obtained using the Vickers indentation configuration can be used to predict the interparticulate contact configuration is potentially more problematic, since the rounding of contacts that occurs as they are pressed together makes the latter resemble the mutual indentation of spheres. As mentioned earlier, spherical indentations of different depths are not geometrically similar. Therefore, one might expect the apparent hardness of compaction to vary as a function of relative density as the interparticulate indentation geometry changes. In practice, one would expect this problem to be significant only at higher relative densities. The models described below are limited to compaction at low relative densities where the contact geometry changes very slowly as a function of interparticle penetration. The hardness literature suggests that this approximation is appropriate in many situations.

3. Random Crystal Orientation in-Die

Recalling the equations describing deformation kinetics, it can be seen that the rate constants contain two geometric factors, α (an orientation factor), and **b** (the Burgers vector for crystal deformation). These are required if one is to characterize the deformation properties of a crystalline material because crystal deformation is anisotropic. By ignoring these factors we essentially are assuming that the material in the die is a randomly packed, polycrystalline assembly. The disadvantage of in-die measurements, then, is the fact that we lose specific information about the nature of crystal deformation and calculate an average "deformability." As long as the packing remains random, this assumption is acceptable. However, if compaction causes crystal reorientation (as is known to occur with anisotropic materials), then the kinetic information will be valid *only* for compacts compressed to the same condition.

4. Equations Used to Relate Contact Parameters to Material Parameters

Semiempirical equations were employed to evaluation E, the elastic modulus, K_c, the fracture toughness, and μ, the shear stress. The primary difficulty here is that it is not easy to determine the error associated with them *in any particular case*. Thus, the failure of a model to predict the compaction of any given material could be due *either* to a failure of the model *or* a failure of these equations. Moreover it is more likely that these difficulties will appear with problematic materials—ones whose behavior is *intuitively unexpected.*

Therefore, the most productive way to manage cases of incorrect prediction is to use them to explore the assumptions used, and eventually, to improve both assumptions and models. Rather than viewing it as a failure, and abandoning the data or the project, nonprediction can viewed as an opportunity for growth.

5. Activation Parameters

We assumed that

- Only *one* barrier controls the deformation kinetics of a material.
- The barrier is *symmetrical.*
- The stress is *high* enough that deformation due to activation in the reverse direction can be neglected.
- That the kinetics can be described by the Halsey-Eyring three-element model (so that the rate of stress relaxation can be related to the strain rate).

The consequences of these assumptions parallel those for the previous section. Once again, the most productive approach is to recognize that the modeling process is still in its infancy, and that any deviations from the trends predicted must be used to explore the nature of the assumptions and the models, with the eventual goal being the improvement of both.

V. MODELS OF TABLET COMPACTION

The goal of our modeling was to predict the behavior of a large number of particles in an aggregate based on the behavior of an "average" individual particle as the space available to each particle is reduced systematically. The easiest way to do this is to imagine that the particle is confined in a box or *cell* whose volume is reduced step by step. What parameters might be associated with this system?

By considering Fig. 3, the following are possibilities:

1. The relationship between the relative density of the assembly and the fraction of the total volume of the cell occupied by the particle (volume fraction)
2. The manner in which the material responds to the loads applied at its contacts (i.e., reversible and irreversible deformation, fracture)
3. The shape of the particle

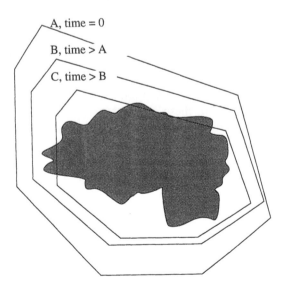

Fig. 3   Conceptual model linking tablet densification with the deformation behavior of a single particle.

4. The shape of the cell
5. The number of contacts that the particle makes with the edges of the cell
6. The number of points at which the particle is loaded externally at any given volume fraction
7. The areas of each of the foregoing contacts as a function of the volume fraction of the system
8. The load sustained at each of the contacts
9. The stress applied to the entire system (i.e., the manner in which the externally applied load is distributed over the surface of the system which can include void space (the cell) or solid space (the particle)

The most helpful thing to do at this stage is to attempt to simplify things as much as possible. The process here is intuitive, but one could think, for example, that mathematical simplicity might dictate that a two-dimensional case—for example, a circular body enclosed within a contracting square—be considered first. This, and many other models were considered and then abandoned as their results did not predict reality. The models described below were not developed in one step. Furthermore the steps given below were not the only attempts made. However they are given in *chronological* order. In other words, it was decided, definitely, that the compaction of brittle and ductile materials would be modeled separately before an interparticulate contact stress was decided.

Finally, I must highlight the degree to which I am indebted to other investigators for their work and insights which preceded mine. In fact, the models that follow can be viewed as a collection of fragments from the vast literature of diverse fields such as metallurgy, materials sciences, chemistry, physical pharmacy, and mathematics that impinge upon solid pharmaceutics. The applicability of information from these other areas essentially is discerned by a simple pattern recognition process. The key, really, is the time spent (much of it "wasted") in exploring the unknown territory.

A. Relationship between the Solid Volume Fraction of the Model and Tablet Relative Density

If the particle/cell system is "average," then its solid volume fraction should be identical with the volume fraction of the tablet.

B. Distinction of "Brittle" and "Ductile" Compaction Behavior

Ductile materials compact in a manner which can be distinguished from the compaction behavior of brittle materials. For example, the form of the plot that results when the tablet relative density is plotted *vs.* the compaction stress according to the Heckel relationship has been shown to be related to the mechanical properties of the particles:

$$\ln \left(\frac{1}{1 - RD} \right) = k_0 + k_H \sigma_p$$

where RD is the relative density of the compact, k_0 is a constant, k_H is the slope of the Heckel plot (inverse mean yield pressure), and σ_p is the punch stress.

A linear Heckel plot is believed to indicate essentially ductile behavior of the particles during compaction, while a curvilinear plot is believed to indicate brittle behavior. Given this fact, and the observation that the compaction of brittle materials is accompanied by considerable particle fracture and asperity crushing,* it would seem appropriate to construct two distinct models of compaction to take this mechanistic difference into account.

"Brittleness" factors have been used extensively in the wear literature [38], and appear to be reasonable criteria to differentiate compaction behavior, based on the facts stated above. The simplest brittleness factor is the ratio of fracture toughness to hardness, K_c/H. This "brittleness index," varying from 10 to 50 for brittle materials, is on the order of unity for semibrittle materials such as sodium chloride. It is two orders of magnitude smaller for metals and polymers, and this divides the materials into two groups (<5 = ductile; >5 = brittle) which correctly reflect the two "coarse" types of Heckel plot as well as the microscopic behavior of the materials during compaction.

Why bother with this differentiation? As some research has shown, most materials are neither purely brittle nor purely ductile, and some Heckel plots are difficult to classify. The reason is that differentiation decreases the number of parameters that must be considered in any given model. In short, it simplifies modeling.

C. An Expression for the Change in the Number of Contacts per Particle During Densification

By referring to Fig. 3 it can be seen that the particle contacts, and is intersected by, the cell at Stage C in the hypothetical densification process. Also, we know that the number of particle-particle contacts increases with increasing density. The simplest way to model such a system is to assume that *whenever the particle contacts the side of its cell, it is actually touching another particle.* Thus, we only need to consider the latter in our discussion of densification.

---

*Whereas ductile particles tend to become rounded and blunted.

A review of the densification literature shows that the total force on a particle is directly proportional to the coordination number which gives the average number of nearest neighbors in a packing of equal-sized particles. Due to particle rearrangement during compaction, the coordination number usually increases during densification. While several relationships between relative density and coordination number (Z) have been described in the literature, the relationship described by Arzt [39] was used in the present work because it describes the densification of randomly packed spheres.

$$Z_{RD<0.85} = 7.3 + 9.5(RD - 0.64)$$
$$Z_{RD\geqslant0.85} = 9.3 + 9.5(RD - 0.85) + 881(RD - 0.85)^3$$

D. An Expression for the Average Area per Contact as a Function of the Particle Geometry and the Average Relative Density of the Compact

1. Ductile Materials

It is assumed that the stress response of crystals of fairly uniform shape resembles that of spheres. While the asperities of the particles initially may be angular, small amounts of deformation will tend to round then. Therefore it is further assumed that on average, the contacts will deform as sections of spheres.

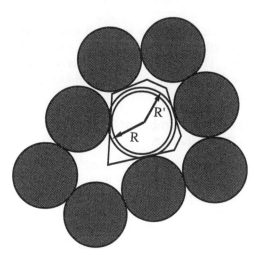

Fig. 4 Exact particle model used to predict the densification of ductile materials.

An elegant model describing the densification of spherical powders was described by Arzt [39]. In this model, a spherical particle with an initial radius of unity (R; Fig. 4) is assumed to grow concentrically within its original cell (40). The cell chosen is termed the "Voronoi cell" and is the set of points closer to a particle's center than to any other particles centers.

As the particle "grows," eventually it will contact and overlap a face of the cell (R', Fig. 4). The volume of the material removed by the contact is assumed to be distributed evenly across the remaining free surface of the particle increasing the apparent particle radius further (R''). By proceeding stepwise through the following equations, one can derive an expression for the average contact area in terms of the relative density:

$$R' = \left(\frac{RD}{RD_i} \right)^{1/3}$$

$$V_{ex} = S_f \frac{R''^3 - R'^3}{3R'^2}$$

$$R^* = R' - 1$$

$$V_{ex} = R^{*2} \left(\frac{\pi Z_i}{3} \right) (2R' + 1) + R^{*3} \left(\frac{\pi C}{12} \right) (3R' + 1)$$

$$S_f = B \, 4\pi R^L - 2\pi Z_i R' R^* - \pi C R' R^{*2}$$

$$A = \frac{\pi}{3ZR'^2} [3Z_i(R''^2 - 1) + C + CR''^2(2R'' - 3)]$$

where RD is the present relative density, RD_i is the initial relative density, S_f is the free surface (surface not contacted by other particles) remaining on sphere, R'' is the present radius of sphere with deposited material covering it, R' is the enlarged radius of sphere *before* material deposition, V_{ex} is the volume of material cut off by a face of the Voronoi cell, Z_i is the initial coordination number, Z is the present coordination number, C is a constant equal to 15.5, and A is the average contact area.

2. Brittle Materials

Most models of densification assume that the particles are spherical. While the first attempt at models to predict the compaction of brittle materials employed this assumption, none of the models were predictive. This led to a deeper reflection on the processes that are observed during compaction. This led to the realization that brittle particles fracture at very low stresses, probably because the load is applied at asperities so that the stress is magnified enormously. Since a brittle material cannot yield to a significant extent, they will fracture instead. What results is two or more *angular* fragments that are displaced from their original positions. This displacement accomodates the volume reduction. Thus, it became obvious that the sphericity assumption was inadequate for brittle particles. However it was not so clear *what shape* would be appropriate. Crystals of many brittle drugs and excipients are acicular (needle-shaped), platy, cubic, etc. and remain that way during compaction. For simplicity it was assumed that the particles are cubic, and that they contact each other so that a cube corners touches the face of another crystal.

For a collection of cubes in a porous assembly, the initial packing will be a function of the *average relative misorientation of the particles*. This assumption has a very interesting consequence: if one assumes that the powder packs to a certain density maximum under zero load, then it implies that the interparticulate friction (and particle surface free energy) will play an important role in determining the entire history of densification. This suggests that this parameter should be added into the model. However for simplicity's sake, this was not done. Instead, one initial set of data were found experimentally (the smallest recordable load and the associated relative density) and used as the initial boundary condition for the model.

When angular bodies contact each other, rotation moments will invariably develop. Therefore, unless the crystals are restricted by mechanical interlocking or extremely high frictional forces, they should rotate relative to one another during compaction. It was assumed that densification produced two effects: (1) crushing (flattening) of the asperities and (2) rotation of the particles. Therefore a relationship between these two effects was required and can be described as follows. As densification proceeds, the area of contact between a surface and a corner of a rectangular parallelepiped, the latter being crushed by contact with the surface, is controlled by their relative orientation, which may be defined by rotations around two axes which pass through the center of the particle. Figure 5 illustrates one such orientation, a tilt of $\theta°$ around the diagonal of a cube face. For this case the height of the crushed asperity (h) is related to the area of contact (A) by the expression

$$A = h^2 \frac{\tan(\theta) + \tan(90 - \theta)}{\sin(\theta)}$$

Since rotation around the diagonal simplifies subsequent calculations and decreases the number of assumptions required in modeling, this geometry was used throughout the modeling process, producing an error of 5% at most.

Rotation is caused by the moment of forces that are applied in a direction away from the center of rotation. This force is balanced by the

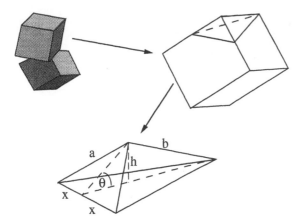

Fig. 5 Geometry of the contact region between two cubic particles.

local stresses at the interparticulate contacts. Therefore densification due to rotation is still controlled by the material properties. The resistance to deformation at the trailing edge will be different from that at the leading edge. At the trailing edge, the rotation will add to the indentation pressure whereas it will be subtracted from the indentation pressure at the leading edge. To formulate this difference in terms of the mechanical properties of the material, we assumed that crushing and the rate of particle rotation are governed by the *constraints* which inhibit the removal of the fragments from the interface. As the particles rotate relative to each other, the angle θ decreases, the asperity crushes further, and the particles can move closer to each other as fragments are displaced Fig. 6. The fragments which are near the side of the asperity which is adjacent to θ in Fig. 6 are subjected to intense shear. Flow occurs laterally, and is therefore relatively unconstrained. At the opposite edge of the asperity, normal stresses inhibit the removal of fragments, therefore the configuration at this location resembles indentation. *The simplest approximation of the relative rates of removal of fragments from each edge of the asperity should then be the ratio of the hardness to shear strength of the material.* This ratio is a function of the von Mises yield criterion and the constraint factor discussed above.

The next step is to relate the interparticulate contact area with the relative density of the compact. This is where the established geometric relationships for spherical particles can be used to advantage. Therefore, a cubic particle of unit volume is assumed to be oriented randomly *within a spherical cell,* many of the latter being packed randomly ($RD_i = 0.60140$). These may be visualized as the spheres in the Arzt model (see Fig. 4)

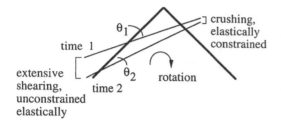

Fig. 6   Illustration of the constraints that controls the rotation and deformation of brittle particles.

arranged within their Voronoi cells. The sphere is assumed to possess a volume of unity when the relative density of the entire assembly is equal to 0.601. The spherical cells provide a geometric reference to describe the densification in terms of particle rotation and asperity crushing.

The particle is oriented within the cell so that it touches the others surrounding it. The actual number of contacts is assumed to be governed by the coordination number for the spherical "cells." During densification, the cube remains the same size but the spherical reference cell contracts. As the reference sphere contracts, the cubic particle must either rotate or be crushed at its asperities to accomodate the decrease in volume. *The debris resulting from crushing is displaced to the void spaces and is assumed to bear no load.** An increase in coordination number with relative density occurs as the asperities of the angular particles impinge upon particles within other cells. The area of contact between two particles and the radius of the cells is related through h, the present height of the crushed asperity, and θ, the present angle between the two particles:

$$\frac{2y}{\sqrt{3}} = \sin(35.26 + \theta)$$

$$\frac{y - h - 0.5}{2} + 0.5 = r$$

Figure 7 illustrates the geometry of the contact and the definitions of the quantities y, h, and r (the radius of the reference sphere).

The relative density at any given instant may be found directly from r by noting that the true volume of the material associated with one refer-

---

* This assumption was encouraged by the experimental observation that a considerable amount of loose debris can be brushed out of the matrix of a tablet that is fractured.

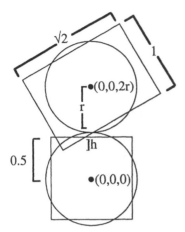

Fig. 7 Geometry which relates the relative orientation of two cubic particles during compaction with the packing of random-close-packed spheres.

ence sphere is always unity (i.e., $V^* = 1$), and that the ratio of the volume of the reference sphere and the total volume is a constant:

$$RD = \frac{V^*}{V_{total}} = \frac{0.601}{V_r}$$

$$V_r = \frac{4\pi r^3}{3}$$

where V^* is the volume of crystal within reference sphere ($=$ unity), V_r is the volume of the reference sphere, and r_r is the radius of reference sphere.

Densification is then modeled as follows:

1. The initial values of a and b are found from the initial value of h. The latter is found from the initial compaction stress and relative density values, the boundary conditions of the model as described above.
2. For a series of increasing a values, each corresponding b value is calculated from the relative orientation of the cells at (RD_i, F_i) and a linear regression slope which is equal to the constraint factor of the material. This is equivalent to assuming that the "relatively unconstrained side" of the asperity grows faster than the "relatively constrained side" of the asperity by a factor which is equal to the ratio of the hardness and the shear strength of the material in question.
3. The corresponding angle of orientation and each increasing crushed asperity height (h) are calculated thus:

$$b - b_i = \sqrt{3}\, CF(a - a_i)$$

$$\theta = \tan^{-1}\left(\frac{a}{b}\right)$$

$$h = a \sin(90 - \theta)$$

E. An Expression for Local Yielding at the Contact

Initially, it was assumed that the local contact stress could be given by the hardness value. Unfortunately, models based on this assumption were not predictive of material behavior. The following analysis yielded local contact stresses that predicted compaction behavior.

1. Hydrostatic Compression

Under hydrostatic loading, two contacting particles experience large normal forces and little shear at their interface. This configuration resembles the mutual indentation of two sharp bodies. For a plastic material, yielding occurs at approximately three times the yield stress of the material (i.e., when the stress is equal to H). In this loading configuration, the most appropriate relationship for local yielding is

$$\sigma_{ct} = H = \frac{F_c}{A}$$

where σ_{ct} is the contact stress, F_c is the contact force, and A is the area of contact.

2. Uniaxial Compression

Uniaxial loading of a mass of particles in a die superimposes shear stresses on the hydrostatic stresses. The extensive shearing that occurs during uniaxial compression increases the probability that the particles will rearrange during compaction. Schwartz and Holland [41] found that the slope of the Heckel plot for iron derived from isostatic compaction was $4.9e^{-8}$ MPa^{-1}, while those quoted by Heckel for iron powders undergoing uniaxial compaction, ranged from $1.1e^{-7}$ to $1.4e^{-7}$ MPa^{-1}. While they attributed this effect to the presence of fines, it is more likely that shear, while not causing densification, facilitates it [42].

3. Ductile Materials

While the contact area in the absence of shear stresses will be governed by the hardness, any tangential movement will enlarge the area of contact if the material is ductile [43]. It is certain that the average contact area will be larger than that which may be characterized by the hardness value, and smaller than that characterized by the shear strength. Therefore, some average proportionality between the shear strength and the contact stress must characterize uniaxial compression at various ratios of punch to die-wall stress. The following simplistic, linear approach was used to calculate the proportionality between the mean contact stress in the die and the shear strength of the material:

1. It was assumed that the proportionality was equal to 1 for orientations where the interparticulate shear stress was greater than the normal contact stress.
2. It was assumed that the proportionality was equal to 5.2 if shear stress was equal to zero.
3. It was assumed that the proportionality varied linearly from 1 to 5.2 between the directions at which the two limiting conditions hold.

For a punch to die-wall stress ratio of 2.5:1 (the average ratio observed during uniaxial compaction of pharmaceutical materials), the average of these values was 3.4. It was expected that this provided a reasonable approximation of the true value given the number of assumptions required to calculate it. Therefore, for ductile materials in uniaxial compression, the contact stress was calculated as follows:

$$\sigma_c = \frac{H}{3.4} = \frac{F_c}{A}$$

The hypothesis that shear facilitates consolidation provides a means of explaining the effect of lubricants and glidants on the densification process. These materials facilitate densification, but weaken the resultant compact. Since very thin layers of lubricants are present on the surfaces of the crystals, they would play a minor role if only normal stresses caused densification.

4. Brittle Materials

As two brittle particles are pressed together, their asperities yield by crushing rather than by flow. Therefore it might seem that the crushing strength of an asperity must be determined. According to Rice [44], however, the compressive failure of a material is limited by plasticity. Therefore, it is probable that the hardness is a good approximation of the contact stress. In fact, it may be that crushing and lateral particle displacement plays a significant role during microindentation itself.

F. An Expression for the Local Stresses at the Contacts in Terms of the Far-Field Stresses

1. Relationship between Uniaxial Compaction Stress and Mean Densification Stress

Die compaction differs from hydrostatic compression since the load is applied from one direction only. For the purposes of the models, it is assumed that densification is caused by the hydrostatic component of the stress within the die, which is smaller than the punch stress, while the shear component of the stress causes only changes in shape [45].

The hydrostatic component of the compaction stress may be calculated as the mean of the principal stresses within the die. This calculation is not trivial, since the effects of friction and the discontinuous nature of the powder compact will cause rotation of the direction of the principal stresses with increasing depth and radius within the compact, and in fact, will probably change from particle to particle. Therefore, to simplify the problem, it will be assumed that the first principal direction is parallel to the axis of the die; the others are perpendicular to this, but otherwise are arbitrarily directed ($\sigma_y = \sigma_z$). The ratio of the die-wall stress (σ_y) to the punch stress (σ_x) has been measured for sodium chloride [46,47] and sucrose [47,48], and appears to be essentially constant for σ_x values between 0 and 197 MPa. The mean stress (MS), assumed to be equal to the hydrostatic component of stress, will be calculated as the mean of these assumed principal stresses:

$$MS = \bar{\sigma} = \frac{\sigma_x + \sigma_y + \sigma_z}{3}$$

2. Relationship between the Mean Compaction Stress and the Average Force at One Contact

Since the hydrostatic stress, approximated by the mean stress in the die, gives rise to densification, it is assumed that the mean compaction stress is distributed over the surface of a sphere enclosing the particle whose deformation is being considered. The average relative density of the contents of the sphere is equal to the present relative density of the compact and the average force on each particle contact region is determined by the coordination number at any given time. The volume of the sphere is calculated as follows.

a. Ductile Materials

It is assumed that the size of the Voronoi cell (or, equivalently, the reference sphere over which the external stress is distributed) is constant, while the particle within it grows. Then the surface area of the reference sphere also remains constant and may be calculated from the initial relative density:

$$RD_i = \frac{V_p}{V_r}$$

$$V_r = \frac{4\pi r_r^3}{3}$$

$$SA_d = 4\pi r_r^2$$

where V_p is the particle volume, V_r is the volume of the reference cell, r_r is the radius of reference cell, and SA_d is the surface area of the external reference cell for a ductile material.

b. Brittle Materials

The volume of the particle of brittle material is assumed to be equal to unity, while the volume of the external reference sphere is assumed to decrease. The geometric reference sphere must contain the particle and sufficient void space so that the average relative density of the contents is equal to the present relative density of the compact. The volume of the sphere over which the mean compaction stress is distributed is equal then to the inverse of the relative density. The surface area of this sphere is found as follows:

$$RD = \frac{V^*}{V_r} = V_r^{-1}$$

$$r_r = (0.75\pi RD)^{\frac{1}{3}}$$

$$SA_b = 4\pi r^2$$

where V^* is the volume of the crystal within the Voronoi cell ($V^* = 1$), and SA_b is the surface area of the reference sphere for a brittle material.

For *both* brittle and ductile materials, the local contact stress then is related to the far-field stresses as follows:

$$MS = \frac{ZF_c}{SA}$$

$$F_c = \sigma_c A$$

where Z is the coordination number, F_c is the force on one contact, σ_c is the contact stress, and A is the area of contact.

G. Predicting Compaction

1. Ductile Materials

The model is applied as follows:

    1. The initial relative density is that of the powder before consolidation begins.

2. A set of increasing relative densities is prescribed.
3. The coordination number for each relative density is calculated.
4. The contact area per asperity is calculated as a function of the relative density.
5. The total force per particle is calculated.
6. The hydrostatic or mean stress is calculated.
7. The punch stress is calculated for each given relative density.

In summary, this procedure calculates the punch stress at each prescribed relative density during uniaxial compaction from the mechanical properties of the ductile crystals being considered. The densification of sodium chloride was predicted using the average ratio of punch stress to die-wall stress found in the literature (i.e. 0.4 [47,48]) and the hardness determined experimentally by microindentation (213 MPa). The predicted and experimental curves coincide [49] within the range of the experimental variability associated with the hardness values and the compaction data (Fig. 8). It is obvious that accuracy of the hardness is crucial to the agreement between theory and experiment. Similar success was found in the prediction of the compaction of potassium bromide and marumerized beads of microcrystalline cellulose [50].

2. Brittle Materials

The following sequence is used to calculate the densification curves.

1. To construct the densification curve we require knowledge of the initial stress applied to permit accurate measurement of the initial relative density.
2. The initial mean stress is calculated from the punch and die-wall stresses.
3. The initial force per contact is calculated by multiplying the mean stress by the initial surface area of a sphere of the same relative density as the compact and dividing by the coordination number at that relative density.
4. The second spherical reference is selected so that it possesses the same volume as the particle ($V^* = 1$) when the relative density of the powder is 0.601 (i.e., "loose random-packed").

The next step involves an iterative process. A radius R_0 which is greater than the radius of the reference sphere is selected arbitrarily to represent the size of the reference cell when the cubes just contact each other. The initial angle θ is found from the length of the cube diagonal ($\sqrt{3}$) and the fact that the reference spheres just touch (i.e., distance from center to center is $2R$). Twice the difference between the arbitrary radius,

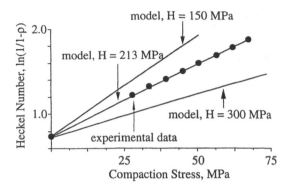

Fig. 8 Comparison of the experimental and predicted Heckel plots for sodium chloride. The limits of error that could occur as a result of error in hardness measurements are shown.

R_0, and that of the reference sphere gives h at the beginning of the compaction. The contact force corresponding to this value of h may then be calculated. This contact force is compared to that calculated in step 3. The length R_0 is varied until the contact forces in step 4 and step 3 are equal. At this point, the initial values of θ and h are set.

    5. The initial values of a and b are found from h in step 4.
    6. The value a is increased in a stepwise manner and then b and h are calculated.
    7. The new contact area, and from this value, the contact stress, are calculated.
    8. The relative density is calculated.
    9. The mean stress may then be calculated from the present coordination number and the surface area of a sphere, the contents of which possess the present relative density (i.e., in effect steps 2 and 3 are carried out in reverse).

While the densification of sucrose (Fig. 9) is predicted by the model to within experimental error, the compaction behavior of acetaminophen (Fig. 10) is not [51]. The lack of fit for the latter material may arise from elastic effects: laminations which were observed in this study but which did not cause the tablet to fail completely could give rise to measured relative densities which are smaller than the true relative densities by a factor related to the volume of the crack. Here is an example of a material for which the simplifications used to faciliate the modeling process are not warranted. The effect of varying the constraint factor is also shown in Fig. 10. The upper predicted curve was calculated using a constraint factor of 3

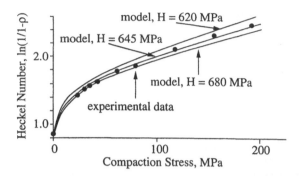

Fig. 9 Comparison of the experimental and predicted Heckel plots for sucrose. The limits of error that could occur as a result of error in hardness measurements are shown.

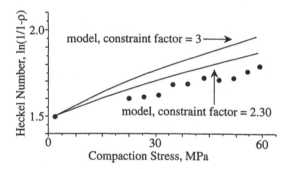

Fig. 10 Comparison of the experimental and predicted Heckel plots for acetaminophen. The lack of prediction may be a result of elasticity effects.

(i.e., the constraint associated with rigid/plastic behavior observed with "ductile" materials). The lower curve, which appears to predict the experimental behavior better, was predicted using a constraint factor which was calculated using the Marsh expression and the ratio of H/E determined experimentally.

VI. MODELS OF TABLET STRESS RELAXATION

Attempts to use linear viscoelasticity theory to predict and explain tableting have limited predictive value because the slope of the stress relaxation plots change with porosity. Although this is far from surprising, it means that the stress relaxation slopes cannot be interpreted because the role of porosity has not been considered explicitly. A successful model must account for the fact that the rate of stress relaxation is a function of porosity.

However deformation kinetic theory may be linked with tablet stress relaxation through a model. This approach was tested first using sodium chloride as a model. The advantage of using this material lies in the fact that its deformation kinetic parameters have been documented previously, which provides a useful test of the model. The following model fulfills these criteria.

A. Assumptions

It is assumed that stress relaxation experiments are carried out at intermediate levels of stress. Under these conditions, much of the particle deformation is concentrated in the interparticle contact regions which in many cases can be assumed to be oriented randomly within the tablet matrix.* Then one need consider only the viscoelastic response at the interparticulate contacts.† It is further assumed that the levels of stress and strain rate are high in these zones so that one mechanism of deformation controls the kinetics as discussed above. Therefore, it is necessary to calculate the parameters of one deformation barrier only. Finally, it is assumed that the yield behavior, or hardness, is independent of the strain history.

B. Compact Model

Whereas tablet compaction arises as a result of the mean hydrostatic stress and shear stress within the die,‡ we can assume that stress relaxation (measured as a function of the force on the punch during a stress relaxation experiment) is largely a function of the changes in punch stress alone. This assumption greatly simplifies our modeling requirements because we need only to find a model that relates the true average interparticle contact area normal to the direction of punch travel to the punch area.

---

* This assumption was also employed in the compaction models.

† At higher relative densities, the entire particle would begin to deform and change shape. Under these conditions, the strain behavior of the entire particle must be considered. One way to do the latter is to employ the Hertz contact equations as differential operators, but this is extremely complex, mathematically. The derivations currently are being undertaken in our group. Hertzian deformation can be visualized by imagining the response of a rubber ball when it is pressed between two flat surfaces.

‡ Which are functions of both the punch pressure and the die-wall stresses.

Table stress relaxation can be linked with the deformation kinetics of single particles by noting the fact that the microindentation text configuration resembles the interparticulate contact configuration. In a microindentation test a load is lowered onto the surface of a test specimen and is allowed to remain there for 10 s. The load and final 10-s contact area are measured and together these measurements permit the hardness to be calculated. In stress relaxation a defined initial load is applied also, however the force is measured not only at the 10-s point, but continually for an extended period. But the significance of the 10-s point should not be lost, for under ideal conditions we know that *the true punch contact stress must equal the hardness then.* Given the hardness and the punch force, one can calculate the true punch/compact contact area immediately. In other words, the contact stress at the 10-s point will equal the hardness, independent of the maximum load applied to the punch. If load is varied, the true 10-s contact area changes proportionally.

Thus it is assumed that compaction serves to change the interparticle contact area, but that once the punch is arrested, there is no further driving force for this process. Instead, the decrease in stress occurs due to the interchange of elastic and plastic strain within the large zone of deformation underlying the contact zone. This assumption is justified by the fact that both experimental and theoretical studies have shown that much of the deformation occurring during the indentation of a viscoelastic material does indeed occur within the crystal bulk, and that the compaction process imposes strain, and not stress, on the compact. Therefore, once the movement of the punches ceases there is decreased impetus for the particles to continue to approach one another which is the driving force for contact area enlargement.

If one makes these assumptions, then the hardness serves as a normalization factor, i.e.,

$$H = \sigma_{t=10\,s}$$

$$A_{cont.,true} = \frac{P_{punch,t=10\,s}}{H}$$

$$\sigma_{t=x} = \frac{P_{punch,t=x}}{A_{cont.,true}}$$

where $A_{cont.,true}$ is the true interparticulate contact area, and $\sigma_{t=x}$ is the apparent interparticulate contact stress at time x. The resultant "true" stress relaxation curve should be independent of the maximum force applied to the punch and relaxation curves at all relative densities should then become *superimposed* (Fig. 11). However, if strain hardening occurs, then the con-

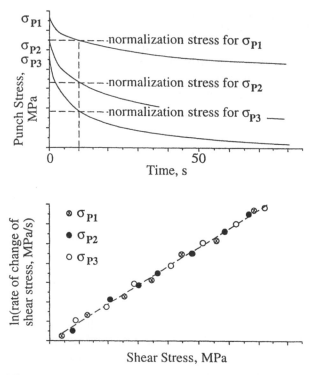

Fig. 11 Using a particle-based model of stress relaxation to normalize the stress relaxation plots determined at different maximum stresses and relative densities. The plots are normalized by assuming that the interparticulate contact area can be calculated 10 s after the beginning of relaxation using the hardness value for the material. If the model is correct, the plots of ln(shear stress rate) versus shear stress at all relative densities should be superimposed.

tact stress/area relationship would no longer be constant and should result in nonoverlap of the normalized relaxation curves.

C. Application of the Models to Evaluate Activation Parameters from Stress Relaxation Experiments

The stress relaxation of sodium chloride compacts were evaluated using this model. When the relaxation curves were normalized and plotted as ln(shear stress rate) *versus* shear stress, they were found to be straight and were superimposed, an observation that supports the assumptions underlying the model.

The activation volume (54 b^3) calculated from the model [23] was of the order of magnitude indicative of a Peierls-Nabarro mechanism [21], literature values for which range from 3.3 to 100 b^3 for this material [52]. The variability in the literature values probably arises from the extreme sensitivity of the deformation of this material to impurities. The activation energy (2.0 × 10^{-19} J) was statistically indistinguishable from the theoretical activation energy which ranges from 1.2 to 3.6 × 10^{-19} J.

The deformation kinetic model did not fare so well when it was applied to the stress relaxation behavior of potassium bromide. For this material, the fact that (1) significant strain hardening occurred and (2) two deformation mechanisms operate in the temperature range studied caused curvilinearity of the activation volume plots and poor rectification by the normalization procedure. That these factors were, indeed, the source of these problems was verified by (1) annealing compacts to remove strain hardening prior to stress relaxation studies, and (2) limiting the test to temperatures in which only one mechanism was operative.

VII. GENERAL CONCLUSIONS

The preceding discussion has provided a rationale for modeling the behavior of compacts using the properties of single crystals and outlined the development of three particular models. In addition, we have highlighted the numbers and types of assumptions that were used, and have described situations in which the models have been both successful and unsuccessful. Both successes *and* failures provide impetus for attempting to predict more complex situations. In the *former case,* one is encouraged to test the limits that can be borne by the assumptions. In the latter case, one is forced to reevaluate both assumptions and the models themselves.

When I model, the primary criterion that I use is an *aesthetic* one. To me, an aesthetically pleasing model is simple, clearly articulated, parsimonious (uses as few parameters as possible), comprehensive (has few exceptions), and it is testable (requiring few assumptions whose validity can be demonstrated experimentally). It might be useful to evaluate the three models discussed above with respect to these criteria.

1.   Simplicity and Parsimony

These models are very simple, both mathematically and conceptually— perhaps too simple, since numerous assumptions were required. The simplicity makes them easier to visualize and to test. Nevertheless, many people have found that the number of equations required in any one model is a bit daunting. The models are also parsimonious. Given the complex nature of the compaction process, it is astounding that one can predict at least an average behavior based on the knowledge of less than five material constants. However the parsimony also detracts from their general applicability. For example, one would want to be able to predict the compaction of brittle materials without making any compacts at all, whereas one is forced to make at least one in-die measurement of relative density and punch force. These facts point to the need for more general models that incorporate parameters such as interparticulate friction and fracture toughness *explicitly.*

2. *Comprehensiveness*

These models are far from comprehensive. For example, one must evaluate the ratio of K_c/H to determine which of the compaction models is most appropriate. However, one must often sacrifice simplicity for comprehensiveness, so in fact, one must seek an optimal balance between these two criteria.

3. *Testability*

Most of the assumptions made are tested quite easily; however, some will require detailed and extensive experimentation. An example of a difficult area of study is the verification of the constraint-limited crushing stress used in the model of compaction for brittle materials. The origin of "hardness" is itself contentious when the material of interest is a brittle one.

The limitations notwithstanding, I believe that the very fact that one can a priori predict the compaction behavior of *some* materials is extremely promising. Perhaps some day, with the use of more comprehensive models, most formulation problems may be predicted and overcome based on the measurement of a few mechanical properties of single crystals.

ACKNOWLEDGMENTS

I am indebted to George Weatherly for his mentorship, and Manny Papadimitropoulos, Meng-Chih Lin, Susan Lum, Xia Guo, Alfred Yu, Mike St. Onge, and David Mount for their excellent input into the research efforts of the group.

SYMBOLS

A	area
a	parameter in brittle model
α	geometrical factor relating active slip system to shear strain direction

A_f, b	activation frequency factor
$\mathbf{b}$	Burgers vector
b	parameter in brittle model
c	crack length
d	distance, diameter
$\Delta E\ (f, b, e)$	activation energies
E	elastic modulus
ϵ	strain
$\dot{\epsilon}$	strain rate
$\dot{\epsilon}_{\text{shear}}$	shear strain rate
F, F_c	force, contact force
H	Vickers hardness, MPa
h	crushed asperity height
K, K_b, K_f	rate constant; rate constant, backward and forward directions, respectively
k	Boltzmann constant, $1.36e^{-23}$ JK^{-1}
K_c	fracture toughness
kH	slope of the Heckel plot
MS	mean compressive stress in-die
ν	Poisson ratio
P	load, N
θ	angle of rotation
R	radius of a reference cell
r	radius
RD, RD_i	relative density, initial relative density
ρ_m	mobile dislocation density
r_r	radius of reference sphere
SA_b, SA_d	surface area of the reference sphere for brittle, ductile materials, respectively
σ	stress
σ	contact stress
S_f	free surface remaining on sphere
σ_f	fracture stress
σ_p	punch stress
T	absolute temperature, K
τ	shear stress
$\dot{\tau}$	shear stress rate
t	time
V_{act}	activation volume
V_{ex}	volume of material cut off by a fact of the Voronoi cell
V_p	particle volume
V_r	volume of the reference cell

W	work
Y	plastic yield stress
y	parameter in brittle model
Z, Z_i	coordination number
Z_r	recovered depth of indentation

REFERENCES

1. A. B. Bangudu, and N. Pilpel, (1985). Effect of composition, moisture and stearic acid on the plasto-elasticity and tableting of paracetamol-microcrystalline cellulose mixtures, *J. Pharm. Pharmacol. 37*: 289.
2. J. R. Britten, and N. Pilpel, (1978). Effects of temperature on the tensile strength of pharmaceutical powders, *J. Pharm. Pharmacol. 30*: 673.
3. J. E. Carless, and S. Leigh, (1974). Compression characteristics of powders: radial die wall pressure transmission and density change, *J. Pharm. Pharmacol. 26*: 289.
4. E. T. Cole, J. E., Rees, and J. A. Hersey, (1975). Relations between compaction data for some crystalline pharmaceutical materials, *Pharm. Acta Helv. 50*: 28.
5. S. T. David, and L. L. Augsburger, (1977). Plastic flow during compression of directly compressible fillers and its effect on tablet strength, *J. Pharm. Sci. 66*: 155.
6. P. York, (1979). A consideration of experimental variables in the analysis of powder compaction behavior, *J. Pharm. Pharmacol. 31*: 244.
7. J. T. Fell, and J. M. Newton, (1971). Assessment of compression characteristics of powders, *J. Pharm. Sci. 60*: 1428.
8. P. J. Rue, and J. E. Rees, (1978). Limitations of the Heckel relation for predicting powder compaction mechanisms, *J. Pharm. Pharmacol. 30*: 642.
9. E. N. Hiestand, and D. P. Smith, (1984). Indices of tableting performance, *Powder Technol. 38*: 145.
10. E. G. Rippie, and D. W. Danielson, (1981). Viscoelastic stress/strain behavior of pharmaceutical tablets: analysis during unloading and post-compressional periods, *J. Pharm. Sci. 70*: 476.
11. J. T. Fell, and J. M. Newton, (1971). Effect of particle size and speed of compaction on density changes in tablets of crystalline and spray-dried lactose, *J. Pharm. Sci. 60*: 1866.
12. R. F. Lammens, T. B. Liem, J. Polderman, and C. J. de Blaey, (1980). Evaluation of force displacement measurements during one-sided compaction in a die, *Powder Technol. 26*: 169.
13. H. H. Harman, (1976). *Modern Factor Analysis*, 3rd ed., University of Chicago Press, Chicago.
14. H. O'Neill, (1967). *Hardness Measurement of Metals and Alloys*, 2nd ed., Chapman and Hall, London.
15. R. Hill, (1950). *The Mathematical Theory of Plasticity*, Clarendon Press, Oxford.

16. A. R. C. Westwood, and N. H. MacMillan, (1973). In *The Science of Hardness Testing and its Research Applications* (J. H. Westbrook and H. Conrads, eds.), ASM, Metals Parks, OH, p. 377.
17. M. C. Shaw, (1973). In *The Science of Hardness Testing and its Research Applications* (J. H. Westbrook and H. Conrad, eds.), ASM, Metals Parks, OH, p. 1.
18. D. Tabor, (1951). *The Hardness of Metals,* Clarendon Press, Oxford.
19. D. M. Marsh, (1964). Plastic flow in glass, *Proc. Roy. Soc. A. 279*: 420.
20. F. A. McClintock, and A. S. Argon, (1965). *Mechanical Behavior of Materials,* Addison-Wesley, Reading, MA., p. 276.
21. A. S. Krausz, and H. Eyring, (1975). *Deformation Kinetics,* Wiley, New York, 1975.
22. W. C. Duncan-Hewitt, (1988). The use of microindentation techniques to assess the ability of pharmaceutical crystals to form strong compacts, Ph.D. thesis. University of Toronto.
23. E. A. Papadimitropoulos, (1990). Post consolidation behaviour of two crystalline materials, sodium chloride and potassium bromide, M.Sc. Phm. thesis, University of Toronto.
24. R. A. Verrall, R. J. Fields, and M. F. Ashby, (1977). Deformation mechanism maps for LiF and NaCl, *J. Am. Cer. Soc. 60*: 211.
25. D. J. Goodman, H. J. Frost, and M. F. Ashby, (1981). The plasticity of polycrystalline ice, *Phil. Mag. A43*: 655.
26. J. C. Conway, Jr., (1986). Determining hardness to elastic modulus ratios using Knoop indentation measurements and a model based on loading and reloading half-cycles, *J. Mat. Sci. Lett. 5*: 2525.
27. E. Breval, and N. H. MacMillan, (1985). Elastic recovery at Vickers hardness impressions, *J. Mat. Sci. Lett., 4*: 741.
28. B. R. Lawn, and T. R. Wilshaw, (1975). *The Fracture of Brittle Solids,* Cambridge University Press, London.
29. B. R. Lawn, and M. V. Swain, (1975). Microfracture beneath point indentations in brittle solids, *J. Mater, Sci. 10*: 113.
30. B. R. Lawn, and T. R. Wilshaw, (1975). Indentation fracture: principles and applications, *J. Mater. Sci. 10*: 1049.
31. M. T. Laugier, (1987). New formula for indentation toughness in ceramics, *J. Mat. Sci. Lett. 6*: 355.
32. B. M. Liaw, A. S. Kobayashi, and A. F. Emery, (1984). In *Deformation of Ceramic Materials II* (R. E. Tressler and R. C. Bradt, eds.), Plenum Press, New York, p. 709.
33. D. B. Marshall, B. R. Lawn, and A. G. Evans, (1982). Elastic/plastic indentation damage in ceramics: the lateral crack system, *J. Am. Cer. Soc. 65*: 561.
34. E. A. Almond, B. Roebuck, and M. G. Gee, (1986). In *Science of Hard Materials* (E. A. Almond, C. A. Brookes, and R. Warren, eds.), Inst. of Physics conference series number 75, Adam Hilger, Boston, p. 155.
35. G. R. Anstis, P. Chantikul, B. R. Lawn, and D. B. Marshall, (1981). A critical evaluation of indentation techniques for measuring fracture toughness. I. Direct crack measurements, *J. Am. Cer. Soc. 64*: 533.

36. G. G. Pisarenko, (1984). In *Advances in Fracture Research (Fracture 84),* Vol. 4 (S. R. Valluri, D. M. R. Taplin, P. Rama Rao, J. F. Knott, and R. Dubey, eds.), Pergamon Press, Oxford, England, p. 2711.
37. A. G. Evans, and E. A. Charles, (1976). Fracture toughness determinations by indentation, *J. Am. Cer. Soc. 59*: 371.
38. T. G. Mathia, and B. Lamy, (1986). Scleromatic characterization of nearly brittle materials, *Wear 108*: 385.
39. E. Arzt, (1982). Densification of metal powders. *Acta Metall. 30*: 1883.
40. J. L. Finney, (1970). Random packings and the structure of simple liquids. I: The geometry of random close packing, *Proc. Roy. Soc. A. 319*: 479.
41. E. G. Schwartz, and A. R. Holland, (1969). Determination of yield criterion for iron powder undergoing compacting, *Int. J. Powder Metall. 5*: 79.
42. J. S. Hardman, and B. A. Lilley, (1972). The influence of shear strain on the compaction of powders, *Proc. 1st Intl. Conf. on the Compaction and Consolidation of Particulate Matter, Brighton, 1972, Powder Technol. Publ. Series #4* (A. S. Goldberg, ed.), p. 115.
43. A. P. Green, (1954). The plastic yielding of metal junctions due to combined shear and pressure, *J. Mech. Phys. Solids. 2*: 197.
44. R. W. Rice, (1971). The compressive strength of ceramics, *Materials Science Research,* Vol. 5, *Ceramics in Severe Environments* (W. W. Krigel and H. Palmour, eds.), Plenum Press, New York, p. 197.
45. K. R. Venkatachari, and R. Raj, (1986). *J. Am. Cer. Soc. 69*: 499.
46. B. A. Obiorah, (1978). Possible prediction of compression characteristics from compression cycle plots, *Int. J. Pharm. 1*: 24.
47. S. Leigh, J. E. Carless, and B. W. Burt, (1967). Compression characteristics of some pharmaceutical materials, *J.Pharm. Sci. 56*: 888.
48. M. P. Summers, R. P. Enever, and J. E. Carless, (1976). The influence of crystal form on the radial stress transmission characteristics of pharmaceutical materials, *J. Pharm. Pharmacol. 28*: 89.
49. W. C. Duncan-Hewitt, and G. C. Weatherly, (1990). Modelling the uniaxial compaction of pharmaceutical powders using the mechanical properties of single crystals. I: Ductile materials, *J. Pharm. Sci. 79*: 147.
50. W. Duncan-Hewitt, (1994). A priori prediction of compact behavior by microcharacterization of the mechanical properties of single particles, AAPS Symposium: Solids: mechanical properties, characterization, and applications, 1994, A.A.P.S., San Diego, p. 33.
51. W. C. Duncan-Hewitt, and G. C. Weatherly, (1990). Modelling the uniaxial compaction of pharmaceutical powders using the mechanical properties of single crystals. II: Brittle materials, *J. Pharm. Sci. 79*: 273.
52. H. J. Frost, and M. F. Ashby, (1982). *Deformation Mechanism Maps,* Pergamon Press, Oxford.

14

Materials for Direct Compaction

Gerad K. Bolhuis

University of Groningen, Groningen, The Netherlands

Zak T. Chowhan

Syntex Research Institute of Pharmaceutical Sciences, Palo Alto, California

I. INTRODUCTION

Compacted tablets are produced from granulations or powder mixtures made by the following general techniques: wet granulation or wet massing combined with tray drying or fluid-bed drying, wet granulation and drying all in one step, dry granulation by roller compaction or slugging, and dry blending (direct compression). Of these, the oldest is the classic wet granulation and drying technique in two steps using tray drying. In spite of enormous improvements in wet granulation techniques (high shear granulation, fluid-bed granulation, extrusion granulation, continuous granulation, and all-in-one granulation), tablet production by direct compression has increased steadily over the years because it offers economic advantages through its elimination of the wet granulation and drying steps. The granulation technique that uses slugging or roller compaction is no longer a method of choice to produce compressed tablets.

II. DIRECT COMPRESSION

The results of a survey conducted in April 1992 in the United States (Shangraw, R. F., June 1992, Land-of-Lakes Conference) of 58 pharmaceutical companies (38 innovator, 14 generic, 8 nonprescription, and 6 vitamin) expressing preferences for the granulation process were surprisingly

419

in favor of the direct compression process [1]. Of the five processes listed in the survey (with 1.0 being the perfect score), the average score for the direct compression process was 1.5, compared to 2.0 for wet massing and fluid-bed drying, 2.5 for wet massing and tray drying, 3.3 for the all-in-one, and 3.6 for the roller compaction. When the respondents were asked about the company policy toward direct compression, 41.4% indicated that direct compression was the method of choice, and 41.1% indicated that they used both direct compression and wet granulation. Only 1.7% of the respondents indicated that they never used direct compression, and 15.5% indicated that the process was not recommended.

Since the introduction of spray-dried lactose in the early sixties as the first excipient specially designed for direct compression [2], other directly compressible excipients, commonly referred to as filler-binders, appeared on the pharmaceutical market (see Table 1). The introduction of the very effective filler-binder microcrystalline cellulose (Avicel PH) in 1964 resulted in an increased interest in the production of tablets by direct compression. It may be assumed that in the United States in the early nineties about 50% of all tablets were manufactured by direct compression. In Europe and the Far East, this percentage was much lower, but the interest in direct compression was increasing.

A. Advantages and Disadvantages of Direct Compression

Compared to wet granulation processes, direct compression offers a number of advantages. It requires fewer unit operations in production that means less equipment and space, lower labor costs, less processing time, and lower energy consumption (see Table 2).

Table 1 Year of Introduction of Some Directly Compressible Filler-Binders

1963	Spray-dried lactose
1964	Microcrystalline cellulose (Avicel PH)
	Anhydrous lactose
	Dicalcium phosphate dihydrate (Emcompress)
	Direct compression starch (STA-Rx 1500)
1967	Spray-crystallized dextrose/maltose (Emdex)
1982	Calcium sulphate dihydrate (Compactrol)
1983	γ-Sorbitol (Neosorb)
1984	Tricalcium phosphate (Tri-Tab)
1988	Ludipress
1990	Cellactose
1991	Modified rice starch (Eratab)
1992	Pharmatose DCL 40

Table 2 Steps in the Production of Tablets via Wet Granulation and by Direct Compression, Respectively

Wet granulation	Direct compression
1. Weighing	1. Weighing
2. Mixing	2. Mixing
3. Moistening	
4. Wet screening	
5. Drying	
6. Dry screening	
7. Admixing disintegrant, lubricant	3. Admixing lubricant
8. Compressing	4. Compressing

The elimination of the wet granulation step increases the stability of drugs that can degrade by moisture and/or heat. Another advantage of direct compression is that the tablets generally disintegrate into primary particles, rather than into granules. The increased surface area for dissolution may result in a fast drug release for some drugs and some drug products. For the majority of tablet formulations, the addition of a disintegrant is necessary in order to obtain a fast tablet disintegration. It has been demonstrated that the efficiency of the disintegrant is strongly dependent on the nature of the filler-binder [3].

The direct compression process also has a number of limitations. Tablets containing a high dose of an active ingredient that has poor compactibility, poor flow properties, and/or low bulk density cannot be prepared by direct compression, because filler-binders have a limited dilution potential, and tablet size and weight are limited. However, if an active ingredient is more compressible and flowable, a greater proportion can be carried successfully by a filler-binder.

The direct compression process generally involves mixing a drug substance with excipients prior to compaction. Because of differences in density of the drug substance particles and excipient particles, direct compression blends are subject to segregation during transfer steps from the mixer to drums, tote bins, hoppers, and so on. The procedure of sampling for analysis must be well defined so that it does not introduce a major error in determining homogeneity of the powder blend. Segregation during the handling of the powder blend before compression and during sampling is a major disadvantage of the direct compression method. Careful consideration must be given to particle size distribution and density of the drug substance and excipients. One way to limit segregation is to match the particle size distribution and the particle density of the drug substance and excipients. However, many low-dose drug substances are reduced in particle size in order to obtain uniform dosage or to obtain a large effective surface area in the dosage form for rapid dissolution. The excipients, especially filler-binders, are chosen so that the powder blend exhibits good flow properties. In some cases, premixing of the fine drug substance particles with large particles of a filler-binder [4] leads to mixing by random adhesion and random mixing (often referred to as ordered mixing) [5]. This phenomenon has been referred to as interactive mixing because of the interparticulate forces which predominate gravity under the right conditions. The primary interparticulate forces are the long-range forces: van der Waals, electrostatic and surface adsorption bonding. In other cases, mixing by spray coating (solvent deposition) may be necessary to avoid segregation and to obtain the homogeneity of the low-dose drug substance in the powder mix.

Generally, the physical and physicomechanical properties of drug substances and excipients in a directly compressible powder mixture need better definition and controls than the materials used in wet granulation. This is because the poor physicomechanical properties of the drug substance are not altered as they are in the wet granulation process; with wet granulation, the solution of the binder (usually a polymer) agglomerates the drug substance and the excipient particles, leaving a thin coat of polymer around the particles and the agglomerated mass. Because of variations in the physical and physicomechanical properties, it has been demonstrated that lot-to-lot variations in directly compressible materials can seriously interfere with tableting properties.

Another disadvantage of direct compression is the high cost of a number of filler-binders (e.g., microcrystalline cellulose and directly compressible starch), as compared to the binders and fillers used in wet granulation. However, the added cost of directly compressible excipients is outweighed by the savings realized by eliminating unit operations such as wet granulation and drying.

B. Requirements for a Directly Compressible Filler-Binder

The most important requirements for a directly compressible filler-binder are listed below.

> High compactibility to ensure that the compacted mass will remain bonded after the release of the compaction pressure. Only a few excipients can be compressed directly without elastic recovery. Most directly compressible filler-binders have undergone physical modification in order to improve tableting properties—mainly compactibility, flowability, and apparent density (see Section II.C).

Compactibility of a filler-binder or blends of filler-binders with other materials is generally tested by plotting tablet crushing strength as a function of the applied load.

Good flowability to ensure that the powder blend flows homogeneously and rapidly and leads to uniform die filling. Although there are different test methods for flowability, the best method is the determination of tablet weight variation under production conditions.

Good blending properties in order to avoid segregation.

Low lubricant sensitivity (see Chapter 16).

Good stability, i.e., the ability to remain unchanged chemically and physically and to remain fully active and effective during storage. Some filler-binders are only stable when stored under certain conditions. Due to hygroscopicity, most sugar-based products and polyols, with the exception of most lactoses, cannot be stored at high humidity conditions [6]. Moreover, the attraction of moisture by hygroscopic filler-binders is often detrimental to the stability of active ingredients in the tablet. Another example of a filler-binder in which stability plays a role is dicalcium phosphate dihydrate, which easily loses its water of hydration when stored above 40°C (See Section V.A).

Inertness, i.e., filler-binders should not accelerate the chemical and physical degradation of the active ingredient or excipients caused by compression or storage conditions.

Compatibility with all substances in any formulation in which it is a part.

Noninterference with the biological availability of active ingredients.

Promotive effect on tablet disintegration (if desired).

Promotive effect on drug release (if desired).

Capability of being reworked without loss of flow or compressibility characteristics.

Worldwide continuous availability.

Batch-to-batch reproducibility of physical and physicomechanical properties (constant quality).

Relatively cost effective.

Not a single excipient fulfills all the optimum requirements. Nevertheless, some products, particularly coprocessed substances, have adequate tableting properties for use as a single filler-binder. It is common to use a combination of two (seldom more) filler-binders in order to obtain a mixture with adequate tableting properties, including good stability and an acceptable cost [7]. It should be realized, however, that the properties of the components of a blend can result in synergistic or antagonistic effects with respect to one or more tableting properties [7]. Such interactions can be visualized by experimental designs, e.g., simplex lattice designs [8,9].

C. Improving Properties for Direct Compression

Although there are a number of drug substances that can be compressed directly into tablets (see Section X), only a few excipients can be compressed directly into tablets without any physical modification. For example, crystalline α-lactose monohydrate 100-mesh can be compressed directly, but the resulting tablet strength is inadequate for any practical purposes. For this reason, special grades of lactose for direct compression have been developed (see Section VII). Yet, lactose 100-mesh is inexpensive and an excellent filler-binder when used in combination with other filler-binders. During recent years, significant progress has been made in producing filler-binders with enhanced physicomechanical properties designed for direct compression into tablets. These improvements were accomplished by using different techniques varying from simple screening to crystal engineering. The most important modifications are listed below.

1. Grinding and/or Sieving

Most directly compressible materials are prepared by crystallization. The crystal size and, in part, the crystal shape are selected by sieving or, in some cases, after grinding. The particle size and shape depend on the grinding process (if ground) or on the mesh size and opening shape of the sieve used (if sieved). Although the purpose of sieving and grinding materials for direct compression is primarily to control flow properties, the compactibility may also alter because of changes in particle properties, such as surface area and possibly surface activation [10].

Crystallized α-lactose monohydrate is either sieved or it is first ground and then sieved in order to make different sieve fractions available to the customers. For direct compression, the unmilled, sieved 100-mesh quality is recommended because of its better flowability. The powdered grades, e.g., lactose 200-mesh or 450-mesh are ground lactoses which have poor flow properties and are intended for use in wet granulation. Dicalcium phosphate dihydrate is commonly milled after crystallization for use in the wet granulation process. For direct compression, however, only the unmilled larger crystal varieties (having at least 25% by weight of particles larger than 125 μm), can be used, because they have better flow and compaction properties [11].

2. Special Crystallization Techniques

The conditions of crystallization determine to a large extent the solid-state properties of directly compressible materials. Controlled crystallization would impart free-flowing properties to excipients and to drug substances, but not necessarily self-binding properties.

If polymorphism exists, the compactibility of the polymorphic forms may be quite different because of the internal arrangement of the molecules within the unit cells of crystals. The forces applied to the sample are not transmitted uniformly and the gliding of the molecules may be more or less difficult, depending on the crystal structure. Plastic deformation may occur depending on the dislocations and slip planes in the crystals. Crystalline substances are subject to such deformations depending on the symmetry within the crystal lattice. The crystal structure that has a greater degree of symmetry will be more prone to deformation on compression and compaction. The symmetry within the crystal structure diminishes in the following order: cubic, hexagonal, tetragonal, rhombohedral, orthorhombic, monoclinic, and triclinic [12]. Thus, cubic crystals such as sodium chloride and potassium chloride can be compacted directly.

α-Lactose monohydrate is obtained by crystallization at temperatures below 93°C. Anhydrous α-lactose is obtained in a stable form by a special thermal dehydration process of sieved α-lactose monohydrate at a carefully controlled temperature and water vapor pressure. β-Lactose is obtained by crystallization from a supersaturated solution at temperatures exceeding 93°C. The compactibility of anhydrous α-lactose is much better than that of the α-lactose monohydrate (see Section VII.B). Cocrystallization of sugars with small concentrations of other materials improves compactibility [13].

The effect of the structure of polycrystalline aggregates on their compactibility has been demonstrated for alumina trihydrate [14]. Mosaic particles with a more disordered structure than radial particles show a much better compactibility. Recrystallization can change the properties for direct compression, as illustrated for potassium chloride [15] and sodium chloride [16]. After recrystallization, the particles are irregular in shape with rounded edges, show a better compactibility, and exhibit reduced friction with punches and dies [17].

Alternative crystallization techniques may improve the tableting properties of directly compressible materials. An example for an active ingredient, naproxen, is given in Section X.

3. Spray Drying

Spray drying involves atomization of an aqueous solution or suspension into a spray, contact between spray and hot air in a drying chamber resulting in moisture evaporation, and recovery of the dried product from the air. Because of the spherical nature of liquid particles after evaporation of water, the resulting spray-dried material consists of porous, spherical agglomerates of solid particles that are fairly uniform in size. The particle size distribution of the spray-dried material is controlled by the atomization process and the type of drying chamber. Fast cooling of a solution and a high rate of crystallization as given by spray technologies produce solids with very imperfect structures containing some amorphous material. Lattice defects imply a high deformability and a good binding tendency of the particles [10]. Moreover, the amorphous component generated by rapid cooling and rapid crystallization acts as a binder [18]. Another advantage of the spray drying technology is the improved flowability caused by the spherical nature of the agglomerated particles.

Several directly compressible materials are produced by spray drying. Lactose was the first pharmaceutical excipient to successfully exemplify the spray-drying technology. The superior binding ability of spray-dried lactose, when compared with α-lactose monohydrate, has been attributed to the amorphous lactose that exhibits plastic flow on compression (see Section VII.D). Examples of other filler-binders prepared by spray drying are Avicel PH (microcrystalline cellulose), TRI-CAFOS S (tricalcium phosphate), Karion Instant (sorbitol), Emdex (cocrystallized dextrose-maltose), different types of directly compressible saccharose, and different types of directly compressible paracetamol.

4. Granulation, Agglomeration, and Coating

Granulation and agglomeration represent the transformation of small, cohesive, poorly flowable powders into a flowable and directly compressible form. Granulation results in nearly spherical particles with relatively high bulk density and strength. Agglomeration, on the other hand, leads to irregularly shaped porous particles with relatively low bulk density and strength. When primary particles have binding properties of their own, the addition of a binder is not necessary. For example, agglomerated lactose (Tablettose) is prepared without the addition of a binder. When primary particles do not have binding properties, a binder is added to aid in the granule formation. The presence of small particles in the agglomerates improves compactibility (see Section VII.E). Granulation of powdered cellulose or starch improves the flowability but increases the lubricant sensitivity to binding (see Sections III.B and IV.C, respectively). Granulation or coating greatly increases the compactibility of ascorbic acid (Section X.A).

5. Pregelatination

Compressible starch is produced by partial hydrolyzation of corn starch. Free amylopectin improves binding properties, whereas free amylose gives a product better disintegration properties than its starting material (see Section IV.C).

6. Dehydration

Increased binding properties of α-lactose monohydrate occur with increasing thermal or chemical dehydration of the solid [19]. Desiccation with methanol gives a much steeper increase than thermal treatment. During dehydration, a gradual transition within each particle from α-lactose monohydrate into anhydrous α-lactose is observed. The anhydrous product has not only much better compactibility but also better flow properties than α-lactose monohydrate (Section VII.B). Change in binding properties of the hydrates by thermal or chemical dehydration has been reported for a number of other materials. These include dextrose monohydrate (see Fig. 1), citric acid monohydrate, calcium sulphate dihydrate, calcium monohydrogen phosphate dihydrate, and dicalcium phosphate dihydrate [20].

7. Hybridization

The hybridization process makes use of the basic principle of interactive powder mixtures. The core powder and the fine powder are first mixed in a preselected ratio. The fine powder adheres to the surface of the core pow-

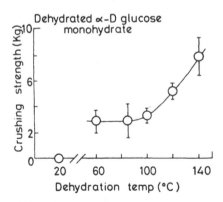

Fig. 1 Crushing strength of tablets, compressed from fully dehydrated *d*-glucose monohydrate, treated at different temperatures. (From Ref. 20; reproduced with permission of the copyright owner.)

der due to static electric charge. The two powders form an "ordered mixture." Without further processing, however, the outer powder could become detached from its carrier. This is prevented by a second step called hybridization which disperses the powders and adds mechanical/thermal energy to embed and/or film the outer "ordered" layer into the core particle. The hybrid powder of six drugs (oxyphenbutazone, prednisolone, theophylline, indomethacin, phenacetin, and aspirin) with potato starch used as a core material immediately after production differed in their structure from interactive mixtures [21]. With the hybrid powders the drug was spread on the surface of the core particles by friction and collision that occurred in the dry process, but with the interactive mixtures, the drug simply adhered as intact particles to the surface of the diluent particles. Although this new technique has not been applied to direct compression of pharmaceuticals, it has a great potential for future application and is a real advance in powder modification technology.

III. CELLULOSE

Cellulose forms the backbone of many excipients used in marketed drug products. Cellulose is built up of repeating glucose units joined by β-1,4-glucosidic bonds. Numerous investigations show that in native cellulose, numerous anhydroglucose units are combined to form a single cellulose molecule. Chain-length determination is very difficult, and average values are given, referred to as the average degree of polymerization (DP).

Pharmaceutical grades of cellulose are obtained by either mechanical or chemical processing, or by both. Attached to carbon atoms 2, 3, and 6 of each anhydroglucose unit is a hydroxyl group capable of undergoing chemical reactions such as esterification or etherification. The reaction of hydroxyl groups varies for steric reasons. Products can be prepared in which one, two, or three hydroxyl groups per anhydroglucose unit have undergone reaction.

The amount of partial esterification or etherification is given for each substituent separately as the degree of substitution (DS). It ranges from 0 to 3 and is equivalent to the average of converted hydroxyl groups per anhydroglucose unit.

Pure cellulose can be ground mechanically with additional treatment by hydrochloric acid. The resulting powder is cellulose powder or microcrystalline cellulose. A survey—with a uniform text in three languages—was made by four pharmacopeias (USP/NF, BP, EP, and JP), with each pharmacopeia board contacting trade associations and excipient manufacturers. Microcrystalline cellulose is among the top 10 excipients and is among the first 3 excipient monographs selected by the USP along with the powdered cellulose for harmonization of excipient standards and test methods with European Pharmacopeia and Japanese Pharmacopeia. The top 25 excipients identified by the survey include cellulose derivatives (13 excipients) (USP undertook to revise and harmonize 17, EP to harmonize 11, and JP to harmonize 9). Only microcrystalline cellulose and powdered cellulose will be discussed

A. Microcrystalline Cellulose

In the survey conducted in April 1992 [1], microcrystalline cellulose was rated as the best filler-binder choice, scoring 1.7 (with 1.0 being the perfect score). The reasons for preferences ranged from solubility, cost, tradition, compatibility, supply, compactibility, handling, and physiological inertness.

Microcrystalline cellulose is described in the NF as a purified, partially depolymerized cellulose prepared by treating α-cellulose, obtained as a pulp from fibrous plant material with mineral acids. The cellulose fibers in the starting material are composed of millions of microfibers. In the microfibers, two different regions can be distinguished: a paracrystalline region, which is an amorphous and flexible mass of cellulose chains, and a crystalline region, which is composed of tight bundles of cellulose chains in a rigid linear arrangement [22]. As an effect of controlled hydrolysis, the amorphous fraction has largely been removed, yielding aggregates of the more crystalline portions of cellulose fibers. After purification by filtration and spray drying, dry, porous microcrystals are obtained. By controlling the atomization and drying conditions, particle size distribution can be varied [23, 24]. Microcrystalline cellulose occurs as a white odorless, tasteless, crystalline powder composed of porous particles or as an agglomerated product.

Microcrystalline cellulose was marketed first in 1964 by the FMC Corporation under the name Avicel PH in four different particle size grades, each with different properties [25]. As of 1992, Avicel PH is available in seven types (see Table 3). Other brands of microcrystalline cellulose are also available on the pharmaceutical market. Most important are Emcocel 50M (comparable with Avicel PH 101) and Emcocel 90M (comparable with Avicel PH 102).

Other qualities of microcrystalline cellulose are available from the Far East and South America. The equivalence of microcrystalline cellulose obtained from different suppliers is discussed later in this section.

Microcrystalline cellulose is one of the most used filler-binders in direct compression. Its popularity in direct compression is because of its extremely good binding properties as a dry binder. It also works as a disintegrant and a lubricant and has a high dilution potential in direct com-

Table 3 Avicel PH Microcrystalline
Cellulose Types

Grade	Mean particle size (μm)	Moisture content (%)
PH-101	50	5
PH-102	100	5
PH-102SCG	120	5
PH-103	50	3
PH-105	20	5
PH-112	100	1.5
PH-200	200	5

Source: FMC Corp., Philadelphia, PA.

pression formulations. In addition to its use in direct compression formulations, microcrystalline cellulose is used as a diluent in tablets prepared by
wet granulation, as a filler for capsules, and for the production of spheres.
    Scanning electron micrographs show that Avicel PH 101 has a
matchstick-like or rodlike structure, whereas Avicel PH 102 is a mixture of
primary particles and agglomerates (see Fig. 2a) [26,25]. The primary particles are composed of fibrils with a radius of 10–15 nm with a hollow axis of
about 2 nm. These intraparticular pores account for 90% of the total surface area [25]. Measurement of the interparticulate porosity by mercury
porosimetry, liquid penetration techniques, and scanning electron microscopy demonstrate that the modal pore radius decreases logarithmically
with an increase in compaction pressure. A theoretical zero porosity would
be reached at a pressure of 300 MPa [27]. The intraparticulate pore size
distribution shows no change at 20 MPa, indicating that the internal pores
do not collapse [28,27].
    Conflicting results of the specific surface area of microcrystalline
cellulose have been reported. Using nitrogen adsorption and the BET
equation, the apparent specific surface area of Avicel PH 101 values
between 1.0 and 1.5 m²/g have been reported [29–31]. Marshall and
Sixsmith [25] concluded that a specific surface area of 1.3 m²/g, obtained
by mercury porosimetry, represents the external interparticle surface
area. Water vapor sorption with application of the BET equation has led
to values of 138 and 150 m²/g [29,30]. These values were ascribed to be
the internal surface area. However, Zografi et al. [32] showed that these
high values do not reflect a true surface area but are caused by penetration of water vapor into the amorphous parts of cellulose and interaction
with individual anhydroglucose units. Using nitrogen and krypton adsorp-

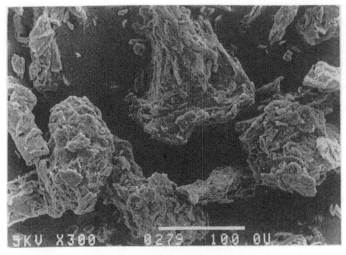

a

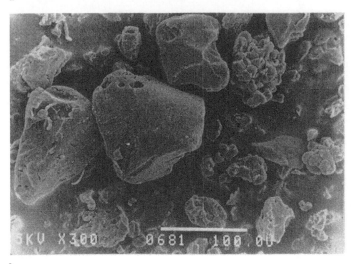

b

Fig. 2 Scanning electron micrographs of directly compressible filler-binders: a = microcrystalline cellulose (Avicel PH 102); b = compressible starch (STARCH 1500); c = modified rice starch (Eratab); d = dicalcium phosphate dihydrate (Emcompress); e = γ-sorbitol (Karion Instant); f = γ-sorbitol (Karion Instant).

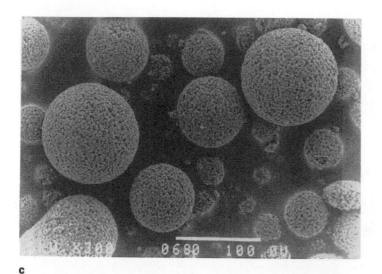

c

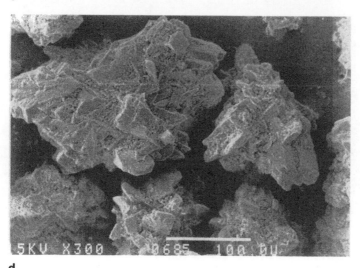

d

Fig. 2 Continued

tion following several methods of sample pretreatment, they found a specific surface area of 1.3 m²/g with no evidence of extensive microporosity. The crystallinity of Avicel PH 101 determined by X-ray diffraction [29,31] and infrared measurement [29], is about 63%. For Emcocel 50M, the crystallinity is somewhat greater [31].

At room temperature and relative humidity between 40% and 50%, Avicel PH 101 has a moisture content of about 5–6% [32]. Water vapor

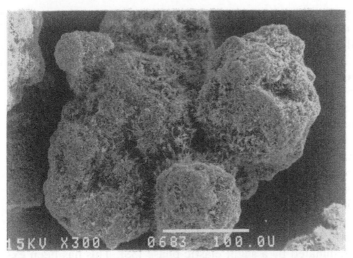

e

f

sorption occurs by the same basic mechanism as in other types of cellulose and starches. This can be described by the sorption of one tightly bound water molecule at an anhydroglucose unit, followed by a second less tightly bound water molecule, with further additional layers of bulk water. In contrast to starch, water is only sorbed in the noncrystalline regions of microcrystalline cellulose [32].

Microcrystalline cellulose exhibits low bulk densities, with values ranging from 0.32 to 0.45 g/cm^3 [24].

The flow properties of Avicel PH 101 are poor [26,33–35] (see Table 4), as explained by the more or less long-drawn, matchstick-like form of the particles, the particle size distribution, and the low bulk density [26, 34]. Although Avicel PH 102 has better flow properties because it is a mixture of agglomerates and primary particles, its flowability is only moderate. The flowability of Avicel PH 103 and Avicel PH 105 is less than that of Avicel PH 101 and has been ascribed to differences in moisture content or particle size distribution [36]. FMC has introduced Avicel PH 200 with a mean particle size of 200 μm in order to improve its flow in direct compression formulations [37]. The improved flow properties of Avicel PH 200 are attributed to spherical aggregates and larger overall particle size distribution. Some drawbacks of these large particle size microcrystalline cellulose are higher lubricant sensitivity and lower carrier capacity.

During compaction, microcrystalline cellulose is thought to undergo stress relief deformation by several mechanisms. At low compression forces, stress relief is dominated by a slight elastic phase [38]. This has been explained by its hollow microfibrillar structure [39]. At higher force, there is either further deformation [40] or permanent deformation by nonspecific plastic flow [41]. Following stress relaxation studies, David and Augsburger

Table 4   Flow Properties of a Series of Filler-Binders, Lubricated with 0.5% Magnesium Stearate

Filler-binder	Flow class (flows through vessel nr:)	Variation coefficient of tablet weight (%)
Avicel PH 101	not through 5	3.2
Avicel PH 102	2	1.0
Elcema G250	2	0.35
STARCH 1500	not through 5	2.2
Emcompress	1	0.9
Compactrol	2	0.3
Emdex	2	0.6
Neosorb 20/60	2	0.5
α-Lactose monohydrate 100-mesh	1	0.4
Tablettose	1	1.2
Spray-dried lactose (Pharmatose DCL 11)	1	0.4
Anhydrous β-lactose (Pharmatose DCL 21)	1	0.45
Anhydrous α-lactose (Pharmatose DCL 30)	1	0.25

Source: Adapted from Ref. 6; reproduced with permission of the copyright owner.

[42] suggested that plastic flow is an important factor affecting the compressibility of microcrystalline cellulose.

Plastic deformation is facilitated by the presence of slip planes, dislocations, and the small size (a few thousand by a few hundred Ångström units) of the individual microcrystals. The hydrolysis portion of the production process is responsible for the formation of the slip planes and plastic deformation of the particles on a microscale. The spray-drying operation is responsible for the deformability of microcrystalline cellulose on a macroscale [23]. The plastic behavior of microcrystalline cellulose is deduced from force-displacement curves [43], Heckel plots [44,45], and creep analysis [46]. Using mercury porosimetry, Vromans et al. [47] show that there is no alteration in the quantity of small pores in tablets compressed from microcrystalline cellulose, which indicates plastic deformation under pressure. The plasticity of microcrystalline cellulose increases with increasing compressing force [48], which is accompanied by a decrease in viscoelasticity [49]. When microcrystalline cellulose tablets are prepared by wet granulation, there is less plastic deformation in comparison with tablets prepared by direct compression [50]. Force versus displacement plots on diametral compression to fracture indicates that Avicel shows stress relief by time-dependent yielding [51]. The yield pressure increases with punch velocity, because of a reduction in the amount of plastic deformation caused by the time-dependent nature of plastic flow [49,52]. This increase in mean yield pressure at increasing compression speeds is reflected by the high strain rate sensitivity (SRS) value for microcrystalline cellulose [52]. Armstrong and Palfrey [53] show that the reduction in crushing strength at increasing compression speed is caused by an increased porosity of the compacted powder bed.

Strong binding properties are caused by hydrogen bonds between hydrogen groups on the plastically deformed, adjacent cellulose particles [41]. The presence of hydrogen bonds is demonstrated by Hüttenrauch [54]. The hydrogen bonds on the extremely large surface area are brought into close contact during plastic deformation. This is the reason for the extremely good compactibility of microcrystalline cellulose, better than that of any other directly compressible filler-binder [26, 55, 56]. The exceptionally good binding properties of microcrystalline cellulose are reflected by its extremely high Bonding Index, whereas its Brittle Fracture Index is extremely low [57].

The compactibility of microcrystalline cellulose depends on its moisture content. It has been suggested that, at its equilibrium moisture content of 5%, most of the water will be within the porous structure of microcrystalline cellulose and a large proportion of this bound moisture is expected to be hydrogen-bonded to small bits of cellulose within the particle.

During plastic deformation, the moisture within the pores should act as an internal lubricant and facilitate slippage and flow within the individual microcrystals [30]. The presence of an optimum amount of water will prevent elastic recovery by forming bonds though hydrogen bond bridges [58].

The compactibility of microcrystalline cellulose decreases with a reduction of its moisture content [58]. The strongest compacts are produced when microcrystalline cellulose contains 7.3% moisture. However, at a lower moisture content, the compacts capped when compressed at 163 MN/m². The capping tendency of microcrystalline cellulose-based formulation is reduced by increasing the moisture content from 3.2% to 6.1% [59]. This is attributed to the strengthening of interparticle binding forces and the reduction of elastic recovery by lowering of the yield point. The effect of water on the compactibility of microcrystalline cellulose is strongly dependent on its thermodynamic state in the excipient, and is similar to the effect of water on the compactibility of starch [60]. Details on starch are given in Section IV.

The particle size of microcrystalline cellulose has only a small effect on its compactibility [44,45]. A maximum compactibility has been found for the 80–100 μm fraction [61].

As an effect of their plastic behavior, cellulose and starches are sensitive to mixing with lubricants [62]. Because of the extremely high compactibility of unlubricated microcrystalline cellulose, softening has practical significance only after relatively long mixing times (Fig. 3). Alternative lubricants, such as hydrogenated vegetable oil (Lubritab) or sodium stearyl fumarate (Pruv), have much less effect on the tablet strength following mixing [46]. More information of the effect of mixing with lubricants on binding properties of filler-binders is given in Chapter 16.

Microcrystalline cellulose has a high dilution potential, defined as the ability of a given quantity of an excipient to bind a specified amount of an active ingredient into an acceptable tablet [23]. This high dilution potential is attributed to low bulk density that impacts high covering power and to broad particle-size distribution that allows optimum packing density.

In a study on the equivalence of different brands of microcrystalline cellulose, two types of Emcocel were compared with comparable types of Avicel PH. Although there were some differences between the physical properties of the two materials, their compressibilities were rather similar [31,63]. When used as a diluent in wet granulation, differences were reported in cohesiveness at different moisture contents between Avicel PH 101 and Emcocel 50M [50]. In other studies, with the exception of Emcocel, large differences in tableting properties existed between the Avicel PH and the microcrystalline cellulose NF from different suppliers, in spite of equivalence in physical properties [64]. For three types of microcrystalline cellulose from different sources, the differences increased after reworking [65].

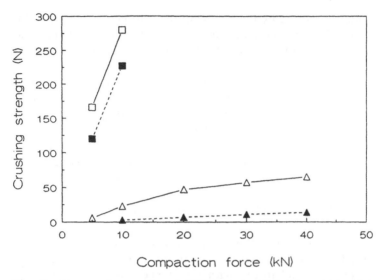

Fig. 3 Compaction profiles of (☐,■) Avicel PH 102 and (△,▲) Elcema G250, respectively, both unlubricated (open symbols) and lubricated with 1% magnesium stearate (closed symbols). 500-mg, 13-mm tablets, hydraulic press, contact time 5.1 s.

In addition to binding properties, microcrystalline cellulose has lubricating and disintegration properties [23]. A mixture of microcrystalline cellulose with up to 40% spray-dried lactose can be compressed without the addition of a lubricant [66]. The lubricating properties are attributed to the very low residual die wall pressure, as an effect of stress relief by viscoelastic relaxation within the die, as well as to the very low coefficient of friction [67,68,57]. From radial movement studies of the outer die wall on compression and recompression of microcrystalline cellulose compacts, Travers and Cox [51] concluded that the self-lubricating properties of microcrystalline cellulose tablets may be due to their elastic properties. With no shear yielding or failure, they regain their original dimensions when the axial force is removed and little or no "locked in" stress remains.

Disintegration of microcrystalline cellulose tablets is attributed to the penetration of water into the hydrophilic tablet matrix by means of capillary action of the pores and by a subsequent disruption of the hydrogen bonds [66]. Penetration measurements demonstrate that microcrystalline cellulose tablets exhibit extremely fast aqueous penetration. The widening of the pores by the breaking of the hydrogen bonds result in a ratio between water uptake and original pore volume up to 20 [69]. The lack of disintegration in fluids with a low dielectric constant confirms the theory that disintegration is caused by the breaking of hydrogen bonds [55]. An increase in compaction load produces an increase in the disintegration time of tablets as an effect of the decreased water penetration into the tablets [66,34,70].

Plain microcrystalline cellulose tablets [41,71] or tablets containing microcrystalline cellulose as a filler-binder [72] softened and had a tendency to swell when aged under increased relative humidity. This moisture uptake is probably caused by high intraparticulate porosity [25]. The rate of moisture uptake is directly proportional to the relative density of plain microcrystalline tablets [72]. The absorbed moisture causes a disruption of the hydrogen bonding and hence a decrease in tablet strength [58]. Although it has been reported that the softening of tablets was reversible on removal of the high humidity [41], other results do not confirm this phenomenon [72].

Because of high cost, poor flow properties, and low bulk density, microcrystalline cellulose is not used solely as a primary filler-binder in a tablet formulation, but is instead mixed with an inexpensive filler with good flowability such as α-lactose monohydrate 100-mesh or dicalcium phosphate dihydrate [73,9,74]. Recent work reveals that blends of microcrystalline cellulose with α-lactose monohydrate 100-mesh, spray-dried lactose, agglomerated lactose or Cellactose can be compressed without problems on a high-speed rotary press with a forced-feed equipment [75,76]. The compactibility of Avicel PH 102/α-lactose monohydrate 100-mesh blends 1:3 and Avicel PH 101/α-lactose monohydrate 1:9, respectively, are disappointing, however, when compared with the compactibility of other filler-binders (see Fig. 4).

B. Powdered Cellulose

Powdered cellulose NF is purified, mechanically disintegrated cellulose, prepared by processing α-cellulose obtained as a pulp from fibrous plant materials. Powdered cellulose is also known as microfine cellulose or cellulose flocs. Degussa Co. markets powdered cellulose under the name Elcema in powder form (P050, P100), in fibrous form (F150), and in granular form (G250 and G400). A well-known type is Elcema G250, consisting of granules prepared from the P100 quality, without the addition of a binder, and with a mean particle size of about 250 μm. Solka Floc from Brown Co., Berlin NH, is available in several grades (Solka Floc BW series) with mean particle sizes between 70 and 90 μm, and in a granular form (Particulate Solka Floc) with a mean particle size of 239 μm.

When it is not in an agglomerated form, powdered cellulose has poor flow properties, so it is an unsuitable direct compression excipient when used alone [77,78]. Although the flow properties of Elcema P050 or P100

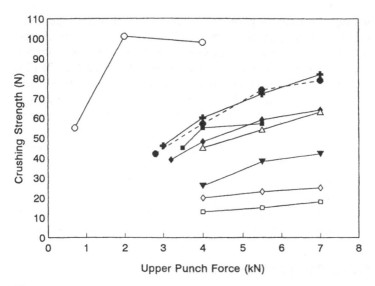

Fig. 4 Compaction profiles of various filler-binders, lubricated with 0.5% magnesium stearate. 60-mg, 5-mm biconcave tablets, Fette P1000 rotary press; rotation speed 30 rpm. (○) Karion Instant, (+) Cellactose, (●) Fast Flo Lactose, (◆) Pharmatose DCL 40, (△) Pharmatose DCL 11, (■) Pharmatose DCL 21, (▼) Tablettose, (◇) α-lactose monohydrate 100-mesh + Avicel PH 102 3:1, (□) α-lactose monohydrate 100-mesh + Avicel PH 101 9:1. (From Ref. 76; reproduced with permission of the copyright owner.)

can be improved by granulation, the lubricant sensitivity increases as well and is related to the bulk density [78]. For the flow properties of the granulated form Elcema G250, conflicting results have been published [26, 78]. During mixing, Elcema G250 particles break, forming irregularly shaped particles and exhibiting decreased flow properties [79]. Solka Floc particulate form has adequate flow properties [77].

Although powdered cellulose has binding properties, the binding is inferior to that of microcrystalline cellulose (see Fig. 3) because of differences in the manufacturing process [23].

Powdered cellulose deforms plastically, as shown by considerable stress relaxation following compaction [80,81] or by the change in the stress-density profile using different dwell times of compression [81,82]. Similar to microcrystalline cellulose, hydrogen bonds are formed between adjacent, plastically deformed cellulose particles, although scanning electron micrographs show that mechanical interlocking may also play a role [79]. The lower compactibility when compared with microcrystalline cellulose is caused by a lack of slip planes and dislocations in the cellulose granules.

The lubricant sensitivity of powdered cellulose is strongly dependent on the size and form of the particles. The binding properties of the granulated forms (Elcema G250 and Elcema G400) show a dramatic decrease after mixing with lubricants such as magnesium stearate (see Fig. 3) [26, 83]. Similar results have been reported for particulate Solka Floc. The powdered and fibrous Elcema qualities are less sensitive to magnesium stearate because of poor flow properties that retard or impart a formation of a lubricant film during mixing [78].

Variations in compactibility have been found between two batches of Elcema G250, even though both batches conformed to the manufacturer's specifications [84]. The variations were thought to be caused by differences in strain rate sensitivity.

Powdered cellulose has self-lubricating properties [26] and requires lower levels of lubricants as compared with other excipients, such as sugars.

IV. STARCH AND STARCH DERIVATIVES

Starches and their derivatives are among the most widely used excipients in drug products. Modified starches include those in which the grain has been more or less completely split, such as pregelatinized starch and esterified and etherified starches (known in the pharmaceutical field as carboxymethyl starches and hydroxyethyl starches) and granulated or agglomerated starches (referred to as "modified starches" in Section IV.C).

A. Native Starches

Starch from commonly used sources consists of two polysacharides, amylose and amylopectin, that are based on a glucose monomer. Amylose is a linear polymer and represents approximately 27% by weight, while amylopectin has a branched structure and represents about 73% by weight. The amylose component resembles cellulose, except for the stereochemical configuration of linkage between monomer units. Starch is insoluble in water because the two polymers (amylose and amylopectin) are intermolecularly associated in the crystal lattice.

Starch is an important excipient for tablet formulations. It is widely used as a disintegrant, diluent, and as a binder in the form of starch paste that is used in the wet granulation process.

Although unlubricated native starches have good compression characteristics, their poor flow properties and their high lubricant sensitivity make them less suitable for use as excipients in tablets prepared by direct compression.

Although it is generally accepted that starch undergoes plastic deformation under pressure [85], the tendency of starches to total and pure plastic deformation seems to be dependent on particle size and particle shape distributions [86,87]. According to the values of yield pressure, corn starch is most prone to plastic flow with only small elastic recovery. Potato starch also flows plastically with ease, whereas barley and wheat starches are more elastic.

The compactibility of starches is dependent on the equilibrium moisture content, which is dependent on the relative humidity of the atmosphere under which the powders are stored. Maximum tablet strength is obtained at 60–70% relative humidity, corresponding to equilibrium moisture content of about 10% w/w [88]. It is generally accepted that water sorption onto starches (just as on cellulose) occurs in different stages. At first, water becomes tightly bound to anhydroglucose units throughout the starch grain until a 1:1 water:anhydroglucose unit stoichiometry (11.1% w/w water) is obtained. Between 1:1 and 2:1 stoichiometry, the water molecules become less tightly bound to the water molecules already bound to anhydroglucose units. Water absorbed at a stoichiometry larger than 2:1 is even less tightly bound and has the properties of bulk water [60]. Water absorbed within the starch particles influences the compaction properties by affecting the degree of viscoelasticity. Water levels below 1:1 stoichiometry reduce the compactibility of starch. Less tightly bound water (stoichiometry between 1:1 and 2:1) is needed to provide plasticity to the starch system, whereas bulk water decreases the ability to form bonds between the starch particles, probably due to the formation of a water film.

In a study about tableting properties of potato, corn, wheat, and barley starch, corn starch was best in compactibility, whereas potato starch was the best with respect to flowability [86]. The latter is, however, extremely susceptible to mixing with magnesium stearate [88]. This high lubricant sensitivity is an effect of the plastic behavior of starch products under compression (see Chapter 16). In another comparative evaluation of several native starches, evidence showed that rice starch had better compaction properties and worse flow properties when compared to corn, potato, and tapioca starches. Moreover, the binding capacity of rice starch was insensitive to mixing with magnesium stearate [[88]. The lack of lubricant sensitivity of rice starch is caused by its poor flowability, which impairs the formation of a lubricant film over the particles during mixing with a lubricant. These physicomechanical properties are attributed to the fine particle size of rice starch as compared to other starches.

Kwan and Milosovich [89] evaluated a pharmaceutical grade of amylose as a filler-binder for direct compression. The amylose used in their study had good compaction properties and was free flowing, self-lubricating, and self-disintegrating.

A major drawback in using amylose as an excipient in direct compaction is its extreme sensitivity to mixing with lubricants, which is caused by plastic deformation of amylose during compression [83,62] (see Chapter 16). A pharmaceutical grade of amylose is no longer commercially available.

In order to improve tableting properties for direct compression, starch can be physically modified in different ways. For example, special pregelatination process improves tableting properties (see Section III.B). Another possibility for improvement of tableting properties of starch is granulation [88]. Although granulation with 2.5% starch paste of the same kind of starch improves flowability, tablets from granulated potato, tapioca, and maize starch have a lower tablet crushing strength than tablets prepared from starches without wet granulation with starch paste (see Table 5). However, wet granulation of rice starch with starch paste improves both flow and compaction properties [88]. Because of their better flow properties, starch granulations are more susceptible to mixing with magnesium stearate than starches prepared without wet granulation. After mixing starches for 30 min with 0.5% magnesium stearate in a Turbula mixer at 90 rpm, only tablets prepared with rice starch granules show

Table 5 Tablet Strength and Magnesium Stearate Susceptibility of Starches as such and Starch Granulations after Storage at 19°C and 44% RH

Starch type	Strength		Strength reduction ratio
	Unlubricated[a]	Lubricated[b]	
Starch as such			
Potato	68 ± 10	0	0
Tapioca	70 ± 4	39 ± 4	0.6
Maize	56 ± 8	30 ± 7	0.5
Rice	102 ± 20	104 ± 11	1.0
Granulations			
Potato	47 ± 3	<10	<0.2
Tapioca	54 ± 2	<10	<0.2
Maize	27 ± 4	<10	<0.4
Rice	139 ± 11	50 ± 8	0.4

[a]Crushing strength (N), unlubricated, compression force 20 kN (except rice starch as such: 10 kN) (stored at 19 ± 1°C and 45 ± 5% relative humidity).
[b]Crushing strength (N), lubricated (0.5% magnesium stearate, mixing time 30 min), compression force 20 kN (except rice starch as such: 10 kN) stored at 19 ± 1°C and 45 ± 5% relative humidity).
Source: From Ref. 90; reproduced with permission of the copyright owner.

sufficient crushing strength (see Table 5) [90]. It has been demonstrated that the lubricant sensitivity of starch granules is dependent, among other factors, on properties of the granules such as bulk density and flowability, which are determined by the method of wet granulation [78]. Commercially available starch granulations are described in Section IV.C.

B. Compressible Starch

Pregelatinized starch NF is starch that has been chemically or mechanically processed to rupture all or part of the granules in the presence of water, and subsequently dried. Although pregelatinized starch is primarily a binder in wet granulation, it can be modified to make it compressible and flowable in character. A special pregelatinized starch for direct compression, commonly called *directly compressible starch* or *compressible starch,* is partially hydrolyzed corn starch, available as *STARCH 1500* from Colorcon (the product was originally marketed as STA-Rx 1500 by A.E. Staley Co., Decatur, IL). Chemically, compressible starch does not differ from starch USP.

STARCH 1500 is prepared by subjecting corn starch to physical compression or shear stress in high-moisture conditions causing an increase in temperature and a partial gelatinization of some of the starch. The product consists of both individual starch grains and aggregates of starch grains bonded to the hydrolyzed starch [91] (see Fig. 2b). During the manufacturing process, some of the hydrogen bonding between amylose and amylopectin are partially ruptured, so that the product contains 5% free amylose, 15% free amylopectin, and 80% unmodified starch [92]. The free amylopectin provides cold-water solubility and aids the binding properties, whereas the free amylose and unmodified starch are responsible for disintegration properties. In addition to binding and disintegration properties, STARCH 1500 has lubrication properties of its own. STARCH 1500 in dry form mixed with a drug substance and other excipients can be compressed directly, or if a wet granulation procedure is used it can serve as a binder [93]. It is also used in dry form as a filler for capsule formulations.

Although STARCH 1500 has been claimed to be a free-flowing material with better flow properties than corn starch, the flow properties are poor when compared with other filler-binders because of the large specific surface of the powder, resulting in strong cohesion between particles [26] (see Table 4). The fluidity can be improved by the addition of 0.25% colloidal silica. An average particle size for STARCH 1500 cannot be given, because the range of particle sizes is very large and the granule shape varies [94]. Moreover, it was demonstrated that the particle size distribution can change dramatically during handling, so it is advisable to determine particle size distribution, compressibility (through determination of bulk and tapped density), and surface area before processing [65].

Manudhane et al. [95] and Sakr et al. [96] showed that STARCH 1500 has many advantages over starch USP with respect to binding, disintegration, and dissolution properties. Because STARCH 1500 has lubricating properties of its own, it can be compressed without the addition of a lubricant. These self-lubricating properties are caused by the same factors as discussed for microcrystalline cellulose [51]. The binding properties of unlubricated STARCH 1500 are rather good (see Fig. 5). When mixed with even small amounts of other products, the presence of a lubricant is necessary.

Density-stress (Heckel plots) and stress-relaxation studies show that STARCH 1500 exhibits extensive, yet slow, plastic deformation during compression. This agrees with microscopic observations after compaction, in which discrete primary particles with approximately the same size as before compaction could be identified, and the reported reduction of the specific surface after compaction [97]. Changes in contact time during compaction, therefore, have a marked effect on tablet properties [81, 98, 53]. STARCH

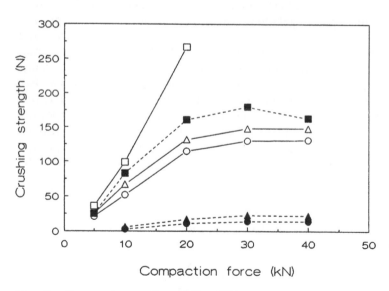

Fig. 5 Compaction profiles of three different starch products, both unlubricated (open symbols) and lubricated with 1% magnesium stearate (closed symbols). (□,■) modified rice starch (Eratab); (△,▲) modified corn starch (Sepistab ST 200); (○,●) compressible starch (STARCH 1500). 500-mg, 13-mm tablets, hydraulic press, contact time 5.1 s.

1500 has a slower initial relaxation than anhydrous lactose, but after prolonged time periods the total relaxation of STARCH 1500 is much greater than that of anhydrous lactose. By measuring density-stress (Heckel) profiles for different excipients and various contact times, an increase in tablet density after prolongation of contact time was in the rank order of STARCH 1500 > Elcema G250 > anhydrous lactose > Emcompress (dicalcium phosphate dihydrate) [81].

As compared with other plastically deforming materials, the strength of STARCH 1500 tablets is low. This effect may be because plastic deformation is too slow to produce adequate interparticle binding during rapid compression. In addition, during compaction at high strain rate, a large proportion of the total deformation will be elastic [81]. When elastic and plastic deformation and interparticles binding occur during compression and when elastic recovery occurs on decompression and ejection, interparticle bonds are not formed rapidly enough to prevent brittle fracture (reducing tablet strength and possibly causing capping) [99].

Contradicting results were published on the effect of particle size distribution of STARCH 1500 and binding properties. McKenna and McCafferty [44] found an increase in tablet crushing strength with a decrease in particle size. This effect was explained by an increased amount of plastic flow and packing with decreasing particle size. Alderborn and Nyström [100] found only a decline in tablet strength with an increase in particle size up to 125 μm; above this size, the tablet strength increased with particle size and leveled off for larger particle size fractions. These changes were attributed to differences in particle shape: particles below 125 μm consisted of fairly smooth primary particles; larger particles exhibited a rougher surface and seemed to consist of very small particles, which were aggregated or fused together. The tableting properties of STARCH 1500 can be improved by combining it with another filler-binder such as microcrystalline cellulose [95] or dicalcium phosphate dihydrate [101].

Because of its plastic behavior under pressure, STARCH 1500 is extremely sensitive to mixing with lubricants (see Fig. 5). After mixing it with 0.5% magnesium stearate in a Turbula mixer at 90 rpm, the tablet crushing strength decreased from 18 kg down to zero within 10 min. For this reason, the use of alkaline stearates should be avoided or kept below 0.25%. When STARCH 1500 is used as a filler-binder in formulations, stearic acid and hydrogenated vegetable oil are acceptable alternatives to magnesium stearate [95]. Additional information about the lubricant sensitivity of filler-binders can be found in Chapter 16.

STARCH 1500 retains its disintegration characteristics despite partial pregelatinization. Although plain starch acts as a disintegrant primarily through wicking and elastic recovery of the deformed grains, STARCH

1500 also acts by swelling in the presence of moisture [91]. This could be why disintegration properties of STARCH 1500 are comparable with or even better than those of plain starch [95]. The dissolution of salicylic acid is faster from tablets containing STARCH 1500 than from tablets containing native starches [102].

C. Modified Starches

Modified rice starch, developed as Era-Tab by Erawan Pharmaceutical Research and Laboratory Co., Ltd. (Bangkok, Thailand), was introduced in 1992. Modified rice starch is produced by physical modification of rice starch and is composed of aggregates of rice starch spherical grains (see Fig. 2c). Modified rice starch is a dry, white, odorless, tasteless powder. The majority of particles have a size range between 75 and 150 μm. Water content is about 11.3% and bulk density is about 0.55 g/cm^3.

The flow properties of modified rice starch, measured as angle of repose and Hausner ratio, are better than those of native rice starch and STARCH 1500 [103]. Modified rice starch has excellent binding properties (see Fig. 5) because of the compactibility of native rice starch (which is better than native corn starch). Like native starches, the compactibility of modified rice starch is dependent on its moisture content.

Mixing modified rice starch with magnesium stearate causes a reduced binding capacity, but the decrease in crushing strength is much smaller than that of STARCH 1500 (see Fig. 5), where zero crushing strength can be reached after mixing for 10 min. The major advantages of modified rice starch, as compared with STARCH 1500, are better flowability and higher tablet crushing strength remaining after lubrication with magnesium stearate.

In tablet formulations, modified rice starch can be used as a single filler-binder or blended with other excipients such as α-lactose 100-mesh or anhydrous β-lactose. Combining with microcrystalline cellulose should be avoided because of the poor flowability of the blend and slow disintegration of the tablets. Data for stored tablets containing model drugs and modified rice starch at 31°C and 75% RH show that modified rice starch can be used in production of tablets for countries with a tropical climate [103].

Modified corn starch is an another example of an agglomerated native starch. Modified corn starch, also referred to as *granulated starch,* is prepared by a patented physical and chemical process from corn starch and is marketed as Sepistab ST200 by Seppic Co., Paris, France. It is made up of roughly spherical granules with an average particle size of 200 μm. Figure 5 shows that the compactibility of Sepistab ST 200, both unlubricated and lubricated with 1% magnesium stearate, is rather similar to the compactibility of STARCH 1500. After lubrication with 1% stearic acid, the compactibility of modified corn starch is comparable with STARCH 1500, but the disintegration properties of modified corn starch are superior [104].

V. INORGANIC SALTS

A. Dicalcium Phosphates

Of the inorganic salts used as directly compressible filler-binders, dicalcium phosphate dihydrate ($CaHPO_4 \cdot 2H_2O$) is the most common. In the United States, the health food sector uses this product in large quantities, and it is estimated that more than 10,000 tons per year is consumed [105]. The calcium phosphates are synthesized by a complicated procedure using phosphoric acid and slaked lime [106,107]. Dicalcium phosphate dihydrate (among other forms) is available in a free-flowing form as Emcompress (formerly Emcompress special) from Mendell, Di-Tab from Stauffer and Rhône-Poulenc, DI-CAFOS from Budenheim, and CalStar from FMC.

In addition to the hydrate form, anhydrous dicalcium phosphate is also used as a directly compressible filler-binder. It is available as Anhydrous Emcompress from Mendell, DI-CAFOS A and DI-CAFOS AN from Budenheim, and A-Tab from Stauffer and Rhône-Poulenc.

Di-Tab is a brand of unmilled dicalcium phosphate dihydrate, whereas Emcompress is a unique form of dicalcium phosphate dihydrate in which particle size distribution is controlled to ensure flowability. According to one investigator, these two products have very similar properties and perform equally well on compression [108], although another investigator has observed some differences [109]. Dicalcium phosphate dihydrate and anhydrous dicalcium phosphate meet the specifications of dibasic calcium phosphate in the USP XXIII. In addition to its use as a directly compressible filler-binder, dicalcium phosphate dihydrate is used as a filler in capsules and in tablets prepared by wet granulation. An advantage of using dicalcium phosphates in tablets for vitamin or mineral supplement is the high calcium and phosphorous content.

Emcompress consists of aggregates of small primary particles (crystallites) (see Fig. 2d). The particle size distribution lies primarily between 75 and 420 μm, with an average particle size of about 130 μm. The primary particles of the aggregates in DI-CAFOS are platelike with a mean size of 2.8 μm. The properties of the dicalcium phosphate dihydrates from other sources for direct compression are very similar to Emcompress [107]. The physical properties of three different calcium phosphates, given in Table 6

Table 6 Physical Properties of Calcium Phosphates

	Dicalcium phosphate Anhydrous	Dicalcium phosphate Dihydrate	Hydroxyapatite
Surface Area (m²/g)	30	2	77
Density (g/cm³)			
True	2.89	2.31	3.15
Bulk (tapped)	0.76	0.94	0.96
Bulk (loose)	0.70	0.83	0.88
Crystal system[a]	Triclinic	Monoclinic	Triclinic

[a]From A. D. F. Toy (1973).
Source: From Ref. 106; reproduced with permission of the copyright owner.

[106], indicate the differences in crystal form, surface area, and densities. The most remarkable difference is in the surface area, which is low for dicalcium phosphate dihydrate when compared to the anhydrous form and even lower when compared to the hydroxyapatite, although bulk and tapped densities do not reflect these differences. This is because of the monoclinic crystal shape of the dihydrate compared to the triclinic shape of the anhydrous form and the hydroxyapatite. Favorable particle size and high density give Emcompress its excellent flow properties [110,26,34] (see Table 4).

Dicalcium phosphate dihydrate is slightly alkaline with a pH of 7.0 to 7.4, which precludes its use with active ingredients that are sensitive to even small amounts of alkali. For example, acidic active ingredients such as ascorbic acid cannot be combined with dicalcium phosphate dihydrate.

During compression, dicalcium phosphate dihydrate undergoes considerable fragmentation as evidenced by photographs [111,97]. The brittle behavior of dicalcium phosphate dihydrate during compaction was shown by scanning electron micrographs of the surface of fracture of a tablet [62]. Heckel plots of Emcompress also exhibit brittle failure during diametrical compression [62,80,82,97]. The brittle nature of Emcompress was also confirmed by stress-relaxation experiments and Heckel plots by Rees and Rue [81]. As expected, Emcompress showed minimal stress relaxation, and Heckel plots showed that increasing the time of material under compression (dwell time), there was no effect on compressibility. Therefore, because of its brittle nature, the tensile strength of dicalcium phosphate dihydrate tablets is hardly affected by an increase in tableting speed [81, 53]. Fragmentation is achieved so rapidly that prolonging exposure to force has no further effect [53].

Duberg and Nyström [97] introduced the strength-isotropy ratio (between axial and radial tensile strength) as a method for the characterization of fragmentation tendency during compaction (see Chapter 16). High strength-isotropy ratios, as found for dicalcium phosphate dihydrate, reflect a large amount of fragmentation during compaction. The fragmentation propensity of dicalcium phosphate dihydrate was also demonstrated by the increase in tablet surface area, measured by permeametry [112] or mercury intrusion porosimetry [47]. Using the latter technique, a comparison of pore volumes at different pore sizes of tablets compressed at two compaction loads showed an increase in the number of small pores during compression (see Fig. 6) [47]. In a detailed evaluation of the compression cycle of Emcompress, using the Heckel equation for both the compression and decompression phases, Duberg and Nyström [113] found that the powder is extensively fragmented even at relatively low loads. The particles continue to fragment at intermediate loads and subsequently appear to undergo mainly plastic deformation, as indicated by a time difference be-

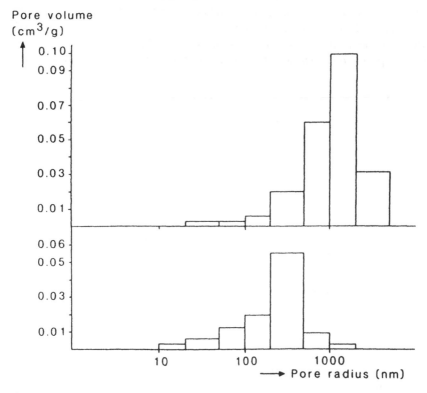

Fig. 6 Comparison of pore volumes at different pore sizes of tablets compressed from dicalcium phosphate dihydrate at a compaction load of 37.8 (upper part) and 302 MPa (lower part), respectively. (From Ref. 47; reproduced with permission of the copyright owner.)

between maximum load and minimum porosity. Using deagglomeration experiments with a Sympatic laser diffraction spectrometer, Herzog [107] showed that particles larger than 63 μm show brittle fracture during compression. On the other hand, particles with a mean diameter of 6 μm do not fragment at all.

As compared with other filler-binders, the binding properties of dicalcium phosphate dihydrate are moderate (see Fig. 7), which is an effect of its brittle nature. Fracture creates a large number of interparticulate contact points and the compaction load per unit area of interparticulate contact is decreased, reducing the strength of the bonds that are produced [114]. The creation of a large number of contact points imply that a comparatively weak type of bonding is involved, e.g., molecular van der Waal's forces [115].

Unlubricated Emcompress tablets are difficult to eject from dies. Free axial compaction of large compacts demonstrate that this effect may be caused by shear failure and rebonding along the shear cones [51]. Investigations on ejection force of dicalcium phosphate dihydrate, carried out on a rotary press, have shown that with the addition of 0.25% magnesium

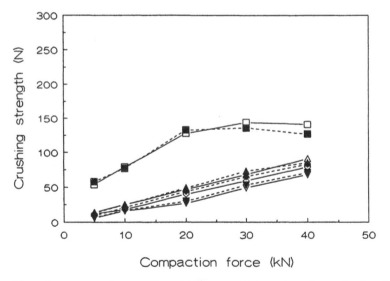

Fig. 7 Compaction profiles of different calcium phosphates, both unlubricated (open symbols) and lubricated with 1% magnesium stearate (closed symbols). ($\square$,$\blacksquare$) tricalcium phosphate (TRI-CAFOS S); ($\triangle$,$\blacktriangle$) dicalcium phosphate dihydrate (Emcompress); ($\diamond$,$\blacklozenge$) dicalcium phosphate dihydrate (DI-CAFOS); ($\triangledown$,$\blacktriangledown$) anhydrous dicalcium phosphate (DI-CAFOS AN). 500-mg, 13-mm tablets, hydraulic press, contact time 5.1 s.

stearate, the ejection force increases significantly with the number of press-ings, indicating insufficient lubrication [116]. With the addition of 0.5 or 1.0% magnesium stearate, the ejection force remains constant even after 500 pressings.

One advantage of dicalcium phosphate dihydrate as a filler-binder is that lubricants such as magnesium stearate have practically no effect on its binding properties (see Fig. 2 in Chapter 16) [83]. This effect is explained by the assumption that clean, lubricant-free surfaces are created by fragmenta-tion of dicalcium phosphate dihydrate during the process of consolidation and compaction [62] (see Chapter 16).

Conflicting results were published regarding the effect of particle size of dicalcium phosphate dihydrate on compressibility of the material. Gillard et al. [117] and Khan and Rhodes [118] found that by using the same compaction pressure, coarse powders rather than fine powders give harder tablets. Large particles are more easily keyed and interlocked than small particles, whereas, because of high lattice energy and high resistance to activation, friction between the particles is less important [10]. On the other hand, Alderborn and Nyström [100] reported that for Emcompress, the tablet strength was almost independent of the particle size. Herzog [107] showed a minimum hardness for tablets compressed from particles of about 200 μm. The increase in tablet hardness with an increase in particle size was attributed to the structure of the larger particles, which are agglom-erates of small primary particles. Because of its brittle behavior, the shape of Emcompress particles has practically no effect on their compaction prop-erties [119].

Most types of anhydrous dicalcium phosphate have a slightly higher compactibility than types of dicalcium phosphate dihydrate [106,107]. The compactibility of DI-CAFOS AN, however, is comparable with the com-pactibility of DI-CAFOS or Emcompress (see Fig. 7). The compactibility of DI-CAFOS A is much lower than that of other types of anhydrous or hydrous dicalcium phosphates. The differences between the compactibilities of DI-CAFOS AN and DI-CAFOS A may be caused by differences in their particle structure; DI-CAFOS A consists of compact spherical particles with a high density (specific surface area by BET = 0.28 m^2/g), whereas DI-CAFOS AN consists of porous particles with a specific surface area of 9.1 m^2/g [107]. The mechanism of consolidation for anhydrous dicalcium phos-phate is similar to that of dihydrate, with the exception of DI-CAFOS A. The latter shows no or insignificant fragmentation during compression, so less binding surface is created as compared with the other products [107].

Because of the more regular particle shape and higher bulk density, the flow properties of DI-CAFOS A are better than those of DI-CAFOS AN, but the latter has sufficient flowability for practical use [107].

When placed in water, dicalcium phosphate dihydrate tablets are rapidly and completely penetrated by the liquid [3]. This is caused by the hydrophilic nature of the filler-binder [120] and the relatively high porosity of the tablets. The small effect of 0.5% magnesium stearate on the water penetration rate, as compared with tablets compressed from other filler-binders, is attributed to the extensive fragmentation of dicalcium phosphate dihydrate during compression [69]. In spite of a fast and complete water penetration, dicalcium phosphate dihydrate tablets do not disintegrate because the excipient is relatively insoluble in water and no disintegration force is developed [121]. This also indicates the type of bonds created on compaction of dicalcium phosphate dihydrate that do not break up when contact with water. The correlation between disintegration time and disintegration force kinetics indicates that, for the disintegration of dicalcium phosphate dihydrate tablets, a disintegrant with an active mechanism such as swelling or disintegration force is needed [121].

Tablets made with dicalcium phosphate dihydrate as a filler-binder may introduce on aging unfavorable changes in the physical properties of the tablets, such as hardness, disintegration time, and drug dissolution time. These are major problems that must be considered when selecting this excipient in direct compression or wet granulated tablets.

Storage in low-humidity conditions of tablets containing 84% dicalcium phosphate dihydrate, 10% starch, 5% naproxen, and 1% magnesium stearate results in an increase in tablet hardness and a decrease in disintegration and dissolution rates [122]. However, on aging under high humidity, hardness, disintegration, and dissolution rates decreased at all initial moistures at which tablets were compacted. The decrease in drug dissolution rate after aging under *low* humidity is because of the limited dissolution and recrystallization of calcium phosphate in the available water in the tablet, whereas the decrease in the dissolution rate on aging under high humidity is ascribed to the expansion and contraction and general opening of the structure of the starch grains and their bonding, via water molecules, to calcium phosphate dihydrate [122]. The idea that the changes in tablet hardness are caused by the disintegrant cannot be excluded because, during storage of disintegrant-free tablets, compressed from dicalcium phosphate dihydrate lubricated with 0.5% magnesium stearate for eight weeks at 20°C and 50% or 85% RH, no change in crushing strength was observed [6].

Delattre and Jaminet [123] saw a decrease in crushing strength of about 40% after storage for one month at 45°C, whereas Horhota et al. [124] found no change in crushing strength and tablet thickness under similar conditions and a decrease in dissolution rate. During storage at

23°C and 44% RH of tablets containing dicalcium phosphate dihydrate with amaranth as a tracer and sodium alginate as a disintegrant, the disintegration and dissolution rates decreased without changes in hardness and tablet weight; during storage at 45°C and 75% RH, however, the tablets showed blotching, substantial weight loss, and complex changes in disintegration and dissolution [125]. The changes at elevated temperatures were attributed to loss of water of hydration; changes at 25°C were thought to be caused by other factors, such as case hardening.

The course of dehydration of dicalcium phosphate dihydrate at elevated temperatures is an extremely complicated process, dependent on water vapor pressure. It was demonstrated that dicalcium phosphate dihydrate easily loses its hydrated water when stored above 40°C, and that this process was irreversible [126,127]. In general this may decrease chemical stability of drugs in tablet formulations containing dicalcium phosphate dihydrate [128,129], possibly after solubilizing the drug in accumulated free water because of dehydration within the drug/excipient system [127].

The disadvantage of dehydration of dicalcium phosphate dihydrate can be avoided by using anhydrous dicalcium phosphate, because this product contains no water that can be lost. Moreover, anhydrous dicalcium phosphate does not absorb water to form the dihydrate.

Emcompress offers the best potential for direct compression when used in combination with microcrystalline cellulose [123] or starch [73]. The crushing strength of tablets prepared from blends of Emcompress with Avicel PH 102 depends on the compression force and the percentage of Avicel [7]. Good compression characteristics can be obtained with 10% to 33% dicalcium phosphate dihydrate and 66% to 90% microcrystalline cellulose with 0.5% magnesium stearate and as little as 0.1% sodium starch glycolate. This combination of excipients can convert up to 20% of a poorly compressible drug substance into a directly compressible formulation [130].

The reworkability of dicalcium phosphate dihydrate–microcrystalline cellulose blends is poor. Stress relaxation studies suggest that reduced plastic flow and work hardening of microcrystalline cellulose is responsible for this behavior [130].

B. Tricalcium Phosphate

Another calcium phosphate for direct compression is tricalcium phosphate (tribasic calcium phosphate USP), available as Tri-Tab (Stauffer, Rhône-Poulenc), Tri-Compress (Mendell), and TRI-CAFOS S (Budenheim). Actually, the commercially available product is hydroxyapatite, $Ca_5(OH)(PO_4)_3$, also commonly and erroneously referred to as basic tricalcium phosphate, $Ca_3(PO_4)_2Ca(OH)_2$, and *tricalcium phosphate* [106]. It is a variable mixture of calcium and phosphates.

Tricalcium phosphate can be used as a filler-binder in direct compression and as a filler in tablets prepared by wet granulation. Its higher calcium load when compared with dicalcium phosphate dihydrate may be advantageous when used in vitamin tablets.

Carstensen and Ertell [106] reviewed the synthesis, physical properties, and chemical properties of tricalcium phosphate. TRI-CAFOS S is probably prepared by spray drying, whereas Tri-Tab is prepared by granulation or a compaction technique [107]. Both products are built up of very fine and porous primary particles with a mean size of 2.2 and 1.5 μm, respectively [107]. Compared with Tri-Tab, TRI-CAFOS S has a much lower bulk density. Some physical properties of tricalcium phosphate (Tri-Tab) are listed in Table 6. Extremely good flow rates were found for Tri-Compress and Tri-Tab and these results were related to the high density of the excipients [131].

TRI-CAFOS S has better compactibility than the different types of DI-CAFOS [105,107]. The compressibility of Tri-Tab is inferior, because of hardness of the agglomerates, which imparts fragmentation [107]. Tricalcium phosphate obeys the Heckel equation only if the particle density of the agglomerated tricalcium phosphate is 1.92 g/cm^3 (rather than the true density of 3.1 g/cm^3 of the individual crystals), and is used to determine the relative density of the compacts [132,133]. Comparing mercury intrusion porosimetry data with nitrogen adsorption data shows that a large proportion of the solid has pores smaller than 60 Å. These micropores, which are attributed to the pore space in the individual crystals, are not compressible in conventional pressure ranges and are therefore not considered as part of the Heckel pore space. The pore space that is lost during compression is between 0.5 and 2 μm in diameter. It was demonstrated that bonding occurs not because of fracture or deformation of the individual crystals, but rather because of the deformation of agglomerates under pressure [134]. This view is supported by the work of Patel et al. [131], which demonstrates that the compactibility of tricalcium phosphate is time dependent. A slow eccentric press yields harder tablets at a given compression force than a much faster rotary press, which is an indication that the mechanism of consolidation is plastic deformation. On the contrary, Herzog [107] shows by Heckel analysis that TRI-CAFOS S and Tri-Tab fragment during compression.

The water content of tricalcium phosphate has a profound effect on its compactibility; its optimum lies between 4% and 5%. Higher or lower values lead to capping and a decreased binding capacity [107].

It has been reported that, in contrast to the dicalcium phosphates, the addition of magnesium stearate has a profound effect on the compactibility of tricalcium phosphate. At 2% magnesium stearate concentration, for instance, tablets cannot be compacted at all [106]. This lubricant sensitivity was assumed to be an indication for the previously noted plastic behavior under pressure. Figure 7 shows however, that TRI-CAFOS S has a low lubricant sensitivity.

A serious drawback of the tricalcium phosphates is their high sticking tendency with dies and punches. This sticking tendency, which results in high ejection forces, is attributed to the fine structure of the primary particles [107].

Another drawback of using tricalcium phosphate as an excipient in tablets is the effect of aging under both low and high humidities on tablet hardness, disintegration time, and dissolution rate. The moisture content of tribasic calcium phosphate–based tablets at the time of compression and the moisture that was gained or lost during aging have a significant effect on the hardness, disintegration time, and dissolution rate of the drug. Some differences and some similarities between the responses of tricalcium phosphate tablets and dibasic calcium phosphate dihydrate tablets are reported [135,122].

In addition to the calcium phosphates, granular, edible bone phosphate was developed by Lensfield as a directly compressible filler-binder under the name Lenphos.

C. Calcium Sulfate Dihydrate

Another inorganic salt used for direct compression but not commonly used in pharmaceuticals is a specially processed grade of calcium sulphate dihydrate NF marketed (among others) as Compactrol by Mendell. Calcium sulfate dihydrate is a commodity item and is produced in commercial quantities for the construction industry. The pharmaceutical grade is an inexpensive excipient used as a filler, also known as terra alba. Compactrol has good flow properties (see Table 4) but poor compactibility [6]. Physical stability studies of tablets prepared with Compactrol, show that tablet hardness decreases at accelerated storage conditions, but that disintegration time and drug dissolution rates are hardly affected [136].

VI. POLYOLS

A. Sorbitol

Sorbitol was discovered in the 19th century in the fruit of the mountain ash tree and has become a major industrial sugar alcohol used in the food and pharmaceutical industries. Chemically, it is an isomer of mannitol. The most significant differences between sorbitol and mannitol are their hygroscopicity and aqueous solubility; sorbitol is hygroscopic at relative humidities above 65%, whereas mannitol is nonhygroscopic, and the aqueous solubility of sorbitol is higher than that of mannitol. According to the monograph in the USP XXII, sorbitol may contain small amounts of other polyhydric alcohols.

Chemically, sorbitol is closely related to glucose, which can be obtained from starch or sugar. It is commercially produced by high pressure catalytic hydrogenation of glucose in the presence of Raney nickel. Different forms of sorbitol are manufactured depending on crystallization and purification conditions, and these forms are available commercially. Pharmaceutical-grade sorbitol is available in several different physical forms from various suppliers. Shangraw et al. [137] showed that there were considerable differences in shape and structure of different sorbitol products resulting in different compaction behavior.

Sorbitol exists in four different crystalline polymorphic forms (α-, β-, γ-, and δ-sorbitol) and one amorphous form [138]. Different sorbitol samples including commercial products were investigated and it was found that the most stable form, γ-sorbitol, had the best compaction properties but showed a longer time for disintegration and dissolution [138,139]. The compactibility of γ-sorbitol was dependent on the particle structure produced by the manufacturing process [138].

γ-Sorbitol is prepared by cooling a melt and seeding it with sorbitol crystals, or by spray drying a sorbitol solution [140]. A product prepared from a melt is available as Neosorb DC in different particle size fractions from Roquette Frères. A spray-dried product containing more than 90% γ-sorbitol is marketed by Merck as Sorbitol Instant (a special pharmaceutical grade is Karion Instant). Neosorb 20/60 DC is a product containing large, nearly spherical particles with smooth surfaces [141,6], whereas Sorbitol Instant consists of agglomerates of loosely packed, randomly oriented, interwoven, filamentous crystals (see Figs. 2e and 2f).

Comparative evaluation of two types of γ-sorbitol [141] showed that the tableting properties were strongly dependent on particle structure, particle-size distribution, and bulk density of the products. Spray-dried sorbitol (Karion Instant) could be compressed into much harder tablets than the γ-sorbitol prepared from a melt (Neosorb 20/60).

Adding magnesium stearate (up to 2%) to both types of γ-sorbitol and blending them together did not markedly affect the hardness of the tablets. At high compression loads, a sintering effect leading to glittering of the tablets was observed when different types of γ-sorbitol were used. The outer surface and the cross section of the tablets compressed at low and high compression forces were examined under the scanning electron microscope, and showed that the individual particles, visible after compaction at low forces, were completely sintered, forming a smooth surface at high compaction forces. It was hypothesized that the phenomenon was caused by plastic deformation of the material under load.

Schmidt and Vortisch [142] also studied the influence of the manufacturing method on compactibility of eight different types of commercially available sorbitol. All products showed different tableting properties, and these differences could not be explained because of the differences in physical properties such as particle size, bulk and tapped density, and polymorphism. The differences in compaction behavior were caused by the method of manufacturing resulting in differences in particle shape and surface properties. Spray drying results in irregular and porous particles that show high compactibility. Sorbitol products prepared from a melt or by crystallization from a solution have lower compactibility.

The compaction profiles obtained after slow compression on a hydraulic press in Fig. 8 show that Karion Instant had very good binding properties and, at relatively low compaction forces maximum tablet hardness was reached. Excellent compactibility was confirmed by Deurloo et al. [76] during compaction of a blend of Karion Instant with 1% magnesium stearate on a high-speed (Fette P1000) rotary press (see Fig. 4). However,

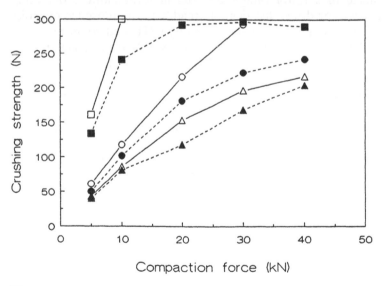

Fig. 8 Compaction profiles of different kinds of sugars and polyols, both unlubricated (open symbols) and lubricated with 1% magnesium stearate (closed symbols). (□,■) γ-sorbitol (Karion Instant); (○,●) Emdex; (△,▲) compressible sugar (Comprima F-gm). 500-mg, 13-mm tables, hydraulic press, contact time 5.1 s.

the author observed moderate flowing of the powder, with sticking to the punches, and capping of the tablets at high compaction forces.

The two types of γ-sorbitol available commercially form ordered mixes (powder mixtures where fine drug substance particles randomly adhere to the large particles of the excipient) with different B vitamins [143]. A large amount of the drug substance particles can adhere to the irregular particles of Karion Instant as opposed to fewer particles adhering to the more regular particles of Neosorb 20/60. For six different antibiotics, Sorbitol Instant had greater adhesion capacity and binding strength than crystalline sorbitol [144].

Direct compression grades of sorbitol can be used for the production of lozenges, chewable tablets, and disintegrating tablets; examples of the formulations (including a number of the so-called problem drugs) can be found in the literature, e.g., Basedow et al. [145], Schmidt [140], and Darrouzet [146]. For disintegrating tablets, a superdisintegrant with a high swelling power, such as croscarmellose sodium, must be used. The use of crospovidone appeared to change tablet properties during aging, especially under high relative humidity conditions [145].

The hygroscopicity of sorbitol limits its use in tablet formulations. A high percentage of sorbitol cannot be used in formulations containing hygroscopic drugs or drugs that degrade readily in the presence of moisture. During storage at 50% relative humidity (RH) and room temperature, the crushing strength of tablets prepared with Neosorb 20/60 did not change. However, during storage at 85% RH, the tablets liquified [6]. Karion Instant tablets increased their weight after aging up to 65% RH, and the weight gain was higher than for tablets compressed from anhydrous β-lactose (Pharmatose DCL 21) or compressible sugar (Di-Pac), but less than the tablets compressed from crystalline sorbitol [145]. The lower hygroscopic behavior of Karion Instant tablets compared with tablets made with crystalline sorbitol was attributed to the smooth tablet surface, which may seal the tablet core against atmospheric moisture to some degree.

During storage, sorbitol tablets can become harder. This effect is caused by dissolution and recrystallization of sorbitol during aging, forming a reinforcing network throughout the tablet core [147]. The inclusion of pregelatinized starch in sorbitol tablets can prevent recrystallization and an increase in tablet crushing strength [147].

B. Mannitol

Mannitol is a polyol isomer of sorbitol and the commercial products are prepared by catalytic reduction of different sugars. It is often obtained with its isomer, sorbitol, from which it is isolated. Mannitol is commonly used in tablets prepared by the wet granulation process. Mannitol and sorbitol have negative heat of solution and sweet taste that make them the most useful excipients for lozenges and chewable tablets. Mannitol is less hygroscopic than sorbitol and the lower hygroscopicity of mannitol is an advantage over sorbitol. A disadvantage of using mannitol in formulations is its higher cost.

Because of its poor flowability and binding properties, unmodified mannitol cannot be used for tablet production by direct compression. For this reason, the product has been modified to improve its tableting characteristics. Kanig [148] prepared a directly compressible mannitol by the spray congealing of a melt. This was possible because of the exceptionally high heat stability of mannitol, as compared with other sugars. It was also possible to produce eutectic blends of mannitol with other carbohydrates by fusion and spray congealing or recrystallization. All modified mannitol products have excellent flow and compression characteristics.

Debord et al. [149] studied the compactibility of different polymorphic forms of mannitol. Four polymorphic forms were characterized, the α-form, the β-form, the δ-form, and an unidentified form. Among the (anonymous) commercial products, the α-form, the β-form, and the unidentified form were characterized. The α-form had the best compactibility. Particle shape had profound effect on compactibility and the most compressible form was the granulated powder. Under compression stress, no polymorphic transition was observed.

Sangekar et al. [73] reported that tablets compressed from granular mannitol picked up less moisture after exposure to high humidity than tablets compressed from anhydrous Celutab, granular sorbitol, or direct compression sugar. This was attributed to the lower hygroscopic nature of mannitol.

Special directly compressible forms are also available and include granular mannitol from ICI and Atomergic Chemetals Corp., and Pearlitol from Roquette. Pearlitol is a granulated mannitol with excellent flow and compaction properties [150] and is available in three grades: Pearlitol FG (average diameter about 250 μm), Pearlitol MG (about 360 μm), and Pearlitol GG2 (about 520 μm). The compactibility of Pearlitol MG, lubricated with 1% magnesium stearate on a Fette P1000 rotary press, is similar to the compactibility of lactose Fast Flo [146].

VII. LACTOSE

Lactose is a natural disaccharide, produced from cow's milk that contains about 4.6% lactose, corresponding to about 38% of its dry solids. Chemically, lactose consists of one galactose unit and one glucose unit. It exists in two isomeric forms, α-lactose and β-lactose, and can be either crystalline or amorphous. Crystalline α-lactose occurs in the monohydrate and the anhydrous forms. The crystalline β-lactose exists in the anhydrous form only. Pure amorphous lactose is not available commercially and is generally present in modified forms of lactose in varying amounts. The modified forms of lactose may contain varying amounts of water depending on the amorphous content.

Crude α-lactose monohydrate is produced from casein or cheese whey by evaporation and crystallization. For the production of pharmaceutical-grade lactose, the crude product is subsequently purified by recrystallization. Depending on the temperature of crystallization, different types of lactose can be obtained (see Fig. 9); below 93.5°C, lactose precipitates as α-lactose monohydrate, and above 93.5°C as anhydrous β-lactose. α-Lactose monohydrate can be processed further either by dehydration into the anhydrous α-lactose form, or by spray drying into the spray-dried form, or by

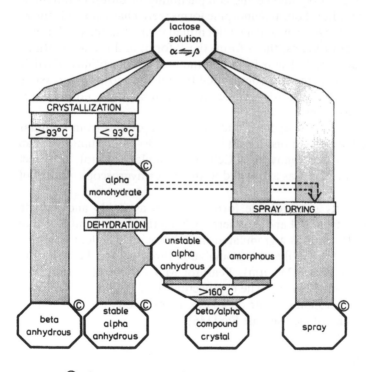

Ⓒ Commercial types of lactose

Fig. 9 Schematic representation of the production of different types of lactose. (From Ref. 227; reproduced with permission of the copyright owner.)

granulation into the agglomerated form. All different types of lactose have different compaction properties.

A. α-Lactose Monohydrate

Lactose is probably the most widely used diluent in tablet formulations, and the most common form used is α-lactose monohydrate, often referred to as *lactose,* or *hydrous lactose,* or *regular lactose.* It is generally used in powdered (ground) form as a filler for tablets, prepared by means of a wet granulation technique. In direct compression, coarse, regular grade, and sieved crystalline fractions of α-lactose monohydrate (particularly the 100-mesh quality) are used because of their good flowability.

α-Lactose monohydrate contains one mole of water per mole of lactose; i.e., it contains 5% water of crystallization. The total moisture content is about 5.2; about 0.2% is free moisture. Sieved lactose is available in mesh sizes ranging from 45/50 to 325. The 100-mesh lactose has a particle size distribution that is almost similar to that of a product previously described as lactose EFK (extra fine crystalline) and microcrystalline lactose in French publications. Both products have an excellent flowability and a high packing density because of the regular form of the particles (see Fig. 10a and Table 4) [110,26,6].

When compared with other filler-binders, α-lactose monohydrate exhibits relatively poor binding properties (see Fig. 11). Experiments based on Heckel and other stress-density equations [151–153], stress relaxation studies, and particle size measurements after compaction [154] showed that α-lactose monohydrate consolidates mainly by fragmentation. In other studies, consolidation by fragmentation and plastic deformation were described [155,156,52,45]. The fragmentation behavior was confirmed by the fast attainment of an ultimate porosity after multiple compressions [67] and by the relative low strain rate sensitivity [52]. Because of its higher yield pressure, Roberts and Rowe [52] concluded that crystalline α-lactose monohydrate was higher in brittleness than spray-dried lactose and anhydrous β-lactose, respectively, and this was attributed to the more angular nature of α-lactose monohydrate as compared with the aggregated collection of smaller crystals of the directly compressible lactose. Roberts and Rowe [45] demonstrated that for powdered grades of α-lactose monohydrate, an increase in particle size resulted in an increase in relative density for a given applied force. This may be caused by the occupation of the interparticulate voids between larger crystals by small particles. During compaction, fragmentation of the larger crystals and filling of the remaining interstices led to a further increase of the relative density. Furthermore, for smaller particle size material, the increased frictional forces associated with particle

a

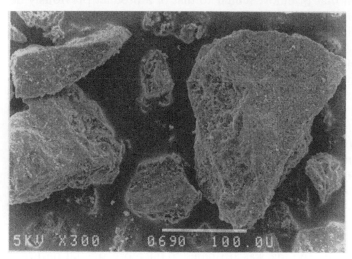

b

Fig. 10   Scanning electron micrographs of directly compressible materials: a = α-lactose monohydrate 100-mesh (Pharmatose 100 M); b = anhydrous β-lactose (Pharmatose DCL 21); c = spray-dried lactose (Pharmatose DCL 11); d = dextrate (Emdex); e = Cellactose; f = ascorbic acid for direct compression C-97.

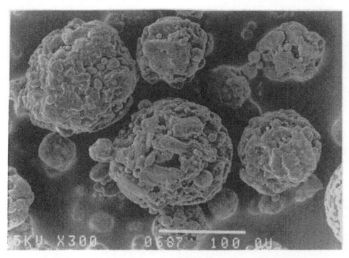

c

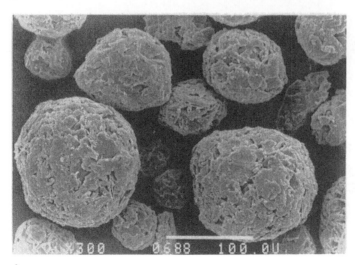

d

sliding can restrict densification. The more uniform size crystalline α-lactose monohydrate showed a higher rate of densification than the powdered grades, which may be caused by the high stress concentrations at contact points, as a consequence of the angularity of the crystals, from which cracks can easily develop. The fractured particles filled the interparticle voids as the fracturing process proceeded during the application of compression force [45].

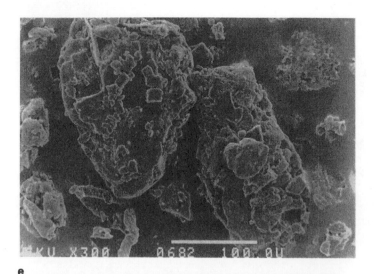

e

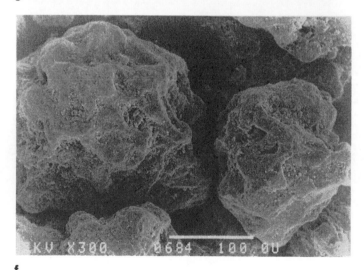

f

Fig. 10 Continued

Duberg and Nyström [97] evaluated different techniques for the evaluation of fragmentation tendency during compaction of α-lactose monohydrate and other materials. From the results of scanning electron microscopy, surface area measurements of starting powder and tablets (krypton adsorption), isotropic ratio, and Heckel plot analysis, it was concluded that the particles of α-lactose monohydrate compacted by plastic deformation

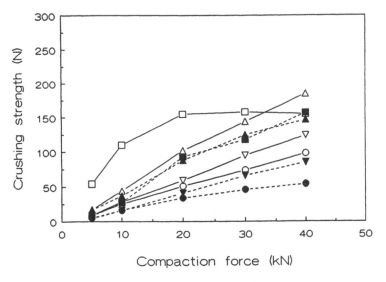

Fig. 11 Compaction profiles of different types of lactose, both unlubricated (open symbols) and lubricated with 1% magnesium stearate (closed symbols). (□,■) spray-dried lactose (Pharmatose DCL 11); (△,▲) anhydrous β-lactose (Pharmatose DCL 21); (▽,▼) agglomerated lactose (Tablettose); (○,●) α-lactose monohydrate 100-mesh. 500-mg, 13-mm tablets, hydraulic press, contact time 5.1 s.

and fragmentation, although the increase in tablet surface area and the isotropic ratio showed that the particles deformed to a large extent by fragmentation [157,97]. Further work at Uppsala University showed that the fragmentation propensity of α-lactose monohydrate could be rank ordered as intermediate using a permeametry technique (see Fig. 12) [112]. A similar linear increase in tablet pore surface area with an increase in compaction force was found using mercury intrusion porosimetry for all types of crystalline lactose (α-lactose monohydrate, anhydrous β-lactose, and anhydrous α-lactose) [47]. Assuming a proportionality between the change in pore surface area and the actual binding surface area, the authors found a unique linear relationship between crushing strength and the pore surface area of tablets, compressed from different types of crystalline lactose at different forces (see Fig. 13) [47]. The unique relationship was valid for the whole range of particle sizes (32–400 μm) of α-lactose monohydrate [158]. The proportionality between crushing strength of brittle lactose tablets and the internal specific surface area was elucidated by a theoretical model [159]. Assuming that a tablet is made up of spherical isometric particles and that the strength of all types of crystalline lactose tablets is caused by van der

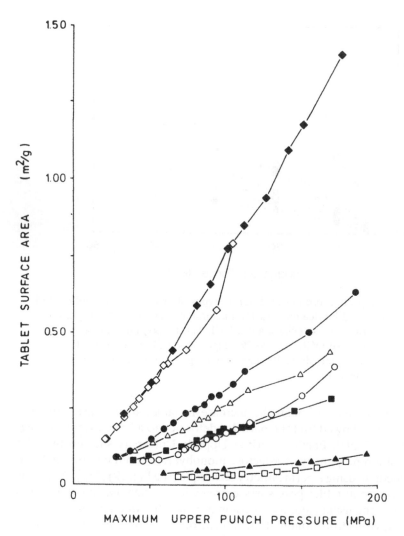

Fig. 12 Specific surface area of sodium chloride (□), sodium bicarbonate (▲), saccharose (○), lactose (●), sodium citrate (■), ascorbic acid (△), para-cetamol (◇), and Emcompress (◆) tablets as a function of compaction pressure. (From Ref. 112; reproduced with permission of the copyright owner).

Waals dispersion forces acting at the coordination points of the particles, a proportionality between tensile strength and internal specific surface of the tablet is obtained.

From the proportionality between crushing strength and tablet pore surface area, it was concluded that fragmentation was the predominant

Crushing strength

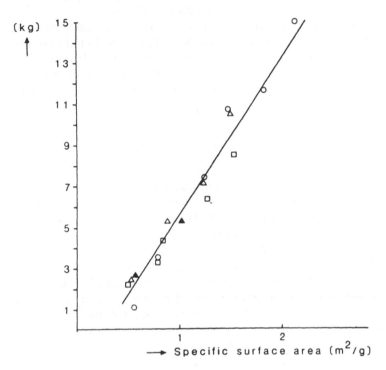

Fig. 13 Crushing strength versus specific surface area for tablets compressed from different types of crystalline lactose: α-lactose monohydrate ($\square$), anhydrous α-lactose ($\bigcirc$), roller-dried β-lactose ($\triangle$), crystalline β-lactose ($\blacktriangle$). (From Ref. 47; reproduced with permission of the copyright owner.)

mechanism of consolidation for all types of crystalline lactose and that the presence of water of crystallization, or the α/β ratio, or the degree of crystallinity had no influence on the binding mechanism [47]. A relation between particle texture and compactibility was also found. Using a gas adsorption apparatus to determine the specific surface areas of crystalline lactose powders, Vromans et al. [160] demonstrated that the binding capacity was directly related to the powder surface area. Thus, the compactibility of crystalline lactose appears to depend on particle texture, which is determined by the manufacturing conditions. Slow crystallization of lactose produces single crystals with low powder surface area and poor compaction properties (e.g., α-lactose monohydrate). Rapid crystallization by dehydration (see Section VII.B) or roller drying (see Section VII.C) results in aggregates of microcrystals (anhydrous lactose). Due to their higher powder surface area, these irregular crystals have better binding capacity compared with the regular crystals of α-lactose monohydrate.

The effect of crystallization conditions of α-lactose monohydrate on its tableting properties was shown by Staniforth [161]. He found that lactose spherulites recovered from a crystallizing solution containing 80% ethanol had the optimum combination of physical characteristics. The improved compression characteristics were attributed to the large number of fine crystallites that formed each spherulite. The spherulites readily broke off and fractured to produce a large amount of fresh surfaces and bonded readily.

The strength of tablets compressed from α-lactose monohydrate increases with a decrease in particle size of the excipient [162,163,158]. Fell and Newton [164] prepared tablets from three particle size fractions of crystalline lactose at two compaction speeds. Calculation of the densification because of particle rearrangement by the Heckel, and Cooper and Eaton methods showed that the smaller fraction underwent the largest rearrangement. Vromans et al. [163] showed that compression of small particles resulted in tablets with increased thickness, compared with compacts obtained from large particles. The observed differences were caused by the higher porosity of tablets compressed from the small particles. For a number of sieve fractions of unmilled crystalline α-lactose monohydrate, fragmentation diminished with decreasing initial particle size, which may increase the role of plastic and elastic deformation [165,158]. However, initial differences in particle surface area did not decrease the fragmentation, so that even larger surface areas were available for bonding from small particles, resulting in stronger compacts [158].

With a decrease in particle size, yield pressure increased and the strain rate sensitivity index (SRS index) decreased [153], which pointed to a reduction in the amount of fragmentation [45]. The transition from a brittle crystal to a purely ductile material was thought to occur at a medium particle size of approximately 20 μm [45]. The change in the mechanism of consolidation at a decreasing particle size of α-lactose monohydrate was confirmed by Vromans et al. [18] by demonstrating that particle fractions of 1–8 μm and 8–16 μm did not meet the unique proportionality between crushing strength and tablet pore surface area, which exists for larger particles of α-lactose monohydrate.

Riepma et al. [166] demonstrated that tablets compacted from α-lactose monohydrate showed a small time-dependent moisture uptake when exposed to an ambient humid atmosphere (45% RH). Moisture sorption was found to reach a plateau within 10 min after compaction, which was accompanied by a decrease in both crushing strength and specific BET surface area of the tablets. Subsequent storage of the tablets in a dry atmosphere resulted in an increase in tablet crushing strength but not in a change in specific surface area. The tablets showed no moisture sorption and no change in crushing strength or specific surface area when transferred immediately after ejection from the die into a dry atmosphere. These results indicate that contact points between lactose particles in a tablet dissolved when exposed to a humid atmosphere and recrystallized on exposure to lower relative humidity conditions. It was suggested that the cause of irreversible decrease in specific surface area of the tablets on exposure to high relative humidity conditions was blockage of the very narrow pores in the tablets by sorbed moisture.

Bolhuis et al. [6] showed that the strength of α-lactose monohydrate tablets, initially measured 30 min after ejection from the die, did not change during storage for eight weeks at 20°C/50% RH or at 20°C/85% RH.

Tablets compressed from α-lactose monohydrate without a lubricant, disintegrate very quickly in water as a result of rapid liquid uptake and fast dissolution of the bonds [167]. The presence of a hydrophobic lubricant has a strong inhibiting effect on water penetration and hence on disintegration time. This effect can easily be counteracted by the addition of microcrystalline cellulose [7] or a high-swelling disintegrant such as sodium starch glycolate or croscarmellose sodium [3,168].

In practice, α-lactose monohydrate 100-mesh is often combined with microcrystalline cellulose [7,74]. This combination results in a strong synergistic effect on disintegration time, whereas the crushing strength increases proportionally to the percentage of microcrystalline cellulose [7]. Therefore, the combination of lactose 100-mesh and microcrystalline cellulose is one of the more popular blends in direct compression. Using a systematic optimization method according to a simplex lattice design, optimum tablet formulations were obtained from α-lactose monohydrate 100-mesh in combination with microcrystalline cellulose, anhydrous β-lactose, or spray-dried lactose [9,169].

B. Anhydrous α-Lactose

The binding capacity of α-lactose monohydrate is increased dramatically by thermal or chemical dehydration of the crystals [19]. During treatment, a gradual transition within each particle from the hydrous to the anhydrous form was observed. The thermally dehydrated product showed strongly increased binding properties with decreasing water content, whereas desiccation by methanol gave a steeper increase in the crushing strength (Fig. 14) [19].

During dehydration, α-lactose monohydrate changes from single crystals into aggregates of anhydrous α-lactose particles. In contrast to the large

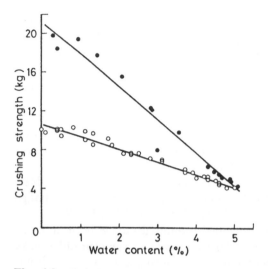

Fig. 14   Relation between crushing strength and percentage water content of, respectively, thermally dehydrated (○) and methanol desiccated (●) α-lactose monohydrate samples, compressed (15000 N) into 500-mg (13 mm) tablets. (From Ref. 19; reproduced with permission of the copyright owner.)

differences in compactibility between α-lactose monohydrate and anhydrous α-lactose, compaction of the two types of lactose resulted in tables with almost equal porosities at the same compaction load, resulting in superimposed Heckel plots [19,47]. Large differences were found between the pore volume distributions of the two types of tablets. Mercury porosity measurements showed that tablets from anhydrous α-lactose exhibited a greater number of pores, much smaller in size, than α-lactose monohydrate. The steeper increase in crushing strength with compaction load and a corresponding increase in pore surface area of tablets lead to the increased fragmentation during compression of the anhydrous product. This effect explains the large increase in binding capacity with increasing dehydration [19,47]. Wong et al. [170] compared the deformation characteristics of single crystals of anhydrous α-lactose and α-lactose monohydrate. Indentation hardness testing showed that the anhydrous crystals were much softer, less elastic, and had a lower resilience than the monohydrate crystals. The anhydrous crystals were less anisotropic than the monohydrate crystals. Stress-strain data of the single crystals and photographs of single crystal deformations showed that the anhydrous crystals withstood a lower maximum load, exhibited lower displacement prior to destructive cracking, and thus required less work to break than the corresponding monohydrate crystals. The monohydrate crystals underwent much more pronounced splitting and fragmentation (spalling) than the anhydrous crystals, which tended to crush by gradual localized cracking at the point of contact. It was concluded that α-lactose monohydrate crystals are hard, elastic, brittle, and strong; compression initiates the progression of large cracks that result in the breaking of both small and large fragments from the crystals. The anhydrous crystals, which are softer, weaker, and less elastic, undergo brittle fracture much more readily and at lower stresses than the monohydrate. The reduced anisotropy of the anhydrous crystals was explained by the removal of water of crystallization, resulting in a partial disruption of the crystalline order. This disruption explains the difference in degree and nature of the fragmentation mechanisms of anhydrous α-lactose and α-lactose monohydrate.

A disadvantage of using anhydrous α-lactose in direct compression formulations is the relatively slow disintegration of the tablets. When in contact with water, the tablets dissolve during the disintegration process. This effect, which is caused by poor water penetration into the tablets, is caused by a combination of small pore diameters and precipitation of dissolved anhydrous α-lactose in the course of the water penetration process into α-lactose monohydrate [167].

Anhydrous α-lactose was manufactured and marketed by DMV as Pharmatose DCL 30. In spite of the excellent flowability (see Table 4) and binding properties [6], the product was withdrawn from the market because relatively slow disintegration turned out to be a major disadvantage.

C. Anhydrous β-Lactose

Another form of lactose that was especially designed for direct compression is anhydrous β-lactose, also referred to as *anhydrous lactose*. In the USP-NF, it is included in the monograph on lactose. The commercial products consist of agglomerated extremely fine crystals, produced by roller drying of a solution of pharmaceutical-grade α-lactose monohydrate followed by subsequent comminution and sieving [137]. Commercial products contain about 80% anhydrous β-lactose; the remaining material is anhydrous α-lactose. The water content is less than 0.5%. Roller-dried anhydrous β-lactose is marketed as Anhydrous DT Lactose by Sheffield Chemicals and as Pharmatose DCL 21 by DMV (see Fig. 10b).

Batuyios [171] investigated a tablet formulation containing 89.25% anhydrous lactose, 10% starch USP, and 0.75% calcium stearate. The tablet formulation performed well on a high-speed tablet press with low weight variation. The tablet parameters of the formulation such as hardness, friability, disintegration, and drug dissolution were excellent. Other studies suggested that the flow properties of anhydrous β-lactose were less than optimum [33,26], fair [172], and good [6,173]. The differences may be caused by the differences in particle size distribution of different materials.

The compaction profiles in Fig. 11 (hydraulic press) and Fig. 4 (rotary press) show that anhydrous β-lactose (Pharmatose DCL 21) has excellent compaction properties and a low lubricant sensitivity. In an examination of six lactose-based materials, anhydrous β-lactose had the best compaction properties [172].

The good compactibility was attributed to the particle structure, resulting in a large surface for bonding [26]. Based on the small deformation before failure during diametrical compression testing of anhydrous β-lactose, Reese and Rue [80] concluded that anhydrous β-lactose is a brittle material. This was confirmed by the relatively small stress relaxation and the absence of a change in nonrecoverable deformation (NRD) with a change in platten rate [81]. The lower yield pressure and the higher strain rate sensitivity of anhydrous β-lactose, as compared with α-lactose monohydrate, showed that anhydrous β-lactose was less brittle than α-lactose monohydrate [52].

Using mercury porosimetry, Vromans et al. [47] showed that the pore size distribution of tablets compressed from roller-dried β-lactose was strongly dependent on compaction load. The shift from larger to smaller pores with increasing compaction force was attributed to the fragmentation of the particles during compaction. In addition, the crushing strength was related to the specific pore surface area (see Fig. 13) and the β-lactose exhibited the same relationship as other crystalline lactoses. The fragmentation propensity of roller-dried β-lactose is related to the morphology of the particles, which in turn is affected by the method of manufacturing (see Section VII.A).

The presence of about 20% anhydrous α-lactose causes a prolonged disintegration time, when tablets from anhydrous β-lactose are compressed at moderate to high compression forces. This phenomenon is caused by a combination of small pore diameters and precipitation of dissolved anhydrous α-lactose in the course of the water penetration process into α-lactose monohydrate [167]. The disintegration time can be enhanced by the incorporation of a disintegrant such as crospovidone, which enables a rapid progress of the water penetration front in the tablets [168].

Anhydrous β-lactose is not hygroscopic. Immediately after ejection, tablets compressed from anhydrous β-lactose show a similar moisture uptake as tablets containing α-lactose monohydrate (see Section VII.A), resulting in a decrease in crushing strength and specific tablet surface area within 10 min after ejection [166]. Storage for eight weeks at 20°C/85% RH resulted in a small increase in hardness [6]. Due to the low moisture content, anhydrous β-lactose is an ideal excipient for moisture-sensitive drugs.

D. Spray-Dried Lactose

Spray-dried lactose was the first product made in a special form for direct compression. In addition to microcrystalline cellulose, it played a paramount role in the early acceptance of direct compression as a valuable tableting technique. In spite of the improved binding and flow properties as compared with α-lactose monohydrate [2], the early quality of spray-dried lactose had a number of problems, particularly discoloration, caused by the presence of contaminants in the mother liquor (mainly 5-(hydroxymethyl)-2-furaldehyde) [174] and the reaction with primary amines [175].

Spray-dried lactose quality was improved in order to overcome the problem of discoloration, and compaction properties were also improved. Although contaminants causing a browning reaction are removed before spray drying, the reaction between all types of lactose and primary amines still remains a possibility. The improved products, marketed since 1973, are prepared by spray drying a concentrated suspension of small lactose crystals in water [176]. In 1995, spray-dried lactose consists mainly of spherical particles containing microcrystals of α-lactose monohydrate, glued together with amorphous lactose (glass) (see Fig. 10c). In contrast to the earlier products that contained a fairly high proportion of large crystals of α-lactose monohydrate because of partial crystallization before spray drying [26,177–179] the later materials contain only a few small individual crystals of α-lactose monohydrate (see Fig. 10c). The improved products contain a higher percentage of amorphous lactose, which is responsible for their better compactibility. Amorphous lactose is the result of quick dehydration of the dissolved lactose in the suspension used in spray drying. Spray-dried lactose has a moisture content of about 5%; a small part of this is free moisture; the remainder is water of crystallization of α-lactose monohydrate. The spray-dried lactose available commercially include Fast Flo Lactose (Foremost Foods Co.), Pharmatose DCL 11 (DMV), and Zeparox (Lactochem).

Because spherical droplets of lactose suspension are spray dried, the dry material produced by the process is spherical in nature (see Fig. 10c), which accounts for excellent flow properties. The binding properties of spray-dried lactose are good (see Fig. 11 for Pharmatose DCL 11), although different spray-dried products from different suppliers have been reported to behave differently [172]. In a comparative evaluation of several directly compressible filler-binders on a high-speed rotary press, Deurloo et al. [76] concluded that the tableting properties of Fast Flo lactose and

Pharmatose DCL 11 were better than those of the other products (see Fig. 4). Spray-dried lactose seems to be less influenced by the rate of compression than other types of lactose [172,76].

It was suggested that spray-dried lactose deforms mainly by brittle fracture, which was supported by the superimposed Heckel plots for two different contact times [44]. The yield pressure was lower than that of equally sized α-lactose monohydrate particles, which would favor plastic deformation [164]. On the other hand, Hiestand and Smith [57], using different tableting indices, reported that spray-dried lactose is not brittle at all, but deforms mainly by plastic deformation.

Previous work pointed out that a decrease in particle size increased the compactibility of spray-dried lactose [162,164,44]. Above a pressure of 140 MPa, capping and lamination began to occur. McKenna and McCafferty [44] attributed the higher strength of tablets prepared from small particles to an increase in cohesive and frictional forces and to a change in particle shape, with decreasing particle size of the spray-dried lactose. Scanning electron micrographs showed that the large particles consisted mainly of α-lactose monohydrate crystals, whereas the small particles were spherical in shape. It should be remembered, however, that the later grades of spray-dried lactose contain only few individual crystals of α-lactose monohydrate. Thus, the effect of the particle size on compactibility, as described by Fell and Newton [162,164] and McKenna and McCafferty [44] may not be used with the improved spray-dried lactose.

Because α-lactose monohydrate has only poor binding properties, the amorphous part of the excipient must be responsible for the increase in compactibility [162,173]. For this reason, the tableting properties of pure amorphous lactose were investigated [180]. Fully amorphous lactose prepared by spray drying exhibited an increasing compactibility after water sorption, and the compact thickness decreased simultaneously (see Fig. 15). Using mercury porosimetry data, it was calculated that, in contrast to crystalline lactose products, the tablet pore surface area did not change with increasing compaction forces. From these observations, and from the high lubricant sensitivity, it was concluded that fully amorphous lactose deforms mainly by plastic flow.

Commercially available spray-dried lactose products contain about 15–20% amorphous lactose and about 80–85% α-lactose monohydrate. In contrast to fully amorphous lactose, the compactibility of the spray-dried lactose was not affected by moisture uptake, and the lubricant sensitivity was lower [180]. The pore surface area increased with increasing compaction load. From these results, it was concluded that the consolidation of spray-dried lactose is determined by the fragmentation of the crystalline α-lactose

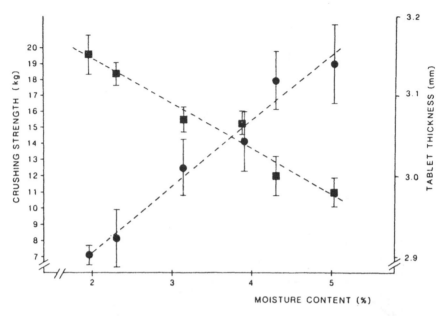

Fig. 15 Compact strength (●) and thickness (■) of tablets of amorphous lactose, compressed at a compaction load of 75 MPa, as a function of the moisture content. (From Ref. 180; reproduced with permission of the copyright owner.)

monohydrate, whereas the binding is largely determined by the amorphous lactose [180]. Further investigations by Vromans et al. [18] with a number of experimental spray-dried lactose products containing different amounts of amorphous lactose demonstrated that the compactibility of the samples was a function of the primary particle size of α-lactose monohydrate and the amount of amorphous lactose (see Fig. 16). Only about 15% of amorphous lactose was enough to increase the compactibility of fine particles of α-lactose monohydrate. The amorphous lactose (glass) was considered to form a binding layer on the particle surface.

    In early publications, it was mentioned that spray-dried lactose showed a browning reaction following storage at elevated temperatures [2]. However, the improved qualities that are produced now exhibit no browning reactions. The physical stability is limited, however, particularly when the product is stored at relative humidities over 50%. This effect caused by the conversion of the amorphous lactose to α-lactose monohydrate was found to depend strongly on the structure and the composition of the product used [6].

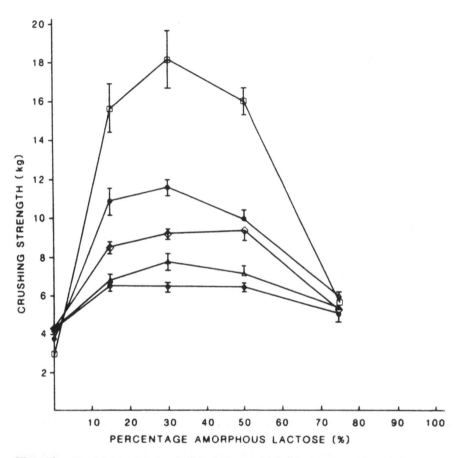

Fig. 16 Crushing strength of tablets of spray-dried lactose samples, compressed at a compaction pressure of 75 MPa. The water content of the amorphous lactose is about 2%. Starting materials: (□) 1–8 μm; (●) 8–16 μm; (◇) 16–24 μm; (△) 24–32 μm; and (○) 32–45 μm. (From Ref. 18; reproduced with permission of the copyright owner.)

E. Agglomerated Lactose

The moderate binding properties of α-lactose monohydrate can be improved by conversion into a granulated form [137]. Tablettose, produced in a fluid-bed granulator-drier by Meggle [75] is made up almost entirely of aggregated crystals of α-lactose monohydrate; it contains no amorphous lactose. Another commercially available agglomerated lactose is Pharmatose DCL 15 from DMV, also available as Dilactose from Freund. The flow properties of Tablettose are good but they are somewhat negatively influenced by the broad particle size distribution and high percentage of fines [6]. Tablettose is resistant to attrition under conditions of low shear blending in a planetary mixer [46].

The binding properties of Tablettose are better than those of α-lactose monohydrate 100-mesh, but not as good as those of spray-dried lactose or anhydrous β-lactose (see Fig. 11 and 4).

Zuurman et al. [181] prepared granulations from different α-lactose monohydrate and roller-dried β-lactose powders by wet granulation, using different techniques with only water as a binder, or by slugging. The authors showed that the compactibility of the granule fractions is dependent on their bulk density and the type of lactose used. Generally, with an increase of the bulk density, the compactibility of a granule fraction decreases (see Fig. 17). It was concluded that with an increase in the total porosity of the granule powder bed the deformation potential increases. A high deformation potential, i.e., a high compactibility, can be obtained by using a granulation procedure, such as fluid-bed granulation, in which

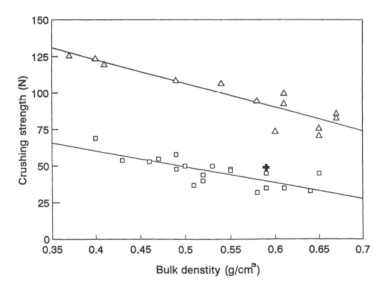

Fig. 17 Crushing strength of tablets compressed from different lactose granule fractions (125–250 μm; 212–425 μm) vs. the bulk density of the granulations before compression. The granules were prepared from different α-lactose monohydrate (□) or anhydrous β-lactose powders (△) by various wet granulation techniques with only water as a binder and by dry granulation, respectively. The figure includes the relationship between the bulk density of Tablettose and the crushing strength of Tablettose tablets (+). Tablets: 500 mg, 13 mm, compacted at 20 kN on a hydraulic press.

granulations with a low bulk density are produced [181]. This means that the increased compactibility of Tablettose, as compared with α-lactose monohydrate 100-mesh must be attributed to the granular texture, which enhances the fragmentation potential (see Tablettose in Fig. 17).

VIII. OTHER SUGARS

A. Compressible Sugar

Sucrose is a nonreducing disaccharide composed of D-glucosyl and D-fructosyl moities and occurs almost universally in all components of practically every existing phenarogam. However, major production sources are sugar beets, sugar cane, and to a lesser extent sorghum and sugar maple. The latter two sources normally provide sucrose containing syrups. As an item of commerce, crystalline sucrose represents the highest-volume organic compound produced worldwide in practically pure state (>99.5%). The only possible exception to this statement is ethylene oxide. In the form of a syrup, it is used extensively in tablets prepared by the wet granulation process.

Large crystals of sucrose provide good flow properties and can be used for direct compression, but the compactibility is poor. The compaction properties of pure sucrose (saccharose) were studied by Hardman and Lilley [151,182] using scanning electron microscopy, surface-area measurements, and mercury porosimetry. Large frictional forces prevented efficient compaction, even when large shear stresses were applied. Although extensive fragmentation took place, the fragments tended to remain in heaps between larger particles rather than to move into the pore space. Fragmentation during compaction was confirmed by stress-density measurements [98] using different techniques (Heckel plots, scanning electron microscopy, strength reduction ratio, and isotropic ratio measurements). Duberg and Nyström [97] concluded that crystalline sucrose behaved as an intermediate between plastic deformation and complete fragmentation, which corresponds to the intermediate brittle fracture index (BFI) [57]. As an effect of the high degree of fragmentation, the strength of sucrose tablets was independent of the particle size [100].

Because of its poor compactibility, sucrose is commonly used in a modified form that makes it more efficient for direct compression. The modified form is known as *Compressible Sugar NF XVII*. Several products fall into this category and are composed of sucrose containing starch, malto-dextrin, or invert sugar.

Compressible sugar is available from different sources. NuTab from Ingredient Technology contains sucrose, about 4% invert sugar, and small percentages of corn starch and magnesium stearate. It is prepared by a roller compaction process and then broken up to the desired particle range. NuTab is available in medium and coarse grades. Di-Pac from Amstar consists of 97% sucrose and 3% modified dextrins and is prepared by a cocrystallization process. Each granule is composed of hundreds of sucrose microcrystals, glued together by the dextrins. Destab Sugar from Desmo Chemical Corp. consists of agglomerates of 96% sucrose and 4% invert sugar [137,108]. Comprima from Suiker Unie is prepared by agglomeration in a fluid bed and is available in two types: Comprima F-gm contains 97% sucrose and 3% dextrose; Comprima F-zd contains 97% sucrose and 3% starch. Sugartab from Mendell is a nonpharmacopeial product because the invert sugar content exceeds 3%.

Mendes et al. [183] evaluated NuTab as a chewable filler-binder for direct compression in combination with several active ingredients and 1.0% magnesium stearate. Generally, good tablets could be prepared with NuTab. Shah et al. [108] compared NuTab, Di-Pac, and Destab Sugar. Particle size distributions as well as bulk and tap densities of the products were different. The compressibility of Destab Sugar was better than that of the other compressible sugars. NuTab and Di-Pac showed almost similar compaction profiles. The compression profile of Comprima F-gm in Fig. 8 shows that the product has good binding properties and is relatively insensitive to mixing with lubricants. David and Augsberger [42] investigated the effect of time of compression and the presence of a lubricant on the tablet strength of various materials including Di-Pac. It was concluded that particle fracture played a more dominant role during compaction of Di-Pac or lactose than during compaction of microcrystalline cellulose or compressible starch. However, other authors mentioned that, in contrast to plain sucrose (which consolidated mainly by fragmentation), plastic deformation in combination with brittle fracture played a large role during compaction of compressible sugar [13].

Because of the porous structure and large surface area of cocrystallized sucrose (e.g., Di-Pac) its compactibility is strongly dependent on its moisture content. The equilibrated free moisture content of 0.4% appears optimum, while higher moisture content may produce hard candy or troche tablets [137].

Tablets will soften during exposure to high humidities because of the hygroscopic nature of sucrose [73]. Compressible sugar containing invert sugar (e.g., NuTab) is more hygroscopic than when it contains maltodextrins (e.g., Di-Pac) because of the absence of reducing sugars. Placebo Di-Pac tablets pick up less than 0.1% moisture during storage for 30 days at 25°C/75% RH [13]. Compressible sugar has few incompatibilities, but is incompatible with primary and with many secondary amines. Compressible sugar is commonly used for lozenges and chewable tablets because of the sweet taste and good mouth feel. Because of the high solubility of sucrose, the tablets do not disintegrate but dissolve during disintegration or dissolution testing.

B. Dextrose

Dextrose, or D-glucose, is a monosaccharide commonly produced by acid hydrolysis of starch. It is available in the anhydrous and monohydrate forms.

Duvall et al. [184] compared the performance of a food grade of hydrous and anhydrous dextrose with spray-dried lactose as an excipient in direct compression tablets. The results indicated that the hydrous dextrose can be partly or completely substituted for spray-dried lactose in some formulations. Dextrose was found to give less browning than spray-dried lactose in formulations containing no amines; more browning was observed in the presence of amines. The binding properties of anhydrous dextrose were inadequate to form a compact; only a 1:1 mixture of anhydrous and hydrous dextrose could be tableted directly. Bolhuis and Lerk [26] showed that coarse fractions of dextrose monohydrate can be compressed directly when mixed with 1.0% magnesium stearate. However, the tablets capped at moderate compaction loads. The effect of moisture on the compaction properties of anhydrous dextrose and dextrose monohydrate was examined by Armstrong et al. [185]. An increase in moisture content of anhydrous dextrose produced a corresponding increase in tensile strength and tablet toughness up to a moisture level of 8.9%. At a higher water content level, tablet strength fell sharply (see Fig. 18). The increase in tensile strength was believed to be caused by an increased lubrication effect and recrystallization of dissolved anhydrous dextrose during compaction. The same moisture content resulted in lower tensile strength values when dextrose monohydrate was used compared to the anhydrous dextrose. Any increase in moisture content obtained by exposure to elevated humidities led to a reduction in both tensile strength and toughness. The differences in the compressional properties of the two types of dextrose were attributed to the different mechanism by which water is held, and to differences in the mobility of water. In dextrose monohydrate, the water of crystallization was thought to be locked into the crystal structure, and therefore not available to encourage bond formation, whereas in anhydrous dextrose the moisture should exist in a relatively mobile form. Any increase in moisture content after exposure to a high humidity resulted in a reduction in both tensile strength and toughness.

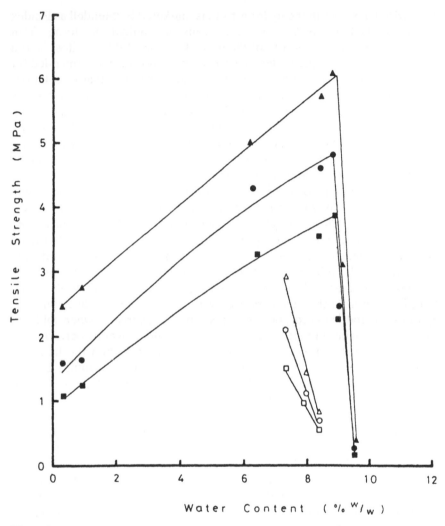

Fig. 18 The relationship between the tensile strength of tablets compressed at 14 (□,■); 19 (○,●) and 25 (△,▲) kN and their water content. Closed symbols: anhydrous dextrose; open symbols: dextrose monohydrate. (From Ref. 185; reproduced with permission of the copyright owner.)

C. Dextrate

The NF XVII describes dextrates as a purified mixture resulting from the controlled enzymatic hydrolysis of starch. It contains between 93.0% and 99.0% dextrose. The product can be hydrous or anhydrous [186]. A spray-crystallized dextrose in the hydrous form is marketed by Mendell as Emdex (formerly available as Celutab in hydrous and anhydrous forms from Penick and Ford). It consists of 90–92% dextrose, 3–5% maltose and a remaining mixture of higher glucose polymers. Emdex is recommended for chewable tablets because of its sweet taste and negative heat of solution, which creates a cooling mouth feel. Emdex is composed of porous spheres (see Fig. 10d) that have excellent flow properties in spite of a sticking together of some particles [26]. Scanning electron micrographs indicate that each spherical granule consists of randomly arranged flat microcrystals bound together by minute amounts of higher saccharides and interspersed with void spaces of various shapes and sizes.

Henderson and Bruno [187] evaluated and compared Celutab with spray-dried and anhydrous lactose. It was concluded that Celutab was superior to the latter two agents as filler-binders for direct compression. Using an instrumented rotary press, Ridgway et al. [188] reported good flow and compaction properties without any sign of capping or difficulty in ejection for both hydrous and anhydrous Celutab, lubricated with 1.5% magnesium stearate. Using an instrumented eccentric press, Bolhuis and Lerk [26] reported that in comparison with a series of other filler-binders, hydrous Celutab showed poor compression characteristics when tableted with 0.5% magnesium stearate. However, excellent overall performance as an excipient for direct compression was exhibited when the lubricant concentration was increased to 1.0%. The compaction profile (Fig. 8) shows that Emdex has good binding properties (somewhat lower than those of Karion Instant) and a small lubricant sensitivity. Consolidation was reported to occur by plastic deformation of the porous particles that deformed along many planes [34,137]. This was confirmed by force-displacement curves for axial and radial displacements [51] and by the linearity of Heckel plots [98].

Because of the large particle size of Emdex, problems with blending were reported. Shangraw [179] advised a preblending of micronized active substances with Emdex before the addition of other excipients. Micronized drug particles lodged in the pores at the surface of Emdex spheres are apparently held in place with sufficient attractive forces to prevent dislodging during the subsequent blending operations.

As an effect of the high solubility of dextrose and the other sugars, tablets prepared from Emdex dissolve, rather than disintegrate, when placed in water [188,26]. As the dissolution takes place mainly from the outer surface, which is initially constant, the disintegration time is not greatly affected by the compaction load [26,34]. When mixing Emdex with microcrystalline cellulose, it should be recognized that opposing forces are playing a significant role in compaction and disintegration of the compact [7].

The large surface area and the presence of polysaccharides make Emdex quite hygroscopic at relative humidities above 75%. Even after storage for eight weeks at room temperature and 50% RH, the crushing strength of tablets decreases dramatically, whereas during storage at 85% RH the tablets liquefy [6].

IX. COPROCESSED PRODUCTS

Excipient mixtures are generally produced to make use of the advantages of each component and to overcome specific disadvantages. The functionality of excipient mixtures is enhanced by a special process by which mixtures are combined. The excipient mixtures used in direct compression have added value compared to simply the physical mixture of excipients. For this reason, ready-to-use blends for direct compression were offered from different suppliers in the past. An example is Emcompress LP, a mixture of Emcompress with a disintegrant. Another mixture, a combination of dicalcium phosphate dihydrate and microcrystalline cellulose was not marketed in spite of a number of advantages when compared to the individual excipients. A major drawback of an excipient mixture is that the ratio of the excipients in a mixture is fixed and in developing a new formulation, a fixed ratio of the excipients may not be an optimum choice for the drug substance and the dose per tablet under development. The choice of filler-binder, disintegrant, lubricant, and their concentrations should be fully investigated with the drug substance under development. The choices depend on optimization of all tablet parameters and consideration for the functionality of each excipient in the formulation and their interactions.

In recent years, a number of excipient mixtures were introduced utilizing particle engineering in the design of combination products for direct compression. It was demonstrated that the use of specially prepared excipient mixtures, defined in this chapter as coprocessed products, can duplicate the advantages of the starting materials while overcoming their respective disadvantages [189]. Examples are Ludipress from BASF, Cellactose from Meggle, and Pharmatose DCL 40 from DMV. A number of other coprocessed products are also available in the United States, including different calcium carbonates and calcium sulphates. They are available from Desmo Ingredient Technology and other suppliers. Most important are the binding and blending properties of the coprocessed excipients, which must be better than those of a physical mixture of the starting materials. Cost is another factor to consider in the selection of combination products.

As of 1995, the drawback of most combination products is their lack of official acceptance in the pharmacopoeias. For this reason, a combination filler-binder will not be accepted by the pharmaceutical industry until it shows significant advantages in the tablet compaction when compared to the physical mixtures of the excipients. On a similar premise, the topic of mixture of excipients was presented in a draft of the NF Admission Policy [190]. It was assigned a priority based on the use of the mixture in dosage forms marketed in the United States, where processing has provided an added value to excipient mixture functionality.

Although spray-crystallized dextrose-maltose (Emdex) and compressible sugar are coprocessed products in reality, they are commonly considered as single components and are official in the USP/NF. For this reason, they are described in the sections on single components.

A. Ludipress

Ludipress is a coprocessed product consisting of 93.4% α-lactose monohydrate, 3.2% polyvinylpyrrolidone (Kollidon 30), and 3.4% crospovidone (Kollidon CL). It contains a filler, a binder, and a disintegrant. It is a free-flowing powder, produced by coating of lactose particles with polyvinylpyrrolidone and crospovidone. The spherical particles are made up of a large number of small crystals with smooth surfaces. The particle size is 10 to 600 μm; no more than 25% of the particles are smaller than 100 μm. The binding properties of Ludipress, both unlubricated and lubricated with 1% magnesium stearate, are good (Fig. 19) and were found to be much better than those of the physical mix [191]. After milling, Ludipress can be recompressed with only a minor loss of binding properties. The dilution potential of Ludipress, with respect to paracetamol, is lower than that of Avicel PH 101, Elcema G250 or Elcema P050 [192]. Although Ludipress contains a disintegrant, the disintegration of tablets takes longer than tablets containing α-lactose monohydrate, anhydrous β-lactose, spray-dried lactose, or Tablettose. The length of disintegration time is attributed to the presence of polyvinylpyrrolidone [172].

B. Cellactose

Cellactose is a coprocessed product produced by the coprocessing and agglomeration of 75% α-lactose monohydrate with 25% cellulose. In this material, the advantageous properties of the two starting materials are combined in a single form to reduce the negative characteristics of the two components. Because of regular particle shape (see Fig. 10e) and favorable particle size distribution, the product has excellent flowability. The flow properties and compressibility of Cellactose are superior to those of physical blends of agglomerated lactose (Tablettose) with either cellulose pow-

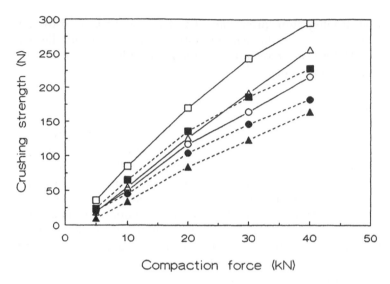

Fig. 19 Compaction profiles of different coprocessed products, both unlubricated (open symbols) and lubricated with 1% magnesium stearate (closed symbols). (□,■) Cellactose; (○,●) Pharmatose DCL 40; (△,▲) Ludipress. 500-mg, 13-mm tablets, hydraulic press, contact time 5.1 s.

der (Elcema P100) or microcrystalline cellulose (Avicel PH 102) [193]. The good compactibility can be attributed to a synergistic effect of consolidation by fragmentation of the lactose, coupled with the concomitant plastic deformation of cellulose [194]. In a comparative evaluation between Cellactose, α-lactose monohydrate, and a mixture of 25% microcrystalline cellulose with 75% dicalcium phosphate dihydrate, Garr and Rubinstein [194] demonstrated that at compression speeds below 330 mm/s, the mean yield pressure increased with the compression speed for both Cellactose and lactose, indicating the dominance of plastic deformation over fragmentation. At higher compression speeds, the yield pressure attained a constant level, indicating that consolidation primarily takes place by fragmentation. In contrast, tablets prepared from the microcrystalline cellulose/dicalcium phosphate dihydrate blend showed no stabilization of the mean yield pressure, suggesting that the presence of microcrystalline cellulose makes plastic deformation the dominating consolidation mechanism. It was also demonstrated that the tablet strength of Cellactose did not diminish with an increasing compression speed. In a comparative evaluation of several filler-binders, compressed on a high-speed rotary press, the tableting properties of Cellactose were found to be excellent (see Fig. 4). Of all the products tested, Cellactose tablets showed the best inscription quality [76].

The good disintegration properties of Cellactose are caused by the presence of cellulose fibers in the macroporous particles. Because cellulose is covered with lactose, the moisture sorption is much lower than that of microcrystalline cellulose alone.

C. Pharmatose DCL 40

Pharmatose DCL 40 is a coprocessed product based on anhydrous β-lactose and anhydrous lactitol. It consists of 95% anhydrous lactose and 5% lactitol. Because of spherical form and favorable particle size distribution, flow properties are very good. Comparing Figs. 19 and 11 shows that the binding properties of Pharmatose DCL 40 are much better than those of all known lactose products. Good tablets can be compressed on a high-speed rotary press (see Fig. 4) [76]. It is claimed that the product has a high dilution potential, better than other commercially available lactose-based products [195]. The water uptake of Pharmatose DCL 40 at increasing humidities is very low.

X. DIRECTLY COMPRESSIBLE ACTIVE SUBSTANCES

Without further processing, only a small number of active substances are directly compressible, in some cases only when used in a particular crystalline or hydrate form [196]. For low-dosage drugs, poor flow and compactibility can be overcome by the addition of suitable excipients. For high-dosage drugs, only a few can be compressed without further processing, after the addition of a suitable lubricant, for example, acetylsalicylic acid and sodium chloride. For adequate flow properties a suitable particle size distribution is required. However, since a vast majority of high-dosage drugs cannot be compressed directly, tablet preparation by wet granulation is the method of choice.

In cases when moisture and heat are detrimental to the stability of the drug (e.g., ascorbic acid) or when compression by wet granulation leads to weak tablets (e.g., paracetamol), special directly compressible forms of the active ingredient have been developed. Compressibility and compactibility of pharmaceutical materials can be modified by crystal engineering and particle design [10,189]. Controlled or alternative crystallization techniques may improve the compactibility and flow properties of active ingredients. Gordon and Chowhan [197] modified naproxen crystals by the spherical crystallization technique. A naproxen/acetone solution was added to water, and the crystals were agglomerated with either hexanol, octanol, or toluene as the nonmiscible solvent. The resulting agglomerates were compact spherical aggregates of plate-shaped crystals with improved compactibility and flow properties. Other examples are found in an excellent review by York [189]. In spite of the potential for improving particle properties by controlling the crystallization process and by using alternative crystallization techniques, the drug products have not yet benefited from this technology on a commercial scale.

A. Ascorbic Acid

Ascorbic acid degrades in the presence of moisture because of an oxidative process leading to biologically active substances. The decomposition is usually associated with browning. For this reason, direct compression can offer advantages in the tableting of ascorbic acid.

Manudhane et al. [198] compressed different types of ascorbic acid in a formulation containing 60% ascorbic acid, 37.5% microcrystalline cellulose, 2% Sterotex or stearic acid, and 0.5% colloidal silica. In contrast to powdered ascorbic acid, which lacked good flow and lubrication properties, satisfactory tablets were prepared using crystalline or coated ascorbic acid. Although medium-size crystals were best from a compression point of view, reflection from the relatively large crystals on the tablet surfaces was distracting and gave the impression of mottling. The coated forms of ascorbic acid, prepared by granulation of powdered ascorbic acid with ethylcellulose in a nonaqueous solvent, were free flowing and had improved stability. It was found, however, that the coated forms required a higher degree of lubrication than the crystalline forms. The use of alkaline stearates, which may accelerate tablet browning, was avoided only when the formulation contained 35–40% microcrystalline cellulose in combination with Sterotex or stearic acid [198]. It was shown by Nyström et al. [157] that dry mixing with methylcellulose powder increased the strength of tablets compressed from crystalline ascorbic acid. The tablet strength was directly related to the degree of surface coating obtained by the binder.

Photomicrographs and permeametry measurements show that ascorbic acid is a brittle material, which fragments markedly during compaction, although its fragmentation is less extensive than that of dicalcium phosphate and paracetamol [157,97,112,199]. The interparticular bonds are weak. The presence of a binder in the coated materials provides stronger bonds between the particles in the tablets, being strong enough to allow stress relaxation and elastic recovery to occur without breakage [157,199]. Because of its brittle nature, good compactibility of ascorbic acid cannot be obtained by preparing coated granular forms by the conventional wet granulation process; the effect is lost when the coated agglomerated crystals break and new surfaces are exposed to fracture during compaction. Granules containing high ascorbic acid content with good compactibility and fluidity can be prepared by fluid-bed granulation of ascorbic acid particles of 100 μm or smaller with aqueous solutions of hydroxypropylcellulose, hydroxypropylmethylcellulose, or starch paste as a binder [200]. The effective coverage with starch paste by the fluid bed granulation technique limits the binder concentration and improves compactibility and stability, as compared with conventional granulation.

Coated ascorbic acid is manufactured by Hoffmann-La Roche under the trade names C-90, C-95, and Ascorbic Acid 98% DC. Micronized ascorbic acid particles granulated with starch paste and lactose is C-90 [201]. The product appears to be extruded and then ground. Another product, C-95, contains 5% excipients, mainly methylcellulose [202]. Ascorbic Acid 98% DC is granulated with methylcellulose. Coated Ascorbic Acid from Takeda is prepared by coating ascorbic acid with 1.2% ethylcellulose [203]. This product is intended to facilitate tableting but is not especially designed for direct compression. Ascorbic acid specially prepared for the direct compression are direct compression C-97 and C-97SF from Takeda (see Fig. 10f). These products are prepared in a fluid bed granulator and are, in fact, granules. C-97 contains 97% ascorbic acid and 3% food starch, whereas starch and sugar-free C-97 SF consists of 97% ascorbic acid and 3% hydroxypropylcellulose. Some physical properties of crystalline ascorbic acid and three forms of directly compressible ascorbic acid are shown in Table 7 [201].

When compressed without a filler-binder, tablets from directly compressible ascorbic acid show capping, lamination, and a tendency to chipping [201]. Technical information from the supplier states that ascorbic acid for direct compression (C-97 or C-97 SF) produces excellent tablets when combined with 5% microcrystalline cellulose or corn starch and 1–1.5% magnesium stearate [203]. In this case, the moisture content is so low that there is no difference between magnesium stearate and stearic acid, and the deleterious effect of magnesium stearate on color stability is suppressed.

Table 7 Physical Properties of l-Ascorbic Acid Forms

Type	Particle shape	Average particle size (μm)	Angle of repose (°)	Packed bulk density (g/cm³)
C97	Medium granules	422	37	0.76
C97 SF	Medium granules	346	41	0.73
C Roche	Large spherical granules	430	48	0.74
C Merck (crystalline)	Cubic crystals	208	55	1.00

Source: From Ref. 201; reproduced with permission of the copyright owner.

The addition of higher amounts of corn starch influences color stability when magnesium stearate is used as a lubricant [204]. On the other hand, in a comparative evaluation of C-97 and C-97SF from Takeda and C-90 from Roche, tablets containing 5% Avicel PH 101, 5% STARCH 1500, or 5% Emdex showed capping, lamination, and chipping. When these excipients were increased to the 10% level, good tablets were formed without any defects [201].

In addition to ascorbic acid, sodium ascorbate is available in directly compressible forms: sodium ascorbate granular (Roche) and sodium ascorbate for direct compression, SA-99 (Takeda). SA-99 contains 99% sodium ascorbate USP and 1% food starch. During conventional granulation, sodium ascorbate discolors rapidly because of moisture and other additives. SA-99 is produced in a fluid-bed granulator with an automatic moisture control and a carefully selected binder, which minimizes discoloration of sodium ascorbate during granulation [205]. Because compactibility of sodium ascorbate decreases with an increase in particle size, the particle size of the starting material for SA-99 is small. Combinations of ascorbic acid and sodium ascorbate reduce the acid taste of vitamin C tablets, an important consideration in the manufacture of chewable tablets.

In 1992, Takeda developed Calcium Ascorbate for Direct Compression (C-CAL-97). It consists of approximately 97% calcium ascorbate and is prepared with hydropropylmethylcellulose and tartaric acid as excipients. It acts as a source of ascorbate and calcium.

The critical compaction behavior of mixtures of C-97, C-CAL-97 and spray-dried sorbitol for direct compression (Karion Instant) were studied by Konkel and Mielck [206,207], using upper-punch power and work with a modified Weibull function and the Heckel equation.

B. Paracetamol

Paracetamol is a high-dose drug that has poor flow properties and poor compactibility. The crystalline and powdered forms of paracetamol produce tablets that are soft or that show capping [208,209]. During compaction, paracetamol fragments intensively. The brittle behavior was confirmed by photomicrography [157], permeametry measurements [112, 199], density/stress profiles [98,12], and by the poor conversion of axial to radial pressure [210,211]. After withdrawal of the pressure, a considerable elastic recovery was measured, which will be responsible for the breakage of interparticulate bonds leading to capping as the tablet was ejected [209–211]. The large number of contact points created, together with weak binding properties, imply that a comparatively weak type of bonding is involved, e.g., van der Waals forces [115]. The difference in tablet strength between paracetamol and another fragmenting material that binds by van der Waals forces (e.g., dicalcium phosphate dihydrate) was attributed to the elastic nature of paracetamol. It was shown that the elastic recovery of dicalcium phosphate dihydrate is very small [115]. From a calculation of the elastic coefficient (defined as the fraction of indentation that rebounds elastically on removal of the load), it has been concluded that paracetamol, in the powdered form and in the directly compressible form, has poor compression properties because of the high elastic recovery [38].

The tableting properties of paracetamol can be improved by dry or wet coating with a binder, which modifies its plastic and elastic characteristics. Although capping and lamination of paracetamol powder can be reduced by admixing Avicel PH 101 [212], the flow properties of such a blend are inadequate [213]. Admixing methylcellulose powder or other dry binders with crystalline paracetamol result in an increase in the binding properties of the latter, whereas the tablet strength is directly related to the degree of surface coating obtained by the binder [157,199].

The function of the binder is to form stronger bonds between particles in the tablet. The bonds should be strong enough to allow stress relaxation and elastic recovery to occur without breakage. The larger the number of binder particles and surface coverage, the stronger is the resulting tablet [157]. During dry mixing paracetamol with binders, surface coverage depends on particle size and deagglomeration of the binder as well as adhesion to the paracetamol particles. Commonly used binders are added in solution because solution binders are more effective in paracetamol than dry binders [214].

By characterization of tablet strength in axial and radial direction, it was demonstrated that the capping tendency of paracetamol granulations (prepared by wet granulation) decreases with an increase in the concentration of gelatin from 2% to 4% [215]. Coated paracetamol prepared by spray drying deforms more plastically than the granules made by the wet-massing and drying technique, because of a high surface concentration of the binder. This greater plastic deformation leads to greater binder-binder contact, resulting in an increased bonding and production of harder tablets [216,217].

A directly compressible paracetamol (Paracetamol D.C., Graesser Salicylates) prepared by spray drying with 4% gelatin hydrolysate, failed by brittle fracture, as evidenced by axial loading of free compacts, axial and radial displacement versus compaction force plots, and indentation measurements [51]. The dominance of fragmentation was concluded from the insignificant change of the yield pressure with increasing punch velocity [52]. Coating or granulation of paracetamol with binders that eliminated capping, resulted in an increase in residual die wall pressure and a decrease in elastic recovery [209,52]. Using a capping index (C_i)—defined as the gradient of elastic recovery versus residual die wall pressure—Krycer et al. [218] showed the difference in capping tendency between crystalline and granulated paracetamol. The higher residual die wall pressure and smaller axial recovery for a paracetamol granulation compared to the crystals was ascribed to the interference of binder films with the bonding between the paracetamol particles and the binder films that yield under elastic recovery to give relaxation at the bond [210,211]. Sheikh Salem et al. [219] also stated that a binder confers on coated paracetamol a much greater plasticity than paracetamol alone, as measured by stress relaxation.

Commercially available paracetamol for direct compression includes Compap (Mallinckrodt) [220]. Compap contains 90% paracetamol and 10% partially pregelatinized starch and is probably prepared by spray drying to give good compactibility, flowability, and a unique particle size distribution. Compap WSE is manufactured using 95% paracetamol and 5% water soluble excipients. Ready-to-use products are Compap L and Compac coarse L from Mallinckrodt. They contain 90% paracetamol and 10% excipients, which are pregelatinized starch, povidone, crospovidone (not in Compac coarse L), and stearic acid. Directly compressible combination products containing paracetamol are Compap CPM (90% paracetamol plus 0.55% chlorpheniramine maleate) and Codacet-60 (76% paracetamol plus 15% codeine phosphate).

C. Other Directly Compressible Active Ingredients

As *acetylsalicylic acid* has good compaction properties and good flowability, it can be compressed directly. It densifies principally by plastic deformation but undergoes some fragmentation [12]. Crystalline or granular products with good flow properties are offered by different suppliers; the notable ones are Bayer and Rhône-Poulenc.

Without prior treatment, *thiamine hydrochloride* has very poor flow and compaction properties. It needs at least 79% of a directly compressible vehicle to yield adequate compression properties [221]. Special types of thiamine for direct compression are available from Takeda and Hoffmann–La Roche. Thiamine hydrochloride for direct compression, TH-97 and *thiamine mononitrate* for direct compression, TM-97 (Takeda), are prepared in a fluidized-bed granulator. They contain approximately 97% thiamine hydrochloride or thiamine mononitrate, and approximately 3% hydroxypropylcellulose. Both are white granular powders with good flow and binding properties. The characteristic odor is suppressed by processing, compared with uncoated thiamine salts. Thiamine Mononitrate 98% DC (Hoffmann–La Roche) consists of granules prepared with 2% methylcellulose.

Pyridoxine hydrochloride for direct compression (B6-97) from Takeda is 97% pyridoxine hydrochloride and 3% cellulose. Similar to the other directly compressible vitamins marketed by Takeda, it is prepared in a fluidized-bed granulator. Pyridoxine Hydrochloride 98% DC from Hoffmann–La Roche consists of a granulation prepared with 2% methylcellulose.

Riboflavin High Flow 95 is a directly compressible vitamin B_2 from Takeda. It consists of 95% riboflavin and about 5% mannitol. Riboflavin 98% DC from Hoffman–La Roche is prepared by granulation with 2% methylcellulose.

A directly compressible *nicotinamide* (vitamin PP) is available from Hoffmann–La Roche as Nicotinamide 100% DC.

Dry vitamin E 50%, Type SD (Hoffmann–La Roche) is a free-flowing powder, containing 500 I.U. of vitamin E per gram in the form of *dl*-α-tocopherol acetate finely dispersed in a matrix of hydrolyzed gelatin. The individual particles are coated with small amounts of silicic acid.

Vitamin A–acetate for direct compression is a free-flowing, light yellow powder of spherical particles composed of vitamin A–acetate oil in droplets of 1 to 2 microns distributed in a modified food starch-coated matrix of gelatin and sucrose, which also contains butylated hydroxy toluene as an antioxidant. This product is marketed as vitamin A–acetate 500 by Dano Chemo A/S, Ballerup, Denmark. The mean yield pressure derived from Heckel analysis is independent of compression speed, in the range 24–60 mm/s, which has been attributed to fragmentation as a predominant mechanism of consolidation. At higher compression speeds, an increase in yield pressure is ascribed to predominantly plastic flow [222]. This plastic deformation is time dependent, as indicated by the decrease in tablet tensile strength with an increase in compression speed. From the changes in tablet tensile strength and relative densities with increasing compaction speeds, the presence of a mixed mechanism of consolidation of vitamin A–acetate 500 is hypothesized [222].

Directly compressible *niacinamide ascorbate* (Romax) from Roche is a biologically complex ascorbic acid and niacinamide prepared by granulation. It is free of any excipients [223].

Calcium lactate for direct compression is available as a granular trihydrate and as a granular pentahydrate form. It is marketed as Puracal by PURAC biochem, Gorinchem, the Netherlands. The compressibility of the pentahydrate is equal or even better than that of the trihydrate [224].

Directly compressible *ibuprofen* is marketed as DCI-63 and DCI-90 by Mallinckrodt, St. Louis, Missouri. DCI-63 is a ready-to-compress blend, containing 63% ibuprofen and 37% of the USP/NF grade excipients (corn starch, sodium croscarmellose, microcrystalline cellulose, polyvinylpyrrolidone, stearic acid, calcium stearate, methylparaben, and propylparaben). DCI-90 contains 90% ibuprofen and 10% excipients (polyvinylpyrrolidone and corn starch). It needs additional binders for adequate compaction [225]. The Desmo Corporation markets a number of pregranulated and coated active ingredients for example, Ferrous Fumarate 90 [226].

REFERENCES

1. R. F. Shangraw and D. A. Demarest, *Pharm. Tech. 17(1)*: 32 (1993).
2. W. C. Gunsel and L. Lachman, *J. Pharm. Sci. 52*: 178 (1963).
3. H. V. Van Kamp, G. K. Bolhuis, A. H. De Boer, C. F. Lerk, and L. Lie-A-Huen, *Pharm. Acta Helv. 61*: 22 (1986).
4. H. Egermann, *Sci. Pharm. 47*: 25 (1979).
5. M. Westerberg, B. Jonsson, and C. Nyström, *Int. J. Pharm., 28*: 23 (1986).
6. G. K. Bolhuis, G. Reichman, C. F. Lerk, H. V. Van Kamp, and K. Zuurman, *Drug Dev. Ind. Pharm. 11*: 1657 (1985).
7. C. F. Lerk, G. K. Bolhuis, and A. H. De Boer, *Pharm. Weekblad 109*: 945 (1974).
8. R. Huisman, H. V. Van Kamp, J. W. Weyland, D. A. Doornbos, G. K. Bolhuis, and C. F. Lerk, *Pharm. Weekblad Sci. 6*: 185 (1984).
9. H. V. Van Kamp, G. K. Bolhuis, and C. F. Lerk, *Pharm. Weekblad Sci. 9*: 265 (1987).
10. R. Hüttenrauch, *Pharm. Ind. 45*: 435 (1983).
11. P. R. Sheth, and J. H. Wiley, Calcium Phosphates in Tablet Compressing, U.S. Patent 3,134,719.
12. P. Humbert-Droz, R. Gurny, D. Mordier, and E. Doelker, *Int. J. Pharm. Tech. Prod. Mfr. 4(2)*: 29 (1983).
13. A. B. Rizzuto, A. C. Chen, and M. F. Veign, *Pharm. Technol., 8(9)*: 32 (1984).
14. S. Veesler, R. Boistelle, A. Delacourte, J. C. Guyot, and A. M. Guyot-Hermann, *Drug Dev. Ind. Pharm. 18*: 539 (1992).
15. J. Lazarus and L. Lachman, *J. Pharm. Sci. 55*: 1121 (1966).
16. Y. Hammouda, H. M. El-Banna, and A. G. Eshra, *Pharm. Ind. 37*: 361 (1975).
17. E. Shotton and B. A. Obiorah, *J. Pharm. Pharmacol. 25 Suppl.*: 37P (1973).
18. H. Vromans, G. K. Bolhuis, C. F. Lerk, H. Van de Biggelaar, and H. Bosch, *Int. J. Pharm. 35*: 29 (1987).
19. C. F. Lerk, A. C. Andreae, A. H. De Boer, G. K. Bolhuis, K. Zuurman, P. De Hoog, K. D. Kussendrager, and J. Van Leverink, *J. Pharm. Pharmacol. 35*: 747 (1983).
20. C. F. Lerk, K. Zuurman, and K. Kussendrager, *J. Pharm. Pharmacol. 36*: 399 (1984).
21. T. Ishizaka, H. Honda, Y. Kikuchi, K. Ono, T. Katano, and M. Koishi, *J. Pharm. Pharmacol. 41*: 361 (1989).

22. J. T. Wallace, in *Encyclopedia of Pharmaceutical Technology*, Vol. 2 (J. E. Swarbrick and J. C. Boylan, eds.), Marcel Dekker, New York, 1990, p. 321.
23. R. L. Lamberson, and G. E. Raynor, *Man. Chem. Aerosol News 47(6)*: 55 (1976).
24. D. Sixsmith, *Man. Chem. Aerosol News 47(8)*: 27 (1976).
25. K. Marshall, and D. Sixsmith, *Drug Dev. Comm. 1*: 51 (1974/75).
26. G. K. Bolhuis, and C. F. Lerk, *Pharm. Weekblad 108*: 469 (1973).
27. D. Sixsmith, *J. Pharm. Pharmacol. 29*: 33 (1977).
28. K. Marshall and D. Sixsmith, *J. Pharm. Pharmacol. 27 Suppl.*: 53P (1975).
29. Y. Nakai, E. Fukuoka, S. Nakajima, and J. Hasegawa, *Chem. Pharm. Bull. 25*: 96 (1977).
30. R. G. Hollenbeck, G. E. Peck, and D. O. Kildsig, *J. Pharm. Sci. 67*: 1599 (1978).
31. P. Pesonen, and P. Paronen, *Drug Dev. Ind. Pharm. 12*: 2091 (1986).
32. G. Zografi, M. J. Kontny, A. Y. S. Yang, and G. S. Brenner, *Int. J. Pharm. 18*: 99 (1984).
33. E. J. Mendell, *Man. Chem. Aerosol News 43(5)*: 43 (1972).
34. K. A. Khan, and C. T. Rhodes, *Pharm. Acta Helv. 51*: 23 (1976).
35. E. M. Rudnic, R. Chilamkurti, and C. T. Rhodes, *Drug Dev. Ind. Pharm. 6*: 279 (1980).
36. K. Marshall and D. Sixsmith, *J. Pharm. Pharmacol. 28*: 770 (1976).
37. D. F. Erkoboni, C. I. Patel, D. K. Mehra, and T. A. Wheatley, Compressional and tabletting performance of an experimental grade large particle size microcrystalline cellulose, FMC Report, Princeton NJ, 1990.
38. M. E. Aulton, H. G. Tebby, and P. J. P. White, *J. Pharm. Pharmacol. 26 Suppl.*: 59P (1974).
39. K. Marshall, D. Sixsmith, and N. G. Stanley-Wood, *J. Pharm. Pharmacol. 24 Suppl.*: 138P (1972).
40. R. Hüttenrauch and J. Jacob, *Pharmazie 25*: 630 (1970).
41. G. E. Reier, and R. F. Shangraw, *J. Pharm. Sci. 55*: 510 (1966).
42. S. T. David, and L. L. Augsburger, *J. Pharm. Sci. 66*: 155 (1977).
43. A. Stamm, and C. Mathis, *Sci. Techn. Pharm. 5*: 245 (1976).
44. A. McKenna, and D. F. McCafferty, *J. Pharm. Pharmacol. 34*: 347 (1982).
45. R. J. Roberts and R. C. Rowe, *J. Pharm. Pharmacol. 38*: 567 (1986).
46. J. N. Staniforth, A. R. Baichwal, H. Ahmed, C. I. Patel, and J. P. Hart, *Man. Chem. 58(6)*: 20 (1987).
47. H. Vromans, A. H. De Boer, G. K. Bolhuis, C. F. Lerk, K. D. Kussendrager, and H. Bosch, *Pharm. Weekblad Sci. 7*: 186 (1985).
48. K. Marshall and D. Sixsmith, *J. Pharm. Pharmacol. 28*: 770 (1976).
49. D. N. Travers, M. Çelik, and T. C. Buttery, *Drug Dev. Ind. Pharm. 9*: 139 (1983).
50. J. N. Staniforth, J. P. Hart, P. W. S. Heng, and A. R. Baichwal, in *Proc. 6th Pharm. Techn. Conf. Canterbury*, Vol. III, 1987, p. 98.
51. D. N. Travers and M. Cox, *Drug Dev. Ind. Pharm. 4*: 157 (1978).
52. R. J. Roberts and R. C. Rowe, *J. Pharm. Pharmacol. 37*: 377 (1985).

53. N. A. Armstrong and L. P. Palfrey, *J. Pharm. Pharmacol. 41*: 149 (1989).
54. R. Hüttenrauch, *Pharmazie 26*: 645 (1971).
55. G. M. Enezian, *Prod. Prob. Pharm. 23*: 185 (1968).
56. A. Stamm and C. Mathis, *Sci. Techn. Pharm. 6*: 65 (1977).
57. E. N. Hiestand and D. P. Smith, *Powder Technol. 38*: 145 (1984).
58. K. A. Khan, P. Musikabhumma, and J. P. Warr, *Drug Dev. Ind. Pharm. 7*: 525 (1981).
59. A. Ritter and H. B. Sucker, *Pharm. Technol. 3*: 24 (1980).
60. G. Zografi and M. J. Kontny, *Pharm. Res. 3*: 187 (1986).
61. R. Hüttenrauch, J. Jacob, and B. Zöbisch, *Pharmazie 27*: 416 (1972).
62. A. H. De Boer, G. K. Bolhuis, and C. F. Lerk, *Powder Technol. 20*: 75 (1978).
63. J. A. Plaizier-Vercammen, A. Bourgeois, and L. De Boeck, *Drug Dev. Ind. Pharm. 17*: 763 (1991).
64. E. Doelker, D. Mordier, H. Iten, and P. Humbert-Droz, *Drug Dev. Ind. Pharm. 13*: 1847 (1987).
65. H. G. Brittain, C. J. Sachs, and K. Fiorelli, *Pharm. Technol. 15(10)*: 38 (1991).
66. C. D. Fox, M. D. Richman, G. E. Reier, and R. F. Shangraw, *Drug Cosm. Ind. 92*: 161 (1963).
67. N. A. Armstrong, N. M. A. H. Abourida, and L. Krijgsman, *J. Pharm. Pharmacol. 34*: 9 (1982).
68. J. E. Rees, and P. J. Rue, *J. Pharm. Pharmacol. 29 Suppl.*: 37P (1977).
69. C. F. Lerk, G. K. Bolhuis, and A. H. De Boer, *J. Pharm. Sci. 68*: 205 (1979).
70. D. Ganderton, and A. B. Selkirk, *J. Pharm. Pharmacol. 22*: 345 (1970).
71. M. A. Shah, and R. G. Wilson, *J. Pharm. Sci. 57*: 18 (1968).
72. H. Nyqvist, and M. Nicklasson. *Int. J. Pharm. Prod. Mfr. 4(3)*: 67 (1983).
73. S. A. Sangekar, M. Sarli, and P. R. Sheth, *J. Pharm. Sci. 61*: 939 (1972).
74. L. Delattre and F. Jaminet, *Labo-Pharma-Probl. Techn. 23*: 1021 (1975).
75. E. Nürnberg and S. Ritsert, *Techn. Nachr. Fette, 92.1*: 1 (1992).
76. M. J. M. Deurloo, J. P. J. M. Peeters, G. K. Bolhuis, and P. De Haan, *Pharm. Weekblad Sci. 14 Suppl.*: F37 (1992).
77. J. H. Shukla, S. Z. Masih, and R. W. Mendes, *Drug Dev. Ind. Pharm. 6*: 161 (1980).
78. C. E. Bos, H. Vromans, and C. F. Lerk, *Int. J. Pharm. 67*: 39 (1991).
79. P. Paronen, and M. Juslin, *Acta Pharm. Fenn. 92*: 187 (1983).
80. J. E. Rees and P. J. Rue, *Drug Dev. Ind. Pharm. 4*: 131 (1978).
81. J. E. Rees and P. J. Rue, *J. Pharm. Pharmacol. 30*: 601 (1978).
82. P. J. Rue and J. E. Rees, *J. Pharm. Pharmacol. 30*: 642 (1978).
83. G. K. Bolhuis, C. F. Lerk, H. T. Zijlstra, and A. H. De Boer, *Pharm Weekblad 110*: 317 (1975).
84. J. E. Rees and P. J. Rue, *J. Pharm. Pharmacol. 30 Suppl.*: 25P (1978).
85. C. Führer, E. Nickel, and F. Thiel, *Acta Pharm. Technol. 21*: 149 (1975).
86. M. Juslin, P. Kahela, P. Paronen, and L. Turakka, *Acta Pharm. Fenn. 90*: 83 (1981).
87. P. Paronen and M. Juslin, *J. Pharm. Pharmacol. 35*: 627 (1984).

88. C. E. Bos, G. K. Bolhuis, H. Van Doorne, and C. F. Lerk, *Pharm. Weekblad Sci. 9*: 274 (1987).

89. K. C. Kwan and G. Milosovich, *J. Pharm. Sci. 55*: 340 (1966).

90. C. E. Bos, *Tropical Tablets. The Development of Tablet Formulations for Use in Tropical Countries*, Krips Repro B. V., Meppel, 1990, p. 42.

91. R. F. Shangraw, J. W. Wallace, and F. M. Bowers, *Pharm. Technol. 5(10)*: 44 (1981).

92. Technical Bulletin ST/C/001. Colorcon, Inc., West Point, PA. (1994).

93. J. B. Schwartz, E. T. Martin, and E. J. Dehner, *J. Pharm. Sci. 64*: 328 (1975).

94. D. E. Wurster, G. E. Peck, and D. O. Kildsig, *Drug Dev. Ind. Pharm. 8*: 343 (1982).

95. K. S. Manudhane, A. M. Contractor, H. Y. Kim, and R. F. Shangraw, *J. Pharm. Sci. 58*: 616 (1969).

96. A. M. Sakr, H. M. Elsabbagh, and K. M. Emara, *Arch. Pharm. Chem. Sci. 2*: 14 (1974).

97. M. Duberg and C. Nyström, *Acta Pharm. Suec. 19*: 421 (1982).

98. P. Humbert-Droz, D. Mordier, and E. Doelker, *Pharm. Acta Helv. 57*: 136 (1982).

99. J. E. Rees, *Acta Pharm. Suec. 18*: 68 (1981).

100. G. Alderborn and C. Nyström, *Acta Pharm. Suec. 19*: 381 (1982).

101. A. Panaggio, C. T. Rhodes, and J. B. Schwartz, *Pharm. Acta Helv. 59*: 37 (1984).

102. T. W. Underwood and D. E. Cadwallader, *J. Pharm. Sci. 61*: 239 (1972).

103. C. E. Bos, G. K. Bolhuis, C. F. Lerk, and C. C. A. Duineveld, *Drug Dev. Ind. Pharm. 18*: 93 (1992).

104. F. Rodriguez, E. Stenger, and G. Trouve, in *Proc. 11th Pharm. Techn. Conf. Manchester*, Vol. II, 1992, p. 181.

105. E. Fischer, Calcium phosphate excipients in the pharmaceutical industry, Proc. CPhI-Exhibition, Milan, 1991.

106. J. T. Carstensen and C. Ertell, *Drug Dev. Ind. Pharm. 16*: 1121 (1990).

107. R. Herzog, Calciumphosphate in der Tablettierung, Thesis, University of Tübingen, Germany, 1991.

108. M. N. Shah, M. A. Carroll, L. G. Miller, *Pharm. Technol. 7(2)*: 45 (1983).

109. R. N. Chilamkurti, C. T. Rhodes, and J. B. Schwartz, *Drug Dev. Ind. Pharm. 8*: 63 (1982).

110. J. Gillard, L. Delattre, F. Jaminet, and M. Roland, *J. Pharm. Belg. 27*: 713 (1972).

111. K. A. Khan and C. T. Rhodes, *Pharm. Acta Helv. 47*: 594 (1972).

112. G. Alderborn, K. Pasanen, and C. Nyström, *Int. J. Pharm. 23*: 79 (1985).

113. M. Duberg and C. Nyström, *Powder Technol. 46*: 67 (1986).

114. J. E. Rees, *Acta Pharm. Suec. 18*: 68 (1981).

115. M. Duberg and C. Nyström, *Int. J. Pharm. Tech. Prod. Mfr., 6(2)*: 17 (1985).

116. P. C. Schmidt and U. Tenter, *Pharm. Ind. 51*: 183 (1989).

117. J. Gillard, A. François, Y. Hermans, and M. Roland, *J. Pharm. Belg. 28*: 395 (1973).

118. K. A. Khan and C. T. Rhodes, *Pharm. Acta Helv. 49*: 258 (1974).
119. L. W. Wong, N. Pilpel, and S. Ingham, *J. Pharm. Pharmacol. 40 Suppl.*: 69P (1988).
120. C. F. Lerk, A. J. M. Schoonen, and J. T. Fell, *J. Pharm. Sci. 65*: 843 (1976).
121. C. Caramella, P. Colombo, U. Conte, F. Ferrari, A. La Manna, H. V. Van Kamp, and G. K. Bolhuis, *Drug Dev. Ind. Pharm. 12*: 1749 (1986).
122. Z. T. Chowhan, *J. Pharm. Pharmacol. 32*: 10 (1981).
123. L. Delattre and F. Jaminet, *Pharm. Acta Helv. 49*: 108 (1974).
124. S. T. Horhota, J. Burgio, L. Lonski, and C. T. Rhodes, *J. Pharm. Sci. 65*: 1746 (1976).
125. J. M. Lausier, C-W. Chiang, H. A. Zompa, and C. T. Rhodes, *J. Pharm. Sci. 66*: 1636 (1977).
126. J. T. Carstensen, *Drug Dev. Ind. Pharm. 14*: 1927 (1988).
127. P. De Haan, C. Kroon, and A. P. Sam, *Drug Dev. Ind. Pharm. 16*: 2031 (1990).
128. D. H. Shah and A. S. Arambulo, *Drug Dev. Comm. 1*: 495 (1974/75).
129. J. B. Mielck and H. Rabach, *Acta Pharm. Technol. 30*: 33 (1984).
130. J. I. Wells and J. R. Langridge, *Int. J. Pharm. Tech. Prod. Mfr. 2(2)*: 1 (1981).
131. N. K. Patel, B. R. Patel, F. M. Plakogiannis, and G. E. Reier, *Drug Dev. Ind. Pharm. 13*: 2693 (1987).
132. X. P. Hou and J. T. Carstensen, *J. Pharm. Sci. 74*: 466 (1985).
133. X. P. Hou and J. T. Carstensen, *Int. J. Pharm. 25*: 207 (1985).
134. J. T. Carstensen and X. P. Hou, *Powder Technol. 42*: 153 (1985).
135. Z. T. Chowhan and A. A. Amaro, *Drug Dev. Ind. Pharm. 5*: 545 (1979).
136. H. M. Elsabbagh, M. H. Elshaboury, and H. M. Abdel-Aleem, *Drug Dev. Ind. Pharm. 11*: 1947 (1985).
137. R. F. Shangraw, J. W. Wallace, and F. M. Bowers, *Pharm. Technol. 5(9)*: 69 (1981).
138. J. W. Du Ross, *Pharm. Technol. 8(9)*: 42 (1984).
139. A. M. Guyot-Hermann, D. Leblanc, and M. Draguet-Brughmans, *Drug Dev. Ind. Pharm. 11*: 551 (1985).
140. P. C. Schmidt, *Acta Pharm. Technol. 30*: 302 (1984).
141. P. C. Schmidt, *Pharm. Technol. 7(11)*: 65 (1983).
142. P. C. Schmidt and W. Vortisch, *Pharm. Ind. 49*: 495 (1987).
143. P. C. Schmidt and K. Benke, *Pharm. Ind. 46*: 193 (1984).
144. I. Nikolakakis and J. M. Newton, *J. Pharm. Pharmacol. 41*: 145 (1989).
145. A. M. Basedow, G. A. Möschl, and P. C. Schmidt, *Drug Dev. Ind. Pharm. 12*: 2061 (1986).
146. H. Darrouzet, *Proc. Pharma Tagung*, ZDS-Fachschule der Deutschen Süsswarenindustrie, Solingen, 1985.
147. J. D. Davis, *Drug Cosmet. Ind. 128(1)*: 38 (1981).
148. J. L. Kanig, *J. Pharm. Sci. 53*: 188 (1964).
149. B. Debord, C. Lefebre, A. M. Guyot-Hermann, J. Hubert, R. Bouche, and J. C. Guyot, in *Proc. 6th Pharm. Tech. Conf.*, Canterbury, Vol. II, 1987, p. 263.

150. Technical Bulletin, Broch. Pear. 1/05/92, Roquette Frères, Lestrem, France, 1992.
151. J. S. Hardman and B. A. Lilley, *Nature 228*: 353 (1970).
152. J. A. Hersey, E. T. Cole, and J. E. Rees, *Proc. First Int. Conf. Compaction and Consolidation of Particulate Matter,* The Powder Advisory Centre, London, 1973, p. 165.
153. P. York, *J. Pharm. Pharmacol. 30*: 6 (1978).
154. E. T. Cole, J. E. Rees, and J. A. Hersey, *Pharm. Acta Helv. 50*: 28 (1975).
155. J. A. Hersey and J. E. Rees, *Nature 230*: 96 (1971).
156. Z. T. Chowhan and Y. P. Chow, *Int. J. Pharm. 5*: 139 (1980).
157. C. Nyström, J. Mazur, and J. Sjögren, *Int. J. Pharm. 10*: 209 (1982).
158. A. H. De Boer, H. Vromans, C. F. Lerk, G. K. Bolhuis, K. D. Kussendrager, and H. Bosch, *Pharm. Weekblad, Sci. 8*: 145 (1986).
159. H. Leuenberger, J. D. Bonny, C. F. Lerk, and H. Vromans, *Int. J. Pharm. 52*: 91 (1989).
160. H. Vromans, G. K. Bolhuis, C. F. Lerk, and K. D. Kussendrager, *Int. J. Pharm. 39*: 207 (1987).
161. J. N. Staniforth, *Int. J. Tech. Prod. Mfr. 5(1)*: 1 (1984).
162. J. T. Fell and J. M. Newton, *J. Pharm. Pharmacol. 20*: 657 (1968).
163. H. Vromans, A. H. De Boer, G. K. Bolhuis, C. F. Lerk, and K. D. Kussendrager, *Acta Pharm. Suec. 22*: 163 (1985).
164. J. T. Fell and J. M. Newton, *J. Pharm. Sci. 60*: 1866 (1971).
165. G. Alderborn and C. Nyström, *Powder Technol. 44*: 37 (1985).
166. K. A. Riepma, B. G. Dekker, and C. F. Lerk, *Int. J. Pharm. 87*: 149 (1992).
167. H. V. Van Kamp, G. K. Bolhuis, K. D. Kussendrager, and C. F. Lerk, *Int. J. Pharm. 28*: 229 (1986).
168. H. V. Van Kamp, G. K. Bolhuis, and C. F. Lerk, *Acta Pharm. Suec. 23*: 217 (1986).
169. H. V. Van Kamp, G. K. Bolhuis, and C. F. Lerk, *Acta Pharm. Technol. 34*: 11 (1988).
170. D. Y. T. Wong, P. Wright, and M. E. Aulton, *Drug Dev. Ind. Pharm. 14*: 2109 (1988).
171. N. H. Batuyios, *J. Pharm. Sci. 55*: 727 (1966).
172. M. Whiteman and R. J. Yarwood, *Drug Dev. Ind. Pharm. 14*: 1023 (1988).
173. M. Nicklasson and H. Nyqvist, *Int. J. Pharm. Tech. Prod. Mfr. 3(4)*: 115 (1982).
174. C. A. Brownley and L. Lachman, *J. Pharm. Sci. 53*: 452 (1964).
175. R. A. Castello and A. M. Mattocks, *J. Pharm. Sci. 51*: 106 (1962).
176. K. Kussendrager, P. De Hoog, and J. Van Leverink, *Acta Pharm. Suec. 18*: 94 (1981).
177. O. Alpar, J. A. Hersey, and E. Shotton, *J. Pharm. Pharmacol. 22 Suppl*: 1S (1970).
178. J. T. Fell, *Pharm. Weekblad, 111*: 681 (1976).
179. R. F. Shangraw, *Drug Cosm. Ind. 122(6)*: 68 (1978).

180. H. Vromans, G. K. Bolhuis, C. F. Lerk, K. D. Kussendrager, and H. Bosch, *Acta Pharm. Suec. 23*: 231 (1986).
181. K. Zuurman, K. A. Riepma, G. K. Bolhuis, H. Vromans, and C. F. Lerk, *Int. J. Pharm.*, 102: 1 (1994).
182. J. S. Hardman and B. A. Lilley, in *Proc. First Int. Conf. Compaction and Consolidation of Particulate Matter,* The Powder Advisory Centre, London, 1973, p. 115.
183. R. W. Mendes, M. R. Gupta, I. A. Katz, and J. A. O'Neil, *Drug Cosm. Ind. 115(12)*: 42 (1974).
184. R. N. Duvall, K. T. Koshy, and R. E. Dashiell, *J. Pharm. Sci. 54*: 1196 (1965).
185. A. N. Armstrong, A. Patel, and T. M. Jones, *Drug Dev. Ind. Pharm. 12*: 1885 (1986).
186. The United States Pharmacopeia XXII, The National Formulary XVII, United States Pharmacopeial Convention, Inc., Rockville, MD, 1990.
187. N. L. Henderson and A. J. Bruno, *J. Pharm. Sci. 59*: 1336 (1970).
188. K. Ridgway, C. Lazarou, and E. E. Thorpe, *J. Pharm. Pharmacol. 24*: 265 (1972).
189. P. York, *Drug Dev. Ind. Pharm. 18*: 677 (1992).
190. Z. T. Chowhan, *Pharmacopeial Forum* 1943 (1991).
191. Technical Bulletin. Ludipress[R], a new direct tableting auxiliary, BASF Aktiengesellschaft, Ludwigshafen, Germany, 1993.
192. T. Baykara, G. Duman, K. S. Özsener, S. Ordu, and B. Özates, *Drug Dev. Ind. Pharm.*, 17: 2359 (1991).
193. D. Reimerdes, and E. H. Reimerdes, *Pharmaceutical Manufacturing International,* (M. S. Barber, ed.), Sterling, London, 1990.
194. J. S. M. Garr, and M. H. Rubinstein, *Pharm. Technol. Int. 3(1)*: 24 (1991).
195. Technical Bulletin DMV, Veghel, the Netherlands, 1993.
196. R. Hüttenrauch and U. Schmeiss, *Pharmazie 23*: 473 (1968).
197. M. S. Gordon and Z. T. Chowhan, *Drug Dev. Ind. Pharm. 16*: 1279 (1990).
198. K. S. Manudhane, C. E. Hynniman, and R. F. Shangraw, *Pharm. Acta Helv. 43*: 257 (1968).
199. C. Nyström and M. Glazer, *Int. J. Pharm. 23*: 255 (1985).
200. N. Kitamori, T. Makino, K. Hemmi, and H. Mima, *Man. Chem. Aerosol News 50(5)*: 54 (1979).
201. S. I. Saleh, and A. Stamm, *S.T.P. Pharma 4(1)*: 10 (1988).
202. R. F. Shangraw, in *Encyclopedia of Pharmaceutical Technology,* Vol. 4 (J. Swarbrick and J. C. Boylan, eds.), Marcel Dekker, New York, 1991.
203. Technical Bulletin P-007-B (91A), Takeda Chemical Industries, Ltd., Osaka/Tokyo, Japan, 1991.
204. Technical Bulletin TB-002 (86G), Takeda Chemical Industries, Ltd., Osaka/Tokyo, Japan, 1986.
205. N. Kitamori, K. Hemmi, M. Maeno, and H. Mima, *Pharm. Technol. 6(10)*: 56 (1982).
206. P. Konkel and J. B. Mielck, *Pharm. Technol. Int. 4(3)*: 62 (1992).

207. P. Konkel and J. B. Mielck, *Pharm. Technol. Int. 4(4)*: 28 (1992).
208. B. A. Oboriah and E. Shotton, *J. Pharm. Pharmacol. 28*: 629 (1976).
209. J. E. Carless and S. Leigh, *J. Pharm. Pharmacol. 26*: 289 (1974).
210. E. Shotton and B. A. Obiorah, *J. Pharm. Sci. 64*: 1213 (1975).
211. B. A. Obiorah, *Int. J. Pharm. 1*: 249 (1978).
212. A. B. Bangudu and N. Pilpel, *J. Pharm. Pharmacol. 36*: 717 (1984).
213. F. Reyss-Brion, D. Chulia, and A. Vérain, *Labo-Pharma-Probl. Tech. 32*: 266 (1984).
214. I. Krycer, D. G. Pope, and J. A. Hersey, *Powder Technol. 34*: 39 (1983).
215. G. Alderborn and C. Nyström, *Acta Pharm. Suec. 21*: 1 (1984).
216. P. Rue, H. Seager, I. Burt, J. Ryder, S. Murray and N. Beal, *J. Pharm. Pharmacol. 31 Suppl.*: 73P (1979).
217. H. Seager, I. Burt, J. Ryder, P. Rue, S. Murray, N. Beal, and J. K. Warrack, *Int. J. Pharm. Prod. Mfr. 1(1)*: 36 (1979).
218. I. Krycer, D. G. Pope, and J. A. Hersey, *J. Pharm. Pharmacol. 34*: 802 (1982).
219. M. Sheikh Salem, J. T. Fell, H. N. Alkaysi, and N. A. Muhsin, *Acta Pharm. Technol. 30*: 312 (1984).
220. A. M. Salpekar, S. R. Freebersyser, and D. A. Robinson, U.S. Patent 4,600,579 (1986).
221. O. K. Udeala and S. A. S. Aly, *Drug Dev. Ind. Pharm. 14*: 499 (1988).
222. J. S. M. Garr and M. H. Rubinstein, *Drug Dev. Ind. Pharm. 18*: 1111 (1992).
223. Technical Bulletin HHN 0455PF/192, Special agent Romax[R], Hoffmann-La Roche, Nutley, NJ, 1992.
224. Technical bulletin A1/92-1/CL233, PURAC biochem, Gorinchem, the Netherlands, 1992.
225. Technical Bulletin 5/25/89, Mallinckrodt, Chesterfield, MO.
226. Anonymous. *Man. Chem. Aerosol News 54(8)*: 30 (1983).
227. H. Vromans, *Studies on Consolidation and Compaction Properties of Lactose,* Van Denderen, Groningen, 1987, p. 11.

15

Compaction Properties of Binary Mixtures

John T. Fell

University of Manchester, Manchester, United Kingdom

I. INTRODUCTION

The complexity of the compaction process has led to the majority of fundamental studies being carried out on single materials. As all tablets consist of more than one material, the prediction of the compaction properties of mixtures from those of the individual components is of obvious interest. For simple mixtures (as opposed to granulated materials), the practical interest of this subject lies in the area of direct compression. In addition to acquiring knowledge on the influence of active ingredients on compaction, blending of two or more excipients may optimize properties.

There are two starting points for such studies. One is to attempt to examine the problem theoretically and to compare predictions with practical results. The second is to study the compaction properties of mixtures of well characterized materials with a view to finding common trends.

This chapter discusses work that has been carried out on the compaction properties of binary mixtures. Only "simple" mixtures will be discussed. Ordered and "interactive" mixtures such as those with magnesium stearate are not dealt with in this chapter. The two main areas for research workers in this field have been the influence of mixture composition on pressure-density relationships (in some form) and on tablet strength. Although the two are interrelated and often have been addressed in a single work, for convenience they will be discussed separately in this chapter.

II. TABLET STRENGTH

In most of the work reported, there is an implicit assumption that the strength of a compact is a reflection of the bonding that has occurred during compaction. This is not unreasonable, although other factors may be involved. For a discussion on the measurement of tablet strength, see Chapter 7.

Most authors who have carried out investigations in this area were interested in the relation between tablet strength and the relative proportions of the two components. Table 1 lists the results of some investigations in terms of the type of relationship observed. The relationships have been divided into five types, shown diagrammatically in Fig. 1.

The results given in the table illustrate the complexity of formulating a unified theory to explain the strength of tablets of binary mixtures. There are no obvious trends in the data and some results are contradictory (e.g., [5,6]). This problem is not uncommon in work related to compaction. Particle size, dimensions of the punch and die, compaction speed, use of added materials (e.g., lubricants), batch of material, dwell time, and other factors can all influence the result obtained. Comparison of results between different workers who use different experimental conditions is, therefore, often impossible. Nonetheless, some proposals to explain the results may be offered.

In a binary mixture consisting of components A and B, three types of bonds will occur after compaction; A–A, A–B, B–B. The relative magnitudes of these bonds may determine the pattern of results obtained. Newton et al. [2] found that mixtures of dicalcium phosphate dihydrate and phenacetin produced tablets of higher strength than either individual component alone. The results are shown in Fig. 2 and illustrate the effect is similar at several compaction pressures. A possible explanation is that dicalcium phosphate dihydrate-phenacetin bonds are stronger than those between the individual materials. However, there were also large porosity differences between tablets containing phenacetin and those containing dicalcium phosphate dihydrate. Porosity enhancement will be returned to later. Newton et al. [2] also point out that addition of a second material may involve the prevention of crack propagation during strength testing. Enhancement of strength in this way would not involve bonding between materials but may add further to the complexity of the problem. Such strength enhancement is well known in materials science (e.g., Gordon [12]) but has not been investigated further in tableting. The experiments of Newton et al. [2] were carried out with magnesium stearate added to the binary mixtures as a lubricant. Magnesium stearate is known for its ability to reduce bonding, especially between plastic materials [13] and its presence may have influenced the result obtained. Cook and Summers [5],

Table 1 Studies on the Relationship Between Tablet Strength and Mixture Composition

Authors (reference)	Value measured	Materials	Compaction type (if known)	Result
Fell and Newton [1]	Tensile strength	Lactoses	Brittle	a
Newton et al. [2]	Tensile strength	Dicalcium phosphate dihydrate/ Phenacetin	Brittle/?	e
Sheikh Salem and Fell [3]	Tensile strength	Sodium chloride/ lactose	Plastic/brittle	b, d[a]
Leuenberger [4]	Tensile strength	Anhydrous lactose/ sucrose	Brittle/brittle	a
	Tensile strength	Lactose monohydrate/ sucrose	Brittle/brittle	d
Cook and Summers [5]	Tensile strength	Aspirin/Emcompress	Plastic/brittle	e
Jetzer [6]	Max. tensile strength[c]	Aspirin/Emcompress	Plastic/brittle	b
	Max. tensile strength[c]	Aspirin/metamizol	Plastic/brittle	a
	Max. tensile strength[c]	Aspirin/caffeine	Plastic/plastic	a
Leuenberger [7]	Max. tensile strength[c]	Sodium stearate/ caffeine	Plastic/brittle	b
Rubinstein and Jackson [8]	Tensile strength	Sodium chloride/ polyethylene	Plastic/plastic	a[b]
Vromans and Lerk [9]	Crushing strength	β-Lactose/cellulose	Brittle/plastic	e
	Crushing strength	Dicalcium phosphate dihydrate/dried cellulose	Brittle/plastic	e
	Crushing strength	Anhydrous/lactose/ dried cellulose	Brittle/plastic	b/d
	Crushing strength	Amylose/dried cellulose	Plastic/plastic	b
Riepma et al. [10]	Crushing strength	Lactoses	Brittle/brittle	a
Rubinstein and Garr [11]	Tensile strength	Dicalcium phosphate dihydrate/ microcrystalline cellulose	Brittle/plastic	e

[a]Result observed depended on pressure, particle size and compaction speed.
[b]Only up to 2% added polyethylene.
[c]Max. tensile strength is an extrapolated value at zero porosity.

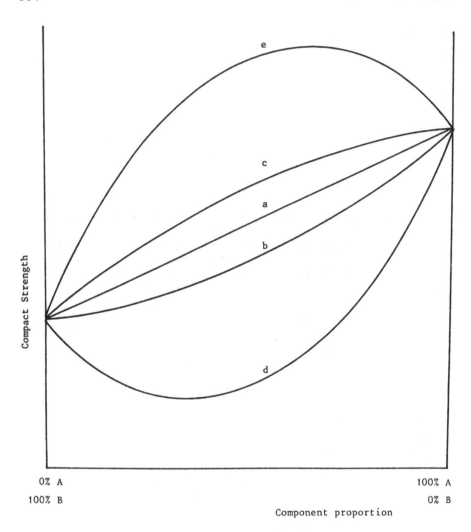

Fig. 1 Observed relationships between component proportion and compact strength. a = linear relationship; and c and b = nonlinear relationship intermediate between either component; and e and d = nonlinear relationship with values above or below either individual.

however, showed a similar enhancement for aspirin-dicalcium phosphate dihydrate mixtures compacted without magnesium stearate. An explanation for this result could not be found in an examination of porosity, lower punch work, or elastic recovery, as these showed almost linear relationships with component percentage (Figs. 3 and 4).

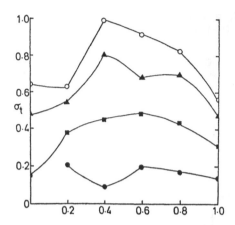

Fig. 2 Tensile strength of tablets (σ_t) prepared at ● 20, ■ 40, ▲ 60, and ○ 70 MN/m² as a function of dicalcium phosphate-phenacetin composition. (From Ref. 2, with permission.)

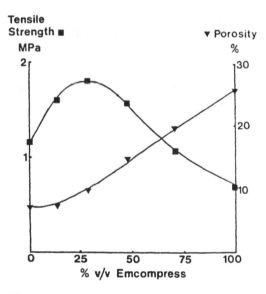

Fig. 3 The relation between tensile strength and porosity against % v/v Emcompress in Emcompress/aspirin compacts. (From Ref. 5, with permission.)

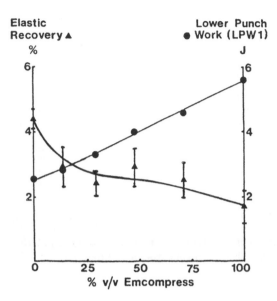

Fig. 4   The relation between elastic recovery and lower punch work against %
v/v Emcompress in Emcompress/aspirin compacts. (From Ref. 5, with permission.)

A further example of strength enhancement is given by Vromans and
Lerk [9]. Mixtures of roller dried β-lactose and cellulose (Sanacel 90)
exhibited the relationships illustrated in Figs. 5 and 6. Here, it is apparent
that the strength increases are related to increases in densification as the
second component is added. Similar enhancements with the addition of
cellulose were obtained for other materials which densified poorly when
compacted alone.

Fell and Newton [1] and Riepma et al. [10] have reported linear relationships between strength and component proportion in the compaction of
mixtures of different types of lactose. Here, similarities in the compaction
properties and bonding mechanisms are assumed to account for this linearity. Riepma et al. [10] demonstrated that all their results could be accounted
for by a unique relationship between the crushing strength and the specific
tablet surface area (Fig. 7). Linear relationships have also been found for
mixtures of anhydrous lactose and sucrose [4], aspirin and metamizol, and
aspirin and caffeine [6]. In the first and second of these cases, the compaction
mechanisms (brittle, plastic) are similar, but a general conclusion that tabletting of materials with similar compaction mechanisms leads to a simple
predictable relationship cannot be drawn (see Table 1).

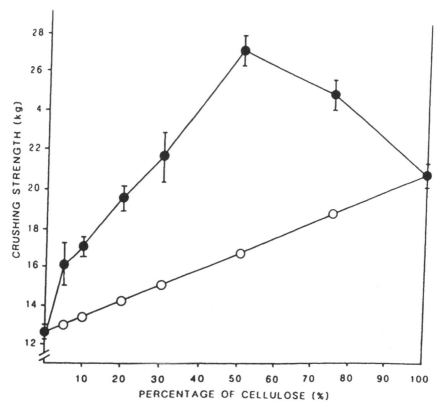

Fig. 5 Measured (●) and calculated (○) data of the crushing strength of tablets compressed at 150 MPa of mixtures of roller-dried β-lactose and cellulose (Sanacel 90) in different proportions. (From Ref. 9, with permission.)

Sheikh-Salem and Fell [3] examined mixtures of sodium chloride (plastic) and lactose (brittle). They found that the addition of small quantities of lactose to sodium chloride caused a marked reduction in tensile strength (Fig. 8). The pattern of results was similar for different compaction pressures and particle sizes. Because lactose undergoes brittle fracture, the number of lactose-lactose bonds in a compact will be considerably greater than the original composition of the mixture would suggest. These are much weaker than sodium chloride bonds as the original strength of the compacts indicates. If it is assumed that lactose and sodium chloride do not bond, a qualitative explanation of the results is forthcoming.

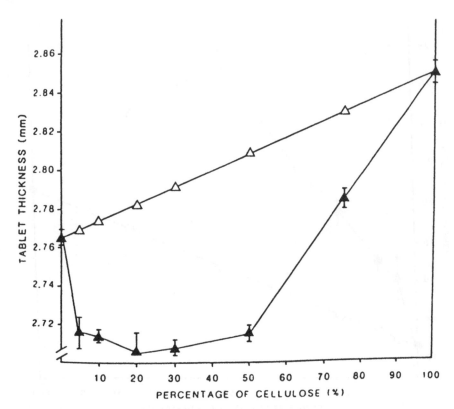

Fig. 6   Measured (▲) and calculated (△) data of the thickness of tablets compressed at 150 MPa of mixtures of roller-dried β-lactose and cellulose (Sanacel 90) in different proportions. (From Ref. 9, with permission.)

Leuenberger [4] has derived an expression to account for the compressibility and compactibility of powder systems based on an examination of bonding and nonbonding contact points in a compact. Compressibility is defined as the ability of a powder to reduce in volume under pressure, and compactibility is the ability of a powder to attain a specific strength. This approach has been applied to binary mixtures in other papers [6,7]. In its final form, the expression is

$$P = P_{max}(1 - \exp(-\gamma\sigma_c\rho_r))$$ (1)

where

P = deformation hardness (measured by indentation)
P_{max} = deformation hardness at $\sigma_c \rightarrow \infty$
σ_c = applied compression force

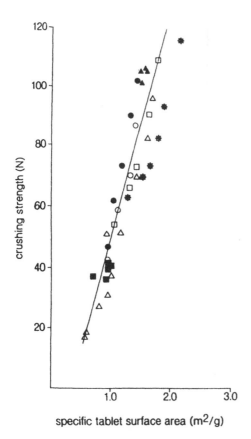

Fig. 7 Crushing strength of tablets compressed from binary mixtures of crystalline lactoses as a function of the specific tablet surface area. (●) α-Lactose monohydrate and anhydrous α-lactose; (□) α-lactose monohydrate and roller dried β-lactose; (■) α-lactose monohydrate and crystalline β-lactose; (○) anhydrous α-lactose and crystalline β-lactose; (▲) anhydrous α-lactose and roller dried β-lactose; (△, ✳) roller dried β-lactose and crystalline β-lactose. (From Ref. 10, with permission.)

γ = proportionality factor, the compression susceptibility
ρ_r = relative density

P and P_{max} may be replaced by the tensile strength. The expression can be extended to include binary mixtures by assuming that the values P_{max} and γ are additive or, if an interaction occurs, by including an interaction term in the equation. Agreement between theory and experiment was achieved for binary mixtures of anhydrous lactose and sucrose using simple

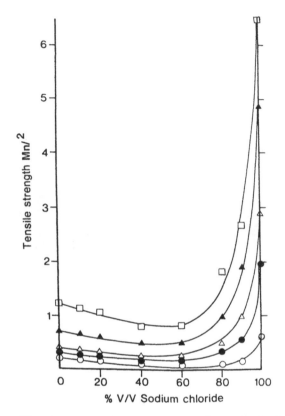

Fig. 8   The relation between tensile strength of tablets prepared at different compaction pressures and mixture composition: $\bigcirc$ = 38.7 MN/m², $\bullet$ = 58.1 MN/m², $\triangle$ = 77.4 MN/m², $\blacktriangle$ = 116.4 MN/m², $\square$ = 193.6 MN/m². (From Ref. 3, with permission.)

additivity of the compression and compaction indices. For sucrose and lactose monohydrate mixtures, an interaction term needed to be included. Using this approach, Jetzer [6] examined tablets of several binary mixtures and concluded that interactions were most likely to occur with mixtures of components with similar compaction mechanisms.

The interaction terms can be related to the relative magnitudes of the adhesive and cohesive forces in the mixture. A negative interaction term indicates that the cohesive forces dominate the adhesive forces. Leuenberger [7] made an analogy between the binary mixtures that he investigated, which showed negative interactions, and regular solutions, in which the cohesive forces between like molecules dominate. Hence, the interaction could be written in terms of the solubility parameters of the materials and predictions made. The solubility parameter is a measure of the cohesive forces between molecules and is therefore likely to be related to bonding in solids. In a similar way, estimates of the Hamaker constant for the van der Waals forces in materials should give an indication of the bonding in compacts.

Luangtana-anan and Fell [14] calculated Hamaker constants from contact angle measurements on several materials. They showed that the tensile strengths of compacts could be largely accounted for by van der Waals forces, by observing the reduction in strength in environments of different dielectric constants. Extrapolation of such an approach to binary mixtures, however, proved unsuccessful [15].

III. PRESSURE-DENSITY RELATIONSHIPS

There are many ways of examining the relationship between compact density and applied pressure. These are reviewed in Chapter 3. This section examines the influence of binary mixtures on such relationships.

Jones [16] considered that the relative volumes at various pressures were not generally additive unless the particle size distribution and particle shape were very similar. He quotes work taken from Bal'shin to illustrate this. The pressing curve for a 50:50 mixture of two iron powders is different from that obtained by assuming the volumes to be additive.

Despite this Ramaswamy et al. [17] suggested that each component of a mixture did behave independently on compaction so that on the application of pressure, volume (V) is a simple additive property for components A and B.

$$V_{mix} = V_A + V_B \quad \text{where } V \text{ is compact volume} \tag{2}$$

$$\frac{v_{mix}}{1 - \epsilon_{mix}} = \frac{v_A}{1 - \epsilon_A} + \frac{v_B}{1 - \epsilon_B} \tag{3}$$

where ϵ is the porosity and v is the volume of material.

$$\epsilon_{mix} = (\epsilon_A X_A + \epsilon_B X_B - \epsilon_A \epsilon_B - \epsilon_A \epsilon_B)/[1 + X_A(\epsilon_A - \epsilon_B) - \epsilon_A] \tag{4}$$

where X_A is the fraction of component a and X_B is the fraction of component B.

On approximation this gives

$$\epsilon_{mix} = (\epsilon_A - \epsilon_B)X_A + \epsilon_B \tag{5}$$

Plots of mixture composition against porosity were linear, justifying the assumption of independence of compression (Fig. 9). This was found to be true for binary mixtures of the following materials: naphthalene-phthalic

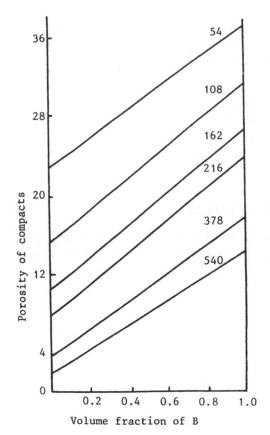

Fig. 9   The relation between porosity and mixture composition for naphthalene (A), phthalic anhydride (B) mixtures: compaction pressure, kg/cm². (From Ref. 17, with permission.)

anhydride, naphthalene-copper sulfate, copper sulfate-calcite, soapstone-calcity, and barytes-calcite. For all the materials, the porosity was linearly related to the log of the compaction pressure, P:

$$\epsilon = K - n \log P \qquad (6)$$

where K and n are constants for a given material.

Therefore, the porosities of compacts of mixtures could be calculated by combining Eqs. (5) and (6) to give

$$\epsilon_{\text{mix}} = X_A[(K_A - K_B) - \log P(n_A - n_B)] + K_B - n_B \log P \qquad (7)$$

This study is interesting in that the materials represent both organic and inorganic substances that possibly compact by different mechanisms and yet a simple additive relationship was observed in each case. It contrasts markedly with the work of Vromans and Lerk [9]. These authors observed enhanced densification (over that predicted from a simple interpolation) of some pharmaceutical excipients when mixed with several types of cellulose. The poorer the densification of the base material, the greater was the enhancement from that predicted. This led to the increase in crushing strength discussed earlier. It is suggested that the cellulose reduces particle interactions and hence consolidation is facilitated.

Sheikh-Salem and Fell [3] examined the consolidation of lactose, sodium chloride mixtures using the Heckel analysis (see Chapter 3). The relationship between K, the slope of the Heckel plots, and mixture composition was not simple and was particle size dependent (Fig. 10). The authors argued that the different compaction mechanisms of the two materials was responsible for this. Lactose compacts by fragmentation and in the early stages of compaction this is a more efficient process. In the later stages, the plasticity of the sodium chloride leads to more efficient compaction. The early fragmentation of lactose may slow down the deformation of sodium chloride as pores are filled more rapidly by fragmentation. Sodium chloride may reduce the fragmentation of lactose by preventing the critical force for fracture from being achieved. The relative effects of these processes, which will be particle size dependent, will result in the different relationships of Fig. 10.

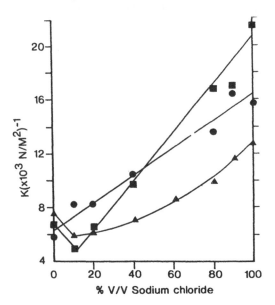

Fig. 10 The relation between K (obtained from Heckel plots) and mixture composition: ■ = 125–150 μm, ● = 45–75 μm, ▲ = less than 45 μm. (From Ref. 3, with permission.)

Rubinstein and Garr [11] working with microcrystalline cellulose and dicalcium phosphate calculated the mean yield pressures of the mixtures from Heckel plots and showed that this was directly related to the proportions of the components. A different, but still direct, relationship was found at a wide range of rates of force application. When the extent of particle rearrangement was examined, this showed a maximum with the addition of 25% microcrystalline cellulose to the dicalcium phosphate dihydrate. This is possibly due to a decrease in the frictional and cohesive forces between particles as suggested by Vromans and Lerk [9].

The examination of pressure-density relationships for mixtures using the compression susceptibility term of Leuenberger [4] (Eq. (1)) leads to the conclusion that no single simple trends that are material unspecific can be observed [4,6].

IV. CONCLUSIONS

From the point of view of designing formulations, it would be valuable to be able to conclude from the foregoing review, that the compaction properties of binary mixtures are predictable. This, however, is clearly not the case. Some work has shown simple relationships between component proportion and some measured compaction parameter [1,4,10,17], but, in the main, results more complex than this have been obtained. The use of dicalcium-phosphate dihydrate has been shown to lead to strength enhancement [2,5,11] possibly due to its high tendency to fragment. However, other fragmenting materials give rise to opposite results [3]. The complexity of the system and the large influence of experimental variables means that simple theoretical approaches are likely to be grossly misleading. The work reviewed here suggests that a single unified theory to account for the compaction of binary mixtures is not yet possible. Experiments on the compaction of well defined materials of simple reproducible shape and size distribution may aid in understanding the behavior of mixtures during compaction.

REFERENCES

1. J. T. Fell and J. M. Newton, *J. Pharm. Pharmac. 22*: 247 (1970).
2. J. M. Newton, D. T. Cook, and C. E. Hollebon, *J. Pharm. Pharmac. 29*: 247 (1977).
3. M. Sheikh-Salem and J. T. Fell, *Int. J. Pharm. Tech. Prod. Manuf. 2*: 19 (1981).
4. H. Leuenberger, *Int. J. Pharm. 12*: 41 (1982).
5. G. D. Cook and M. T. Summers, *J. Pharm. Pharmac. 37*: 29 (1985).
6. W. E. Jetzer, *Int. J. Pharm. 31*: 201 (1986).

7. H. Leuenberger, *Int. J. Pharm. 27*: 127 (1985).
8. M. H. Rubinstein and I. M. Jackson, *Int. J. Pharm. 36*: 99 (1987).
9. H. Vromans and C. F. Lerk, *Int. J. Pharm. 46*: 183 (1988).
10. K. A. Riepma, C. F. Lerk, A. H. de Boer, G. K. Bolhuis, and K. D. Kussendrager, *Int. J. Pharm. 66*: 47 (1990).
11. M. H. Rubinstein and J. S. M. Garr, *Int. J. Pharm. 73*: 75 (1991).
12. J. E. Gordon, *The Sciences of Strong Materials*, Penguin Book, Harmondsworth, p. 98, 1973.
13. G. K. Bolhuis, C. F. Lerk, H. T. Zijlstra, and A. H. de Boer, *Pharm. Weekblad. 110* (1975).
14. M. Luangtana-anan and J. T. Fell, *Int. J. Pharm. 60*: 197 (1990).
15. M. Luangtana-anan, The role of surface free energy in the compaction of powders, Ph.D. thesis, University of Manchester, 1988.
16. W. D. Jones, *Fundamental Principles of Powder Metallurgy*, Edward Arnold, London. pp. 288–289, 1960.
17. C. M. Ramaswamy, Y. B. G. Varma, and D. Venkateswarlo, *Chem. Eng. 1.* 168 (1970).

16

Lubricant Sensitivity

Gerad K. Bolhuis

University of Groningen, Groningen, The Netherlands

Arne W. Hölzer

Astra Hässle AB, Mölndal, Sweden

I. INTRODUCTION

Lubricants are commonly included in tablet formulations in order to reduce die wall friction during both compaction and ejection of the tablet. Their presence, however, may cause undesirable changes in tablet properties. Shotton and Lewis [1] investigated the effect of magnesium stearate on the diametrical crushing strength of tablets of crystalline materials and of two simple granulations without binders. They found that the lubricant decreased the strength of all tablets, but at most for the crystalline materials. The softening of tablets by lubricants has been previously reported by Strickland et al. [2], who observed that magnesium stearate and other lubricants, added as dry powder to granules, appeared to adhere to and form a coat around individual granules.

II. EFFECT OF LUBRICANTS ON TABLET CRUSHING STRENGTH

Bolhuis et al. [3] showed that magnesium stearate forms an adsorbed lubricant film around host particles during the mixing process (Fig. 1). This lubricant film interferes with the bonding properties of the host particles by acting as a physical barrier. This can be seen when the tablet crushing strength is plotted as a function of the logarithm of the mixing time with the

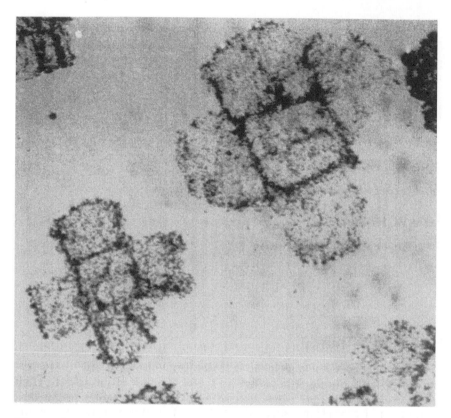

Figure 1 Unfolded magnesium stearate envelopes on water, isolated from sodium chloride crystals, after mixing with magnesium stearate (approx. 30× and 60×). (From Ref. 3; reproduced with permission of the copyright holder.)

lubricant. Figure 2 shows that the effect depends strongly on the material used [4]. For microcrystalline cellulose (Avicel PH 102), powdered cellulose (Elcema G250), compressible starch (STARCH 1500) and calcium sulphate dihydrate (Compactrol) a strong decrease in crushing strength can be seen with an increase in mixing time with the lubricant. On the other hand, the bonding properties of the two kinds of dicalcium phosphate dihydrate are not affected by mixing with magnesium stearate.

In addition to the decreased bonding properties, the addition of hydrophobic lubricants causes increased disintegration times and decreased dissolution rates [1,5–11].

The magnitude of the effect of lubricants on crushing strength is dependent on a large number of factors, e.g., the nature and properties of

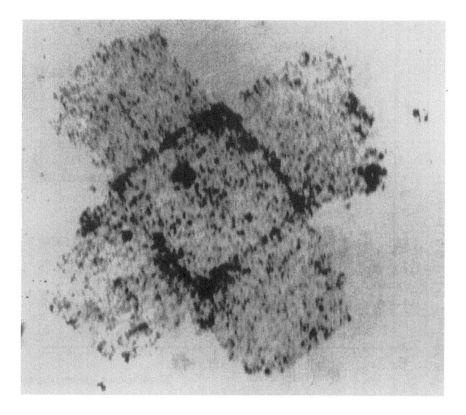

Figure 1 (*continued*)

the lubricant, the nature and properties of the other tablet ingredients and the processing conditions. These factors affect the formation of a lubricant film during the mixing process and/or the adventures of the film during compaction and this will be discussed in the following sections.

III. FILM FORMATION OF LUBRICANTS DURING MIXING

When a powdered lubricant is added to a tableting blend and submitted to a mixing action it is distributed either as a free fraction or, when the lubricant is prone to deagglomeration and a following delamination, as a surface film on the base/carrier material. Prolonged mixing time will transfer more lubricant from the free fraction to the surface film. The phenomenon of decreased crushing strength with an increase in mixing time of tablet ingredients with lubricants is caused by the formation of this lubricant film, which interferes with the binding of the particles [3]. The decrease in crushing strength has

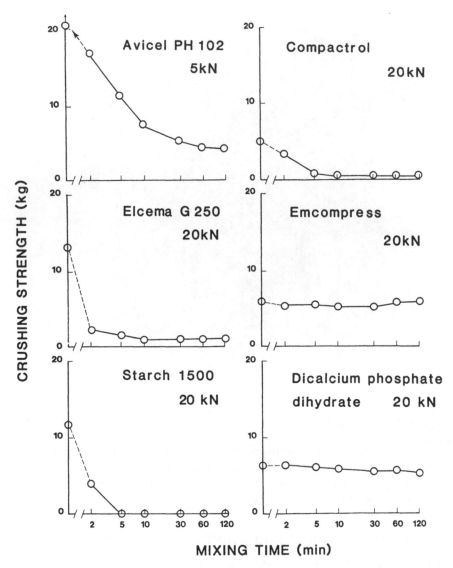

Figure 2 Crushing strength versus lubricant mixing time (log scale) for tablets compressed from filler-binders with 0.5% magnesium stearate. (From Ref. 4; reproduced with permission of the copyright holder.)

been attributed to weaker bonds after compression between lubricant-lubricant molecules rather than strong excipient-excipient bonds [12].

It has been demonstrated by means of X-ray examination that magnesium stearate can be extended by means of a slight pressure to a continuous film, composed of parallel-oriented molecules [13]. Apparently, during a common mixing process, the shear forces are sufficient to shear off molecular layers from magnesium stearate particles and form a film upon the host particles. It has been postulated that film formation is developed in accordance with the Langmuir pattern and can be regarded as an ordered mixing process [6,14–16].

In an attempt to assess film formation and understand the causes of deleterious effects of magnesium stearate, a number of investigations have concentrated on studying the lubricant distribution on host particles and conflicting statements are reported regarding the nature of the lubricant film.

Bolhuis et al. [3] calculated the quantity of magnesium stearate needed to cover different host particles with a monoparticulate and a monomolecular layer, respectively of the lubricant. For amylose, a mono-particulate layer of magnesium stearate will be formed when about 1–2% magnesium stearate is present. For a monomolecular layer, only about 0.05% magnesium stearate was calculated to be sufficient. Tablets of amylose admixed with magnesium stearate with a concentration lower than needed for a monoparticulate layer but larger than for a monomolecular layer gave zero tablet strength (Table 1). It was concluded that magnesium stearate forms a monomolecular lubricant film around the amylose particles.

This hypothesis was proved by mixing amylose with a magnesium stearate *tablet*. The crushing strength of tablets compressed from the

Table 1   Percentage of Magnesium Stearate (specific surface area by weight 15.6 m²/g) Theoretically Required to Cover a Particulate Solid with a Monoparticulate and with a Monomolecular Layer, Respectively.

| | | Percentage of magnesium stearate | | |
| | | Monoparticulate layer | | Monomolecular layer |
Particulate solid	Specific surface A_W (m²/g)	W_{MP}(spheric)	W_{MP}(flat)	W_{MM}
Amylose V	0.097	1.9	0.9	0.025
Crystalline sodium chloride	0.010	0.2	0.1	0.003
Powdered sodium chloride	0.115	2.2	1.1	0.03

Source: From Ref. 3; reproduced with permission of the copyright holder.

blends, where samples were taken at different mixing times, showed that zero crushing strength was attained after a weight loss of the magnesium stearate tablet, corresponding to a concentration in the amylose tablets of 0.04%. This concentration was found to be of the same order of magnitude as that required to coat the host particles with a monomolecular layer.

In a theoretical consideration, Shah and Mlodozeniec [6] suggested that the degree and extent of surface coverage of a substrate particle by lubricants can be described by at least three different mechanisms: (a) adsorption or surface contact adhesions; (b) diffusion or solid penetration, which includes mechanical interlocking; and (c) delamination or deagglomeration of the lubricating agent to form a film coating (usually discontinuous) on the substrate particles. According to the mechanism of boundary lubrication [2], solid lubricants such as magnesium stearate are adsorbed on the surface of substrate particles and form a uniform surface-adsorbed film in a manner similar to a Langmuir-type adsorption [6]. The authors assumed that magnesium stearate particles first adsorb on the surface of host particles, followed by a uniform distribution over the substrate surface as an effect of delamination or deagglomeration of magnesium stearate particles resulting from shear effects upon continued mixing. Diffusion or solids penetration appeared to play a minor role in lubricant spreading during mixing. A continuous coating was thought not to occur during mixing, but only during the compression process [6].

Until now, it was not possible to see lubricant films on host particles by means of scanning electron microscopy (SEM), but isolated films could be detected by transmission electron microscopy [3]. Several attempts have been made to detect lubricant films by chemical or physical methods for a better understanding of the nature of these films; most of these investigations have relied, however, on indirect techniques.

Lerk and Bolhuis [17] measured contact angles of sodium chloride/magnesium stearate blends. For binary mixtures of small particle sized powders a linear relationship exists between the cosine of the contact angle of the mixed system and the proportion of the components. On this basis, the addition of small amounts of magnesium stearate with a contact angle of 121° to crystalline sodium chloride with a contact angle of 28° would be expected to increase the contact angle of the blend by at most a few degrees. The observed increase of the contact angle of sodium chloride up to 121° after mixing with 1% magnesium stearate indicates the existence of a lubricant film [17].

Another indirect technique to determine surface coverage with lubricant is the measurement of the dissolution rate of host particles. Based on the idea that a hydrophobic lubricant decreases the effective drug/solvent interfacial area and thereby decreases the release of the drug, Nicklasson and Brodin [18] calculated the effective surface area coated by lubricants by comparing the initial rates of release of drugs from nondisintegrating drug/lubricant disks. The observed data were explained by comparing the results with SEM pictures. The areas were found to be dependent on the type and amount of lubricant used and the size of the host particles. The influence of mixing time on the dissolution properties were found to be small: 1 min seemed to be enough to distribute the lubricant uniformly. Using 5% magnesium stearate or sodium stearyl fumarate (Pruv), up to 60% of the disk area was coated; with 5% of an 1:1 mixture of talc and stearic acid, the coated area was about 20%. It must be emphasized, however, that film formation of magnesium stearate on the surface of a tablet is not directly related to the film formation of the lubricant mixing stage since the tablet surface film can be influenced by the spreading out of the lubricant at the ejection of the tablet, the affinity of the lubricant to the tablet surface and the die wall, respectively, and the migration of magnesium stearate in the tablet mass during compaction [19]. The film formation of magnesium stearate on the surface of acetylsalicylic acid particles has been investigated by applying a flow-through dissolution technique [15]. The film formation on the surface of the host particles increased by increasing mixing time. Dependent on the specific surface area of the magnesium stearate quality used, a maximum coverage of about 70% was obtained after mixing between 10 and 30 mins.

Rowe [20] determined the extent of the surface coverage of films of magnesium stearate and calcium stearate, respectively upon tablet surfaces with a film-coating/tablet adhesion technique. The results were very similar to those obtained by the rotating disk technique [18], although the substrate materials were widely different.

While most information has accumulated on the physical characteristics of lubricant films, only a few studies have focused on the chemical structure of the *in situ* film, primarily using elemental microanalysis techniques.

A direct technique to examine visually the distribution of magnesium stearate on host particles is possible by SEM in combination with energy-dispersive X-ray microanalysis (EDAX). Using this technique, Lerk and Bolhuis could not make X-ray maps for magnesium because the low concentration of magnesium produced a Mg-K peak, which was drowned in the background radiation [17]. More successful were attempts with higher magnesium stearate concentrations where the formation of a magnesium stearate film on a substrate by EDAX analysis has been confirmed visually [21]. The thickness of the film was found to be variable, and even after 60-min mixing time magnesium stearate crystals were found in the blend. More recently, however, Roblot-Treupel and Puisieux [22] confirmed the preferential location of magnesium stearate in the cavities of host particles and the regularization of the surface provided by the lubricant. Lerk and Sucker [23,24] found likewise that part of magnesium stearate was trapped in cavities of the granules.

Hussain et al. [16] estimated the percentage surface coverage of sodium chloride by different magnesium stearate concentrations from the EDAX data. Film formation by lubricants from different manufacturers was examined, and results suggested similarity in mechanism but different degree of host surface coverage for equivalent mixing conditions. Depending on the magnesium stearate used, the infinite surface coverage varied between 4.5% and 15%. From the fact that the EDAX data fitted the Langmuir type of adsorption isotherm, it was suggested that a molecular film was first formed, which, on further blending was followed up by a gradual buildup of a particulate film. This particulate layer may have been initiated at gross defect points on the host particle surface. A simple and direct correlation between lubricant surface area and surface coverage of the host particles was found. Noteworthy is the difference between the estimated infinite surface coverage determined by EDAX, a maximum of 15% [16], and that determined by changes in dissolution rates, 60% [15,18]. It was suggested that in the dissolution technique the free fraction and loose agglomerates of the lubricant may also contribute to the estimate of percentage surface coverage of the host [16].

Another direct method for measuring magnesium stearate film formation is secondary ion mass spectrometry (SIMS). From static SIMS spectra (sampling depth 20 Å) of sodium chloride tablets, containing magnesium stearate, it has been concluded that the lubricant film tends to be patchwise in nature rather than a continuous mono- or multimolecular film [25]. In another study [26], it has been observed that during mixing sodium chloride with magnesium stearate, lubricant film formation appeared to begin instantly, and the maximum rate of film formation occurred during the first few minutes. During prolonged mixing, film formation continued at a much lower rate.

The effect of magnesium stearate on bonding properties depends strongly on the completeness of the film during the bonding stage of the compaction process. The completeness of the film depends on possibilities and velocity of film formation during mixing and the adventures of the film during compaction and consolidation.

The possibility and velocity of film formation is influenced by:

• Nature and properties of the lubricant
• Nature and properties of the host particles

- Specific surface area of the host particles in relation to the concentration of the lubricant
- Presence of other additives in the blend
- Mixing time and intensity
- Type, size and content of the mixer

The adventures of a formed lubricant film depend on (a) the consolidation and compaction behavior of the host particles and perhaps on (b) storage conditions of lubricated products.

IV. EFFECT OF LUBRICANT PROPERTIES ON FILM FORMATION

A. Nature of the Lubricant

The effect of tablet lubricants on bonding properties of host particles depends on the nature of the lubricant used. Lerk et al. [27] compared stearic acid, different metal soaps of stearic acid and polytetrafluoroethylene (PTFE). Stearic acid and the stearates decreased the crushing strength of STARCH 1500 tablets, but the effect of magnesium-, calcium-, or aluminum stearate was larger than that of sodium stearate and stearic acid (Figure 3). It was shown, however, that the results may be influenced by the specific area of the lubricants used. Tablets with the coarse sodium stearate were stronger than tablets with the fine lubricant grade. PTFE did not affect the bonding properties between STARCH 1500 particles, because it has, in contrast to stearic acid and the stearates no laminar structure and hence cannot form a film around the excipient particle. The absence of a lubricant film on sodium chloride particles after intensive mixing with PTFE was confirmed by contact angle measurements [17].

Theoretical calculations of interactions of lubricants and host particles were presented by Rowe [28]. Based on a model involving the integrating of the Lennard-Jones pair potential function to predict the properties of a single- and a two-component system, the intensity of the molecular interactions were predicted for the excipients microcrystalline cellulose and anhydrous lactose and the lubricants magnesium stearate, stearic acid and PTFE, using literature values of their partial solubility parameters. Table 2 shows data for the interaction parameter and strengths of both the adhesive and cohesive interactions within the lubricant for the excipient/lubricant systems studied. For both excipients, the values for the adhesive interactions between excipient and lubricant decrease in the order magnesium stearate > stearic acid > PTFE. Table 2 shows that in the case of magnesium stearate the lubricant-excipient interactions are higher than the lubricant-lubricant

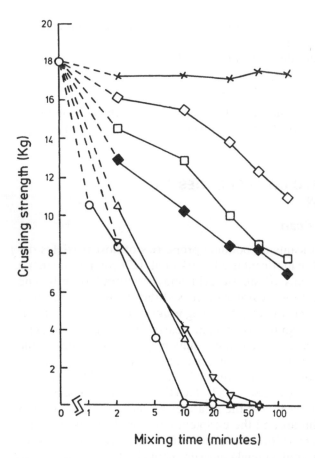

Figure 3   Effect of mixing time on the crushing strength of tablets compressed from blends of STARCH 1500 with 0.5% of various lubricants (×) PTFE; (◇) coarse sodium stearate; (◆) powdered sodium stearate; (□) stearic acid; (▽) aluminium stearate; (△) calcium stearate; (○) magnesium stearate. (From Ref. 27; reproduced with permission of the copyright holder.)

interactions. The strong adhesive interactions explain the formation of a monomolecular magnesium stearate film over the excipient. This will decrease the number of the strong cohesive interactions between the excipient particles causing a decrease in tablet strength. For PTFE the adhesive interactions are less than half the cohesive interactions (Table 2) which indicates that this lubricant will tend to concentrate in the interstitial pores between the excipient particles. As an effect, the excipient-excipient interactions and hence the tablet strength will not significantly decrease. Stearic acid will

Table 2 Interaction Parameters (ϕ) and Strengths of the Adhesive (A-B) and Cohesive (B-B) Interactions for Microcrystalline Cellulose and Anhydrous Lactose Lubricated with Magnesium Stearate, Stearic Acid, and Polytetrafluoroethylene (PTFE)

Excipient A	Lubricant B	ϕ	Strength [adhesive (A-B)]	Interaction (MPa) [cohesive (B-B)]
Microcrystalline cellulose	Magnesium stearate	0.48	85.8	82.8
Microcrystalline cellulose	Stearic acid	0.40	69.2	77.4
Microcrystalline cellulose	PTFE	0.16	22.0	49.0
Anhydrous lactose	Magnesium stearate	0.48	87.1	82.8
Anhydrous lactose	Stearic acid	0.38	65.7	77.4
Anhydrous lactose	PTFE	0.14	19.6	49.0

Source: From Ref. 28; reproduced with permission of the copyright holder.

have properties between the two extremes but closer to magnesium stearate than PTFE. This is consistent with studies of the effect of stearic acid and PTFE on tablet crushing strength [27,29–31].

The effect of a decrease in tablet crushing strength during mixing with a lubricant is not limited to stearates and stearic acid. A significant reduction in crushing strengths has been found during mixing of excipients with hydrogenated oils [29,31–33], glycerides [29,32,33], sodium- and magnesium lauryl sulfate [32], sodium stearyl fumarate [32,34] and polyethylene glycol 4000 [31,33]. On the other hand, just like PTFE, talc and graphite did not affect tablet strength [29]. In other work, however, it has been reported that talc may decrease tablet crushing strength [31].

The effect of mixing time with lubricant on both tablet properties and lubricating properties was studied by Hölzer and co-workers, using a large number of lubricants and different test materials [35,32,36]. Some results from these papers, completed with unpublished data are shown in Figure 4. The figure shows the effect of mixing time with 0.1% lubricant on percentage reduction in tensile strength of sodium chloride tablets as compared to the values of unlubricated sodium chloride tablets. Mixing with lubricant was performed in a 2-L Turbula mixer at 42 rpm (20, 200, and 2000 revolutions, respectively); the tablets were compacted at 150 MPa in an eccentric press. Table 3 lists the physical and chemical properties of the lubricants used, ordered by specific surface area. Figure 4 shows that a prolonged mixing time generally increases the reduction in tensile strength but that the magnitude of the effect depends on the lubricant used. Lubricants with a large surface area, e.g., the magnesium stearates give a high reduction in tensile strength even when admixing the

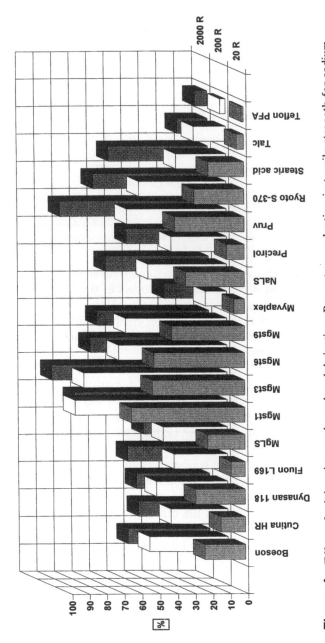

Figure 4 Effect of mixing time on boundary lubrication. Percentage reduction in tensile strength for sodium chloride tablets containing 0.1% lubricant.

Table 3 Physical and Chemical Properties of Lubricants Used in Trials with Sodium Chloride, Ordered by Specific Surface

Lubricant	Description	Area (m^2/g)	Density (g/cm^3)	Melting point (°C)
Mg stearate 1	Salt of fatty acid	6.29	1.13	143
Mg stearate 3	Salt of fatty acid	4.91	1.12	—
Mg stearate 6	Salt of fatty acid	4.52	1.08	134–136
Dynasan 118	Triglyceride	4.46	1.08	71
Talc	Silicate	3.93	3.05	—
Na stearyl fumarate	Salt of fatty acid	2.72	1.14	225
Mg stearate 9	Salt of fatty acid	2.45	1.14	140–154
Fluon L169	Polymer, PTFE	2.06	2.43	>300
Mg lauryl sulfate	Alkyl sulfate	0.88	1.26	241
Na lauryl sulfate	Alkyl sulfate	0.62	1.21	189
Cutina HR	Triglyceride	0.60	1.04	86
Ryoto S-370	Sugar ester	0.54	1.14	71
Boeson VP	Triglyceride	0.34	1.04	>48
Teflon PFA powder	Polymer, PTFE	0.29	2.48	>300
Précirol	Triglyceride	0.26	1.06	63
Myvaplex 600	Triglyceride	0.11	1.02	50–60
Stearic acid	Fatty acid	0.09	1.00	68

lubricant as short as 20 revolutions. The triglycerides seem to have a smaller negative effect on tensile strength than the stearates. For two PTFE qualities a different behavior can be seen: Fluon particles are spread out during mixing which reduces the tensile strength; Teflon PFA has hardly any effect on tablet strength even after prolonged mixing.

B. Concentration and Specific Surface Area of the Lubricant

In addition to the nature of origin, the concentration and the specific surface area of the lubricant determine the effect on tablet crushing strength. Using low magnesium stearate concentrations without changing the mixing conditions, film formation will be slower and hence the decrease of crushing strength will be smaller for a given mixing time, as compared to higher lubricant concentrations [3,34]. For amylose particles it was shown that the effect of mixing with magnesium stearate on tablet strength was controlled by the mixing time rather then by the lubricant concentration, providing that the concentration is high enough for the formation of a monomolecular film [3]. A similar effect may be expected when the mean particle size of the lubricant particles increases: Mixing excipients with large magnesium stearate particles will slow down the film formation process, as compared to mixing with small lubricant particles. This may be the reason that for a fixed mixing time, the effect of granulated magnesium stearate on crushing strength was found to be lower than that of powdered magnesium stearate [1]. Sufficient fine-powdered lubricant was transferred to the surface of the base granulate particles to reduce the strength of the bond. This idea was used by Johansson [37,38] who compared magnesium stearate in a granular form with powdered magnesium stearate as a lubricant, with respect to both effect on crushing strength and lubricating properties. Under ideal circumstances, an agglomerate of fine primary lubricant particles should be strong enough to withstand the forces at the lubricant mixing, but during compaction the granules and the released primary particles should be spread out due to the shearing action at the die and punch surfaces. For a fixed mixing time, the granular magnesium stearate showed lubricating properties which are comparable to powdered lubricant without negatively affecting tablet properties when present in concentrations above 1% [37]. This effect is supported by the observation that, compared with powdered magnesium stearate, the granular form gives a lower final surface coverage of host particles [15]. Increasing the particle size of the magnesium stearate granulation increased the amount of lubricant required to obtain lubrication similar to powdered magnesium stearate [39]. The negative effect of the granular lubricant on tablet crushing strength increased, however, by a longer mixing time or when the batch size was increased [40]. These effects were explained, respectively by attrition of the lubricant granules after long mixing and by higher shear forces, due to the larger masses involved.

It is often considered that the surface area of the lubricant is the most critical parameter of the material with respect to the deleterious effects on tablet properties. Hölzer and Sjögren [35] found a quantitative relationship between the tensile strength of sodium chloride or lactose tablets and the surface area of sodium stearyl fumarate. A correlation between the lubricant surface area and coverage on the base material after a mixing period of as short as 1 min has been reported too [15]. For three batches of magnesium stearate differing in morphology, particle size, bulk density, and specific surface area, used in such amounts to develop equivalent lubricating areas, the effect on tablet crushing strength was almost identical [41]. From experiments with two different types of magnesium stearate, blended with hydrochlorothiazide as a drug model, Colombo and Carli [42] concluded that on the basis of intrinsic dissolution rate, contact angle and ejection work data the crystalline structure of the lubricant has no marked influence on its film-forming properties and on the characteristics of the resulting mixtures.

On the other hand, they showed that the fraction of hydrochlorothiazide coated with lubricant was linearly dependent on their specific surface area.

In a study on batch to batch variations of seven commercial magnesium stearates, the differences in effect on tablet properties could be explained to a great extent by differences in surface areas and spreading out the lubricant particles during the short mixing process, both for sodium chloride and anhydrous lactose tablets [36]. From these studies [36,41,42], it can be concluded that lubricant surface area seems to be a critical parameter. In another study, however, using different kinds of magnesium stearate with a different specific surface area no direct correlation between the lubricant surface area and surface coverage of sodium chloride particles could be found [16]. Although surface area may play a role during the initial mixing stages, on further mixing the delamination propensity for film formation and the inherent material properties of magnesium stearate as well as host powder surface will influence the total percentage surface coverage [16].

C. Morphology and Crystal Modification of the Lubricant

For magnesium stearate, it is well documented that large batch variations exist with respect to both chemical and physical properties [36]. In addition to chemical differences, structural and crystalline characteristics are likely to be important criteria in determining the rate and extent of surface coverage [43–45].

SIMS analysis indicated a profound difference between commercial magnesium stearates and a high-purity magnesium stearate with respect to film generation; the commercial samples covered the host particles to a much greater extent than the pure product [26]. The poor film formation of the pure product was attributed to its crystal structure and shearing strength; the more crystalline and high-purity lubricant seems to be more resistant to shearing during the mixing process than the less crystalline commercial materials [46]. Magnesium stearate consisting of needles shows, in contrast to magnesium stearate consisting of platelets, a poor distribution over host particles [47,48].

V. EFFECT OF HOST MATERIAL PROPERTIES ON FILM FORMATION

A prerequisite for the formation of a lubricant film on host particles is a distribution of lubricant particles among the host particles. This means that particle size and flow properties of host particles influence the rate of lubricant film formation. It has been found that the effect of mixing time on tablet crushing strength decreased with increasing particle size of microcrystalline cellulose particles [49]. This effect was attributed to the larger shear forces in a mixer, created by the larger particles. Moreover, the flow properties of host particles may have a large effect on the rate of lubricant film formation. When the flow properties of the host particles are extremely poor, the distribution of the lubricant particles and the consequent formation of a lubricant film during mixing will be a very slow process. Bos et al. [50] compared rice starch with extremely bad flow properties with potato starch which has reasonable flow properties. It can be seen from Table 4 that the crushing strength of rice starch tablets were, in contrast to potato starch tablets, not affected by mixing for 30 min with 0.5% magnesium stearate. In this study strength reduction, defined as the lubricant sensitivity ratio (LSR) was used as a quantitative measure to express the sensitivity to mixing with a lubricant of tableting materials. The LSR is the ratio between the decrease in crushing strength of tablets, due to mixing with a lubricant and the crushing strength of unlubricated tablets:

$$\text{LSR} = \frac{Cs_u - Cs_l}{Cs_u}$$

where Cs_u and Cs_l are the crushing strengths of tablets prepared without and with a lubricant, respectively. Prolonged mixing of rice starch for several hours with the lubricant caused a marked decrease of the lubricant sensitivity ratio. When the rice starch was granulated (Table 4), the improved flow properties resulted in an increased lubricant sensitivity.

Table 4 Flow Properties, Tablet Strength, and Magnesium Stearate Susceptibility of Different Starches and Starch Granulations

	Hausner ratio[a]	$Cs_u{}^{b}$ (N)	Cs_l (N)	LSR
Potato starch	1.20	68 ± 10	0	1.0
Tapioca starch	1.33	70 ± 4	39 ± 4	0.4
Maize starch	1.38	56 ± 8	30 ± 7	0.5
Rice starch	1.38	102 ± 20	104 ± 11	0.0
Potato starch granulation	1.18	47 ± 3	<10	>0.8
Tapioca starch granulation	1.25	54 ± 2	<10	>0.8
Maize starch granulation	1.21	27 ± 4	<10	>0.6
Rice starch granulation	1.21	139 ± 11	50 ± 8	0.6

[a] Unlubricated.
[b] Cs_u = crushing strength of unlubricated tablets; Cs_l = crushing strength of lubricated tablets; LSR = lubricant sensitivity ratio.
Source: Adapted from Ref. 50.

In a later study, Vromans et al. [51] showed that for different types of lactose, the sensitivity to lubrication was related to the bulk density of the powder. Although bulk density is a secondary parameter, depending on fundamental properties such as true density, particle size, shape, texture, and surface roughness, the authors proposed some theoretical considerations. Firstly, a low bulk density is an indication for poor flowability of a powder, which might delay or even prevent the formation of a lubricant film during the mixing process. Secondly, a lower bulk density will result in a larger contribution to particle rearrangement and consequently higher friction during consolidation. This could disturb an already formed lubricant film and enhance bond formation. A similar relationship between bulk density and lubricant sensitivity ratio was found for granulations based on native starches or on modified celluloses, but the relationships were different for the different investigated materials [52]. For the granulations, the flowability of the particulate system was thought to be the predominant mechanism in the sensitivity to lubrication with magnesium stearate.

The consolidation and compaction characteristics of host particles are known to have considerable influence on their susceptibility to lubrication [12,53]. De Boer et al. [12] illustrated that the sensitivity of tablet excipients to magnesium stearate depends on the compression behavior and of the bonding mechanism of the material. The bonding properties of brittle materials like dicalcium phosphate dihydrate (Fig. 2) and anhydrous β-lactose (Fig. 5) were hardly influenced by lubrication. The phenomenon was explained by the assumption that clean, lubricant-free surfaces are created by fragmentation of the particles during consolidation of the particle system. On the other hand, a maximum effect of magnesium stearate was found for excipients that undergo complete plastic deformation without any fragmentation under compression and are bonded by cohesion, such as starch and some starch derivatives (for instance STARCH 1500; see Fig. 2).

Most tablet excipients behave, however, in a manner intermediate between complete plastic deformation and complete brittle fracture, so that the lubricant sensitivity depends on the extent of fracture of the particles during compression. This should be the reason that for many materials the crushing strength decreases not to zero but to a minimal level after mixing with magnesium stearate (Fig. 5).

De Boer et al. [12] illustrated the effect of magnesium stearate on excipients with a different consolidation behavior by means of scanning electron micrographs. The upper surface of an unlubricated amylose tablet (Fig. 6a) shows that the particles are plastically deformed, but keep their individuality. When compaction was performed after 1 h mixing with 0.05% magnesium stearate, the particles are plastically deformed too, but

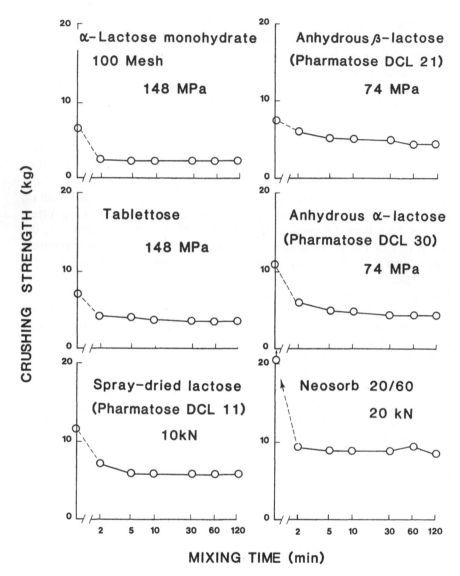

Figure 5 Crushing strength versus lubricant mixing time (log scale) for tablets compressed from filler-binders with 0.5% magnesium stearate. (From Ref. 4; reproduced with permission of the copyright holder.)

the magnesium stearate film on the amylose particles has prevented bonding (Fig. 6b). Similar results were found for crystalline sodium chloride, a plastically deforming material. Comparison of the cross section of lubricated and unlubricated sodium chloride tablets (Figs. 6c,d) shows that the failure occurs mainly across the particles for plain sodium chloride tablets, but around the particles for tablets compressed from blends containing magnesium stearate. No differences could be seen between micrographs of surfaces of dicalcium phosphate dihydrate tablets without and with magnesium stearate, respectively, because of the extensive fragmentation behavior of the excipient.

In recent work, Riepma et al. [54] showed for different materials that the lubricant sensitivity was not always related to the degree of fragmentation during compaction. In contrast to dicalcium phosphate dihydrate, both α-lactose monohydrate and sodium citrate exhibited a considerable reduction in tablet strength upon lubrication, whereas it is known that intensive fragmentation occurs during consolidation [54,55]. Photomicrographs of fractures of lubricated tablets of α-lactose monohydrate and of sodium citrate showed that the tensile failure occurred around the interfaces between the original crystals [54]. Realizing that these surfaces are coated with magnesium stearate, it was suggested that a three-dimensional matrix of magnesium stearate is sustained during compression of the particulate system. On the other hand, on photomicrographs of fractures of lubricated tablets of dicalcium phosphate dihydrate, no original crystals could be distinguished. This means that addition of the lubricant did not lead to weak bonds between the original particles. Based on these results, the authors presented two different models for the elucidation of the lubricant sensitivity of brittle materials [54]. Figure 7 shows the noncoherent matrix model. A coherent network of magnesium stearate (A), created by dry mixing the excipient with the lubricant, is interrupted by fragmentation and consolidation of the particulate system (B). The fat lines in the figure represent the magnesium stearate film upon the surface of the original particles. As a consequence, the effect of a lubricant on the compactibility of an excipient will be limited by the created lubricant-free surfaces. Figure 8 shows the coherent matrix model. A coherent network of magnesium stearate (A), created by dry mixing the excipient with the lubricant, is sustained during the process of consolidation (B). Fragmentation occurs within the areas surrounded by the lubricant. Consequently, the strength of a compact is principally determined by the structure of a magnesium stearate matrix, created during the process of mixing the excipient with the lubricant.

The formation of a coherent lubricant matrix is strongly dependent on the surface texture of the host particles [4,24]. Lerk and Sucker [24]

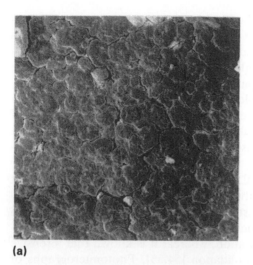

(a)

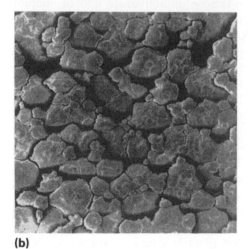

(b)

Figure 6   (a) Micrograph of the upper surface of a tablet compressed from Amylose V (× 150 approx.) (b) Micrograph of the upper surface of a tablet compressed from a mixture of Amylose V with 0.05% magnesium stearate (× 150 approx.) (c) Micrograph of the surface of fracture of a tablet compressed from crystalline sodium chloride (× 150 approx.) (d) Micrograph of the surface of fracture of a tablet compressed from a mixture of crystalline sodium chloride with 0.1% magnesium stearate (× 150 approx.) (From Ref. 12; reproduced with permission of the copyright holder.)

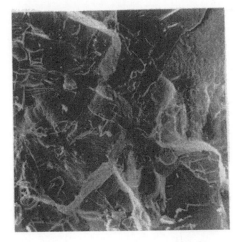

(c)

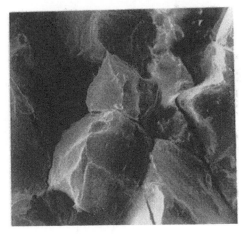

(d)

Figure 6 *(continued)*

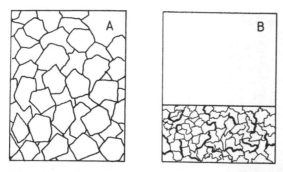

Figure 7 Noncoherent matrix model. (A) and (B) as described in the text. (From Ref. 54; reproduced with permission of the copyright holder.)

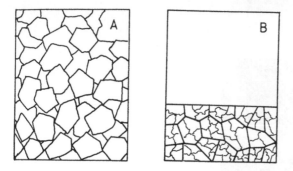

Figure 8 Coherent matrix model. (A) and (B) as described in the text. (From Ref. 54; reproduced with permission of the copyright holder.)

showed that during the mixing process of irregular-shaped granule particles magnesium stearate formed a discontinuous layer around the particles of the granular excipient. Part of the magnesium stearate was trapped into the asperities and cavities and was therefore not available for the formation of a magnesium stearate film. Evidently no coherent matrix of magnesium stearate was formed. In order to assure the presence of a complete lubricant film, in two studies excipients were coated with carboxylic acid in diethyl ether [51,54]. Table 5 shows totally different lubricant sensitivity ratios for three different excipients when dry-blended with magnesium stearate but almost equal lubricant sensitivities on liquid coating with decanoic acid. This result endorses the determining effect of the presence of a coherent or a noncoherent matrix of magnesium stearate within a particulate system. The very irregular texture of dicalcium phosphate dihydrate particles, as compared to the relatively smooth crystal surfaces of

Table 5 Effect of Lubrication with Magnesium Stearate (0.5%) and Decanoic Acid (0.5%), Respectively, on the Strength of Tablets Compacted at 20 kN from a Fraction (250–300 μm) of Several Materials

	Magnesium stearate[a]			Decanoic acid		
Material	Cs_u (N)	Cs_l (N)	LSR	Cs_u (N)	Cs_l (N)	LSR
α-Lactose monohydrate	42	24	0.43	42	21	0.50
Sodium citrate	23	5	0.78	23	11	0.52
Dicalcium phosphate dihydrate	32	29	0.09	32	16	0.50

[a] Cs_u = crushing strength of unlubricated tablets; Cs_l = crushing strength of lubricated tablets; LSR = lubricant sensitivity ratio
Source: From Ref. 54; reproduced with permission of the copyright holder.

α-lactose monohydrate and sodium citrate, prevents the formation of a continuous film on the blending with magnesium stearate and explains the small lubricant sensitivity of this excipient on direct compression. It can hence be concluded that the susceptibility of a material to lubricants like magnesium stearate is a complex function of a number of factors including surface area, surface texture, flowability, mixing properties, and consolidation behavior.

VI. EFFECT OF PROCESS CONDITIONS ON FILM FORMATION

The extent of film formation and hence the negative effect of lubricants on tablet crushing strength depends strongly on mixing time (Fig. 2) and the mixing procedure [3,6]. The influence of magnesium stearate on tablet strength has been investigated using different types of lab-scale mixers [27,56,57]. It was concluded that not only mixing time but also mixing intensity plays a large role [27]. When production-scale volumes are used, mixing and shearing intensity increase. As the rate at which an ordered mix is formed increases significantly when the batch size is increased [58], it may be expected that in large industrial mixers, the shear forces, affecting the migration of magnesium stearate from magnesium stearate particles to excipient particles, will be much greater than in lab-scale mixers. Therefore, it may be expected that the formation of a film of magnesium stearate while the tablet ingredients are being mixed with the lubricant proceeds faster in production-scale mixers and will depend on the type of the mixer used and its rotation speed. This means that the tablet properties also will depend on the type, size, load, and rotation speed of the mixer. Few studies have been made, however, of the effect on tablet properties of scaling up of the lubricant mixing process. Johansson [40] studied the effect of mixing with magnesium stearate on tablet strength using three different double cone mixers containing 5, 25, and 80 kg material, respectively. The negative effect of the lubricant increased with batch size, which effect was attributed to the higher shear forces in the tumbling mixer with the larger masses. As in tumbling mixers the energy required for the formation of an ordered mixture is mainly provided by shear forces within the mixture, larger masses promote the deagglomeration of lubricant particles and the formation of an ordered mixture.

Bolhuis et al. [59] mixed a lactose/microcrystalline cellulose test formulation with 0.5% magnesium stearate in seven different mixers, operating at different mixing velocities (Table 6). The decrease in crushing strength of tablets, compressed from the blends occurred much more quickly in production-scale mixers than in the lab-scale mixers when they operated at the same rotation speed. The critical mixing time, i.e., the time required for a decrease of the crushing strength from 180 N (unlubricated tablets) down to an arbitrarily minimal level of 60 N varied between 1.5 and 30 min. For the production-scale mixers, the decrease in tablet crushing strength as an effect of lubricant admixing depended more on the rotation speed than on the type or size of the mixer or the size of the load. This can be seen by comparing different production scale mixers operating

Table 6 Sensitivity of the Test Formulation, When Mixed in Various Mixers at Various Speeds, to the Addition of the Lubricant[a]

Mixer	Capacity (L)	Rotation speed (rpm)	Crushing strength half-life (min)	Critical mixing time (min)
Turbula	2	45	1.0	10.5
		90	<1.0	1.5
Cubic	13	20	8.0	30.0
		60	2.7	17.0
Drum	45	10	1.0	9.0
Planetary	90	25	1.0	4.0
		42	<1.0	2.0
Planetary	200	26	2.0	3.8
Planetary	900	10	3.3	7.9
V-shaped	1000	22	1.5	3.6

[a] The sensitivity is reflected both in the crushing strength half-life and in the critical mixing time.
Source: From Ref. 59; reproduced with permission of the copyright holder.

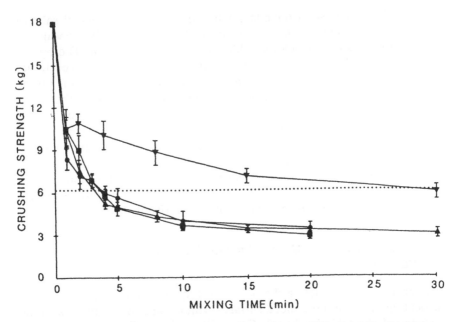

Figure 9 Effect of lubricant mixing time on the crushing strength of tablets compressed from the test formulation. (▼) Formulation mixed in a 13-L cubic mixer at 20 rpm; (●) formulation mixed in a 90-L planetary mixer at 25 rpm; (■) formulation mixed in a 200-L planetary mixer at 26 rpm; (▲) formulation mixed in a 1000-L V-shaped mixer at 22 rpm. (From Ref. 59; reproduced with permission of the copyright holder.)

with speeds between 20 and 26 rpm (Fig. 9). It was found that a Turbula mixer, operating at high rotation speeds can be used to predict the effect of lubricant admixing on tablet crushing strength in production scale mixers. This is important in preformulation work, where lab-scale mixers are commonly used to determine the lubricant sensitivity of tableting blends.

VII. EFFECT OF LUBRICANT FILM FORMATION ON LUBRICATION PROPERTIES

A. Boundary Lubrication

Effective lubricants such as magnesium stearate and sodium stearyl fumarate are examples of boundary lubricants. The theory of Bowden and Tabor shows that effective lubrication is provided only if the lubricant can react with the surface of the die wall to form a layer with strong lateral adhesion [2,60,61]. Solid boundary lubricants such as the metal salts of stearic acid, e.g., magnesium stearate form a layered film at the die wall and on punch surfaces on which the tablet slides during ejection. The hydrophobic carbon chains are directed outward from the die surface. Tablet lubricants act as a friction lowering material at the die wall building up the boundary shearing and gliding protection layer and/or as an antiadherent to prevent adhesion, sticking and binding of the tablet material to the tooling surfaces as well as other machine parts. Moreover, lubricants can act as a glidant that promotes the flow of the granulation within the hopper and die cavity. The next sections will show that variation in mixing time with lubricants will not only affect tablet properties such as strength, disintegration, and drug dissolution, as shown in first part of this chapter, but will also have an effect on the lubricating properties of the lubricant.

B. Effect of Lubricant Mixing Time on Tablet/Die Wall Friction

In pharmaceutical tableting the desired effect of a lubricant is to have a lubricating layer on those metal surfaces which come into contact with the particles of the tableting blend. Generally, only small amounts of tablet lubricants are needed to lower the friction forces effectively [34]. The lubrication and friction properties have been measured with the aid of force transmission, by the determination of friction forces or friction coefficients, by acoustic emission and by electric power consumption measurements in order to optimize the amount of lubricant and to minimize the negative effects of lubricants [2,32,60,62–65]. The ejection force, calculated per unit contact area between the tablet and the die wall is recommended as a good measure of friction during tableting, but has the disadvantage that it depends on the compaction load [47,66].

To obtain a measure, which for most materials is independent of load and/or dimension, the friction coefficients at the maximum compaction (μ_1 = ratio of axial force difference (FD) and die wall force at compression maximum (DWF)) and at ejection (μ_2 = ratio of ejection force (EJF) and radial force at ejection (DWFE)) have been calculated [63,32,45,67,68]. It was shown that the ejection force EJF is due to variation in friction coefficient and not due to variation in radial forces at ejection DWFE [32].

1. Strength of the Lubricant Boundary Layer

In a classical series of papers Strickland et al. showed the mechanism of action and evaluation of 70 materials as tablet lubricants [60]. After coating the die wall with boundary type lubricants, the ejection force was measured during tableting of eight successive tablets of a nonlubricated sodium bicarbonate test granulation. The ejection force increased from 6 kg up to 8, 10, 20, 43, 65, 88, and 103 kg when the die was prelubricated with solid magnesium stearate. The strength of the lubricant boundary layer formed by magnesium stearate, conditioning the die wall was also investigated by Hölzer and Sjögren [63]. First unlubricated sodium chloride tablets were compacted at 115 MPa upper-punch pressure. The friction coefficient μ_1 measured was 1.4 (see Fig. 10). Then material lubricated with 1.0% magnesium stearate was tabletted. The polar parts of the boundary lubricant molecules adhered to the metal surface, forming a resistant layer (a boundary lubricant film) on the die wall. The friction coefficient was reduced and reached a constant value of 0.3. When changing to unlubricated sodium chloride once more, the friction coefficient increased again, but more than 30 tablets were needed before the starting value was reached. This result shows that the boundary lubricant film on the die wall is very resistant to abrasion by sodium chloride.

2. Effect of Mixing Time on Ejection Force

Müller et al. [62] have shown the influence of mixing time on the remaining force and ejection force of tablets. Two brittle tablet masses, granulatum simplex and crystalline lactose, were mixed with different magnesium stearates, used in concentrations between 0.2 and 1%. Minimum values for both remaining and ejection forces were obtained after 2–5 min mixing in a Turbula mixer. Prolonged mixing times had a negligible effect on the lubricating properties.

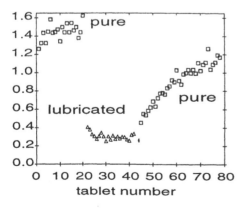

Figure 10 Strength of boundary lubricant film. Friction coefficient (μ_1) at compaction for consecutive sodium chloride tablets.

Ragnarsson et al. [34] studied the influence of mixing on the lubricating properties of magnesium stearate. The ejection force of tablets compressed from blends of sodium chloride, anhydrous β-lactose or calcium citrate granulate with 0.1–2.5% magnesium stearate were measured in an eccentric press. Magnesium stearate reduced the ejection force, and this effect was related to the concentration but not to the mixing time except at the lowest concentration of 0.1%. Generally, it seems possible to get a good lubrication even when the lubricant is poorly spread out in the mixture. This means that for the lubricating effect a short mixing time is sufficient when admixing the common used standard concentrations 0.25 to 0.50 w/w% magnesium stearate to tableting mixtures or granulations.

The effect of mixing time with lubricant on both tablet properties and lubricating properties was studied by Hölzer and co-workers, using a large number of lubricants and different test materials [66,69,35,63,32,36,70]. Some results from these papers, completed with unpublished data are shown in Fig. 11. The figure shows the effect of mixing time (20, 200, or 2000 revolutions) with 0.1% lubricant on the ejection force per unit contact area (EJF/A) of sodium chloride tablets, compacted at 150 MPa in an eccentric press. Mixing with lubricant was performed in a 2-L Turbula mixer at 42 rpm. Table 3 lists the physical and chemical properties of the lubricants used, ordered by specific surface area. Figure 11 shows that a prolonged mixing time generally increases the lubrication efficiency of 0.1% lubricant, expressed as reduction in EJF/A, but that the magnitude of the effect depends on the lubricant used. Lubricants with a large surface area, e.g., the magnesium stearates give a high reduction in EJF/A even when admixing the lubricant as short as 20 revolutions. The EJF/A for tablets with lubricants containing agglomerates; e.g., the triglycerides Boeson VP, Cutina HR, Dynasan 118, Myvaplex, and Precirol show a great influence of the mixing time. The lubricant particles are spread out efficiently and the sodium chloride crystals are covered much more after a longer mixing time. A similar effect was found for stearic acid; the EJF/A values were reduced 40–60% due to the increased surface coverage. Fluon particles are also spread out efficiently and the ejection forces are reduced 20% up to 50%. For another PTFE quality, Teflon PFA, no lubricant effect could be seen, even not at prolonged mixing.

Johansson studied the influence of the lubricant mixing time on the lubrication and tablet properties of four different tablet masses, using magnesium stearate in either powdered or granular form as lubricant [37–40]. The EJF/A was used for estimation of the lubrication effect. The adhesion to the punch faces was measured by visual inspection. Tablets with granular magnesium stearate showed values of EJF/A comparable with those of tablets with powdered lubricants at high concentrations (2–5%). Powdered

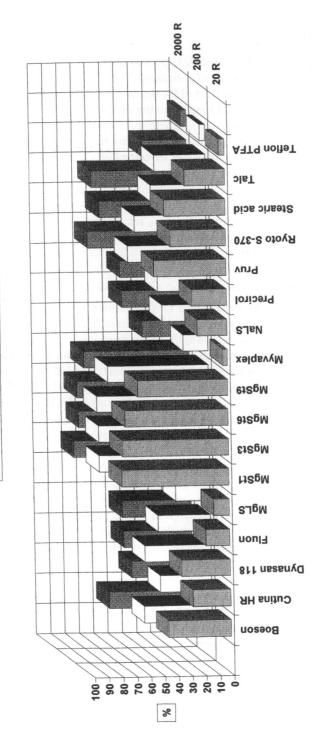

Figure 11 Effect of mixing time on boundary lubrication. Percentage reduction in ejection force per unit contact area for sodium chloride tablets containing 0.1% lubricant.

545

lubricant was better than granular in preventing adhesion to the punch faces at low concentration. The lubricating effect of the powdered lubricant was found to be almost unaffected by the mixing time, while the granular lubricant required somewhat longer mixing times to obtain optimal lubrication properties. Increasing the particle size of the magnesium stearate granulation increased the amount of lubricant required to obtain lubrication similar to powdered magnesium stearate [39]. Variation in the specific surface area of the starting material could be masked by using them in granular form [39]. The results show that film formation is not a prerequisite for good lubrication. The granular magnesium stearate mixtures formed not a film as strong and not as fat as the powdered magnesium stearate lubricant but gave efficient lubrication on the die wall due to the boundary lubrication effect.

3. Effect of Mixing Time on Friction Coefficients

Hölzer and Sjögren [63] measured DWFE simultaneously with EJF. The ratio EJF/DWFE was expressed as the dynamic friction coefficient during ejection μ_2. In further work, the authors measured the effect of mixing time with 0.1% of different lubricants (see Table 3) on the reduction of the friction coefficient μ_2 of sodium chloride tablets, compacted at 150 MPa in an eccentric press [32]. Mixing with lubricant was performed in a 2-L Turbula mixer at 42 rpm. Figure 12 shows that the highest reduction in friction coefficient μ_2 was obtained by the magnesium stearates, followed by sodium stearyl fumarate (Pruv), stearic palmitic sugar ester (Ryoto S-370), stearic acid, and glyceryl palmitic stearate (Boeson VP). PTFE in a coarse grade (Teflon PFA) had very little effect as a lubricant. The effect of mixing time on friction coefficients at ejection (μ_2) in Fig. 12 shows similar results as obtained for the EJF/A values in Fig. 11. The μ_2 values are reduced even more when the mixing is increased because the ejection forces are reduced more than the radial die wall force during ejection. By measuring both die wall forces and ejection forces during ejection one obtains the possibility to distinguish if the friction is lowered by the lubricant.

Using another technique for the measurement of the dynamic friction coefficient at ejection, μ_2, Kikuta and Kitamori [67] found likewise that an increase in mixing time with the lubricant gave lower friction coefficients. Moreover, adhesion to the punches, estimated as the intercept in the correlation to Coulomb's law, decreased on prolonged mixing time [67].

Comparing Figs. 4, 11, and 12 shows that the negative effect of lubricants on tablet crushing strength should be correlated with their positive friction lowering effects. Hölzer and Sjögren [32] showed for mixtures of sodium chloride with 0.1% of different lubricants that the percentage

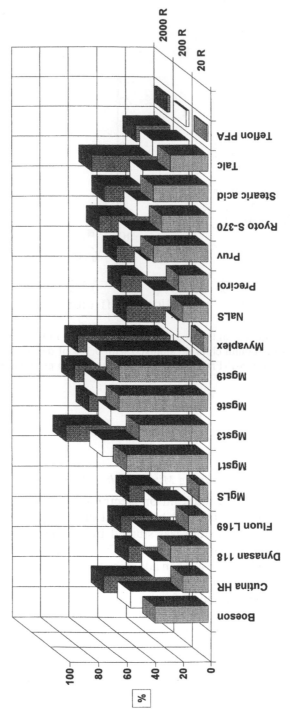

Figure 12 Effect of mixing time on boundary lubrication. Percentage reduction in friction coefficient at ejection (μ_2) for sodium chloride tablets containing 0.1% lubricant.

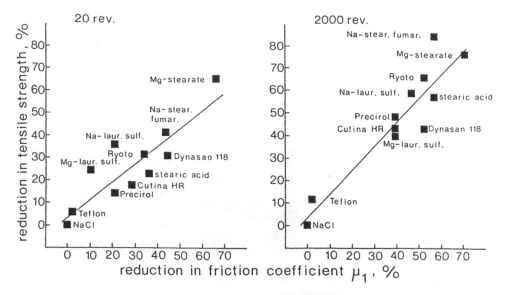

Figure 13 Reduction in tensile strength vs. reduction in friction coefficient (μ_1) for tablets of sodium chloride admixed with 0.1% of different lubricants for 20 and 2000 revolutions. (From Ref. 32; reproduced with permission of the copyright holder.)

reduction in tensile strength was related to the reduction in friction coefficient μ_1 (Fig. 13).

4. Power Consumption during Mixing with Lubricant

The influence of a solid lubricant on electric power consumption during mixing and the effect of addition of colloidal silica on particle friction of a directly compressible mixture was studied by Schrank-Junghäni and co-workers in a series of papers [65,71,72]. The electric power consumption was reduced from 80 to 40 W when admixing 0.2% magnesium stearate to a lactose granulation for about 1 min in a planetary mixer (see Fig. 14). Measuring the interparticulate friction allows comparison of the wall and internal friction and provides a classification of different lubricants [72].

For stearic acid higher concentrations and longer mixing times are required to reach the same effect on power consumption as for magnesium stearate mixing, because magnesium stearate forms the boundary layer on particulate solids much faster than stearic acid. This can, to some extent, be explained by differences in particle surface area. The results show that power consumption is a valuable tool for the determination of the mini-

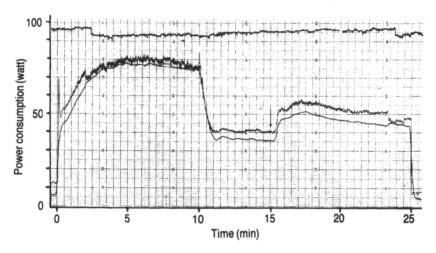

Figure 14 Plot of power consumption versus time for a directly compressible mixture with 0.2% magnesium stearate and 0.2% colloidal silica added 5 min later. (From Ref. 71; reproduced with permission of the copyright holder.)

mum concentration and mixing times required with lubricants, so that the negative effect on tablet properties can be minimized.

C. Effect of Lubricant Mixing Time on Adhesion

A more severe problem than friction is the adhesion of tablet material on machine parts and tooling, causing sticking and picking on punch and die surfaces. Often a much higher concentration of the lubricant is necessary to reduce the adhesion then to reduce the friction to a minimum level. Methods for studying the antiadhesive properties are (1) measuring the adhesion force at the die wall, (2) measuring the push off force from the lower punch surface and (3) by visual inspection of the punch surfaces [64]. Only a few studies describe the effect of mixing time with lubricant on the antiadhesive properties.

Hölzer and Sjögren [35] found that the adhesion-reducing effect of lubricants can not only be increased by an increasing concentration of the lubricant, but also by prolonged mixing. Arbitrary chosen scores for picking and sticking decreased for anhydrous lactose tablets produced with different concentrations magnesium stearate or sodium stearyl fumarate when the mixing time was increased from 20 to 2000 revolutions in a 2-L cubic mixer at 42 rpm. This effect was attributed to an increased surface coverage and film forming of the boundary lubricant on prolonged mixing [35].

VIII. EFFECT OF THIRD COMPONENTS ON THE FILM FORMATION OF LUBRICANTS

The film formation of lubricants during mixing can be influenced by third components. Simultaneous mixing of excipient particles with magnesium stearate and colloidal silica (Aerosil 200) can significantly suppress the negative effect of the lubricant on the bonding properties (Fig. 15). A larger effect was obtained when the host particles were blended with colloi-

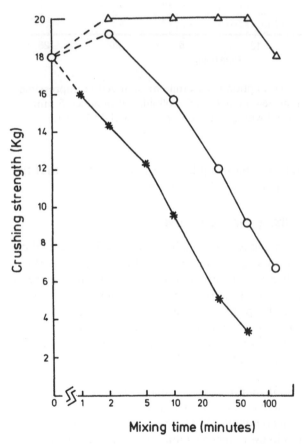

Figure 15 Effect of mixing time on the crushing strength of tablets compressed from blends of STARCH 1500 with magnesium stearate or with magnesium stearate and Aerosil 200. (✳) 0.1% magnesium stearate; (○) 0.1% magnesium stearate + 0.1% Aerosil 200; (△) 0.1% magnesium stearate + 0.4% Aerosil 200 (From Ref. 27; reproduced with permission of the copyright holder.)

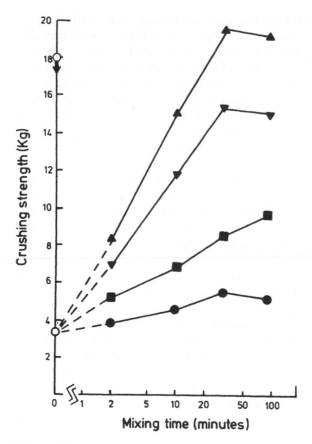

Figure 16 Effect of mixing with Aerosil 200 on the crushing strength of tablets compressed from previously mixed (1 h) blends of STARCH 1500 and 0.1%
magnesium stearate. (●) 0.1%; (■) 0.2%; (▼) 0.3%; (▲) 0.4% Aerosil 200.
(From Ref. 27; reproduced with permission of the copyright holder.)

dal silica prior to the addition of magnesium stearate [27,73]. The addition
of colloidal silica after previous mixing of the excipient with magnesium
stearate (Fig. 16) may even restore the bonding properties, when the ratio
between colloidal silica and magnesium stearate is 4 to 1 [27]. Even low,
commonly used concentrations of colloidal silica (0.2%) can suppress the
deleterious effect of 0.5% magnesium stearate on tablet bonding to a certain extent when the excipient particles were first mixed with the glidant
and consequently for a short time with the lubricant.

The interaction between magnesium stearate and colloidal silica has
been elucidated by the measurement of contact angles and electron micro-

probe analysis (EDAX) [17] and by dissolution measurements [17,19]. It was shown that a magnesium stearate film upon sodium chloride particles can be stripped from the substitute during mixing with fourfold the amount of colloidal silica under the formation of separate colloidal silica/magnesium stearate spheres. If colloidal silica is admixed together with the lubricant, the surface coverage with lubricant is reduced [19]. The effect of premixing of host particles with low concentrations of colloidal silica has been elucidated by competitive inhibition of magnesium stearate molecular layers at the adhesion sites which are occupied by colloidal silica particles [74].

The intensity of the interactions between the components of the ternary powder system microcrystalline cellulose, magnesium stearate, and colloidal silica have been predicted using literature and calculated values for their partial solubility parameters [75]. These show that there is a greater interaction between magnesium stearate and colloidal silica than between magnesium stearate and microcrystalline cellulose, although both interactions are greater than the cohesive interaction within the lubricant itself. Moreover, the interaction between colloidal silica and microcrystalline cellulose was found to be high, although less than the cohesive interactions within colloidal silica itself. From these results it can be predicted that in a ternary system the microcrystalline cellulose will be preferentially coated by colloidal silica and that the majority of the magnesium stearate will be enrobed by the colloidal silica. On compaction at high shear forces, shearing of the enrobed magnesium stearate will occur first, followed by shearing at the microcrystalline cellulose–colloidal silica interface. Although the model is something crude and oversimplified, the theoretical considerations are consistent with the previously found "protecting" effect of colloidal silica with respect to lubricant sensitivity [75].

The interaction between magnesium stearate and the glidant colloidal silica does not only effect the bonding properties of an excipient but also affects disintegration time [34], drug dissolution rate [17] and lubricating properties [34,19,15,71,76] Lerk et al. [27] showed that the ejection force of tablets, compressed from a compressible starch/lactose blend 1:1, lubricated with 0.5% magnesium stearate was not affected by the addition of 0.2% colloidal silica. However, when the concentration of colloidal silica is equal to that of magnesium stearate or even higher, the glidant will have a significant effect on the lubrication properties. Ragnarsson et al. [34] studied the effect of admixing colloidal silica on the tableting properties of sodium chloride, lubricated with magnesium stearate. Mixing time, sequence of mixing and the concentrations were varied (see Table 7). From the blends, tablets were compacted at 200 MPa maximum upper punch pressure. Mixing with 0.1% magnesium stearate and 0.5% colloidal silica simultaneously or mixing with 0.1% magnesium stearate after previous

Table 7 Properties of Tablets Compressed from Sodium Chloride, Sodium Chloride/Magnesium Stearate Blends, Sodium Chloride/Colloidal Silica Blends, and Sodium Chloride/Magnesium Stearate/Colloidal Silica Blends, Respectively

Magnesium stearate (%)	Colloidal silica (%)	Ejection force (kN cm^{-2})	Porosity (%)	Tensile strength (MPa)	Disintegration time (min)
—	—	1.31	5.0	1.3	4.2
0.1[a]	—	0.18	3.6	0.5	8.8
0.5[a]	—	0.09	3.5	<0.1	19.0
—	0.5[a]	1.28	6.1	1.7	5.7
0.1[a]	0.5[a]	0.88	4.9	1.2	8.1
0.5[a]	0.5[a]	0.27	4.2	0.6	19.5
0.1[a]	0.5[b]	0.43	4.8	0.6	9.0
0.1[a]	0.5[c]	0.43	4.8	0.6	8.6
0.1[b]	0.5[a]	0.71	4.8	1.2	7.7

[a] Mixed with sodium chloride, 25 min.
[b] Admixed to mixture a, another 25 min.
[c] Admixed to mixture a, another 50 min.
Source: From Ref. 34; reproduced with permission of the copyright holder.

mixing with 0.5% colloidal silica restored the bonding properties, but increased at the same the ejection force as compared with tablets containing 0.1% magnesium stearate only. Mixing with 0.5% colloidal silica after previous mixing with 0.1% magnesium stearate or mixing with 0.5% magnesium stearate and 0.5% colloidal silica simultaneously gave rather low ejection forces but could not restore the bonding properties. This means that the positive effect on strength may be counteracted by a negative effect on friction lowering properties. Similar results were reported by Schrank-Junghäni et al. [71] using power consumption measurements (see Fig. 14). Addition of 0.2% colloidal silica to a direct compression blend, premixed for 5 min with 0.2% magnesium stearate results in an increase in power consumption from 40 up to 55 W. This indicates increased interparticle friction and the friction toward the mixer vessel walls.

The addition of colloidal silica on both surface coverage with magnesium stearate and tablet ejection force of an acetylsalicylic acid granulation, lubricated with magnesium stearate was investigated by Johansson and Nicklasson [15,19]. A flow-through technique was used for calculating the surface coverage of the lubricant. When 1% colloidal silica was admixed after previous mixing with 1% magnesium stearate the surface coverage was not affected, but the ejection force/area increased twice. If 1% colloidal silica and 1% magnesium stearate were mixed simultaneously with the granulation, both surface coverage and ejection force/area were affected. The authors concluded that colloidal silica imparts primarily with the free fraction of magnesium stearate which is partly withdrawn from further coverage of the base material as well as from the lubrication of the die wall. Moreover, it was concluded that the lubricating effect of magnesium stearate is not directly related to the coverage on the surface of the base material.

In contrast to colloidal silica, little has been reported on the interaction between talc and magnesium stearate. In a series of papers Staniforth and Ahmed [73,76,77] showed that, in contrast to colloidal silica, 2% talc cannot restore the bonding properties of microcrystalline cellulose, mixed with 0.5% magnesium stearate (Fig. 17). A preblend of talc and magnesium stearate caused even a larger decrease of the work of failure than magnesium stearate alone. It was concluded that talc becomes coated with magnesium stearate particles, which promotes dispersion of the lubricant, whereas colloidal silica itself coats magnesium stearate particles, preventing the formation of a lubricant film on the base material [73]. The ejection force of the tablets, which fell from 700 to 275 N after the addition of 0.5% magnesium stearate, increased to values between 350 and 420 N when 2% talc or 2% colloidal silica were present as a third component. These results indicate that components such as talc and colloidal silica can modify the influence of magnesium stearate on lubrication and compactibility independently [76].

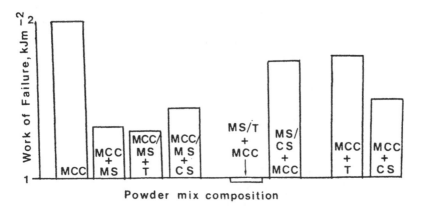

Figure 17   Relationship between powder mixing conditions and tablet work of failure. MCC = microcrystalline cellulose; MS = magnesium stearate; T = talc; CS = colloidal silica. (From Ref. 73; reproduced with permission of the copyright holder.)

Lerk and Sucker [23,24] showed that the addition of small amounts of talc forces magnesium stearate out of cavities of a material during the mixing process and hence promotes film formation of magnesium stearate. As the amount of magnesium stearate consumed for film formation is no longer available in the asperities to reduce the friction during the ejection of a compact, a decreased lubrication efficiency was found. Consequently, the interaction between talc and magnesium stearate is deleterious for both bonding properties and lubrication efficiency.

IX. LIMITING LUBRICANT SENSITIVITY

There are different ways to limit the deteriorating effect of magnesium stearate on tablet properties, without affecting the lubrication properties to a large extent.

1. Undoubtedly the best method is *omitting a lubricant* in a tablet formulation and applying alternative lubrication methods, mostly involving modifications to tablet machines. Recently reported techniques are the addition of exact amount of a suitable lubricant directly onto the punch and die surfaces by electrostatic deposition [77] or by the PKB method (Presskammerbeschichtung) immediately after tablet ejection [78]. Other techniques are using bonded PTFE die linings or dies made from steel with lubricant inclusions [79] or cycling compressions of a carrier formulation first containing a lubricant, to create a lubricant film on the die wall, followed by compression of the unlubricated formulation [80]. All these methods have been received with limited success or are extremely expensive.

2. The lubricant concentration should not be higher than is strictly necessary. The *minimum* concentration required with lubricants can be determined by means of power consumption measurement [65,71,72]. Magnesium stearate reduces the electrical power consumption of a mixer, even when used in very small concentrations. The optimum concentration can also be calculated by means of external friction measurements [81] or based on surface area of both lubricant and host particles [82]. Hölzer and Sjögren found for anhydrous β-lactose that the surface ratio between the lubricant and the host particles should be greater than 10–15% [83].

3. The choice of *alternative lubricants* seems an attractive solution of the problem of the deleterious effect of magnesium stearate on the bonding properties of tablet excipients. Several comparative evaluations have been carried out between such materials and magnesium stearate, considering both lubrication efficiency and effect on tablet properties [33,84,35, 29,32,85]. It has been demonstrated for different stearates, hydrogenated vegetable oils, glycerides, sugar esters, PTFE, talc and graphite that the

time-dependent reduction in tablet crushing strength is coupled with the lubrication efficiency (see also Fig. 13) [29,32]. Generally, magnesium stearate and calcium stearate gave the lowest ejection force or the highest reduction in friction coefficient μ_1, but caused the largest reduction in tablet strength with an increase in mixing time. Good overall properties were found for glycerides such as glyceryl palmitostearate (Précirol) and Boeson VP [33,84,29], hydrogenated vegetable oils e.g., hydrogenated cotton seed oil (Sterotex) [84,29], hydrogenated vegetable oil (Lubritab) in combination with an antiadherent [85], and stearic acid [33]. All these alternative lubricants are effective in concentrations of 1% or more.

Sodium stearyl fumarate (Pruv) has been suggested as a suitable lubricant in tableting. It has been claimed not to have the disadvantages of magnesium stearate in respect of tablet properties under which tablet strength [86]. Later it has been shown that sodium stearyl fumarate reduced the friction and the adhesion to about the same degree as magnesium stearate but had also about the same influence on tablet strength and disintegration [35]. In some formulations, however, sodium stearyl fumarate appears to be less sensitive to processing variables and less hydrophobic than magnesium stearate [70]. In formulations where magnesium stearate causes problems, sodium stearyl fumarate may be a good alternative [83].

The results from the experiments with sodium chloride tablets (Table 3, Figs. 4, 11, and 12), completed with effect on disintegration time (Fig. 18) can serve as a directive for the choice of alternative lubricants. The negative effects of lubricant on tablet tensile strength and disintegration time are well correlated to the positive effect on lubrication. However, the negative properties are more pronounced when the mixing time is increased. It can be seen that the lauryl sulfates had more effect on tablet strength than on disintegration time; on the other hand, the triglycerides increased disintegration time more than they reduced tablet strength. For adhesive excipients such as anhydrous β-lactose, the antiadhesive properties of lubricants should be taken into account too. Good antiadherents are magnesium stearate, magnesium lauryl sulfate, sodium stearyl fumarate (Pruv) and talc in combination with magnesium stearate.

4. Vezin et al. [87] showed for microcrystalline cellulose that loss of tablet strength arising from lubricant *overmixing* can be substantially reduced by careful independent adjustment of main and precompression forces. This effect was attributed to the greater separation in time of the two distinct compaction events, providing a longer interval in which time-dependent effects such as stress relaxation and escape of air may occur.

5. Another possibility is *changing the mixing procedure*. Drugs and excipients should be mixed without a lubricant first. After the addition of a lubricant, the mixing should be continued for a short period. It has been

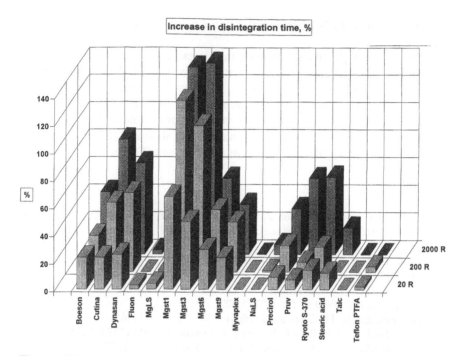

Figure 18 Effect of mixing time on boundary lubrication. Percentage increase in disintegration time for sodium chloride tablets containing 0.1% lubricant.

demonstrated that a short mixing time, resulting in a poor distribution of magnesium stearate, did not impair its lubricating efficiency [34]. A similar effect as a short mixing time can be reached by using a coarser magnesium stearate or magnesium stearate granules, as proposed by Johansson [37–40].

6. The lubricant sensitivity can also be decreased by a proper *choice of tablet excipients*. As mentioned, the effect of magnesium stearate depends on the nature of the excipient used. The largest reduction of crushing strength can be expected for tablets containing starch or cellulose products (see Fig. 2), the smallest effects will be found for tablets containing brittle materials such as dicalcium phosphate dihydrate or anhydrous β-lactose (see Fig. 5). It should be realized, however, that most tablet formulations are mixtures of materials with very different compaction properties and in practice it is difficult to anticipate the compressional characteristics of a tableting mixture. For example, dicalcium phosphate dihydrate which is regarded as a brittle material can behave as a plastic material when it is

mixed with only 10% of microcrystalline cellulose and 1% magnesium stearate [84].

7. Another possibility to limit the deleterious effect of magnesium stearate on bonding properties is premixing with *colloidal silica* (Aerosil 200). The glidant Aerosil can delay the film formation of magnesium stearate when tablet ingredients are mixed with Aerosil first and then with magnesium stearate. When the concentration of colloidal silica is low as compared with the concentration of magnesium stearate, the effect on the lubrication efficiency of magnesium stearate is limited [27].

REFERENCES

1. E. Shotton, and C. J. Lewis, *J. Pharm. Pharmacol. 16* (Suppl.): 111T (1964).
2. W. A. Strickland, E. Nelson, L. W. Busse, and T. Higuchi, *J. Am. Pharm. Assoc. Sci. Ed. 45*: 51 (1956).
3. G. K. Bolhuis, C. F. Lerk, H. T. Zijlstra, and A. H. De Boer, *Pharm. Weekbl. 110*: 317 (1975).
4. G. K. Bolhuis, G. Reichman, C. F. Lerk, H. V. Van Kamp, and K. Zuurman, *Drug Dev. Ind. Pharm. 11*: 1657 (1985).
5. G. Levy, and R. H. Gumtow, *J. Pharm. Sci. 52*: 1139 (1963).
6. A. C. Shah, and A. R. Mlodozeniec, *J. Pharm. Sci. 66*: 1377 (1977).
7. K. S. Murthy, and J. C. Samyn, *J. Pharm. Sci. 66*: 1215 (1977).
8. G. K. Bolhuis, A. J. Smallenbroek, and C. F. Lerk, *J. Pharm. Sci. 70*: 1328 (1981).
9. C. F. Lerk, G. K. Bolhuis, A. J. Smallenbroek, and K. Zuurman, *Pharm. Acta Helv. 57*: 282 (1982).
10. T. A. Iranloye, and E. L. Parrott, *J. Pharm. Sci. 67*: 535 (1978).
11. A. Soininen, and P. Kuusivuori, *Acta Pharm. Fenn. 89*: 215 (1980).
12. A. H. De Boer, G. K. Bolhuis, and C. F. Lerk, *Powder Techn. 20*: 394 (1978).
13. B. W. Müller, *Pharm. Ind. 38*: 394 (1976).
14. J. N. Staniforth, *J. Pharm. Pharmacol. 39*: 329 (1987).
15. M. E. Johansson, and M. Nicklasson, *J. Pharm. Pharmacol. 38*: 51 (1986).
16. M. S. H. Hussain, P. York, and P. Timmins, *Int. J. Pharm. 42*: 89 (1988).
17. C. F. Lerk, and G. K. Bolhuis, *Pharm. Acta Helv. 52*: 39 (1977).
18. M. Nicklasson, and A. Brodin, *Acta Pharm. Suec. 19*: 99 (1982).
19. M. E. Johansson and M. Nicklasson, *Pharmaceutical Technology: Tableting Technology*, Vol. 1 (M. H. Rubinstein, ed.), Ellis Horwood, Chichester, 1987, p. 43.
20. R. C. Rowe, *Acta Pharm. Suec. 20*: 77 (1983).
21. K. Pintye-Hódi, I. Tóth, and M. Kata, *Pharm. Acta Helv. 56*: 320 (1981).
22. L. Roblot-Treupel, and F. Puisieux, *Int. J. Pharm. 31*: 131 (1986).
23. P. C. Lerk, and H. Sucker, *Acta Pharm. Technol. 34*: 68 (1988).
24. P. C. Lerk, and H. Sucker, *Acta Pharm. Technol. 34*: 72 (1988).
25. M. C. Davies, A. Brown, and J. M. Newton, *J. Pharm. Pharmacol. 39* (Suppl.): 122P (1987).

26. M. S. H. Hussain, P. York, P. Timmins, and P. Humphrey, *Powder Technol.* *60*: 39 (1990).
27. C. F. Lerk, G. K. Bolhuis, and S. S. Smedema, *Pharm. Acta Helv.* *52*: 33 (1977).
28. R. C. Rowe, *Int. J. Pharm.* *41*: 223 (1988).
29. G. K. Bolhuis, C. F. Lerk, and P. Broersma, *Drug Dev. Ind. Pharm.* *6*: 15 (1980).
30. O. Alpar, J. J. Deer, J. A. Hersey, and E. Shotton, *J. Pharm. Pharmacol. 21* (Suppl.): 6S (1969).
31. P. J. Jarosz, and E. L. Parrott, *Drug Dev. Ind. Pharm.* *10*: 259 (1984).
32. A. W. Hölzer, and J. Sjögren, *Acta Pharm. Suec. 18*: 139 (1981).
33. L. Delattre, J. Gillard, F. Jaminet, and M. Roland, *J. Pharm. Belg. 31*: 497 (1976).
34. G. Ragnarsson, A. W. Hölzer, and J. Sjögren, *Int. J. Pharm. 3*: 127 (1979).
35. A. W. Hölzer, and J. Sjögren, *Int J. Pharm. 2*: 145 (1979).
36. A. W. Hölzer, *Labo-Pharma Probl. Techn. 32* (338): 28 (1984).
37. M. E. Johansson, *Int. J. Pharm. 21*: 307 (1984).
38. M. E. Johansson, *Acta Pharm. Suec. 22*: 343 (1985).
39. M. E. Johansson, *J. Pharm. Pharmacol. 37*: 681 (1985).
40. M. E. Johansson, *Acta Pharm. Technol. 32*: 39 (1986).
41. C. Frattini, and L. Simioni, *Drug Dev. Ind. Pharm. 10*: 1117 (1984).
42. I. Colombo, and F. Carli, *Il Farmaco Ed. Pr. 39*: 329 (1984).
43. B. W. Müller, *Proc. 1st Int. Conf. Pharm. Techn*, Vol. IV, Paris, pp. 134–141 (1977).
44. T. A. Miller, and P. York, *Int. J. Pharm. 23*: 55 (1985).
45. T. A. Miller, and P. York, *Powder Technol. 44*: 219 (1985).
46. T. A. Miller, P. York, and T. M. Jones, *J. Pharm. Pharmacol. 35*: 42P (1983).
47. K. J. Steffens, Die physikalischen Eigenschaften von Magnesiumstearat und ihr Einfluss auf das tribologische Verhalten bei der Tablettierung, Thesis, Marburg, 1978.
48. G. K. Bolhuis, *Enkele aspecten van de formulering en bereiding van tabletten met direct comprimeerbare vulbindmiddelen*, Krips Repro, Meppel, 1978, p. 76.
49. J. G. Van der Watt, *Int. J. Pharm. 36*: 51 (1987).
50. C. E. Bos, G. K. Bolhuis, H. Van Doorne, and C. F. Lerk, *Pharm. Weekbl. Sci. Ed. 9*: 274 (1987).
51. H. Vromans, G. K. Bolhuis, and C. F. Lerk, *Powder Technol. 54*: 39 (1988).
52. C. E. Bos, H. Vromans, and C. F. Lerk, *Int. J. Pharm. 67*: 39 (1991).
53. H. Egermann, *Sci. Pharm. 46*: 137 (1978).
54. K. A. Riepma, H. Vromans, and K. Zuurman, *Int. J. Pharm. 97*: 195 (1993).
55. M. Duberg, and C. Nyström, *Acta Pharm. Suec. 19*: 421 (1982).
56. J. Bossert, and A. Stamm, *Drug Dev. Ind. Pharm. 6*: 573 (1980).
57. L. Roblot, F. Puisieux, and D. Duchêne, *Labo-Pharma Probl. Techn.* 31: 843 (1983).
58. K. Malmqvist, and C. Nyström, *Acta Pharm. Suec. 21*: 21 (1984).
59. G. K. Bolhuis, S. W. De Jong, H. V. Van Kamp, and H. Dettmers, *Pharm. Technol. 11*(3): 36 (1987).

60. W. A. Strickland, T. Higuchi, and L. W. Busse, *J. Am. Pharm. Ass.*, *Sci. Ed.* *49*: 35 (1960).

61. F. P. Bowden and D. D. Tabor, *Friction and Lubrication of Solids.* Paperback edition. Clarendon Press, Oxford, 1986.

62. B. W. Müller, K. J. Steffens, and P. H. List, *Pharm Ind.* *44*: 636 (1982).

63. A. W. Hölzer, and J. Sjögren, *Int. J. Pharm.* *7*: 269 (1981).

64. T. A. Miller, and P. York, *Int. J. Pharm.* *41*: 1 (1988).

65. H. Junghäni, H. P. Bier, and H. Sucker, *Pharm Ind.* *43*: 1015 (1981).

66. A. W. Hölzer, and J. Sjögren, *Drug Dev. Ind. Pharm.* *3*: 23 (1977).

67. J. Kikuta, and N. Kitamori, *Powder Technol.* *35*: 195 (1983).

68. J. Kikuta, and N. Kitamori, *Drug Dev. Ind. Pharm.* *11*: 845 (1985).

69. A. W. Hölzer, and J. Sjögren, *Acta Pharm. Suec.* *15*: 59 (1978).

70. A. W. Hölzer, *Acta Pharm. Suec.* *24*: 71 (1987).

71. H. Schrank-Junghäni, H. P. Bier, and H. Sucker, *Pharm. Technol.* *7(9)*: 71 (1983).

72. H. Schrank-Junghäni, H. P. Bier, and H. Sucker, *Acta Pharm. Techn.* *30*: 224 (1984).

73. J. N. Staniforth, and H. A. Ahmed, *J. Pharm.* *38* (Suppl.): 50P (1986).

74. G. K. Bolhuis, and C. F. Lerk, *J. Pharm. Pharmacol.* *33*: 790 (1981).

75. R. C. Rowe, *Int. J. Pharm.* *45*: 259 (1988).

76. J. N. Staniforth, and H. A. Ahmed, *J. Pharm. Pharmacol.* *39* (Suppl.): 68P (1987).

77. J. N. Staniforth, S. Cryer, H. A. Ahmed, and S. P. Davies, *Drug Dev. Ind. Pharm.* *15*: 2265 (1989).

78. P. Gruber, V. I. Gläsel, W. Klingelhöller, und Th. Liske, *Pharm. Ind.* *50*: 839 (1988).

79. J. A. Hersey, *Aus. J. Pharm. Sci.* *1*: 76 (1972).

80. B. J. Lael, H. Irvington, and P. J. Pinto, U.S. Pat. 3.042.531 (1962).

81. Y. Fukumori, and J. T. Carstensen, *Int. J. Pharm. Tech. Prod. Mfr.* *4(4)*: 1 (1983).

82. F. J. Bavitz, and P. K. Shiromani, *Drug Dev. Ind. Pharm.* *12*: 2481 (1986).

83. A. W. Hölzer, *Acta Pharm. Suec.* *18*: 72 (1981).

84. A. Stamm, A. Kleinknecht, and D. Bobbé, *Labo-Pharma Probl. Techn.* *25*: 215 (1977).

85. J. N. Staniforth, *Drug Dev. Ind. Pharm.* *13*: 1141 (1987).

86. N. O. Lindberg, *Acta Pharm. Suec.* *9*: 207 (1972).

87. W. R. Vezin, K. A. Khan, and H. M. Pang, *J. Pharm. Pharmacol.* *35*: 555 (1983).

17

The Development and Optimization of Tablet Formulations Using Mathematical Methods

Fridrun Podczeck

School of Pharmacy, University of London, London, United Kingdom

I. INTRODUCTION

The degree of development in the area of natural sciences is characterized by the ability to integrate theoretical principles in the practical work, and to assess and predict the results of a manual process [1]. The dosage form development in the classical trial-and-error procedure does not reflect the degree of development in the field of natural sciences and neglects the possibilities which the modern computer technique offers to a research pharmacist. A dosage form development using trial and error can be characterized as gambling [2], which never guarantees identification of the true optimum of the formulation conditions, and a rather second-class formulation conditions, and a rather second-class formulation will be provided as an optimal drug product.

   The development of tablet formulations, which release the incorporated drug substance(s) in a way that allows an optimal effect, is still a tedious and complicated process. The application of computer science may help if the research pharmacist was able to model the problems and dominant laws, which are related to the problems of development, comprehensively and to implement development strategies based on computer techniques. In this way, the actual work could be reduced to the creative part of finding an optimal tablet formulation. The understanding of the complicated relationships between tablet manufacture and formulation and the evaluation of the information given based on computer applications leads to a more scientific way of dosage form development. A series of possible formulation principles can be created in a comparatively short time, and their objective comparison provides a straightforward optimal solution.

There are many different ways of applying mathematical methods in tablet formulation discussed in the literature, and the following notes review some of them with the aim to show the improvement of the methods used from the early 1970s to the current day.

II. UNIVARIATE ANALYSIS OF VARIANCE (ANOVA) WITH MORE THAN ONE INDEPENDENT VARIABLE (FACTORIAL DESIGN)

Simple ANOVA is a parametric test of significance to compare mean values of one dependent variable in more than two sets of measurements. These measurements are differentiated by their systematic different appearance of at least one independent variable [3]. Application requires a normal distribution of the data in each data set, where the variances of each set are not significant different [4].

Three estimates of variance are needed to test whether the mean values of each data set are similar or significant different [5]:

1. The total variance, i.e., the estimate of the variance of all data without considering the grouping of them.
2. The variance in the sets, which represents the average estimate based on the deviations of all single measurements from the related mean value of the data set.
3. The variance between the sets, which is an estimate based on the mean values of the sets.

If the sets of data are similar, then all three calculations represent an estimate of the variance of the entire population. Otherwise, the three estimates reflect the influence of the factors chosen to group the data into the sets in different ways. Hence, they are indicators of the variation of the independent variable and the measuring errors, which will be used in the test of significance [3].

Having more than only one independent variable, e.g., in a factorial design, the ANOVA gives a test of significance for each independent variable and for their interactions. In this respect, a special calculation technique of ANOVA was presented by Yates [6], which allows the evaluation of interaction tables to provide a tool for their interpretation. More often, computerized ANOVA procedures are used [7].

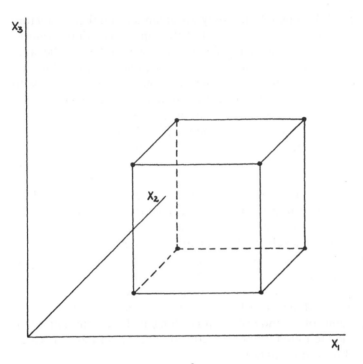

Figure 1 Factorial space of a 2^3-design (eight experiments). X_i = independent variable (factor); ●, experimental point.

There are different ways to set up the experimental plan for a factorial design [8], and the ways of analyzing the data depend on these experimental details. In the simplest case, two or more independent variables (factors n) are tested at different levels (f). For a so-called fully factorial design the number of experiments equals f^n. Figure 1 shows the factorial space covered by a design of three independent variables at two levels (2^3-design = 8 experiments). The number of levels of each factor can raise the number of experiments into an excessive range. For example, a 3^2-design involves 9 experiments, and a 4^2-design consists of 16 experiments. The addition of one factor (3^3- and 4^3-designs) leads to 27 and 256 experiments respectively. Hence, only designs with 2 (e.g., [9–12]) or a maximum of three levels (e.g., [13–15]) are commonly used to minimize the number of experiments. The number of factors mainly varies between 2 and 4, but extraordinary large designs of 6 factors [16–18] are also reported.

In the first instance it appears that the factorial design combines a series of useful properties without having major related disadvantages.

Theoretically, it will be possible to study an unlimited number of factors and their interactions on the same object at the same time. Furthermore, the calculation procedure is relatively simple, and even in the unbelievable case of not having a computer or statistical software package available, the calculations could be done manually. The problems, however, arise with the interpretation of the results. The explanation and the balancing of the contents of the different main and interaction effects, which are "statistically significant," are not always easily identified. Often, main effects and interaction terms are discussed separately, as if they gave different information. If, however, interactions of factors are significant, then the factors themselves have no real main effect, even if the ANOVA certifies them as significant. The degree of the effectivity of the factors always depends in such a case on the quantitative value of the accompanying factors. This is often not considered in the conclusions drawn.

Designs using two levels imply a linear dependence of the response variable on the factors tested. An increase to three levels may also allow the handling of quadratic relationships, but to fit the results to an exact nonlinear function, five or more levels would be needed. Nevertheless, many authors have used the data of a 3^n-factorial design to calculate regression model equations of second order, and as long as these equations were used carefully, and their limitations were considered, this strategy provided a useful tool in dosage formulation.

Having three levels of each factor increases the number of experiments of a fully factorial design tremendously. Hence, special designs were developed, which reduce the number of experiments to a considerable number, but still provide three points for each factor included. Such a design is the "central composite design" (CCD) [19]. The number of experiments will be calculated as follows: for a number of n factors an ordinary 2^n design has to be created. A center point experiment will be added, which sets the level of each factor at the center of the related geometrical figure. A number of $2n$ star point experiments follows. Geometrically, the star points are positioned on the coordinate axes of the factorial design space in a distance of α to the center point. Hence, the total number of experiments in a CCD is $2^n + 2n + 1$ (see Fig. 2). The CCD is not as stiff as the ordinary factorial design, because many additional points could be added along the coordinate axes. For an ordinary factorial design, the levels of the factors have to be fixed in advance, and some preliminary experiments are needed to guarantee, that the factorial space covers a range that includes the final dosage form. By adding CCD points, however, small corrections of the model space are possible, if the final dosage form point is not too far outside the basic factorial space.

Another popular method in dosage form design is the "simplex lattice design" (SLD) [20,21]. The factorial space described is an equilateral trian-

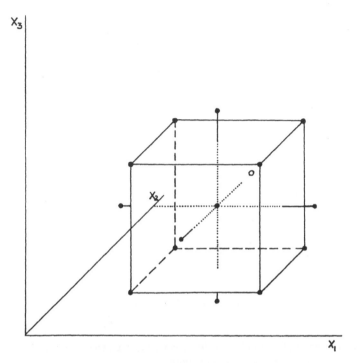

Figure 2 Central composite design with three factors (number of experiments = $2^3 + 2 \cdot 3 + 1 = 15$). X_i = independent variable (factor); •, experimental point.

gle (see Fig. 3). Having three excipients in a formulation, the proportions always add up to unity. Hence, there are three edge points to study (only one excipient used), three surface points have to be included, where always only two excipient are used (commonly in a half-to-half concentration), and finally a mid point of a mixture of all three components at a concentration of a third each completes the test set. The SLD is a very rough design and should only be used as a preliminary study.

In tabletting, factorial design is an established method for many years. Fell and Newton [22] investigated the production conditions of spray-dried lactose in terms of the compaction properties of the spray products as early as 1971. Moldenhauer et al. [23–25] have used this technique in 1978 to investigate the use of hydrophilic matrix components for prolonged-release tablets. Podczeck [26] studied the effect of excipients and tabletting pressure in basic tabletting mixtures in 1984. In 1988 Zubair et al. [27] used factorial design to quantify the influence of filler quality and quantity in tabletting of paracetamol. Even in this decade, factorial design

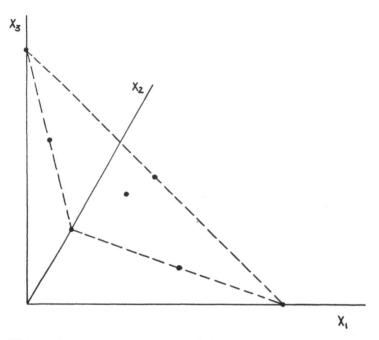

Figure 3 Factorial space of a simplex-lattice design (seven experiments). X_i = independent variable (factor); •, experimental point.

was used to investigate the influence of excipients in tablet and capsule formulation [28]. A CCD design to describe the influence of maize starch and tabletting pressure has been used by Leuenberger [29], who also reports the use of a CCD with five factors to optimize another tablet formulation [29]. González [30] set up a SLD to calculate the effect of three different excipients on the tablet strength of a formulation.

For the purpose of developing a computer-aided dosage form design of tablet formulations using multivariate statistics and linear optimization procedures, Podczeck [31] deducted a "Center of gravity design" (CGD), which keeps the advantages of the CCD, but reduces the number of experiments further. The experiments start with a midpoint, which should be in the expected factorial space of the final tablet formulation. From this midpoint ("the center of gravity"), at least four points have to be placed along each coordinate axis in a way, that the resulting geometric space becomes as large as possible, but includes only meaningful experiments (total number of experiments: $4n + 1$). For example, the Aerosil content of a direct tabletting formulation will rarely exceed 3%, because the bulk volume of a higher content in a mixture is too large to fit into the limits of a die space of a tablet machine. Hence, the center of gravity of Aerosil could be 1.5%, and the four points along the related coordinate axis could be 0.5%, 1.0%, 2.0%, and 2.5%. Table 1 and Fig. 4 show a CGD for a direct tabletting mixture. If some of the coordinate points were not placed far enough from the center of gravity, further points can be added at any time and at any position of the geometric space created. In the first instance, the basic design assumes that the effect of a factor is similar at any level of the other factors. If the multivariate analysis of the data indicates an interaction between two factors, a second or even third factor axis can be placed at different levels of the related factor by parallel distortion of the basic axis along the related factor axis, which allows a check as to the validity of the basic assumption. Because only relevant interations are studied, the number of experiments is still far lower than for a fully factorial design.

In any case, the design leads to a regression model, which can be used for simulation or optimization purposes. Podczeck and Wenzel [32–34] have used two CGDs to demonstrate the use of multivariate statistics in describing the influence of formulation factors on powder and tablet proper-

Table 1 Center of Gravity Design for a Direct Tabletting Problem

No.	A [%]	S [%]	M [%]	P [%]
1	0.5	10.0	0.3	200
2	1.0	10.0	0.3	200
3G	**1.5**	**10.0**	**0.3**	**200**
4	2.0	10.0	0.3	200
5	2.5	10.0	0.3	200
6	1.5	0.0	0.3	200
7	1.5	5.0	0.3	200
8	1.5	15.0	0.3	200
9	1.5	20.0	0.3	200
10	1.5	10.0	0.1	200
11	1.5	10.0	0.2	200
12	1.5	10.0	0.4	200
13	1.5	10.0	0.5	200
14	1.5	10.0	0.3	50
15	1.5	10.0	0.3	100
16	1.5	10.0	0.3	300
17	1.5	10.0	0.3	350

A = aerosil; S = starch; M = magnesium stearate; P = tabletting pressure; G = center of gravity.

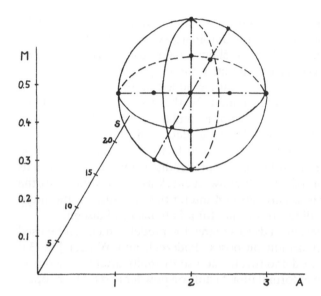

Figure 4 Factorial space of a center of gravity design (values given in Table 1 except factor 4 "tabletting pressure"). A = aerosil content [%]; S = starch content [%]; M = magnesium stearate content [%]; •, experimental point.

ties of direct tabletting basic mixtures, finally calculating a multivariate regression model.

III. MULTIVARIATE ANALYSIS OF VARIANCE (MANOVA) AND DISCRIMINANT ANALYSIS

The application of ANOVA, allows the influence of formulation factors on only one dependent variable to be measured. The experimental factors, however, predict normally a series of dosage form properties, e.g., in tabletting the tensile strength and the disintegration at the same time. In many of the papers mentioned previously, therefore, the ANOVA was repeated several times, until each formulation property was studied. The relationships between these properties are neglected, although it is well known, that, e.g., tensile strength and tablet disintegration are related properties.

The MANOVA approach draws attention to the simultaneous observation of several dosage form properties, without the need of a general change of the statistical designs described above. The MANOVA provides a complete overview of the relationships and dependencies, which are included in the data. It should be pointed out that the results of a MANOVA cannot be attributed to the results of single ANOVA tests. It may happen that dosage form properties, which appeared unrelated to the factors in the ANOVA test, include very important information for the MANOVA. Other properties, which appeared greatly effected by ANOVA approach, suddenly lose their importance, mainly because the information included is similar to other properties, which stay in the model [35].

Basically, the data now have to be multinormal distributed, where the multinormal distribution is the probability distribution of more than one normal distributed variable. The arithmetic mean of the univariate statistics is replaced by the mean vector, and the variance becomes a variance-covariance matrix because of the combination of the variables. The principle of splitting the total variance into parts as in the ANOVA test is kept, but the analysis deals with matrices instead of single numbers [3].

Multivariate analyses have their own multivariate test criteria, which are mainly Hotelling's T^2-test [36] and Wilks's Λ-test [37,38]. Unfortunately, there is no exact zero distribution, except for some special cases. Hence, tests of significance are only possible by approximation of the multivariate test criterion usually onto the F distribution. An approximation on χ^2 is only useful for a large number of observations and a wide set of dependent variables [39].

The calculation of multivariate test criteria allows the separation of important dosage form properties from those properties which do not contribute to the information hidden in the data material. Multivariate test procedures select the optimal set of variables normally by maximization of the test criterion [35]. Sometimes, a manufacturing process cannot be described with data, which are measured on the same object, alone. In tabletting, for example, the flow properties of the powder blend are as important as the properties of the tablets, but they cannot be measured on the same object. Hence, either an independent MANOVA has to be undertaken for both groups of properties, or it must be proved, whether or not the two groups can be combined. The complete tabletting process can be optimized with one mathematical model only for a combination. The MANOVA offers a "test of affinity" in such a case. Using the different test criteria, a coefficient of affinity can be calculated, which allows the judgment as to whether or not a combination of the properties is possible. The user of the MANOVA, however, must guarantee that such a combination is meaningful, because the statistical test is a pure numerical procedure, which detects relationships of numerical values and not of implications of properties measured.

The use of the MANOVA approach provides a lot of useful information about the data structure and the influence of the formulation factors on the dosage form behavior. However, it is lacking in terms of its presentation, and an understanding of the results may be very difficult. Hence, MANOVA tests should be combined with discriminant analysis.

Discriminant analysis consists of a set of techniques, which are designed to describe and to group individuals, which are characterized by a large number of variables [40]. The discriminant analysis transfers the extracted effects and relationships from the basic multidimensional space into a smaller dimensional space, which makes the interpretation of the results easier. The mathematical procedures are divided into a linear and a nonlinear discriminant analysis. The advantage of the nonlinear procedure is that nonlinear discriminant variables are uncorrelated, and the first variable includes the most important information of the data. Any further discriminant variable has less importance compared with its predecessor. Hence, the first two discriminant variables are often used to create two-dimensional drawings, the "canonical drawings," which enhance the relationships and differences of a grouped data material. The influence of different factors on the multivariate properties of, e.g., a formulation problem becomes "visible."

Both MANOVA approach and the discriminant analysis are described in detail by Anderson [41] and by Morrison [42].

Podczeck [31] has used MANOVA/discriminant analysis (nonlinear) to investigate several CGDs for direct tabletting mixtures. The simplest CGD comprised an analysis of the effect of the concentration of starch (0, 5, 10, 15, 20, 40%) and talc (0, 1, 2, 3, 5, 10%) in a tabletting mixture. Three independent sets of data were measured to characterize as much of the tabletting process as possible: flow properties of the dry powder blends, force-displacement parameters of the mixtures, and properties of the tablets. The MANOVA answered three key questions:

1. The dependent variables (e.g., tablet properties) in the final data set should not correlate too well. High correlations make it impossible to change one property without a proportional change in the other correlated property. Hence, an optimization of the formulation would become critical.
2. From a theoretical point of view, the three sets of properties measured were not obtained on the same object, and hence it should be tested whether or not a combination of the groups is possible.
3. The use of optimization techniques, which was the aim of the design, will only be possible if the different experiments of the design mainly lead to statistically different results. In other words, the levels chosen for each factor should affect the tablet properties significantly.

From the initial set of 20 formulation properties, the MANOVA selected 10 as important. This remaining set of properties did not consist of high correlated variables, even if the diametral crushing strength, the friability, and the disintegration time were still included. The correlation matrix did not indicate any significant correlation. Other important properties selected were for example the mean powder flow rate and the packing density of the tablets. A comparison of the multivariate test criteria of the complete set of data and the data for each independent group of dosage form properties (test of affinity) indicated, that a combination of all properties regardless of their observation source (powder, tablet) was possible. The final multivariate test criterion was highly significant (approx. $F = 10.58; f_1 = \infty; f_2 = 101; F_{\infty,100,0.05} = 1.28$). The experiments of the CGD lead to statistically different results, and the influence of the factors studied could be measured numerically. Starch and talc were identified to be interacting variables.

The discriminant analysis reduced the information included in the data to a two-dimensional problem. On basis of the two significant discriminant variables, each single measurement was reordered into the experiment, which consisted of the most similar results of the properties measured. Not all single measurements, however, were reordered into the original experimental group. The error of discrimination can be used as an indicator of the reproducibility of the experiments. In this study, it was shown that an increase in starch concentration decreases the reproducibility of the experiments. Hence, the amount of starch in the final tabletting mixture should be as low as possible. From the canonical drawing, it could be seen that the effect of talc on the formulation properties is nearly as strong as the influence of starch on the important properties left (see Fig. 5). The overlap of the discriminant classes of the different factors is characteristic for an interaction of the factors.

MANOVA and discriminant analysis were used in a similar way by Podczeck and Wenzel [33] to study two CGDs for tablet formulation.

IV. DATA STRUCTURE, ENHANCED USING PRINCIPAL COMPONENT ANALYSIS (PCA), CLUSTER ANALYSIS (CA), AND CANONICAL ANALYSIS

The PCA allows a descriptive summary of a set of N observations with p numeric variables. It reduces the variety of variables into p' principal components (PC; $p' \leq p$), which reflect to a great proportion the original information content [43]. The new PCs are uncorrelated, and hence they may replace an original set of highly correlated variables advantageously [44]. Mathematically, the eigenvalues ("latent roots") of the variance-covariance matrix of the data have to be calculated, and normally, $p' = p$

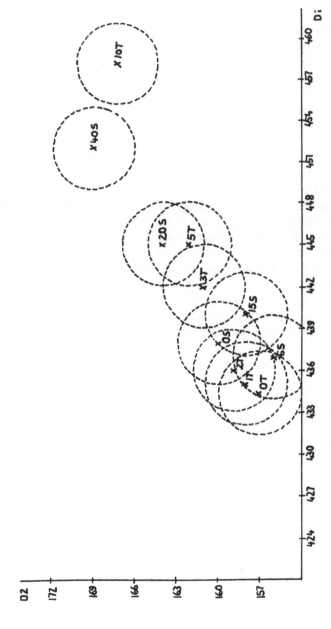

Figure 5 Canonical drawing based on the first two nonlinear discriminant variables (see text for details of the experiment). X, midpoint of a discriminant class; circle (- - -), radius of a discriminant class; D_i = discriminant variable; 0T, 1T, 2T, 3T, 5T, 10T, concentration of talc [%]; 0S, 5S, 15S, 20S, 40S, concentration of starch [%].

principal components result, which in total represent 100% of the original information given. The singular case of having only $p' < p$ PCs (the last eigenvalues are zero) is rather uncommon, and in such a case it would be useful to check whether N was smaller than p. The observation number should be considerably larger than the number of variables. The quasi-singular case is far more important. This occurs when the relationships between the original variables are not really linear or when the relationships are blurred by excessive measuring errors. The quasi-singularity is characterized by the fact that the last PCs have very small eigenvalues. Obviously, these PCs do not provide any more relevant information about the data, but they reflect the variability of the original variables caused by inaccurate measurements [44].

In tabletting, several models are used to describe the deformation process. Most of them give several numbers, which should be discussed together, when interpretations are made. Often this is difficult, because such data are not well organized. One such model is the registration of force-displacement curves (FDC), as mainly developed by Führer et al. [45], Hilmann and Fuchs [46], and Moldenhauer et al. [47]. Commonly, several values are discussed: the network $[E_2]$; the elastic recovery energy $[E_3]$; an energy model constant $[E_{max}]$, which equals the maximum energy a substance could absorb and store, if it behaved like a steel spring without any elasticity; the plasticity $[PL = E_2/(E_2 + E_3)]$ [48]; and an energy ratio $[EV = E_2/E_3(E_{max} - E_2 - E_3)]$ [49]. Table 2 lists these values for 20 drug substances, determined on an eccentric tablet press using an operating speed of 1 rpm. The PCA was used to highlight the internal structure of the data and to find out what is the main information of the data.

Normally, the results of a PCA are firstly listed giving the eigenvalues λ_i of the variance-covariance matrix, the percentage of the information extracted by each PC, R, and the cumulative information extracted, C. The following values were calculated from the data in Table 2:

PC	λ_i	R [%]	C [%]
1	3.373	67.5	67.5
2	1.126	22.5	90.0
3	0.327	6.5	96.5
4	0.168	3.4	99.9
5	0.006	0.1	100.0

Especially the last PC (i.e., PC 5) has a very small eigenvalue, which indicates a quasi-singularity. According to the correlation matrix, only the relationship between E_3 and PL was completely linear ($r = 0.942$). All

Table 2 Force-Displacement Parameters of Pure Drug Substances

Substance	E_{max}[J]	E_2 [%]	E_3 [%]	PL [%]	EV
Acetylsalicylic acid	3.87	18.3	16.8	52.1	0.017
Quinidine hydrogensulf.	5.48	43.1	9.9	81.4	0.100
Quinidine sulfate	5.63	36.6	3.0	92.4	0.208
Codeine phosphate	7.06	32.6	8.5	79.3	0.066
Caffeine	5.17	33.2	13.2	71.7	0.048
Sodium salicylate	2.95	11.0	36.3	23.2	0.006
Papaverine HCl	4.80	37.3	12.0	74.9	0.067
Paracetamol	4.95	20.6	13.3	60.9	0.024
Phenacetin	6.15	25.2	11.3	68.8	0.036
Phenobarbitone	5.10	27.1	14.4	65.1	0.033
Phenylbutazone	6.88	26.2	9.3	73.9	0.044
Pholedrin sulfate	3.14	41.1	11.4	78.2	0.076
Propipocaine HCl	5.37	24.7	13.1	64.9	0.031
Propyphenazone	7.45	27.8	10.7	72.2	0.042
Theophylline	5.60	25.5	12.2	67.5	0.034
Aminophylline	3.84	33.1	16.7	66.5	0.040
DL-Ephedrine HCl	4.81	22.2	13.7	61.9	0.025
Potassium chloride	4.54	34.7	10.2	77.3	0.062
Phenytoin	7.59	17.8	14.9	54.4	0.018
Propranolol HCl	4.92	35.9	5.3	87.1	0.119

Tabletting speed 1 rpm; tabletting pressure 86 ± 1 MPa; tablet weight 200.0 mg; punches 11 mm flat; ($\bar{x}$; $n = 5$; average standard deviation 1%)
Source: Unpublished data material, undertaken at the Martin-Luther University Halle/Saale (Germany) by F. Podczeck

other correlation coefficients were in a range between 0.7 and 0.8 or even below. Hence, the quasi-singularity is a result of nonlinearity between the data. In such a case the small PCs have to be neglected, and only the first two PCs (i.e., PCs 1 and 2) are indicated as significant. The coefficient matrix has, therefore, the following structure:

	PC 1	PC 2	Communality
E_{max}	0.330	0.924	0.963
E_2 [%]	0.859	−0.364	0.871
E_3 [%]	−0.926	−0.250	0.920
PL [%]	0.990	0.007	0.980
EV	0.830	−0.277	0.766

The value of the communality weighs the importance of the variable, and the value of 0.766 for EV suggests that this value is comparatively unimportant. The first PC can be characterized as the "PC of the energy proportions." The greater the value of E_3, the smaller is the value of PL, and taking the absolute values of the coefficients into account, PL and E_3 reflect the information of a FDC in terms of the energy consumption of the substance comprehensively. The second PC could be called the E_{max} – PC, because it only reflects the model constant as a main factor (coefficient $0.924 >>$ than all other coefficients). In the theory of FDCs the E_{max} plays a key role, because the percentages of E_2, E_3 and the constants PL and EV are based on this value. Hence, the way of estimating E_{max} is mainly responsible for all other conclusions drawn from FDCs.

The calculation of PCs provides the ability to draw the "global property FDC" of every drug substance in a two-dimensional graph using the first two PCs (Fig. 6). Comparisons of the multivariate data become possible, and the drug substances can be classified. It becomes quite clear that sodium salicylate (far left) is an odd substance, which has very different compaction properties. Acetylsalicylic acid and phenytoin are still distinctly different from the other substances (PC1 coordinate about 55). Comparing the actual values, it can be seen that the relative network E_2 is below 20%. On the far right are four substances—either quinidine salt, propranolol, and pholedrin—which provide comparatively large EV values above 0.075. The average drug substance (cloud of points in the middle) has a relative E_2 below 30%, but some substances (codeine, potassium chloride, papaverine, caffeine, aminophylline) show an E_2 value between 30% and 38%. Their points are more right-shifted from the average cloud.

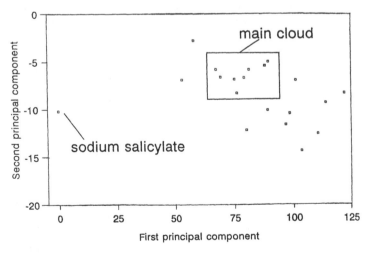

Figure 6 Differences of the force-displacement curves of 20 drug substances, characterized by their principal components.

As in the example given, the approach of PCA is more commonly applied to the study of drug substance or excipient properties in terms of their applicability for a dosage form design [50,51]. It has, however, also been used to quantify the effects of granulation liquids and granulation procedures on granule and tablet properties [52,53].

Cluster analysis (CA) [37] is a tool for detecting similarities in data. The aim of the analysis is to group a set of objects, which is characterized by a number of variables, into clusters ("classes") of similar objects. There are different types of classification in use, e.g., partitioning and hierarchical classification. The mathematical procedure comprises more-or-less iterative ways to find the clusters. Depending on the type of classification different measures of homogeneity, heterogeneity, and quality of classification are in use. In the case of a "partitioning with average linkage" the indices of homogeneity quantify the similarity of the objects connected in one cluster by calculating the distance between each member of the class. The indices of heterogeneity, however, measure the distance between pairs of cluster taking the average distance of all objects into account. The quality of the whole classification depends on the indices chosen and reflects the total success, i.e., to classify the objects into most homogenic clusters, which are most heterogenic to each other.

For the characterization of the densification process during tabletting mainly two models are used: the Heckel graph [54,55] and the Lüdde-Kawakita equation [56,57]. The results of these models are often not completely interpretable, and many authors have tried to obtain more information by creating derived values such as (a) the yield strength σ [58] as a measure for the resistance of a substance to deformation, (b) the density of the compact B that is reached without bond formation during compression [59] or its equivalent D_B [60], which represents the relative part of the densification process taken place without any bond formation, and (c) the specific yield strength π [58], which marks the tabletting pressure, below which only elastic deformation occurs. Table 3 summarizes the original Heckel and Lüdde-Kawakita constants as well as the derived values for 20 drug substances studied.

A CA (partitioning with average linkage) was applied to identify similar drug substances, i.e., substances, which have similar densification properties regarding the multivariate description of this feature given in Table 3. The number of clusters was optimized empirically by varying the cluster number required. Allowing only two clusters, class 1 included 19 drug substances, whereas class 2 comprised only pholedrin sulfate. Pholedrin sulfate sticks out because of its large value of K_p (394.2 MPa). The increase to three or four clusters did not provide significant additional information, but a classification into five clusters provided useful similarities. Pholedrin sulfate

Table 3 Heckel and Lüdde-Kawakita Constants of 20 Drug Substances

Substance	K_p [MPa]	A	σ [MPa]	B [g cm^{-3}]	D_B [%]	a [%]	b [MPa^{-1}]	π [MPa]
Acetylsalicylic acid	71.2	2.06	23.7	1.33	35.4	34.42	0.022	23.86
Quinidine hydrogens	188.0	0.85	62.7	0.50	28.1	52.21	0.022	49.66
Quinidine sulfate	229.8	1.21	76.6	0.93	46.1	52.72	0.025	44.60
Codeine phosphate	49.4	1.08	16.5	0.83	44.2	55.84	0.022	57.48
Caffeine	105.4	1.32	35.1	0.94	41.8	49.43	0.020	48.87
Sodium salicylate	162.4	1.28	54.1	1.03	49.9	56.06	0.021	60.75
Papaverine HCl	211.1	1.03	70.4	0.60	29.3	42.08	0.021	34.60
Paracetamol	185.6	1.38	61.9	0.79	30.3	37.55	0.018	33.40
Phenacetin	218.7	1.23	72.9	0.70	29.9	37.75	0.022	27.56
Phenobarbitone	92.7	1.27	30.9	0.89	40.3	51.99	0.025	43.32
Phenylbutazone	85.7	1.37	28.6	1.14	53.7	61.43	0.021	75.84
Pholedrin sulfate	394.2	1.01	131.4	0.64	32.5	60.86	0.022	70.68
Propipocaine HCl	55.1	0.98	18.4	0.66	35.4	64.33	0.037	48.74
Propyphenazone	116.9	0.78	39.0	0.60	37.4	74.96	0.026	115.14
Theophylline	97.9	1.54	32.6	0.98	35.6	44.05	0.020	39.36
Aminophylline	55.4	0.99	18.5	0.58	29.4	65.84	0.031	62.17
DL-Ephedrine HCl	133.0	1.68	44.3	1.18	42.1	55.10	0.021	58.44
Potassium chloride	46.1	0.81	15.4	0.16	7.9	55.05	0.028	43.74
Phenytoin	221.2	1.22	73.7	0.85	39.5	57.05	0.015	88.55
Propranolol HCl	34.5	1.27	11.5	0.65	25.9	45.55	0.021	39.84

K_p = mean yield pressure; A = constant associated with densification due to particle rearrangement; σ = yield strength; B = compact density reached without bond formation; D_B = relative amount of B on the whole densification process; a, b=Lüdde-Kawakita constants; π = specific yield strength.
Source: Unpublished data material, undertaken at the Martin-Luther University Halle/Saale (Germany) by F. Podczeck.

is still a separate class. Phenylbutazone and propylphenazone build a second class, which is characterized by an average class performance of $K_p = 101.3$ MPa, $\sigma = 33.8$ MPa, $D_B = 45.6\%$, $a = 68.2\%$, and $\pi = 95.5$ MPa. The π value is especially large, and the class could be characterized as "distinctive elastic substances." A next class comprises quinidine sulfate, papaverine HCl, and phenacetin. As average class performance, a large value of K_p (219.9 MPa) and σ (73.3 MPa) and a comparatively low value for the possible volume reduction ($a = 44.2\%$) were separated. The class could be classified as "ductile substances with little compressibility." A further class included quinidine hydrogensulfate, sodium salicylate, paracetamol, DL-ephedrine HCl, and phenytoin. For this group of substances, all average model parameters were of average height. The last class included the remaining nine substances, which were characterized by a very low K_p value (67.5 MPa) and yield strength ($\sigma = 22.5$ MPa). This class could be described as "plastic substances."

Canonical analysis is a tool for quantification of relationships between two groups of variables, e.g., the physical/pharmaceutical properties of drug substances and the properties of tablet formulations which include those substances. Mathematically, it is related to the PCA, but the mathematical algorithms are applied to both groups of variables using the overall variance-covariance matrix. The equivalents to the PCs are now the canonical variables [61,62], but a set of additional measures [63] are needed for the interpretation. A summary of the most useful algorithms, measures, and their meanings is given in [64]. Because of the complexity of the analysis, its use to date is limited, but there are some papers in the field of tabletting [33,64–66] which report the benefits of applying canonical analysis.

The flow properties of a powder blend determine whether or not it is suitable for use in direct tabletting. The batch-to-batch variability of the excipients and drug substances included in such a powder blend may change the flow properties considerably, and very sensitive mixtures may sometimes cause problems in the tabletting process. Hence, the influence of the material properties on the powder mixture properties should be investigated. Canonical analysis is one possible test method.

To demonstrate the use of canonical analysis for the purpose described, a basic direct tabletting mixture composed of 0.4% magnesium stearate, 9.0% starch, and 90.6% microcrystalline cellulose was mixed with 15 different drug substances in concentrations of 30%; 40%; 50%; 60%, and 70% of drug under standardized conditions. The drug substances were characterized using a wide range of different physical and pharmaceutical properties, for which the experimental details and results are not listed here. The powder mixtures were characterized by their bulk volume, Hausner's ratio, and the flowability measured on a powder flowmeter. The latter measurements described the powder flow as a mean flow rate and their standard deviation per 50-g mixture passing a copper/nickel funnel of defined dimensions. Canonical analysis was used to detect the important drug substance properties, which if changed may affect the powder mixture behavior. Every measurement was undertaken in five replicates, and the analysis is based on the single values instead of the average values. In this way, the observation number was increased to allow canonical analysis to be applied. The total observation number was, therefore 375, and the group of the drug substance properties consisted of $q = 32$ different variables. In the group of the powder mixture properties, Hausner's ratio and the standard deviation of the flow rate had to be excluded because of large intercorrelations, and a preliminary MANOVA had indicated their dispensability. Hence, the number of variables in this goup was $p = 2$ only. According to Thorndike [67], the observation number for a canonical analysis has to be in the interval of $(p + q) < N < (p + q)^2 + 50$. This requirement was fulfilled.

In a first step, the canonical variables, which are uncorrelated like the principal components, must be calculated. There are maximal p canonical variables determinable. The canonical corelations quantify the degree of relationship between the information in the two groups of variables, now extracted and transposed into the canonical variables. For the problem described above the canonical correlations were $r_1 = 0.900$ and $r_2 = 0.669$. These measures, however, do not tell anything about the significance of the relationship measured. The multivariate test criterion Λ ($\Lambda = 0.105$) and its approximation onto the F distribution ($F = 22.36$; $f_1 = 64$; $f_2 = 686$; $F_{tab} = 1.39$) proved that the drug substance properties influence the powder mixture behavior. Hence, the measures of redundance could be deduced from the structure matrices. They quantify the part of the whole variance of one variable group that can be explained by the information given in the other variable group. In this example, only the value for the powder mixture properties is interesting ($g = 0.732$), but the analysis delivers the measures of redundance for both groups. Here, a redundancy of 73.2% indicates that the powder mixture properties as a whole are predicted by the drug substance properties, but this does not mean that both mixture properties are equally dependent on the drug substance properties, nor that all drug substance properties have an effect on the powder mixture behavior. Hence, the canonical analysis provides the interranging communalities. From these values, the final effect structure can be read. It was found that, first, both the bulk volume V and the mean powder flow rate fr are similar influenced by the drug substance properties ($d_V = 0.754$; $d_{fr} = 0.710$). In canonical analysis these values are regarded to be high ($0 < d < 1$), and both mixture properties appear highly related to the drug substance properties. On the other hand, only a few drug substance properties exceeded a critical value of $d = 0.1$ [68]. These are the following parameters:

Kawakita const. a	$d = 0.633$
Bulk volume	$d = 0.587$
Hausner's ratio	$d = 0.523$
Powder porosity	$d = 0.498$
Powder flow rate	$d = 0.388$
Specific surface area	$d = 0.346$
Angle of shearing resistance (linear shear cell)	$d = 0.201$

In a direct tabletting mixture, the flow properties of the incorporated drug substance influence the flowability of the mixture as demonstrated here (Hausner's ratio, powder flow rate of the pure drug, angle of shearing resistance). Every component of the mixture influences the bulk volume of the assembly. In this respect, it might be surprising that the mean particle size and the particle size range are not identified as important drug substance properties. However, the specific surface area was identified. This value correlates to a certain extent with the particle size distribution. Summarizing the results of the canonical analysis, it would be useful to set up fixed limits for the drug substance properties identified as important to guarantee that a batch-to-batch variability does not distort the applicability of a formulation from being directly tablettable into a useless one and hence could be used in the development of a tablet formulation.

V. MATHEMATICAL OPTIMIZATION

Many people write or talk about the "optimization of a formulation," but many of them use this terminology even if the so-called optimum has been identified by trial and error. Even the use of a factorial design is not primary an optimization, because in the first instance the analysis of the design experiments identifies the main effects and interactions of the factors only. The experiments, however, can be used advantageously in optimization procedures. In fact, optimization is a mathematical method to search for and find the "optimum," which is the most advantageous state of the system studied [69]. Mathematical optimization techniques should be applied, when the optimal formulation and the optimal process variables to manufacture a dosage form are needed.

There is a wide range of optimization techniques available. Every method is based on the quantitative measuring of the structure of the system, which is to be observed, by a mathematical model, and the quantitative description of the optimal point, the purpose of the investigation. Hence, all methods require, that the variable or variables to be optimized are related quantitatively to their predictor variables, and that the function describing this relationship is consistent over the whole n-dimensional space of the predictor variables. The nature of the mathematical function leads to the division of the optimization techniques into those which are linear and those which are nonlinear.

The actual method chosen furthermore depends on how many dependent variables are to be optimized. In tabletting, for example, there are always many dependent variables such as tensile strength, disintegration time or tablet height to consider. Unfortunately, they are mostly contradictory; i.e., the "optimum" has to be a suitable compromise between all tablet properties included. The most common optimization techniques allow the optimization of only one variable at the same time. Hence the variable to optimize is often an artificial composition of all dependent variables, and the way of building this variable becomes the key problem [70]. Vector optimization [71,72], on the other hand, allows the use of all dependent variables at the same time without the need of combining them into a single variable. There are as yet, however, few software systems available, because this kind of optimization technique is still in development.

The linear optimization is based on the condition that the dependent variable(s) and all restrictions, which define the actual space confined by the predictor variables, are linearly dependent on the predictor variables [73]. The mathematical problem then consists of finding a solution in between that space that either maximizes or minimizes the dependent variable(s). Linear optimization techniques only work with restrictions, which limit the p-dimensional Euclidean space (R^p; p = number of predictor variables), so that a finite space G results, in which the optimal point will be separated from the range of possible solutions. G is convex because of the linearity of the restrictions. In a convex space two random points can be connected by a line, which is positioned completely in the space. Any point that cannot be a midpoint of such a line will be called an "extreme point." If the linear optimization problem (LOP) was composed of more than two predictor variables, G becomes a polyeder. The optimal solution of the problem is positioned either at an edge point of G or at a surface line of G. Except for the special case of the optimal surface line, the optimal solution is always a single point. The edge points of G are the "basic points" of the LOP and the optimal point is the "efficient point" of the LOP. The most common mathematical procedure in linear optimization is Dantzig's simplex method [74].

Nonlinear optimization techniques are based on the fact that any nonlinear function has at least one maximum or minimum, which can be determined numerically by differential calculus. In practice, however, this becomes more difficult, the more complex is the function used to describe the relationship between the predictor variables and the dependent variable(s). Usually, iterative methods have to be used.

A special category of optimization techniques for a single dependent variable is called "hill climbing." Here, the way of calculation can be described as climbing to the top of a hill, which is the maximum of all possible results of the function used. One of these methods is the Nelder-Mead method [69]. An nonregular simplex is fitted onto the optimization function by expansion, contraction, or reflection. The "nonregular simplex" is a geometric figure of a fixed number of experimental points, which is distorted along the space of the predictor variables following special mathematical rules (see Fig. 7). The geometrical figure of the simplex is a triangle in the case of two predictor variables, and a polyeder for three or more

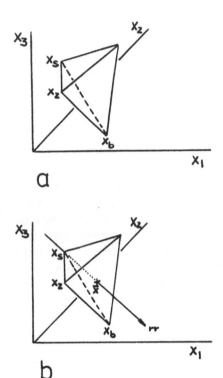

Figure 7 Definition of a "simplex" and simplex transformation in the Nelder-Mead method. X_i = independent variable; X_s = worst point of the simplex; x_b = best point of the simplex; x_z = second best point of the simplex; $\bar{x}$ = center of gravity of the simplex; x_r = simplex point calculated for reflection; x_e = simplex point calculated for expansion; x_k = simplex point calculated for contraction; rr = direction of reflection: (a) start simplex for 3D-LOP; (b) determination of the center of gravity of the simplex and the direction of reflection; (c) expansion of the simplex in the direction of reflection after reflection ($x_r = \bar{x} + \alpha(\bar{x} - x_s)$); (d) simplex after expansion ($x_e = \bar{x} + \gamma(x_r - \bar{x})$); (e) simplex contraction ($x_k = \bar{x} + \beta(x_r - \bar{x})$); (f) simplex contraction ($x_k = \bar{x} + \beta(x_s - \bar{x})$).

predictor variables of the LOP, because at least one point more than the number of predictor variables is always needed in the LOP. At each new simplex point, the theoretical value of the dependent variable has to be calculated using the function that describes the relationship between the set of predictor variables and the dependent variable. The direction of movement (simplex transformation) depends on the result calculated for the new simplex point. If the new point is a "better" solution, than the direction of transformation of the simplex will be kept. Otherwise, the direction of

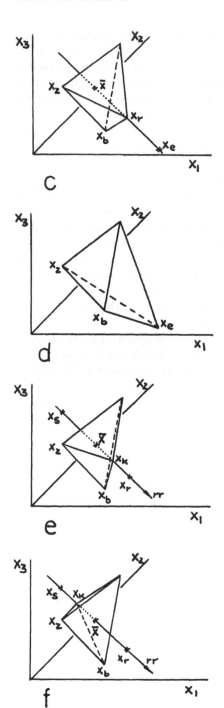

transformation will be changed or even inverted. The mechanism stops when every transformation in any possible direction leads to a less satisfying result. Because this does not involve differential calculus, the method is suitable for linear functions as well as for nonlinear, e.g., quadratic functions. The Nelder-Mead method can be used for a simultaneous optimization (based on a model equation) and a sequential optimization (after a few preliminary experiments to set up the simplex each new experiment is calculated by a simplex transformation, and the next simplex transformation is based on the experimental result).

Another group of techniques belonging to the hill-climbing methods uses the gradient of the mathematical function and requires at least a quadratic term, because they are based on differential calculus and work without considering restrictions. The two favorite methods are named after Fletcher and Powell and Fletcher and Reeves [69]. At the minimum of the function, the gradient is perpendicular to the function, which will be used as the stopping criterion. For functions with more than one predictor variable, the differential calculus provides a gradient vector (first derivate of the function) and the Hesse matrix (second derivate of the function), and the problems can only be solved iteratively.

With the exception of the vector optimization, the method of Lagrange multiplicators [69] is one of the most flexible methods in treating restrictions. The methods using Lagrange functions solve a LOP with restrictions by determining the limiting value of a series of solutions of special constructed auxilliary problems without restrictions [75]. Again, only one dependent variable can be optimized, but the other interesting variables can be included into the problem by setting up real restrictions, which control the optimization process. Furthermore, the original concept can be transferred to nonlinear optimization problems using nonlinear restrictions [75] if the convexity of the space G is still guaranteed.

The basic Lagrange function

$$F(x,\lambda) = f(x) + \lambda_i g_i(x) \qquad i = 1(1), m \qquad (1)$$

where $F(x,\lambda)$ = the Lagrange function, $f(x) \rightarrow$ min! $(x \in G \subset R^p)$, λ_i = Lagrange multiplicator, $g_i(x)$ = ith restriction, m = number of restrictions, has a local minimum x' if a surrounding $U(x')$ exists, which is defined as [75]

$$f(x) \geq f(x') \qquad x \in [G \cap U(x')], \quad x \neq x' \qquad (2)$$

For example, the very simple LOP for a tablet formulation could consist of the following requirements: the property to be minimized shall be the disintegration time D, the tensile strength T shall be above 1.5 MPa m^{-2}, and the tablet height H shall equal 3 mm. Using a statistical design, which included three independent variables X_i, the following regression equations could be found:

$$D = bo_1 + b_{11}X_1 + b_{12}X_2 + b_{13}X_3 \tag{3a}$$
$$T = bo_2 + b_{21}X_1 - b_{22}X_2 - b_{23}X_3 \tag{3b}$$
$$H = bo_3 + b_{31}X_1 + b_{32}X_2 + b_{33}X_3 \tag{3c}$$

D will be minimized, and hence Eqs. (3b) and (3c) have to be transformed into restrictions, where S_i = limit of the variable:

$$S_1 > bo_2 + b_{21}X_1 - b_{22}X_2 - b_{23}X_3 \tag{4a}$$

(transformation into equation)

$$S_1 = bo_2 + b_{21}X_1 - b_{22}X_2 - b_{23}X_3 - q^2 \tag{4b}$$
$$S_2 = bo_3 + b_{31}X_1 + b_{32}X_2 + b_{33}X_3 \tag{5}$$

(transformation into final restrictions)

$$0 = bo_2 + b_{21}X_1 - b_{22}X_2 - b_{23}X_3 - q^2 - S_1 \tag{6a}$$
$$0 = bo_3 + b_{31}X_1 + b_{32}X_2 + b_{33}X_3 - S_2 \tag{6b}$$

Hence, the Lagrange function would now have the following structure:

$$\begin{aligned} F(x,\lambda,q) = \ & bo_1 + b_{11}X_1 + b_{12}X_2 + b_{13}X_3 \\ & + \lambda_1(bo_2 + b_{21}X_1 - b_{22}X_2 - b_{23}X_3 - q^2 - S_1) \\ & + \lambda_2(bo_3 + b_{31}X_1 + b_{32}X_2 + b_{33}X_3 - S_2) \end{aligned} \tag{7}$$

Differential calculus has to be used to transform the Lagrange function into a system of equations to be solved:

$$\frac{\partial F}{\partial X_1} = b_{11} + b_{21}\lambda_1 + b_{31}\lambda_2 = 0 \tag{8a}$$

$$\frac{\partial F}{\partial X_2} = b_{12} - b_{22}\lambda_1 + b_{32}\lambda_2 = 0 \tag{8b}$$

$$\frac{\partial F}{\partial X_3} = b_{13} - b_{23}\lambda_1 + b_{33}\lambda_2 = 0 \tag{8c}$$

$$\frac{\partial F}{\partial \lambda_1} = bo_2 + b_{21}X_1 - b_{22}X_2 - b_{23}X_3 - q^2 - S_1 = 0 \tag{8d}$$

$$\frac{\partial F}{\partial \lambda_2} = bo_3 + b_{31}X_1 + b_{32}X_2 + b_{33}X_3 - S_2 = 0 \tag{8e}$$

$$\frac{\partial F}{\partial q} = -2q\lambda_1 = 0 \tag{8f}$$

The equation system consists of six unknown values $(X_1, X_2, X_3, \lambda_1, \lambda_2, q)$ and six equations. Hence, the problem is exactly defined and can be solved with an iterative procedure.

The example given is very simple, and in tabletting there are far more dependent variables to consider. Furthermore, there should be limits set for the X values, e.g., a maximum tabletting pressure or a maximum lubricant level. A very complex problem is reported in [31] listing the complete system of equations.

Mathematical optimization has been used in tabletting many times, e.g., in [28,76–82]. The spectrum of methods used covers all types of methods discussed above. The experiments to be undertaken should follow a system such as given in a statistical design [20], which guarantees that the model equations deduced from the results describe the relationship between the variables comprehensively. In this respect, the most complicated model does not always appear to be the best one, and a careful residual analysis should be done before a regression model is used in optimization.

VI. EXPERT SYSTEMS (EXPS)

People have always dreamed about machines or computer systems which make decisions for them, especially when they had to solve a difficult problem or when the decision was related to a financial or other risk. In the field of artificial intelligence, such tasks are handled by knowledge manipulation using EXPS.

EXPS are programs which include a wide range of knowledge in a restricted domain. The knowledge is manipulated by inferential reasoning to solve a problem in a way which a human expert could have chosen [83].

Software engineers praise the advantages of EXPS, but the EXPS are never an "overall problem solver" which replaces the human expert as soon as the development has finished. EXPS only consist of a comprehensive knowledge, if all information about the domain described was included, and if the knowledge base was permanently updated. EXPS for specific problems cannot be transferred, because the way of tackling these problems varies from user to user. For example, the development and manufacture of tablets differs from company to company, and there are clear disagreements in the politics (excipients and equipment used, different manufacturing procedures such as direct compression or granulation). The EXPS certainly document the way a decision was made, but is the decision right? Usually, the knowledge base consists of facts and rules which had been used before to solve a problem, and the new problem is solved by comparing past and present. It may be that there is no comparable case reported. Many EXPS try to give an answer by "creating" such a case combining the most suitable rules instead of admitting the limitations of the system [84]. Hence, the solution recommended by an EXPS should always be checked for its credibility, which means that EXPS should be used by human experts to make their work easier. Using EXPS, experts can use their own knowledge comprehensively, and they may spend a lot more time on creative work than before.

The main part of an EXPS is its knowledge base, which consists of facts and rules. Facts comprise knowledge in the form of descriptions such as "PARACETAMOL: Solubility 19 g L^{-1}," "CAFFEINE: Deformation mechanism ductile," "THEOPHYLLIN Dose (rapid release) 100 mg," "MICROCRYSTALLINE CELLULOSE: hygroscopic," or in form of definitions such as "MAGNESIUM STEARATE: lubricant / 0.1–0.5%." Rules are descriptors, which will be used to combine the facts. The most common form of rules is the "production rule" [84]. There are passive semiactive and active rules requiring differentiation. "IF a drug substance has an amino group THEN it is incompatible to lactose ($P = 80\%$)" is a passive rule, which does not lead to an action by itself. It gives, however, a further piece of knowledge about the substance, which can be used in another rule. The value in parentheses indicates the probability that the rule is true. Probability values are needed to value the chance that the final solution is successful. The calculation of the final probability can be done using Bayesian theorem [83]. A semiactive rule "fires" if a decision criterion was true, e.g., "IF the drug substance is brittle THEN use a basic tabletting mixture, which includes microcrystalline cellulose as a filler ($P = 85\%$)." An active rule also tests limits, such as "IF the bulk volume of the substance is between 2 and 4 cm^3 g^{-1} THEN use tabletting basic mixture 1 ($P = 75\%$)." Semiactive and active rules lead to a decision, which brings the program step by step to a final conclusion.

Knowledge must be found and transformed into facts and rules. In general, there are three main ways to get the knowledge. The first way is classical knowledge engineering. Experts will be questioned by the knowledge engineer, who tries to find out what the expert knows and how he decides on a problem. The knowledge engineer then tries to copy the knowledge and the way of problem solving. Tabletting, however, is a very complicated process, which still cannot be described comprehensively. Therefore, other ways of getting the knowledge, which have to be done mainly by the experts themselves, must be included. For example, collecting "past knowledge" from the literature will help to develop the knowledge base, but even then in tabletting an EXPS would still not work. Hence, the development of EXPS for tablet formulation requires an experimental component of knowledge acquisition. Experiments done in the past can be systematized, or new experiments can be undertaken, preferably using a statistical design. In both cases the results need a mathematical treatment, which should include modern multivariate statistical methods such as PCA, CA, and canonical analysis. In this way, the knowledge formed into rules is based on a rational background, which leads to higher-quality EXPS.

EXPS contain a problem-solving component which consists of an inference engine and a control module. The latter regulates the contact between the knowledge base, the inference engine, and the user (task component). The most important part of the problem-solving component, however, is the inference engine. In a rule-based system, the inference engine combines suitable rules until all information, which is available to solve a problem, has been included in the problem-solving process. There are four main strategies to combine the rules: Backward chaining with depth-first search, backward chaining with breadth-first search, forward chaining with depth-first search, and forward chaining with breadth-first search [85]. Backward chaining and forward chaining are the inference mechanisms. Backward chaining implies a possible solution and checks the rules to prove this idea true or wrong, whereas forward chaining checks the rules for their agreement with start information, and the addition of the rule-provided information leads finally to a conclusion. Depth-first and breadth-first search characterize the way, in which the logical rule tree will be checked, branch after branch, until an answer has been given, or along all branches at the same time.

EXPS are implemented at computers either using EXPS shells or by programming in a higher programming language (C or Pascal) or artificial intelligence language (Prolog, Lisp). The use of an EXPS shell reduces the programming effort tremendously, because task, explanation, and inference components are prebuilt, and normally only the knowledge base has to be fitted into the system. However, this may restrict the flexibility of the system, because the knowledge base has to be constructed in the way required by the shell. Programming languages certainly prolong the implementation time, but they allow an exact copy to be formed of the basic idea of the system and the knowledge base.

In tabletting, there are two types of EXPS in use: diagnosis and formulation EXPS. Murray [86] described a diagnosis EXPS, which can be supplied with new Korsch high-speed rotary tablet presses. In the first instance it is designed to discover and remove weak points and difficulties running the machine for a special formulation. It helps to adjust the optimal machine parameters for the manufacturing process. Furthermore, it includes tools for machine maintenance and training, which amplifies the area of application.

In the literature, there are three formulation EXPS for tabletting described [87–89], but the published number certainly does not reflect the true number of existing systems. Many companies prefer not to mention the existence of such systems, and some of the published papers give hardly any detail about the facts and rules included, to ensure that competing companies will not get easy access to the formulation of the original products. Hence, the systems can only be compared in terms of their basic principles, and phrases like "more advanced" than others or "in no way complete" as used in [88] to gather more approval are inappropriate.

Basically, the systems use two different principles. The first principle [87,88] is that the user enters information about the drug substance (physical and pharmaceutical properties), the probable dose, and ideas about the preferred tablet dimensions and weight. The system recommends a possible set of excipients according to the drug substance properties and a basic quantitative composition of them. It also provides the proposed properties of the formulation. Then the user has to perform experiments comparable to a sequential optimization; i.e., after experimental check of the basic solution the results will be reentered into the system and further rules are used to adjust the formulation parameters until the tablet formulation matches the requirements asked by the user completely. This principle reflects the formulation policy of large companies.

The second principle [89] reflects more the policy of smaller companies, which often use so-called basic mixtures as standard tabletting excipients. The drug substance will be mixed with the basic mixture to a defined amount, and a standardized manufacturing method will be used. It needs, however, the identification of which basic mixture from a given set matches the drug substance properties best, and in which concentration the basic mixture and the drug substance must be mixed. Again, the system requires, first, the input of a series of drug substance properties. A set of general rules is used to determine the best basic mixture (either granulation or one of three direct compression basic mixtures). A separate set of facts and rules, derived using multivariate statistical methods, exists for each basic mixture to determine the most suitable concentration of the drug substance and the accompanying tabletting pressure. Because tablet properties such as tensile strength or disintegration time are contradictory to each other, a mathematical optimization procedure is added to provide the user with a final solution. The system also calculates the probability with which the recommended final formulation will be successful when compared to a set of requirements that a tablet formulation has to fulfill. The solution given is already optimized, and no further experiments are needed.

None of the systems involve detailed rules about the mixing of the powder blends or a complete analysis of machine parameters during the tabletting, but these subprocesses influence the final success of the formulation. Hence, none of the systems is comprehensive in terms of a description of the complete tabletting formulation process and scaling up. This is, however, to be viewed in a positive way, because it allows the development of the systems permanently introducing new strategies both in formulation and in knowledge acquisition. The development of EXPS is a dynamic process, which will never end, because there is always an increase in knowledge about the domain covered.

REFERENCES

1. R. Hüttenrauch, *Acta Pharm. Technol.*, *Suppl. 6*: 55 (1978).
2. H. Leuenberger, P. Guitard, and H. Sucker, *Pharm. unserer Zeit. 5*: 65 (1976).
3. W. R. Glaser, *Varianzanalyse*, Gustav Fischer Verlag, Stuttgart, New York, pp. 17, 63, 274–276, 1978.
4. E. Weber, *Grundriß der biologischen Statistik*, 8th ed., Gustav Fischer Verlag, Jena, p. 264, 1980.
5. J. Adam, *Einführung in die medizinische Statistik*, 3rd ed., Verlag Volk & Gesundheit, Berlin, p. 216, 1971.
6. F. Yates, *The Design and Analysis of Factorial Experiments*, Imperial Bureau of Soil Science, London, p. 4, 1937.
7. O. L. Davies, *The Design and Analysis of Industrial Experiments*, Oliver & Boyd, London, p. 263, 1960.
8. G. Stetsko, *Drug Dev. Ind. Pharm. 12*: 1109 (1986).
9. L. Öner, and M. J. Groves, *Pharm. Res. 10*: 621 (1993).
10. M. Chariot, J. Francès, G. A. Lewis, D. Mathieu, R. Phan Tan Luu, and H. N. E. Stevens, *Drug Dev. Ind. Pharm. 13*: 1639 (1987).
11. L. Hasznos, I. Langer, and M. Gyarmathy, *Drug Dev. Ind. Pharm. 18*: 409 (1992).
12. J. F. Pinto, G. Buckton, and J. M. Newton, *Int. J. Pharm. 83*: 187 (1992).
13. I. Khattab, A. Menon, and A. Sakr, *J. Pharm. Pharmacol. 45*: 687 (1993).
14. P. Merkku, and J. Yliruusi, *Eur. J. Pharm. Biopharm. 39*: 75 (1993).
15. P. Merkku, O. Antikainen, and J. Yliruusi, *Eur. J. Pharm. Biopharm. 39*: 112 (1993).
16. L. Hellén, J. Yliruusi, P. Merkku, and E. Kristoffersson, *Int. J. Pharm. 96*: 197 (1993).
17. L. Hellén, J. Yliruusi, and E. Kristoffersson, *Int. J. Pharm. 96*: 205 (1993).
18. L. Hellén, and J. Yliruusi, *Int. J. Pharm. 96*: 217 (1993).
19. G. P. E. Box, and K. B. Wilson, *J. R. Stat. Soc. Ser. B 13*: 1 (1951).
20. R. Huisman, H. V. van Kamp, J. W. Weyland, D. A. Doornbos, G. K. Bolhuis, and C. F. Lerk, *Pharm. Weekbl. Sci. Ed. 6*: 185 (1984).
21. O. Reer, and B. W. Müller, *Eur. J. Pharm. Biopharm. 39*: 105 (1993).
22. J. T. Fell, and J. M. Newton, *Pharm. Acta Helv. 46*: 441 (1971).

23. H. Moldenhauer, H.-J. Loh, and H. Kala, *Pharmazie 33*: 69 (1978).
24. H. Moldenhauer, and H.-J. Loh, *Pharmazie 33*: 216 (1978).
25. H. Moldenhauer, H.-J. Loh, and H. Kala, *Pharmazie, 33*: 349 (1978).
26. F. Podczeck, *Beiträge zum Einsatz mathematische Verfahren für die Rezepturentwicklung fester Peroralia*, Diplomarbeit, Martin-Luther-Univ. Halle-Wittenberg, Halle/Saale, pp. 25–28, 1984.
27. S. Zubair, S. Esezobo, and N. Pilpel, *J. Pharm. Pharmacol. 40*: 278 (1988).
28. B. Iskandarani, J. H. Clair, P. Patel, P. K. Shiromani, and R. E. Dempski, *Drug. Dev. Ind. Pharm., 19*: 2089 (1993).
29. H. Leuenberger, in *Pharmazeutische Technologie* (H. Sucker, P. Fuchs, and P. Speiser, eds.), Georg Thieme Verlag, Stuttgart, New York, pp. 137, 139, 1991.
30. A. G. González, *Int. J. Pharm. 97*: 149 (1993).
31. F. Podczeck, *Beiträge zur rechnergestützten Entwicklung fester Peroralia*, Thesis (Dr. rer. nat.), Martin-Luther-Univ. Halle-Wittenberg, Halle/Saale, pp. 54, 59, 113, 1986.
32. F. Podczeck, and U. Wenzel, *Pharm. Ind. 52*: 230 (1990).
33. F. Podczeck, and U. Wenzel, *Pharm. Ind. 52*: 348 (1990).
34. F. Podczeck, and U. Wenzel, *Pharm. Ind. 52*: 496 (1990).
35. H. Ahrens, and J. Läuter, *Mehrdimensionale Varianzanalyse*, 2nd ed., Akademie-Verlag, Berlin, pp. 2–3, 72, 1981.
36. H. Hotelling, in *Proc. Second Berkeley Symp. Mathematical Statistics and Probability*, Berkeley, CA, pp. 23–41, 1951.
37. J. Hartung, and B. Elpelt, *Multivariate Statistik*, R. Oldenbourg Verlag, Munich, Vienna, pp. 175, 443, 1984.
38. S. S. Wilks, *Biometrika 24*: 471 (1932).
39. H. Gaensslen, and W. Schubö, *Einfache und komplexe statistische Analyse*, 2nd ed., E. Reinhardt Verlag, Munich, Basel, p. 176, 1976.
40. W. J. Krzanowski, *J. Am. Stat. Ass. 70*: 782 (1975).
41. T. W. Anderson, *An Introduction to Multivariate Statistical Analysis*, Wiley, New York, pp. 215–221, 1958.
42. D. F. Morrison, *Multivariate Statistical Methods,* McGraw-Hill, New York, pp. 170–245, 1976.
43. L. Lebart, A. Morineau, and J.-P. Fenelon, *Statistische Datenanalyse*, Akademie Verlag, Berlin, pp. 10–11, 1984.
44. B. Flury, and H. Riedwyl, *Angewandte multivariate Statistik*, Gustav Fischer Verlag, Stuttgart, New York, pp. 112–113, 124, 1983.
45. C. Führer, G. Bayraktar-Alpmen, and M. Schmidt, *Acta Pharm. Technol. 23*: 215 (1977).
46. J. Hilmann, and P. Fuchs, *Pharm. Ind. 39*: 72 (1977).
47. H. Moldenhauer, H. Kala, G. Zessin, and M. Dittgen, *Pharmazie 35*: 714 (1980).
48. A. Stamm, and C. Mathis, *Acta Pharm. Technol. 22* Suppl. (1): 7 (1976).
49. M. Dürr, D. Hanssen, and H. Harwalik, *Pharm. Ind. 34*: 905 (1972).
50. N.-O. Lindberg, and T. Lundstedt, in *Proc. 12th Pharmaceutical Technology Conf.*, Elsinore, pp. 37–67, 1993.
51. A. Hagman, and S. Jacobsson, *Drug Dev. Ind. Pharm. 16*: 2527 (1990).

52. L. Benkerrour, D. Duchéne, F. Puisiéux, and J. Maccario, *Int. J. Pharm. 19*: 27 (1984).
53. T. Schofield, J. F. Bavitz, C. M. Lei, L. Oppenheimer, and P. K. Shiromani, *Drug Dev. Ind. Pharm. 17*: 959 (1991).
54. R. W. Heckel, *Trans. Metall. Soc. AIME 221*: 671 (1961).
55. R. W. Heckel, *Trans. Metall. Soc. AIME 221*: 1001 (1961).
56. K. H. Lüdde, and K. Kawakita, *Pharmazie 21*: 393 (1966).
57. K. Kawakita, and K. H. Lüdde, *Powder Technol. 4*: 61 (1970/71).
58. J. Hinsch, *Anwendbarkeit einiger Preßgleichungen auf die Tablettierung polymorpher Modifikationen des Tolbutamid*, Thesis (Dr. rer. nat.), Univ. Hamburg, Jesteburg, pp. 19, 24, 1983.
59. P. Humbert-Dróz, D. Mordier, and E. Doelker, *Pharm. Acta Helv. 57*: 136 (1982).
60. R. J. Roberts, and R. C. Rowe, *J. Pharm. Pharmacol. 38*: 567 (1986).
61. H. Hotelling, *J. Educ. Psych. 26*: 139 (1935).
62. H. Hotelling, *Biometrika 28*: 321 (1936).
63. R. M. Thorndike, *Multiv. Beh. Res. 12*: 75 (1977).
64. F. Podczeck, G. Merkel, and P. Révész, *Int. J. Pharm. 97*: 15 (1993).
65. F. Podczeck, and U. Wenzel, *Pharmazie 44*: 468 (1989).
66. F. Podczeck, and U. Wenzel, *Pharm. Ind. 51*: 524 (1989).
67. R. M. Thorndike, *Correlation Procedures for Research*, Gardener Press, New York, 1978.
68. M. Röhr, *Kanonische Korrelationsanalyse*, Akademie Verlag, Berlin, p. 84, 1987.
69. U. Hoffmann, and H. Hofmann, *Einführung in die Optimierung*, Chemie Verlag GmbH, Weinheim, pp. 1, 127–129, 132–143, 150, 1971.
70. B. Zierenberg, and H. Stricker, *Pharm. Ind. 43*: 777 (1981).
71. J. H. de Boer, A. K. Smilde, and D. A. Doornbos, *Acta Pharm. Technol. 34*: 140 (1988).
72. L. Baert, H. Vermeersch, J. P. Remon, J. Smeyers-Verbeke, and D. L. Massart, *Int. J. Pharm. 96*: 225 (1993).
73. E. Seiffart, and K. Manteuffel, *Lineare Optimierung*, BSB Teubner Verlagsgesellschaft, Leipzig, p. 10, 1974.
74. G. B. Dantzig, A. Orden, and P. Wolfe, *Pacific J. Math. 5*: 183 (1955).
75. Ch. Großmann, and A. Kaplan, *Strafmethoden und modifizierte Lagrange-Funktionen in der nichtlinearen Optimierung*, BSB Teubner Verlagsgesellschaft, Leipzig, pp. 3, 125, 16–17, 1979.
76. A. Devay, J. Uderszky, and I. Rácz, *Acta Pharm. Technol. 30*: 239 (1984).
77. J. B. Schwartz, J. R. Flamholz, and R. H. Press, *J. Pharm. Sci. 62*: 1165 (1973).
78. J. B. Schwartz, J. R. Flamholz, and R. H. Press, J. *Pharm. Sci. 62*: 1518 (1973).
79. N. Benkaddour, L. Bonnét, F. Rodriguez, and R. Rouffiac, *Labo-Pharma Probl. Technol. 32*: 270 (1984).
80. D. E. Fonner, J. R. Buck, and G. S. Banker, *J. Pharm. Sci. 59*: 1587 (1970).

81. F. Podczeck, and U. Wenzel, *Pharm. Ind. 52*: 627 (1990).

82. G. Zierenberg, A. R. Gupte, H. Harwalik, and H. Stricker, *Pharm. Ind. 44*: 741 (1982).

83. A. Hart, *Knowledge Acquisition for Expert Systems*, 2nd ed., McGraw-Hill, New York, pp. 21, 92, 1992.

84. C. I. Doukidis, and E. A. Whitley, *Developing Expert Systems*, Chartwell-Bratt, pp. 8, 11, 1988.

85. F. K. Y. Lai, in *Encyclopedia of Pharmaceutical Technology* (J. Swarbrick and J. C. Boylan, eds.), Vol. 5, pp. 361–378, Marcel Dekker, New York, 1992.

86. F. Murray, *Pharm. Technol. 3*: 100 (1989).

87. K. V. Ramini, M. R. Patel, and S. K. Patel, *Interfaces 22*: 101 (1992).

88. R. C. Rowe, and N. G. Upjohn, *Pharm. Technol. Int. Bipharm. 5*: 46 (1993).

89. F. Podczeck, in *Proc. 11th Pharmaceutical Technology Conf.* Manchester, pp. 240–264, 1992.

Index

[Dicalcium phosphate dihydrate]
  brands, 429-430, 436
  calculation of energy of formation, 93
  calculation of net work, 89
  compression properties, 448-450, 503
  excipient, 447-453
  lubricant sensitivity, 451, 520
  physical characterization, 147
  preparation, 424
  surface area, 20
  tablet surface area, 200, 206-207
  volume reduction, 18, 193
Die wall friction (*see* Friction)
Dielectric constants
  importance for interparticulate
    bonding, 29, 511
  of liquids used for compaction, 29-35
Dirac delta function, 107
Direct compaction/compression
  advantages-disadvantages, 420-422
  drugs in D.C. forms, 486-493
  excipients, 422-486
  particle engineering methods, 424-
    428, 486-487
  tablet production by, 419-420
Direct tensile strength (*see* Axial tensile
  strength)
Directly compactible fillers/binders
  list of, 420
  methods for improving processing
    properties, 424-428
  requirements, 422-423
Discharging, 11-12
Disintegration time, percolation
  thresholds for, 138
Displacement vs. time curve (*see* Load
  vs. time curve)
Double compression, in network
  calculation, 87
Double layer tablet, 44-48
  measurement of axial tensile strength,
  44
Double torsion testing, 309-310
Drug dissolution
  of caffeine, 159
  as described by percolation theory and
    fractal geometry, 154-162
  in relation to pore distribution, 157
Drugs in directly compactible forms,
  486-493

Dry binder sensitivity, 344-346
Ductile material (*see also* Plastic
  deformation)
  effect on direct tensile testing, 166

Ejection force, 542-546
Elastic deformation (*see also* Volume
  reduction mechanisms *and*
  Elasticity)
  definition, *viii*, 284
  effect on tablet strength, 377
  evaluation of elastic strain, 233
  evaluation from Heckel plot, 70
  evaluation from indentation testing and
    yield pressure, 102, 292-294, 389
  evaluation from load vs. time plot,
    100-101
  negative effect of, 48-51
  in network calculation, 81, 83-88
  stiffness, 181, 284
Elastic modulus (*see* Young's modulus)
Elasticity, 104 (*see also* Elastic
  deformation)
  definition of, 104
  effect of punch velocity, 114-115
  Hooke's law, 104, 283
  modulus, 102, 104, 283-294, 387, 389
  storage compliance, 114
  storage modulus, 114
  Young's modulus, 106, 283-294
Electrostatic charge
  discharging, 11-12
  effect on mixing behavior, 12-13
  effect on powder flow behavior, 10-12
  permanent, 6-7
  triboelectrostatic, 3-6
Electrostatic forces, 21-22
Electrostatic interactions, 2-7
Emcompress (*see* Dicalcium phosphate
  dihydrate)
Energy consumption during milling, 255
Energy dispersive X-ray microanalysis
  (EDAX), 523-524
Energy of formation
  definition of, 93
  relation to tablet strength, 93
Energy ratio, 82
Excipient mixture, 483-486
Excipients, 419-500
Expert system, 586-590

Pharmaceutical Powder Compaction Technology

DRUGS AND THE PHARMACEUTICAL SCIENCES

A Series of Textbooks and Monographs

edited by

James Swarbrick
AAI, Inc.
Wilmington, North Carolina

ADDITIONAL VOLUMES IN PREPARATION

Pharmaceutical Powder Compaction Technology

edited by
Göran Alderborn
Christer Nyström

Uppsala University
Uppsala, Sweden

informa
healthcare

New York   London

Informa Healthcare USA, Inc.
52 Vanderbilt Avenue
New York, NY 10017

Library of Congress Cataloging-in-Publication Data

Pharmaceutical powder compaction technology.
    edited by Goran Alderborn and Christer Nystrom.
    p. ; cm. -- (Drugs and the pharmaceutical sciences ; v. 71)
    Includes bibliographical references and index.
    ISBN-13: 978-0-8247-9376-0 (alk. paper); ISBN-10: 0-8247-9376-5 (alk. paper)
    1. Tablets (Medicine). I. Alderborn, Goran.
  II. Nystrom, Christer.
  III. Series.

RS201.T2P465 1995
615'.19--dc20   96-24571

Visit the Informa Web site at
www.informa.com

and the Informa Healthcare Web site at
www.informahealthcare.com

Preface

Oral administration is the dominant method of delivering drugs to the human systemic blood circulation because of its safety and simplicity. Thus, great interest has been focused within pharmaceutical science on the design of oral dosage forms with optimal therapeutic properties. The prevailing oral dosage form today is the tablet due to its elegance. Tablets of various types and biopharmaceutical properties—from conventional, disintegrating tablets, to advanced modified release systems—exist, but their common denominator is the way in which they are formed, i.e., powder compaction. Physical and technological aspects of this process, from a pharmaceutical point of view, are the theme of this book.

The complexity of the compaction process—what at first sight seems to be a simple mechanical operation—was recognized early. Problems still exist in large-scale production of tablets, such as low tablet strength, capping, limited use of direct compression, and sensitivity to batch variability of starting materials. Moreover, the use of basic physical data in formulation work in order to predict tableting behavior of particles such as compressibility (ability to reduce in volume) and compactibility (ability to cohere into compacts) is limited. Thus, tablet formulation must still be based to a large extent on empirical knowledge rather than on scientific theory.

An improved theoretical understanding of the compaction process will enable a more rational approach to the formulation of tablets. However, the investments in research on the physics of the compaction process have, in relative terms, been limited in universities and the pharmaceutical industry. In spite of this, a large number of publications on the theme of the formation of tablets by compaction exist today in the pharmaceutical literature. This literature can be broadly classified into three categories: (1) reports on specific formulations and their compactibility and on formulation solutions to compaction-related problems, (2) studies on mechanisms of and theories for the compression and the compaction of pharmaceutical powders (such studies also include articles dealing with the development and evaluation of methods for theoretical studies), and (3) evaluation, with recognized methods and theories, of the compression and compaction behavior of pharmaceutical tableting excipients.

In the older literature, publications were focused mainly on the practical aspects of the preparation of tablets. However, since the late 1940s, articles focused on the theoretical aspects of the compaction process have been presented in the pharmaceutical scientific literature. As a consequence of the growing interest in directly compactable formulations, new excipients with improved tableting performance have been developed and the compaction characteristics of these have been the object of scientific studies. Despite this growing literature on the physics and technology of powder compaction, the interest in bringing together the accumulated knowledge in the form of comprehensive reference works has hitherto been limited. It is thus a great pleasure for the editors of this volume to present a book on theoretical and practical aspects of the process of forming compacts by powder compression. This is, to our knowledge, the first book devoted entirely to this theme. It has been made possible by the contribution of chapters from researchers throughout Europe and North America. To achieve the high level needed, only recognized scientists, representing academia or the pharmaceutical industry, have been involved, and each contributor has been encouraged to focus on his or her field of expertise. The role of the editors has been to primarily select topics and authors for the contributions and to find a suitable structure for the book. The consequence of this is that different concepts and beliefs in the field of powder compaction are presented and discussed in the book, and we have not attempted to hide this diversity. This diversity reflects the complexity of studying and establishing theories for the handling and processing of "real" materials. Moreover, there are also different traditions with respect to the nomenclature used in the discussion on powder compaction, and this inconsistency among researchers in this respect is also reflected in this book. The editors allowed each author to use terms in accordance with his or her tradition. However, to improve the stringency in the use of the nomenclature for the future, a short list of definitions follows this preface.

During the preparation of this book, some topics within the area of pharmaceutical powder compaction have not been dealt with as separate chapters, as they are not covered extensively in the literature. Examples of such topics are energy aspects of the formation of tablets, physical instability in compacts during storage, and mathematical expressions for the tensile strength of compacts. However, these topics are discussed and references are given in some of the chapters of this book.

Although great progress in the theoretical understanding of the compaction process has been made since the late 1940s, the need for further research is obvious. It is our hope that this volume can contribute to and stimulate such intellectually challenging research.

We are very grateful to Marcel Dekker, Inc., for taking the initiative to prepare a book on pharmaceutical powder compaction technology. We express our sincere appreciation especially to Sandra Beberman and Ted Allen, for pleasant cooperation during the preparation of this book, for their qualified contributions, and for their support and patience with us in our role as editors.

We are also very grateful to all contributors to this volume, for their positive attitude to share their expertise in the field of powder compaction and for the time and effort taken to write articles of high quality. Without their collaboration and contributions, the writing of this book would never have been accomplished.

Finally, we would like to thank Mrs. Eva Nises-Ahlgren for qualified administrative work in connection with the preparation of this book.

Göran Alderborn
Christer Nyström

Nomenclature

Below are the definitions of some terms commonly used in relation to powder compaction. It should be noted that within this book the terms are not used strictly in accordance with these definitions, as some authors have used the terms differently. The list is narrow in that it presents only the terms that seem to cause the most confusion in discussions of powder compaction. Less ambiguous terms, describing mechanical properties of and bonding mechanisms between particles, are defined within the individual chapters and in other books on material science and chemical engineering.

Compactibility   The ability of a powder bed to cohere into or to form a compact. Usually described in terms of tablet strength as a function of applied compaction stress.

Compaction   The transformation of a powder into a coherent specimen of defined shape by powder compression.

Compressibility   The ability of a powder bed to be compressed (be reduced in volume) due to the application of a given stress.

Compression   The reduction in volume of a powder bed due to the application of a stress, e.g., loading or vibration.

Consolidation   Mostly used synonymously with compaction. The term has also been used to describe compression of powders.

Elastic deformation of particles   Time-independent, recoverable deformation of a particle. Deformation occurs parallel to a contraction of the particle.

Hardness   The resistance of a specimen against penetration into the surface of the specimen.

Particle deformation   The change in shape of a particle during compression. Can be quantified with some shape factor for the particle as a function of applied stress during compression.

Plastic deformation of particles   Time-independent, permanent deformation of a particle. Degree of deformation is thus controlled by the applied stress and independent of the time of loading. Deformation occurs without a change in particle volume.

Particle fragmentation   The fracturing of a particle into a number of smaller, discrete fragments during compression. Can be quantified as the change in particle size or particle surface area with applied stress during compression.

Time-dependent deformation of particles   Degree of deformation of a particle is controlled by the applied stress and the time of loading.

Viscoelastic deformation of particles   Time-dependent recoverable deformation of a particle.

Viscous deformation of particles   Time-dependent permanent deformation of a particle.

Contents

V. COMPACTION PROPERTIES OF EXCIPIENTS AND POWDER MIXTURES

Contributors

Göran Alderborn, Ph.D.   Division of Pharmaceutics, Uppsala University, Uppsala, Sweden

N. Anthony Armstrong, Ph.D.   Welsh School of Pharmacy, University of Wales, Cardiff, United Kingdom

Gerad K. Bolhuis, Ph.D.   Department of Pharmaceutical Technology and Biopharmacy, University of Groningen, Groningen, The Netherlands

Jean-Daniel Bonny, Ph.D.   School of Pharmacy, University of Basel, Basel, Switzerland

Zak T. Chowhan, Ph.D.   Department of Formulation Development, Syntex Research Institute of Pharmaceutical Sciences, Palo Alto, California

Peter N. Davies, Ph.D.*   School of Pharmacy, University of London, London, United Kingdom

---

Current affiliation: Roche Products, Welwyn Garden City, Hertfordshire, United Kingdom

Wendy C. Duncan-Hewitt, Ph.D. Faculty of Pharmacy, University of Toronto, Toronto, Ontario, Canada

John T. Fell, Ph.D. Department of Pharmacy, University of Manchester, Manchester, United Kingdom

Claus Führer, Ph.D. Institut für Pharmazeutische Technologie der Technischen Universität Braunschweig, Braunschweig, Germany

Everett N. Hiestand, Ph.D.* Upjohn Company, Kalamazoo, Michigan

Arne W. Hölzer, Ph.D. Astra Hässle AB, Mölndal, Sweden

Jukka Ilkka, M.Sc. Department of Pharmaceutical Technology, University of Kuopio, Kuopio, Finland

Per-Gunnar Karehill, Ph.D. Astra Hässle AB, Mölndal, Sweden

Ruth Leu, Ph.D. School of Pharmacy, University of Basel, Basel, Switzerland

Hans Leuenberger, Ph.D. School of Pharmacy, University of Basel, Basel, Switzerland

Fritz Müller, Ph.D. Pharmazeutisches Institut der Universität Bonn, Bonn, Germany

J. Michael Newton, Ph.D. School of Pharmacy, University of London, London, United Kingdom

Christer Nyström, Ph.D. Division of Pharmaceutics, Uppsala University, Uppsala, Sweden

Petteri Paronen, Ph.D. Department of Pharmaceutical Technology, University of Kuopio, Kuopio, Finland

Fridrun Podczeck, Ph.D. School of Pharmacy, University of London, London, United Kingdom

---

*Retired. Residing in Galesburg, Michigan

Gert Ragnarsson, Ph.D.* Pharmacia Biopharmaceuticals, Stockholm, Sweden

Ron J. Roberts, Ph.D. Zeneca Pharmaceuticals, Macclesfield, Cheshire, United Kingdom

Ray C. Rowe, Ph.D. Zeneca Pharmaceuticals, Macclesfield, Cheshire, United Kingdom

Martin Wikberg, Ph.D. Kabi Pharmacia Therapeutics Uppsala, Uppsala, Sweden

*Current affiliation: Astra Draco AB, Lund, Sweden

1

Interparticulate Attraction Mechanisms

Claus Führer

Institut für Pharmazeutische Technologie der Technischen Universität Braunschweig, Braunschweig, Germany

I. INTRODUCTION

A deciding factor in the technology of all dispersed systems is the influence of interparticulate attractions (see also [15,16]). Examples are the influence of the interparticulate attraction on the stability of emulsions and suspensions or on the flowability and compactibility of powders. The interparticular forces producing these attractions are a function of the nature of the particles and the interparticular distances. Normally the deciding properties are located on the particle surface or at particle interfaces.

In the following these considerations are applied to various phenomena in powder technology. A powder can be described as a special case of a dispersed system solid in gas, where the solid particles remain in contact. Because of this the powder may also be regarded to understand as a more or less bicontinuous system. The permanent interparticular contact is caused by external forces such as gravity and by interparticular interactions. Interparticular attractions between faces of the same material are called cohesion forces, independent of the nature of the interactions, whereas the attractions between faces of different materials are called adhesion forces [1,2]. Both depend on the same principles, and the distinction between them is thus sometimes very arbitrary. The adhesivity (resp. cohesivity) of a powder bed can be measured indirectly with a shearing cell (see also [3]).

Shearing forces may overcome the forces of interaction. Since the forces of interaction normally have their origin on the particle surfaces, whereas the shearing forces mostly depend on the mass of the particles, the phenomena become more significant with decrease of the particle diameter. Depending on the intensity of the mechanical treatment and on the special properties of the powder bed, the resulting irreversible deformations can be classified into three groups:

1. The contact areas change during the deformation from particle to particle, preserving the coherency of the solid phase in its totality. In this case one observes the typical flow properties of a powder bed.
2. A second possibility is the fracture of the powder bed into two or more parts. This corresponds to a fracture of a solid body. In this case the powder does not flow, it breaks.
3. If the solid phase is completely dispersed in the surrounding gas with no or only few interparticular contacts, the system has lost the typical properties of a powder and represents an aerosol.

Compression of a powder bed with partial elimination of the gas phase enhances the enlargement of the interparticular contact areas and thus the strength of the powder bed, which may finally be transformed in this way to a solid body (e.g., a tablet).

II. DIFFERENT TYPES OF INTERPARTICULATE ATTRACTIONS

The interparticular forces are roughly classified according to their origin as electrostatic or molecular interactions. For further classifications see also [4]. This classification is also arbitrary, and the names of the two types are not quite correct. Corresponding to this classification one understands electrostatic interactions as interactions of relatively highly charged particles or highly charged faces, being active over a relatively long distance. Molecular interactions normally also have the nature of electrostatic phenomena, but with small charges or polarizations on molecules or especially only on some atomes of the molecules, which are responsible for the valencies of secondary bonds. Because of the limited charge they have only a very short distance of activity (see also [5]).

A. Electrostatic Interactions

Two small single particles 1 and 2 with a distance r between them much larger than their diameters and with the electrostatic charges q_1 and q_2 interact with force F_{12}:

$$\frac{F_{12} = q_1 q_2}{4\pi\epsilon_0 r^2}$$

where ϵ_0 is the permitivity constant. The force is negative and attractive if the two charges have opposite signs. If both charges are positive or if both are negative, the force is repulsive. In the case of an assembly of particles with different charges, the equation holds for every pair of particles (Fig. 1).

Contrary to chemical linkages, the electrostatic interactions cannot be saturated. The total influence of the assembly on particle 1 can be expressed by the vector equation (see also Fig. 1)

$$F_{1i} = F_{12} + F_{1a} + F_{14t} + \ldots$$

Such equations can be used in an aerosol where the distances between particles are relatively large. In a powder bed the particles touch each other, and the electrostatic force influencing one particle is thus affected by interactions with the close surrounding particles. Because of the small distances or even direct contacts, the resulting forces can be relatively high. If the attractions are of the order of the weight of the particles or even higher, the powder cannot flow.

1. Triboelectrostatic Charge

A turbulent movement of a powder, especially in mixers, causes very frequent collisions of the particles mutually and with the surfaces of the vessel and the tools. With each collision normally an interfacial charge transfer of electrons occurs (Fig. 2a). Because of the intensive movement of the material causing the charge transfer, it seems sensible to nominate this event as a

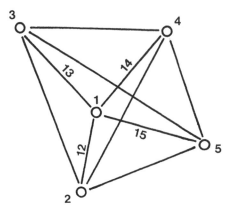

Fig. 1 Electrostatic interactions in an assembly of particles.

triboelectrostatic charge transfer. The direction of this charge transfer is given by the difference of the electron affinity of the two touching materials in the contact area at the moment of collision. The respective electron affinity depends not only on the chemical state of the materials but also on the crystallographic state, including the type of the lattice and distribution of lattice defects in the surface. But it depends also on the special state of the material under the influence of the high-energy transfer concentrated in the touching regions at the moment of collision.

The charge remains on the points of the transition if the respective contact area breaks under the influence of the further movement of the powder bed (Fig. 2b) and the particles separate again. Subsequent collisions of the charged particles have extremely low probabilities to find just the same points on the surface for the new interparticular contacts. Consequently, it is assumed that every or nearly every collision causes a new interparticular charge transfer (Fig. 2c). With each collision a new charge will be transfered. If the charges have no mobility within the surface, the process continues. Consequently, the density of the charged spots on the surface rises. Also the probability rises that an already charged spot collides with the surface of a particle. Thus the charge transfer will be hindered more and more.

Particles with highly polar surfaces have, under normal conditions, a water sorption layer which is able to distribute the transfered charge over the surface. The thicker this layer, the higher is its conductivity and the more easily is this charge distribution effected on the surface. The distribution becomes relatively homogeneous on the surface of an amorphous

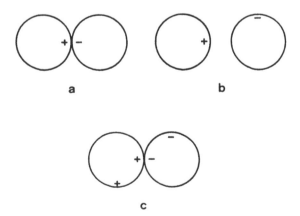

Fig. 2 Formation of the triboelectrostatic charge (schematically). a, First collision; b, separation; c, second collision.

spherical particle (Fig. 3a). But in the case of crystalline nonspherical particles, especially of milled crystalls with pointed edges, extremely high values of charge will be concentrated in the tops of the edges (Fig. 3b).

A relatively homogeneous charge distribution over the surface of the whole particle hinders more charge transfers. The process may reach an equilibrium. But with nonhomogeneous charge distributions, as schematicaly demonstrated in Fig. 3b, the total charge of the particle, especially in the region of edges, may rise to extreme values.

Mostly the surrounding gas is not indifferent [6]. In a humid climate the polar substances normally have a multimolecular water sorption layer and have a relatively frequent exchange of the water molecules of the particle surface with those of the atmosphere. Consequently, the charge is not merely distributed over the surface, but it can also leave the particle. Therefore, the total charge of the particles will be relatively low compared with materials without polar groups in the surface and no water sorption layer. In this case the total charge of each particle may rise to very high values depending on the intensity of the movement.

This shows that mixing two different powdered materials results in a disproportion of the charge within the powder bed. The two materials

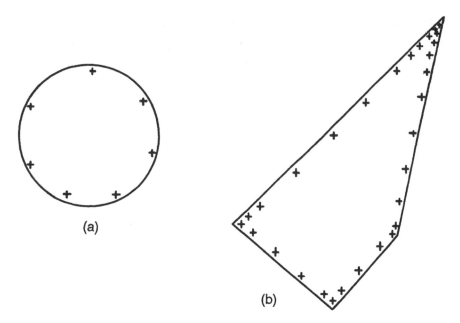

(a)

(b)

Fig. 3 Charge distribution under the influence of conductivity in the surface. a) Amorph spherical particle, b) milled crystal.

attract each other because of their opposite polarities. Depending on the composition of the mixture and the intensity of the movement, one can observe an increase in the bulk density and a decrease of the flowability.

In case of uniform particles of the same material a powder bed charges in a uniform sense with intensive movements, whereas the machine takes over the opposite charge. A similar effect can be observed if large amounts of air pass the powder bed, as for instance in a fluid bed. The powder bed and the air have opposite polarities. The machine usually has a certain conductivity, and the charged air leaves the system. Because of this the charge on the powder, compared with the ground potential, may rise to some 10,000 volts. In extreme dangerous cases, especially with nonpolar fine powders of large specific surface area, violent dust explosions may occur.

Powders charged with a uniform polarity are characterized by low bulk densities because of their elevated interparticular repulsion forces.

The charge distribution in a powder bed of a mixture allows ordered mixtures to be obtained and in this way to overcome the statistical optimal homogeneity. A uniformly charged powder allows, because of the elevated interparticular repulsions, sticking to be avoided in flow procedures. But charged powders do have the great disadvantage of bad charge reproducibility and bad stability.

2.   Permanent Electrostatic Charge on the Particle Surface

The permanent electrostatic charge is not induced by the mechanical treatment of the powder or the interaction of the particles but is a consequence of the surface structure of the powder. This charge is immobile and can only be suppressed by oppositely charged materials.

Unlike amorphous materials, the surface of a crystalline material is not uniform. Each face cuts the crystallographic lattice of the particle and represents a special case of a crystallographic plan of the lattice. The face, consequently, has its own structure, depending to its orientation against the lattice. It contains only some special atoms or atomic groups of the material in a state of order. Hence, each face has its own energy level, its own surface or interface tension, and, with the presence of polar functional groups, its own polarity.

a.   Ionized Surfaces

Due to the special chemical and crystallographic state of the surface of solid particles, surfaces can be charged, even if each particle in total is neutral. This is possible especially if the material itself is a salt or is coated by a ionized tensid.

With crystalline salts the state of order gives theoretically the possibility of the formation of crystallographic faces with high positive or high negative charge. Depending on the conditions of crystallization, it is to be expected that one can find neutral faces on the surface of the same particle along with positive or negative charged faces. With sodium chloride the octaeder faces (111) are positively or negatively charged, depending on the last layer of the ions forming the crystal surface (sodium or chloride). Such faces have a very high energy level and may exist only under certain conditions.

The surfaces of precipitates of salts are occupied initially by those ions which have been in stoichiometric surplus. Thus, a precipitation of silver chloride, for instance, yields particles with a surface occupied by chloride ions, if after the precipitation the mother solvent still has a certain cloride concentration. In contrast to a surplus of silver ions the precipitate has an outer layer of silver ions. In the first case the surface of the particle is negatively charged, whereas in the second case the surface is positively charged.

Amorphous materials also may have uniform charged ions in their surface which are responsible for electrostatic attractions or repulsions. A significant example is the attraction of the particles in a mixture of ion-exchange resins of opposite type. In a regenerated mixed-bed ion-exchange column, one observes a relatively highly ordered mixture, where the anion-active and the cation-active particles are in alternating places. This demonstrates the possibility of higher homogeneities in charged mixtures than the statistical optimum.

The coat of ionized tensid molecules around a particle of relatively low polarity has an orientation where the polar groups of the tensid molecules are located in the surface and the opposite nonpolar part covers the outer layer of the particle. If the particles are completely surrounded by such surface-active molecules, neighboring particles will come under strong interparticular repulsion forces. Thus the tensid has a disagglomerating effect on the powder.

b. Presence of Polar Functional Groups in the Surface

A crystalline material with highly ordered functional polar groups can also be the reason for a permanent charge on the surface. But the field intensity surrounding these groups is too low to cause electrostatic attraction or repulsion forces to influence decisively the mechanical properties of a powder bed.

B. Molecular Interactions

Highly charged particles and faces give interactions over long distances. The molecular interactions are also of an electrostatic nature, but they have only a very limited range. They are secondary bonds or van der Waals attractions that can be classified as ion-dipole, dipole-dipole, or van der Waals–London interactions.

Dry particles with no water sorption layer allow only very few molecular interparticular interactions. To form a molecular interaction, a short interatomic distance and a certain steric arrangement between the respective functional groups are required. Because of the roughness of the particle surfaces, the probability of such an interparticular arrangement is extremely low. When two bodies are placed in contact, there are, if any, very few functional groups which are by accident in that suitable opposite position to allow interparticular linkages (Fig. 4a). These few interactions are much too weak to resist the usual mechanical stress that makes a powder flow. Only small forces are needed to overcome the few linkages of interaction. Thus, even polar substances with a relatively high concentration of free secondary dipole-dipole valencies in the surface often have very low coherency in the dry state, which means good flow properties. The enlargement of the microscopic contact areas is proportional to the normal force, because the contact regions deform plastically under the great stress that develops in these regions (Fig. 4b).

In a humid atmosphere the surfaces of polar materials adsorb water. The thickness of the sorption layer depends on the polarity of the surface

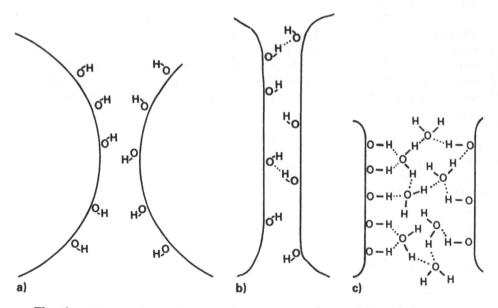

Fig. 4 Molecular interactions of polar substances: a) dry surface, b) dry surface after plastic deformation, c) with water sorption layer.

and the humidity of the atmosphere. The water molecules are linked to the surface and among themselves by hydrogen bonds. But the surface layer does not have the properties of a liquid. Information on the structure of the sorption layers is given by the arrangement of the polar groups in the surface of the particles. Thus even with a multimolecular layer the water molecules are not able to form the normal cluster arrangement. With the sorption layer the particle gets a new highly polar outer surface. The polar groups in this face have, contrary to the groups of the dry particles, a high flexibility, rising with the thickness of the sorption layer (Fig. 4c).

If two such particles come together, the water sorption layers might possibly interact over a high concentration of hydrogen bonds (Fig. 4c). The interparticular layer may be deformed, and finally the two particles have a joint water sorption layer. The result is a strong interparticular attraction. In consequence of this consideration one can understand the strong dependence of the coherency of polar materials on the humidity of the surrounding atmosphere.

C. Capillary Forces

A wetted powder contains, in addition to the sorption water, water as a separate phase, with the typical properties of a liquid. If the angle of contact between the water and the surface is small, a limited amount of water takes a preferred place in the interparticular contact regions with the formation of a meniscus corresponding to Fig. 5. A simple explanation of

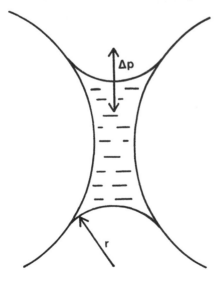

Fig. 5   Capillary forces between two particles.

the capillary forces may be given by the following: The concave deformation of the liquid surface requires a pressure difference between the atmosphere and the liquid, corresponding to

$$\Delta p = \frac{2\sigma}{r}$$

where Δp is the pressure difference, σ is the surface tension of the liquid phase, and r is the radius of the concave deformation. The pressure in the liquid phase must be less than the atmospheric level. Consequently, one can observe an interparticular attraction. The high surface tension of pure water gives strong attractions. On the other hand, on adding a surfactant the attractions become weak.

The attraction is influenced by the enlargement of the interparticular contact area and the radius of the meniscus as well as the surface tension. The radius of the meniscus depends not only on the angle of contact between the water and the solid surface but also, because of the more or less round form of the particle, on the amount of water between the particles. Very small amounts of water produce a surface with a concavity of small radius but of small interparticular contact area. Large amounts have large contact areas but also large radii of the meniscus. Since the functions of the enlargement of the contact area and the radius of meniscus are different, the attraction forces reach a maximum with a certain amount of the wetting liquid. This corresponds to the fact that the wet granulation requires a certain amount of water to give optimal mechanical properties in the mass, corresponding obviously to the maximum dilatant rheological behavior.

The importance of the surface tension in this effect demonstrates that all ingrediants lower the surface tension of the water.

III. CONSEQUENCES FOR PRACTICAL PHARMACEUTICAL TECHNOLOGY

In powder technology the interparticular interactions are important for the flow properties and bulk density of a powder, the mixing behavior, granulating, and tabletting or capsule filling. Only some of these will be discussed.

A. Flow Properties, Bulk Density

Interparticulate attraction in any case makes the flow properties worse. On the other hand, electrostatic repulsions may facilitate the flow of a powder bed. For the rheology of powders see [7–10].

Despite this, triboelectrostatic charges should be avoided, especially for powders provided for direct compression to tablets or for capsule filling. Both the tablet and capsule-filling machine regulate the dose volumetrically. In order to have tablets or capsules with a sufficiently constant mass, the bulk density of the powder must be as constant as possible during the whole fabrication of the batch. The bulk density of powders is very sensitive to electrostatic charge. The triboelectrostatic charge is not reproducible, does not distribute homogeneously within a batch of a powder, and changes rapidly during storage and the operation of tablet compression. Because of this the triboelectrostatic charge influences in a negative way the uniformity of the bulk density and with this the uniformity of dosage.

It is in many cases not possible to exclude absolutely any charge transfer by special measures during the handling of a powder (e.g., in milling, mixing, or drying operations), but there are various ways to diminish the triboelectrostatic charge transfer or to discharge the material.

1. The powder should have a sufficiently polar surface and a water sorption layer; i.e., it should have been handled or stored under the influence of air of sufficient humidity (e.g., 70% rh). If the material contains polar groups due to its chemical state, these groups should be present in the surface. With crystalline materials the crystallization conditions often allow variation of the polarity of the surface. A certain aid is given by the rough rule that the more polar the medium of crystallization, the more polar is the crystal surface. The temperature of crystallization also sometimes has an influence: the lower the temperature, the higher the polar character of the medium and the higher the polarity of the crystallizate.

2. The polarity of the surface can be enhanced by small concentrations of excipients (e.g., Aerosil), but also in the extreme case of nonpolar surfaces, with amphiphilic substances of the O/W type. They distribute on the surface coating the material with a more or less statistical orientation. In this way a certain concentration of polar groups in the surface may obtained. The typical O/W orientation of, e.g., emulsions are not to expected because for powders the air as gas does not attract the polar groups of the amphiphilic molecules.

3. Triboelectrostatic charged materials can be discharged by treatment with ionized air, such as air that has passed an electric field of high voltage. A second possibility is treatment with humid air, but many substances do not allow this. Both methods do not change permanent charges on the surface, and their efficiency is very limited.

4. All charges discussed, which are located in the particle surface-independent of their nature—can be compensated by the addition of amphiphilic excipients. The optimal excipient must be found empirically by experiment. The aim is to find a substance whose polar groups have an intensive affinity for the respective charged groups. In this case they take orientation where the polar groups are located on the original particle

surface and the nonpolar groups form the outer surface. Very often one choses substances with W/O character. These are more universal, whereas the more selective O/W surfactants may have a much better efficiency.

5. The possibility also exists of coating the particles with Aerosil in order to isolate more or less charged regions of the surface, against direct contact with the respective regions of other particles.

These considerations show that the normally used excipients in powder technology (e.g., lubricants) diminish the negative influence of the charges. But they demonstrate also that after the treatment of a powder with these excipients, intensive powder movements must be avoided. For the influence of interparticular attraction on the fluidization of a powder bed see [11–13].

B. Mixing Behavior

The mixing operation is, in the case of mutually indifferent substances and where the components have the same physical properties, a pure entropical operation. The optimal homogeneity of such mixtures is given by a pure statistical distribution of the components. In case of attractions or repulsions within the material this distribution can be disturbed.

A two-component system where one component has a positive charge and the other a negative occurs because of the interparticular attraction of the two different components, which causes better homogeneities than the statistical one. This effect is supported because of the repulsion between particles of the same component. In the case of a 1:1 mixture the material tends to an alternating state of order. It is sometimes useful to compare these states of order with the state of order of a crystalline substance. To do this the term "particle crystalline state" is proposed. The state of order represented under this definition is a more or less polycrystalline one. In the case where one component has a very high concentration and the other a low, the low-concentration component is separated and surrounded by the high-concentration one. In a statistical distribution the probability always exist that two or even more particles of the low-concentration substance are together. The charged material represents something like a solution. Between the two extremes—the particle crystalline state and the solution—homogeneous transitions exist.

It is evident that in any case the quality of a two-component mixture improves with the charge disproportion. The discussed disadvantages of triboelectrostatic charges as badly reproducible and unstable are not very important for the improvement of homogeneity.

Multicomponent systems are much more complicated. Already, in a pure statistical state, each component in the mixture has its own quality of distribution. Different homogeneities exist for one and the same mixture, depending on the respective questioned components. The influence of charges on the quality of such a mixture depends on the type of charge distribution. Consequently, for some components the effect is less than for others. One can also imagine that a multicomponent mixture of two or more components becomes worse under the influence of a triboelectrostatic charge.

C. Granulation

After having produced a mixture of a suitable quality, the aim of granulation is to agglomerate the particles to produce new particles of enhanced size. Each particle of the granulate should represent the whole composition of the mixture. For this purpose the interparticular attraction needed for the agglomeration must be uniform throughout the whole mass, independent of the character of the components.

The type of interparticular attraction used for this purpose must be relatively nonspecific with regard to the properties of the surfaces of the different materials. This is possible with interactions within the water sorption layers or with capillary effects. The water sorption layer is primarily used for dry granulations as the compaction, whereas the wet granulation uses the capillary forces. It is evident that both methods requires polar surfaces of all components. The distribution of the different components in terms of the particle sizes in a granulate is often a mirror of the polarity of the materials. Excipients used for granulating and tabletting are normally highly polar in their surfaces, whereas the active drug substances very often are much less polar. Consequently they have a much smaller tendency to agglomerate. Because of this one often finds preferred the active drug substances in the smallest particles of the granulate.

A good wettability (i.e., a small angle of contact) is a precondition for agglomeration with capillary forces. To enhance the wettability of the active drug substances, one sometimes adds surface-active substances. Since the agglomeration by wet granulation also depends on the surface tension of the granulation liquid, it may happen that the particle size distribution of the granulate shifts toward smaller particle sizes because of a poorer agglomeration tendency in the whole mixture. But often it is possible to find a suitable compromise between the wettability differences of all components and the tendency of the whole mass to agglomerate.

With dry granulation, one presses the powder (e.g., in a compactor), thus increasing the interparticular contact areas. It is in any case to be assumed that to some extent plastic deformations will occur. The interparticular linkages are qualitatively the same as those one observes in tablet compression. Since for most tablets adhesion and cohesion forces are mostly required to ensure short disintegration times, one expects with dry granulation methods that these forces are responsible for the strength of the granulate.

Consequently one has to require that the surface of all components be polar with a multimolecular water sorption layer. Nonpolar-active drug substances can be included in the compacted mass but do not actively participate in the solidification. Because of the water sorption layer required, the mass should be in equilibrium with a humid atmosphere of about 60–70% rh.

IV. SOLID BRIDGES

With mechanical treatment—tabletting or dry granulation—and with wet granulation, the interparticular interfaces of the contact areas within a powder bed may disappear by sintering or by recrystallization [14]. Thus, the particles grow together and lose their individuality. The material of these areas passes by diffusion over the contact interface, resulting in a more or less weak transition of the composition and structure between the primary particles. The strength of the solid bridges corresponds to the internal strength of the particles. It is not possible by mechanical or other treatments to divide these aggregates into their original particles. To have a separation, a fracture of the solid bridge must occur. Thus, the resulting particles are not identical to the original. This is the significant difference of the solid bridges to the above-mentioned interparticular interactions. Agglomerates formed by the latter may be separated into their primary particles.

Solid bridges are always to be expected if the material has, over a limited time period, a certain mobility in the contact areas. This mobility may be caused by a fused state under the influence of heat or a dissolved state under the influence of a multimolecular solvent sorption layer, preferably water.

With mechanical treatment, elevated energy levels in the interparticulate contact areas occur, which may overcome the molecular interactions. This occurs at first in the preferred crystallographic gliding or fracture planes but with higher levels also in the stronger planes. Because of the broad energy distribution in the stressed substance, regions with energy levels sufficient to destroy the crystallographic state exist, i.e., to transfer the material into an amorphous liquid state. Extremes of those energy levels are the hot spots, i.e., very small regions where the energy of the material corresponds to extreme temperatures being able even to decompose the molecules.

The consequence of the broad energy distribution and the steep gradients of energy within the material is that the energy level may change very fast. The half-life of the hot spots is very short. Passing the maximum energy level, the material solidifies and may recrystallize during the following decrease of energy. Naturally, such solid bridges have a high concentration of lattice defects or remain even in the amorphous state.

A special case of solid bridges is the interparticular interactions with strong binders. One assumes that the material of the particles and the binder diffuse across the interfaces. Thus, the interface becomes a more or less broad and diffuse region. This consideration demonstrate a relationship of this type of interaction to the normal interparticular interaction without binder.

REFERENCES

1. A. Martin, J. Swarbrick, and A. Cammarata, *Physical Pharmacy*, Lea & Febiger, Philadelphia, PA, 1983.
2. O. Molerus, *Particle Technol.* (Conference), Amsterdam NL, June 3–5, 1980, p. 932.
3. W. Schütz and U. H. Schubert, *Chem. Eng. Tech. 48*: 567 (1976).
4. H. Rumpf, *Chem. Eng. Tech. 30*: 144 (1958).
5. E. N. Hiestand, *J. Pharm. Sci. 55*: 1325 (1966).
6. K. Rietema, J. Boonstra, G. Schenk, and A. H. M. Verkooyen, *Particle Technol.* (Conference), Amsterdam NL, 1980, p. 981.
7. T. Y. Chem, W. P. Walawender, and L. T. Fan, *Powder Technol. 22*: 89 (1979).
8. J. S. M. Botterill and B. B. Abdal-Halim, *Powder Technol. 23*: 67 (1979).
9. S. Stemerding and G. W. J. Wes, *Chem. Weekblad, 75*: 8 (1979).
10. O. Molerus, *Powder Technol. 20*: 161 (1978).
11. S. M. P. Musters and K. Rietema, *Powder Technol. 18*: 239 (1977).
12. S. M. P. Musters and K. Rietema, *Powder Technol. 18*: 249 (1977).
13. K. Schügerl, *Fluidisation*, (J.F. Davidson and D. Harrison, eds.), Academic Press, London, 1971, Chap. 6.
14. C. Führer, *Labo-Pharma-Probl. Tech.* (1977), 25(269), 759–62.
15. 2nd World Congress Particle Technology, Kyoto, Japan 1990 Society of Powder Technology, Japan; VDI, Germany; Fine Particle Soc., USA.
16. K. Iinoya, J. K. Beddow, and G. Jimbo, *Powder Technology*, Hemisphere, Washington, DC, 1981.

2

The Importance of Intermolecular Bonding Forces and the Concept of Bonding Surface Area

Christer Nyström

Uppsala University, Uppsala, Sweden

Per-Gunnar Karehill

Astra Hässle AB, Mölndal, Sweden

I. INTRODUCTION

A. Stages in the Compaction Process

When a force is applied on a powder bed consisting of more or less nonporous particles, a number of mechanisms become involved in the transformation of the powder into a porous, coherent compact with a well-defined shape. A compaction process is normally described by a number of sequencial phases [1–3]. Initially, the particles in the die are rearranged resulting in a closer packing structure. At a certain load, the packing characteristics of the particles or a high interparticulate friction between particles will prevent any further interparticulate movement. The subsequent reduction of compact volume is therefore accompanied by elastic and plastic deformation of the initial particles [4–7]. Elastic deformation is a reversible, while the plastic is an irreversible deformation of the whole or a part of a particle. For many materials these particles are then fragmented. Fragmentation can be defined as a dividing up of a particle into a number of smaller, discrete parts [8,9]. The particle fragment will then normally find new positions, which will further decrease the compact volume. When the applied pressure is further increased, the smaller particles formed could again undergo deformation [3]. Thus, a single particle may pass through one or several of these processes several times during a compression. As a consequence of the compression of the powder, particle surfaces are brought into close proximity to each other and interparticulate attraction or bonds will be formed.

The volume reduction processes consume energy (endothermal processes) and will normally increase the amount of particle surface area capable of forming interparticulate attraction forces. Bond formation is however, an exothermal process, thereby releasing energy [10]. During ejection, when the load is reduced, many materials produce laminated (capped) compacts or results in compacts of pronounced low strength. These observations indicate the importance of the elastic component of tableting materials.

To summarize, the following processes are involved in the compaction of a powder:

1. Particle rearrangement
2. Elastic deformation of particles
3. Plastic deformation of particles
4. Fragmentation of particles
5. Formation of interparticulate bonds

Examples of materials consolidating by plastic deformation are sodium chloride, starch, and microcrystalline cellulose [8,11,12]. Fragmenting materials are, for example, crystalline lactose, sucrose, and Emcompress [8,13–16]. However, all materials possess both an elastic and a plastic component. The volume reduction mechanism which will dominate for a specific material is dependent on factors such as temperature and compaction rate. Lower temperatures and faster loading [17,18] during compression will generally facilitate consolidation by fragmentation. Pharmaceutical materials normally consolidate by more than one of these mechanisms [3], which emphasizes the need for adequate characterization techniques.

B. Physical Description of a Tablet

The axial compaction of pharmaceutical powders results in anisotrope and inhomogeneous compacts or tablets; i.e. a tablet shows varying values of some characteristics, porosity, density (e.g. [19]), bonding, mechanical strength in different directions and parts (Fig. 1).

For normal compaction pressures, not exceeding 300–500 MPa, the final compact porosity is 1% to 25%, depending on the powder compressibility. Two extreme models could be used to describe the distribution of

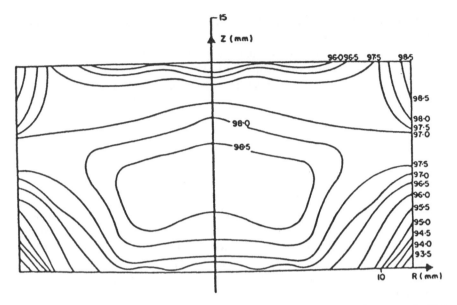

Fig. 1 Density variation in the cross section of a die compact. (From Ref. 19).

A Swiss Cheese   A particulate dispersion in air

Fig. 2 Models for describing the physical structure of a pharmaceutical tablet.

this gas phase (Fig. 2). First, the air could be regarded as a disperse phase of individual units incorporated in a solid continuous phase (like a Swiss cheese). Then the pores are to be considered as intra particulate pores in a large particle (the tablet). Second, the air could be regarded as a continuous phase, in which solid particulate units are dispersed. In this case, the individual solid particles are separated by some distance and the tablet contains continous pores like a loose powder plug. Then the tablet can be penetrated by a flowing medium and characterized on, e.g., permeability properties [20] and permeametry surface area.

Which of these extreme models that are closest to a correct description is largely related to the degree of compression and to the nature of the dominating bond type. If solid bridges easily can be formed due to melting, the first model may be relevant. This could be the case for some polymeric materials with a low melting temperature. However, for common tabletting materials strong evidence has been presented supporting the second model, as will be discussed below.

C. Primary and Secondary Factors for Tablet Strength

Two factors could be regarded as primary factors for the compactability of powders [21–27]: the dominating bond mechanism and the surface area over which these bonds are active. Owing to considerable experimental difficulties, these factors have not been evaluated in any detail for pharmaceutical materials. Instead, more indirect, secondary factors are normally studied and used for correlations with tablet strength. Such secondary factors are particle shape, surface texture and particle size. The importance of volume reduction mechanisms (i.e., elastic deformation, plastic deformation and particle fragmentation) have also been studied in detail.

D. Bonding Surface Area

The term *bonding surface area* is often defined as the effective surface area taking part in the interparticulate attraction. In the case of solid bridges, the term corresponds to the true interparticulate contact area, while for intermolecular forces the term is more difficult to define. It can seldom be estimated from direct measurements of the surface area of the starting material. This is especially obvious for strongly fragmenting materials [28]. Furthermore, in practice, many powders possess, in addition to their external visible surface area, an internal surface area. This internal surface area is small for dense crystalline solids such as sodium chloride, but in porous bodies such as microcrystalline cellulose and Emcompress the internal surface area may be considerably greater than the external surface area [8].

Thus the bonding surface area is a function of several secondary factors [29]. Apart from the complex origin of a bonding surface area, this property is also difficult to define exactly. Consequently, experimental determinations are rare in the literature. Instead of a direct measure of the bonding surface area, the secondary factors listed in Table 1 have been measured and used for the correlation to tablet strength [29].

Of special importance for the final bonding surface is probably the particle elasticity of the materials. This property is normally not measured, but the axial elastic recovery of the tablet is determined. Extensive

Table 1 Factors Influencing the Surface Area of Tablet Particles and the Bonding Surface Area in Tablets

Tablet particle surface area		Bonding surface area	
Before compaction	After compaction	During compaction	After Compaction
Particle size	Particle size	Particle size	Particle size
Particle shape	Particle shape	Particle shape	Particle shape
	Fragmentation	Fragmentation	Fragmentation
		Plastic deformation	Plastic deformation
		Elastic deformation	Elastic deformation
			Elastic recovery
			Friction properties
			Bond strength

particle elasticity could cause a drastic decrease in tablet strength, due to the breakage of interparticulate bonds, thereby reducing the bonding surface area [3].

E. Bonding Mechanisms

The general bonding mechanisms co- or adhering particles have been classified by Rumpf to be of mainly five types [30]:

1. Solid bridges (sintering, melting, crystallization, chemical reactions, and hardened binders)
2. Bonding due to movable liquids (capillary and surface tension forces)
3. Non-freely-movable binder bridges (viscous binders and adsorption layers)
4. Attractions between solid particles (molecular and electrostatic forces)
5. Shape-related bonding (mechanical interlocking)

This general classification has been widely accepted in the literature. In the case of compaction of dry, crystalline powders, it has been suggested that the mechanisms of importance could be restricted to classes 1 and 4 [15] and perhaps class 5 [31]. However, it cannot be excluded that the presence of liquids in a compact might be of significance for the tablet strength [27,32]. It can be discussed, though, if the change in tablet strength is due to an effect of liquid on the compressibility of the powder or on the nature of the particle-particle interactions.

However, the dominating bond types adhering particles together in

compression of dry powders could for simplicity be limited to three types
[31] (Table 2):

1. Solid bridges (due to, e.g., melting)
2. Distance attraction forces (intermolecular forces)
3. Mechanical interlocking (between irregularly shaped particles)

The first type corresponds to strong bonds, where a true contact is established between adjacent particles. The second group could roughly be
described as weaker bonds acting over distances.

The term *intermolecular forces* is used in this chapter as a collective
term for all bonding forces that act between surfaces separated by some
distance. Thus, the term includes van der Waals forces, electrostatic forces,
and hydrogen bonding [33]. The dominant interaction force between solid
surfaces in the van der Waals force of attraction [34–36]. This force operates in vacuum, gas, and liquid environments up to a distance of approximately 100–1000 Å. Hydrogen bonding is predominantly an electrostatic
interaction and may occur either intramolecularly or intermolecularly [37].
These bonds are of special importance for many direct compressible binders such as Avicel, Sta-Rx 1500, and lactose. Electrostatic forces arise
during mixing and compaction due to triboelectric charging. These electrostatic forces are neutralized with time by electrostatic discharging. For
compacts stored at ambient relative humidity or in liquids, this is a relatively fast process owing to the high diffusivity of the charges in the liquid
or the adsorbed liquid layers.

Table 2 Some Specifications of Bonding Mechanisms in Compacted Dry
Powders

Type (−)	Dissociation energy (kcal/mol)	Separation distance at equilibrium (Å)	Maximum attraction distance (Å)
Solid bridges			
Covalent homopolar	50–150	<2 ⎫	
Covalent heteropolar	100–200	⎬	<10
Ionic	100–200	<3 ⎭	
Intermolecular forces			
Hydrogen	2–7 ⎫		
van der Waals	1–10 ⎭	3–4	
Electrostatic	—	—	100–1000
Mechanical interlocking	—	—	—

Solid bridges that contribute to the overall compact strength can be defined as areas of real contact, i.e., contact at an atomic level between adjacent surfaces in the compact. Different types of solid bridges have been proposed in the literature, such as solid bridges due to melting, self-diffusion of atoms between surfaces, and recrystallization of soluble materials in the compacts [38–41]. Solid bridges can be detected by electrical resistivity measurements [42,43]. Electric conductance is found in compacts manufactured from metal powders or polycrystalline materials. Most of the electric conductance arises from valency vacancies and impurities in the crystal [43]. The amount of electricity that travels between the different crystals in solid and liquid bridges is in the ideal case proportional to the area of real contact between the surfaces. Calculations of the contact area between metallic surfaces have indicated that the area of real contact is relatively small compared with the available geometrical surface area [42].

The term *mechanical interlocking* is used to describe the hooking and twisting together of the packed material. It has been claimed [31] that materials bonding predominantly by this mechanism require high compression forces and have low compact strength and an extremely long disintegration time. However, a more limited description of this bonding mechanism is that it is dependent on the shape and surface structure of the particles; i.e., long needle-formed fibers and unregular particles have a higher tendency to hook and twist together during compaction compared with smooth spherical ones.

II. REPORTED MEANS OF ESTIMATING THE DOMINATING BOND MECHANISMS

A. Surface Specific Tablet Strength

Since solid bridges seem to be relatively strong bonds, while intermolecular forces are weaker, though acting over distances, a ratio between the compact strength and the surface area of the starting compound could perhaps be used to distinguish the two bonding mechanisms [23]. The surface specific strength would then give a high value for a material bonding predominantly with solid bridges and a low value for a material bonding with long-range forces as the dominating bond type. In Table 3 the compact strength is expressed in relation to the surface area of the respective materials.

The coarse sodium chloride qualities gave high values, and iron, Sta-Rx 1500, sodium bicarbonate, Avicel PH 101, and sodium chloride <63 μm gave lower values. It is of special interest that the surface specific strength of Avicel values is of the same order as the other materials in this group. This indicates that the high absolute tablet strength of Avicel probably is caused by a high surface area taking part in the bonding.

Table 3 Surface Specific Radial Tensile Strength of Test Materials Compacted at 150 MPa

Material (−)	Radial tensile strength δ_x (kPa)	Powder surface area s (cm^2)	Surface specific tensile strength, δ_x/s (kPa/cm^2)
Sodium chloride 425–500 μm	1150	44[a]	26.1
Sodium chloride 250–355 μm	793	68[a]	11.7
Sodium chloride <63 μm	2990	612[b]	4.9
Iron	1880	1490[b]	1.3
Avicel PH 101	7330	1610[b]	4.6
Sodium bicarbonate 90–150 μm	571	503[b]	1.1
Sta-Rx 1500 90–150 μm	514	358[b]	1.4

[a] Surface area of the amount of material corresponding to a tablet, as measured by gas adsorption.
[b] Surface area as measured by permeametry.

The results indicate that all of the investigated materials are predominantly bonding with long-range forces; however, two of the materials (i.e., sodium chloride 425–500 μm and sodium chloride 250–355 μm) are also bonding with a significant contribution from solid bridges.

B. The Use of Lubricant Films

The effect of mixing sodium chloride, iron, Avicel PH 101, sodium bicarbonate, or Sta-Rx 1500 with small amounts of magnesium stearate on compact strength was investigated [24]. The effect of removing magnesium stearate by soaking the compacts in an organic solvent was also studied in an attempt to regain the initial compact strength [24].

The main effect of the magnesium stearate film on compact strength is to reduce bonding with intermolecular forces. If intermolecular forces were the only bond type present and no fragmentation of the particles or rupture of the lubricant film occurred, a compact strength close to zero would be expected. For materials bonding with solid bridges, relatively high local stresses may be formed within the compact, as suggested for coarse sodium chloride [23]. Such materials ought then to be capable of penetrating the lubricant film prior to the formation of solid bridges. The compact strength should level off at higher concentrations of magnesium stearate, thereby ideally reflecting the relative contribution of solid bridges to the total compact strength.

The proportion of lubricant added to each test material corresponded to values above and below the theoretical amount required to form a monomolecular coat on each material [58].

The plot (Fig. 3) of the remaining strength (%) against the weight of added magnesium stearate in relation to the compound surface area shows that the radial tensile strength decreases for all materials with increasing amounts of magnesium stearate. This is in agreement with earlier reports [15,44], where this effect was explained in terms of the formation of a molecular lubricant film around the compound particles.

At concentrations between 0.2 and 1 $\mu g/cm^2$, the compact strength

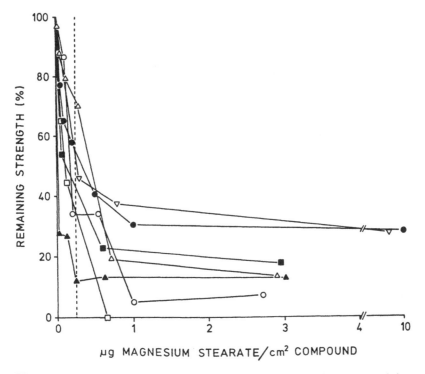

Fig. 3 The effect of surface specific lubricant concentration on remaining strength. (△) Avicel PH 101; (■) iron; (○) sodium chloride <63 μm; (●) sodium chloride 250–355 μm; (▽) sodium chloride 425–500 μm; (▲) sodium bicarbonate; (□) Sta-Rx 1500. (From Ref. 23).

leveled off and reached a constant value for all materials except Sta-Rx 1500. This is in agreement with the lubricant concentration predicted to give a molecular film around the particles on the assumption that magnesium stearate corresponds to a molecular surface area of 400 m^2/g [44]. This is equivalent to 0.25 μg magnesium stearate/cm compound and is denoted by a dotted line in Fig. 3.

The results indicate that a shift in main bond type has taken place; i.e., intermolecular forces have been filtered out by the magnesium stearate film and that the dominating bond type involves solid bridges when the plateau has been reached. If this assumption is correct the relative contribution by solid bridges could be estimated.

The coarser fractions of sodium chloride appear to bond with a relatively high proportion of solid bridges, while for the other materials the contribution is less. However, for Avicel PH 101 an incomplete surface coverage by the lubricant and a limited particle fragmentation can be expected which will influence the interpretation of the results. This assumption was supported by the fact that the constant level found for Avicel PH 101 at 150 MPa was absent when the material was compressed at 35 MPa. Then zero compact strength was obtained. According to Sixsmith [12] there is limited fragmentation tendency below 50 MPa. The level found for Avicel PH 101 at 150 MPa is thus best explained by an incomplete surface coverage with magnesium stearate after compaction due to fragmentation of the Avicel PH 101 particles rather than solid bonds penetrating the surface film.

All lubricated tablets were then tested on soaking by immersing them in a mixture of isoamylalcohol and chloroform (1/1) for seven days. For all materials, soaking the unlubricated compacts resulted in little change in the mechanical strength, indicating that the solvent did not cause any specific interaction, such as dissolution followed by recrystallization. For all lubricated compacts except sodium bicarbonate the soaking procedure resulted in a reduction in the amount of magnesium stearate present in the compacts and with the exception of sodium bicarbonate, the lubricated compacts increased significantly in strength after the soaking (Figs. 4 and 5). Theoretically, if magnesium stearate functions as a filter for intermolecular attraction and thereby reduces the long-range bonding forces, the compact strengths ought to be fully restored after soaking, provided all lubricant is removed and that the proportion and degree of solid bridges have not been altered due to the admixing of the lubricant. However, the soaked compacts never attained the same strength as the unlubricated compacts. The most probable explanation for this is incomplete soaking of the specimens tested. For sodium chloride <63 μm (Fig. 4), the treatment resulted in an almost complete recovery of compact strength.

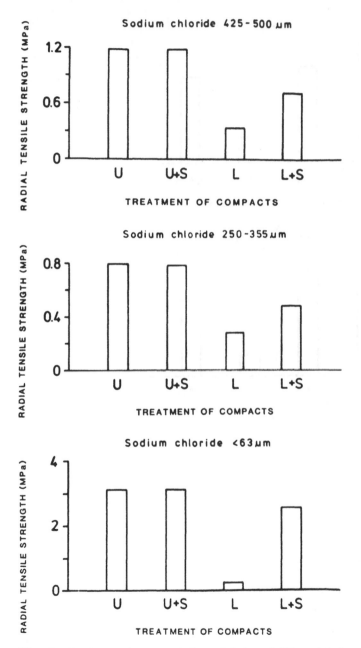

Fig. 4 Radial tensile strength for unlubricated (U), unlubricated plus soaked (U+S), lubricated (L) and lubricated plus soaked (L+S) compacts of sodium chloride. (From Ref. 23).

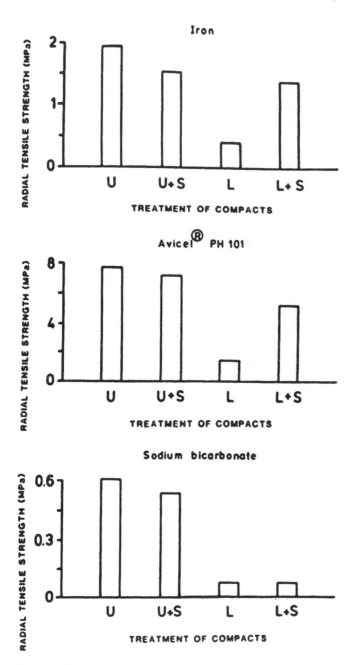

Fig. 5 Radial tensile strength for unlubricated (U), unlubricated plus soaked (U+S), lubricated (L) and lubricated plus soaked (L+S) compacts of iron, Avicel PH 101 and sodium bicarbonate. (From Ref. 23).

The results from the soaking experiments indicate that intermolecular forces, acting over distances, contribute substantially to the strength of the tested compacts. These forces could be filtered out by the addition of a film-forming lubricant. However, this effect seems to be reversible in the sense that removal of the lubricant increases the strength again. Intermolecular forces thus constitute the dominating or only bonding mechanism for the investigated materials. Only for sodium chloride is there substantial evidence for the existence of solid bridges.

C. Powder Compaction in Liquids with Different Dielectric Constants

If a surface is covered with a liquid, the liquid will act as a barrier for the adhesion forces and hence reduce the interaction between surfaces in close proximity. However, if more liquid condenses on the surface, liquid bridges may be formed between surfaces in close contact. These bridges are strong, and could thus increase the compact strength. However, it should be noted that only the interfacial forces at the liquid-gas interface contribute to the bonding force between the grains. As soon as the liquid completely envelops the compact, all capillary bonding force vanishes [38].

Shirvinskii reported that the adhesion forces for a number of different particulate materials (silicon, aluminum, and calcite) adhering to surfaces of silicon, quartz, Teflon, and steel decreased with increasing dielectric constant for liquid binary media or along a curve with a minimum plateau value [45]. The basic idea of this theory is that the forces interacting between surfaces in close contact are considered to be due to a fluctuating electromagnetic field. Because of the quantum mechanical fluctuations, this field is always present in the interior of a material medium, and it also extends beyond its boundaries. This theory is applicable to any material independent of its molecular nature [46,47].

From the Hamaker and Lifshits theories some general conclusions related to the forces interacting in a medium between particles can be drawn:

a. The London–van der Waals forces between two particles of the same material dispersed in a fluid are always attractive, provided there is no marked orientation of the fluid molecules. If the particles are of different composition, the resultant force may be repulsive in its nature [46,48].
b. The London–van der Waals forces between any two bodies in vacuum are always attractive [46,48].
c. A decrease in London–van der Waals forces with an increase in the dielectric constant of the medium could be expected [45,46].

It seems, therefore, reasonable to determine the tablet strength for a number of pharmaceutical materials compressed in liquids with different dielectric constants, and to compare these strength values with the compact strength values obtained in both ambient air and in vacuum. The importance of solid bridges or mechanical interlocking in relation to bonding with intermolecular forces could then be obtained since these latter forces are at a minimum for liquids with dielectric constants of 10 to 20 [45].

1. Apparatus for Compaction in Liquids

Tablets were compressed in liquids in a specially designed apparatus manufactured of steel (Fig. 6) equipped with flat-faced punches of diameter 1.13 cm [24].

The apparatus was filled with liquid to the top of the die. All liquids used in this study were preconditioned with the solid material prior to compaction to obtain a saturated solution. Excess liquid was then removed from the die with a pipette, making it possible to add the powder in dry form. An amount of powder sufficient to give a compact height of approxi-

Fig. 6   Tablet compression apparatus: (1) upper punch, (2) load cell, (3) compression chamber, (4) lower punch, and (5) die. (From Ref. 24).

mately 0.3 cm at 150 MPa pressure was poured into the die. Liquid was then added to the apparatus to completely wet the powder mass. After 1 min the upper punch was inserted in the die and the compression chamber was mounted in the hydraulic press. The upper punch pressure was then raised to 150 MPa over 10 s. This procedure gave two effects. First, the liquid was able to leave the tablets when the load increased without disturbing the consolidation of the compact, and, second, it increased the time for plastic flow in the tablet. Ejection of the tablet from the die was performed in the liquid by turning the die and applying pressure with the hydraulic press on the lower punch. The compact was not exposed to air during the compression and ejection phase since the compression chamber was completely filled with liquid. After ejection the tablets were removed and stored in the saturated liquid for 24 h before the tensile strength was measured. The compaction procedure described was carried out in a ventilated safety box.

Two plastically deforming materials (sodium chloride, Avicel PH101) and one material undergoing extensive fragmentation (lactose) were compressed both in air at ambient conditions and in liquids with different dielectric constants [24].

The use of liquids to reduce bonding with intermolecular forces has one important limitation for compounds easily soluble in water (e.g., sodium chloride and lactose). The solubility of these materials increases drastically with increasing dielectric constant of the test liquids [49]. A high solubility increases the risk that solid bridges and surface properties of the compact can change due to dissolution. This risk is difficult to grade and evaluate. However, for liquids with a dielectric constant below 10 to 20 the solubility is limited and subsequently does not significantly affect the bonding properties of the tested materials [24].

2. Apparatus for Characterization of Radial Tensile Strength in Vacuum

A vacuum tensile strength tester was constructed [24] (Fig. 7). The tester was manufactured from a tube of stainless steel with two movable vacuum tight end plates. On each side of the tube two bellows were attached with vacuum tight welding. Inside the bellows the tablet-crushing device moved freely with a minimum of friction to transmit the force to the tablets in the tester. The force was manually transmitted to the crushing device by a traction wheel and measured with a piezoelectric crystal attached to the bellows. The tester could hold five compacts, stored in a sliding magazine. Prior to each measurement, the tablets were degassed for 24 h or longer at a pressure of less than 10^{-4} mm Hg. For each material, 10 tablets were

Fig. 7 Vacuum tensile strength tester: (1) load cell, (2) traction wheel, (3) end plate, (4) magazine, (5) bellow, and (6) tube. (From Ref. 24).

measured in vacuum and 10 in air in the apparatus. The radial tensile strength was then calculated.

The results (Table 4) showed that all investigated materials increased in radial tensile strength in vacuum. This was explained by the removal of condensed material, primarily water from the particle surfaces in the compact [50]. Similar results have been reported for water absorption on degassed silica compacts [51]. They found that the compact strength decreased proportionally to the surface coverage of water. This suggests that the strength increase in vacuum is primarily caused by an increased surface interaction due to removal of adsorbed water vapor and surface contamination, which act as a filter to reduce bonding with intermolecular forces [24]. A completely clean surface correspond to a dielectric constant of unity (i.e., maximal interaction between surfaces in close contact).

In Fig. 8 the compact strength in liquids is compared with the radial tensile strength in vacuum presented, and in Fig. 9 the corresponding remaining strength values (%) are given. Since the tensile strength of compacts prepared in air can be regarded as a starting point before transfer takes place to a vacuum or a liquid environment, the compact strength measured under ambient conditions is denoted by a different symbol (X) in the figure.

Table 4 Radial Tensile Strength Measured under Ambient Conditions and in Vacuum of Compacts Prepared at 150 MPa

Material (−)	Tensile strength (MPa)[a]		Increase in tensile strength (MPa)
	Ambient	Vacuum	
Avicel PH101	6.37 (0.81)	8.08 (0.73)	1.71
Sodium chloride, coarse	1.01 (0.13)	1.28 (0.20)	0.27
Sodium chloride, fine	4.13 (0.56)	5.00 (0.56)	0.87
Lactose, coarse	0.73 (0.08)	0.80 (0.10)	0.07
Lactose, fine	1.03 (0.22)	1.40 (0.21)	0.37

[a] Standard deviations are given in parentheses.

All materials decreased with increasing dielectric constant in the interval tested. This is similar to the results obtained above with small amounts of magnesium stearate [23]. The formation of a plateau was suggested to be typical for a material bonding with at least two different bond types, say solid bridges and intermolecular forces [21,23]. When a stable plateau is obtained, the compact strength will in the ideal case be determined by solid bridges or mechanical interlocking alone.

It seems therefore that the results [24] for both size fractions of sodium chloride support the existence of solid bridges for this material. A similar result was obtained for coarse lactose compacts which also formed a plateau. Although lactose is described as a brittle material, the participation of plastic flow in the densification process has been reported [3]. A limited contribution of bonding with solid bridges was therefore regarded as a possible explanation for the plateau for coarse lactose. For the fine fraction of lactose, no coherent compacts were formed when the liquid dielectric constant exceeded approximately seven, indicating that intermolecular forces is the only bonding mechanism for this size fraction.

Avicel PH 101 showed a continuous decrease in tablet strength down to zero with increasing dielectric constant. Thus bonding with intermolecular forces seems to be the dominating bonding mechanism for this material.

In a recent study [52] also the effect of compaction pressure on the relative contribution of solid bridges was tested for sodium chloride (Fig. 10). In these experiments only one liquid (butanol) was used with a dielectric constant of 17.8. The results demonstrate that the development of solid bridges is increased with compaction pressure.

Also earlier some studies have been reported [53,54], where the concept of using liquids for filtering out, predominantly the intermolecular

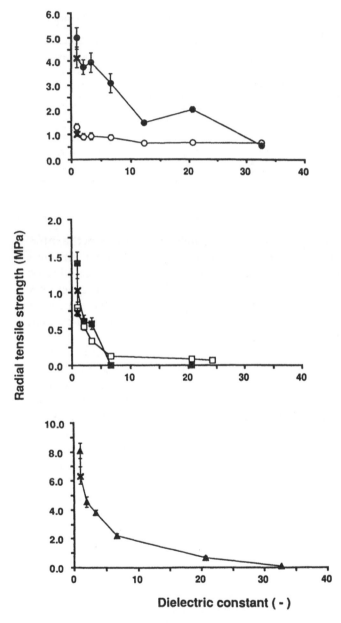

Fig. 8 The effect of the dielectric constant of the test liquid on the compact strength. (○) sodium chloride coarse; (●) sodium chloride fine; (□) lactose coarse; (■) lactose fine; (▲) Avicel PH101, compressed at 150 MPa. Error bars represent confidence intervals of the mean for 95% probability. For some of the result, precision is better than can be denoted in figure. (From Ref. 24).

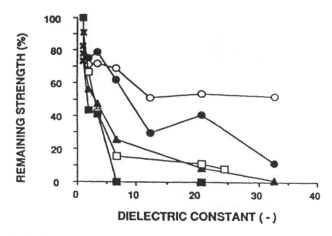

Fig. 9 The remaining compact strength in liquids compared with the radial tensile strength in vacuum. Compact strength in ambient surrounding is marked with X. Symbols as in Fig. 8. (From Ref. 24).

forces, was applied. However, in those reports, the compacts were prepared in conventional manner, i.e., in a dry state, and then subsequently soaked in different organic liquids. Although, this technique might be applicable in cases where relatively porous tablets, without any distinct peripheral high-density "seal," are at hand, it has been established in our laboratory to be an unsuitable technique for most pharmaceutical materials. The main problem is the limited possibility for liquids to penetrate the compact and to reach the interior of the tablet.

III. REPORTED MEANS OF ESTIMATING BONDING SURFACE AREA

A. Relations between Tablet Surface Area and Bonding Surface Area

For many materials it has been possible to relate an improved tablet strength to, e.g., a decrease in particle size or a change to more irregularly shaped particles. It has also been claimed that materials which deform plastically bind better than materials undergoing elastic deformation or extensive particle fragmentation [55,56]. Plastic deformation is thus believed to be of special importance for the formation of a large bonding surface area (e.g., [31]).

Some fundamental studies in the field of metallurgy [42] have indicated that the surface area taking part in the attraction between compact

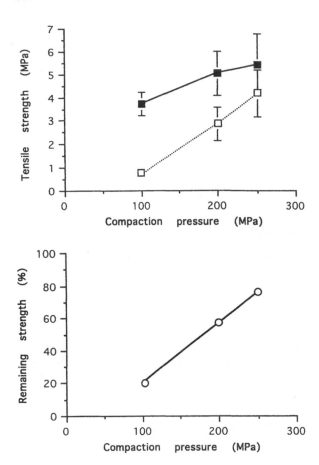

Fig. 10 Upper graph: Effect of the compaction pressure on the compact
strength of sodium chloride 20–40 μm. Compaction under ambient conditions ϵ
= 1 (■), compaction in butanol ϵ = 17.8 (□). Lower graph: The remaining
compact strength of sodium chloride 20–40 μm in butanol compared with the
compact strength under ambient conditions.

particles is relatively small, being only a minor fraction of the geometrical
surface area available. Studies on pharmaceutical materials, using gas adsorption techniques, have, however, suggested that larger surface areas are
involved [11,28,57,58]. In these studies, a decrease in tablet surface area
with compaction load has been regarded as a reflection of the surface area
utilized for the bonding between particles.

　　With few exceptions, the relations between compaction load and measured tablet surface are a presented in the literature are relatively complex

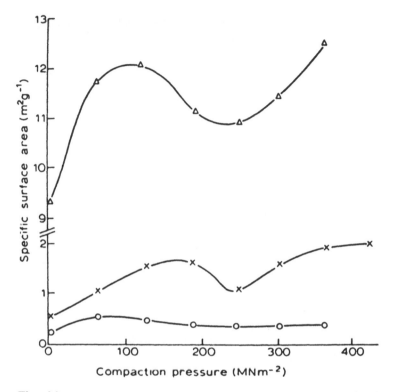

Fig. 11 Specific surface area—compaction pressure profiles of sodium chloride (○), phenacetin (×) and magnesium carbonate (△). (From Ref. 57).

(e.g., [57]) (Fig. 11). Normally, an initial increase in tablet surface area with compaction load is recorded. This has been explained by a fragmentation of the compressed particles, resulting in the formation of new surfaces. With increasing compaction load, more dense compacts will be formed, bringing particle surface areas into closer proximity. The surface area available for the gas molecules will then be dependent upon the penetration capacity of the technique, i.e., mainly the size of the gas molecules used [21]. For gases like nitrogen and krypton, that are normally used in these kind of studies, the molecules will only reach a fraction of the total tablet surface area. Since the distance between solid surfaces needed for the development of bonds could be substantially smaller than the size of the gas molecules, only a part of the "nonavailable" surface area may be utilized for bonding [11]. Additionally, the use of porous materials and the possibility for pores and cracks to be closed during compression may further complicate the evaluation of a bonding surface area [8,20].

Using *gas adsorption* data the following relation has been applied to estimate the surface area utilized for bonding:

$$S_T = S_P + S_F - S_B \tag{1}$$

where

S_T = tablet surface area as measured by gas adsorption
S_P = surface area of uncompressed particles
S_F = new surface area created by fragmentation during compression
S_B = surface area consumed for bonding

From the discussion above it is, however, obvious that such calculations involve too many uncontrolled factors, especially if porous, granulated materials are tested. These difficulties thus make the conclusion that larger surface areas may be involved in pharmaceutical tablets highly uncertain.

If, however, solid nonporous materials undergoing volume reduction mainly by plastic deformation without any tendency toward fragmentation are used, some kind of proportional relation between compact surface area and bonding surface area ought to be obtained. For iron powder and coarse particulate sodium chloride, two materials belonging to this group, this has been investigated at our laboratory [21].

The characteristics of both iron and sodium chloride compacts, compressed at increasing loads, are presented in Fig. 12. The compact strength increased approximately linearly for both materials tested. The minimum load at which coherent compacts could be obtained was about 100 and 50 MPa for iron and sodium chloride respectively. The slope of the linear part of the strength-pressure profile was approximately three times higher for iron than for sodium chloride, indicating that the iron powder was bonding with a stronger bond type or that the surface area utilized for bonding was higher. Assuming that the major bond types involved are unchanged with increasing pressure, it seems reasonable that an increase in strength is accompanied by a proportional increase in bonding surface area. If the permeametry and gas adsorption techniques are capable of detecting such an increase in bonding surface area, results would be expected showing a decrease in compact surface area after 100 and 50 MPa for the two materials respectively. However, for the iron compacts, the surface areas obtained by both permeametry and gas adsorption (Fig. 12) seem to be fairly constant with increasing pressure. For permeametry, even a slight increase was observed. It could, however, be questioned whether this is an artefact or not. In an earlier study [20], it was shown that measurements on powders compressed at relatively high loads could result in an overestimation of compact surface area. The surface areas measured by gas adsorption showed an appreciable variation, especially at high loads, making it diffi-

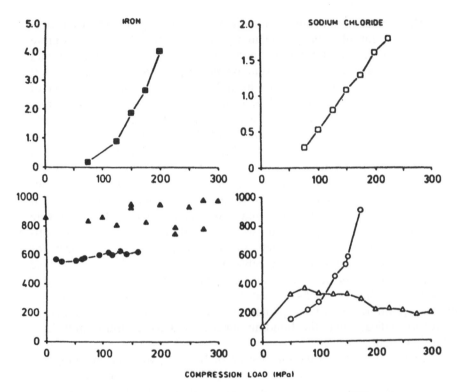

Fig. 12 The effect of compression load on compact strength (☐) and compact surface area as measured by gas adsorption (△) and permeametry (○), for iron (closed symbols) and sodium chloride (open symbols). (From Ref. 21).

cult to draw firm conclusions. The result could not be interpreted as giving a significant decrease in compact surface area with compression load, after 100 MPa. The relatively small difference in surface area obtained by the two techniques investigated supports the idea that the iron particles were essentially nonporous.

The lack of increase in permeametry surface area with increased compression load indicates that the iron particles were not significantly fragmenting during compression. The data obtained for the iron powder therefore indicate that the surface area utilized for interparticulate attraction is equal to or less than the precision of the surface area methods tested. Since it is reasonable to assume that these methods will tend to overestimate the bonding surface area [11] as discussed in the introduction, the results imply that the bonding surface area in the iron compacts is very small [42].

The surface area profiles for sodium chloride (Fig. 12) show a different pattern. Initially, the permeametry surface area shows a moderate increase, whereafter an extensive increase is observed. Although the data obtained at higher pressures presumably correspond to an overestimation of the surface area [20], the results indicate that volume reduction of sodium chloride is accompanied by some tendency toward particle fragmentation. The surface area as measured by gas adsorption shows initially a higher increase with compression load than the corresponding data obtained by permeametry. This probably reflects the fact that cracks and pores were formed during compression. After approximately 75 to 100 MPa, the surface area decreases with an increase in compression load. This decrease in gas adsorption surface area could then be a reflection of an increase in bonding surface area. This means that sodium chloride particles bond with a significantly weaker bond type than the iron particles, resulting in weaker compacts but utilizing a larger surface area for interparticulate attraction, which subsequently is reflected in the decrease of gas adsorption surface area. This decrease could alternatively be interpreted as a sealing of cracks created. Considering the profiles obtained for the iron powder, the latter explanation seems more probable. The results therefore indicate that the bonding surface area for sodium chloride is relatively small.

In an attempt to vary the degree of bonding without changing the compression load and subsequently not affecting the fragmentation tendency or formation of cracks, mixtures of sodium chloride and varying amounts of magnesium stearate were compressed at 150 MPa. The effect of increasing additions of the lubricant on both compact strength and surface area for sodium chloride is presented in Fig. 13.

Using minute additions, the compact strength decreased strongly with increasing magnesium stearate concentrations. This is in agreement with results reported earlier [8,15,23,59], where this effect was explained in terms of the formation of a lubricant film around the compound particles. However, at additions in excess of approximately 0.005 wt.%, the compact strength leveled off and even the use of a relatively high concentration (0.05 wt.%) gave no further reduction in strength. Similar profiles have been obtained for silica compacts [51], where only a small amount of adsorbed water decreases the compact strength considerably, whereafter the strength remained fairly constant when the adsorption of water was further increased. The interpretation of the data obtained is as discussed that two different bond types are involved in the interparticulate attraction [23,24].

One bonding mechanism (e.g., molecular forces of van der Waals type) is very sensitive for surface changes, while the other mechanism (e.g., solid bridges due to ionic bonding) could be established even in the presence of a lubricant film by, due to penetration by point irregularities. When a total surface coverage of sodium chloride particles by magnesium stearate is obtained, the latter bond type alone would determine the compact strength.

As is evident from Fig. 13, the admixture of the different lubricant amounts did not affect the powder surface area as characterized by the gas adsorption technique. However, the gas adsorption surface area of the corresponding compacts showed a decrease with increased quantities of lubricant. This effect was not expected, considering the obtained decrease in compact strength, which was supposed to expose the surface area used for bonding [28]. The result could probably be explained by a reduction in particle fragmentation and crack formation in the systems containing lubricant additions. Similar observations have been reported earlier [57]. The results therefore indicate that the surface area used for bonding is relatively small, and masked by the surface area changes due to fragmentation and crack formation.

The results obtained demonstrate the problems involved when investigating the compaction behavior of pharmaceutical materials. Although sodium chloride in pharmaceutical studies (e.g., [8,15,57]) normally represents a nearly ideal model substance (nonporous and consolidating by plastic deformation), it is evident from the results that sodium chloride must be classified as a complex substance with respect to its compaction behavior. In contrast to the iron powder, sodium chloride particles undergo such changes during compression (fragmentation and formation of cracks and pores) that a simple monitoring of changes in surface area with pressure (Fig. 12) cannot unambiguously be used for the evaluation of bonding surface area. Considering the even more complex nature of most pharmaceutical materials used in some reported studies [8,11,28,57,58], it is questionable whether the interpretations suggested, purporting relatively high bonding surface areas [11,28,57,58], are justified solely on the basis of previously published experimental data.

The results in this study indicate that the surface area taking part in the bonding between particles for sodium chloride, as well as for iron powder, is small in relation to the geometrical surface area available. As discussed, it seems reasonable that the major bonding is facilitated by long-range forces (e.g., molecular forces of van der Waals type), thereby explaining why the two surface area techniques are not capable of detecting any substantial change in bonding surface area with changes in the different parameters tested.

Thus, a tablet which can be described as a dispersion of solid particles in gas phase where the individual particles are separated by some distance

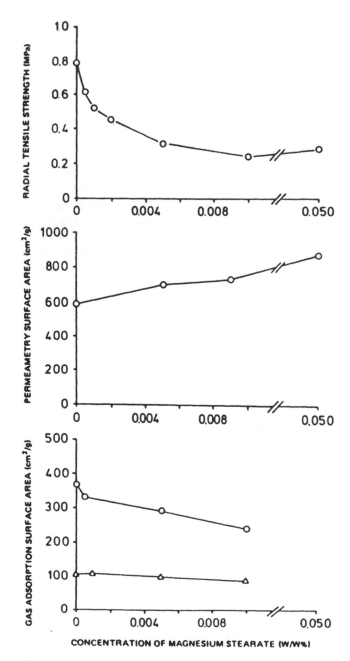

Fig. 13 The effect of lubricant concentration on compact strength and surface area for sodium chloride, compressed at 150 MPa. (From Ref. 21).

cannot probably be assayed on bonding surface area in any direct sense by surface area methods. The possibility of obtaining a proportional relation between measured tablet surface area and the fraction of surface area participating in bonding is then probably limited to relatively simple test systems. So far, such relations have only been reported to a limited extent, e.g., for lactoses undergoing fragmentation during compression [59]. In these studies, a unique relation was claimed to exist between compact strength and surface area as measured by mercury porosimetry (Fig. 14). It was also suggested that the obtained results indicated that the same bonding mechanism must be active in all types and size fractions of the lactoses tested. However, it could be discussed whether mercury porosimetry is the best method to reflect the surface area of tablet particles that potentially could

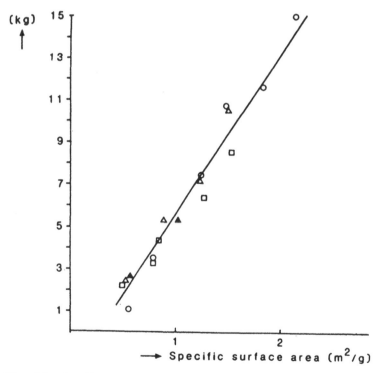

Fig. 14 Crushing strength versus specific surface area for tablets compressed from different types of crystalline lactose; α-lactose monohydrate ($\square$), anhydrous α-lactose ($\bigcirc$), roller-dried β-lactose ($\triangle$), crystalline β-lactose ($\blacktriangle$). (From Ref. 59).

participate in bonding since mercury porosimetry also monitors the surface area of intraparticulate pores. Another approach is to use a permeametric technique. Several studies have demonstrated the usefulness of this technique for the characterization of external surface areas in tablets (e.g., [9,20,60,61]).

B. The Importance of Volume Reduction Mechanisms for Bonding Surface Area

The effective amount of surface area available for bonding is dependent on several material properties [29]. Both the particle characteristics of the starting material and the changes caused by the volume reduction will be determining factors. The problems of measuring and defining a bonding surface area have resulted in a great interest for more indirect, secondary parameters (Table 1). In this section, the influence of volume reduction mechanisms on bonding surface and compact strength will be discussed.

To study the influence of volume reduction mechanisms on flattening of the surface roughness and the creation of bonding surface areas, a model system was developed [26] consisting of two layers that are compressed to form a single tablet (Fig. 15). The first layer was precompressed at a compaction pressure of 25–200 MPa. Powder material was then added on top of the first layer, and the lower punch was adjusted to give a compaction pressure of 200 MPa for the double-layer tablet. The tablet strength was characterized by measuring the axial tensile strength according to Nyström et al. [62].

The axial tensile strength for double layer tablets of all materials are presented in Figs. 16 and 17. For all materials tested, an increase in pressure on the first layer of the double-layer tablet resulted in a decrease in the axial strength. All the plastically deforming materials (sodium chloride, Sta-Rx

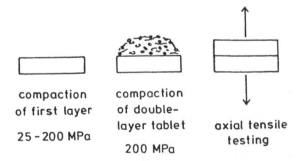

compaction
of first layer

25 - 200 MPa

compaction
of double-
layer tablet

200 MPa

axial tensile
testing

Fig. 15 Double-layer tablet technique for studying the influence of volume reduction behavior on bonding surface area and tablet strength.

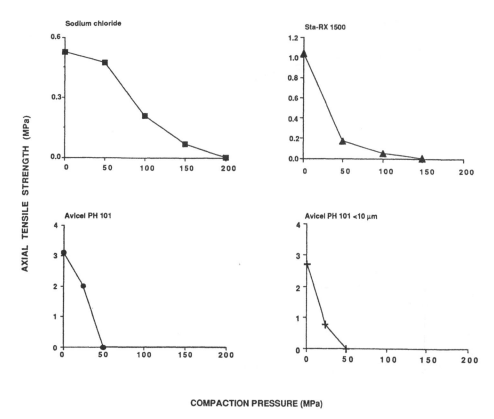

COMPACTION PRESSURE (MPa)

Fig. 16 The effect of compaction pressure of the first portion on the axial tensile strength of the double-layer compact compressed at 200 MPa for plastically deforming materials. (From Ref. 26).

1500, Avicel PH 101, and Avicel PH 101<10 μm) failed in the contact zone between the first and the second layer of the double-layer tablet, indicating that the bonding strength in the contact zone generally was lower than in the intervidual layers. The materials consolidating mainly by fragmentation failed generally in the first layer of the double-layer tablet, indicating that the bonding strength between the two layers was higher than that of the individual layers.

In Fig. 18, the percentage decrease in strength is plotted against the compaction pressure of the first layer of the double-layer tablet for all materials tested. Avicel PH 101 and Sta-Rx1500 showed the greatest sensitivity followed by sodium chloride. The materials undergoing extensive fragmentation (i.e., Emcompress, lactose, and sucrose) are relatively insen-

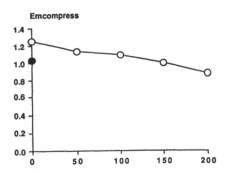

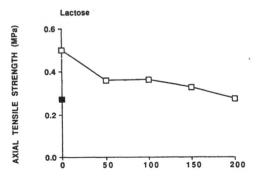

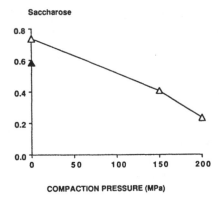

COMPACTION PRESSURE (MPa)

Fig. 17 The effect of compaction pressure of the first portion on the axial tensile strength of the double-layer compact compressed at 200 MPa for materials undergoing volume reduction mainly by fragmentation. The strength of a single tablet compressed two times at 200 MPa is denoted by a filled symbol on the y-axis. (From Ref. 26).

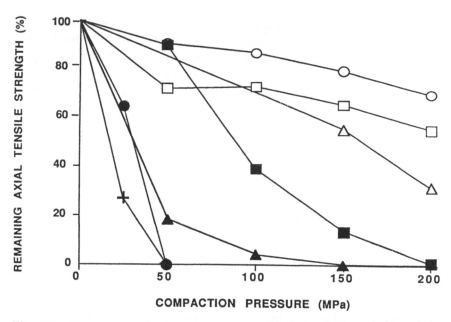

Fig. 18 The remaining axial tensile strength with the increase in compaction pressure of the first portion of the double-layer tablet for all materials tested. Symbols as in Figs. 16–17. (From Ref. 26).

sitive to an increase in compaction pressure compared to the plastically deforming materials. Thus volume reduction by fragmentation seems to be a more efficient means of producing larger surface areas that would promote bonding between particles in the compacts [26,59]. This is especially valid for materials bonding with intermolecular attraction forces—probably the majority of pharmaceutical compounds and excipients.

The so-called plastically deforming materials used in this study were all very sensitive to a decrease in the surface roughness of the first layer of the double-layer tablet; i.e., the plastic deformation of these materials was inadequate for the development of large zones of intimate contact between the layers. It is debatable whether the term *plastic deformation* should be applied to all less fragmenting materials possessing varying degrees of plasticity. Many of the commonly used amorphous tablet binders with pronounced plastic deformation may provide an effective means of creating large interparticulate attraction surface areas. For most of the plastically deforming materials, possessing a moderate plasticity and bonding with intermolecular forces, it seems that a high external specific surface area is a prerequisite for high compactibility. This is best achieved by using fine particle sizes or qualities with high surface roughness, as reported earlier for Sta-Rx 1500 [14] and for sodium bicarbonate [63]. It has also been suggested that the high compactability of Avicel PH 101 is due to the large external specific surface area of the irregular particles [23].

A large surface area and an irregular particle shape will probably promote all bonding mechanisms discussed. Fragmenting materials normally bonding by intermolecular forces [29] do not seem to be severely affected by an increase in compression load since the particles in the second layer can develop large bonding surface areas not only by particle rearrangement together with a limited plastic deformation but also by extensive fragmentation. For the plastically deforming materials used in this study, the initially high surface roughness is reduced after compression, with a subsequent reduction in intermolecular forces, the development of solid bridges (in, e.g., sodium chloride), and mechanical interlocking.

IV. CONCLUSIONS

In this chapter, several pharmaceutical materials have been reported on volume reduction behavior, dominating bond mechanism, and, to a limited extent, bonding surface area. The following general conclusions have been suggested.

1. Intermolecular forces constitute the dominating bond mechanism for pharmaceutical materials.
2. A proportional relation between compact surface area and bonding surface area could in some cases be established.
3. The following material properties will in principle favor a high compact strength:
    *   limited elastic deformation
    *   high compact surface area
        fine particulate starting materials
        highly fragmenting starting materials
        starting materials possessing high surface roughness
    *   extreme plastic deformation (e.g., amorphous binders)

A. Classification of Tabletting Materials

It is believed that for materials undergoing extensive fragmentation, a large number of interparticulate contact points are created. The compaction load per unit area of such contact points will thereby be low. This indicates that mainly relatively weak attraction forces, acting over distances, can be formed. However, due to the large number of bonds or attraction zones, a relatively strong compact could be formed. For less fragmenting materials, a smaller number of contact points are formed, which then would result in strong compacts only if relatively strong attraction forces, such as solid bridges, could be developed. This is probably the situation for compacts of coarse crystalline sodium chloride. If the material exhibits extensive plastic deformability, such as many amorphous binder materials, the number of weak distance forces would probably be much higher and thereby contribute significantly to the compact strength. Also materials having a rough surface texture ought to be capable of forming a relatively large number of weak distance forces, in spite of the fact that such a material does not fragment extensively. If the powder being compressed consists of particles with both a rough surface texture and a pronounced plastic deformability, compacts of extremely high mechanical strength ought to be obtained, due to the large number of weak distance attractive forces developed. Microcrystalline cellulose (e.g., Avicel) could be an example of such a material. A suggestion for how materials can be classified according to their compactability and its dependence on volume reduction and bonding properties is presented in Table 5.

An important parameter that will influence the final compact strength is the proportion of elastic deformation and, consequently, elastic recovery that will take place after compaction. By definition, so-called plastically deforming materials show little elastic deformation. Thus for fragmenting materials it seems important to distinguish between those where the smaller particles formed undergo mainly plastic deformation or elastic deformation. Most of the materials intended for direct compression have in common a minute elastic behavior.

High tablet strength is thus primarily produced by materials possessing a low elastic component during consolidation and having a high bonding surface area that could develop intermolecular forces. To this group of materials belong fine particulate materials, milled materials, and strongly fragmenting materials such as granulations. Excipients belonging to this group may produce a large bonding surface area in various ways; e.g., the material can be highly fragmenting, very plastically deforming, or exhibit a pronounced surface roughness.

B. Advantages of Using Fragmenting Materials

Some possible advantages for the use of materials with a high fragmentation tendency are presented in Table 6.

Several studies have shown that a pronounced material fragmentation will result in a reduced dependence on initial particle size, surface shape and texture, additions and load rate for the compact strength. In this

Table 5 Primary and Secondary Factors Affecting Compact Strength for
Pharmaceutical Materials

	Low tablet strength	High tablet strength	
Dominating volume reduction mechanism	Small number of weak attractions (low bonding surface area and bonding predominantly with intermolecular distance forces)	Large number of weak attractions (high bonding surface area and bonding predominantly with intermolecular distance forces)	Small number of strong attractions (small bonding surface area and bonding both with intermolecular distance forces and solid bridges)
Plastic deformation	Not capable of forming solid bridges e.g., sodium bicarbonate	Very fine particulate qualities e.g., fine particulate sodium chloride	Capable of forming solid bridges e.g., coarse particulate sodium chloride
		Pronounced surface roughness e.g., Avicel, Sta-Rx 1500 and milled qualities of materials	
		Very plastically deformable e.g., amorphous binders	
Fragmentation and plastic deformation	Nonexisting	With a small elastic component e.g., lactose Emcompress	
Fragmentation and elastic deformation	With a large elastic component e.g., phenacetin paracetamol		

Table 6 Possible Advantages with Fragmenting Materials

Less sensitive for variations in particle size and shape of starting materials
Less sensitive for admixture of lubricants
Less sensitive for load rate
Less prone to undergo postcompaction strength changes by stress relaxation
High compactability, provided the elastic component of smaller particles (fragments) formed is minute

chapter it has also been suggested that fragmenting materials in general will result in a high bonding surface area and compact strength, provided the elastic component of the material is limited.

REFERENCES

1. D. Train, *Inst. Chem. Eng. 35*: 258 (1957).
2. J. T. Carstensen, *Solid Pharmaceutics: Mechanical Properties and Rate Phenomena*, Academic Press, New York, 1980.
3. M. Duberg and C. Nyström, *Powder Technol. 46*: 67 (1986).
4. I. Krycer and D. G. Pope, *Int. J. Pharm. Technol. Prod. Manuf. 3*: 93 (1982).
5. C. Führer and J. Ghadially, *Acta Pharm. Suec. 3*: 201 (1966).
6. C. Führer, E. Nickel, and F. Thiel, *Acta Pharm. Technol. 21*: 149 (1975).
7. C. Führer, *Acta Pharm. Technol. 6*: 129 (1978).
8. M. Duberg and C. Nyström, *Acta Pharm. Suec. 19*: 421 (1982).
9. G. Alderborn, K. Pasanen, and C. Nyström, *Int. J. Pharm. 23*: 79 (1985).
10. D. P. Coffin-Beach and R. G. Hollenbeck, *Int. J. Pharm. 17*: 313 (1983).
11. J. S. Hardman and B. A. Lilley, *Nature 228*: 353 (1970).
12. D. Sixsmith, *J. Pharm. Pharmacol. 34*: 345 (1982).
13. E. T. Cole, J. E. Rees, and J. A. Hersey, *Pharm. Acta, Helv. 50*: 28 (1975).
14. G. Alderborn and C. Nyström, *Acta Pharm. Suec. 19*: 381 (1982).
15. A. H. de Boer, G. K. Bolhuis, and C. F. Lerk, *Powder Technol. 25*: 75 (1978).
16. A. McKenna and D. F. McCafferty, *J. Pharm. Pharmacol. 34*: 347 (1982).
17. R. J. Roberts and R. C. Rowe, *J. Pharm. Pharmacol. 37*: 377 (1985).
18. R. J. Roberts and R. C. Rowe, *J. Pharm. Pharmacol. 38*: 567 (1986).
19. A. Kandeil, M. C. De Malherbe, S. Critchley, and M. Dokainish, *Powder Technol. 17*: 253 (1977).
20. G. Alderborn, M. Duberg, and C. Nyström, *Powder Technol. 41*: 49 (1985).
21. C. Nyström and P. G. Karehill, *Powder Technol. 47*: 201 (1986).
22. Z. T. Chowhan and Y. P. Chow, *Int. J. Pharm. Technol. Prod. Manuf. 2*: 29 (1981).
23. P. G. Karehill, E. Börjesson, M. Glazer, G. Alderborn, and C. Nyström, *Drug. Dev. Ind. Pharm. 19*: 2143 (1993).
24. P. G. Karehill and C. Nyström, *Int. J. Pharm. 61*: 251 (1990).
25. P. G. Karehill and C. Nyström, *Int. J. Pharm. 64*: 27 (1990).

26. P. G. Karehill, M. Glazer, and C. Nyström, *Int. J. Pharm. 64*: 35 (1990).
27. J. J. Benbow, in *Enlargement and Compaction of Particulate Solids* (N. G. Stanley-Wood, ed.), Butterworths, London, 1983, p. 171.
28. N. G. Stanley-Wood and M. S. Shubair, *Powder Technol. 25*: 57 (1980).
29. M. Duberg and C. Nyström, *Int. J. Pharm. Technol. Prod. Manuf. 6*: 17 (1985).
30. H. Rumpf, *Chem. Eng. Tech. 30*: 144 (1958).
31. C. Führer, *Lab. Pharma. Probl. Technol. 269*: 759 (1977).
32. C. Ahlneck and G. Alderborn, *Int. J. Pharm. 54*: 131 (1989).
33. J. N. Israelachvili, *Intermolecular and Surface Forces*, Academic Press, London, 1985, p. 21.
34. B. V. Derjaguin, *Sci. Am. 203*: 47 (1960).
35. B. V. Derjaguin, I. I. Abrikosova, and E. M. Lifshitz, *Quart. Rev. Chem. Soc. 10*: 295 (1956).
36. J. N. Israelachvili and D. Tabor, *Prog. Surf. Membr. Sci. 7*: 1 (1973).
37. J. N. Israelachvili, *Intermolecular and Surface Forces,* Academic Press, London, 1985, p. 98.
38. H. Rumpf, in *Agglomeration* (W. A. Knepper, eds.), Interscience, New York, 1962, p. 379.
39. G. R. B. Down and J. N. McMullen, *Powder. Technol. 42*: 169 (1985).
40. A. G. Mitchell and G. R. B. Down, *Int. J. Pharm. 22*: 337 (1984).
41. C. Ahlneck and G. Alderborn, *Int. J. Pharm. 56*: 143 (1989).
42. F. P. Bowden and D. Tabor, *The Friction and Lubrication of Solids*, Oxford University Press, New York, 1950, p. 25.
43. R. P. Bhatia and N. G. Lordi, *J. Pharm. Sci. 68*: 222 (1979).
44. G. K. Bolhuis, C. F. Lerk, H. T. Zijlstra, and A. H. de Boer, *Pharm. Weekblad 110*: 317 (1975).
45. A. E. Shirvinskii, V. A. Malov, and I. S. Lavrov, *Colloid J. U.S.S.R. 46*: 867 (1984).
46. I. E. Dzyaloshinskii, E. M. Lifshitz, and L. P. Pitaevskii, *Adv. Phys. 10*: 165 (1961).
47. J. N. Israelachvili, *Intermolecular and Surface Forces*, Academic Press, London, 1985, p. 115.
48. H. C. Hamaker, *Physica 4*: 1058 (1937).
49. J. N. Israelachvili, *Intermolecular and Surface Forces*, Academic Press, London, 1985, p. 30.
50. H. K. Sartor, *Proc. Int. Fachtagung Komprimate* 2, Solingen, 1978.
51. D. Dollimore and G. R. Heal, *J. Appl. Chem, 11*: 459 (1961).
52. Å. Svensson, H. Olsson, and C. Nyström, in preparation.
53. M. Luangtana-anan and J. T. Fell, *Int. J. Pharm. 14*: 197 (1990).
54. D. R. Fraser, Powder Advisory Center, London, 149, 1973.
55. G. Milosovich, *Drug Cosmet. Ind. 92*: 557 (1963).
56. J. E. Rees, *Acta Pharm. Suec. 18*: 68 (1981).
57. N. A. Armstrong and R. F. Haines-Nutt, *Powder Technol. 9*: 287 (1974).
58. T. Higuchi, N. A. Rao, L. W. Busse, and J. V. Swintosky, *J. Am. Pharm. Assoc. Sci. Ed. 42*: 194 (1953).

59. H. Vromans, A. H. de Boer, G. K. Bolhuis, C. F. Lerk, and K. D. Kussendrager, *Pharm. Weekblad Sci. Ed.* 7: 186 (1985).
60. G. Alderborn and C. Nyström, *Powder Technol.* 44: 37 (1985).
61. C. Nyström and M. Glazer, *Int. J. Pharm.* 23: 255 (1985).
62. C. Nyström, W. Alex, and K. Malmqvist, *Acta Pharm. Suec.* 14: 317 (1977).
63. G. Alderborn, E. Börjesson, M. Glazer, and C. Nyström, *Acta Pharm. Suec.* 64: 31 (1988).

3

Porosity–Pressure Functions

Petteri Paronen and Jukka Ilkka

University of Kuopio, Kuopio, Finland

I. INTRODUCTION

The powder column is a heterogeneous system consisting of solid particulate material and air. Air can exist both between particles (interparticulate voids) and inside particles (intraparticulate voids). The physical nature of a powder column is different from that of a solid body, as powder can flow and have rheological properties typical of liquids. On the other hand, permanent deformation (plastic flow), reversible deformation (elasticity), and brittle fracturing of particles typical phenomena for solid bodies occur in powders. Thus, the behavior of powders in pharmaceutical processes, e.g., during compression, is often very complicated. In die compaction of powders, materials are subjected to compressive forces which lead to a volume reduction of the powder column. A volume is reduced by decreases in the inter- and intraparticulate pore space. The process of volume reduction is generally divided into different stages: die filling, rearrangement of particles, deformation by elastic changes, permanent deformation by plastic flow, or particle failure by brittle fracturing. The measuring of porosity changes as a function of the compression pressure is a method widely used in describing the compaction processes of powders.

Porosity is a function of the voids in a powder column, and in general all pore space is considered, including both inter- and intraparticulate voids. For porosity measurements, the dimensions and weight of a powder column (i.e., apparent density) and the particle density (referred to often as true density) of the solid material should be known. The porosity, ϵ, can be expressed by the equation

$$\epsilon = 1 - \frac{\rho_A}{\rho_T} \tag{1}$$

where ρ_A is the apparent density of a powder column and ρ_T is the particle density of the compressed material. The value of ρ_A/ρ_T, also referred as D, is regarded as the relative density or the packing fraction, which describes the solid fraction of a porous powder column. A value for the applied pressure, P, while loading a powder column under pressure is a function of the compressional force, F, and the punch tip area, A:

$$P = \frac{F}{A} \tag{2}$$

Instrumentation of tablet presses with force and displacement transducers enables a data acquisition for porosity-pressure analysis. Several equations have been proposed for describing the relationship between the porosity of a powder column and the applied pressure [1–3].

II. DETERMINATION OF POROSITY AND PRESSURE

Measurements for the relationship between porosity and pressure have usually been made during one-sided, uniaxial compressions. In such studies, instrumented eccentric tablet presses were utilized, as well as hydraulic and electrical universal testing machines, and high-speed rotary presses with double-ended compression were seldom used.

Compressive force is currently measured by strain gauges or piezoelectric transducers mounted directly on the compressing parts of the press. After proper calibration reliable force measurements can be obtained. Acceptable measurements of accuracy and precision of force are possible with modern strain gauges and amplifiers.

For an accurate determination of porosity, the height and diameter of a powder column must be reliably measured. Any looseness or elastic deformation of the compression devices' parts should be evaluated to allow an accurate measurement of the true displacement. The effects of machine deformation may be diminished by mounting the displacement transducers as close as possible to the compressing punch tip. Furthermore, the connection of a transducer to both the upper and lower punches makes the measurement even more accurate, e.g., by preventing a summation of the nonlinearity of discrete transducers. Several studies have been published on the mounting, calibration, and validation of measuring systems [4–7].

Besides the measuring transducers and their mounting also the amplifiers and analogue digital converters also affect the accuracy and precision of measurements [4,5]. The resolution of an A/D converter could have a remarkable effect on the accuracy of measuring systems. Typically a 12-bit conversion is used, which is enough for obtaining measurements for fully describing a compression process [8]. Smoothing of the original measurement data (e.g., using polynomial fitting) is not needed if the resolution of a measuring system is high enough and the electrical noise is removed by filtering.

The weight and the particle density of the compressed material, as well as the bulk density of loose powder, should also be carefully measured. Preweighed powder samples may be used to improve the repeatability of the measurements. The apparent density of a compact is obtained by dividing the weight by the apparent volume of powder compact. Compaction is often considered to begin at the loose packing state described by the bulk density of powder. The bulk density as measured by pouring a known quantity into a graduated vessel may differ considerably from that obtained by a die-filling process. In addition, the apparent density of a powder column, measured at the moment of the first detected force, has been used for calculations instead of separately measured bulk density [9].

Pycnometric methods, especially helium pycnometry, is preferably used for measurements of the particle density. Adsorbed volatile impurities and even absorbed water may cause variations. In this respect the particle density can not always be regarded as a material constant. In addition, even the particle density may change during compression. Changes are probably rare, and difficult to demonstrate. However, polymorphic transitions and changes in crystallinity are typical phenomena related to changes in density. An elastic deformation of a crystal structure could theoretically lead to reversible changes in density. Obviously, unpredictable and temporary changes in density can cause oddities in porosity-pressure analyses [10].

The volume of a powder column, and its corresponding porosity, can be measured either during compression or after ejection of the compact from a die. Instrumentation of a tablet press enables continuous monitoring of the powder column height during the volume reduction process. This method is referred as "in die" or "at pressure." In the "ejected tablet" or "at zero pressure" method, the dimensions of a compact are measured after ejection. The maximum compression pressure observed is usually taken to correspond to an achieved volume reduction or a certain porosity. A single compression provides only one measurement as a function of

compression pressure. A disadvantage of this method is the need to perform several compressions at different pressure levels. Also an "at pressure" method is possible, where only the maximum compression pressure
and the corresponding porosity of a powder column in the die are taken for
further analysis. Consequently, compressions at different pressure levels
are also needed in this method.

The porosity-pressure relationship is greatly influenced by the
method in which the dimensions of a powder column or a compact are
measured. When pharmaceutical powders are subjected to compression,
part of volume reduction in a powder column may occur by reversible,
elastic densification which then cause measurable changes in compact dimensions, at the decompression phase and even after ejection. Elastic
recovery of a compact has even been used to indicate the elastic properties
and bonding of a compressed material. However, the mechanical properties of a solid material may differ from those of a powder column consisting
of particulate material and air.

III. POROSITY-PRESSURE EQUATIONS

Numerous mathematical models describing the change of relative density
in a powder column as a function of the applied pressure have been derived
and adopted from other fields of industry for research in pharmaceutical
compression processes [1–3]. Three equations have been widely applied to
pharmaceutical purposes, namely Heckel (also called as Athy-Heckel),
Kawakita, and Cooper-Eaton. These equations will be considered in this
chapter.

A. Heckel Equation

Heckel [11,12] introduced an equation for the densification phenomenon
following the first-order kinetics. The equation is

$$\ln \frac{1}{1 - D} = kP + A \tag{3}$$

where k and A are constants obtained from the slope and intercept of the
plot $\ln(1/(1 - D))$ versus P, respectively, D is the relative density of a
powder column at the pressure P (Fig. 1a).

A is an intercept which is extrapolated from the linear part of the
Heckel plot. As the plots were curved at low pressures, Heckel related the
constant A to processes of volume reduction which have taken place by (1)
die filling and (2) particle rearrangement before deformation and bonding

of the discrete particles [11,12]. Densification of a powder by die filling can be expressed as

$$\ln \frac{1}{1 - D_0} \tag{4}$$

where D_0 is the relative density of a powder column at resting pressure, and usually derived from the bulk density. The combined effect of die filling and particle rearrangement at low pressures can be described by the equation

$$A = \ln \frac{1}{1 - D_0} + B \tag{5}$$

where B describes a volume reduction purely by particle rearrangement. Relative densities corresponding the processes above are D_A, which includes both die filling and particle rearrangement, and D_B, which describes only the extent of particle rearrangement.

The relative densities can be related by the equation

$$D_A = D_0 + D_B \tag{6}$$

and D_A may be calculated from

$$A = \ln \frac{1}{1 - D_A} \tag{7}$$

In his original work Heckel studied the densification of metal powder [11]. The slope, k, of the Heckel plot was intended to give a measure of the plasticity of a compressed material. Consequently, greater slopes indicated a greater degree of plasticity of material. The slope was also related to the yield strength, Y, of the material by the equation

$$k = \frac{1}{3} Y \tag{8}$$

Hersey and Rees [13] later defined the reciprocal of k to be the mean yield pressure, P_Y, in order to study whether the fragmentation of particles was the predominant compaction mechanism of powders.

The constants of the Heckel equation are commonly determined by the linear regression analysis by using the least-squares method. However, two difficulties are associated with this procedure. First, correct selection of a linear region of the function, and, second, the different weight values of measurement points on the logarithmic scale. In using the first and second derivatives of the function, the linear region can be correctly se-

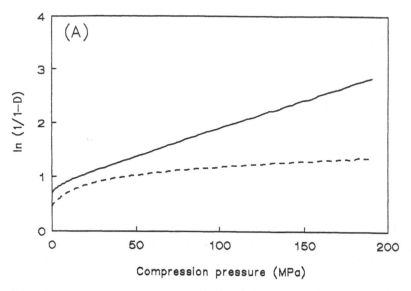

Fig. 1 Graphical presentation of the Heckel (A), Kawakita (B), and Cooper-Eaton (C) equations. The same measurement data obtained by the "at pressure" method for pregelatinized starch (Starch 1500) [——] and dicalcium phosphate dihydrate (Emcompress) [– – –] are plotted in each figure.

lected [9,14]. In the strictly linear regions of the plot the first derivative is constant and the second derivative is zero. The other possibility is to select a range of measurement points where the linear regression coefficient is higher than that decided for the threshold value. These methods are better than straight-line calculations over a constant pressure range without ensuring the actual linearity of the plot. However, if the method of constant range is used the regression coefficient should be reported, and in some cases deductions from the behavior of a compressed material can even be done with a variation in linearity.

The linearity calculations performed by using the least-squares method are based on the assumption that the function's values have an equal weight on the scale used. This is true only on an arithmetic scale but not on logarithmic scales. The problem of different weights could be avoided by using iterative techniques and nonlinear fitting for the minimizing function values. The most common methods used for iterative minimization of nonlinear equations is the simplex technique, which is often based on the Gaussian algorithm. By using commercial software, the proper mathematical procedures for calculations can easily be determined and performed.

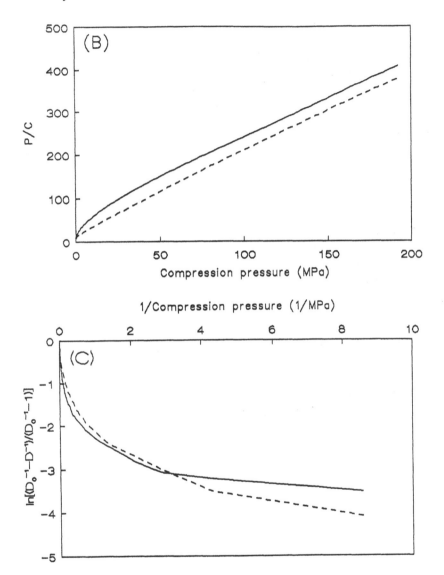

B. Kawakita Equation

Kawakita [15] introduced an equation describing the relationship between the volume reduction of a powder column and the applied pressure. The Kawakita equation is written as

$$C = \frac{V_0 - V}{V_0} = \frac{abP}{1 + bP} \tag{9}$$

where

$$C = \text{degree of volume reduction}$$
$$V_0 = \text{initial volume}$$
$$V = \text{volume of a powder column under the applied pressure } P$$
$$a, b = \text{constants characteristic to powder being compressed}$$

In addition, C can be described by means of the density and porosity as

$$C = 1 - \frac{\rho_0}{\rho_P} = \frac{\epsilon_0 - \epsilon_P}{1 - \epsilon_P} \tag{10}$$

where

$$\rho_0 = \text{bulk density}$$
$$\rho_P = \text{apparent density at pressure } P$$
$$\epsilon_0 = \text{porosity at the bulk state}$$
$$\epsilon_P = \text{porosity at pressure } P$$

Equation (9) can also be rearranged in linear form as

$$\frac{P}{C} = \frac{P}{a} + \frac{1}{ab} \tag{11}$$

From the graphical presentation of P/C versus P, the constants may be evaluated (Fig. 1b). The constant a is given as a reciprocal of the slope from the linear part of the plot and is equivalent to the value of C at infinitely high pressures:

$$C_\infty = \frac{V_0 - V_\infty}{V_0} = a \tag{12}$$

An intercept at the P/C-axis, extrapolated from the linear region of the plot, gives a value for $1/ab$.

The constant a gives an indication of the maximum volume reduction available and is considered to describe the compressibility of a powder, while b is considered to describe an inclination toward volume reduction. However, the actual physical meaning of the constants a and b have been in question [1,16].

Kawakita and co-workers [1,16] have also applied the equation in describing the volume reduction on tapping and vibrating processes. For the former application, the pressure, P, is replaced by the tapping number, N, and for latter by the vibration time, T. For these applications the meaning of the constants a and b become physically more significant than in Eq. [11], since the experimental conditions in tapping and vibrating processes are strictly independent of the powder tested. In die compaction a material may influence the development of the applied pressure.

C. Cooper-Eaton Equation

The Cooper-Eaton equation is [17]

$$\frac{1/D_0 - 1/D}{1/D_0 - 1} = a_1 \exp\left(-\frac{k_1}{P}\right) + a_2 \exp\left(-\frac{k_2}{P}\right) \tag{13}$$

where

D_0 = relative density at zero pressure or the bulk density divided by the particle density

D = relative density at pressure P, or the apparent density of a powder column divided by the particle density

a_1, a_2 = dimensionless constants that indicate the fraction of the theoretical maximal densification which could be achieved by filling voids of the same size (a_1) and of a smaller size (a_2) than the actual particles

The most probable pressures at which the respective densification process would occur are described by k_1 and k_2.

Cooper and Eaton [17] considered the compaction of powders to take place in two stages. First by the filling of voids of the same or larger size than the particles, where the original particles are moved and rearranged during this stage. At the second stage, the voids smaller than the original particle size are filled due to particle deformation. This phase may proceed by elastic deformation, plastic flow, or fragmentation of the compressed particles.

According to Cooper and Easton [17] densification can be described by a biexponential equation. If a nonporous powder column is produced under infinite pressure, then the sum of a_1 and a_2 is unity. If the sum is greater than unity, then a nonporous compact can be achieved at lower pressures. If the sum is less than unity, then other processes operating before complete volume reduction is achieved are indicated.

The Cooper-Eaton plots are typically biphasic linear plots, at least for hard and monodisperse metal and ceramic powders [17]. The measurements are presented graphically by the equation (Fig. 1c)

$$\ln\left(\frac{1/D_0 - 1/D}{1/D_0 - 1}\right) \text{ versus } \frac{1}{P} \tag{14}$$

where a and $a_1 + a_2$ can be determined from the ordinate intercepts of the first and the second linear regions, respectively, while k_1 and k_2 can be determined from the slopes of these two linear regions.

Cooper and Eaton [17] have studied the applicability of their equation with metal and ceramic powders. With relatively hard materials and monodisperse powders their equation can clearly distinguish the two putative densification stages, where $a_1 + a_2$ is close to unity with these kinds of materials. The studies with relatively soft materials and polydisperse pharmaceutical powders have pointed out that the densification stages are not always so clearly distinguishable [18–22]. Also the deviation of $a_1 + a_2$ from unity has been more extensive, and values both over and under unity have been reported. Thus it seems that the Cooper-Eaton equation is not as suitable for soft polydisperse powders as it is for hard monodisperse powders. This may result from the densification of the powder column by several simultaneous mechanisms. Thus, it might be impossible to notice the totally separated rearrangement and deformation stages. This is often supported by the poor linear regressions of Cooper-Eaton plots with pharmaceutical materials. On the other hand, the application of linear regression analysis may be questionable, again due to the logarithmic scale of function values and reciprocals of the compression pressure.

The main advantage of the Cooper-Eaton equation is, however, the possibility to accurately study the initial stages of volume reduction, i.e., measurement points at large $1/P$ values. Thus, information can be obtained from the effects of particle surface properties, shape, and size of the densification of powder columns.

So far, there is a lack of a universal compaction equation which would be capable of describing the whole volume reduction process. Most equations seem to be limited to certain conditions in which they are applicable, commonly within a certain range of pressure or porosity. Also, the susceptibility of the equations to experimental variables differ. In the Kawakita equation the constant C is highly dependent on the initial packing state. For example, it is obvious that varying die filling methods or dies with different diameter provide a different values for constants of the Kawakita equation [23]. Also, for the Cooper-Eaton equation, the initial packing state must be determined. Another difficulty with the Cooper-Eaton equation is in measuring the constant a_1, which is highly dependent on the pressure range chosen for the determination [22,23]. In the Heckel equation, the constant k (or $1/k$) is not dependent on the initial packing of the powder column, since the constant measured after the particle rearrangement phase [23]. Moreover the Kawakita equation is valid at low pressure and large to intermediate porosity, whereas the Heckel equation gives the best fit at intermediate to high pressure and low porosity [24]. However, different opinions on the applicability of the equations have been presented, and often the extremes on both pressure and porosity have been found to give a poor fit [2,3,22–24].

IV. APPLICATIONS OF POROSITY-PRESSURE FUNCTIONS

In pharmaceutical tablet compression the ability of the powdered material to form coherent compacts by inter- and intraparticulate bonding is of prime importance. The known compression properties of materials to be tabletted give a basic indication of the tablettability and may allow prediction of the properties of the formulation. With the porosity-pressure functions, a relationship may be created between tablettability or compact formation and the volume reduction properties of powders. In this chapter the applications of the porosity-pressure functions in pharmaceutics are reviewed. The published reports deal mainly with the utilization of the Heckel equation. Thus only a few reports are available for the other equations.

A. Shapes of Heckel Plots

Three types of volume reduction mechanisms of pharmaceutical powders have been distinguished by using the Heckel equation [13,18]. The types are referred as A, B, and C (Fig. 2). Materials were categorized by compressing different particle size fractions of various powders. In type A, size fractions had different initial packing fractions and the plots remained parallel as the compression pressure was increased. In type B, the plots were slightly curved at the initial stages of compaction and later became coincidental. In type C, the plots had an initial steep linear part after which they became coincidental with only trivial volume reduction. Generally, type A behavior was related to the densification by plastic flow, preceded by particle rearrangement. In type B, powder densification occurs by fragmentation of the particles. Differences in initial packing have no effect on further densification as the initial structure of a powder column is completely destroyed by fragmentation of the particles. Type C densification occurs by plastic flow but no initial particle rearrangement is observed.

The effects of experimental variables on the Heckel plots have been studied quite intensively. Rue and Rees [25] and York [26] have published critical notes about the limitations of Heckel plots used for predicting the compaction mechanisms. Rue and Rees [25] pointed out that caution is necessary when materials are classified according to changes in the Heckel plots with different particle sizes. The predominant compaction mechanism may change with the particle size. Transitions from brittle to ductile behavior, and even vice versa, have been described [27–29]. The effect of compression time was also described by Rue and Rees [25]. Using a microfine cellulose powder, an increased volume reduction was observed with increased compression time. This was shown to indicate deformation by plastic flow. In contrast, no increase in volume reduction as a function of

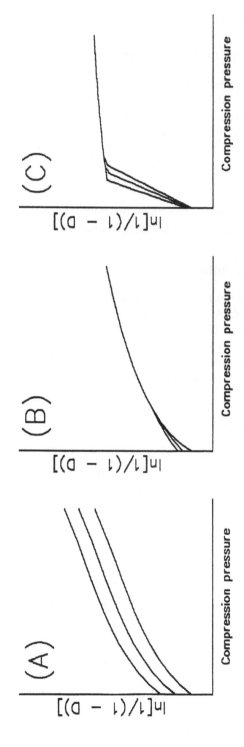

Fig. 2 Different types of compression behavior distinguished by the Heckel equation.

contact time was observed for dicalcium phosphate, this being characteristic of brittle materials. Rue and Rees [25] also proposed measuring the area under the Heckel plot (AUC) to quantify the amount of plastic deformation during compaction.

York [26] reviewed several studies where the densification behavior of crystalline lactose was evaluated and he pointed out that the general form of the Heckel plots was similar in all cases though the numerical values describing the compression process were dependent on experimental variables. York [26] also demonstrated a dramatic effect of die size on the densification behavior with microfine cellulose powder. York [26] has listed several variables such as the state and type of lubrication, rate of compaction, mode of die filling, contact time, dimensions of tools, and techniques used to measure compact dimensions, all which are necessary to take into account in tabletting studies.

Ragnarsson and Sjögren [30,31] found that the die wall friction between powder particles and wall surfaces had a clear effect on the Heckel plots of pharmaceutical powders. The upper punch pressure P in Heckel equation was noticed to be friction dependent and the mean of upper and lower punch pressures may be used instead [30]. In general, the state of lubrication, particle interactions and friction had only minor effects on Heckel plots and calculated parameters such as yield pressure, assuming the mean punch pressure was used instead of the upper punch pressure [31].

Duberg and Nyström [32] have used the initial curvature of Heckel plots as an indication of particle fragmentation. Also the correlation coefficient describing the linearity of the Heckel plots have been used for the same purpose. Thus, nonlinear plots may indicate fragmentation, and linear plots the deformation by plastic flow. Also poor packing of powder due to cohesiveness, small particle size and irregular particle shape may lead to initial curvature in a Heckel plot. In most cases, however, the rearrangement of the relatively regular particles occurs already at low pressures. Thus measurements of the nonlinearity could be a useful tool for categorizing pharmaceutical materials. This is especially true if fragmentation of multicrystalline particles or aggregates of primary particles are concerned, since they are intensively fragmented to smaller particles already at low compression pressures and this results in considerable volume reduction. At higher pressures the volume reduction becomes more difficult, and is expressed by nonlinear Heckel plots. On the other hand, the densification due to plastic flow and elastic deformation of particles, and the volume reduction of a powder column proceed both steadily, since according to Heckel function the porosity reduction occurs exponentially.

The compaction of granulated materials might be considered even more complicated than the compaction of discrete, solid primary particles. Deformation of aggregates or agglomerates is proposed to give an additional mechanism in the compaction of powder columns [33]. In literature this process has gained limited attention. Obviously, a distinction between void volumes associated with a whole powder column, single granule, and primary particle is complicated. Thus Heckel plots, for example, have been found to be difficult to interpret in terms of the physical meaning of yield pressure for granulated materials [34].

A novel treatment of porosity-pressure measurements based on the Heckel equation was proposed by Carstensen and Hou [35] and Carstensen et al. [36]. At infinite pressure, a density of a powder column was considered to approach a value different from the particle density. This was due to the assumption that only pores and voids having a diameter larger than a threshold value, as measured by mercury porosimetry, were considered to take part in volume reduction. As a consequence, a totally nonporous state at infinite pressure is not achieved.

B.   Constants of Heckel Equation

In several reports it has been pointed out that the mean yield pressure, determined by the Heckel equation, cannot be taken as the actual yield point of a compressed material [19,22]. The practical value of the mean yield pressure is that it gives a general impression from the deformation tendency of a powder column. In addition, the mean yield pressure values determined by using "at pressure" and "at zero pressure" methods are often very different. This is due to the elastic deformation of a powder column considered only in the "at pressure" method.

Particle shape affects the deformation properties of a powder column. Regular particles tend to move more easily at the initial stages of compression [37]. In the case of ductile materials, irregular particles have been noticed to deform more easily than regular particles [38], and the mean yield pressure values were smaller for irregular particles.

The irregular shape and surface roughness of the particles may support the plastic flow, as deformation begins in surface asperities due to mechanical shear forces. On the other hand, the irregular shape of primary particles may reflect the existence of crystal defects which facilitate the plastic flow. The particle shape of mainly fragmenting materials did not have a remarkable effect on the yield pressure values [38]. However, for irregular particles greater fragmentation propensity measured by air permeability was observed.

Forbes et al. [39] noticed that elongated particles from a series of *para*-aminosalicylate salts had smaller yield pressures than spherical ones.

They concluded that the orientation of elongated particles and therefore the orientation of crystal planes in a powder column was more uniform and preferable for deformation. Thus the applied pressure affects more effectively on the crystal planes of the nonspherical particles.

Roberts and Rowe [9] have studied the effect of punch velocity on compaction properties of pharmaceutical powders by using a compaction simulator. The velocity was found to have a considerable effect on the Heckel plots, and on the constants derived from the equation. However, changes in the plots were greatly dependent on the material being compressed. For materials which were known to deform by plastic flow, an increase was found in the mean yield pressure as the punch velocity increased. This was concluded to result from either a change from ductile to brittle behavior, or a reduction in the amount of plastic deformation due to the time-dependent nature of plastic flow. For microcrystalline cellulose and maize starch the latter mechanism was considered to be more likely, whereas for mannitol and sodium chloride an increase in brittle behavior was suspected. No major change in the mean yield pressure was found as punch velocity was increased for materials deforming mainly by fragmentation.

Roberts and Rowe [9] introduced a strain-rate sensitivity (SRS) index which was calculated by using the mean yield pressure values from compressions with high and low velocities. The equation for the SRS index is

$$SRS = \frac{P_{Y2} - P_{Y1}}{P_{Y2}} \times 100 \tag{15}$$

where P_{Y2} and P_{Y1} are the main yield pressures at the velocities of 300 mm/s and 0.033 mm/s, respectively. According to Roberts and Rowe [9], it is possible to use the SRS index in ranking materials. Materials which are strain-rate sensitive (high SRS index) tend to deform by plastic flow. In further studies Roberts and Rowe [27,28] showed the importance of the relationship between particle size and compression behavior of paticulate materials. For example, a transition for lactose from a brittle to a ductile material was determined to occur at a median particle size of about 20 microns. In another example of a drug material a transition in the deformation mechanism of a phthalazine derivative, from a ductile to a brittle fracture, was observed as the particle size was reduced.

Using the areas under the Heckel plots (AUC), as suggested by Rue and Rees [25], McKenna and McCafferty [40] evaluated the effects of particle size and contact time on the densification of pharmaceutical powders. They concluded that a greater amount of plastic flow (larger AUC) was associated with the smaller particle size of modified starch. A compaction mechanism of spray-dried lactose was found to be brittle fracturing and independent of particle size fraction. These results from lactose contradict those of Roberts and Rowe [27]. A comparison between these two studies is, however, difficult due to the different methods used.

Duberg and Nyström [41] and Paronen [42] have attempted to evaluate the elastic behavior of powder compacts from the decompression phase of the Heckel plots (Fig. 3). The reciprocal of the slope, calculated from the downward portion of the plot, can be utilized as the yield pressure of fast elastic deformation [22,42]. This parameter gives an indication of the fast elastic behavior of a compact. In comparing the Heckel plots, obtained by using both the "at pressure" and the "ejected tablet" methods, conclusions from the total elastic recovery can be drawn. By subtracting the slope

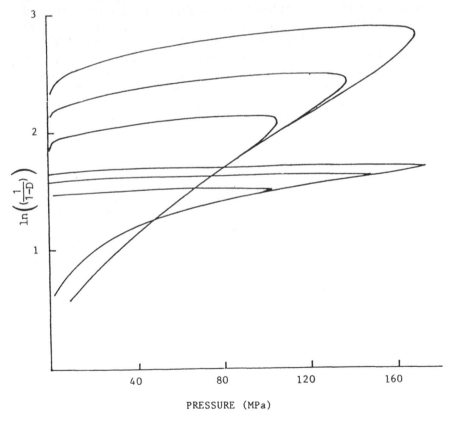

Fig. 3 Heckel plots for microcrystalline cellulose (upper curves) and dicalcium phosphate dihydrate (lower curves) obtained by "at pressure" method using three different compression pressures. (From Ref. 14.)

obtained from the "ejected tablet" method from that obtained from the "at pressure" method, and taking the reciprocal of that value, it is possible to get a parameter that describes the tendency of a compact to recover elastically [22].

C. Other Applications

Most of the reports on porosity-pressure functions are derived from single-component powders. Reports concerning the compressional behavior of multicomponent powder mixtures are, however, more rare [19,43–50]. The determination of the porosity-pressure relationships, and their respective constants, have been widely applied to the industrial R&D of tablets. It would be very useful to be able to predict the compressional behavior of multicomponent powders from the measured behavior of their individual components, though contradicting results for this possibility have been published. Humbert-Droz et al. [44] have noticed a linear relationship between the mean yield pressure and the proportions of the mixture's components. However, Ilkka and Paronen [49] noticed that although the packing fraction values determined from Heckel plots were in a linear relationship with the mixture composition (Fig. 4), there were systematic deviation from the linearity in the mean yield pressures, both positive and negative (Fig. 5). They assumed that the compression behavior of a powder mixture was often dominated by one of the mixture's components.

The applicability of the porosity-pressure functions to powder compaction can be evaluated under extreme conditions, e.g., compressing metal powders or polymeric materials under exceptionally high pressures with fast and slow loading. Page and Warpenius [51] noticed that the equations are differentially sensitive to the densification close to the particle density or infinite density of the material. Due to the logarithmic scale, the Heckel equation is much more inaccurate under these conditions than the Kawakita equation. The Kawakita and Cooper-Eaton equations can be generally found to fit better to whole compression data than the Heckel equation. The reason for this is the scaling and the biphasic nature of Kawakita and Cooper-Eaton equations, respectively.

V. CONCLUSIONS

Porosity-pressure functions have been successfully applied to obtain basic understanding of the compression behavior of particulate materials. The volume reduction process of a powder column is considered as a continuum in which the proportion of air between and inside the solid particles is decreased according to theoretical or empirical equations. Thus, a powder

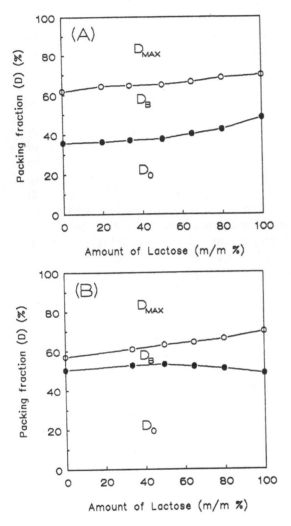

Fig. 4 The packing fractions (%) for the binary mixtures of (A) crystalline lactose and dicalcium phosphate dihydrate, and (B) crystalline lactose and modified starch. Densification due to the die filling D_0 (●) and particle rearrangement D_B (○), respectively. The sections noted with D_0 and D_B correspond the relative part of the densification of powder mixtures achieved by the respective mechanisms. At D_{MAX} a densification of powder column is obtained mainly by deformation mechanisms. Packing fraction of 100% corresponds a nonporous state, and it is not usually achieved during compression. (Data adapted from Ref. 49).

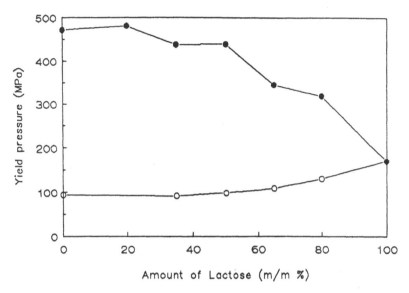

Fig. 5 The mean yield pressures for the binary mixtures of (●) crystalline lactose and dicalcium phosphate dihydrate, and (○) crystalline lactose and modified starch, obtained by "at pressure" method. (Data adapted from Ref. 49).

column is primarily seen as a body having macroscopic dimensions. On the other hand, microscopic factors, e.g., various solid-state, surface, and interface properties, are not directly included into these functions. Currently, especially the Heckel equation has been used to relate the behavior of a powder column to the fundamental properties of the compressed material. However, it is interesting to note that in many of the distinguished compaction studies for pharmaceutical powders, deviations from Heckel's theory have been utilized in obtaining additional information. So, most often the Heckel equation is strictly speaking invalid on most stages of compaction of pharmaceutical powders. It might be advantageous to utilize other porosity-pressure functions and methods of material science contemporaneously with the Heckel equation.

REFERENCES

1. K. Kawakita, and K. H. Lüdde, *Powder Technol. 4*: 61 (1970/71).
2. P. J. James, *Powder Metall. Int. 4*: 82, 145, 193 (1972).
3. V. A. Belousov, V. P. Fedin, *Pharm. Chem. J. 12*: 263 (1978).
4. R. F. Lammens, J. Polderman, and C. J. DeBlaey, *Int. J. Pharm. Tech. Prod. Mfr. 1*: 26 (1979).

5. R. F. Lammens, J. Polderman, C. J. DeBlaey, and N. A. Armstrong, *Int. J. Pharm. Tech. Prod. Mfr. 1*: 26 (1980).
6. M. J. Juslin and P. Paronen, *J. Pharm. Pharmacol. 32*: 796 (1980).
7. A. Y. K. Ho, J. F. Barker, J. Spence, and T. M. Jones, *J. Pharm. Pharmacol. 31*: 471 (1979).
8. P. Ridgway Watt, *Tablet Machine Instrumentation: Principles and Practice*, Ellis Horwood, Chichester, 1988, p. 203.
9. R. J. Roberts and R. C. Rowe, *J. Pharm. Pharmacol. 37*: 377 (1985).
10. H. K. Chan and E. Doelker, *Drug Dev. Ind. Pharm. 11*: 315 (1985).
11. R. W. Heckel, *Trans. Metal. Soc. AIME. 221*: 671 (1961).
12. R. W. Heckel, *Trans. Metal. Soc. AIME. 221*: 1001 (1961).
13. J. A. Hersey and J. Rees, *Nature PS. 230*: 96 (1971).
14. P. Paronen, *Using the Heckel Equation in the Compression Studies of Pharmaceuticals*, Proc. 4th Int. Conf. Pharmaceutical Technology, Paris, pp. 301–307 (1986).
15. K. Kawakita and Y. Tsutsumi, *Bull. Chem. Soc. Japan 39*: 1364 (1966).
16. M. Yamashiro, Y. Yuasa, and K. Kawakita, *Powder Technol. 34*: 225 (1983).
17. A. R. Cooper and L. E. Eaton, *J. Am. Ceramic Soc. 45*: 97 (1966).
18. P. York and N. Pilpel, *J. Pharm. Pharmacol. 25*: 1 (1973).
19. T. R. R. Kurup and N. Pilpel, *Powder Technol. 19*: 147 (1978).
20. N. A. Armstrong and F. S. S. Morton, *Pharm. Weekbl. Sci Ed. 1*: 234 (1979).
21. Z. T. Chowhan and Y. P. Chow, *Int. J. Pharm. 5*: 139 (1980).
22. P. Paronen and M. J. Juslin, *J. Pharm. Pharmacol. 35*: 627 (1983).
23. M. Sheikh-Salem and J. T. Fell, *J. Pharm. Pharmacol. 33*: 491 (1981).
24. R. Ramberger and A. Burger, *Powder Technol. 43*: 1 (1985).
25. J. Rue and J. E. Rees, *J. Pharm. Pharmacol. 30*: 642 (1978).
26. P. York, *J. Pharm. Pharmacol. 31:* 244 (1979).
27. R. J. Roberts and R. C. Rowe, *J. Pharm. Pharmacol. 38*: 567 (1986).
28. R. J. Roberts and R. C. Rowe, *Int. J. Pharm. 36*: 205 (1987).
29. R. J. Roberts, R. C. Rowe, and K. Kendall, *Chem. Eng. Sci. 44*: 1647 (1989).
30. G. Ragnarsson and J. Sjögren, *Acta Pharm. Suec. 21*: 141 (1984).
31. G. Ragnarsson and J. Sjögren, *J. Pharm. Pharmacol. 37*: 145 (1985).
32. M. Duberg and C. Nyström, *Acta Pharm. Suec. 19*: 421 (1982).
33. J. Van der Zwan and C. A. M. Siskens, *Powder Technol. 33*: 43 (1982).
34. M. Wikberg and G. Alderborn, *Int. J. Pharm. 62*: 229 (1990).
35. J. T. Carstensen and X-P. Hou, *Powder Technol. 42*: 153 (1985).
36. J. T. Carstensen, J. M. Geoffroy, and C. Dellamonica, *Powder Technol. 62*: 119 (1990).
37. P. York, *J. Pharm. Pharmacol. 30*: 6 (1978).
38. L. W. Wong and N. Pilpel, *Int. J. Pharm. 59*: 145 (1990).
39. R. T. Forbes, P. York, and R. Davidson, *Compaction Behavior within a Salt Series: Salts of p-Aminosalicylic Acid*, Proc. 10th Pharmaceutical Technology Conf., Vol. 1., Bologna, pp. 181–197, 1991.
40. A. McKenna and D. F. McCafferty, *J. Pharm. Pharmacol. 34*: 347 (1982).
41. M. Duberg and C. Nyström, *Powder Technol. 46*: 67 (1986).

42. P. Paronen, in *Pharmaceutical Technology: Tabletting Technology*, Vol. 1. (M. H. Rubinstein, ed.), Ellis Horwood, Chichester, 1987, p. 139.
43. M. Sheikh-Salem and J. T. Fell, *Int. J. Pharm. Tech. Prod. Mfr. 2*: 19 (1981).
44. P. Humbert-Droz, D. Mordier, and E. Doelker, *Acta Pharm. Technol. 29*: 69 (1983).
45. M. Duberg and C. Nyström, *Int. J. Pharm. Tech. Prod. Mfr. 6*: 17 (1985).
46. H. Vromans and C. F. Lerk, *Int. J. Pharm. 46*: 183 (1988).
47. H. C. M. Yu, M. H. Rubinstein, I. M. Jackson, and H. M. Elsabbagh, *Drug Dev. Ind. Pharm. 15*: 801 (1989).
48. J. S. M. Garr and M. H. Rubinstein, *Int. J. Pharm. 73*: 75 (1991).
49. J. Ilkka and P. Paronen, *Predictability of Compressional Behavior of Powder Mixtures Using Heckel Treatment*, Proc. 10th Pharmaceutical Technology Conf., Vol. 1., Bologna, pp. 225–238, 1991.
50. J. S. M. Garr and M. H. Rubinstein, *Int. J. Pharm. 82*: 71. (1992).
51. N. W. Page and M. K. Warpenius, *Powder Technol. 61*: 87 (1990).

4

Force-Displacement and Network Measurements

Gert Ragnarsson*

Pharmacia Biopharmaceuticals, Stockholm, Sweden

I. FORCE-DISPLACEMENT MEASUREMENTS IN COMPRESSION STUDIES

Force-displacement measurement has been one of the most popular methods for studying the compression process during tabletting. Force-displacement curves are obtained from measurements of punch force and displacement. Energy is needed for the compression of materials and formation of strong compacts. It seems logical to correlate the properties of the compact with the energy input rather than the compression pressure. The force-displacement profiles as such may also be useful as a material characteristic in preformulation work or for detecting batch-to-batch variations in the compression properties of materials.

The major reason for the interest is, however, the assumption that it should be possible to correlate the energy input, or work of compression, with the deformation and tablet-forming properties of materials. In many of the studies, the work of compression has been calculated as the total area under the upper punch force versus upper punch displacement curve [1,2], although it was realized in very early studies [3] that this value will include the work to overcome die wall friction.

A number of different methods have been suggested for evaluating and

---
**Current affiliation:* Astra Draco AB, Lund, Sweden

interpreting force-displacement measurement. The possibility for obtaining data from a variety of materials by carrying out large numbers of accurate studies have improved considerably during the last 10 to 15 years due to the development and improvement of the equipment and sampling technique.

This chapter presents a general description of force-displacement measurements plus discusses its possibilities in the evaluation of the compaction process and material properties. No attempt is made to present a full review of the literature.

II. INSTRUMENTATION

Force-displacement measurements requires accurate recording of upper punch force and displacement as well as lower punch force. The latter merits special discussion and will be treated later. It is possible, theoretically, to gather adequate data by using any of the different types of instrumented machines discussed in the literature (i.e., hydraulic presses, excenter machines, rotary presses, and compaction simulators). The compaction simulator meets the requirements of an ideal testing device as it allows measurements under dynamic conditions simulating normal tabletting and is suitable for small amounts of testing materials. Although studies have been carried out with compaction simulators [4,5] as well as rotary presses [6], most studies and theoretical discussions are based on experiments using single-punch excenter machines.

The instrumentation of tablet machines in general will not be discussed in this chapter. There are, however, several aspects of special importance in force-displacement measurements. The sampling procedure has to be sufficiently sensitive and accurate to register the whole compression cycle and the data sampling system should be able to handle a considerable amount of data. This is achieved, for example, by using high-quality force transducers, amplifiers, inductive displacement transducers and a AD converter connected to a suitable computer [7].

Attempts have been made to eliminate some registration error by using complex calibration methods [8,9]. The need for such complicated and tedious calibrations has been reduced by improved measuring devices and sampling equipment (amplifiers, AD converters, etc.). Nonetheless, the true accuracy of the system should be determined and special attention should be paid to error sources such as electrical noise generators.

Figure 1 illustrates the improvement in the data registration quality by a careful search for and elimination of electrical noise generators, primarily different earth connections [7]. The figure may also serve as a typical example of a force-displacement registration in a single-punch excenter press.

To make accurate force-displacement measurements, it is necessary

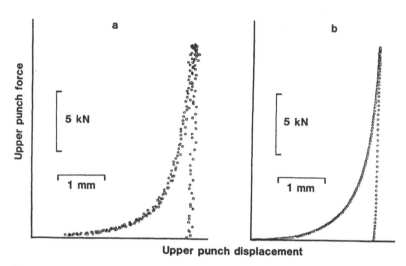

Fig. 1 Typical upper punch force versus upper punch displacement plots illustrating the improvement (b) in the force-displacement registration by elimination of electrical noise generators. (From Ref. 7).

to consider and carefully compensate for the deformation of machine parts such as punches and punch holders. Preferably, the deformation at normal machine rate should be measured. This can be obtained by compressing a flat steel disc between the punches, e.g., a 4–5-mm-thick disc, with a diameter slightly smaller than the die diameter and consisting of a steel quality that is insignificantly deformed within the pressure range of the tablet machine. The deformation during the compression can be represented by a mathematical function. If the deformation is linear, the deformation of the machine parts will be directly proportional to the compaction force, facilitating recalculation of the apparent displacement readings.

It is also advisable to study the displacement registration during the decompression phase. A hysteresis between the displacement during compression and decompression indicates that machine parts undergo time-dependent deformation at normal machine rate. Hysteresis may also indicate time dependent deformation of force transducers (such as piezoelectric force transducers). As will be discussed in more detail below, the area under the force-displacement curve during the decompression phase, i.e., the expansion work, is often used as a measure of the elastic deformation of the tested material. Failure to compensate for the deformation of machine parts may have a considerable impact on the accuracy of the calculated expansion work.

Furthermore, a time-dependent deformation means that the calibration should be carried out at the machine rate actually used in the study and at different pressure levels. When increasing the pressure, the upper punch velocity at impact with the steel disc, or the test material, should increase [10]. This may influence the slope of the force versus displacement line as well as the hysteresis during decompression. When running a real test, compensation should be made for the deformation of the measurement equipment during both compression and decompression, including the hysteresis, at the actual pressure level.

III. QUANTITATIVE ANALYSIS OF FORCE-DISPLACEMENT PROFILES

Figure 2 shows a schematical plot of upper punch force versus upper punch displacement. The work of compression, sometimes also called gross work input or upper punch work, is represented by the total area *ABC*. (The origin represents the point where the upper punch gets in contact with the material in the die. The areas E_1–E_3 will be used in the discussion to follow.)

Some of this work, or energy, is recovered during decompression as

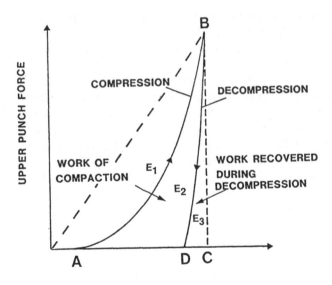

Fig. 2 Schematical plot of upper punch force versus upper punch displacement during compression and decompression including different areas (E_1–E_3) used in the characterization of the compression process.

work performed on the upper punch, which is represented by the area *DBC*. The area *ABD* should represent the apparent net work used in the formation of the compact and the work needed to overcome die wall friction. ("Apparent net work" is a more appropriate term than "net work" since the expansion is not complete.) As mentioned in the introduction, these areas have been related to the deformation properties of tested materials as well as their binding properties. A simplified description of the compression process identifies three components; elastic deformation, plastic deformation and fragmentation, as illustrated in Fig. 3.

The work needed to deform an *elastic* material will be completely recovered during the decompression phase and there will be no net work used in the formation of a compact. On the other hand, *plastic* deformation, with or without *fragmentation*, yields a net work input. Materials that are irreversibly deformed to a large extent should be expected to give a large work input.

The absolute values of the areas in Fig. 2 will of course be dependent on the force level. In a number of papers [11–14] it has been shown that the upper punch force versus displacement generally follows a hyperbolic form and can be described by hyperbolic constants. The constants should be independent of the pressure level and attempts have been made to correlate them with the properties of different mixtures [14].

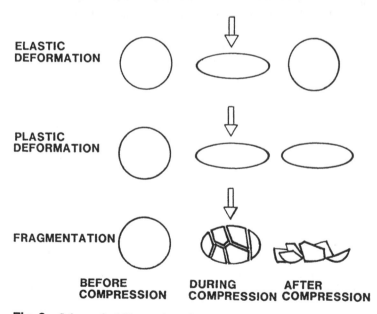

Fig. 3 Schematical illustration of processes that take place during compression.

A more common approach has been to use ratios between different areas in the force displacement plot. Dürr et al. [15] used the areas E_1, E_2, and E_3 (see Fig. 2) and the sum of these areas ($E_{max} = E_1 + E_2 + E_3$) to characterize the compressibility of powders and compacts. According to the authors' experience, E_1 should be as small as possible and the ratios ($E_2 + E_3)/E_1$ and E_2/E_3 should be as large as possible. An energy ratio (Energieverhältnis) was defined as $E_{2\%}/E_{3\%}$ $E_{1\%}$. To make the value E_2 pressure-independent they used the equation $E_2 = E_2\infty$ $P/(b + P)$, where P is the force, b a constant and $E_2\infty$ the energy transferred to the tablet at a infinitely large pressure. They suggested that a small b (indicating a fast increase in E_2 at increasing pressure) and a large $E_2\infty$ are favorable. In an attempt to reduce the influence of die wall friction, they used the lower punch force in the calculations.

Stamm and Mathis [16] used similar methods to calculate plasticity constants. Plasticity (P1) was calculated as $100E_2/E_2 + E_3$; i.e., the ratio of the work used to form the compact to the total work input. A high P1-value indicates that a large part of the energy input is utilized in irreversible deformation of the material. As the measure is not pressure independent, they characterized a variety of materials by compressing a series of tablets with different tablet hardness. Like Dürr et al. they used lower punch values. They ranked a large number of excipients. The "plasticity" values ranged for example from 58.2 (mannitol) to 94.1 (Avicel PH 101).

Doelker et al. [17,18] used the same plasticity coefficient as Stamm and Mathis but called it PL1 and defined it as the ratio between apparent net work input and the "lower punch work." The apparent net input is equal to the upper punch work minus the expansion work and is thus equivalent to E_2 in the equations above. The "lower punch work" is the area under the lower punch versus upper punch displacement curve. The term "lower punch work" is unfortunate since the lower punch cannot produce any physical work as long as it is stationary. The main reason to use lower punch data in this and the previous studies is, however, that these data have been suggested as being independent of die wall friction [19]. The validity of this approach will be discussed later on.

A second plasticity parameter, PL2, was calculated as the net input-to-upper-punch-work ratio. To calculate the net input they used lower punch values and a double-compression technique suggested by de Blaey and Polderman [19,20]. De Blaey and Polderman assumed that only elastic deformation will take place during a second compression of the tablet. According to this assumption, the compression work during the second compression should represent true expansion work and thus make possible improved calculation of the actual net input used for permanent deformation and particle bonding.

The authors stated that incorporation of a binder gave higher values for each kind of work, i.e., increased the resistance to deformation, and gave higher plasticity coefficients. They suggested that a comparatively greater proportion of energy is consumed by tablets made from granulations. They also concluded that tablet strength was related to the net work input. Furthermore rank orders and not the magnitude of mechanical strength, can be predicted because "the work taken up by the tablet is not a direct indication of the work used for the formation of bonds and of the bonding capacity of the components."

In a large number of studies, force-displacement measurements have been used to characterize the compression properties and to show correlations between net work values, or alternatively different work ratios, and tablet strength [2,5,21–38]. Most studies concluded that improved tablet strength was associated with a large degree of plasticity.

There are sound reasons to consider the accuracy and relevance of the findings reported in many force-displacement studies. As discussed in the instrumentation part of this chapter, it is difficult to measure expansion work. Small errors in the compensation for the deformation of machine parts will have a large impact. The influence of such errors will be very large if ratios between the expansion work and measures of the compaction work such as E_2/E_3 are used. E_3 usually assumes a value which is only a fraction of E_2. Thus dividing a large value by a small one, itself measured with considerable uncertainty, runs the risk of considerable error. Accurate expansion work measurement is made more difficult still by time-dependent, incomplete tablet expansion within the die and die wall friction.

An in depth review of the literature reveals that different techniques and assumptions have been used in the calculation of net work input as well as in the interpretation of the results. It is thus justified to discuss the net work concept in more detail.

IV. CALCULATION OF NET WORK FROM FORCE-DISPLACEMENT PROFILES

De Blaey, Polderman, and co-workers have proposed a definition of the different energy-consuming steps during compression and introduced methods to calculate the net work input. According to their definition [19], work is consumed during compression

    a. For arriving at the closest possible proximity of the particles of the granulate
    b. By friction between the particles
    c. By friction with the die wall

   d. By plastic deformation
   e. By elastic deformation

(Similar lists have been presented by others, such as Führer and co-workers.) It may also be argued that energy used for fragmentation and energy released due to reduced surface energy during bonding (see Sect. VI) should be included when discussing the overall energy balance. This list above, however, been of fundamental importance in the attempts to define net work. The authors further assumed that the steps a and b, which can be regarded as particle rearrangement and interparticulate friction, can be neglected. The work used to overcome die wall friction (c) and the work recovered during decompression (e) should, on the other hand, be subtracted from the gross work input to form a net work (NETW) input. They consequently assumed that the calculated NETW represented the energy used for plastic deformation and bond formation. The NETW calculation should thus be very informative in characterising a material provided that these assumptions are correct.

As discussed above, we first have to subtract the work needed to overcome friction (FW) and the expansion work (EXPW) from the total work consumption (see Fig. 4).

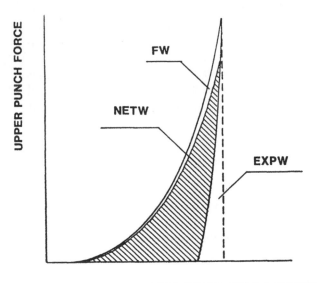

Fig. 4 Force-displacement plot illustrating the net work (NETW), work of friction (FW) and work recovered during expansion (expansion work, EXPW).

De Blaey and Polderman [19] have suggested that the work of die wall friction may be calculated according to Eq. (I) in Fig. 5. This method was criticized by Järvinen and Juslin [39], who tried to derive a work of friction based on the movement of the particles in contact with the die wall and the force acting on the particles rather than the force and displacement of the upper punch. They derived Eq. (II) in Fig. 5.

Equation (I) in Fig. 5 implies that the FW should be equal to the area limited by the curves D_SA and D_SB in Fig. 5, i.e., the area D_S-A-B. The lower punch reading should consequently be independent of die wall friction and useful as force measure in NETW calculations.

Invoking the second equation results in values approximately half of those obtained by de Blaey and Polderman (at low or moderate friction levels).

The assumption that the frictional force should not coincide with the movement of the upper punch appears reasonable. Only particles in the upper layer of the compact are capable of moving the same distance as the upper punch. Particles at the lower punch surface will be stationary.

The method by de Blaey and Polderman has been the more widely used as discussed previously. The validity of the two methods has been the

I. $FW=\int_{D_S}^{D_M} (UPF-LPF)\, dD$

     i.e. area D_S-A-B

II. $FW=\int_{D_S}^{D_M} \left\{ UPF-(UPF-LPF)/\ln(UPF/LPF) \right\} dD$

     $=\int_{D_S}^{D_M} (UPF)\, dD -\int_{D_S}^{D_M} \left\{ (UPF-LPF)/\ln(UPF/LPF) \right\} dD$

     i.e. area D_S-A-C

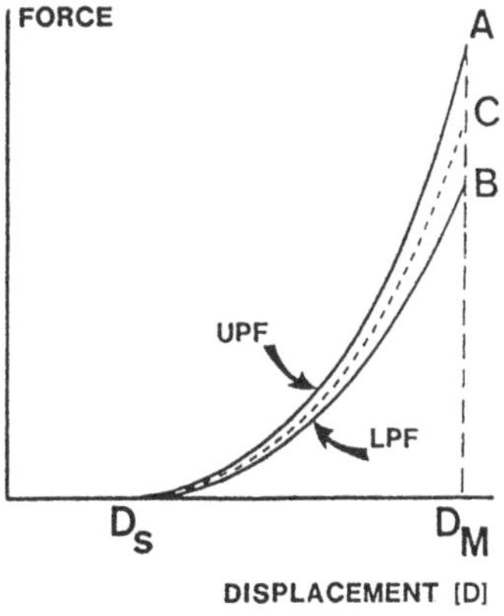

Fig. 5 Calculation of work needed to overcome die wall friction (FW) according to I (de Blaey and Polderman [19]), II (Järvinen and Juslin [39]), illustrated with a schematical force-displacement plot. The following abbreviations have been used: D, the displacement of the upper punch, measured relative to the lower punch; D_S, the point at which the force rises from zero; D_M, the maximum displacement of the upper punch; UPF, the upper punch force; LPF, the lower punch force.

subject of some discussion, however, [40, 41] since neither method had
been confirmed empirically.

A recent series of experiments was carried out to resolve this controversy [34] by holding constant all experimental factors except the die wall
friction. Under these conditions true net work should be constant even
though the friction varies, provided that FW is calculated correctly.

The experiments were carried out using a die prelubricated by compressing a number of tablets containing a large excess (approx 25%) of
magnesium stearate. It is known that magnesium stearate forms a lubricant
layer that can resist a number of repeated compressions [42]. When a series
of tablets of an unlubricated material are compressed in a prelubricated
die, the lubricant layer will gradually be worn off, resulting in a gradual
increase in die wall friction. This is illustrated in Fig. 6.

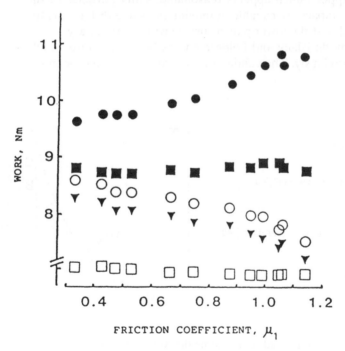

Fig. 6 Work of compression versus friction coefficients of consecutive anhydrous lactose tablets compressed in a prelubricated die. ●, upper punch work;
○, "lower punch work"; □, expansion work; ▼ net work (NETW) when work
of friction (FW) is calculated according to de Blaey and Polderman [19]; ■,
NETW when FW is calculated according to Järvinen and Juslin [39]. (Adapted
from Ref. 34.)

As can be expected, friction increased as the lubricant wore off and increased the total work input. The Järvinen and Juslin method to calculate FW resulted in an approximately constant NETW (filled squares in Fig. 6). The de Blaey and Polderman method produced a decreasing NETW value. Thus the latter method appears to overestimate the influence of die wall friction, which seems theoretically sound.

As predicted by Järvinen and Juslin, the FW values were close to half of those calculated according to Eq. (I) in Fig. 5. This was found to be true not only at low friction levels but over the whole friction range, even at friction levels where further compressing was impossible. As a practical consequence, the mean of the upper punch force (UPF) and the lower punch force (LPF); i.e., the mean punch force (MPF) can be regarded as a compression force measure that is independent of die wall friction. The NETW may thus be calculated by a simple integration of the MPF versus upper punch displacement plot and subtraction of the expansion work (EXPW). The usefulness of MPF as a general friction independent measure of compaction force has been demonstrated in other studies [43,44] and is gaining wider acceptance [5].

The axial expansion of a tablet within a die is imcomplete during the decompression phase which makes the measurement of the EXPW difficult. For example, comparison of the tablet height after ejection and in the end of the decompression phase, have shown that only about 65% of the total axial expansion of anhydrous lactose tablets took place in the die [7]. This value was obtained with a well-lubricated die and decreased to about 40% at high die wall friction.

De Blaey and Polderman suggested [19] that the EXPW should be measured by using a double-compression technique. The tablet will continue its expansion in the die when the first compression is completed and the area under the decompression curve (EXPW$_1$) should thus underestimate the true expansion work. They suggested that the upper punch work during a second compression should represent the true EXPW. A problem with the double-compression technique is that a second compression may not only include elastic deformation but also some plastic deformation [45]. In addition, the properties of the compact may change at repeated compression [46].

When the friction is high, neither the single- nor the double-compression technique appears to give the complete EXPW due to incomplete expansion in axial direction. At low friction, obtained by prelubrication of the die or lubrication of the test material, only small differences between the two methods have been obtained [19].

When all these aspects are considered there appears to be little justification in using the more complicated double-compression technique. I recommend that force-displacement measurements are carried out using the single-compression technique in prelubricated dies giving low die wall friction without interference with the bonding properties of the test material.

The calculated EXPW will underestimate the true expansion work, not only due to incomplete expansion in axial direction. The tablets will also expand in radial direction after ejection. It appears, nevertheless, reasonable to subtract the calculated EXPW in NETW calculation. The error in the NETW calculation due to an underestimated EXPW should in general be small. Empirically, the EXPW tends to be only a few percentage of the total compression work for pharmaceutical materials (e.g., 0.5–10% of the NETW values for a variety of materials in [7]).

The importance of the criticism put forward in this chapter regarding established methods to calculate FW and EXPW is dependent on the aim of the studies, experimental conditions and the general accuracy of the measurement system. It is, for example, reasonable to believe that data registration limitations have affected results and conclusions more than the choice of calculation methods in some of the early studies.

Of more fundamental interest are the basic assumptions about the different processes taking part in powder compaction and their contribution to the NETW.

V. INFLUENCE OF INTERPARTICULATE FRICTION AND BONDING ON CALCULATED NET WORK

As discussed previously, the introduction of the NETW calculation (i.e., [20,21,24]) was an attempt to calculate the work used for plastic deformation and for bond formation. While it was considered necessary to subtract FW and EXPW from the total work input, the work needed for particle rearrangement and interparticulate friction were assumed to be negligible. Further deformation, requiring a net work input, should reflect the inherent deformation properties of the test material.

In a crystal, plastic deformation or fragmentation takes place when the intermolecular forces are exceeded. Schematically, the plastic deformation can be described as taking place along slip planes inside the crystals as, e.g., reviewed by Moldenhauer et al. [47].

A compact is deformed plastically if it undergoes irreversible deformation by plastic flow or fragmentation of the particles (crystals) in the compact and shows little elastic recovery during decompression. It may thus be reasonable to distinguish between plastic deformation of a crystal and of a compact consisting of a large number of particles, unless it can clearly be demonstrated that interactions between the particles play an insignificant role in the compression event.

Attempts have been made to vary the interaction of interparticulate friction and bonding among the particles in a compact with minimal effect on the composition by varying the particle size, the lubrication, and the moisture content [35]. The influence on both NETW and Heckel plots were studied. The Heckel equation, which will be discussed in more detail elsewhere in this book, is one of the experimental powder compression equations that relates applied pressure to the volume change during compression.

For example by adding a small amount of a lubricant, 0.5% magnesium stearate to dicalcium phosphate, Emcompress, a small decrease in the tablet strength was obtained in addition to reducing the interparticulate friction. The result was a small reduction in the NETW (Fig. 7). The reduction in the NETW was not due to changes in the EXPW, which was supported by measurements of the total axial expansion of the tab-

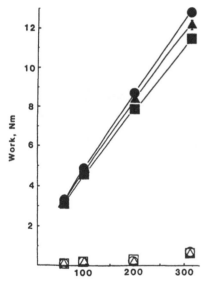

AXIAL ELASTIC RECOVERY, %			
MPP	EC	EC1	EC30
55	2.2	2.1	2.1
96	3.1	3.0	3.0
198	2.4	2.3	2.4
313	3.3	3.4	3.5

Mean punch pressure (MPP), MPa

	Emcompress (EC)	Emcompress + 0.5% Magnesium stearate Mix. time 1 min. (EC1) Mix. time 30 min. (EC)30	
Net work	●	▲	■
Expansion work	○	△	□

Fig. 7 Net work and expansion work when compressing dibasic calcium phosphate dihydrate, Emcompress, with or without small amounts of magnesium stearate (0.5%) admixed for 1 and 30 minutes. (Adapted from Ref. 35.)

lets. As a precaution die wall friction was kept low by using prelubricated dies. Mean punch pressure was used as a measure of the compaction pressure.

The bonding properties were drastically reduced by adding magnesium stearate to starch, Sta-Rx, known to be very sensitive to lubricants. Total elimination occurred after mixing for 30 min (Fig. 8). As can be seen in Fig. 8 this reduced particle to particle interaction resulted in a significant reduction in the NETW. It seems very unlikely that a small amount of a lubricant and its admixing time should affect the ability of the individual units of the starting material to deform plastically. In contrast to earlier suggestions, interparticulate effects during compaction appears to significantly affect the NETW. Tests on other materials verified that decreased particle to particle interaction, obtained by changing the surface properties

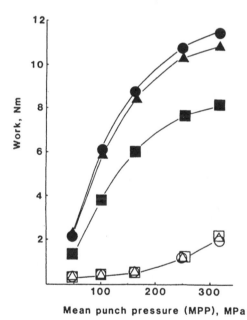

AXIAL ELASTIC RECOVERY, %			
MPP	SX	SX1	SX30
48	17.0	16.9	–
103	16.1	16.4	–
164	15.9	16.0	–
251	17.8	17.8	–
316	19.3	19.6	–

	STA-RX (SX)	STA-RX + 0.5% Magnesium stearate Mix. time 1 min. (SX1)	Mix. time 30 min. (SX30)
Net work	●	▲	■
Expansion work	○	△	□

Fig. 8 Net work and expansion work when compressing corn starch, Sta-RX 1500, with or without small amounts of magnesium stearate (0.5%) admixed for 1 and 30 min. (Adapted from Ref. 35.)

of particles in the compact will reduce the resistance to consolidation and thereby lower the NETW.

There is no simple correlation between the NETW and the deformation properties of a material since the NETW is also substantially affected by the particle to particle interaction. The experiment with lubricated Sta-RX suggests that about 30% of the NETW may be due to such interactions!

Is it possible that particle-to-particle interaction can be this large? Our standard model assumes that the force needed to overcome particle-to-particle interactions is small compared with the force needed to deform particles in the compact. Yet the results of the lubricated Sta-RX experiment appear to violate this assumption.

Let us have another look at the Sta-RX and magnesium stearate experiment, now presented as force displacement curves (Fig. 9). We see how the resistance to compression is actually reduced, especially during the initial part of the compression. The reduction in *force* needed to compress

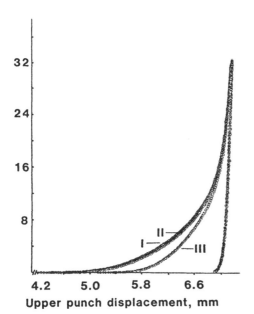

Fig. 9 Force-displacement curves of Sta-RX compressed with or without small amounts of magnesium stearate. I, pure Sta-RX; II, 0.5% magnesium stearate admixed for 1 min; III, 0.5% magnesium stearate admixed for 30 min. (Adapted from Ref. 35.)

the lubricated material is indeed small, but takes place over a long distance. The result is a substantial reduction in NETW.

The effect of interparticulate friction and bonding may perhaps be explained with the aid of a simplified figure (Fig. 10). When the particles within the die are compressed, irregular rearrangement takes place. In the beginning of the process there will be both particles in mutual contact and interparticulate voidage.

Further particle rearrangement can reduce the porosity of the compact, registered as a punch displacement, without particle fragmentation or plastic deformation. Such rearrangement will, however, probably be counteracted by high bonding or friction forces in the contact points between particles. Increased friction and bonding may thereby increase the work needed to compress the material independent of any changes in the deformation mechanism. Any change in the surface properties that promotes bonding should consequently give an increase in the NETW due to higher total resistance to deformation of the large number of particles that forms a

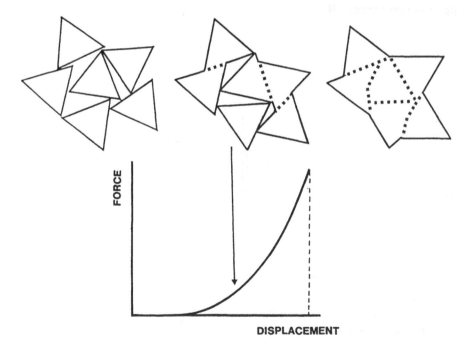

Fig. 10 Schematical figure illustrating the influence of interparticulate friction and bonding on force-displacement plots. Bonding or friction forces in contract points (dots) are suggested to have a significant effect on the work needed for further compression (see text).

compact. If this theory is correct, it is obvious that the importance of the deformation mechanism has been overestimated in many of the earlier NETW measurements while the effect of surface properties, bonding mechanism and bonding strength has been underestimated.

VI. ENERGY BALANCE DURING COMPACTION

Apart from the attempts to correlate force-displacement data with the tablet-forming properties of materials, such curves may be used to study the energy balance in tabletting. All energy used to compress a material will be released as heat if no changes in the energy content of the material takes place. The work of compaction, Wc, will then be equal to the heat released during compaction, Qc. i.e.,

$Ec = Wc - Qc$
Ec = energy change during compression
Wc = work done on the powder (work of compression)
Qc = heat released by the system
(Adapted from Coffin-Beach and Hollenbeck [48])

Führer and Parmentier [13] estimated that about 90% of the work of compression was released as heat.

Coffin-Beach and Hollenbeck [48] have more recently made very interesting studies of the energy balance with the aid of a highly sensitive compression calorimeter. By measuring the work of compression during compaction at a constant pressure, they found that the energy released as heat was *larger* than the energy of compression for all materials tested. They measured only the effect during the compression phase and not during decompression but efforts were made to compensate for energy changes associated with the deformation of machine parts.

The extent to which the heat released *exceeded* the work of compression was termed the *energy of formation* as it was assumed that this energy was equal to the reduction in surface energy due to bonding. For example, microcrystalline cellulose, Avicel, gave a high energy of formation while dicalcium phosphate, Di-Tab, known to fragment to a large extent during compression, gave considerably lower values. It was further suggested that fracture and bonding balanced each other at forces below approximately 10,000 N for dicalcium phosphate, while particle recombine and bond at higher pressures resulting in increased *energy of formation*. The *energy of formation* correlated with the tensile strength for each material but it appears not to be a simple general correlation.

As suggested by the authors, these type of measurements may not only give quantitative evaluation of the energetics of the interparticulate interactions due to bonding but possibly also some information about the true particle area involved in bonding.

Simultaneous measurements of compression work and heat released during compression should be rather tricky, however. During compression the punches will deform elastically. They will thereby act as springs, as very roughly illustrated in Fig. 11. The upper punch is drawn as a relaxed spring at the beginning of compression and as a compressed spring at maximum pressure (the total deformation of machine parts has for example been shown to be over 200 μm at a pressure of 250 MPa using 11.3-mm punches in an excenter machine [7]).

Let us assume, for simplicity, that all elastic deformation of the equipment is caused by the upper punch. The true deformation of the test material, D_I, can then easily be obtained by subtracting the known elastic deformation of the punch making it possible to calculate the compression work Wc_I. Unfortunately, it take some time to measure the heat released and it appears necessary to stop the compaction process at a suitable pressure level. During the heat measurement the material tends toward further deformation, as the deformation process in general is a time-dependent process and the material is under pressure of the elastically deformed punch. The deformation of the material generates additional work of com-

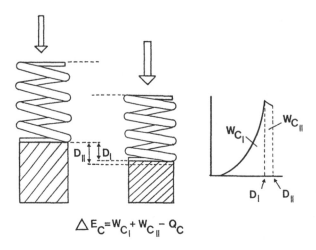

$$\triangle E_C = W_{C_I} + W_{C_{II}} - Q_C$$

Fig. 11 Schematic illustrating the problems of measuring energy change during compression (ΔE_C) under static conditions. The punch, which is elastically deformed, and illustrated as a spring in the figure, will keep the material under pressure and may cause further plastic deformation ($D_{II} - D_I$). This will cause additional compaction work (Wc_{II}) which should be included in the energy balance where Qc is equal to heat released by the system.

pression which should be included in the energy balance equation above. It may be difficult to measure this work as the displacement transducer is mounted some distance from the punch tip.

Figure 11 shows punch deformation. The punch may expand the distance $D_{II} - D_I$ without giving any visible increase in displacement. The deformation should, however, give a decrease in the force reading. By using equipment capable of keeping the punch force exactly constant by a constant slow compression, governed by the deformation rate of the test material, it might be possible to measure the total compression work, i.e., including Wc_{II}. If this extra compression work is not considered, the energy of formation of materials such as cellulose with a very-time-dependent deformation [49] should be overestimated when compared with a material with no or little time-dependent deformation such as dicalcium phosphate [49].

Coffin-Beach and Hollenbeck used the static method in a tablet press with a slow rate of compression. Sufficient stress relaxation may have occurred during the compression to give an acceptable calculation of the total work input (personal communication with the authors).

VII. SUMMARY AND CONCLUSION

Force-displacement measurements have received much attention, especially during the 1970s and early 1980s. The basis for this interest has mainly been the proposed possibility to characterize the compression and deformation properties of materials and to obtain predictive data. Modern data registration and computer systems have given us increasingly accurate raw data. However, there remain pitfalls and problems in the measurement techniques, as well as with the evaluation of obtained data, which keep force-displacement and NETW registration from being uncomplicated tools.

Based on published data it appears reasonable to conclude that the NETW may be useful for detecting batch-to-batch variations in the compaction properties of materials due to its high sensitivity to *both* inter- and intraparticulate properties, good reproducibility and low dependence upon die wall conditions.

Results obtained during the last century indicate that the predominant assumption that the influence of particulate interaction during compression can be neglected is incorrect. Additional methods will be needed [35] to determine whether interlot variations in the NETW consumption during compaction are due to altered surface properties or intraparticulate changes (e.g., polymorphism and lattice defects). This should be kept in mind when considering the predictive value of NETW measurements.

A natural area where NETW calculations should be useful is in the evaluation of crystallographic changes during compaction. In these cases it appears obvious that the work, or energy, used during compression is of far more interest than the maximum force or pressure used in the process. This possibility has been mainly overlooked except for a few studies [50].

Simultaneous measurement of work of compression and heat released during compression is an interesting approach that may benefit from new and very sensitive measurement devices. The technique will not only increase our knowledge about thermal effects in compaction but may, theoretically, yield information about surface energy reduction due to bonding and thus indicate the true bonding surface in tablet formation. It would be of particular interest to see a comparison between the results of such studies and the results from some of the new approaches to measuring the true bonding area which are discussed elsewhere in this book.

REFERENCES

1. H. Heins, W. Ott, and C. Führer, *Pharm. Ind. 31*: 155 (1969).
2. N. A. Armstrong and F. S. S. Morton, *J. Powder Bulk Solids Technol. 1*: 32 (1977).
3. E. Nelson, L. W. Busse, and T. Higuchi, *J. Am. Pharm. Assoc. Sci. Ed. 44*: 223 (1955).
4. S. D. Bateman, The effect of compression speed on the properties of compacts, Ph.D. thesis, School of Pharmacy, Liverpool Polytechnic, 1988.
5. J. S. M. Garr, and M. H. Rubinstein. *Int. J. Pharm. 64*: 223 (1990).
6. R. J. Oates, and A. G. Mitchell, *J. Pharm. Pharmacol. 41*: 517 (1989).
7. G. Ragnarsson, Evaluation of tabletting properties in preformulation and early formulation work, Ph.D. thesis, Uppsala University, 1985.
8. R. F. Lammens, J. Poldermen, and C. J. de Blaey, *Int. J. Pharm. Tech. Prod. Mfr. 1*: 26 (1979).
9. R. F. Lammens, J. Polderman, C. J. de Blaey and N. A. Armstrong, *Int. J. Pharm. Tech. Prod. Mfr. 1*: 26 (1980).
10. P. Colombo, U. Conte, C. Caramella, A. La Manna, J. C. Guyot, A. Delacourte, B. Devise, M. Traisnel, and M. Boniface, *Il Farmaco - Ed.Pr. 33*: 531 (1978).
11. C. Führer, (1965). *Dtsch. Apoth. Ztg. 105*: 1150 (1965).
12. W. Parmentier, Untersuchungen zur Gesetzmässigkeit des Kraftferlaufs bei der Tablettierung, Diss., der Technischen Universität Carolo-Wilhelmina zu Braunschweig, 1974.
13. C. Führer, and W. Parmentier, *Acta Pharm. Technol. 23*: 205 (1977).
14. C. Führer, G. Bayraktar-Alpmen, and M. Schmidt, *Acta Pharm. Technol. 23*: 215 (1977).
15. M. Dürr, D. Hansen, and H. Harwalik, *Pharma. Ind. 34*: 905 (1972).
16. A. Stamm and C. Mathis, *Acta Pharm. Technol. Suppl. 1*: 7 (1976).

17. E. Doelker, *Pharm. Acta Helv. 53*(*6*): 182 (1978).
18. E. Doelker, R. Gurny, and D. Mordier, *Acta Pharm. Technol. 26*: 155 (1980).
19. C. J. de Blaey, and J. Polderman, *Pharm. Weekbl. 105*: 241 (1970).
20. C. J. de Blaey, and J. Polderman, *Pharm. Weekbl. 106*: 57 (1971).
21. C. J. de Blaey, M. C. B. van Oudtshoorn, and J. Polderman, *Pharm Weekbl. 106*: 589 (1971).
22. C. J. de Blaey, A. B. Weekers-Andersen, and J. Polderman, *Pharm. Weekbl. 106*: 893 (1971).
23. J. T. Fell, and J. M. Newton, *J. Pharm. Sci. 60*: 1428 (1971).
24. J. Polderman, and C. J. de Blaey, *Farm. Aikak. 80*: 111 (1971).
25. J. Gillard and M. Roland, *Pharm. Acta Helv. 52*: 154 (1971).
26. M. Dürr, *Acta Pharm. Technol. 22*: 185 (1976).
27. Erhardt, and E. Schindler, *Pharm. Ind. 41*: 1213 (1979).
28. Erhardt, and E. Schindler, *Pharm Ind. 42*: 1213 (1980).
29. G. Ragnarsson, and J. Sjögren, *Int. J. Pharm. 12*: 163 (1982).
30. I. Krycer, D. G. Pope, and J. A. Hersey, *Int. J. Pharm. 12*: 113 (1982).
31. I. Krycer, D. G. Pope, and J. A. Hersey, *J. Pharm. Pharmacol. 34*: 802 (1982).
32. I. Krycer, and D. G. Pope, *Powder Technol. 34*: 39 (1983).
33. I. Krycer, and D. G. Pope, *Powder Technol. 34*: 53 (1983).
34. G. Ragnarsson and J. Sjögren, *J. Pharm. Pharmacol. 35*: 201 (1983).
35. G. Ragnarsson, and J. Sjögren, *J. Pharm. Pharmacol. 37*: 145 (1985).
36. N. Kaneniwa, K. Imagawa, and M. Otsuka, *Chem. Pharm. Bull. 32*: 4986 (1984).
37. T. Cutt, J. T. Fell, J. R. Rue, and M. S. Spring, *Int. J. Pharm. 49*: 157 (1989).
38. M. Otsuka, T. Matsumoto, and N. Kaneniwa, *J. Pharm. Pharmacol. 41*: 665 (1989).
39. M. J. Järvinen, and M. J. Juslin, *Farm. Aikak. 83*: 1 (1974).
40. R. F. Lammens, T. B. Liem, J. Polderman, and C. J. de Blaey, *Powder Technol. 26*: 169 (1980).
41. M. J. Järvinen and M. J. Justin, *Powder Technol. 28*: 115 (1981).
42. A. W. Hölzer, and J. Sjögren, *Int. J. Pharm. 3*: 221 (1981).
43. G. Ragnarsson, and J. Sjögren, *Int. J. Pharm. 16*: 349 (1983).
44. G. Ragnarsson, and J. Sjögren, (1984). *Acta Pharm. Suec. 21*: 141 (1984).
45. I. Krycer, D. G. Pope, and J. A. Hersey, *Drug Dev. Ind. Pharm. 8*: 307 (1982).
46. N. A. Armstrong, N. M. A. H. Abourida, and L. Krijgsman, *J. Pharm. Pharmacol. 34*: 9 (1982).
47. H. Moldenhauer, H. Kala, G. Zessin, and M. Dittzen, *Pharmazie 35*: 714 (1980).
48. D. P. Coffin-Beach, and R. G. Hollenbeck, (1983). *Int. J. Pharm. 17*: 313 (1983).
49. J. E. Rees and P. J. Rue, *J. Pharm. Pharmac. 30*: 601 (1978).
50. M. Otsuk, T. Matsumoto, and N. Kaneniwa, *J. Pharm. Pharmacol. 41*: 665 (1989).

5

Viscoelastic Models

Fritz Müller

Pharmazeutisches Institut der Universität Bonn, Bonn, Germany

I. INTRODUCTION

A comparison of load (force, stress) versus time (σ/t) and displacement (strain) versus time curves (ϵ/t) often shows that their maxima do not occur simultaneously (Fig. 1). The load reaches its maximum first and decreases to a significantly lower value when the upper and lower punches are in the closest position (Schierstedt [1], Schierstedt and Müller [2], Caspar [3], Müller and Caspar [4], Ho and Jones [5]). This effect can also be noticed in the time independent load versus displacement-curves (σ/ϵ) where the coordinates of the maximal load and maximal displacement are different [1, 2, 6, Fig. 2].

Experiments involving various agents and excipients show that the described time shift is typical for the behavior of compressed materials (Müller and Caspar [4]). Macromolecular substances, among them important tabletting materials such as starches and celluloses, comply with this empirical experience (Cole, Rees, and Hersey [7]). First observations in this field have been reported by David and Augsburger [8], Wiederkehr [9], and Rees and Rue [10]. These experiments were performed in a static manner, for example by stopping the machine at its lowest point, the dead center position of the upper punch and observing the time-dependent change of the load. These phenomena can only be explained by the influence of the upper punch proceeding into the already compacted powder

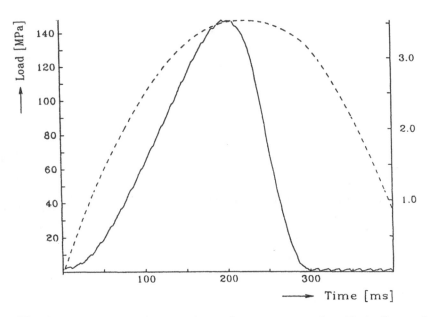

Fig. 1 Load vs. time (solid line) and displacement vs. time (dashed), raw data for Starch.

bed where the resistance diminishes, giving way due to the characteristics of the substance, alternatively there could be entrapped air which will escape as a function of the gas pressure in the die, as supposed by Casahoursat et al. [11,12]. Continuum mechanics assigns this type of phenomena to elasticity, viscosity, viscoelasticity, or viscoplasticity.

Furthermore, the curves indicate, that the upper punch has not yet reached the original position with no load, when the load reaches zero again at the point of punch lift-off. The difference between the initial and final displacement can serve as a measure of plastic deformation.

Elastic deformation is displayed by the symmetry of the load versus time curve or by the declining section of the load versus displacement curves. Using appropriate parameters corresponding to the shape of the load versus time curves faciliates the evaluation of tablet mass deformation. Emschermann [13] and Emschermann and Müller [14] investigated various tablet excipients using a single-punch machine. The authors subdivided the area under the load versus time curve into two sections: one from the origin to the maximum of compression force and the other one from there to the end of the process. The ratio of the two areas represents a measure of curve symmetry. It allows us to discern substances, that

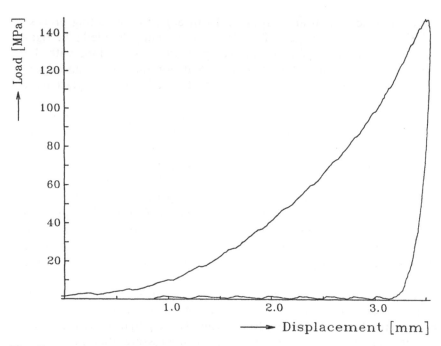

Fig. 2 Load vs. displacement, raw data for Starch.

undergo plastic rather than elastic deformation, from substances which behave vice versa. An ideally elastic material should produce a symmetrical curve and an area ratio of unity. Tenter [15] and Schmidt et al. [16] applied this area evaluation principle to load versus time curves of a rotary press in some papers. Since pinpointing the compaction maximum is difficult in this case due to the trapezoidal formed curve, they calculated the point of subdivision to the time coordinate of the center of gravity under the curve. Subsequently, depending on the mass deformation, the area ratio now assumed values below and above unity. More recently the authors divided the area into four sections in order to get more information [17]. Dietrich and Mielck [18] introduced sensitive quantitative parameters to give an interpretation of the shape of a modified Weibull function to get time-independent results. All attempts to obtain the influence of time on the slope of a Heckel-plot [19] as strain rate sensivity (Roberts and Rowe [20]) are efforts to characterize the viscoelastic properties in this indirect manner.

One can expect that all properties are superimposed upon each other, so that components that arise dynamically through the tabletting process should be isolated part by part. Therefore, a thoroughly detailed investigation of force relaxation phenomena, creep characteristics along with the development of mechanical models and calculation methods is an important field of theory in tabletting to predict the load versus displacement curve shape, or vice versa.

For example, the Brinell value is suitable for judging the plastification characteristics of a tablet. It involves measuring the imprint caused by a round object that is pressed against the upper or lower surface of a tablet with a given force (Jetzer, Leuenberger, and Sucker [21]).

The load versus displacement curves show that tablet materials are neither ideally elastic, nor ideally viscous (or plastic). The behavior of pharmaceutically used substances lies between these features. The physical description of these properties is based on the theory of continuum mechanics. David and Augsburger [8] first applied this kind of theoretical models on tablets. For this purpose they loaded tablets in a static way (i.e., using constant stress or strain).

Once pharmaceutical powder is compressed to test tablets with certain dimensions, as is typical in the field of technical material testing, various methods can be applied. Roberts and Rowe [20] offered a relationship between indentation hardness and yield pressure that provides a method for calculating the elasticity of compacted substances. To ensure compatibility, the authors suggested extrapolating elasticity moduli for porous samples to a state of zero porosity.

An important subject within the theory of tabletting is the exact investigation of the phenomena of stress relaxation and creep compliance, especially the development of mechanical models and mathematical methods to predict load versus displacement curves. In this way it should be possible to get information about the physics of flow properties during the compaction of materials.

Müller and Caspar [4, 22] undertook high resolution load versus displacement measurements using a single-punch machine. The authors showed a specific time shift of load and displacement maxima, due to the properties of the substances and drew attention to the fact that relaxation phenomena are also detectable during the dynamic process of compression. They adapted the section up to the load maximum by (static) one-dimensional simple models and found all material features, the height of the compressed tablets and the maximum of the load to be important. Only the particle size was found not to be relevant. The parameters were poorly reproducible and sometimes negative.

Sutanto [23] compared various mathematical methods to estimate different viscoelastic parameters using load versus time and displacement versus time curves on single-punch machines. He used an explicit solution of the constitutive equation, the linear creep compliance, to characterize the materials and described the stress relaxation under a dynamic load. These functions were found by deconvolution of the load versus time curves or the displacement versus time curves respectively. The obtained creep compliances showed irregularities such as two maxima which could be a result of two different, but superimposed events like consolidation and creep compliance and made good physical sense only in the first phase of compressing. Subsequently the deconvolution degenerated. Fischer [24] tested the applicability of the deconvolution method and found it not to be satisfactory, even with error-free test data.

Danielson et al. [25] and Rippie and Danielson [26] studied the decompression and the postcompression phase of substances with known differing compression characteristics using a rotary machine. The so-called postcompressing phase is the phase after compression when the tablet remains for some minutes in the die. Using load versus time curves deduced from radial and axial load measurements and with displacement versus time functions deduced mathematically from the rotary machine's geometry, they adjusted three-dimensional viscoelastic models where the reexpansion volume was related to an elastic process and the deformation to viscoelastic behavior. The arithmetic evaluation was resolved with a constitutive equation, i.e., the stress-strain-relation, correlating with the respective model. The parameters determined were found to relate to the maximum compression load and some turned out to be negative (eluding a physical explanation). Mathematical methods for three-dimensional models have been discussed by Morehead [27], who reduced these three dimensions to the cylindrical dimensions of tablets. He introduced in this paper viscoelastic properties in relation to work of compaction.

This short review shows the difficulties to obtain reasonable figures to characterize viscoelastic effects during tabletting. Hence it seems to be important to show the theoretical background of this behavior. Because there are many effects, one depending on another, this chapter may have a more philosophical touch then a technical report should have. Nevertheless it is the only way to understand the information hidden in the displacement versus time and the load versus time curves of a tabletting process.

A. Simple Models Explaining Viscosity and Elasticity

A mathematical procedure is given by adjusting mechanical models to the load versus time and displacement versus time curves in order to quantify material characteristics. Mathematical models incorporating viscous, elastic, and viscoelastic substances make physical sense and are easily understandable. Real existing materials react in a rather complex way. Therefore an attempt to cover all observed reactions with a simple constitutive equation, i.e., a relation of internal and external forces to its properties, is hardly possible. Constitutive equations are suited rather to define "ideal substances," such as the "ideal elastic material," the "ideal viscous fluids," or the "viscoelastic materials." The behavior of real substances can be attributed to these material models after initial and boundary conditions as well as applicable regions have been determined. Under these circumstances one has to consider that in powders the flow properties are bound to unknown borders as given by plasticity and viscosity (yield values).

Cases in which a complete description of a material is unnecessary or inaccessible, other effects, for example thermal can be uncoupled in order to simplify the theory. This would, however, not imply a neglect of the influence of temperature upon mechanical processes. It is expected that the temperature and/or temperature distribution is known or can be investigated separately. As the constitutive equations now hold only static and kinematic variables they are called stress-strain-relationships.

Stress (σ) describes surface forces which act perpendicularly on small surface elements. Strain (ϵ) as simple shear, tension or compression during tabletting describes local deformation related to original dimensions.

1. Elasticity (Spring)

Linear elastic behavior is represented by a spring. When a massless linear spring is pulled apart or compressed, the necessary force (load) is proportional to the relative expansion or compression. The relative change of length ($\Delta l/l_0$) is a consequence of the action of force without time delay and independent of the present state of load. Once a load is removed the spring returns immediately to its original state (l_0). Then Hooke's law is applicable:

$$\sigma = E\epsilon \tag{1}$$

where

σ = stress
ϵ = relative strain $\Delta l/l_0$
E = elasticity modulus

For the practical use of Hooke's law in the field of tabletting one has to consider with the effect of Poisson's law, because in a die the material cannot increase wider and wider in diameter like a free handled body will do (Bauer [28]).

2. Viscosity (Dashpot)

The dashpot exemplifies viscosity clearly. Force is necessary to move a piston through a cylinder filled with viscous fluid. The velocity of motion is proportional to this force. This linear behavior is described by Newton's law:

$$\sigma = F\dot{\epsilon} \tag{2}$$

where

σ = stress
$\dot{\epsilon}$ = $d\epsilon/dt$
F = viscosity modulus

The change in length depends on the time interval. A constant stress σ_0 in an interval Δt results in a strain of ϵ:

$$\epsilon = \frac{\sigma_0}{F}\Delta t \tag{3}$$

Since there is no restoring force, the dashpot remains elongated when the load is removed.

B. Composite Models Explaining Viscoelasticity

Viscoelastics are substances, the behavior of which lies betweeen elastic solids and viscous fluids. Elastic (Hookean) solids and viscous (Newtonian) fluids show great differences in their deformation behavior. Deformed elastic materials return to their original shape, once the load is removed. On the other hand, fluids remain in their state. Furthermore elastic deformation (ϵ) and stress (σ) are directly proportional to one another, whereas a fluid's internal stress (σ) depends on its rate of deformation (ϵ).

1. Maxwell Model

The serial connection of the two basic elements, spring and dashpot, is marked in a way that the entire elongation of such a model is equal to the sum of elongations of its components. As all components are accessible through the same amount of stress (σ) for the serial connection of a spring and a dashpot, out of the so-called Maxwell model, the following elongation results:

$$\epsilon = \epsilon_D + \epsilon_S \tag{4}$$

From $\sigma = E\epsilon_D$ for the spring and $\sigma = F\dot{\epsilon}_S$ for the dashpot follows for a Maxwell model a relationship between force and elongation (stress and strain) with the derivative of ϵ with respect to time:

$$\dot{\epsilon} = \dot{\epsilon}_D + \dot{\epsilon}_S = \frac{\dot{\sigma}}{E} + \frac{\sigma}{F} \tag{5}$$

The standardized constitutive equation as the mathematical stress-strain relationship is as follows:

$$\sigma + p_1\dot{\sigma} = q_1\dot{\epsilon} \tag{6}$$

The microconstants E for the elasticity (Young's) modulus and F for viscosity, which can be easily interpreted physically, are related to the hybrid parameters p and q in the constitutive equation, whose significance is recognizable only in the most simple model, for example in this Maxwell model:

$$p_1 = \frac{F}{E} \quad \text{and} \quad q_1 = F. \tag{7}$$

2.   Kelvin (Voigt) Model

In the Kelvin (Voigt) model the spring and the dashpot are connected in parallel, so the elongation (ϵ) of both parts is equal at any time in the entire model. In such an arrangement the stress σ is distributed in different shares on the spring and the dashpot. Shares of the spring (σ_S) and shares of the dashpot (σ_D) are

$$\sigma_S = E\epsilon; \; \sigma_D = F\dot{\epsilon} \tag{8}$$

and add up to the total stress σ:

$$\sigma = \sigma_S + \sigma_D \tag{9}$$
$$\sigma = E\epsilon + F\dot{\epsilon} \tag{10}$$

With $q_0 = E$ and $q_1 = F$ the constitutive equation of this model becomes

$$\sigma = q_0\epsilon + q_1\dot{\epsilon} \tag{11}$$

The stress (σ) in the Kelvin element is proportional both to the elongation of the spring and to the rate of deformation of the dashpot.

3.   Generalized Models

All higher-order models can be attributed to the two basic combinations: the "simple Kelvin model" and the "simple Maxwell model." The "generalized Kelvin model" results from the serial connection of n Kelvin models, from which one or, at maximum, two may be reduced to a dashpot and/or to a spring. The total extension of such a model is the sum of the elongations of the n-Kelvin models. Writing the constitutive equation of a single Kelvin model,

$$\sigma = E\epsilon + F\dot{\epsilon} \tag{12}$$

in operator form:

$$\sigma = \{E + F\delta_t\}\epsilon \tag{13}$$

(with $\{\delta_t\} \equiv \delta/\delta t$ the linear differential time operator) adding up the ϵ_i of the n elements, when solving for ϵ the result for a solid is as follows:

$$\epsilon = \frac{\sigma}{E_1 + 1/F_1\delta_t} + \frac{\sigma}{E_2 + 1/F_2\delta_t} + \cdots + \frac{\sigma}{E_n + 1/F_n\delta_t} \tag{14}$$

The "generalized Maxwell model" results from parallel connection of n Maxwell models, one or two of which can be reduced to a dashpot and/or to a spring. In this model the n strains σ_i of the single elements add up to the total strain σ:

$$\sigma = \frac{\dot{\epsilon}}{\delta_t/E_1 + 1/F_1} + \frac{\dot{\epsilon}}{\delta_t/E_2 + 1/F_2} + \cdots + \frac{\dot{\epsilon}}{\delta_t/E_n + 1/F_n} \tag{15}$$

A practical problem are the indices of the reduced elements. It would be convenient to use the same index for both reduced elements (Table 1).

Newton's law dictates that a sudden extension of a dashpot at the time $t = 0$ requires an infinite load. The peak of the load that induces such a theoretic expansion is described mathematically with a single step function $\Delta(t)$ and its derivative, the Dirac delta function $\delta(t)$.

The Heaviside unit step function $\Delta(t)$ is divided into two sections, one with $t < 0$ where $\Delta(t) = 0$, and a second with $t > 0$, where $\Delta(t) = 1$. The derivative $d[\Delta(t)/dt] = \Delta'(t) = \delta(t)$ of the single step function, the so-called Dirac delta function, must be therefore

$$\delta(t) = 0 \text{ for } t \neq 0 \tag{16}$$
$$\delta(t) = \infty \qquad \text{for } t = 0 \tag{17}$$

and so the integral becomes unity:

$$\int_{-\infty}^{+\infty} \delta(t)\, dt = \int_{0-}^{0+} \delta(t)\, dt = 1 \tag{18}$$

Table 1   Identification of the First 12 Simple Viscoelastic Models. The indices
for the degenerated elements are always 1.

Fluids	Solids
11: $1p$ model (Newton)	12: $1p$ model (Hooke)

21: $2p$ model (Maxwell)	22: $2p$ model (Kelvin)

31: $3p$ model	32: $3p$ model

41: $4p$ model	42: $4p$ model

51: $5p$ model	52: $5p$ model

61: $6p$ model	62: $6p$ model

Physical experiments can only emulate extension leaps within finite
time intervals. In principle it is acceptable, that these intervals are small
compared to the total observation time, but the integral should be equal to
unity. With a continuous function $f(t)$ and $t > 0$

$$\int_{-\infty}^{t} f(t')[\delta(t')]\, dt' = f(t)[\Delta(t)] \tag{19}$$

with t' as the variable of integration results the stress relaxation for example of a Maxwell model

$$\sigma(t) = E\varepsilon_0 / e^{-Et/F} [\Delta(t)] = E\epsilon_0 e^{-\tau t} [\Delta(t)] \tag{20}$$

with $E/F = \tau$ as the relaxation time or $1/\tau$ as the retardation time and for a Kelvin model:

$$\sigma(t) = E\epsilon_0 [\Delta(t)] + F\epsilon_0 [\delta(t)] \tag{21}$$

The solution for the strain is analogous (Mase [29]).

II. MATHEMATICAL METHODS

A. Construction of Constitutive Equations

The equations for the simple models show how to formulate the unknown stress-strain behavior of a new model. For models from the series of the "generalized Kelvin models" the following statements apply:

1. The elongation ϵ of the Kelvin models adds up to its specific increase for each element.
2. The total strain σ is applied to each element.
3. The following elements can be serially connected:
   a. n Kelvin elements and possibly
   b. one spring as a reduced Kelvin element and/or
   c. one dashpot as a reduced Kelvin element.

1. Microconstants and Hybrid Parameters

For calculations of parameters during tabletting experiments and for the physical understanding of the process, the coherence between the hybrid parameters (p_i and q_i) in the constitutive equations and the microconstants (E_i and F_i) in the models are of great interest. Basically it is always easier to calculate the hybrid parameters than the microconstants than vice versa (Table 2). In the latter case for a $4p$ model up to four solutions for a microconstant can be obtained and their physical validity has to be tested to find a real solution (Sutanto [23], Stahn [30], Sirithunyalug [31]). Due to this calculation and the analog distribution of the elements within the model a simplified method results for all models, which belong to a group with just n elements, where a spring, a dashpot, or a Kelvin model represents an element.

The following terms are an example for the Laplace-transformed figures (23) of Eq. (22):

$$\epsilon = \frac{\sigma}{E_1 + 1/F_1\delta_t} + \frac{\sigma}{E_2 + 1/F_2\delta_t} + \frac{\sigma}{E_3 + 1/F_3\delta_t} \tag{22}$$

Table 2 Hybrid Parameters, Calculated with the Microconstants for Elasticity (E_i) and Viscosity (F_i)

Model	Hybrid parameters

11    $q_1 = F_1$

12    $q_0 = E_1$

21    $p_1 = \dfrac{F_1}{E_1}, q_1 = F_1$

22    $q_0 = E_1, q_1 = F_1$

31    $p_1 = \dfrac{F_1 + F_2}{E_2}, q_1 = F_1, q_2 = \dfrac{F_1 F_2}{E_2}$

32    $p_1 = \dfrac{F_2}{E_1 + E_2}, q_0 = \dfrac{E_1 E_2}{E_1 + E_2}, q_1 = \dfrac{E_1 F_2}{E_1 + E_2}$

41    $p_1 = \dfrac{E_1 F_1 + E_1 F_2 + E_2 F_1}{E_1 E_2}, p_2 = \dfrac{F_1 F_2}{E_1 E_2}, q_1 = F_1, q_2 = \dfrac{F_1 F_2}{E_2}$

42    $p_1 = \dfrac{F_2 + F_3}{E_2 + E_3}, q_0 = \dfrac{E_2 E_3}{E_2 + E_3}, q_1 = \dfrac{E_2 F_3 + E_3 F_2}{E_2 + E_3}, q_2 = \dfrac{F_2 F_3}{E_2 + E_3}$

51    $p_1 = \dfrac{E_2 F_1 + E_2 F_3 + E_3 F_2 + E_3 F_1}{E_2 E_3}, p_2 = \dfrac{F_1 F_2 + F_1 F_3 + F_2 F_3}{E_2 E_3}$

     $q_1 = F_1, q_2 = \dfrac{E_2 F_1 F_3 + E_3 F_1 F_2}{E_2 E_3}, q_3 = \dfrac{F_1 F_2 F_3}{E_2 E_3}$

52    $p_1 = \dfrac{E_1 F_2 + E_1 F_3 + E_2 F_3 + E_3 F_2}{E_1 E_2 + E_1 E_3 + E_2 E_3}, p_2 = \dfrac{F_2 F_3}{E_1 E_2 + E_1 E_3 + E_2 E_3}$

     $q_0 = \dfrac{E_1 E_2 E_3}{E_1 E_2 + E_1 E_3 + E_2 E_3}, q_1 = \dfrac{E_1 E_2 F_3 + E_1 E_3 F_2}{E_1 E_2 + E_1 E_3 + E_2 E_3}$

     $q_2 = \dfrac{E_1 F_2 F_3}{E_1 E_2 + E_1 E_3 + E_2 E_3}$

61    $p_1 = \dfrac{E_1 E_3 F_1 + E_1 E_3 F_2 + E_1 E_2 F_3 + E_1 E_2 F_1 + E_2 E_3 F_1}{E_1 E_2 E_3}$

     $p_2 = \dfrac{E_1 F_1 F_3 + E_1 F_1 F_2 + E_1 F_2 F_3 + E_2 F_1 F_3 + E_3 F_1 F_2}{E_1 E_2 E_3}$

     $p_3 = \dfrac{F_1 F_2 F_3}{E_1 E_2 E_3}, q_1 = F_1, q_2 = \dfrac{E_2 F_1 F_3 + E_3 F_1 F_2}{E_2 E_3}, q_3 = \dfrac{F_1 F_2 F_3}{E_2 E_3}$

62    $p_1 = \dfrac{E_1 F_2 + E_1 F_3 + E_2 F_1 + E_2 F_3 + E_3 F_1 + E_3 F_2}{E_1 E_2 + E_1 E_3 + E_2 E_3}, p_2 = \dfrac{F_1 F_2 + F_1 F_3 + F_2 F_3}{E_1 E_2 + E_1 E_3 + E_2 E_3}$

     $q_0 = \dfrac{E_1 E_2 E_3}{E_1 E_2 + E_1 E_3 + E_2 E_3}, q_1 = \dfrac{E_1 E_3 F_2 + E_1 E_2 F_3 + E_2 E_3 F_1}{E_1 E_2 + E_1 E_3 + E_2 E_3}$

     $q_2 = \dfrac{E_1 F_2 F_3 + E_2 F_1 F_3 + E_3 F_1 F_2}{E_1 E_2 + E_1 E_3 + E_2 E_3}, q_3 = \dfrac{F_1 F_2 F_3}{E_1 E_2 + E_1 E_3 + E_2 E_3}$

for a model with $n = 3$ elements ($6p$ solid).

$$\bar{\epsilon} = \bar{\sigma} \left[\frac{1}{E_1 + sF_1} + \frac{1}{E_2 + sF_2} + \frac{1}{E_3 + sF_3} \right] \tag{23}$$

Simple calculations show the transformed strain-stress relation expressed as a sequence of sums:

$$\bar{\epsilon}(E_1 + F_1)(E_2 + sF_2)(E_3 + sF_3) = \bar{\sigma}[(E_2 + sF_2)(E_3 + sF_3)$$
$$+ (E_1 + sF_1)(E_3 + sF_3) + (E_1 + sF_1)(E_2 + sF_2)] \tag{24}$$

Multiplication and rearrangements of terms in the latter equation lead to a polynomial equation (Sutanto [23], Stahn [30]):

$$\bar{\epsilon}[q'_0 s^0 + q'_1 s^1 + q'_2 s^2 + q'_3 s^3] = \bar{\sigma}[p'_0 s^0 + p'_1 s^1 + p'_2 s^2] \tag{25}$$

It is obvious that by the rules of transformation and by comparison of the coefficients, a parameter p'_k, which is a factor of s^k in the transformed equation and the parameter p'_i in the inverse transformation corresponds to the factor of σ_i. The parameters q'_i are also related as shown in the following generalized equations:

$$\bar{\epsilon} \sum_{k=0}^{m} p'_k s^k = \bar{\sigma} \sum_{k=0}^{n} q'_k s^k \tag{26}$$

$$\sum_{i=0}^{m} p'_i \frac{\partial^i \sigma}{\partial t^i} = \sum_{i=0}^{n} q'_i \frac{\partial^i \epsilon}{\partial t^i} \tag{27}$$

Described in standard form, the number of microconstants of a model (the number of all springs and dashpots) is equal to the number of parameters belonging to the constitutive equation. For "solids," $q_0 \neq 0$, and for "fluids," $q_0 = 0$; for "solids" and "fluids" with spontaneous elastic answer, $q_n = 0$ for the nth element.

Therefore, for each three-element model the constitutive equation can easily be found without mathematical effort, for instance:

$$\bar{\epsilon} \sum_{k=0}^{3} p'_k s^k = \bar{\sigma} \sum_{k=0}^{2} q'_k s^k \tag{28}$$

and in the inverse general transformation:

$$\sum_{i=0}^{2} p'_i \frac{\partial^i \sigma}{\partial t^i} = \sum_{i=0}^{3} q'_i \frac{\partial^i \epsilon}{\partial t^i} \tag{29}$$

For example, in reduced models the parameters p'_i and q'_i, which can be deduced from the $6p$ solid, can be determined by setting spring and/or dashpot constants to zero.

For physical reasons (no negative parameters) there are relations in form of inequalities between the microconstants E_i and F_i. The relations which are a result of the classification save the use of complicated calculations to determine the inequalities (Table 3). If the equalities of a non-reduced Kelvin model with n elements are known, the inequalities of the derived models appear by setting the parameter q_0 and/or q_n to zero according to the scheme.

2. Laplace Transformation

For each element (string, dashpot, Kelvin) the stress-strain relation can be performed separately. The mathematical problem to connect the individual differential equations to the stress-strain equation of the total model can be solved by means of a Laplace transformation.

The solutions for all models are differential equations of the general form:

$$p_0' \, \sigma + p_1' \, \dot{\sigma} + p_2' \, \ddot{\sigma} + \ldots = q_0' \, \epsilon + q_0' \, \epsilon + q_1' \, \dot{\epsilon} + q_2' \, \ddot{\epsilon} + \ldots \quad (30)$$

Table 3 Inequalities of the First 12 Models

Model	Inequalities
11	$q_1 > 0$
12	$q_0 > 0$
21	$p_1, q_1 > 0$
22	$q_0, q_1 > 0$
31	$p_1 q_1 > q_2$
32	$q_1 > p_1 q_0$
41	(1) $p_1^2 > 4p_2$, (2) $p_1 q_1 q_2 > p_2 q_1^2 + q_2^2$
42	(1) $q_1^2 > 4q_0 q_2$, (2) $p_1 q_1 > p_1^2 q_0 + q_2$
51	(1) $q_2^2 > 4q_1 q_3$, (2) $p_1^2 > 4p_2$
52	(1) $q_1^2 > 4q_0 q_2$, (2) $p_1^2 > 4p_2$
61	(1) $q_2^2 > 4q_1 q_3$, (2) $\dfrac{3p_1 p_3 - p_2^2}{3p_3^2} < 0$,
	(3) $\left(\dfrac{1}{27} \dfrac{p_2^3}{p_3^3} - \dfrac{1}{6} \dfrac{p_1 p_2}{p_3^2} + \dfrac{1}{2p_3} \right)^2 + \left(\dfrac{3p_1 p_3 - p_2^2}{9p_3^2} \right)^3 < 0$
62	(1) $4p_2 < p_1^2$, (2) $\dfrac{3q_0 q_2 - q_1^2}{3p_0^2} < 0$,
	(3) $\left(\dfrac{1}{27} \dfrac{q_1^3}{q_0^3} + \dfrac{1}{6} \dfrac{q_1 q_2}{q_0^2} - \dfrac{q_3}{2q_0} \right)^2 + \left(\dfrac{3q_0 q_2 - q_1^2}{9q_0^2} \right)^3 < 0$

If both sides of the differential equation are divided by p_0', the standard form of the constitutive equation is as follows with $p_0 = 1$ as unity:

$$\sigma + p_1 \dot{\sigma} + p_2 \ddot{\sigma} + \ldots = q_0 \epsilon + q_1 \dot{\epsilon} + q_2 \ddot{\epsilon} + \ldots \tag{31}$$

Terms of stress and strain are combined in the most generalized form:

$$\sum_{i=0}^{m} p_i \frac{\partial^i \sigma}{\partial t^i} = \sum_{i=0}^{n} q_i \frac{\partial^i \epsilon}{\partial t^i} \tag{32}$$

Flügge [32] stated that the number m of the stress terms has to be smaller or equal to the number n of strain terms; otherwise the differential equations show a behavior of the substance different from linear viscoelastic materials.

B. Complex Modulus and Complex Compliance (Dynamic Modulus and Dynamic Compliance)

Besides the work with viscoelastic models, the calculation of their parameters and physical interpretation of their components, for example in the standard experiments, there is another more common method for experimental exploration with a different kind of mathematical description of the models. This method was introduced to pharmaceutics primarily for the characterization of gels or ointments.

The experimental approach is to apply an oscillating (for example, a sinusoidal) stress to the sample and to observe the strain as a response (or vice versa) which may be altered in phase and amplitude. With a loading stress, which is very similar to the tabletting load function of an excenter:

$$\sigma(t) = \sigma_{max} \sin(\omega t) \tag{33}$$

there will be a resulting strain response

$$\epsilon(t) = \epsilon_{max} \sin(\omega t + \delta) \tag{34}$$

The angular velocities ω of stress and strain are the same, but within the strain response, there is a lagtime like the maximum differences in load and displacement curves, defined by the loss angle δ as a phase shift. A large field for experimental alterations is opened, because it is easy to apply an alternating displacement to a sample and obtain a load as a response. For example, in the field of tabletting, experiments can be made with the first compression or repeated compressions within the same die. Using the whole tablet in a compaction simulator as often as possible studying the resulting strain could be another way. But the common way, for instance, for a plastic material is to apply an oscillatory strain to one end of a rod and to measure the torque on the other end, which means the rod is exposed to alternating positive and negative stress. In the case of compacting a tablet in a die, only the first quarter of the first oscillation is relevant to describe the whole oscillation. For the experimental estimation of the parameters on the other hand an equilibrium between elastic and viscous movements is necessary. Therefore, for a correct experiment one needs several full oscillations:

$$E^* = \frac{\sigma_{max}}{\epsilon_{max}} \quad \text{absolute dynamic modulus} \tag{35}$$

$$G' = \frac{\sigma_{max} \cos \delta}{\epsilon_{max}} \quad \text{storage modulus (elasticity)} \tag{36}$$

$$G'' = \frac{\partial_{max} \sin \delta}{\epsilon_{max}} \quad \text{loss modulus (viscosity)} \tag{37}$$

$$J' = \frac{\epsilon_{max} \cos \delta}{\sigma_{max}} \quad \text{storage compliance (elasticity)} \tag{38}$$

$$J'' = \frac{\epsilon_{max} \sin \delta}{\sigma_{max}} \quad \text{loss compliance (viscosity)} \tag{39}$$

$$\eta' = \frac{G''}{\omega} \quad \text{real part of viscosity} \tag{40}$$

$$\eta'' = \frac{G'}{\omega} \quad \text{complex part of viscosity} \tag{41}$$

This kind of description is a consequence of the generalized formulation in complex form of stress and strain which always leads to two solutions in real and imaginary parts.

The nomenclature concerning storage and loss moduli indicates the close vicinity to problems of energy. A short theoretical introduction is given by Pipkin [33]. A first attempt to characterize the work of tabletting using viscoelastic parameters was done by Morehead [27].

Very important for discussing tabletting problems is the change of storage and loss moduli with the change of the frequency of the compression device, that is the change with respect to the speed of the punch within the die. When the frequency increases, the loss modulus G'' may be lowered and, vice versa, the storage modulus G' increases. Computing the numbers of a simple model like a Kelvin or Maxwell model, one can see the different influence of a free spring, a free dashpot, or a Kelvin element (Table 4).

Table 4 Interaction of Viscous and Elastic Features in Model, Depending on the Speed of Deformation

	Spring	Dashpot	Maxwell	Kelvin	3p fluid	3p Solid
G'	$E \rightarrow$ const	$0 \rightarrow$ const	$0 \rightarrow E$	$E \rightarrow$ const	$0 \rightarrow$ max	min $\rightarrow E_1$
G''	$0 \rightarrow$ const	$0 \rightarrow$ max	$0 <<$ max $>> 0$	$0 \rightarrow$ max	min $\rightarrow$ max	min $<$ max $>$
δ	$0 \rightarrow$ const	$\pi/2 \rightarrow$ const	$\pi/2 \rightarrow 0$	$0 \rightarrow \pi/2$	$\pi/2 >$ min $< \pi/2$	$0 < \pi/2 > 0$
η'	$0 \rightarrow$ const	$F \rightarrow$ const	$F \rightarrow 0$	$F \rightarrow$ const	max $\rightarrow$ min	max $\rightarrow$ min
σ_{max}	const	$<<$ max	$<$ max	$<<$ max	min $\rightarrow$ max	min $\rightarrow$ max

In a spring model the stress is always directly proportional to the strain, so the displacement versus time and strain versus time curves respectively do not change with altering frequencies. This means that the loss angle δ is zero and therefore the loss modulus G'' and the loss compliance J'' are zero too, because sin δ is zero. On the other side, cos δ is unity, so the storage modulus G' and the storage compliance J' are important to characterize the elastic properties of a model.

The behavior of a dashpot is also easy to understand. In this case the resulting stress is always proportional to the rate of deformation, so at any frequency the loss angle is $\pi/2$. That means sin δ is at its maximum and cos δ is zero. The characteristics are the loss modulus G'' or the loss compliance J''. Because in a dashpot the resulting stress is proportional to a given frequency of deformation the maximum load will increase with ascending radian frequency.

Both elementary models are the base for composite models like Kelvin, Maxwell, or higher-order models. To obtain in these models the relation of strain and stress, a series of different frequencies must be used to estimate the loss and storage moduli. With a plot of G', G'', J', J'' or η' versus frequency the viscoelastic properties become evident. A reasonable approach is to use log/log plots to get simple relations if they exist.

In a Maxwell model (Fig. 3) the storage modulus G' will approach Young's modulus E for the spring in the model with higher speed (higher frequencies) of tabletting. With lower speed the viscosity η' approaches the viscosity modulus, which is the same as the microconstant F of the dashpot in the model. The interaction between elasticity and viscosity in a Maxwell model becomes evident in the loss modulus G'', which has a maximum at a medium value of the frequency. On the other hand, the loss angle δ decreases from $\pi/2$ at lowest frequency to zero at high frequencies and the maximum of the load will rise.

In the Kelvin model (Fig. 4), the dashpot and the spring are connected in parallel. So they must always be in the same position, but the

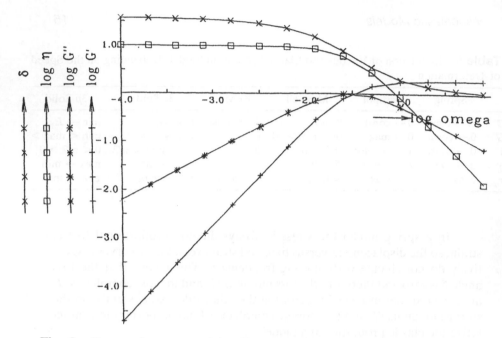

Fig. 3 Change of storage modulus G', loss modulus G'', viscosity η', and loss angle δ of a Maxwell model with time (theoretic data with $E = 2$ and $F = 10$).

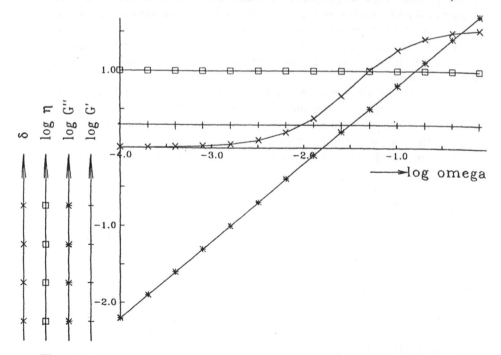

Fig. 4 Change of storage modulus G', loss modulus G'', viscosity η', and loss angle δ of a Kelvin model with time (theoretic data with $E = 2$, $F = 10$).

stress in each element is different and the storage modulus G' representing elasticity and η' representing viscosity do not vary with altering frequencies. But the load needed in this example to obtain the same elongation as at a low speed is in this example 25 times higher, and the loss modulus G'' will rise also. The loss angle δ changes from zero at low speed to $\pi/2$ at the highest frequency, and the maximum load rises to a significantly higher value.

In $3p$ models the loss angle δ shows a minimum or a maximum in contrast to the simpler models which have been discussed. G'' rises in the $3p$ fluid and in the $3p$ solid we see a maximum.

All these relations show on the one hand the possibilities to obtain information and on the other hand the difficulties to apply the theory to the process of powder compaction. The main problem is the nonequilibrium in the first quarter of the oscillation in a tabletting process. Correct values could only be estimated in the following full oscillations when elastic and viscous properties are in equilibrium and the strain is alternating with the positive and negative parts of the theoretical oscillations as shown in Figs. 5 and 6 for a Maxwell model. Here it seems to be very important to look at the first quarter of the oscillation which is highly influenced by the initial conditions.

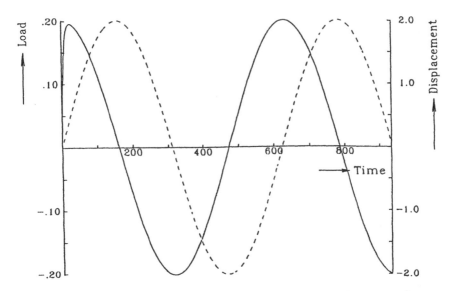

Fig. 5 Load vs. time (solid line) and displacement vs. time (dashed line) curves for a Maxwell model (theoretic data with $E = 2$, $F = 10$) at slow speed.

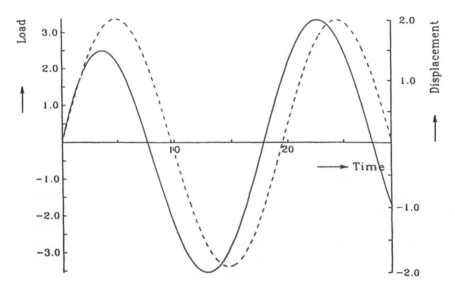

Fig. 6 Load vs. time (solid line) and displacement vs. time (dashed line) curves for a Maxwell model (theoretic data with $E = 2$ and $F = 10$) at high speed.

C. Explicit Solutions

1. Linear Creep Compliance

The strain-stress behavior of different materials is completely described by their constitutive equations. But the physical features of materials can better be understood by application and physical interpretation of a standard test: The explicit solutions, which were derived from the constitutive equations, characterize these materials as a function of time in a creep compliance with constant stress or a stress relaxation test with constant strain. In the standard experiment one of the variables is constant, so the derivatives vanish and the explicit solutions are simple.

As we only consider the linear viscoelastic materials, the change in the relative extension ϵ is proportional to the constant stress σ_0 and can be described as follows:

$$\epsilon(t) = \sigma_0 J(t) \tag{42}$$

The function $J(t)$, the linear creep compliance,

$$J(t) = \frac{\epsilon(t)}{\sigma_0} \tag{43}$$

describes the elongation or compression of a material under unitary (constant) load. The diagram of individual behavior of material, the creep compliance function, is equivalent to the mathematical description with constitutive equations with a constant stress (Table 5). To get a specific answer to an analysis of viscoelastic materials the creep compliance must be estimated as a response function and should be fitted by a physically obvious model.

Table 5 Creep Compliance, Calculated with Microconstants (E_i and F_i)

Model	Creep compliance
11	$\dfrac{t}{F_1}$
12	$\dfrac{1}{E_1}$
21	$\dfrac{1}{E_1} + \dfrac{t}{F_1}$
22	$\dfrac{1}{E_2}(1 - e^{-(E_2/F_2)t})$
31	$\dfrac{1}{E_2}(1 - e^{-(E_2/F_2)t}) + \dfrac{t}{F_1}$
32	$\dfrac{1}{E_1} + \dfrac{1}{E_2}(1 - e^{-(E_2/F_2)t})$
41	$\dfrac{1}{E_1} + \dfrac{1}{E_2}(1 - e^{-(E_2/F_2)t}) + \dfrac{t}{F_1}$
42	$\dfrac{1}{E_2}(1 - e^{-(E_2/F_2)t}) + \dfrac{1}{E_3}(1 - e^{-(E_3/F_3)t})$
51	$\dfrac{1}{E_2}(1 - e^{-(E_2/F_2)t}) + \dfrac{1}{E_3}(1 - e^{-(E_3/F_3)t}) + \dfrac{t}{F_1}$
52	$\dfrac{1}{E_1} + \dfrac{1}{E_2}(1 - e^{-(E_2/F_2)t}) + \dfrac{1}{E_3}(1 - e^{-(E_3/F_3)t})$
61	$\dfrac{1}{E_1} + \dfrac{1}{E_2}(1 - e^{-(E_2/F_2)t}) + \dfrac{1}{E_3}(1 - e^{-(E_3/F_3)t}) + \dfrac{t}{F_1}$
62	$\dfrac{1}{E_2}(1 - e^{-(E_2/F_2)t}) + \dfrac{1}{E_3}(1 - e^{-(E_3/F_3)t}) + \dfrac{1}{E_4}(1 - e^{-(E_4/F_4)t})$

Barry [34] estimated viscosity, elasticity, and several serial Kelvin elements from the creep compliance by numerical or graphical methods using the standardized experiment with constant stress. The creep compliance shows a very common solution for the generalized Kelvin model with two reduced elements:

$$J(t) = \frac{1}{E_1} + \frac{1}{F_1}t + \frac{1}{E_i}\sum_{i=2}^{n}\left[1 + e^{-(F_i/E_i)t}\right]$$ (44)

Under a tensile stress real materials will expand and under pressure they will be compressed. This behavior corresponds with the monotonously increasing course of the creep compliance. Therefore the microconstants E_i and F_i are always positive. The rules of conversion show that also the hybrid parameters can only assume positive values. These consequences do not become evident if one only deals with differential equations. Among the microconstants there are relationships existing in form of inequations (Table 3), which must hold for physically meaningful models. Under the given conditions of the standard experiment the creep compliance is the explicit solution of the constitutive equation. The solution of the differential equations can be obtained in most cases by means of Laplace transformation.

2. Linear Relaxation Function

The second explicit solution of the differential equation describes the behavior during a relaxation experiment. The stress relaxation function $G(t)$ gives the response of a stressed (compressed or expanded) material as a function of time, related to a given constant strain. The solution of the constitutive equation is determined in the same way as that of the creep compliance, but it is more complicated to calculate its parameters (Table 6). This applies also for the transformed function $\overline{G}(s)$ according to Laplace. Physically, in the time-dependent domain, no simple relation between the creep compliance and the relaxation function can be found. But after a Laplace transformation the functions have been converted to the very simple relation (Flügge [32]):

$$\overline{J}(s)\overline{G}(s) = s^{-2}$$ (45)

In the stress relaxation functions of different models no simple and clear relations can be found as in the functions of the creep compliance. This is because the relaxation of force in more complex models, parts of elongation ϵ_i of the springs, dashpots, and Kelvin elements equalize differently in time and direction, until a forceless condition is reached. Using the hybrid parameters, an arrangement of terms of the stress relaxation function can be found, which allows one to formulate the stress relaxation functions of the different models with one, two, or three elements (Stahn [30]). But

Table 6 Stress Relaxation Function, Calculated with Hybrid Parameters (p_i and q_i)

Model	Stress relaxation function
11	$q_1 \delta(t)$
12	q_0
21	$\dfrac{q_1}{p_1} e^{-t/p_1}$
22	$q_0 + q_1 \delta(t)$
31	$\dfrac{q_2}{p_1} \delta(t) + \dfrac{1}{p_1}\left(q_1 - \dfrac{q_2}{p_1}\right) e^{-t/p_1}$
32	$\dfrac{q_1}{p_1} e - t/p_1 + q_0(1 - e^{-t/p_1})$
41	$\dfrac{1}{\sqrt{p_1^2 - 4p_2}}\left((q_1 + q_2\alpha)e^{\alpha t} - (q_1 + q_2\beta)e^{\beta t}\right)$
42	$q_0 + \dfrac{q_2}{p_1} \delta(t) - \left(q_0 - \dfrac{q_1}{p_1} + \dfrac{q_2}{p_1^2}\right) e^{-t/p_1}$
51	$\dfrac{q_3}{p_2} d(t) + \dfrac{1}{\sqrt{p_1^2 - 4p_2}}\left((q_1 + q_2\alpha + q_3\alpha^2)e^{\alpha t} - \left(q_1 + q_2\beta + q_3\beta^2\right)e^{\beta t}\right)$
52	$q_0 + \dfrac{1}{\sqrt{p_1^2 - 4p_2}}\left((\dfrac{q_0}{\alpha} + q_1 + q_2\alpha)e^{\alpha t} - \left(\dfrac{q_0}{\beta} + q_1 + q_2\beta\right)e^{\beta t}\right)$
61	no solution
62	$q_0 + \dfrac{q_3}{p_2} \delta(t) + \dfrac{1}{\sqrt{p_1^2 - 4p_2}}\left(\left(\dfrac{q_0}{\alpha} + q_1 + q_2\alpha + q_3\alpha^2\right)e\alpha t - \left(\dfrac{q_0}{\beta} + q_1 + q_2\beta + q_3\beta^2\right)e\beta t\right)$
	with $\alpha, \beta = \dfrac{-p_1 \pm \sqrt{p_1^2 - 4p_2}}{2p_2}$

both relations, creep compliance as well as stress relaxation function, hold at the same point and the same time.

D. Estimation of Parameters from Compaction Dynamics

1. *Constitutive Equation*

Differential equations like the constitutive equation are usually solved explicitly resulting in the creep compliance or the stress relaxation function.

In this case it is possible to provide a physical interpretation for the relation between the variables. If one only wants the numbers of the parameters, for example to compare different methods of tablet compaction, it is not necessary to know the complex solution. Therefore the constitutive equation was reassembled to get 100 to 300 numerical equations related from all the 100–300 data of a load versus time and displacement versus time curves:

$$\sigma_i = -p_1\dot{\sigma}_i - p_2\ddot{\sigma}_i - \ldots + q_0\epsilon_i + q_1\dot{\epsilon}_i + q_2\ddot{\epsilon}_i + \ldots \qquad (46)$$

The displacement ϵ_i, the speed $\dot{\epsilon}_i$, the acceleration $\ddot{\epsilon}_i$, the change of load $\dot{\sigma}_i$ and the second $\ddot{\sigma}_i$ or higher derivatives with respect to time of the ith data point must be known at every time. If all these values are known at each of the 100 to 300 data points and an estimate of the experimental load σ_i^* at this point, the sum of squares becomes

$$\sum_{i=1}^{n} (\sigma_i - \sigma_i^*)^2 \equiv \min \qquad (47)$$

which can be minimized by the method of maximum likelihood or similar methods.

The *Numerical Recipes* [35] contain very helpful source modules in Fortran or Pascal to find the numerical solution of the constitutive equation. The main problem is to find the smoothing function for load and displacement and their first to third derivatives with sufficient accuracy. The numerical value of the calculated load will be added from up to five terms in the constitutive equation and must be subtracted from the estimated load. In order to find adequately fine calculated data points, we used polynomial smoothing algorithms, which can easily be derivated in double precision (Sutanto [23]).

The advantage of this procedure is that it is not necessary to find accurate boundary and initial conditions as for the explicit solutions (i.e., Dirac delta function). The disadvantage is the high precision required by the polynomials. So we used polynomials of 9th to 15th degree. Once programmed, they allowed to get the significance of the polynomial in a first run and to get the best model out of every combination up to four- (sometimes even six-) parameter models in a second run with, e.g., by multiple linear or nonlinear regression. Other functions for load or strain like sums of sine and cosine or exponential functions from a Fourier transformation may be feasible for smoothing too. In all cases it should be easy to obtain the derivatives.

To find satisfying solutions for the 4p to 6p models, the calculated parameters must be tested for their significance: If the parameters are greater than 1000, the reciprocal will be so small that the calculated effect of the element will vanish (Sirithunyalug [31]). A small residual amount may remain and will add to another (reduced) element, so that another uncertainty will arise.

2. Explicit Solutions of the Constitutive Equation

a. Superposition Principle

For the mathematical analysis of the displacement versus time or stress versus time behavior it is important that the explicit solutions, i.e., the creep compliance and/or the stress relaxation function, which describe the material in a static situation, must be applicable to characterize the dynamic of the process of powder compaction for example in a single-punch machine. Therefore the Boltzmann superposition principle is used. The linear behavior of an elastic or viscoelastic material is determined by the total effect of a sum of causes, which equals the sum of the single effects of a single cause, but at consecutive times.

For example, a compaction curve from $t = t_0$ to $t = t_n$ is a response to a sum of n changing-of-stress events in single steps and can be described as follows:

$$\epsilon_n = \sigma_0 J(t_n - t_0) + \sigma_1 J(t_n - t_1) + \sigma_2 J(t_n - t_2) + \ldots + \sigma_{n-1} J(t_n - t_{n-1}) \ (48)$$

In general,

$$\epsilon_n = \sum_{i=1}^{n} \sigma_{i-1} J(t_n - t_{i-1}) \tag{49}$$

If the complete stress history of a material is known from minus infinity to the time t_0, the creep compliance or the stress relaxation curves can be calculated from the load versus time and displacement versus time curves. But it is much easier to get a solution if the initial conditions start at zero stress and at zero strain. Figure 7 shows, how a load curve is built up from single steps of a discrete creep compliance ($4p$ fluid) corresponding to the displacement curve. In the range of decreasing load, steps of negative load also are superimposed.

In earlier publications of Rippie [26] and Danielson [25], Neuhaus [36] and Sutanto [23] the sections of increasing and decreasing load were treated separately. A simple example shows that the direction of the load is of no consequence for the mathematical execution of superposition (Stahn [30]).

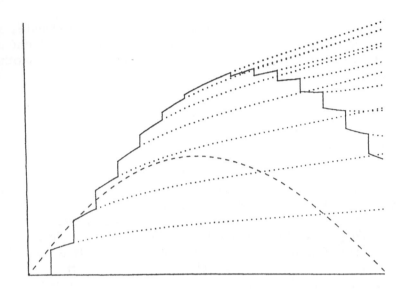

Time

Fig. 7 Schematic construction of displacement vs. time curve for a 4p fluid with a load vs. time curve (dashed line) and 15 consecutive creep compliances (dotted lines). Envelope curve as a first approximation (solid line).

b. Hereditary Integral

When the entire process is divided into many small steps, this leads finally to an integral, which describes the history of the powder compacting process. The hereditary integral is defined by the equation

$$\epsilon(t) = \sigma(0)J(t) + \int_0^t J(t - t')\frac{d\sigma'}{dt'}\,dt' \tag{50}$$

The explicit solution for tabletting problems becomes possible with a substitution of $J(t - t')$ and $d\sigma'$ and then following the rules of integration of products.

$J(0)$, the initial condition of the creep compliance (for example, free spring with a spontaneous answer), and for $J(t)$ the terms of the creep compliance will be used. The equation for the 4p fluid (Rieger [37]) with the derivative of the creep compliance for a 4p fluid (in brackets) is as follows:

$$\epsilon(t) = 1/E_1\sigma(t) + \int_0^t \sigma(t')\left[\frac{1}{F_1} + \frac{F_2}{E_2^2}e - (F_2/E_2)(t - t')\right]dt' \tag{51}$$

The calculation of the load curve $\sigma(t)$ is performed by an analog transformation. The integration can be done for example according to Romberg's method.

c. Convolution Processes

Principally in

$$\epsilon(t) = \sigma_0 J(t) + \int_0^t \sigma(t)' J(t - t') \, dt' \tag{52}$$

the integral is a convolution operation (53), whereby the load appears as an input function, the creep compliance as a response function, and the displacement curve as the output function. According to Stepanek [38] in

$$\sigma(t) * J(t) = \epsilon(t) \tag{53}$$

where the asterisk stands for the convolution operator. A similar notation is used to find the inverse function

$$J(t) = \epsilon(t) \, {}^*_* \, \sigma(t) \tag{54}$$

where ${}^*_*$ represents the deconvolution operator. Therefore (53) is the displacement curve and (54) is the creep compliance.

Deconvolution. There are different ways to find the solution of $\epsilon(t)$ by deconvolution: with help of the Laplace transformation, by development of time series (FFT), or by a numerical method. Here we use the numerical step-by-step method to solve the deconvolution problem. The cumulative displacement curve is the output function obtained from the response function as the creep compliance and the load curve as the input function, split to fractional intervals. As the equation for the displacement is calculated by means of repeated creep compliances, the deconvolution is calculated vice versa. The creep compliance itself always keeps the same shape after each step of deconvolution of load. Therefore,

$$t_0 \qquad J(t) = 0 \qquad \text{resp. } \frac{1}{E} \text{ with an elastic response}$$

$$t_1 \qquad J(t) = \epsilon_1/\sigma_0$$

$$t_2 \qquad J(t) = (\epsilon_2 - J_1 \, d\sigma_1)/\sigma_0$$

$$t_3 \qquad J(t) = (\epsilon_3 - (J_1 \, d\sigma_2 + J_2 \, d\sigma_1))/\sigma_0$$

$$\vdots \qquad \vdots$$

$$t_n \qquad J(t) = (\epsilon_n - \sum_{i=1}^{n-1} J_i d\sigma_{n-i})/\sigma_0 \tag{55}$$

This formula is not difficult to program, but load and displacement have to be continuous functions and the time intervals must be small and constant. Therefore experimental data have to be smoothed and divided to equal intervals. As long as the data of displacement and load are only interpolated, a fifth- to ninth-degree polynomial is used to obtain the graphical construction of the creep compliance $J(t)$. The shape of the resulting curve with the height and the position of the maximum is dependent on the substance and the maximum load. Very often the curves cannot be interpreted because they are distorted by oscillations and degenerate. Reasons for this include limited accuracy of the numerical transaction and the uncertainty of the initial conditions. In case of tabletting the load versus time and the displacement versus time curves are not ideal, because the consolidation, reducing the volume, adds to the viscoelastic features. So the deconvolution can never produce an ideal creep compliance. The parameters of the exposed creep compliance are found by a Simplex process or by regression using the method of the steepest slope. The advantage of the method of deconvolution is the "once only" single exposure of the creep compliance, which is then subject to the regression to get the parameters of the chosen model.

Convolution. The method of convolution is more reliable. Here a first guess of a parameter sequence for the creep compliance of the given model is estimated. The entire displacement function is calculated with the parameters of an assumed creep-compliance function. By iteration with the Simplex algorithm improved parameters are searched until the result (sum of squares) is satisfactory. Before a new cycle with a new set of parameters for the creep compliance starts, it has to be checked, whether the new parameters will meet the inequalities. The calculation will be stopped, if the variation of the parameters does not improve the fit of the curve.

If the calculation converges—this should be the case after 300–800 steps of variation—the iteration is stopped. Another advantage of convolution is the possibility to stay within the physically feasible region of the parameter space with its equalities and inequalities discussed above. If the load maximum is exceeded, the steps of the load become negative. The alterations of the displacement curves change their directions and are taken from those parts of the displacement which emerged from earlier positive steps of load and which increase even further (Fig. 7). The best possible model can be found by repetition of the calculations with different candidates for the model. It is very helpful to follow the overlays visually on the screen. The smaller the steps of load and the greater the number of overlays, the smoother the envelope curve will be, and the better it will also fit the Dirac delta function. The number of steps is chosen according to the number of datapoints (200–300) except for rough calculations. The first result shows that the calculated load versus time and displacement versus time curves reproduce the real initial situation in a qualitatively and quantitatively correct way: the load maximum is located in front of the displacement curve maximum, the extent of the time shift results from the choice of the model and the parameters of the substance. In the same way the load curve is adapted, according to the displacement curve. The corresponding displacement curves are calculated from the load curves by a shifted addition of the estimated stress relaxation function curves. To check the correct program function, alternate calculations of load and displacement curves should be performed with different initial parameters. The advantage compared to the deconvolution is the invariance of the given creep compliance, because in this case the stable function will be used to find the "not ideal" load curve as similar as possible.

III. REMARKS CONCERNING PRACTICAL APPLICATIONS

The numerical methods are difficult, extremely expensive in programming and different for each element. The main problem however is the numerical stability of the regression methods. Therefore it is a good practice to prove the numeric stability with a set of well-defined, known theoretical data (Sirithunyalug [31]).

For the use of theoretical numbers there must be a given function to simulate the strain or the stress respectively. This function should be easy to transform into the Laplace domain. This could be for example a quadratic polynomial, a sum of two or three exponential terms, a simple or composite sine function, just to get a similar behavior as a displacement (resp. load) curve. The function must however have enough derivatives to meet the chosen model. The most natural approximation is the calculated function with the geometry of the press. The next step is to choose a model to balance either the stresses or the strains of the single elements and then to transform the equations to Laplace. Now it is possible to calculate the theoretical stress (resp. strain) function with the given set of parameters (the microconstants of the different elements). It is convenient to use the data to prove the smoothing function and to find the best approximation to the parameters, which are now well known. In this point every inequation must be satisfied and all parameters must be positive! The next criterion is the variance of the calculated parameters. The results must be in a reasonable order of magnitude. But the main problem lies in a personal decision: The resolution of the numerical calculation of the constitutive equation is based on the degree of the polynomial. So it is recommended to use polynomials of every power of the polynomial from 3 to 15 to find the best approximation of the well-known parameters and a reasonable sum of squares (there is a minimum), besides the above-mentioned properties of the inequalities. With this feeling for the accuracy the same procedure should be chosen in the main experiment.

With the criterion for the regression being the sum of squares one should be aware that, e.g., the constitutive equation is composed of a sum of two to six single terms, each of which is a different product. Hence, there exist billions of variations of terms, but only one meets the true value. That means the shape of the curve should be characteristic for the model which could be achieved by a high number of data points, a typical segment of the curve and in the case of convolution as the most effective method by simultaneous calculation of the added sums of squares from the creep compliance and the stress relaxation function, because both functions are "true" at all points and times.

Practical investigations for estimation of viscoelastic properties should be validated with materials that display the characteristics expected in the experiments. Using convolution or integral methods, one needs the initial condition, which is the most difficult prerequisite for a solution of a model. In practice the viscoelastic behavior is overlapped with the consolidation phase of the materials in the die. Both events are power consuming. We always failed to find this point in tabletting experiments. The best approach is a precompression to a distinct load and restart the machine to get the final compaction and decompression phase. On rotary machines we used the precompression station [39]. Another approximation is to substitute the displacement with the corresponding analogon such as the relative porosity like in the experiments by Heckel. In this case the dimension of the displacement will be lost and the parameters become relative (Heikamp [39]).

Stahn [30] validated the experimental methods using a sintered poly-ethylene powder. In this way he obtained well-defined initial conditions and compressed without a die, avoiding the irregularities produced according to Poisson's law.

In Figs. 5 and 6 we see how the type of a model can change its behavior with the frequency, that is with the punch velocity (Müller [40]). So the Maxwell model converts from a single spring to a real Maxwell model and at the highest speed to a simple dashpot (Table 4).

It seems important to find the best model at a given speed which shows viscoelastic properties out of all possible combinations. If the chosen model is too complex, then it will degenerate. A single dashpot or a single spring assume such big numbers that in a calculation of, for instance, the creep compliance, there will be no effect of these elements because the reciprocals of the parameters are nearly zero and eliminate these terms. If there are too many Kelvin elements, this has two consequences: First, a huge parameter of the Kelvin spring brings the whole element to zero, or a huge parameter of the Kelvin dashpot eliminates the exponential term in the creep compliance. In this case only the effect of the spring is important. If both parameters are large and of the same magnitude, a free dashpot will occur (Sirithunyalug [31]).

The equation of Hooke is theoretically unlimited for a massless linear spring and is valid only in one dimension. Real elastic materials show a constant ratio in the change in length and diameter during their pull and push strain in their elastic section. In contrast, during compaction the expansion in diameter will be confined by the die. For these reasons the pure linear models of the theory of viscoelasticity cannot be perfect, as long the model is not three-dimensional (Bauer [28]).

On the other hand, the region of validity for the ideal viscoelastic behavior will probably be very small, because there are only some microns in the compaction bed where such effects are evident. Therefore, one needs an excellent resolution of the electronic displacement registration chain.

Analyses of the creep compliance, the relaxation functions or the constitutive equations of the tabletting processes require an exact synchronization of measurement for displacement and load. Thus the dependence from speed and acceleration becomes evident. The correction of the elasticity of the punch and of other parts of the machine is estimated by a punch-to-punch compression [1,2], which should be used for correction of every point of displacement in relation to the stress. The real displacement with respect to the time can be calculated from the corrected displacement and load measurements in the punch-to-punch compression. Therefore the assembled electronic modules and amplifiers to get and transmit the signals need special care. A simple test for good synchronization is a sudden stroke with a hammer to the upper side of the excenter and the time delay in the registration device between the first occurrence of stress and strain (Bauer [28]).

The investigations must discriminate elastic, viscous, and viscoelastic components during the phase of stress and release. In the present state of knowledge no one can expect to get precise results from the experiments. But it is easy to do "forward calculations" to realize the puzzle of stress and strain.

Theoretical investigations with oscillating load show new difficulties depending on the tabletting rate (Müller [40]): While elastic properties always remain proportional to the load, in contrast viscous needs a maximal load to hold the displacement constant (as in a displacement versus time curve) and viscoelastic (Maxwell or Kelvin model) materials give different pictures depending on the speed of tabletting as shown in Figs. 3 to 6.

In real experiments one will never see the theoretical punch upset to calculate l_0 in the sense of viscoelasticity. It is only possible to compare results within the same powder mass, for example at different speeds or of different maximal loads.

The only way to avoid all these difficulties seems to be the use of a well-equipped compaction simulator with precompression and an oscillating load to characterize the materials in the equilibrium state of viscous and elastic properties at different frequencies to determine the material constants. But even in this manner there remains the nightmare of more or less entrapped air. Casahoursat and co-workers [11,12] stopped the press and observed the stress while the displacement was varied by a crank of the lower punch. The authors defined the exponential decay of stress as the effect of escaping air.

So far as there is no better chance, we recommend the most primitive method, suggested by [1,4,22,40]: To use the time delay between load and displacement as a characteristic and try to find a function out of (35) to (41) to extrapolate the history prior to the load maximum, just to overcome the consolidation phase.

NOTATION

δ	loss angle
$\delta(t)$	Dirac delta function
$\{\delta^{(i)}\}$	linear differential time operator
$\Delta(t)$	Heaviside step function
ϵ	strain (displacement) ($\Delta l/l_0$) in the time domain
ϵ_0	initial strain (displacement)
ϵ_{max}	maximal strain (displacement) in oscillation experiments
ϵ_S	strain of a spring ($\Delta l/l_0$)
ϵ_D	strain of a dashpot ($\Delta l/l_0$)
$\overline{\epsilon}$	strain in the Laplace domain
$\dot{\epsilon}$	derivative of strain with respect to time
$\ddot{\epsilon}$	second derivative of strain with respect to time
E^*	absolute (complex) dynamic modulus
E_i	$i = 1$ to n, elasticity (Young's) modulus in the 1st to nth element; in reduced elements, $i = 1$ always
F_i	$i = 1$ to n, viscosity modulus in the 1st to nth element; in reduced element, $i = 1$ always
$G(t)$	stress relaxation function in the time domain
$\overline{G}(s)$	stress relaxation function in the Laplace domain
G'	storage modulus (real component)
G''	loss modulus (imaginary component)
$J(t)$	creep compliance in the time domain
$\overline{J}(s)$	Creep compliance in the Laplace domain
J'	storage compliance (real component)

J''	loss compliance (imaginary component)
η'	real component of viscosity
η''	complex component of viscosity
p_i'	absolute hybrid constants related to stress
p_i	relative hybrid constants related to stress
q_i'	absolute hybrid constants related to strain
q_i	relative hybrid constants related to strain
s	Laplace variable
σ	stress (load) in the time domain
σ_0	initial stress (load)
σ_{max}	maximal stress (load) in oscillation experiments
$\bar{\sigma}$	stress in the Laplace domain
σ_S	stress (load) of the spring
σ_F	stress (load) of the dashpot
$\dot{\sigma}$	derivative of stress with respect to time
$\ddot{\sigma}$	second derivative of stress with respect to time
t	time
t'	variable for integration
τ	relaxation time
$1/\tau$	retardation time
ω	radiant frequency

REFERENCES

1. D. Schierstedt, Rückdehnung der Tabletten während der Kompression, Diss. rer. nat., Bonn, 1982.
2. D. Schierstedt and F. Müller, *Pharm. Int. 44*: 932 (1982).
3. U. Caspar, Viskoelastische Phänomene während der Tablettierung, Diss. rer. nat., Bonn, 1983.
4. F. Müller and U. Caspar, *Pharm. Ind. 46*: 1049 (1984).
5. A. Y. K. Ho and T. M. Jones, *J. Pharm. Pharmacol. Suppl. 40*: 75 (1988).
6. N. M. A. H. Abourida, Ph.D. thesis, University of Wales, 1980.
7. E. T. Cole, J. E. Rees, and J. A. Hersey, *Pharm. Acta Helv. 50*: 28 (1975).
8. S. T. David and L. L. Augsburger, *J. Pharm. Sci. 66*: 155 (1977).
9. C. Wiederkehr von Vincenz, Instrumentierung und Einsatz einer Rundlauf-Tablettenpresse zur Beurteilung von pharmazeutischen Pressmaterialien, Diss., ETH Zürich, 1979.
10. J. E. Rees and P. J. Rue, *J. Pharm. Pharmacol. 30*: 601 (1978).
11. L. Casahoursat, G. Lemagnen, and D. Larrouture, *Drug Dev. Ind. Pharm. 14(15–17)*: 2179 (1988).
12. L. Casahoursat, G. Lemagnen, D. Larrouture, and A. Etienne. *Study of the Visco-elastic Behaviour of Some Pharmaceutical Powders*, II Congresso Int. diciencas farmaceuticas, Barcelona, 1987.

13. B. Emschermann, Korrelation von Presskraft und Tabletteneigenschaften, Diss. rer. nat., Bonn, 1978.
14. B. Emschermann and F. Müller, *Pharm. Ind. 43*: 191 (1981).
15. U. Tenter, Preßkraft- und Weg-Zeit-Charactreristik von Rundlaufpressen. Diss. rer. nat., Marburg, 1986.
16. P. C. Schmidt, S. Ebel, H. Koch, T. Profitlich, and U. Tenter. *Pharm. Ind. 50*: 1409 (1988).
17. P. C. Schmidt and H. Koch, *Pharm. Ind. 53*: 508 (1991).
18. R. Dietrich and J. B. Mielck, *Pharm. Ind. 46*: 863 (1984)
19. R. W. Heckel, *Trans. Metall. Soc. 221*: 1001 (1961).
20. R. J. Roberts and R. C. Rowe, *Int. J. Pharm. 37*: 15 (1987).
21. W. Jetzer, H. Leuenberger, and H. Sucker, *Pharm. Technol. 7(4)*: 33 (1983).
22. U. Caspar and F. Müller, *Pharm. Acta Helv. 59*: 329 (1984).
23. L. Sutanto, Berechnung von Parametern zur Beschreibung der Tablettierung, Diss. rer. nat., Bonn, 1986.
24. W. Fischer, Berechnung von Parametern zur Viskoelastizität bei der Tablettierung, Diss. rer. nat., Bonn, 1989.
25. D. W. Danielson, W. T. Morehead, and E. G. Rippie, *J. Pharm. Sci. 72*: 342 (1983).
26. E. G. Rippie and D. W. Danielson, *J. Pharm. Sci. 70*: 476 (1981).
27. W. T. Morehead, *Drug Dev. Ind. Pharm. 18(6,7)*: 659 (1992).
28. A. Bauer, Untersuchungen zur Prozeßdatengewinnung, Viskoelastizität und Struktur von Tabletten, Diss. rer. nat., Bonn, 1990.
29. G. E. Mase, *Theory and Problems of Continuum Mechanics*, McGraw-Hill, 1970.
30. P. Stahn, Modellrechnungen zur Parametrisierung von Tablettierkurven an Polyethylenproben, Diss. rer. nat., Bonn, 1990.
31. J. Sirithunyalug, Validierung der Berechnungsoperationen zur Bestimmung viskoelastischer Eigenschaften von Tabletten. Diss. rer. nat., Bonn, 1992.
32. W. Flügge, *Viscoelasticity*, 2nd ed., Springer-Verlag, Berlin, Heidelberg, New York, 1975.
33. A. C. Pipkin, *Lectures on Viscoelasticity Theory*, Springer-Verlag, New York, Heidelberg, Berlin, 1972.
34. B. W. Barry, in *Advances in Pharmaceutical Sciences*, Bean e.a., Vol. 4, Academic Press, London, 1974.
35. W. Press, B. P. Flannery, S. A. Teukolsky, and W. T. Vetterling, *Numerical Recipes*, 2nd ed., Cambridge University Press, 1986.
36. J. Neuhaus, Viskoelastische Phänomene beim Tablettieren auf Rundläuferpressen, Diss. rer. nat., Bonn, 1985.
37. H. Rieger, personal communication. RHRZ Universität Bonn.
38. E. Stepanek, *Praktische Analyse linearer Systeme durch Faltungsoperationen*, Akad. Verlagsges. Geest & Portig K.G., Leipzig, 1976.
39. H. P. Heikamp, Bestimmung viskoelastischer Eigenschaften von Pulvermischungen auf einer Rundläufertablettenpresse, Diss. rer. nat., Bonn, 1990.
40. A. Müller, Probleme bei der Auswertung von Weg-Spauuwigs-Daten an Hanol viskoelastischer Modelle bei der Tablettierung. Diss. rer. nat., Bonn, 1994.

6

Application of Percolation Theory and Fractal Geometry to Tablet Compaction

Hans Leuenberger, Ruth Leu, and Jean-Daniel Bonny

School of Pharmacy, University of Basel, Basel, Switzerland

I. SHORT INTRODUCTION TO PERCOLATION THEORY AND FRACTAL GEOMETRY

Percolation theory [1] and fractal geometry [2] represent novel powerful concepts which cover a wide range of applications in pharmaceutical technology [3]. Both concepts provide new insights into the physics of tablet compaction and the properties of compacts [4–12].

A. Percolation Theory

Different types of percolation can be distinguished: random-site, random-bond, random-site-bond, correlate chain, etc. Generally, percolation theory deals with the number and properties of clusters [1]. A percolation system is considered to consist of sites in an infinitely large real or virtual lattice. Applying the principles of random-site percolation to a particulate system, a cluster may be considered as a single particle or a group of similar particles which occupy bordering sites in the particulate system (see Figs. 1a,b). In the case of bond percolation, a group of particles is considered to belong to the same cluster only when bonds are formed between neighboring particles.

In random-bond percolation, the bond probability and bond strength between different components can play an important role. The bond proba-

P = 0,50

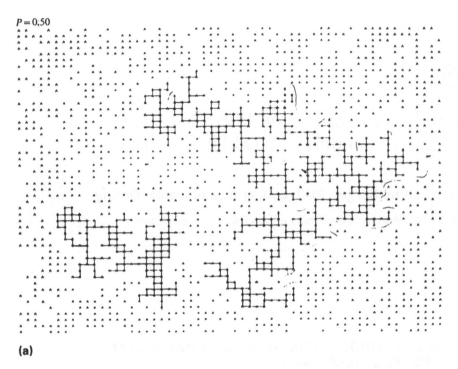

(a)

Fig. 1 (a) Example for percolation on a square lattice for $p = 0.50$ [1]. Occupied sites are shown with asterisk; empty sites are ignored. Two clusters are marked by lines. (b) Example for percolation on a square lattice for $p = 0.6$ [1]. The "infinite cluster" is marked by lines.

bility p_b can assume values between 0 and 1. When $p_b = 1$, all possible bonds are formed and the tablet strength is at its maximum; i.e., a tablet should show maximal strength at zero porosity when all bonds are formed. In order to form a stable compact it is necessary that the bonds percolate to form an "infinite" cluster within the ensemble of powder particles filled in a die and put under compressional stress. Tablet formation can be imagined as a combination of site and bond percolation phenomena. It is evident that for a bond percolation process the existence of an infinite cluster of occupied sites in a lattice is a prerequisite. Figure 2 shows the phase diagram of a site-bond percolation phenomenon [13].

Site percolation is an important model of a binary mixture consisting of two different materials. In the three-dimensional case, two percolation thresholds, p_c, can be defined: a lower threshold, p_{c1}, where one of the components just begins to percolate, and a second, upper percolation

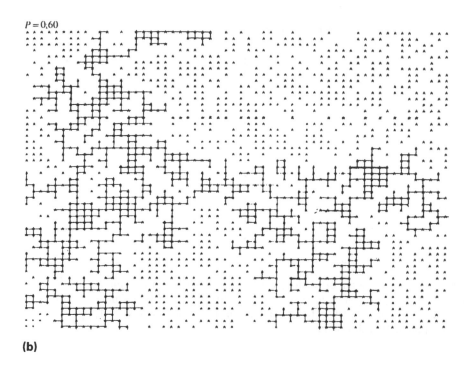

$P = 0.60$

(b)

threshold, p_{c2}, where the other component ceases to have an infinite cluster. Between the two thresholds the two components form two interpenetrating percolating networks. Below the lower or above the upper percolation threshold, the clusters of the corresponding components are finite and isolated. Thus, in site percolation of a binary powder mixture, p_c corresponds to a critical concentration ratio of the two components. From emulsion systems these concentrations are well known where oil-in-water or water-in-oil emulsions can be prepared exclusively.

Table 1 shows critical volume-to-volume ratios for well-defined geometrical packing of monosized spherical particles. The critical volume-to-volume ratios depend on the type of percolation and the type of lattice. In the case of real powder systems the geometrical packing is a function of the particle size, the particle size distribution, and the shape of the particles.

As different types of packing of monosized spherical particles show different porosities, a powder system which has porosity ϵ can be represented in an idealized manner as an ensemble of monosized spheres having hypothetical mean diameter x and a mean coordination number k corresponding to hypothetical (idealized) geometrical packing. Table 2 shows

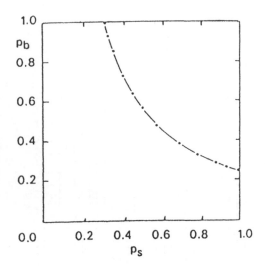

Fig. 2 Phase diagram of random site-bond percolation in the simple cubic lattice; p_s = site probability, p_b = bond probability (Monte Carlo simulation [13]). When $p_s = 1$, $p_b = 0.249$ = bond percolation threshold; when $p_b = 1$, $p_s = 0.312$ = site percolation threshold.

Table 1 Selected Percolation Thresholds
for Three-Dimensional Lattices

Lattice type	Site	Bond
Diamond	0.428	0.388
Simple cubic	0.312	0.249
Body-centered cubic	0.245	0.179
Face-centered cubic	0.198	0.119

Source: From Ref. 1.

Table 2 Coordination Numbers of Isometric
Spherical Particles for Different Packing Structures

Lattice type	Coordination number	Porosity
Diamond	4	0.66
Simple cubic	6	0.48
Body-centered cubic	8	0.32
Face-centered cubic	12	0.26

Source: From Ref. 19.

the coordination number of isometric spherical particles of different packing structures.

Using the simplified model of powder systems mentioned, the following equation was developed [14]:

$$k = \frac{\pi}{\epsilon} \tag{1}$$

for porosities in the range $0.25 < \epsilon < 0.5$. This equation is a rough estimate and does not hold for compacts, where usually $\epsilon < 0.25$.

At a percolation threshold some property of a system may change abruptly or may suddenly become evident. Such an effect starts to occur close to p_c and is usually called a critical phenomenon. As an example, the electrical conductivity of a tablet consisting of copper powder mixed with Al_2O_3 powder may be cited. The tablet conducts electricity only if the copper particles form an "infinite" cluster within the tablet, spanning the tablet in all three dimensions.

In case of a pharmaceutical tablet consisting of an active drug substance and excipients, the principle of function is not the electric conductivity and the tablet usually does not consist of a binary powder system compressed. However, often also in case of a complex tablet composition the system can be reduced to a type of binary powder system dividing the drug and excipients, etc., involved in two classes of function, such as material which is swelling or is easily dissolved in water. Thus, in case of a mixture KCl-StaRX 1500 cornstarch the two percolation thresholds expected are well recognized as a function of the disintegration time of the tablet [3] (see Fig. 3).

B. Fractal Geometry

Fractal geometry is related to the principle of self-similarity; i.e., the geometrical shape is kept identical independent of the scale, magnification, or power of resolution [2]. In practice the range of self-similarity may, however, sometimes be limited to only a few orders of magnitude.

A typical case of fractal geometry is the so-called Coastline of Britain Problem [2]: the length of the coastline is continuously increasing with increasing power of resolution, i.e., with a smaller yardstick to measure the length. Thus a log-log plot of the length of the coastline as a function of the length of the yardstick to perform a polygon approximation yields a straight line with slope $1-D_1$, where D_1 is equal to the fractal dimension of the coastline. Coastlines with perfect self-similarity can also be constructed mathematically (see Fig. 4). The fractal dimension of a coastline is thus in between the Euclidean dimension 1 for a straight line and 2 for a surface.

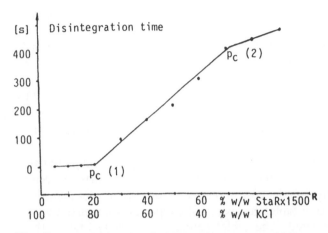

Fig. 3 Percolation thresholds for the compacted binary mixture KC1-StaRX 1'500; tablet property: disintegration time [3].

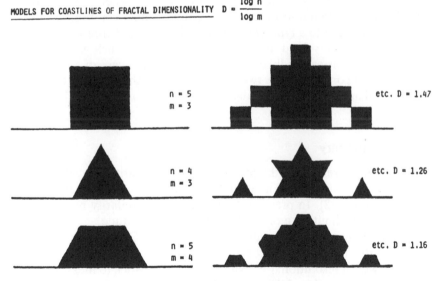

Fig. 4 Different self-similar structures as coastline models; m = theoretical number of equal parts of the unit length projected on a straight line, n = theoretical number of equal parts of unit length describing the coastline structure [5].

It is also fruitful to imagine a fractal surface dimension D_s describing the roughness of a surface. Such a description again includes the prerequisite of a self-similar shape independent of the scale. In this respect the introduction of a surface fractal is very advantageous in powder technology: the result of a measurement of the specific surface of a powder is, as is well known, dependent on the power of resolution of the apparatus (e.g., Blaine, mercury intrusion porosimetry, nitrogen gas adsorption BET method, etc.). Thus the result of a measurement with a chosen method is not able to describe adequately the roughness of the surface. However, if this roughness shows at least within a certain range an approximative self-similarity, it is possible to describe the surface by indicating a value for the specific surface and a value for the surface fractal D_s. Consequently, it is possible to know the specific surface data for different powers of resolution applying a log-log plot of the specific surface as a function of the yardstick length, describing the power of resolution, where the slope of the resulting straight line is equal to $2\text{-}D_s$. As the slope is negative—i.e., the surface is larger for a smaller yardstick length—the value for D_s is between 2 and 3.

In case of porous material it is also possible to define a volume fractal. This concept is based on the fact that as a function of the power of resolution to detect a pore volume or pore size the void volume is increased. It is evident that in a practical case the porosity of a material attains a limiting value; i.e., the self-similarity principle is only valid within a limited range. Based on a mathematical self-similar model of pores—i.e., a Menger sponge (see Fig. 5)—the relationship between the accessible void space (sum of pore volumes) and the power of resolution of the pore size was established [9]. For this purpose the mercury intrusion porosimetry is the method of choice as the pore volume; i.e., the void space of a tablet is filled with mercury as a function of the mercury intrusion pressure, which is related to the accessible pore size. It is, however, necessary to keep in mind that as a consequence of percolation theory not all of the pores are accessible in the same way and some of them are not accessible at all; i.e., there may be closed pores present and pores of increased size may be hidden behind pores of smaller size, a fact that is responsible for the hysteresis loop between filling up and draining off the mercury from the void space within the tablet. These reservations have to be taken into account when the volume fractal dimension D_v of a porous network is determined.

In case of a porous network the solid fraction—i.e., relative density $\rho_r = 1 - \epsilon$ of the tablet (determined according to the pore volume fraction ϵ (d) filled up by mercury)—as a function of the pore diameter d is related to the volume fractal D_v as follows:

$$\log \rho_r = (3 - D_v)\log d + c \tag{2}$$

with c = constant.

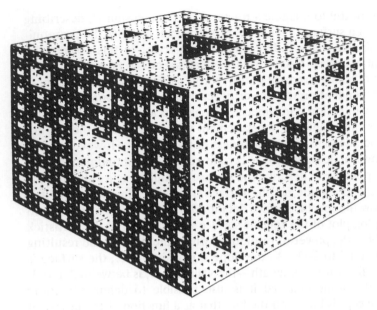

Fig. 5 Menger sponge with fractal dimension of 2.727 (idealized three-dimensional network of a pore system) [2].

It is a unique property of the Menger sponge that its surface and volume fractals D_s and D_v are identical and equal to 2.727 = log 20/log 3.

On the other hand, in case of an agglomerate or aggregate of the size L consisting of identical primary particles of diameter δ, the following relationship holds:

$$\log \rho_r = (3 - D_v)\log\left(\frac{\delta}{L}\right) + c \qquad (3)$$

with c = constant.

A volume fraction ϕ can be attributed to such an aggregate. The above equation plays an important role in the gelification of, e.g., silica particles, which is by nature a percolation process. It has to be taken into account, however, that secondary aggregates of volume fraction ϕ (fractal blobs) and size L and not the primary individual silica particles of diameter δ are the percolating units. Such frail structures may have fractal dimensions below D_v = 2. In case of Aerosil 200 aggregates a fractal dimension D_v = 1.77 was determined [15]. Thus, depending on the structure of an aggregate, the range of D_v overlaps the range of linear and surface fractals, introduced in a first step to trigger the imagination. This is not a contradiction, as the definitions of linear, surface, and volume fractals are arbitrarily related to the Euclidean dimensions to which we are better accustomed. In fact, the electron micrograph of an Aerosil aggregate shows a chainlike structure leading to a fractal dimension of 1.77 as mentioned. Other types may have a fractal dimension close to 2 or even 3. Well known by the work of Mandelbrot [2], the concept of fractal geometry has numerous applications in other fields. In this chapter the concept of fractal geometry is treated only in respect to the physics of tablet compaction and the resulting tablet properties.

II. THE FORMATION OF A TABLET [10]

Filling of the die: For simplicity, it is imagined that the volume of the die is spanned by a three-dimensional lattice. The lattice spacing is assumed to be of the order of a molecular diameter. Thus granules represent clusters of primary particles and primary particles are considered as clusters of molecules. After pouring the particles/granules to be compacted into the die, the lattice sites are either empty forming pores or occupied by molecules forming clusters with site occupation probability p_s. This site occupation probability is equal to the relative density $\rho_r = 1 - \epsilon =$ porosity.

A. Loose Powder Compacts

To fill hard gelatine capsules, one of the principles consists in performing a loose powder compact as a unit dosage form. As only relatively low compressional force is applied, no brittle fracture or plastic flow is expected. However, at the relative density ρ_r in the range of $\rho_p \leq \rho_r \leq \rho_t$ ($\rho_p =$ poured, $\rho_t =$ tapped relative density) bonds are already formed at the contact points throughout the powder bed. This process can be considered as a bond percolation problem. For a sufficient strength of this powder bed only a weak compressional stress σ_c is needed.

In this simplified model it is assumed that the particle size of the powdered substances is sufficiently small to form a cohesive compact under a weak compressional stress. In practice the residual moisture content and the individual substance specific capacity to form weaker or stronger bonds have to be taken into account. Thus a bond percolation threshold ρ_0 is expected in the range $\rho_p \leq \rho_0 \leq \rho_t$.

B. Dense Powder Compacts

1. Pharmaceutical Tablets

Tablets represent the majority of solid dosage forms on the pharmaceutical market. This position is due to its elegance and convenience in application.

Thus, a tablet usually has, among other properties, smooth surfaces, low friability, and sufficient strength (e.g., tensile strength or deformation hardness). For the production of such tablets the compressional stress σ_c needs to be important enough to induce plastic flow and/or brittle fracture of the primary granules or particles, i.e., to produce simultaneously new surfaces and bonds in this dense powder compact.

2. Process of Uniaxial Compression

The usual tableting machines work according to the principle of uniaxial compression. Thus the upper and lower surface areas of the tablet remain constant during the compression process; the thickness of the tablet is reduced with application of the compressional stress σ_z in the z-direction. Because of the initial high porosity of the powder bed, the radial transmission σ_r of the main stress cannot be calculated easily.

In the following simplified model of uniaxial compression, the radial stress σ_r need not be specified explicitly. However, a lateral displacement or rearrangement of particles occupying former pore sites is allowed. The compression process is now studied starting from a loose powder compact with a relative density ρ_p and a moving upper punch. Again a three-dimensional lattice with lattice spacing of molecular diameter spanning the die volume is imagined. During the compression the number of sites to be occupied is constantly reduced and the material (particles, granules, i.e., cluster of molecules) is available to occupy remaining sites. According to the principle of uniaxial compression the mean particle-particle separation distance is more reduced in the z-direction than in the lateral directions. Thus it can be assumed that in the beginning a one-dimensional bond percolation (i.e., a chain of molecules) is responsible for the stress transmission from the upper to the lower punch due to the repulsion forces of the electron shell (Born repulsion forces). After the rearrangement of the particles/granules at a relative density $\rho_r = \rho_r{}^*$ an important buildup of stress occurs as particles/granules can no longer be displaced easily. At this relative density $\rho_r{}^*$ the compact can be considered as the "first" dense tablet.

On a molecular scale the molecules react in a first approximation as a hard-core spheres model and span as an infinite cluster the die. This situation is typical for a site percolation process. Above the percolation threshold $\rho_c = \rho_r{}^*$ still-empty lattice sites can be occupied due to brittle fracture and/or plastic flow of particles. Thus at higher compressional stress one can imagine that new bonds are formed with a certain bond formation probability p_b and sites are occupied with a site occupation probability p_s, typical for a site-bond percolation phenomenon. Figure 2 represents the phase diagram of a site-bond percolation process. Thus, due to the complex situation during the formation of a tablet, no sharp percolation threshold is expected.

3. Stress Transmission in the Die

It is well known that the compressional stress is transmitted from the upper punch to the lower punch by means of particle-particle contact in the powder bed. Thus, one may expect that the stress is conducted similar to electric current. As a consequence it is of interest to measure at the same time the stress and the electric current transmitted. This experimental work was realized by Ehrburger et al. [15], using as a conducting material different types of carbon black and for comparison the electrical insulating silica particles Aerosil 200. The physical characterization of the material tested is compiled in Table 3.

Silica was chosen because this material is often used to study the gelation process which can be adequately described by percolation theory. From nonlinear regression analysis the parameters of the following power laws were determined:

$$C = C_0(\rho_r - \rho_c)^t \tag{4}$$

$$\sigma_c = \sigma_0(\rho_r - \rho_0)^\tau \tag{5}$$

Where

C = conductivity $[(\text{ohm cm})^{-1}]$
C_0 = scaling factor
σ_c = compressional stress transmitted
σ_0 = scaling factor
ρ_c, ρ_0 = percolation thresholds
t, τ = experimentally determined scaling exponents, expected to equal the electrical conductivity coefficient $\mu = 2$.

Table 3 Physical Characterization of Aerosil 200 and Different Types of Carbon Black [15]

Material	BET surface (m²/g)	Density (g/cm³)			Fractal dimension	
		True	Poured	Tapped		
Aerosil 200	202	2.2	0.016	0.022	1.77	±0.05
Types of carbon black						
Noir d'acétylène	80	1.87	0.027	0.041	1.99	±0.10
TB #4500	57	1.83	0.043	0.066	1.76	±0.12
TB #5500	206	1.84	0.025	0.043	1.83	±0.13
Sterling FT	15	1.85	0.27	0.40	3.0	

Table 4 Stress Transmission and Conductivity of Aerosil 200 and Different
Types of Carbon Black [15]

Material	Stress transmission			Conductivity		
	ρ_0	τ	ρ_{max}	ρ_c	t	ρ_{max}
Aerosil 200	0.025	1.5±0.1	0.06	—	—	—
Types of carbon black						
Noir d'acétylène	0.032	2.9±0.1	>0.24	0.024	1.9±0.1	≥0.24
TB #4500	0.050	2.2±0.1	0.15	0.040	1.8±0.1	≥0.27
TB #5500	0.033	2.1±0.1	0.10	0.019	1.8±0.1	>0.2
Sterling FT	0.27	3.9±0.2	>0.56	0.27	3.4±0.2	>0.56

The results of these investigations are compiled in Table 4. The authors [15] concluded that both the stress transmission and the conductivity
follow the power laws of percolation. The value of ρ_{max} should indicate the
range of relative densities ρ_r: $\rho_0 < \rho_r < \rho_{max}$ and $\rho_c < \rho_r < \rho_{max}$ where this
power law is still valid. Table 4 shows rather low values for the percolation
thresholds ρ_0 and ρ_c. This fact can be explained that in the process of
percolation secondary agglomerates (aggregates of size L) consisting of
primary carbon black or silica particles (of size δ) are responsible for the
stress transmission. Taking into account the fractal geometry of these aggregates, i.e., the volume fractal D_v, Ehrburger et al. [15] obtained a good
estimate for the ratio L/δ, using the following equation based on the percolation threshold ρ_0:

$$\rho_0 = 0.17\left(\frac{L}{\delta}\right)^{D_v-3} \tag{6}$$

The values calculated for L/δ were in good agreement with estimates
obtained from independent experiments (BET and porosimetry measurements).

III. TABLET PROPERTIES

A number of tablet properties are directly or indirectly related to the
relative density ρ_r of the compact. According to percolation theory the
following relationship holds for the tablet property X close to the percolation threshold p_c:

$$X = S(p - p_c)^q \tag{7}$$

Where

X = tablet property
p = percolation probability
p_c = percolation threshold
S = scaling factor
q = exponent

In case of a tablet property X the values of S and q are not known a priori. In addition, the meaning of p and p_c has to be identified individually for each property X. In the case of site percolation the percolation probability p is identical to the relative density ρ_r as mentioned earlier. For obvious reasons one can expect that the tensile strength σ_t and the deformation hardness of P are related to the relative density ρ_r in accordance with the percolation law (Eq. (7)).

Unfortunately, in a practical case only the experimental values of σ_t, P, and ρ_r of the tablet are known. In this respect, it is important to take into account the properties of the power law (Eq. (7)) derived from percolation theory: (1) the relationship holds close to the percolation threshold but it is unknown in general how close; (2) the a priori unknown percolation threshold p_c and the critical exponent q are related. Thus there is a flip-flop effect between p_c and q, a low p_c value is related to a high q value, and vice versa. As a consequence, the data evaluation based on nonlinear regression analysis to determine S, p_c, and q may become very tedious or even impossible.

In selected cases such as percolation in a Bethe lattice, the percolation exponent q is equal to unity. A list of selected exponents for 2, 3, and infinite dimensions (i.e., Bethe lattice) known from first principles is compiled in Table 5 and cited from Stauffer [1]. Details should be read there concerning the relevant equation for the property of the system described.

Table 5 Percolation Exponents for Two Dimensions, Three Dimensions, and in the Bethe Lattice and the Corresponding Quantity

Exponent	Dimension			Quantity/property
	2	3	Bethe	
α	−2/3	−0.6	−1	Total number of clusters
β	5/36	0.4	1	Strength of infinite network
γ	43/18	1.8	1	Mean size of finite clusters
ν	4/3	0.9	1/2	Correlation length
μ	1.3	2.0	3	Conductivity

In case of the tablet properties such as tensile strength σ_t and deformation hardness P no meaningful results can be obtained without additional expertise. It was a rewarding endeavor [10] to combine the following two equations derived earlier [16] with the well-known Heckel equation [17]:

$$\sigma_t = \sigma_{t\ max}(1 - e^{-\gamma_t \sigma_c \rho_r}) \tag{8}$$

$$P = P_{max}(1 - e^{-\gamma \sigma_c \rho_r}) \tag{9}$$

Where

$$\sigma_{t\ max} = \text{maximum tensile strength at } \epsilon \rightarrow 0$$
$$P_{max} = \text{maximum deformation hardness at } \epsilon \rightarrow 0$$
$$\sigma_c = \text{compressional stress}$$
$$\gamma, \gamma_t = \text{compression susceptibility}$$

and

$$\ln \frac{1}{1-\rho_r} = a + b\sigma_c \tag{10}$$

Where a, b = constants specific to the particulate material compressed and σ_c = compressional stress. This derivation does not take into account the exact value of ρ_r (usually between 0.6 and 1) in the exponents of Eqs. (8) and (9). The combination of Eqs. (8) and (10) and Eqs. (9) and (10) yielded the following general relationships:

$$\sigma_t = \frac{\sigma_{t\ max}}{1 - \rho_c}(\rho_r - \rho_c) \tag{11}$$

$$P = \frac{P_{max}}{1 - \rho_c}(\rho_r - \rho_c) \tag{12}$$

with ρ_c = critical relative density (percolation threshold).

It is evident that Eqs. (11) and (12) are formally identical with the fundamental law of percolation theory (Eq. (7)):

$$\sigma_t = S'(\rho_r - \rho_c) \qquad \text{with } S' = \frac{\sigma_{t\ max}}{1 - \rho_c} \tag{13}$$

$$P = S(\rho_r - \rho_c) \qquad \text{with } S = \frac{P_{max}}{1 - \rho_c} \tag{14}$$

and with the exponent $q = 1$ corresponding to a percolation in a Bethe lattice [1].

The general relationships (11), (12) can be specified, on the one hand, for the formation of loose compacts and, on the other hand, for the formation of dense compacts.

For loose compacts, i.e., at a low-pressure range, ρ_c equals the bond percolation threshold ρ_0. P_{max} and $\sigma_{t\,max}$ of the low-pressure range do not correspond to the maximal possible deformation hardness or tensile strength respectively at the relative density $\rho_r = 1$, but describe the strength of the substance specific particle-particle interaction at low relative densities, where the primary particles have not yet lost their identity. The scaling factors S and S' are a measure of the strength of this interaction. In this range the compact can still be separated into the original particles.

For the formation of dense compacts, i.e., at a median pressure range, ρ_c equals the site percolation threshold ρ_r^*. In this case P_{max} and σ_{tmax} correspond to the maximum deformation hardness or tensile strength respectively of the substance at $\rho_r \to 1$.

At higher relative densities of the tablet the pore network may no longer form an infinite cluster. Thus another percolation threshold p_c has to be expected for $\rho_r = \rho_\pi$. It is evident that this threshold is important for, e.g., the disintegration time. Due to the complexity of the tablet formation again no sharp percolation threshold is expected at ρ_π.

The experimental methods and the data evaluation are described in detail in a recent paper [10]. For the physical characterization of the materials used see Table 6. The experimentally determined Heckel plot is approximated by two linear sections: a linear section for low and another for median pressures. The percolation thresholds ρ_0 and ρ_r^* are calculated on the basis of the intercepts of these two linear sections of the Heckel plot. It

Table 6 Physical Characterization of the Substances Used

	Densities					Mean particle size (μm)
	True (g/mL)	Poured (g/mL)	ρ_p	Tapped (g/mL)	ρ_t	
Avicel PH 102 FMC 2843	1.58	0.325	0.206	0.439	0.278	97.0
Caffeine anhydrous Sandoz 88828	1.45	0.323	0.223	0.417	0.288	55.0
Emcompress CaHPO$_4$·2H$_2$O Ed. Mendell & Co. E27B2	2.77	0.714	0.258	0.870	0.314	106.0
Lactose α-monohydrate DMV 171780	1.54	0.562	0.365	0.735	0.477	53.2
PEG 10'000 Hoechst 605331	1.23	0.568	0.462	0.719	0.585	135.8
Sta-RX 1'500 Sandoz 86823	1.50	0.606	0.404	0.741	0.494	68.1

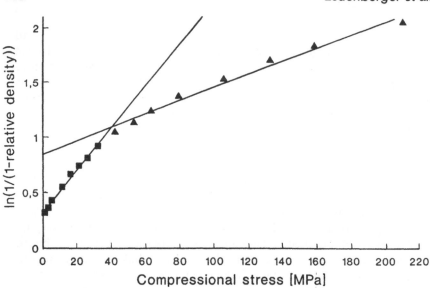

Fig. 6 Heckel plot of microcrystalline cellulose (Avicel) in order to determine the percolation thresholds for loose and dense compacts (ρ_0 and $\rho_r{}^*$).

is evident that the Heckel equation is an approximation of the pressure-density profile and the application of two linear sections leads to a better fit of the relationship found in reality (see Figs. 6–8). According to Eqs. (13) and (14) σ_t and P are plotted against the relative density and linearized by two regression lines for the same ranges as in the Heckel plot (see Figs. 9–14). In Table 7 the experimentally determined percolation thresholds from the Heckel plot (Eq. (10)), from the plot of tensile strength against relative density (Eq. (13)), and from the plot of deformation hardness against relative density (Eq. (14)) are summarized. Table 8 shows the values of S, S' and the resulting values for $\sigma_{t\,max}$ and P_{max} at the median pressure range. The values of $\sigma_{t\,max}$, P_{max}, γ, and γ_t according to Eqs. (8) and (9) are compiled in Table 9 and the corresponding plots are shown in Figs. 15 and 16. The comparison of the results and estimated standard deviations indicates that Eqs. (13) and (14) offer in general more reliable estimates for the $\sigma_{t\,max}$ and P_{max} values than Eqs. (8) and (9). However, the squared correlation coefficients of the evaluation according to Eqs. (13) and (14) are lower than according to Eqs. (8) and (9). This is due to the fact that the linear model is not as flexible as the exponential one. That is the reason why it can happen that the exponential model can lead to unreasonable values of the model parameters with large standard errors in spite of the good

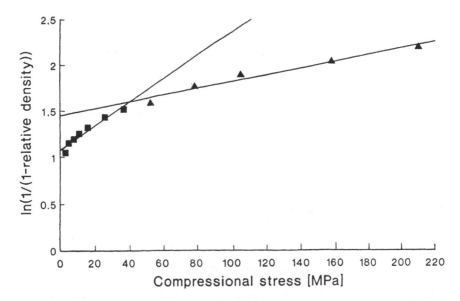

Fig. 7 Heckel plot of lactose in order to determine the percolation thresholds for loose and dense compacts ($\rho_{0 \text{ and }} \rho_r{}^*$).

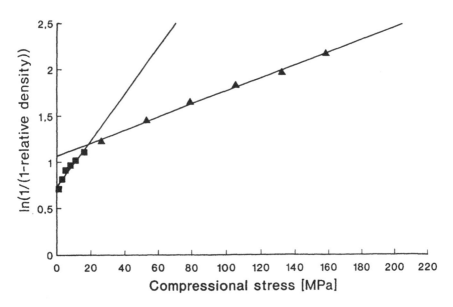

Fig. 8 Heckel plot of caffeine in order to determine the percolation thresholds for loose and dense compacts (ρ_0 and $\rho_r{}^*$).

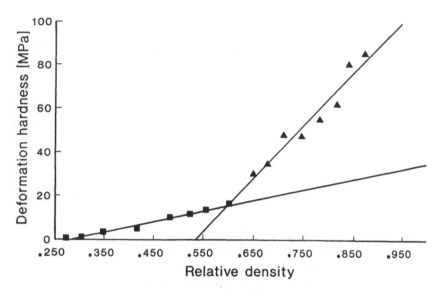

Fig. 9 Tablet property: deformation hardness of Avicel compacts as a function of relative density according to Eq. (14).

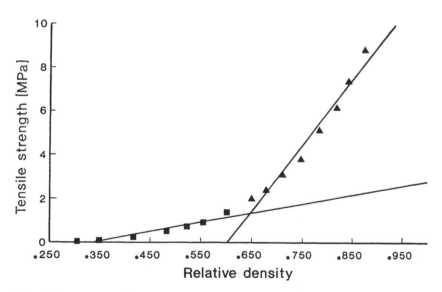

Fig. 10 Tablet property: tensile strength of Avicel compacts as a function of relative density according to Eq. (13).

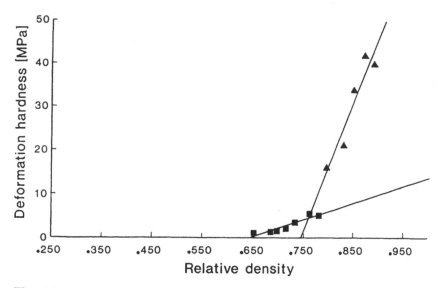

Fig. 11 Tablet property: deformation hardness of lactose compacts as a function of relative density according to Eq. (14).

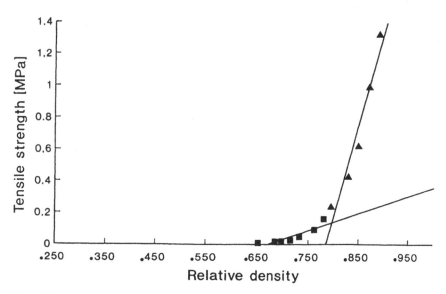

Fig. 12 Tablet property: tensile strength of lactose compacts as a function of relative density according to Eq. (13).

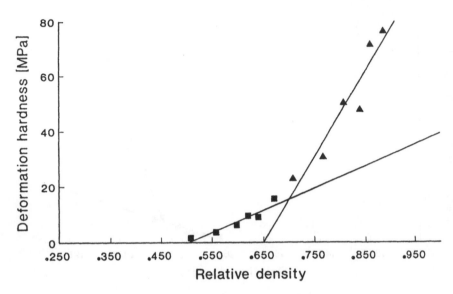

Fig. 13   Tablet property: deformation hardness of caffeine compacts as a function of relative density according to Eq. (14).

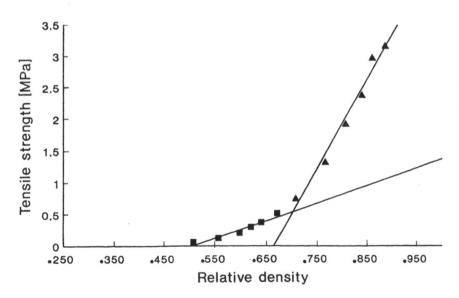

Fig. 14   Tablet property: tensile strength of caffeine compacts as a function of relative density according to Eq. (13).

Table 7 Comparison of Values for ρ_c Experimentally Determined According to Eqs. (10), (13), and (14) for (a) Loose and (b) Dense Compacts

(a) Loose compacts (i.e., low-pressure range $\rho_c = \rho_0$)					
Heckel plot Eq. (10)		Tensile strength Eq. (13)		Deformation hardness Eq. (14)	
$\rho_0 \pm s$	r^2	$\rho_0 \pm s$	r^2	$\rho_0 \pm s$	r^2
Avicel	0.275 ± 0.013	0.922	0.333 ± 0.023	0.911	0.278 ± 0.011	0.980
Caffeine	0.521 ± 0.028	0.939	0.464 ± 0.014	0.946	0.500 ± 0.010	0.952
Emcompress	0.429 ± 0.025	0.924	0.454 ± 0.016	0.854	0.448 ± 0.020	0.801
Lactose	0.660 ± 0.026	0.945	0.659 ± 0.013	0.820	0.646 ± 0.015	0.841
PEG 10'000	0.591 ± 0.028	0.966	0.644 ± 0.016	0.897	0.576 ± 0.008	0.978
Sta-RX 1'500	0.506 ± 0.015	0.976	0.521 ± 0.010	0.917	0.499 ± 0.012	0.927

(b) Dense compacts (i.e., median-pressure range $\rho_c = \rho_r{}^*$)					
$\rho_r{}^* \pm s$	r^2	$\rho_r{}^* \pm s$	r^2	$\rho_r{}^* \pm s$	r^2
Avicel	0.572 ± 0.039	0.982	0.602 ± 0.013	0.968	0.532 ± 0.025	0.941
Caffeine	0.657 ± 0.023	0.996	0.703 ± 0.019	0.914	0.650 ± 0.028	0.901
Emcompress	0.566 ± 0.006	0.996	0.581 ± 0.008	0.965	0.586 ± 0.018	0.808
Lactose	0.765 ± 0.052	0.966	0.775 ± 0.010	0.930	0.757 ± 0.011	0.934
PEG 10'000	0.807 ± 0.147	0.943	0.813 ± 0.011	0.969	0.823 ± 0.033	0.745
Sta-RX 1'500	0.619 ± 0.067	0.982	0.688 ± 0.014	0.976	0.595 ± 0.070	0.820

squared correlation coefficients. In the linear model the data are fitted by a straight line which results in poorer fits but at the same time more reasonable values for the parameters P_{max} and $\sigma_{t\,max}$.

The description of other tablet properties such as disintegration time [3], dissolution rate, etc., according to percolation theory are limited as long as the theoretical models are not established to allow more to be known about the respective percolation exponents. However, in the special case of a matrix-type slow release system [8,9] it is possible to apply simultaneously the concept of percolation theory and fractal geometry. This is the topic of the next section.

IV. DRUG DISSOLUTION FROM A MATRIX-TYPE CONTROLLED RELEASE SYSTEM

A. Ants in a Labyrinth and Drug Dissolution Kinetics

Molecules of an active substance, which are enclosed in a matrix-type controlled release system, may be called ants in a labyrinth [1] trying to

Table 8 S- and S'-values for Low- and Median-Pressure Ranges and P_{max}- and $\sigma_{t\,max}$-Values Resulting from the S- and S'-Values for Median Pressures According to Eqs. (13) and (14)

| | Tensile strength (Eq. (13)) | | |
| | Low-pressure range | Median-pressure range | |
	$S' \pm s$ (MPa)	$S' \pm s$ (MPa)	$\sigma_{t\,max} \pm s$ (MPa)
Avicel	4.199 ± 0.586	30.232 ± 2.261	12.023 ± 0.564
Caffeine	18.924 ± 2.253	2.443 ± 0.376	0.726 ± 0.075
Emcompress	0.467 ± 0.086	3.451 ± 0.379	1.446 ± 0.135
Lactose	0.705 ± 0.165	10.364 ± 1.421	2.332 ± 0.240
PEG 10'000	1.621 ± 0.316	14.322 ± 1.475	2.688 ± 0.141
Sta-RX 1'500	1.142 ± 0.172	12.167 ± 1.359	3.807 ± 0.282

| | Deformation hardness (Eq. (14)) | | |
| | Low-pressure range | Median-pressure range | |
	$S \pm s$ (MPa)	$S \pm s$ (MPa)	$P_{max} \pm s$ (MPa)
Avicel	48.17 ± 2.77	239.22 ± 24.55	111.85 ± 6.12
Caffeine	99.16 ± 11.19	309.34 ± 51.33	108.39 ± 10.19
Emcompress	21.60 ± 4.82	262.17 ± 73.77	108.63 ± 26.20
Lactose	40.72 ± 8.85	333.41 ± 44.34	81.17 ± 7.50
PEG 10'000	18.45 ± 1.25	139.80 ± 47.18	24.72 ± 4.50
Sta-RX 1'500	32.03 ± 4.49	74.69 ± 24.71	30.25 ± 5.11

escape from an ordered or disordered network of connected pores. For site occupation probabilities p_s far above the percolation threshold p_c, the random walk distance R of such an ant is related to time t as follows: $R^2 = Dt$, where D is the diffusivity. In a more general form this diffusion law can be expressed as $R \propto t^k$ with $k = 0.5$ for p_s above p_c. For p_s far below p_c, R approaches a constant for large times, i.e., $k = 0$. Right at the critical point the value of k ranges between these two extremes and is about 0.2 in three dimensions. This process is called anomalous diffusion.

From these first principles one can conclude that there is at least one percolation threshold, i.e., p_{c1}, where the active drug is completely encapsulated by the water-insoluble matrix substance and where the usual square-root-of-time law for the dissolution kinetics is no longer valid. As in general, in a three-dimensional system two percolation thresholds can be expected, an experiment was set up to elucidate this phenomenon. For

Table 9 Evaluation of the Data According to Eqs. (8) and (9) over the Whole Pressure Range

	Tensile strength (Eq. (8))		
	$\sigma_{t\,max} \pm s$ (MPa)	$\gamma_t \times 10^{-3} \pm s$ (MPa^{-1})	r^2
Avicel	14.44 ± 0.59	5.22 ± 0.30	0.999
Caffeine	4.32 ± 0.34	9.61 ± 1.26	0.995
Emcompress	5.89 ± 12.95	0.57 ± 1.29	0.995
Lactose	494.53 ± 28849	0.01 ± 0.80	0.994
PEG 10'000	3.34 ± 0.39	11.93 ± 2.15	0.992
Sta-RX 1'500	5.26 ± 0.70	3.98 ± 0.79	0.990
	Deformation hardness (Eq. (9))		
	$P_{max} \pm s$ (MPa)	$\gamma \times 10^{-3} \pm s$ (MPa^{-1})	r^2
Avicel	93.22 ± 5.65	12.34 ± 1.44	0.984
Caffeine	94.66 ± 15.64	10.99 ± 3.22	0.968
Emcompress	725.14 ± 10187	0.32 ± 4.60	0.944
Lactose	58.56 ± 12.15	7.45 ± 2.51	0.958
PEG 10'000	37.06 ± 17.73	8.85 ± 5.95	0.924
Sta-RX 1'500	20.38 ± 1.44	19.93 ± 4.20	0.945

this purpose, a highly water-soluble model drug (caffeine anhydrous) and a plastic water-insoluble matrix substance (ethyl cellulose) were chosen. The materials and methods used are described in detail by Bonny and Leuenberger [12]. Here, only the theoretical background and conclusions are summarized. In this experiment the drug content was varied from 10% to 100% (w/w), and the drug dissolution from one flat side of the tablets was studied. In order to measure this intrinsic dissolution rate the tablet was fixed into a paraffin matrix to leave only one side accessible for the dissolution medium (distilled water).

For low drug concentrations, i.e., low porosity of the matrix, most of the drug is encapsulated by the plastic matrix and the release is incomplete. At the lower percolation threshold p_{c1} the drug particles begin to form a connective network within the matrix and according to the theoretical considerations the diffusion should be anomalous. At the upper percolation threshold p_{c2} the particles which should form a matrix start to get isolated within the drug particles and the tablet disintegrates.

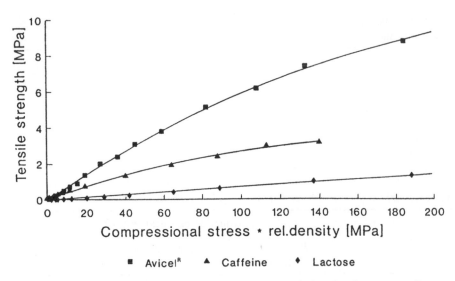

Fig. 15 Plot of the tensile strength σ_t versus the relative density ρ_r according to Eq. (8) for Avicel, caffeine, and lactose.

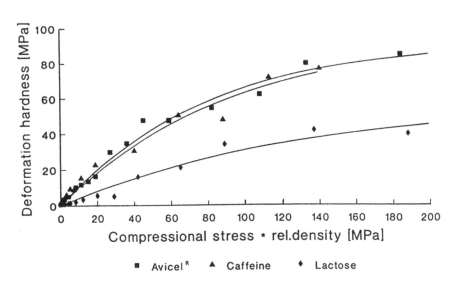

Fig. 16 Plot of the deformation hardness P_{max} versus the relative density ρ_r according to Eq. (9) for Avicel, caffeine, and lactose.

The concentration of the drug particles within the matrix can be expressed as site occupation probability p_s. The amount of drug $Q(t)$ released from one tablet surface after the time t is proportional to t^k and the exponent k depends on the percolation probability p_s:

Case 1: $p_s < p_{c1}$. Only the few particles connected to the tablet surface can be dissolved and $Q(t)$ reaches a constant value.

Case 2: $p_s \approx p_{c1}$. Anomalous diffusion with $k \approx 0.2$ in three dimensions [1].

Case 3: $p_{c1} < p_s < p_{c2}$. Normal matrix-controlled diffusion with $k = 0.5$.

Case 4: $p_{c2} < p_s$. Zero-order kinetics with $k = 1$.

Between the two percolation thresholds p_{c1} and p_{c2} the particles of drug and matrix substance form a bicoherent system; i.e., the drug release matches the well-known square-root-of-time law of Higuchi [18] for porous matrices:

$$Q(t) = \sqrt{\frac{D_0 \epsilon}{\tau}} \, C_s (2A - \epsilon C_s)t = \sqrt{DC_s(2A - \epsilon C_s)t} \qquad (15)$$

where

Q = cumulative amount of drug released per unit exposed area
D_0 = diffusion coefficient of drug in permeating fluid
ϵ = total porosity of empty matrix
τ = tortuosity of matrix
C_s = solubility of drug in permeating fluid
A = concentration of dispersed drug in tablet
D = apparent or observed diffusion coefficient

Close to the percolation threshold the observed diffusion coefficient obeys a scaling law which can be written as [1]

$$D \propto (p_s - p_{c1})^\mu \qquad (16)$$

where

p_s = site occupation probability
p_{c1} = critical percolation probability (lower percolation threshold)
μ = conductivity exponent = 2.0 in three dimensions [1]

In the case of a porous matrix p_s can be expressed by the total porosity ϵ of the empty matrix and p_{c1} corresponds to a critical porosity ϵ_c, where the pore network just begins to span the whole matrix. Equation (16) can then be written as

$$D \propto (\epsilon - \epsilon_c)^\mu \qquad \text{or} \qquad D = \kappa D_0(\epsilon - \epsilon_c)^\mu \qquad (17)$$

with $\kappa D_0 =$ scaling factor.

In the cases where the dissolution kinetics are in agreement with Eq. (15) the dissolution data can be linearized by plotting $Q(t)$ versus $\sqrt{t}$ giving a regression line with the slope b,

$$b = \sqrt{DC_s(2A - \epsilon C_s)} \tag{18}$$

which leads to

$$D = \frac{b^2}{C_s(2A - \epsilon C_s)} \tag{19}$$

Combining Eqs. (17) and (19) and assuming $\mu = 2.0$ results in

$$\frac{b^2}{C_s(2A - \epsilon C_s)} = \kappa D_0(\epsilon - \epsilon_c)^{2.0} \tag{20}$$

or

$$\beta = \frac{b}{\sqrt{2A - \epsilon C_s}} = \sqrt{\kappa D_0 C_s}(\epsilon - \epsilon_c) \tag{21}$$

i.e., the tablet property β is determined by a linear relationship of ϵ,

$$\beta = c(\epsilon - \epsilon_c) = -c\epsilon_c + c\epsilon \tag{22}$$

where the constant c equals $\sqrt{\kappa D_0 C_s}$.

By the help of Eq. (22) ϵ_c can easily be calculated by using a nonlinear, or even a linear, regression analysis, giving a slope of c and an intercept of $-c\epsilon_c$.

The results of the intrinsic dissolution test of the tablets with the different caffeine loadings are plotted in Fig. 17. A change in dissolution kinetics can be assumed between 70% and 80% of caffeine loading. In order to test this assumption the release data are evaluated according to the model $Q(t) = a + b\sqrt{t}$ by a simple linear regression to clarify for which loadings the square-root-of-time law is fulfilled. To analyze the diffusion mechanism the data are also evaluated according to $Q(t) = a' + b't^k$ by a nonlinear least square fit. The results are compiled in Table 10.

Comparing the squared correlation coefficients of the $\sqrt{t}$ evaluation, there is a clear decrease in the grade of correlation for caffeine loadings higher than 70%; i.e., the drug loadings from 80% to 100% caffeine are no longer in good agreement with the $\sqrt{t}$ law. For low drug concentrations only a small amount of drug is released and the dissolution curve runs nearly parallel to the abscissa. In these cases the correlation coefficient cannot be used as an indicator for compliance with the model.

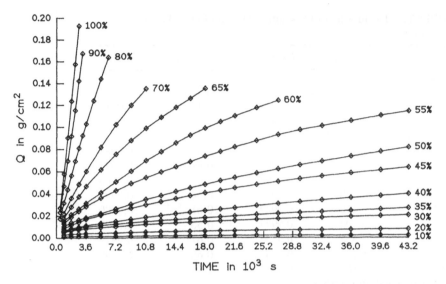

Fig. 17 Cumulative amount $Q(t)$ of caffeine released per unit area as a function of time for tablets with caffeine loadings between 10% (w/w) and 100% (w/w).

The estimation of k according to the model $Q(t) = a' + b't^k$ yields values for k between 0.17 and 1.09. For 35% to 55% of caffeine the exponent k ranges between 0.41 and 0.61, which is in good agreement with the $\sqrt{t}$ kinetics with $k = 0.5$. For higher loadings there is a clear change from $\sqrt{t}$ kinetics to zero-order kinetics with $k = 1$. Both evaluations show that the upper percolation threshold p_{c2} lies between 70% and 80% of caffeine.

For the quantitative determination of the lower percolation threshold p_{c1}, i.e., the critical porosity ϵ_c, Eqs. (21) and (22) are used. The needed data are summarized in Table 11.

The initial porosity ϵ_0 before leaching is calculated from the apparent volume V_{tot} and the true volume V_t of the tablet constituents:

$$\epsilon_0 = \frac{V_{tot} - V_t}{V_{tot}} \qquad (23)$$

ϵ_d is the porosity corresponding to the volume occupied by the drug substance in the matrix and is calculated as follows:

$$\epsilon_d = \frac{m_d}{\rho_d V_{tot}} \qquad (24)$$

Table 10 Evaluation of Dissolution Data and Percolation
Thresholds

Drug content	$Q(t) = a + b\sqrt{t}$		$Q(t) = a' + b't^k$	
	b	r^2	k	r^2
10	0.006	0.9858	0.17	0.9615
20	0.027	0.9975	0.27	0.9932
Lower percolation threshold expected				
30	0.085	0.9989	0.37	0.9983
35	0.116	0.9994	0.41	0.9973
40	0.189	0.9954	0.52	0.9963
45	0.320	0.9987	0.54	0.9992
50	0.415	0.9929	0.61	0.9980
55	0.593	0.9989	0.56	0.9979
60	0.858	0.9936	0.67	0.9979
65	1.15	0.9983	0.66	0.9991
70	1.56	0.9941	0.74	0.9996
Upper percolation threshold expected				
80	2.55	0.9863	0.84	0.9998
90	4.11	0.9752	1.02	0.9991
100	5.38	0.9784	1.09	>0.9999

Drug content in % (w/w)
b = slope in 10^{-3}g cm^{-2} s$^{-1/2}$
k = dimensionless exponent
r^2 = squared correlation coefficient

with m_d = total amount of drug present in the tablet and ρ_d = true density
of drug. ϵ is the total porosity of the empty matrix:

$$\epsilon = \epsilon_0 + \epsilon_d \tag{25}$$

D is calculated according to Eq. (19), and the tablet property β according
to Eq. (21).

   For estimating ϵ_c with the help of Eq (22), the data for ϵ and β for
35% to 55% of caffeine are used, because in this range there is the best
agreement with the normal diffusion law, i.e., $\sqrt{t}$ kinetics with $k = 0.5$ (see
Table 10). The nonlinear regression yields $\epsilon_c = 0.35 \pm 0.01$ and $c = (2.30 \pm
0.15) \cdot 10^{-3}$ g$^{1/2}$ cm$^{-1/2}$ s$^{-1/2}$. A linear regression analysis leads to the same
result. The critical porosity of 0.35 corresponds to a caffeine content of
about 28% (w/w). Figure 18 shows the plot of β versus ϵ where the point of
intersection with the abscissa just indicates ϵ_c.

Table 11 Calculation of D and the Tablet Property β

Drug content	ϵ_0	ϵ_d	ϵ	A	b	D	β
10	0.134	0.078	0.212	0.110	0.006	0.0045	0.013
20	0.128	0.158	0.286	0.225	0.027	0.0437	0.041
30	0.121	0.242	0.363	0.344	0.085	0.282	0.104
35	0.118	0.285	0.403	0.405	0.116	0.446	0.130
40	0.116	0.328	0.444	0.466	0.189	1.03	0.198
45	0.109	0.375	0.484	0.532	0.320	2.58	0.313
50	0.110	0.418	0.528	0.594	0.415	3.88	0.384
55	0.106	0.465	0.571	0.660	0.593	7.13	0.520
60	0.103	0.512	0.615	0.727	0.858	13.5	0.717
65	0.098	0.562	0.660	0.798	1.15	22.2	0.918
70	0.099	0.608	0.707	0.863	1.56	37.7	1.197
80	0.092	0.708	0.800	1.006	2.55	86.4	1.811
90	0.088	0.811	0.899	1.151	4.11	196	2.729
100	0.080	0.920	1.000	1.306	5.38	296	3.353

Drug content in % (w/w)
ϵ_0 = initial tablet porosity
ϵ_d = porosity due to drug content
ϵ = total porosity of matrix
A = concentration of dispersed drug in tablet in g cm^{-3}
b = slope in 10^{-3} g cm^{-2} s$^{-1/2}$
D = apparent diffusion coefficient in 10^{-6} cm^2 s^{-1}
β = tablet property in 10^{-3} g$^{1/2}$ cm$^{-1/2}$ s$^{-1/2}$

In Table 1 selected percolation thresholds [1] for three-dimensional lattices (as volume-to-volume ratios) and in Table 2 the corresponding coordination numbers [19] for isometric spherical particles are compiled. Comparing the experimentally determined percolation threshold of 0.35 (volume-to-volume ratio) with the theoretical values shows good agreement with the simple cubic lattice, which has a site percolation threshold of 0.312. In the tablet a brittle and a plastic substance of different grain size are compacted together so that the postulation of a lattice composed of isometric spheres is only a rough estimate, but it points out that the magnitude of ϵ_c is reasonable.

B. Fractal Dimension of the Pore System of a Matrix-Type Slow Release System

If the matrix-type slow release dosage form of the preceeding chapter is removed from the dissolution medium after, e.g., a maximum of 60%

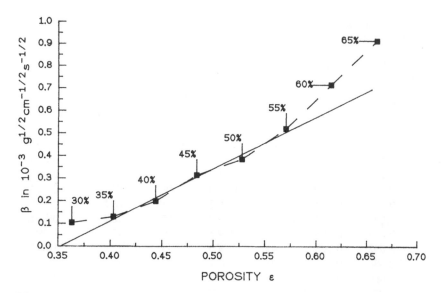

Fig. 18 Tablet property β as a function of porosity ϵ (power law) according to percolation theory.

(w/w) of drug dissolved to guarantee the physical stability of the remaining carcass, the open pore system left can be analyzed by mercury intrusion porosimetry to determine the fractal dimension. It is evident that the pore structure depends on the particle size distribution of the brittle model drug caffeine anhydrous originally embedded in the plastic ethyl cellulose matrix. The volume fractals [9] of leached tablets with an initial caffeine content of 50% (w/w) range between 2.67 and 2.84, depending on the particle size distribution of the water-soluble drug. The system which contained caffeine with a broad particle size distribution (125–355 μm) yielding a fractal volume dimension of 2.734 is closest to the dimension of the Menger sponge ($D_v = 2.727$).

C. Fractal Dimension of the Porous Network of a Fast Disintegrating Tablet

In the thesis of Luy [20] tablets were produced from granules which showed a very high porosity and internal surface. Those granules were obtained with a novel process technology: vacuum fluidized-bed spray granulation [21,22]. Independent of the granule size distribution the resulting tablets released 100% of the soluble drug (solubility in water < 0.01%) within 5 min. The disintegration time was smaller than 1 min. The fractal dimension of the porous network ranged from 2.82 to 2.88.

V. CONCLUSIONS

The concept of percolation theory and fractal geometry allowed new insights into the physics of tablet compaction and the properties of the tablets. The results attained so far are promising and should stimulate further research in this field.

LIST OF SYMBOLS

A	concentration of the dispersed drug in the tablet
b	slope of the regression line in the plot of $Q(t)$ versus $\sqrt{t}$
C_s	solubility of the drug in the permeating fluid
D	diffusivity or apparent diffusion coefficient
D_0	diffusion coefficient of the drug in the permeating fluid
D_1	linear fractal dimension
D_s	surface fractal dimension
D_v	volume fractal dimension
k	mean coordination number
L	size of agglomerate or aggregate
p	percolation probability
p_b	bond formation probability
p_c	percolation threshold
p_{c1}	lower percolation threshold
p_{c2}	upper percolation threshold
P	deformation hardness (Brinell hardness)
$P_{\max}$	maximum deformation hardness at $\epsilon \to 0$
p_s	site occupation probability
q	scaling exponent
Q	cumulative amount of drug released per unit exposed area
S	scaling factor
S'	scaling factor
t	scaling exponent
β	tablet property as defined in Eq. (21)
δ	diameter of particle; size of primary particle
ϵ	total porosity of the empty matrix
ϵ_o	initial porosity
ϵ_c	critical porosity of the matrix
μ	conductivity exponent
ρ_c	percolation threshold
$\rho_{\max}$	maximal relative density where the percolation power law still is valid
ρ_p	poured relative density

ρ_r relative density
ρ_π percolation threshold of the pore network
ρ_r^* site percolation threshold, relative density of the "first" dense tablet
ρ_t tapped relative density
ρ_0 percolation threshold
σ_t tensile strength
$\sigma_{t\,max}$ maximum tensile strength at $\epsilon \to 0$
τ scaling exponent; tortuosity of the matrix

REFERENCES

1. D. Stauffer, *Introduction to Percolation Theory*, Taylor and Francis, London and Philadelphia, 1985.
2. B. B. Mandelbrot, *The Fractal Geometry of Nature*, Freeman, San Francisco, 1982.
3. H. Leuenberger, B. D. Rohera, and C. Haas, *Int. J. Pharm. 38*: 109 (1987).
4. L. E. Holman, and H. Leuenberger, *Int. J. Pharm. 46*: 35 (1988).
5. H. Leuenberger, L. E. Holman, M. Usteri, and S. Winzap, *Pharm. Acta Helv. 64*: 34 (1989).
6. D. Blattner, M. Kolb, and H. Leuenberger, *Pharm. Res. 7*: 113 (1990).
7. L. E. Holman and H. Leuenberger, *Powder Technol. 64*: 233 (1991).
8. H. Leuenberger, J. D. Bonny, and M. Usteri, Proc. Second World Congress Particle Technology, Kyoto, Japan, 1990.
9. M. Usteri, J. D. Bonny, and H. Leuenberger, *Pharm. Acta Helv. 65*: 55 (1990).
10. H. Leuenberger, and R. Leu, *J. Pharm. Sci. 81*: 976 (1992).
11. J. D. Bonny, and H. Leuenberger, Proc. Int. Symp. Control. Rel. Bioact. Mater., 18, Controlled Release Society, Inc., 1991.
12. J. D. Bonny, and H. Leuenberger, *Pharm. Acta Helv. 66*: 160 (1991).
13. D. Stauffer, A. Coniglio, and M. Adam, *Adv. Pol. Sci. 44*: 103 (1982).
14. W. O. Smith, P. D. Foote, and P. F. Busang, *Phys. Rev. 34*: 1272 (1929).
15. F. Ehrburger, S. Misono, and J. Lahaye, *Conducteurs granulaires, théories, caractéristiques et perspectives*, journée d'études Oct. 10, Paris, Textes de communication, pp. 197–204, 1990.
16. H. Leuenberger, *Int. J. Pharm. 12*: 41 (1982).
17. R. W. Heckel, *Trans. Metall. Soc. AIME, 221*: 671 (1961).
18. T. Higuchi, *J. Pharm. Sci. 52*: 1145 (1963).
19. P. J. Sherrington, and R. Oliver, *Granulation*, Heyden, London, 1981, p. 34.
20. B. Luy, *Vakuum-Wirbelschicht*, Ph.D. thesis, University of Basel, Basel, 1991.
21. B. Luy, P. Hirschfeld, and H. Leuenberger, *Drugs Made in Germany, 32*: 68 (1989).
22. H. Leuenberger, B. Luy, and P. Hirschfeld, Proc. Preworld Congress Particle Technology, Gifu, Japan, 1990.

7

Mechanical Strength

Peter N. Davies* and J. Michael Newton

School of Pharmacy, University of London, London, United Kingdom

I. INTRODUCTION

The determination of the mechanical strength of pharmaceutical compacts is carried out for several reasons:

1. As an in-process control to ensure that tables are sufficiently strong to withstand handling yet reman bioavailable
2. To assist in obtaining a fundamental understanding of compaction mechanisms
3. To aid in the characterization of the mechanical properties of the compacted material

For many years the strength of pharmaceutical compacts has been determined in terms of the force required to fracture a specimen across its diameter. The fracture load obtained is usually reported as a "hardness value." This is an unfortunate use of a term which has a specific meaning in material science, associated with indentation. Mechanical strength would be a better term to use. Such a test does not take into account the mode of failure or the dimensions of the tablet; i.e., the result is not a fundamental property of the compact. An ideal test would allow comparison to be made

---

**Current affiliation*: Roche Products, Welwyn Garden City, Hertfordshire, United Kingdom,

between samples of different shapes or sizes. An essential feature of such a test is that the geometry of the specimens and the loading must be such that a calculable stress state prevails at the section where fracture occurs so that the fracture stress can be readily calculated from the fracture load.

II. DIRECT TENSILE TESTING

Pharmaceutical powder compacts tend to be brittle, that is fracture is not preceded by significant permanent deformation. For this reason the simple tensile specimen is not ideal for these materials and is rarely used. Tensile tests have been performed on brittle materials, such as gypsum, by preparing dumbbell-shaped specimens. The enlarged ends are held in special grips and an axial tensile load applied. The results of such uniaxial tests are open to question. Photoelastic studies show large stress concentrations in the gripped portions of the specimen which can cause fracture elsewhere than in the central cross section. In tests with concrete conglomerates, the fracture usually occurs at the weakest point between the grips and the central cross section [1]. Strength calculations using this testing procedure neglect these stress concentrations. Also, small misalignments in the grips can add a bending component to the applied tensile stress. Ductile materials would correct this problem by plastic flow, without significantly affecting the results. Brittle materials have little or no capacity to flow plastically and the bending stress can seriously lower the measured strength [2]. In the case of gypsum the material can be introduced into the mold while in a fluid state before it hardens, making the preparation of samples relatively simple. The preparation of compacted powder specimens would be problematic, making the method unsuitable for the testing of pharmaceutical compacts.

An axial tensile strength test has been used to determine the tensile strength of round plane-faced tablets [3]. The tablets were fixed to a pair of platens by a cyanoacrylate adhesive and strained in tension until fracture occurred. No analysis of the stress state induced by this method is reported.

III. FLEXURAL TESTS

In the flexural test or bending test a specimen in the form of a parallel beam is subjected to three- or four-point bending and the maximum tensile stress estimated from the load at fracture. A feature of such a test is that under the correct conditions of loading the specimen will be subjected to a pure longitudinal tensile stress along a line on the opposite surface to that on which the load is applied.

In general for a beam subjected to bending the tensile fracture stress, σ_f, can be calculated from the following expressions [4]:

$$\sigma_f = \frac{My_{max}}{I} \tag{1}$$

where

M = bending moment at fracture
y_{max} = transverse coordinate measure in the plane of bending (Fig. 1)
I = second moment of bending

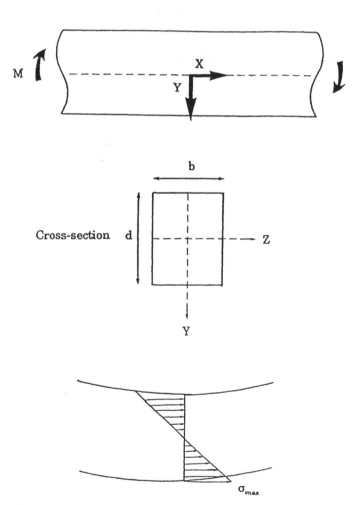

Fig. 1 Diagrammatic representation of the stresses induced by beam bending tests.

For a beam of rectangular cross section, width b, and depth d,

$$I = \frac{bd^3}{12} \tag{2}$$

and

$$y_{\max} = \frac{d}{2} \tag{3}$$

The maximum bending moment, M, is dependent on the loading configuration used. In practice, one of two loading models is used: three-point loading or four-point loading (Fig. 2).

For the symmetrical three-point loading configuration the maximum bending moment is

$$M = \frac{WL}{4} \tag{4}$$

where W is the fracture load and L is the distance between supports. Thus at failure the tensile fracture stress is

$$\sigma_f = \frac{WL}{4} \cdot \frac{d}{2} \cdot \frac{12}{bd^3} \tag{5}$$

leading to

$$\sigma_f = \frac{3WL}{2bd^2} \tag{6}$$

The maximum bending moment obtained using the four-point loading configuration is

$$M = \frac{Wa}{2} \tag{7}$$

thus at failure,

$$\sigma f = \frac{Wa}{2} \cdot \frac{d}{2} \cdot \frac{12}{bd^3} \tag{8}$$

leading to

$$\sigma_f = \frac{3Wa}{bd^2} \tag{9}$$

Modifications for asymmetric cross sections can be readily introduced.

3 point bending

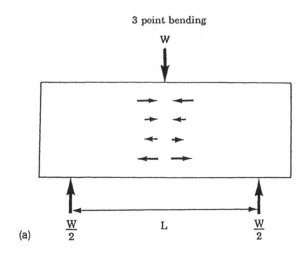

(a)

4 point bending

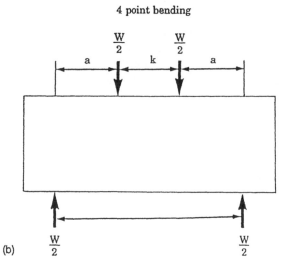

(b)

Fig. 2 Loading configurations for beam bending tests: (a) three-point bending, (b) four-point bending.

Four-point loading is usually considered to be preferable since a region of constant bending moment is obtained between the two inner loading points. The stress distribution in the specimen is nonuniform, varying from zero at the neutral axis to a maximum at the outer edge surface. This accentuates the effects of surface conditions on the measured strength, and results obtained using this test are considerably higher than the true tensile strength [5]. Beams of rectangular cross section prepared from pharmaceutical materials have been tested using this method to characterize the material properties of compacted powders [6, 7]. Rectangular beams are not a conventional tablet shape and the density distributions developed during beam formation are unlikely to be the same as those produced in a right circular cylinder [8].

The three-point flexure test has been applied to plane-faced round tablets [9]. This was a manually operated test, which meant the application of stress was variable and operator dependent. While the design of the test produced a bending moment there was also likely to be a large contribution of shear stresses at failure.

David and Augsburger [10] also used the three-point flexure to measure the tensile strength of round tablets. One problem with such an approach is that the dimensions of the tablets are unsuitable, generally being too stubby to allow a valid application of the equations derived from classical theories of bending; shear stresses would be important in such specimens. It is unlikely that the tensile fracture stresses obtained by diametral compression and flexure will be equal [11]. It is, therefore, surprising that David and Augsburger [10] found results that correlated very well. This could be explained by the fact that the mathematical analysis was incomplete. Stanley and Newton [12] stated that some of the mathematical analysis was based on incorrect assumptions.

Capsule-shaped tablets have been tested using a three-point flexure test [13]. The breaking load was used to express the strength of the compacts and no attempt was made to calculate the tensile stresses. Stanley and Newton [12] calculated the theoretical tensile stresses induced by the three-point flexure tests on capsule specimens. It was concluded that a reasonable approximation of the tensile stresses could be obtained using the equation

$$\sigma_x = \frac{3WI}{2d^2} \left[\frac{6 + 2h_c}{6A_c + bd} \right] \tag{10}$$

where A_c is the area of the curved segment of the capsule, h_c is the height of that segment, and other terms are as in Eqs. (2) and (4).

The flexure test has been used to characterise the strength of compacts prepared from pharmaceutical materials. The test does not, however, lend itself to the testing of circular compacts or thin square compacts where the shear stresses significantly affected the results obtained.

IV. DIAMETRAL COMPRESSION TEST

A. Analytical Solution

A method of testing brittle materials that does not suffer from the previous disadvantages is the diametral compression test. This consists of a simple flat-faced disc specimen which is subjected to two diametrically opposed point loads. The diametral compression test was developed independently at the same time by Barcellos and Carneiro [14] in Brazil and by Akazawa [15] in Japan and is referred to as the "Brazilian disc" or indirect tensile test, the indirect referring to the fact that a tensile fracture is obtained from compressive loading.

The test has been used to measure the tensile strength of concrete [11], coal [5], gypsum [16], and pharmaceutical compacts [17]. A complete analytical solution exists for the stress state induced by this loading configuration.

Assuming plane stress and considering a circular disc with concentrated loads on the diameter, three general equations can be used to express the stress conditions at all points within the disc (Fig. 3) [18]:

$$\sigma_x = \frac{-2P}{\pi t} \left(\frac{(R-y)x^2}{r_1^4} + \frac{(R+y)x^2}{r_2^4} - \frac{1}{D} \right) \tag{11}$$

$$\sigma_y = \frac{-2P}{\pi t} \left(\frac{(R-y)^3}{r_1^4} + \frac{(R+y)^3}{r_2^4} - \frac{1}{D} \right) \tag{12}$$

Considering points between OC on the horizontal X axis where $y = 0$,

$$\tau_{xy} = \frac{2P}{\pi t} \left(\frac{(R-y)^2 x}{r_1^4} - \frac{(R+y)^2 x}{r_2^4} \right) \tag{13}$$

$$r_1 = r_2 = \sqrt{x^2 + R^2} \tag{14}$$

Both stresses vanish at the circumference and reach maximum values at the center. The stresses at the center are

$$\sigma_x = \frac{2P}{\pi t D} \tag{15}$$

$$\sigma_y = \frac{-6P}{\pi t D} \tag{16}$$

Therefore, the compressive strength needs to be at least three times the tensile strength to ensure a tensile failure.

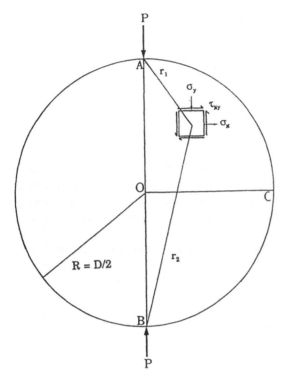

Fig. 3 Diagram of stress distribution across loaded diameter for a cylinder compressed between two line loads.

Along the Y-axis AB where $x = 0$,

$$r_1 = R - y \tag{17}$$

$$r_2 = R + y \tag{18}$$

$$\tau_{xy} = 0 \tag{19}$$

$$\sigma_x = \frac{2P}{\pi t D} \tag{20}$$

$$\sigma_y = \frac{-2P}{\pi t} \left(\frac{2}{D-2y} + \frac{2}{D+2y} - \frac{1}{D} \right) \tag{21}$$

This indicates that tensile failure of the specimen could start at any point along the vertical AB axis due to the even distribution of tensile stresses.

The compressive stress on this axis increases from $\sigma_y = -6P/\pi Dt$ at the center to $\sigma y = \infty$ at the loading points. With a concentrated load, the specimen will fail at the loading points due to the compressive stresses and not in the central part of the specimen due to the tensile stress.

B. Effect of Loading Conditions on the Diametral Compression Test

In practice, the theoretical condition of lines of contact between specimen and platens cannot be achieved; instead the load is distributed over areas of contact. Comparison of Fig. 4a, b shows, that provided the area of contact is small in relation to the diameter of the disc, this only affects the stress distribution near the ends of the loaded diameter and the tensile stress is $2P/\pi Dt$ over a large fraction of the loaded diameter [2].

It has been shown that tensile stresses can be held uniform across a reasonable proportion of the loaded diameter if the width of the contact area does not exceed one-tenth of the specimen diameter [19]. The solution for the tensile stresses can only be used to calculate the tensile strength of a specimen provided it fails in tension. This is characterized by failure along the loaded diameter (Fig. 5). The fracture does not always extend right to the ends of the diameter. A second fracture pattern, the triple-cleft failure has also been identified as being failure in tension [2]. Compressive failure occurs at the specimen surface immediately beneath the loads where the compressive stresses are at a maximum and appears as local crushing. If this crushing is not extensive it may only result in an increase in the area over which the load is applied so that ultimate failure may be in shear or tension.

As shown in Fig. 4 the maximum shear stresses occur beneath the surface. The exact location and magnitude of these stresses depend on the distribution of applied loads. Shear failures start at an angle to the loaded diameter and are followed by secondary failures causing an irregular fracture pattern. Thus, the validity of using the diametral compression test under a given set of conditions to determine a tensile strength from a fracture load can be easily assessed by examining the specimen fragments after failure [20]. The mode of failure is affected by the width of the contact area. Some materials such as concrete [1] and ceramics [2] require relatively soft packing pieces to be placed between the platens and specimens to ensure adequate load distribution and therefore failure in tension. Other specimens such as autoclaved plaster deform sufficiently at the points contact to ensure that failure occurs in tension [16]. The failure is also influenced by the type and extent of padding. The effect of padding can be predicted by mathematical study, although experimental evidence indicates that such predictions are only general at best [1].

The material properties of the platen will modify the stress distribution within the disc. Photoelastic studies using epoxy resin discs showed

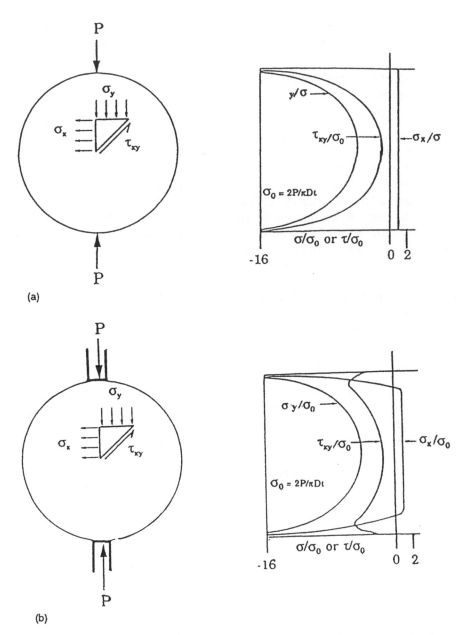

Fig. 4 (a) Diagram of stress distribution across loaded diameter for a cylinder compressed between two line loads (b) Diagram of stress distribution across loaded diameter for a cylinder compressed between two platens with uniform contact pressures.

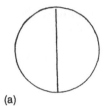
(a)

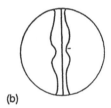
(b)

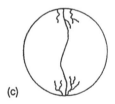
(c)

Fig. 5 Types of failure induced by the diametral compression test: (a) simple tensile failure, (b) triple-cleft failure (tensile failure), (c) failure due to shear at platen edges.

that platens made of different materials such as steel, rubber, and cardboard produced different stress patterns [20]. Experimentally determined fracture strengths indicated that the lower the elastic moduli of the platens or padding the higher was the apparent tensile strength of the disc and the greater the variance of the results. This may be explained by the increased contact area causing a reduction in the volume of material subjected to the maximum tensile stress. Such effects appear to be material and condition dependent as Wright [11] and Fell and Newton [17] reported a decrease of variability with an increase in padding material.

The shape of the platens is also of importance in determining the outcome of the diametral compression test. Concave semicircular platens have been used to induce tensile failure in graphite specimens where formerly only compressive failure occurred. A modified solution for the stresses produced by the platens is required in this case [21].

C. Effect of Loading Rate on Diametral Compression Test

The rate at which the load is applied to specimens can affect the results obtained. Increasing the rate of loading of concrete cylinders resulted in higher observed tensile strengths [1]. Tablets containing lactose and microcrystalline cellulose were tested at loading rates corresponding to crosshead movements of 0.05 to 5 cm min^{-1} [22]. An increase in the loading rate produced a significant increase in the breaking strength, although the standard deviation of replicate values was apparently unaffected. It was also concluded that discrepancies in tensile strength values obtained by different testing instruments may be partially attributed to differences in rates of loading.

The effect of the loading rate on the value of tensile strength obtained for compacts of acetylsalicylic acid and ammonium sulfate has been investigated [23]. Increasing the loading rate produced a decrease in the tensile strength obtained for compacts of both materials. For ammonium sulfate, the change in strength with strain rate was a linear function of the log of the strain rate. The response of a material to loading rate changes appears to be dependent on the mechanism of failure. Variations in strain rate can lead to changes in the failure mode. At low strain rates some materials may fail in a ductile manner. The faster the strain rate the more likely the failure is to be brittle in nature.

D. Tensile Strength Testing of Non-Plane-Faced Circular Compacts

The diametrical compression test has also been used to measure the tensile strength of round convex-faced compacts. The loads required to produce tensile failure of flat-faced and deep convex tablets were compared to produce an empirical equation for tensile strength [24]. The method requires that a linear relation exists between the breaking load and the compaction pressure for the convex tablets and does not allow the isolated determination of the tensile strength of convex-faced tablets.

An equation for the tensile strength of convex-faced tablets was calculated utilizing the central thickness of the tablet [25]. The equation was a correction for the difference in the cross sectional area between a convex-faced and plane-faced specimen which has no theoretical basis in relation to elastic theory.

The stresses induced in convex-faced specimens by the diametral compression test have been examined photoelastically [26]. Formulae were derived that enabled the tensile strength of a convex-faced specimen to be determined from a knowledge of the fracture load and the specimen dimensions. The formulae were verified by comparing the loads required to produce tensile failure of flat-faced and convex-faced gypsum specimens [27]. The tensile strength of a convex specimen may be determined from the fracture load by the equation:

$$\sigma_f = \frac{10P}{\pi D^2} \left(2.84 \frac{t}{D} - 0.126 \frac{t}{C_L} + 3.15 \frac{C_L}{D} + 0.001 \right)^{-1} \quad (22)$$

where

P = fracture load of the convex-faced disc
D = diameter
C_L = cylinder length
t = disc overall thickness

This approach was used to study the influence of the curvature on the mechanical strength of aspirin tablets, when it was established that the "normal" concave punches gave the optimum strength [28].

The diametrical loading of flat-faced discs with a groove across the diameter of one face has been examined photoelastically [29]. The tensile stress distribution was affected by the depth of the groove and also the orientation of the groove with respect to the loaded diameter.

V. LINE LOADING OF RECTANGLES

An analysis of the line loading of rectangles was performed by using both point loading and loading with indentors of different widths [30] (Fig. 6). The analysis showed that in general the average tension σ_x along the loaded line was given by

$$\sigma_x = \frac{2P}{\pi Bt} \quad (23)$$

where B is the length of the line of loading. This expression is independent of the proportions of the rectangle, with the middle portion of the loaded line subjected to approximately uniform normal stress.

The maximum tensile stress is the important quantity in determining failure; for point loading this tensile stress was calculated to rise to a peak

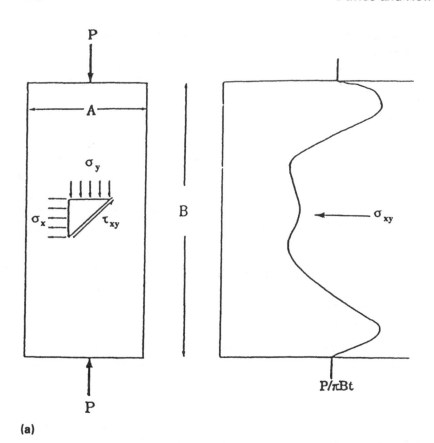

(a)

Fig. 6 (a) Diagram of stress distribution across loaded line for a rectangle ($A/B = 0.4$) compressed between two line loads. (b) Diagram of stress distribution across loaded line for a rectangle ($A/B = 1$) compressed between platens ($a/A = 1$).

value near to the loading platens, before becoming compressive in the immediate vicinity of the indentor (Fig. 6). However, for broad indentors ($a_i/B = 0.1$, where a_i = width of indentor), there was little if any rise in the tensile stress along the loaded line (Fig. 6). The stress remained uniform for a considerable portion of the section before becoming compressive in the loading zone. Hence, for the latter condition, the tensile strength σ_f would be given by

$$\sigma_f = \frac{2P}{\pi Bt} \tag{23'}$$

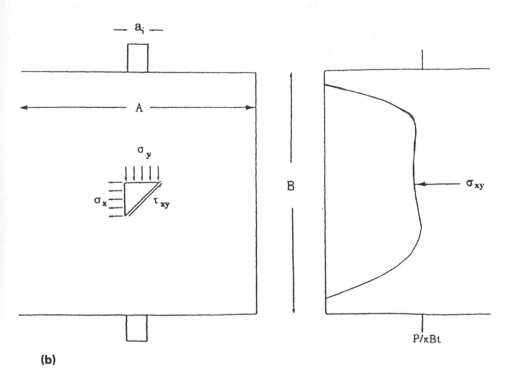

(b)

The tensile strength of concrete cubes was assessed experimentally by using an indentor method [31]. The indentors were used with padding and, under these conditions, good correlation was found between the determined tensile strength of the cubes using Eq. (23′) and the determined tensile strength of cylinders using Eq. (20). Subsequently, a finite element method was used [32] to assess the indentor method. Equation (23′) was shown to be a good approximation of the tensile strength of the material. The indentor test was examined by photoelasticity for squares [5]. The width of the indentor, a_i, was varied and the resulting stress state calculated. As the indentor width was increased (from $a_i/B = 0.1$ to $a_i/B = 0.75$) with the load kept constant, the tensile stress along the loaded line decreased. However, when the applied stress was kept constant the maximum tensile stress reached its greatest value at $a_i/B = 0.4$.

A theoretical solution for indentor loading was attempted by Jordan and Evans [33]. Four approximate solutions for the elastic stresses in a square were obtained. These solutions were compared with the photoelastic analysis of Berenbaum and Brodie [5]. Only one solution was found to be in approximately good agreement with the experimental results; the other theoretical results were markedly different. The inadequacies of theoretical solutions can be attributed to the differences in approximating the boundary conditions.

A different theoretical solution for the indentation test on squares was attempted by Sundara and Chandrashekhara [34]. The stresses along the central loaded axis were evaluated by considering the shape of the pressure distribution under the indentor. Again difficulties were found in the theoretical modeling of the stress distribution.

The photoelastic results of Berenbaum and Brodie [5] are used in preference to the theoretical results above when a reliable value for a tensile strength is required [35]. However, the experimental results of Berenbaum and Brodie [5] and the theoretical solution of Goodier [30] give approximately the same tensile strength for indentor widths in the region $a_i/b = 0.1$ to 0.2. Therefore, when these particular indentor widths are used, Eq. (23') can be used to determine the tensile strength.

Cube-shaped specimens have also been tested by loading across a diagonal (i.e., applying the load to opposite edges rather than faces). Analysis of this test configuration by the finite element method indicated that the maximum tensile stress along the loaded axis is approximately 0.77 of that predicted by Eq. (20) [32].

VI.   LINE LOADING OF ELLIPSES

A theoretical analysis for the stress distribution in an elliptical disc due to concentrated axial loading showed that the tensile stresses were a function of the width-to-length ratio and were at a maximum for a circular disc [36]. The tensile stress along the loaded axis, σ_x, was calculated to be uniform for the majority of the central portion, but rises to a maximum value at the edge of ellipse at the loading point. The stress σ_x was not calculated to become compressive under the loading zone, this being in contrast to the experimentally determined stress distribution of Frocht [37], indicating a limitation of the analysis of Brisbane [36].

A finite element method was used [38] to predict the stress distribution when ellipses and rectangles were loaded along their central axis. Photoelastic results of Phillips and Mantei [39] and the theoretical results of Brisbane [36] and of Goodier [30] were used to demonstrate the validity of the solutions obtained.

The analyses showed that any elliptical disc with a width-to-length ratio of 0.9 to 1.1 would have the entire Y-axis loaded to 95% or more of the maximum tensile stress. This was because of the nonuniform stresses produced along the Y-axis, particularly once the shape of the ellipse departed significantly from that of a circle.

VII. STATISTICAL TREATMENT OF TENSILE STRENGTH DETERMINATIONS

The stress solutions presented for all the aforementioned tensile strength determinations assume the specimen is isotropic. Such an assumption is unlikely to be true for any pharmaceutical compact as studies have shown that there are distinct density distributions within a compact [40]. Furthermore, within a brittle material there are structural and material defects in the form of interstitial cavities, fractured particles, and interparticular boundaries. Since these defects are randomly distributed throughout the material and are of random severity, there is an inherent variability in the strength of nominally identical brittle specimens. This variabilitly requires that tensile strength test results for brittle materials be treated statistically. The Weibull probability distribution is often used for this purpose [41]. An important attribute of the Weibull hypothesis is that the mean fracture stress of a batch of nominally identical compacts is a characteristic of the specimen and not of the material itself and it is size dependent. The larger the specimen the more likely it is that it will contain a flaw of a given severity, and consequently the smaller will be the mean fracture stress of a batch of such specimens. The predicted relationship between the mean fracture stresses and volumes of two batches A and B is

$$\frac{\sigma_{f_A}}{\sigma_{f_B}} = \left(\frac{V_B}{V_A} \right)^{1/m} \tag{24}$$

where V is the specimen volume and m is the Weibull modulus. This equation assumes that the materials of the different batches are physically identical, which is unlikely to be the case for two batches of tablets. Stanley and Newton [42] prepared six sizes of lactose compacts and compared the experimental mean fracture stresses obtained with those predicted by the Weibull distribution function. For each batch the strength variability was satisfactorily represented by the Weibull distribution, although the effect of size could not be predicted by Eq. (24). A similar conclusion was obtained when comparing two sizes of sodium chloride tablets [43].

VIII. MECHANICAL PROPERTIES RELATED TO STIFFNESS: DETERMINATION OF YOUNG'S MODULUS

Tensile strength has been the most common expression used to describe the strength of compacts, but tensile strength values alone are not sufficient to fully characterize a material's mechanical properties. The stress-strain behavior of a material has been used to measure the toughness and stiffness of materials.

Young's modulus, E, defined as the ratio of stresses, σ, to strain, ϵ, describes the stiffness of a material. Young's modulus of Avicel PH-101 has been determined using the four-point flexure test [44]. In such an arrangement the tensile stress and the associated strain are

$$\sigma = \frac{3Wa}{2bd^2} \tag{25}$$

$$\epsilon = \frac{4\delta d}{k^2} \tag{26}$$

where δ is the vertical displacement of the midpoint of the beam and k is the distance between the loading points on the upper surface of the beam. It was demonstrated that Young's modulus increases as the porosity decreases. Determination of Young's moduli for a range of pharmaceutical excipients demonstrated that there is no relationship between Young's modulus and the tensile strength of a material [6]. To enable comparison to be made of Young's modulus values of porous materials. Sprigg's equation has been used to determine empirically Young's modulus at zero porosity [45]:

$$E_S = E_0 \exp(-cp) \tag{27}$$

where

E_0 = Young's modulus at zero porosity
E_S = specimen's Young's modulus value at porosity p
c = material constant

The ability to calculate Young's modulus from the molecular structure via the cohesion energy density was established by Roberts et al. [46] for a series of organic solids. Attempts to predict Young's modulus of binary powder mixtures, however, were less succesful [47], although the approach proposed by Nielsen [48] appeared to be the most appropriate.

IX. FRACTURE MECHANICS

The four-point bending test has been used to characterize the fracture toughness of Avicel PH-101 [7]. For a crack to grow under static loading the stress must be high enough to initiate fracture and the energy released must be at least as much as that required to form the new surfaces. The stress field near the crack tip will be proportional to the general stress in the material and the square root of the crack length and is called the stress intensity factor (K). This is related to the rate of strain-energy release, G, with crack growth by the elastic modulus of the material:

$$K^2 = EG \tag{28}$$

The crack will grow when the stress has been raised sufficiently for K and G to reach their critical values K_{IC} and G_{IC}. K_{IC} can be determined by carrying out four-point bending tests on beam-shaped specimens in which a notch has been made in the surface at the point of maximum tension. The value can be calculated from the dimensions of the beam, the maximum load, and the notch depth when this load is reached.

Similar studies have been performed by York et al. [49], who obtained lower values of K_{IC} for Avicel PH-101 than those reported by Mashadi and Newton [7]. The differences were attributed to Avicel PH-101 being a material with a rising crack growth resistance curve. For such materials the method of crack induction and notch geometry become critical. The fracture mechanics approach has been used to characterize the mechanical properties of microcrystalline cellulose [50].

Roberts and Rowe [51] described a method of determining the critical stress intensity factor using circular compacts into which a radial crack was cut. Values for K_{IC} were obtained by two methods: (1) edge opening in which the cracked compacts were gripped in the jaws of a tensometer and pulled apart and (2) a diametral compression test performed with the crack positioned along the line of loading. Of the two techniques the edge opening was preferred since it gave the most stable crack propagation.

X. WORK OF FAILURE

The diametral compression test has also been used to measure the "toughness" of compacts [52]. In this case the change in length of the loaded diameter was monitored throughout testing. Values were recorded for force and the corresponding change in diameter. The product of these values, the area under a force-displacement curve, was called the work of failure, W_f:

$$W_f = \int P \, dx \tag{29}$$

where dx is the rate of change of distance x.

Tablets with a high work of failure were considered to deform plastically under compressive loading, thus requiring a relatively large platen displacement to produce failure while brittle materials require only a small displacement to produce failure.

Since most specimens fail in tension during the diametral compression test a further expression, the normalized work of failure (NWF), was introduced to convert the applied load, P, to a tensile stress [53]:

$$\text{NWF} = \frac{2}{\pi Dt} \int P \, dx \tag{30}$$

As the specimen fails in tension it might be argued that the increase in the transverse diameter would be a more appropriate measurement to determine the work done. This technique is significant because it demonstrates that materials do behave differently under load, exhibiting different degrees of deformability. This may be due, in part, to the hardness of the material rather than the inherent strength of the compact. Work of failure measurements have been extended to the flexure test [54] and axial tensile strength tests [55].

XI. BRITTLE FRACTURE INDEX

The work-of-failure test has been used to distinguish between ductile and brittle materials. A further method used to measure the brittleness of materials is the brittle fracture index (BFI) [56]. Two sets of square compacts are prepared, one set containing a small circular hole along the axis. The compacts are tested by an indenter test with a platen to compact width ratio of 0.4, and the tensile strengths of the compacts with and without the central hole are compared. Under the test conditions the hole acts as a stress concentrator. Elasticity theory predicts that the stress concentration factor is approximately 3.2 for a hole in an isotropic solid. However, for most pharmaceutical materials the ratio of the tensile strengths of the two types of compact is less than 3. This is believed to be because of the relief of the highly localized stresses by plastic deformation, i.e., the material is not completely brittle. Thus, the ratio of the tensile strength of a compact without a hole, σ_f, to the tensile strength with a hole, σ_{fo}, may indicate its ability to relieve stress by plastic deformation. Based on this the brittle fracture index has been defined:

$$\text{BFI} = 0.5 \left(\frac{\sigma_f}{\sigma_{fo}} - 1 \right) \tag{31}$$

Roberts and Rowe [57] have used the BFI to measure the brittleness of circular compacts using the diametral compression test and have obtained values in good agreement with those of Hiestand and Smith [56].

XII. INDENTATION HARDNESS

While tensile strength measurements describe the global strength of a specimen, indentation hardness describes the "local" plasticity of a material. Hardness may be defined as the resistance of a solid to local permanent deformation, and is usually measured by a nondestructive indentation or scratch test.

The most widely used methods in determining hardness are static indentation methods. These involve the formation of a permanent indentation on the surface of the material to be examined. Normally the diameter of the impression is determined and, from it, the hardness is calculated by means of the formula

$$BHN(Q) = \frac{2W}{\pi D_i (D_i - \sqrt{D_i^2 - d_i^2})} \tag{32}$$

where

W = applied load
D_i = diameter of the spherical indenter
d_i = diameter of the indentation (Fig. 7)

The Brinell hardness number (BHN) is not a constant for a material but depends on the load and diameter of the indenter [58]. The method has been used to determine the hardness at various points across the diameter of aspirin compacts [59] and direct compression excipients [60]. Studies on the surface hardness distribution over tablet faces with different face curvatures indicated that, as the degree of curvature increased, the hardness of the outer portions of the compact increased relative to the center of the compacts [61].

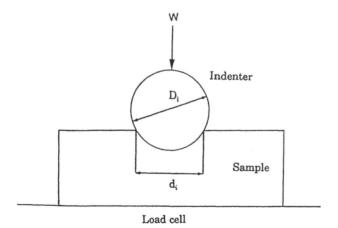

Fig. 7 Diagram of the indentation hardness test.

Hardness tests have been used to measure the elastic recovery. Under load the indenter, radius of curvature r_1, penetrates to a depth h_1 giving an impression of diameter d_i. When the load is removed, elastic recovery occurs and the radius of curvature of the impression increases to r_f, its depth decreasing to h_2. The change $h_1 - h_2 = h$ is a measure of the elasticity. Using this method it is possible to derive a value for the elastic modulus of elasticity [59].

Hardness values of tablets have also been determined using dynamic tests. In dynamic methods, either a pendulum is allowed to strike from a known distance or an indenter is allowed to fall under gravity onto the surface of the test material. The hardness is determined from the rebound height of the pendulum or the volume of the resulting indentation.

The volume of the indentation is directly proportional to the kinetic energy of the indenter. It therefore implies that the material offers an average pressure of resistance to the indenter equal to the ratio

$$\frac{\text{Energy of impact}}{\text{Volume of indentation}}$$

This ratio has the dimensions of pressure and is sometimes referred to as the dynamic hardness number. Hiestand et al. [35] have used the dynamic method to characterize a range of pharmaceutical materials. The indentation hardness, Q, was divided by the tensile strength of the tablet to form a bonding index;

$$\text{Bonding index} = \frac{\text{Indentation hardness}}{\text{Tensile strength}}$$

The objective of the bonding index is to estimate the survival, during decompression of areas of true contact that were established at maximum compression stress.

There are several disadvantages associated with the measurement of the indentation hardness of a specimen. Pharmaceutical compacts consist of particles compacted with voids between them. It is very difficult to decide whether an indentation has been made into a particle, a void or a combination of the two, raising doubts as to what the indentation figure is describing. Furthermore, if the elastic recovery is time dependent, it is difficult to assess when the recovery measurement should be made.

XIII. PRACTICAL STRENGTH TESTING

The choice of method used to determine the mechanical strength of pharmaceutical compacts should be related to the information required. If required for in-process control, it is preferable to provide a system which is reproducible and accurate and which possesses sufficient sensitivity to identify differences between tablets. Several types of instrument are available. Unfortunately, most simply crush the tablet and do not allow the assessment of the mode of failure of the tablet. Such devices will be very insensitive and hence may be apparently reproducible. The variable mode of failure could, however, allow such variations which are not directly related to the properties of the tablet. Simple crushing of tablets of complex shapes can be misleading. If good controlled breaking with a consistent known mode of failure can be induced, the need for complex evaluation of poor-quality systems, as described by Bavitz et al. [62], would be eliminated. There are instruments which can produce this consistent mode of failure and are adaptable for different shapes. When applied to the determination of a fundamental understanding of compaction mechanisms or characterization of the mechanical properties of materials, then only tests which are fundamentally sound should be applied. Even here there are still problems. To undertake tests, the powders have to be compacted. The process of preparation is influenced by the mechanical properties of the material. Hence the final test procedure is an assessment of the mechanical properties of a specimen compounded by its method of formation. For example, the tensile strength of different quantities of lactose at a range of formation pressures was found not to be independent of the sample weight [63]. Subsequent characterization of the tablet formation process by the area under the pressure/time curve rather than the maximum formation value provided a better comparison of tensile strength values for the different quantities of material [64].

There does not appear as yet to be any single method of material characterization which adequately describes the ability of material to form tablets and hence a range of properties, as suggested by York [65], would appear to be more appropriate.

SYMBOLS

a	horizontal distance between upper and lower loading points in four-point bending test.
a_i	indenter width
A	width of rectangle
A_c	area of the curved segment of a capsule-shaped compact
b	width of beam
B	length of the line of loading of rectangle specimen
BFI	brittle fracture index
BHN	Brinell hardness number

c	constant
C_L	cylinder length
d	depth of rectangular beam
d_c	contact width
d_i	indentation diameter
dx	change of distance x
D	diameter of disc specimen
D_i	indenter diameter
E	Young's modulus
E_0	Young's modulus at zero porosity
E_S	Young's modulus of specimen
G	strain energy release
G_{IC}	critical strain energy release
h_c	height of the curved segment of a capsule-shaped specimen
h_1	initial depth of penetration of indenter
h_2	final depth of penetration of indenter
h	$h_1 - h_2$
I	second moment of bending
k	distance between the loading points on upper surface of beam
K	stress intensity factor
K_{IC}	critical stress intensity factor
L	distance between lower supports in the three-point bending test
m	Weibull modulus
M	bending moment
NWF	normalized work of failure
p	porosity
P	applied load
Q	deformation hardness
r_1 r_2	distance from loading points to point within disc specimen
r	initial radius of indenter
r_f	final radius of curvature of indentation
R	radius of disc specimen
t	thickness of disc specimen
V_A	volume of specimen A
V_B	volume of specimen B
W	applied load, tensile test, indentation test
W_f	work of Failure
x	distance along x axis

y_{max}	transverse coordinate measure in the plane of bending
y	distance along y axis
δ	vertical displacement of the midpoint of beam during beam bending tests
ϵ	strain
σ_f	tensile failure stress
σ_{fO}	tensile fracture of specimen with a hole
σ_{fA}	tensile strength of specimen A
σ_{fB}	tensile strength of specimen B
σ_x	tensile strength
σ_y	compressive stress
$\tau_{x,y}$	shear stress
ν	Poisson's ratio

REFERENCES

1. N. B. Mitchell, *Mater. Res. Stand. 1*: 780–788 (1961).
2. A. Rudnick, A. R. Hunter, and F. C. Holden, *Mater. Res. Stand. 3*: 283–289 (1963).
3. C. Nyström, A. Wulf, and K. Malmquist, *Acta Pharm. Suec. 14*: 317–320 (1977).
4. P. Stanley, Post-Graduate School on Production Processes in Tablet Manufacture, School of Pharmacy, University of London.
5. R. Berenbaum, and I. Brodie, *Brit. J. Appl. Phys. 10*: 281–287 (1959).
6. M. S. Church and J. W. Kennerley, *J. Pharm. Pharmacol. 35*: Suppl. 43P (1983).
7. A. B. Mashadi and J. M. Newton, *J. Pharm. Pharmacol. 39*: 961–965 (1987).
8. B. Charlton and J. M. Newton, *Powder Technol. 41*: 123–134 (1985).
9. C. J. Endicott, W. Lowenthal, and H. M. Gross, *J. Pharm. Sci. 50*: 343–346 (1961).
10. S. T. David, and L. L. Augsburger, *J. Pharm. Sci. 63*: 933–936 (1974).
11. P. J. F. Wright, *Mag. Concrete Res. 7*: 87–96 (1955).
12. P. Stanley and J. M. Newton, *J. Pharm. Pharmacol. 32*: 852–854 (1980).
13. G. Gold, R. N. Duvall, and B. T. Palermo, *J. Pharm. Sci. 69*: 384–386 (1980).
14. F. L. L. B. Barcellos and A. Carneiro, *R.I.L.E.M. Bull.* No. 13: 97–123 (1953).
15. T. Akazawa, *R.I.L.E.M. Bull.* No. 16: 11–23 (1953).
16. R. Earnshaw and D. C. Smith, *Austral. Dental J. 11*: 415 (1966).
17. J. T. Fell, and J. M. Newton *J. Pharm. Sci. 59*: 688–691 (1970).
18. J. P. Den Hartog, *Advanced Strength of Materials*, McGraw-Hill, New York, 1952.
19. R. Peltier, *R.I.L.E.M. Bull.* No. 19: 33–74 (1954).
20. E. Addinall and P. Hackett, *Civil Eng. Pub. Works Rev. 59*: 1250–1253 (1964).

21. H. Awaji and S. Sato, *J. Eng. Mater. Tech. 101*: 139–147 (1979).
22. J. E. Rees, J. A. Hersey, and E. T. Cole, *J. Pharm. Pharmacol. 22*: Suppl. 65S–69S (1970).
23. J. M. Newton, S. Ingham, and O. O. Onabajo, *Acta Pharm. Technol. 32*: 61–62 (1986).
24. J. M. Newton, G. Rowley, and J. T. Fell, *J. Pharm. Pharmacol. 24*: 503–504 (1972).
25. S. Esezebo and N. Pilpel, *J. Pharm. Pharmacol. 28*: 8–16 (1976).
26. K. G. Pitt, J. M. Newton, and P. Stanley, *J. Phys. D: Appl. Phys. 22*: 1114–1127 (1989).
27. K. G. Pitt, J. M. Newton, and P. Stanley, *J. Mater. Sci. 28*: 2723–2728 (1988).
28. K. G. Pitt, J. M. Newton, R. Richardson, and P. Stanley, *J. Pharm. Pharmacol. 41*: 289–292 (1989).
29. J. M. Newton, P. Stanley, and C. S. Tan, *J. Pharm. Pharmacol. 29*: Suppl. 40P. (1977).
30. T. N. Goodier, *Trans A.S.M.E. 54*: 173–183 (1932).
31. S. Nillson, *R.I.L.E.M. Bull.* No. 11: 63–65 (1961).
32. J. P. Davies and D. K. Bose, *A.C.I. J. 8*: 662–669 (1968).
33. D. W. Jordan and I. Evans, *B. J. Appl. Phys. 13*: 75–79 (1962).
34. K. T. Sundara and K. Chandrashekhara, *Brit. J. Appl. Phys. 13*: 501–507 (1962).
35. E. N. Hiestand, J. E. Wells, C. B. Peot, and J. F. Ochs, *J. Pharm. Sci. 66*: 510–519 (1977).
36. J. J. Brisbane, *J. Appl. Mech. Trans. A.S.M.E. 85*: 306–309 (1963).
37. M. M. Frocht, *Photoelasticity,* Vol. 2, Wiley, New York, 112–129 (1948).
38. F. J. Appl, *J. Strain Anal. 7*: 178–185 (1972).
39. H. B. Phillips and C. L. Mantei, *Expl. Mech. 9*: 137–139 (1969).
40. D. Train, *J. Pharm. Pharmacol. 8*: 745–761 (1956).
41. W. Weibull, *J. Appl. Mech. 18*: 293–297. (1951).
42. P. Stanley and J. M. Newton, *J. Powder Bulk Solids Technol. 1*: 13–19 (1978).
43. J. W. Kennerley, J. M. Newton, and P. Stanley, *J. Pharm. Pharmacol. 29*: 39P Suppl., (1971).
44. M. S. Church and J. W. Kennerley, *J. Pharm. Pharmacol. 34*: Suppl. 50P (1982).
45. J. C. Kerridge and J. M. Newton, *J. Pharm. Pharmacol. 38*: Suppl. 79P (1986).
46. R. J. Roberts, R. C. Rowe, and P. York, *Powder Technol. 65*, 139–146 (1991).
47. F. Bassam, P. York, R. C. Rowe, and R. J. Roberts, *Powder Technol. 65*: 103–111 (1991).
48. L. E. Nielsen, *J. Appl. Phys. 41*: 4626–4627 (1970).
49. P. York, F. Bassam, R. C. Rowe, and R. J. Roberts, *Int. J. Pharm. 66*: 143–148 (1990).
50. G. F. Podczeck and J. M. Newton, *Die Pharmazie 47*: 462–463 (1992).
51. R. J. Roberts and R. C. Rowe, *Int. J. Pharm. 52*: 213–219 (1989).
52. J. E. Rees and P. J. Rue, *Drug Dev. Ind Pharm. 4*: 131–156 (1978).

53. J. E. Rees, P. J. Rue, and S. C. Richardson, *J. Pharm. Pharmacol.* 2P.: Suppl. 38P (1977).
54. A. E. Moschos and J. E. Rees, *J. Pharm. Pharmacol. 38*: Suppl. 32P (1986).
55. P. J. Jarosz and E. L. Parrott, *J. Pharm. Sci. 71*: 607–614 (1982).
56. E. N. Hiestand and D. P. Smith, *Powder Technol. 38*: 145–149 (1984).
57. R. J. Roberts and R. C. Rowe, *J. Pharm. Pharmacol. 38*: 526–528 (1986).
58. H. Leuenberger and B. D. Rohera, *Pharm. Res. 3*: 12–22 (1986).
59. K. Ridgway, M. E. Aulton, and M. H. Rosser, *J. Pharm. Pharmacol. 22*: Suppl. 70S–78S (1970).
60. M. E. Aulton, H. G. Tebby, and P. J. P. White, *J. Pharm. Pharmacol. 26*: Suppl. 59P–60P (1974).
61. M. E. Aulton and H. G. Tebby, *J. Pharm. Pharmacol. 27*: Suppl. 4P (1975).
62. J. F. Bavitz, N. R. Bohidar, J. I. Karr, F. A. Restaino, *J. Pharm. Sci. 62*: 1520–1524 (1973).
63. J. M. Newton, G. Rowley, J. T. Fell, D. G. Peacock, and K. Ridgway, *J. Pharm. Pharmacol. 23*: 1955–2015 (1971).
64. J. M. Newton and G. Rowley, *J. Pharm. Pharmacol. 24*: 250–257 (1972).
65. P. York, *Drug Dev. Ind. Pharm. 18*: 677–721 (1992).

8

Tablet Surface Area

N. Anthony Armstrong

Welsh School of Pharmacy, University of Wales, Cardiff, United Kingdom

I. INTRODUCTION

The process by which a particulate solid is transformed by the application of pressure to form a coherent compact or tablet can essentially be divided into two stages: consolidation and bond formation.

1. Consolidation. As pressure is applied, the porosity of the bed of particles is reduced. Initially this is achieved by particle rearrangement, for which only a low pressure is required. Subsequently when rearrangement is effectively complete, further consolidation is achieved by particles undergoing fragmentation or deformation, or most probably both fragmentation and deformation in varying degrees, depending on the solid. Thus, for example, microcrystalline cellulose is primarily a deforming material, dicalcium phosphate dihydrate fragments, and lactose holds an intermediate position.

2. Bond formation. After particles, by whatever mechanism, have been brought into a sufficiently close proximity with each other, a coherent compact cannot be formed unless some form of bonding occurs between particles. Fuhrer [1] concluded that three types of bond are applicable to tablets: solid bridges, intermolecular forces, and mechanical interlocking.

Solid bridges have been defined as areas of physical contact between adjacent surfaces. They can occur due to melting followed by resolidification or by dissolution of solid materials followed by recrystallization [2,

3]. Measurements of conductivity of metal powder compacts indicate that only a small proportion of the total surface of particles is in contact in this intimate fashion [4].

If two surfaces are sufficiently close to each other, they will exhibit mutual attraction. It should be noted that unlike solid bridges, the particles need not necessarily be in contact with each other. Intermolecular forces include van der Waal's forces, hydrogen bonding, and electrostatic forces. The first is the most important, and they can be significant at particle separations of up to 10 nm.

The incidence and importance of mechanical interlocking obviously depends on the size and shape of the particles. Smooth spherical particles will have little tendency to interlock, whereas irregularly shaped particles might be expected to do so [5].

II. DETERMINATION OF SURFACE AREA OF COMPACTED POWDER SYSTEMS

The term "surface area" of a particulate solid can be defined in a variety of ways:

1. As the visible or "outer" surface.
2. As the external surface, which is the sum of the surface areas of all the particles, the latter being regarded as a group of nonporous units.
3. As the total area including fine structure and pores within the particles. In a compacted system, there will inevitably be an internal structure.

Surface area is quantified by the calculation of the specific surface area, which is the surface area per gram of solid.

Methods of determination such as those described below differ in the numerical values of surface area which they yield, and hence there is no "correct" figure. Essentially the methods differ in their degree of penetration into porous structure.

A. Methods of Surface Area Determination

A considerable number of methods are available for the determination of surface area of particulate systems. At its simplest, the surface area can be calculated from particle size data, making assumptions of the shape and polydispersity of the powder. However only three methods, namely gas adsorption, mercury porosimetry, and gas permeametry, have been used to any appreciable extent to investigate the surface area of compacted powder systems.

1. Adsorption Methods

These measure the total area of solid which is accessible to the adsorbate, usually a gas such as nitrogen, helium, or krypton. It follows therefore that any internal structure to which the adsorbate has access will be included in the final result. In principle, the method involves determining the quantity of adsorbate which is required to cover all the available surface with a layer one molecule thick. Langmuir [6] postulated that at equilibrium the amount of adsorbate adsorbed was constant at a specific temperature and pressure. He assumed that only a single layer of adsorbate was present and derived the equation

$$\frac{P}{V} = \frac{1}{bV_m} + \frac{P}{V_m} \tag{1}$$

where

V = volume of gas adsorbed at pressure P
V_m = volume required to form a monolayer
b = constant

A plot of P/V against P yields V_m.

However, the adsorbed gas is unlikely to be only one molecule thick. This point was considered by Brunauer, Emmett, and Teller [7], who made three assumptions:

1. Multilayer adsorption occurs, and each layer obeys the Langmuir isotherm.
2. The average heat of adsorption of the second layer is the same as the third and subsequent layers and is equal to the heat of condensation of the liquid.
3. The average heat of adsorption in the first layer is different from that in the second and subsequent layers.

They derived the so-called BET equation:

$$V = \frac{V_m cP}{(P_0 - P)[1 + (c - 1)P/P_0]} \tag{2}$$

where

V = total volume of adsorbed gas on the surface of the adsorbate under conditions of standard temperature and pressure

V_m = volume of adsorbed gas if the entire surface were covered with a layer one molecule thick

P = partial pressure of adsorbate

P_0 = saturation pressure of adsorbate (thus P/P_0 is relative pressure)

c = constant

Equation 2 can be rearranged to

$$\frac{P}{V(P_0 - P)} = \frac{1}{V_m c} \left(\frac{c - 1}{V_m c} \right) \frac{P}{P_0} \tag{3}$$

Thus a graph of V versus P/P_0 gives a straight line where the slope, a, equals $(c - 1)/V_m c$ and the intercept, b, equals $1/V_m c$. Hence $V_m = 1/(a + b)$.

Since the volume which one mole of gas occupies at STP is 22.412×10^3 cm^3, the volume of the monolayer can be converted into moles. One gram-mole of gas at STP contains 6.023×10^{23} molecules (Avogadro's number), so the number of molecules in the monolayer can be calculated. Then if the cross-sectional area of each molecule is known (for example, the cross-sectional area of nitrogen is 16.2×10^{-20} m^2), then the area occupied by the monolayer can be determined.

The underlying theory of the BET approach and alternative treatments has been described by Allen [8]. In general, methods can be divided into two types: static and dynamic. In the former, the adsorbate is stationary, whereas in the latter, adsorption takes place from a moving stream of adsorbate. A number of commercial instruments are available and have been used on tabletted systems, e.g., Sorptometer (Perkin-Elmer) and Quantasorb (Quantachrome). An essential aspect of surface area measurement by this technique is in surface preparation. All solid surfaces are covered with a physically adsorbed film, which must be removed before any quantitative measurements can be made. This is termed "degassing" and is usually achieved by maintaining the solid in vacuo or in a stream of adsorbate, perhaps at elevated temperatures. It naturally follows that the degassing process should not in itself alter the nature of the surface.

2. Mercury Porosimetry

Mercury intrusion methods have been widely used in powder technology for a number of years.

For a nonwetting liquid to rise up a narrow capillary, pressure must be applied. Assuming a capillary with a circular cross section, then the Washburn equation [9] applies:

$$\Delta P - \frac{2\gamma \cos \theta}{r} \tag{4}$$

where

ΔP = increase in pressure
γ = surface tension of liquid
r = radius of capillary
θ = angle of contact between liquid and capillary wall

Mercury is forced into the pores of a powder bed or tablet, and from the volume of mercury utilized and the pressure needed to force the mercury into the pores a distribution of pore sizes can be established.

Assuming cylindrical pores, the surface area S can be calculated from the equation

$$S = 4 \sum_i \left(\frac{V_i}{d_i} \right) \tag{5}$$

where V_i is the volume of mercury needed to fill pores of diameter d_i.

It follows that if the pores are not cylindrical (i.e., circular in cross section), then results can only be comparative. There are two other potential problems with this technique. The first is the compressibility of the solid. If there are pores which do not connect with the surface, then these pores will be deformed when under pressure without making a contribution to the pore size distribution. Furthermore such deformation will also affect the pore size distribution of those pores which are accessible to the mercury. A further complication is the presence of "ink bottle" pores, which have constricted openings into large void volumes. The presence of pores of this shape in a compacted system is highly likely.

3. Permeametry Methods

Consider air flowing through a powder bed under the influence of a pressure drop. The principal resistance to airflow is at the particle surfaces, and so the permeability of the powder bed is an inverse function of its surface area.

If the powder bed is regarded as a system of parallel capillaries whose diameters are dependent on the average particle size, then the flow rate through a single capillary is given by Poiseuille's equation

$$\frac{V}{t} = \frac{\pi r^4 \Delta P}{8L\eta} \tag{6}$$

where

V = volume of air that flows in t sec
r = radius of capillary
ΔP = pressure drop across length L
η = viscosity of air

The average pore diameter in a powder depends on the particle diameter (d) and the degree of consolidation, which is related to the porosity (ϵ). Furthermore in a real powder bed the capillaries are not parallel cylinders of uniform radius and length. They are tortuous, and hence their length is greater than that of the powder bed. Neither are they circular in cross section.

These factors are taken into account by the Kozeny-Carman equation (Eq. (7)), which is the basis of most permeability methods [10].

$$S_K^2 = \frac{\Delta P t A}{k_1 L V \eta} \cdot \frac{\epsilon^3}{(1 - \epsilon)^2} \tag{7}$$

where

A = cross-sectional area of bed
S_K = surface area
k_1 = constant, which includes shape and tortuosity factors, usually 5

It has been pointed out [11] that for very fine powders, the rate of gas flow through the compact is greater than would be predicted by Eq. (7). This is because of slip flow The contribution of molecular slip (S_M) to the surface area is given by

$$S_M = \frac{\Delta P t A}{L V P} \left(\frac{\epsilon^2}{1 - \epsilon} \right) k_2 \sqrt{\frac{RT}{M}} \tag{8}$$

where

R = gas constant
T = absolute temperature
M = molecular mass of air

Hence the total surface area S_V can be calculated by combining Eqs. (7) and (8) to give

$$S_V = \frac{S_M}{2} + \sqrt{\frac{S_M^2}{4} + S_K^2} \tag{9}$$

The applicability of Eq. (7) to the determination of surface area has been discussed by Buechi and Soliva [12], who pointed out some of the assumptions which must be made before this equation can be used. The use of the Kozeny-Carman equation in connection with tablets has been criticized by Selkirk [13], but nevertheless this approach has found considerable application.

III. THE RELATIONSHIP BETWEEN SURFACE AREA AND TABLET PROPERTIES

There is a considerable body of published work on the surface area determination of tablets. A list of solids, tablets of which have been the subject of surface area studies, is given in Table 1, together with the method of surface area determination used and the appropriate reference. However there are very few studies which use more than one method of measurement. Therefore it is convenient to subdivide the topic according to the technique of measurement which was used. Work which uses more than one technique is of particular interest and will be discussed more fully later.

A. Surface Area Measurement in Tablets by Gas Adsorption

In 1952, a series of publications began to emerge from workers at the University of Wisconsin under the direction of Professor T. Higuchi. These papers had a profound effect on the subsequent course of tabletting research. In an early paper in this series [14], the relationship between compression force and the specific surface area of the tablet was examined. Using a static nitrogen adsorption method, the specific surface area of tablets made from sulfathiazole granulated with starch was determined.

The results are shown in Fig. 1. The shape of the curve was explained as follows. At lower forces (below 2500 lb, equivalent to 156 MPa), the net surface area of the tablet rose because granules were breaking down to give smaller particles with a higher specific surface area. Above 2500 lb, closer association of the particles became the more dominant process, even though further granule breakdown was undoubtedly occurring. The particles became so closely associated that the surface between them was not available to molecules of nitrogen.

The suggestion that the shape of the compression pressure-surface area curve could be explained by changes in the particle size within the tablet received experimental confirmation from the work of Armstrong and Haines-Nutt [15], who used a Perkin-Elmer Sorptometer. These workers found that when tablets of magnesium carbonate were placed in water, they rapidly disintegrated into their constituent particles. It therefore was possible to prepare tablets from particles of known size, measure the surface area of the tablets, and then disintegrate them so that the particle size after compression could be determined. Figure 2 shows the weight fractions of particles in the 20–30 μm and 90–100 μm fractions before and after compression. The "before-compression" data is represented by the intercepts of the lines on the ordinate (i.e., at zero pressure). Smaller particles

Table 1   Solids, Tablets of Which Have Been the Subject of Surface Area Study

Ascorbic acid	p	25
Bentonite	ga	20
Calcium phosphate dihydrate (Emcompress)	ga	21
Calcium phosphate dihydrate (Emcompress)	mp	32, 35
Calcium phosphate dihydrate (Emcompress)	p	25
Coal	ga	19
Dextrose monohydrate	ga	16
Iron powder	p	29, 30
α-Lactose (amorphous)	mp	35
α-Lactose (anhydrous)	ga	36, 37, 40
α-Lactose (anhydrous)	mp	32, 36
α-Lactose monohydrate	ga	21, 36, 37, 38, 40
α-Lactose monohydrate	mp	33, 34, 35, 36
α-Lactose monohydrate	p	25, 27
β-Lactose	ga	36, 37, 40
Lactose, spray-dried	mp	32, 35
Magnesium carbonate	ga	15, 17, 18
Magnesium oxide	ga	20
Magnesium trisilicate	ga	20
Microcrystalline cellulose (Avicel PH101)	mp	32
Microcrystalline cellulose (Avicel PH101)	p	30
Paracetamol	ga	16
Paracetamol	p	25
Phenacetin	ga	16
Sodium bicarbonate	p	25, 26, 27, 30
Sodium chloride	ga	21, 23
Sodium chloride	p	23, 25, 27, 29, 30
Sodium citrate	ga	21
Sodium citrate	p	25, 26, 27
Starch 1500 (Sta-Rx)	ga	21
Starch 1500 (Sta-Rx)	p	30
Sucrose	ga	19, 21, 23
Sucrose	p	23
Sulfathiazole/starch	ga	14

ga = surface area determined by gas adsorption; p = permeametry; mp = mercury porosimetry.

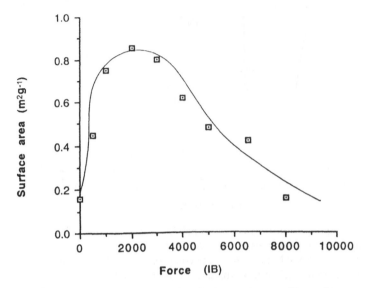

Fig. 1 The effect of compression force on the specific surface area of sulfathiazole tablets. (Reproduced from Reference 14 with permission of the copyright owner, the American Pharmaceutical Association.)

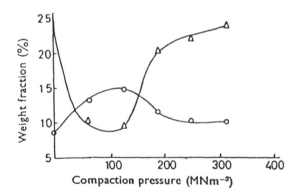

Fig. 2 Percent weight fraction-compression pressure profile of two size fractions of magnesium carbonate; $\bigcirc$, 20–30 μm; $\triangle$, 90–100 μm. (Reproduced from Reference 15 with permission of the copyright holder.)

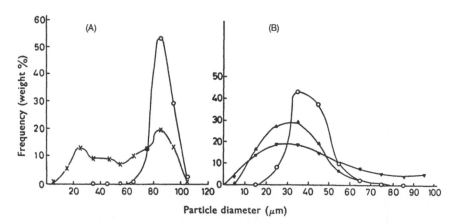

Fig. 3   Particle size distribution of (A) a nominal 73–89 μm fraction; (B) a nominal 33–45 μm fraction of magnesium carbonate; $\bigcirc$, before compression; $\times$, after compression at 62.5 MN m^{-2}; $\blacktriangledown$, after compression at 250 MN m^{-2}. (Reproduced from Reference 15 with permission of the copyright holder.)

show an increase, followed by a decrease, and the reverse is shown by the larger particles. Magnesium carbonate was then fractionated, compressed at either 62.5 MN m^{-2} or 250 MN m^{-2}, and the tablets disintegrated as before. The 75–89 μm fraction showed particle reduction. However at the higher pressure, considerably more "larger" particles were detected after compression than had been present in the original powder (Fig. 3).

Not all particulate materials show the surface area changes described above. For example, Armstrong and Griffiths [16] showed that for phenacetin and dextrose, in addition to the rise and fall in surface area already described, a second increase occurred at pressures greater than 150 kg cm^{-2} (equivalent to 147 MPa) (Fig. 4). It was shown that this was associated with a high value for elastic recovery of the tablets, which in turn resulted in a high degree of lamination or capping [17]. Furthermore it was shown that the magnitude of the surface area changes, and especially the pressure at which maximum surface area was achieved was associated with the presence of lubricating agents, both liquid [16] and solid [18].

Hardman and Lilley [19] used a dynamic method of surface area measurement by gas adsorption, studying sodium chloride (a plastically deforming material), sucrose (a brittle material) and coal powder, which has a considerable pore structure. They measured the surface area of tablets which had been compressed at pressures up to about 200 MPa. For sodium chloride, there was little change in surface area over the whole

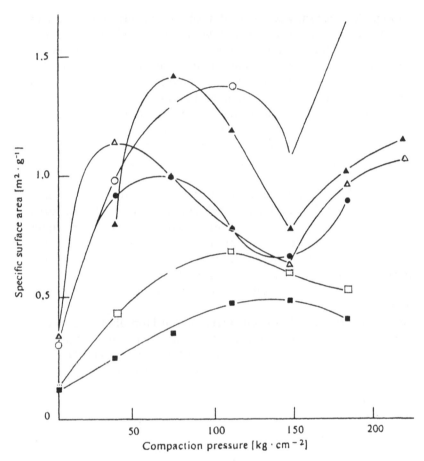

Fig. 4 Surface area-compression pressure profiles of compacts prepared from phenacetin, paracetamol, and dextrose monohydrate: ○, compacts prepared from phenacetin granules, peripheral moisture content 0%; ●, compacts prepared from phenacetin granules, peripheral moisture content 2.5%; □, compacts prepared from paracetamol granules, peripheral moisture content 0%; ■, compacts prepared from paracetamol granules, peripheral moisture content 2.5%; △, compacts prepared from dextrose monohydrate granules, peripheral moisture content 0%; ▲, compacts prepared from dextrose monohydrate granules, peripheral moisture content 6.6%; (Reproduced from Reference 16 with permission of the copyright holder.)

pressure range. Sucrose showed a considerable rise in surface area, attributed to fragmentation, up to about 170 MPa, beyond which the surface area fell. They suggested that this was due to cessation of fracture, accompanied by a small increase in interparticulate contact, some plastic flow also taking place. The surface area of the coal compacts continued to rise over the whole pressure range. Hardman and Lilley attempted to calculate the area of contact between particles by equating this to half the difference in surface area before and after compression. However they pointed out the inability of the nitrogen molecule to penetrate openings of less than 2 nm. Thus surfaces separated by less than this distance would be counted as being in contact.

Further work using nitrogen adsorption was carried out by Stanley-Wood and Johansson using a static method [20]. They attempted to correlate compact density with surface area changes. They found that for magnesium trisilicate and bentonite, surface area fell at pressures up to about 50 MPa and then rose slightly, whereas that for magnesium oxide remained relatively constant over a pressure range from 0 to 500 MPa. They related this difference in behavior to the different consolidation mechanisms of the three solids.

Considerable research has been carried out at Uppsala University on mechanisms by which particles aggregate to form tablets, and surface area measurement has played an important part in this. Duberg and Nystrom [21] evaluated several methods by which particle fragmentation could be assessed. One of these was krypton adsorption, surface area of both powder and tabletted systems being calculated by the BET equation. Tablets were prepared at 165 MPa.

The results they obtained were, at least in part, anomalous. Some substances, such as calcium phosphate dihydrate (Emcompress), which are known to fragment under a compressive load, showed very large increases in surface area after compaction, whereas others, such as starch 1500 (StaRx 1500) and sodium chloride, showed much smaller increases. However, for other substances the change in area was much more difficult to interpret. Duberg and Nyström pointed out that materials with a large number of cracks and pores will give a high value for powder surface area. Furthermore, as particle consolidation increases on compression, that part of the surface area which represents particle contact will be lost. Thus, Duberg and Nyström concluded that gas adsorption was probably an inappropriate method for measuring the degree of fragmentation occurring during compression, and suggested that a technique based on permeametry would be more appropriate. Some of the confusion surrounding their results may stem from the fact that they only measured the surface area of the tablets at one pressure. All previous work had shown that the relationship between compression pressure and surface area is at least parabolic, and the position of the maximum of the parabola is substance dependent. A one-point measurement would not detect this.

Duberg and Nyström concluded that the strength isotropy ratio (the ratio between the axial and radial tensile strengths of the tablet) [22] was a more appropriate means of assessing fragmentation, though they recommended that the scanning electron microscope was essential for a qualitative evaluation of fragmentation.

B. Surface Area Measurement of Tablets by Gas Permeametry

Because of their reservations about measuring tablet surface area by low-temperature gas adsorption, workers at Uppsala University developed a method of measuring surface area by permeametry [23]. Their method was based on the Blaine apparatus [24].

Using sodium chloride and sucrose compressed at pressures between 40 and 300 MPa, they measured surface area by both krypton adsorption and permeametry. The results are shown in Fig. 5. Surface area by

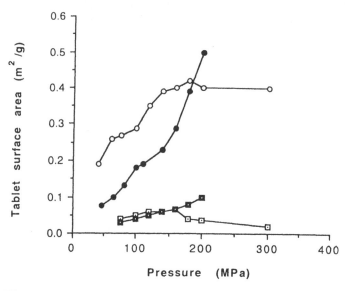

Fig. 5 Weight specific surface area, measured by permeametry (closed symbols) and by the BET method (open symbols) as a function of compaction pressure for sodium chloride (□) and saccharose (○) tablets. (Reproduced from Reference 23 with permission of the copyright holder.)

permeametry increases over the whole of the pressure range for both substances, though the increase for sucrose is much larger. Surface area by gas
adsorption shows the parabolic shape reported by earlier workers [14,15]
with a maximum at about 150 MPa for both substances. Hence, at the
highest pressures studied, there is a considerable difference in magnitude in
the results given by the two methods.

Using the permeametry technique described above, Alderborn et al.
studied the consolidation behavior of a number of substances [25]. They
reasoned that if a solid fragmented under a compressive load, then that
should be reflected in an increase in the surface area. They used two
materials which deformed plastically (sodium chloride and sodium bicarbonate), four which primarily underwent fragmentation (lactose, sodium
citrate, ascorbic acid, and paracetamol) and one (calcium phosphate
dihydrate: Emcompress) which fragmented extensively. Tablets were prepared at a pressure range of 40–200 MPa, and the surface areas of the
resulting tablets were measured.

Surface area increased for all materials, indicating that some fragmentation occurred during compaction. However the shapes of the surface
area–compression pressure curves differed markedly, their slopes being in
the same rank order as their propensity to fragment (Fig. 6).

Alderborn [26] then went on to compare the surface areas of intact
tablets with those which had been deaggregated by gentle manual abrasion, sodium citrate and sodium bicarbonate being used as solids. He
found that, at least at pressures up to 100 MPa, the surface area of the
intact tablets differed little from that of the deaggregated tablets. From
this finding, he concluded that the actual surface area used for bonding
was relatively low.

Alderborn and Nyström [27] investigated if particle size had an effect
on the slope of the surface area compression pressure profile, using sodium
chloride, sodium bicarbonate, lactose, and sodium citrate as test materials,
and five different size fractions of each solid. For all materials and size
fractions, the tablet surface area increased with increasing compression pressure. The magnitude of the increase depended on whether the solid's primary means of consolidation was by fragmentation. For all materials, the
finer size fractions gave a larger absolute increase in surface area, though the
reverse was true if the relative surface area change was considered.

If all the lines were regarded as rectilinear, then the values of the
slopes of these lines could be regarded as a quantitative relationship between surface area and compresson pressure, at least up to pressures of
about 125 MPa. Alderborn and Nyström then plotted the slopes of these
lines against the surface area of the powders (Fig. 7), expressing the slopes

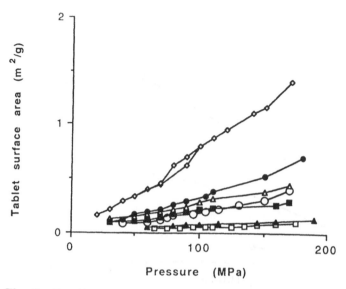

Fig. 6 Specific surface area of sodium chloride (▢), sodium bicarbonate (▲), saccharose (◯), lactose (●), sodium citrate (■), ascorbic acid (△), paracetamol (◇), and Emcompress (◆), as a function of compaction pressure. (Reproduced from Reference 25 with permission of the copyright holder.)

of these lines as a measure of the rate of increase of tablet surface area with pressure. Thus

$$\frac{ds}{dP} = kS^n \tag{10}$$

where k is a constant describing the fragmentation propensity of the solid. They found a value for n of $\frac{2}{3}$ was most appropriate. Thus over the range of pressures 0 to P, with corresponding surface areas of S_1 and S_2, then

$$3(S_2^{\frac{1}{3}} - S_1^{\frac{1}{3}}) = kP \tag{11}$$

They drew attention to the similarity between this equation and that of Bond [28], which relates the energy consumed during size reduction to the particle size of the product.

Attempts have been made using surface area measurement to calculate the extent of the actual area of contact between particles in a tablet [29]. These workers used nonfragmenting particles (iron and sodium chloride) and measured surface area by both permeametry and krypton adsorp-

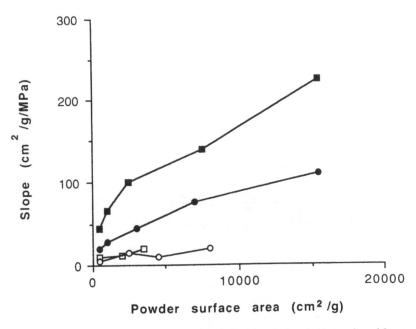

Fig. 7 Effect of powder surface area on the absolute increase in tablet surface area with pressure (i.e., the slope from a plot of tablet surface area as a function of compaction pressure (0–125 MPa)): sodium chloride (○), sodium bicarbonate (□), sodium citrate (●), lactose (■). (Reproduced from Reference 27 with permission of the copyright holder.)

tion. Since no significant change in surface area in the iron particles could be attributed to bonding, it was concluded that the bonding surface of the particles was very small and below the level of detection of the technique.

Sodium chloride showed a different pattern, A fall in surface area measured by gas adsorption was noted at higher pressures, but as the permeability surface area continued to increase, it was concluded that sodium chloride also utilized only a small portion of its surface area for bonding.

Karehill et al. [30] also studied the relationship between the surface area of the solid and the compact strength, using plastically deforming materials—sodium chloride, sodium bicarbonate, starch 1500, microcrystalline cellulose, and iron powder. They concluded that only a very small fraction of the total available surface was used to form solid bridges between adjacent particles, and that tablet strength must be due to relatively long-range forces. The strength of tablets made from some of these substances increased when stored for two days, but no changes in surface area were detected which could account for this change [31].

C. Surface Area Measurement of Tablets by Porosimetry

Another body of work utilizing surface area measurements for research on powders and tablets is that carried out by Professor Lerk and colleagues from the University of Groningen in The Netherlands. This work has been principally concerned with varieties of lactose. Surface area, usually described as pore surface area, was most frequently determined by mercury porosimetry, though gas adsorption was sometimes used.

Vromans et al. found a linear relationship between surface area and compaction load for α-lactose monohydrate and anhydrous α-lactose [32]. Calcium phosphate dihydrate (Emcompress) showed a similar effect. The authors claimed that microcrystalline cellulose did not show an increase in surface area with increased load, but critical examination of their data shows that the latter statement may not be totally valid. Notwithstanding this, their work supports that of other workers [25] in that for fragmenting substances, surface area increased markedly as the load increased, the slope of the line being proportional to the fragmentation propensity of the solid. From this it was concluded that the degree of fragmentation of α-lactose monohydrate was less than that of anhydrous α-lactose.

As tablet strength also increases with compression pressure, it follows that crushing strength must increase with surface area. This is shown in Fig. 8. The significant point about this graph is that all points lie on the same straight line, indicating that tablet strength is related to surface area and hence the degree of fragmentation. Furthermore it would appear that neither the presence of water in the α-lactose monohydrate nor the ratio between α and β lactose has any influence on the tablet strength, suggesting that the same binding mechanism applies to all types of crystalline lactose.

Using a series of sieve fractions of α-lactose monohydrate, de Boer et al. [33] studied the relationship between compaction pressure and compact strength. A nonlinear relationship was obtained, with clearly defined maxima, but stronger tablets were obtained from smaller particles at any given pressure. When the surface area of all the tablets prepared in this study was measured, it was found that a good rectilinear relationship was obtained between surface area and compact strength for all the original size fractions.

In contrast to the behavior of crystalline lactose monohydrate, the pore surface area of amorphous lactose did not increase with compaction pressure. It was therefore inferred that amorphous lactose deformed plastically rather than undergoing consolidation by fragmentation [34]. However,

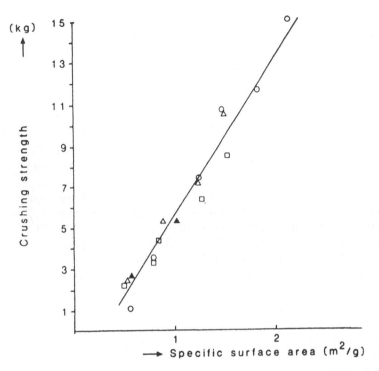

Fig. 8 Crushing strength versus specific surface area for tablets compressed from different types of crystalline lactose: α-lactose monohydrate (□); anhydrous α-lactose (○); roller-dried β-lactose (△); crystalline β-lactose (▲). (Reproduced from Reference 32 with permission of the copyright holder.)

tablet crushing strength showed an increase with compaction pressure. It was thus concluded that for materials which consolidate by deformation, the tablet strength was not related to surface area, as had been found for fragmenting substances. Spray-dried lactose contains about 15% amorphous lactose, which makes a major contribution to tablet strength despite making little contribution to surface area changes. Since little new surface area is generated on compaction, the strength of tablets made from materials such as amorphous lactose are greatly reduced by the addition of magnesium stearate. Compactibility was, however, increased after the solid had been exposed to water vapor. Again this was interpreted as being characteristic of materials which consolidate by deformation rather than fragmentation.

This study was taken further by preparing samples of spray-dried lactose containing differing proportions of amorphous lactose [35]. It was found that the proportion of amorphous lactose present had little effect on the surface are of the tablets, but had a major influence on the tablet strength. It was suggested that the primary role of amorphous lactose was to act as a binder.

Continuing the work with lactose, Vromans and co-workers studied the compactability of a series of lactoses which differed in respect of size, texture, water content, and α/β ratio [36]. They found that the binding capacity was directly related to the surface area of the starting material, surface area being measured by nitrogen adsorption. As shown in Fig. 9, a

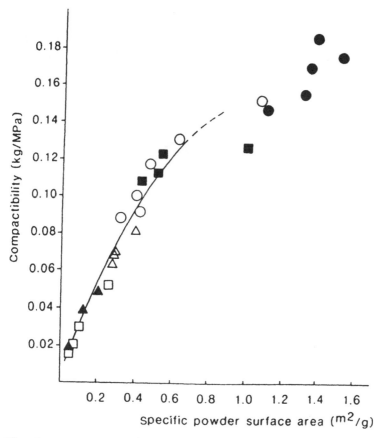

Fig. 9 Compactibility of samples of crystalline lactose plotted versus the powder surface area (S_{N_2}), □, α-lactose monohydrate; ○, anhydrous α-lactose; ●, anhydrous α-lactose; △, roller-dried β-lactose; ▲, crystalline β-lactose; ■, compound crystalline lactose. (Reproduced from Reference 36 with permission of the copyright holder.)

rectilinear relationship was obtained between compactibility and surface area, except for highly porous samples of anhydrous lactose, which showed lower compactibility than expected. Compactibility was defined as the slope of the compression pressure-crushing strength curve.

To avoid capillary condensation (see later), they studied tablets prepared at low pressures (37.5 MPa) in which the incidence of small capillaries would be expected to be lower. They found a good rectilinear relationship, except for the porous materials noted above. Furthermore an excellent straight line was obtained by plotting tablet surface area against powder surface area (Fig. 10). It is of interest to note that the slope of this line is very close to unity.

The properties of tablets containing binary mixtures of various types of lactose were studied by Riepma et al. [37]. Using size fractions of the same size for all materials studied, the rectilinear relationship noted earlier between tablet surface area and tablet crushing strength was reported. To

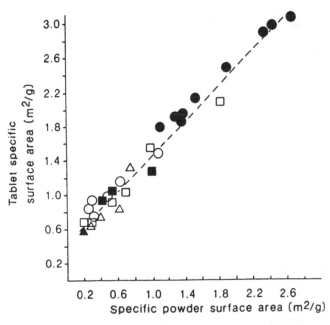

Fig. 10 Specific surface area of tablets compacted at 37.5 MPa versus specific powder surface area of the starting material. Symbols as in Fig. 9. (Reproduced from Reference 36 with permission of the copyright holder.)

avoid capillary condensation, tablets were compressed at a force of only 10 kN (equivalent to about 75 MPa).

Using different size fractions of the same materials, Riepma et al. found again that specific surface and crushing strength were linearly related [38]. The graphs of both surface area against percent fine fraction and tablet strength against fine fraction were very similar, in that they showed a minimum at about 40% fine fraction. The authors explained this by suggesting that the fine fraction possessed less tendency to fragment. A suggestion that smaller particles contain fewer cracks and are therefore less likely to fragment forms part of the Griffith crack theory [39].

As stated earlier, a linear relationship between crushing strength and surface area was found for a wide variety of different types of lactose, and from this it was assumed [32] that the number of bonding sites was proportional to the surface area. Leuenberger et al. [40] derived a theoretical model in which coordination numbers (i.e., the number of isometric spherical particles which form a compact) were calculated. The size of the spheres was obtained from surface area measurements and then by calculating the equivalent diameter of the sphere. The coordination number then depends on the geometrical packing of the particles [41]. Fair agreement was obtained for various types and sizes of lactose, but the authors point out that the calculated diameter of the spheres is very dependent on the method used to measure surface area. Furthermore calculating the equivalent sphere diameter introduces an error, since a comparison of the particle size and surface area data clearly shows that the particles are not spherical.

IV. THE RELATIONSHIP BETWEEN COMPRESSION PRESSURE AND SURFACE AREA IN TABLETTED SYSTEMS

Whether consolidation of a powder bed under pressure takes place by fragmentation or particle deformation, there will be some change in surface area. Then, as the particles assume an even closer relationship, this too will cause surface area changes. It follows that surface area studies may provide important information about the extent and mechanisms of particle consolidation. However as will have been seen from the above discussion, different groups of workers have obtained very different shapes for the graph of surface area against compression pressure, and an explanation of this variation must be sought.

The explanation put forward by the original workers in this area [14] (Fig. 1) is worthy of serious consideration. Three processes which will affect the surface area of the tablet will proceed simultaneously. These are

1. Fragmentation of particles, leading to an increased surface area
2. Increased interparticulate contact, leading to a decrease in surface area
3. Flattening of particle asperities, giving a smoother particle surface and hence a decrease in surface area

The overall shape of the compression pressure-surface area curve is thus determined by the relative importance of these three mechanisms for a given substance at any given pressure. This will be a function of the properties of the substance and in particular its fragmentation propensity. It is to be expected that virtually all particles will undergo some fragmentation either as the result of the imposition of a compressive force or even, at low pressures, by abrasion as the particles slide over one another to give a less porous powder bed, but one which is still composed of discrete particles.

At higher pressures some materials deform significantly, e.g., cellulose derivatives, and a granulating agent, if present, leads to a more deformable product [42].

However, these considerations are complicated by another important factor, namely the method by which the surface area is determined. Each of the three methods of measuring surface area which has been used to examine tablets measures something different. These are the surface to which an adsorbate molecule, e.g., nitrogen, has access, the volume to which pressurized mercury has access, and the surface over which gas flow occurs. Consequently it is not surprising that in the few comparative studies which have been carried out, numerical identity is not achieved. However it is potentially more serious that the overall shape of the curve appears to be dependent on the experimental method used to measure surface area.

Those methods which utilize gas adsorption (see, for example, [14], [16]) usually show a rise in surface area followed by a decrease, whereas the use of permeametry [25] or mercury porosimetry [32] gives a progressive increase in surface area over the whole range of compression pressures studied. Such a divergence in behavior renders more important the few comparative studies which have been carried out, and it is worth examining these in some detail.

Vromans et al. [32] reported a linear relationship between specific surface area as assessed by porosimetry and compression pressure for α-lactose monohydrate and anhydrous lactose, but a curvilinear relationship for β-lactose measured by gas adsorption [36]. Bearing in mind the similarity of behavior of all types of lactose referred to earlier [32], such a difference was considered worthy of further investigation.

Vromans et al. [36] reported that compared to mercury porosimetry results, surface area determined by gas adsorption gave poor reproducibility. The magnitude of the surface area appeared to be related to the degassing and storage conditions the tablets underwent after compression. Tablets examined immediately after compression, i.e., without any pretreatment showed a larger surface area than those which had been subjected to degassing at 120° (Fig. 11). Furthermore the surface area changed after storage for 24 hr, the extent of the change depending on the relative humidity of the storage conditions (Fig. 12). They suggested that this variability was due to capillary condensation, and since the pores in the tablet become progressively smaller as the compression pressure is increased, then the effect of capillary condensation might be expected to be more pronounced in tablets prepared at higher pressures. Hence, doubt must be cast on reported surface area measurements which have used the gas adsorption technique. As Vromans et al. pointed out, it is impossible to prevent some capillary condensation which occurs almost immediately after the tablet is prepared.

These workers also suggested that removing water from the tablet by degassing will have an unavoidable disruptive effect on the tablet structure. Hence for subsequent work using gas adsorption. Vromans and co-workers used tablets compressed at the relatively low pressure of 37.5 MPa, at which capillary condensation would be expected to be low.

The magnitude of the effects of degassing will depend in part on the substance from which the tablet is made. Vromans et al. degassed α-lactose

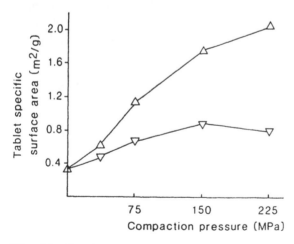

Fig. 11 Specific surface area (S_{N_2}) of tablets compacted from roller-dried β-lactose (100–160 μm) versus the compaction pressure. The tablets were measured immediately after compaction ($\triangle$) or after outgassing at 120°C ($\triangledown$). (Reproduced from Reference 36 with permission of the copyright holder.)

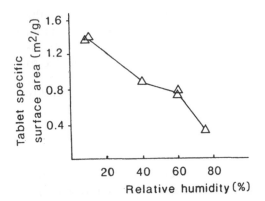

Fig. 12 Specific surface area (S_{N_2}) of tablets compacted from roller-dried β-lactose at 225 MPa plotted in relation to the storage conditions, being the relative humidity at 20°C. (Reproduced from Reference 36 with permission of the copyright holder.)

monohydrate at 120° and an appreciable quantity of lactose might be expected to dissolve and subsequently recrystallize under such circumstances, with obvious effects on the pore structure and surface area of the tablet. Also liberation of the water of crystallization must be considered. However, other workers have used much less soluble solids, e.g., magnesium carbonate [15], and degassed at lower temperatures, so the use of gas adsorption cannot be totally condemned. Nor would the theory of capillary condensation explain the second rise in surface area detected by Armstrong and co-workers [16,18].

Alderborn et al. [23] found similar differences between surface area determinations using gas adsorption and permeametry (see Fig. 5). These authors pointed out that as particles are pressed more closely together, surface area will be lost, due not only to interparticulate contact but also because adsorbate molecules fail to gain access to part of the surface area. Hence, surface area is underestimated. With permeametry, on the other hand, an overestimate may well be obtained, since, as consolidation increases, nonhomogeneity of the compact will also increase. This may well be so, but it is worthy of comment that in all the gas adsorption studies reported, the maximum in the compression pressure surface area curve is always at about 150 MPa, despite the wide range of substances used. Furthermore the two solids used in this work are both freely soluble in water, and so disruption of the particle surfaces might well occur during the degassing process.

REFERENCES

1. C. Fuhrer, *Labo-Pharma Probl. Techn. 25*: 759 (1977).
2. P. York and N. Pilpel, *J. Pharm. Pharmacol. 25*: 1P (1973).
3. W. B. Pietsch, (1969). The strength of agglomerates bound by salt bridges. *Can. J. Chem. Eng. 47*: 403.
4. F. P. Bowden and D. Tabor, *The Friction and Lubrication of Solids*, Oxford University Press, New York, pp. 145–159, 1950.
5. P. J. James, *Powder Metall. 20*: 199 (1977).
6. I. Langmuir, *J. Am. Chem. Soc. 40*: 1361 (1918).
7. S. Brunauer, P. H. Emmett, and E. Teller, *J. Am. Chem. Soc. 60*: 309 (1938).
8. T. Allen, *Particle Size Measurement*, 3rd ed., Chapman and Hall, London, pp. 465–513, 1981.
9. E. W. Washburn, *Phys. Rev. 17*: 374 (1921).
10. B. H. Kaye, *Powder Technol. 1*: 11 (1967).
11. D. T. Wasan, W. Wnek, R. Davies, M. Johnson, and B. H. Kaye, *Powder Technol. 14*: 209 (1976).
12. W. Buechi and M. Soliva, *Powder Technol. 38*: 161 (1983).
13. A. B. Selkirk, *Powder Technol. 43*: 285 (1985).
14. T. Higuchi, A. H. Rao, L. W. Busse, and J. V. Swintosky, *J. Am. Pharm. Assoc. Sci. Ed. 42*: 194 (1953).
15. N. A. Armstrong and R. F. Haines-Nutt, *J. Pharm. Pharmacol. 22*: 8S (1970).
16. N. A. Armstrong and R. V. Griffiths, *Pharm. Acta Helv. 45*: 583 (1970).
17. N. A. Armstrong and R. F. Haines-Nutt, *Powder Technol. 9*: 287 (1974).
18. N. A. Armstrong and R. F. Haines-Nutt, Proc. 1st Int. Conf. on Compaction and Consolidation of Particulate Matter, Brighton, pp. 161–164, 1972.
19. J. S. Hardman and B. A. Lilley, *Proc. R. Soc. London A 333*: 183 (1973).
20. N. G. Stanley-Wood and M. E. Johansson, *Drug Dev. Ind. Pharm. 4*: 69 (1978).
21. N. Duberg and C. Nyström, *Acta Pharm. Suec. 19*: 421 (1982).
22. C. Nyström, W. Alex, and K. Malmquist, *Acta Pharm. Suec. 14*: 317 (1977).
23. G. Alderborn, M. Duberg, and C. Nyström, *Powder Technol. 41*: 49 (1985).
24. R. L. Blaine, *ASTM Bull. 108*: 17 (1943).
25. G. Alderborn, K. Pasanen, and C. Nyström, *Int. J. Pharm. 23*: 79 (1985).
26. G. Alderborn, *Acta Pharm. Suec. 22*: 177 (1985).
27. G. Alderborn and C. Nyström, *Powder Technol. 44*: 37 (1985).
28. F. C. Bond, *Mining Eng. 4*: 484 (1952).
29. C. Nyström and P. G. Karehill, *Powder Technol. 47*: 201 (1986).
30. P. G. Karehill, G. Alderborn, M. Glazer, E. Borjesson, and C. Nyström, Studies on direct compression of tablets. XX. Investigation of bonding mechanisms of some directly compressed materials by addition and removal of magnesium stearate, cited by Karehill, P. G. (1990). Studies in bonding mechanisms and bonding surface area in pharmaceutical compacts. Ph.D. thesis, Uppsala University.
31. P. G. Karehill and C. Nyström, *Int. J. Pharm. 64*: 27 (1990).

32. H. Vromans, A. H. deBoer, G. K. Bolhuis, C. F. Lerk, K. D. Kussendrager, and H. Bosch, *Pharm. Weekbl. Sci. Ed.* 7: 186 (1985).
33. A. H. De Boer, H. Vromans, C. F. Lerk, G. K. Bolhuis, K. D. Kussendrager, and H. Bosch, *Pharm. Weekbl. Sci. Ed.* 8: 145 (1986).
34. H. Vromans, G. K. Bolhuis, C. F. Lerk, K. D. Kussendrager, and H. Bosch, *Acta Pharm. Suec.* 23: 231 (1986).
35. H. Vromans, G. K. Bolhuis, C. F. Lerk, H. van de Biggelaar, and H. Bosch, *Int. J. Pharm.* 35: 29 (1987).
36. H. Vromans, G. K. Bolhuis, C. F. Lerk, and K. D. Kussendrager, *Int. J. Pharm.* 39: 207 (1987).
37. K. A. Riepma, C. F. Lerk, A. H. deBoer, G. K. Bolhuis, and K. D. Kussendrager, *Int. J. Pharm.* 66: 47 (1990).
38. K. A. Riepma, J. Veenstra, A. H. deBoer, G. K. Bolhuis, K. Zuurman, C. F. Lerk, and H. Vromans, *Int. J. Pharm.* 76: 9 (1991).
39. A. A. Griffith, *Philos. Trans. R. Soc. London A221*: 163 (1920).
40. H. Leuenberger, J. D. Bonny, C. F. Lerk, and H. Vromans, *Int. J. Pharm.* 52: 91 (1989).
41. P. J. Sherrington and R. Oliver, *Granulation*, Heyden, London, p. 54, 1981.
42. N. A. Armstrong, N. M. A. H. Abourida, and A. M. Gough, *Pharm. Technol.* 6: 66. (1982).

9

Rationale for and the Measurement of Tableting Indices

Everett N. Hiestand*

Upjohn Company, Kalamazoo, Michigan

I. INTRODUCTION

The impracticability of measuring the properties of individual particles, the real participants of tableting, is a major impediment to obtaining an accurate description of the tableting performance of the powder. The challenge has been to obtain information from the properties of the compact, the "continuum," that will definitively indicate the processing performance of the material. It is believed that the tableting indices[†] developed at the Upjohn Company provide this information.

II. BACKGROUND

A. The Origin of Bond Strength

The forces acting at particle-particle interfaces would be preferred information; but the surface energies and the interaction energies between particles have their origin from the same force. This was demonstrated by using mechanical measurements to estimate solubility parameters [1]. The sur-

---

Retired. Residing at 11,378 East G Avenue, Galesburg, Michigan 49053

[†]This is not a review of all indices proposed. Only the system developed by Upjohn is discussed.

face energies are more accessible than the forces per se. Thus, for convenience, surface energies become the subject of discussion. In general, the surface energy of organic materials are in the 25 to 85 ergs/cm^2 range. Note that the largest is less than four times the smallest value. Clearly, it is impossible to account for the observed variation of bond strength among different materials, variations greater than an order of magnitude (see tensile strengths in Table 1), by considering only the kinds and magnitudes of forces acting. A legendary explanation of the large variation of strength is to assume that the areas of contacts between the particles varies sufficiently to yield these strength differences; however, in Table 1 the tensile strength of compacts at the same solid fraction, i.e., essentially the same interparticle contact area, shows a range greater than 10 to 1. Certainly melting-resolidification could give differences; however, this explanation often collapses when it is recognized that some of the stronger bonding materials decompose before they melt; and melting induced by applied pressure requires the opposite volume change (the Clapeyron-Clausius equation) from that of most organic materials. Furthermore, plastic deformation usually occurs at lower pressures than melting. Clearly, this cannot be a general explanation. While chemical interactions cannot be excluded in all situations, certainly they cannot be invoked to account for the universal adhesion between particles.

Table 1 Indentation Hardness and Tensile Strength Values (kN/cm^2) for Various Materials at Selected Dwell Times[a] for H Values and Time Constant of 15 sec for σ_T; Compact Solid Fraction 0.9

Material	H_0	$H_{1.5}$	H_{30}	H_∞	$\sigma_T \times 10$
Avicel[b]	24.1	6.79	6.01	—	8.1
Sorbitol	40.8	3.13	1.39	0.45	1.9
Lactose-Sp.Dr.	33.4	12.6	10.8	6.53	1.2
Sucrose	47.3	—	8.0	—	1.9
Ibuprofen[c]	12.8	—	2.38	—	0.97
Aspirin	1.40	—	0.57	—	0.29
Caffeine	24.9	12.3	8.00	<3.1	3.59
Phenacetin	3.15	2.87	1.99	—	2.78
Acetaminophen	8.8	7.4	7.4	1.4	0.66
CaSO$_4 \cdot$ 2H$_2$O	23.5	—	14.1	—	1.86
Starch (maize)	10.5	—	2.0	—	0.80

[a] Subscripts indicate approximate dwell time in minutes, except ∞ is 24-hr dwell time value.
[b] PH-101, microcrystalline cellulose.
[c] Lot-to-lot variation regularly observed.

Modern adhesion theories emphasize the role of mechanical properties in determining strength. Co/adhesion strength is the derivative of the maximum incremental work done during the separation of surfaces. The components of the work are (1) the energy needed for the new surfaces formed by the separation, (2) the plastic deformation that occurs as the surfaces separate, and (3) charge separation when the separation is of surfaces containing charged (transfer) sites. Plastic deformation may be either the ductile extension of the entire contact regions or the localized deformation at the interparticle contact perimeters at the receding boundary between two particles. In this process the perimeter is the equivalent of the crack tip region considered in studies of fracture mechanics. Item (3) is the most difficult to estimate since with insulating materials, charge separation is dependent on charge transfer at defect and impurity sites on the crystal surface, which can have dramatic effects even when present at unknown, undetectable levels. However, in most cases, it is believed to be negligible because surface conductivity limits the survival of localized charges.

Universal attraction (except when close enough for Born repulsion) exists because it is the lowest free-energy state and realized by the particles coming together. The particle rearrangement and plastic deformation produced by compression increase the total number of effective contact sites in a cross section where this low-energy state is attained. All of these processes have been included in models of the processes of tablet bonding [2,3].

B. The Challenge for Tableting Indices

The indices are based on a philosophy that is unique only because it is seldom used in pharmaceutical studies, viz. the indices are determined using measurements on specially prepared test specimens (compacts); only limited data are collected during the making of the compact. A compact, by definition, has sufficient rigidity to be handled without uncontrolled changes occurring as a result of the handling. This makes it a preferred form. However, how does one produce equivalent test compacts from any or all materials that need to be evaluated? This must be done without adding excipients that would "dilute" or modify the property one is trying to measure, and must be done without introducing macroscopic flaws into the compact which would distort the test results. To identify a procedure that works for all materials is a very big challenge; and the making of the test specimen becomes a critical part of the evaluation of tableting properties. It is as important as the choice of measurements made on the test specimen.

III. CONSIDERATIONS FOR STRENGTH MEASUREMENTS

Mechanical properties are the response to an applied stress state. Strength is defined as the stress required to produce a permanent change of shape; both plastic yielding and fracture do this. Thus strength measurements may include response to both compressive and tensile loading. Strength measurements provide the simplest method of characterizing mechanical properties.

A. Fracture Strength

While the tensile strength* is an obvious choice for indicating bond strength, there are numerous complicating factors to be considered. Any macroscopic† defect within the test volume can falsify the results and must be avoided. Also, all compacts are Mohr bodies. Thus, the observed strength is dependent on the hydrostatic stress within the compact, i.e., the stress state. Furthermore, nearly all organic materials exhibit some degree of viscoelastic behavior; therefore, the values of strength obtained depend on the rate of strain during the test. One must use equivalent, not identical, strain rates when the materials being compared exhibit different degrees of viscoelastic properties. Meticulous attention is a requisite to reproduce the exact procedure, including the equivalent strain rate, to obviate these potential problems. Also, to reduce the risk of a macroscopic defect distorting the result, the strength test should subject the minimum volume to the maximum stress that produces the fracture, i.e., minimize the volume from which a crack will originate. Chance defects far removed from this "critical" volume would not significantly distort the results. An additional complication arises from the fact that compacts are not of uniform consolidation‡ throughout, and the magnitudes of the density differences within the compact are dependent on the properties of the material. Clearly, it would be best, when more than one measurement is being used in a comparison, for all measurements to depend on the properties of the same region within the compact. For example, one could measure the properties of the central portion of the compact. Also, strength values increase nearly exponentially with the solid fraction, ρ_r; therefore, one must identify any given tensile strength value, σ_T, with the ρ_r value of the compact measured.

---

*Tensile strength and fracture strength are used herein as synonymous because tensile failure of compacts always is a brittle mechanism.

†For this Chapter macroscopic indicates a size much larger than the particles or pores.

‡Consolidation may be expressed as solid fraction or relative density, which is equal to one minus the porosity of the powder bed or compact.

B. Tensile Strength from Compression

The transverse compression of square compacts provides a tensile test that meets more of the requirements discussed above than any other test procedure known to the author. For the square compacts used with this system, the compact is placed between two platens that cover the central 0.4 of the edge-width of the compact; pads are placed between the platens and the compact to reduce the potential for stress concentrations at the edge of the platen initiating fracture (polymeric platens without pads have been used successfully, also). The maximum tensile stress appears only in the central portion of the test specimen [4,5], the same region used for the indentation hardness described later. It is believed that with this procedure the fracture starts in this volume.

An extreme example will illustrate the need for control of the strain rate. What strain rate should be used to test the fracture strength of a very viscoelastic silicone polymer—specifically, the one sold as (in the United States) "Silly Putty"? Clearly, it must be a very high strain rate; otherwise ductile extension occurs. A convenient strain rate to use in routine testing of most materials would be much too low to cause fracture of the silly putty. However, if a test is to be applicable to all materials, a systematic method to arrive at comparable strain rates must be incorporated into the test procedure when viscoelastic materials are used. With the indices testing, this is done by assuming that the rate of stress increase during the tensile test is approximately exponential with strain. Therefore, one can apply a time constant to the loading-to-fracture process. The time constant is set as a fixed elapse of time between when $1/e$, 0.368, of the fracture force was being applied and the time of fracture. The time interval, the time constant, used in the author's laboratory is 15 sec. (The platen movement rate for a single test is constant but with other materials or other solid fractions the constant rate may be different.) The choice of 15 sec is arbitrary but appears to be suitable for nearly all compacts.* Table 2 shows data for sucrose and sorbitol obtained with different time constants and confirms that the strength varies with strain rate. The setting of the test instrument to produce the desired strain rate will be by trial and error; for this one can use the indentation hardness specimens left over from that test. While the 15-sec rate would not cause silly putty to fracture, one does not expect to make tablets from materials with such unusual properties.

---

*More recent work indicates that a time constant of perhaps 5 sec provides better breaks with some materials than does the 15-sec case. It is important that comparisons be made only using data obtained with the same time constant. (10 sec has been used successfully by Professor McGinity at the University of Texas.)

Table 2 Tensile Strength[a] Variation with Time Constant[b]

Material	Solid fraction	Time constant (sec)	Tensile strength (kN/cm^2)
Sorbitol	0.9055 ±0.0014	43.30 ±1.92	0.1745 ±0.0024
	0.9038 ±0.0008	13.40 ±0.48	0.2049 ±0.0038
	0.9037 ±0.0002	4.70 ±0.19	0.2342 ±0.0040
	0.9044 ±0.0010	1.70 ±0.17	0.2782 ±0.0063
	0.9045 ±0.0006	0.80 ±0.02	0.3021 ±0.0059
Sucrose	0.9071 ±0.0004	26.50 ±0.08	0.2043 ±0.0094
	0.9073 ±0.0002	7.70 ±0.30	0.2274 ±0.0179
	0.9069 ±0.0008	3.32 ±0.08	0.2400 ±0.0085
	0.9068 ±0.0003	1.14 ±0.02	0.2481 ±0.0075
	0.9074 ±0.0005	0.572±0.013	0.2608 ±0.0114

[a] These data were for different lots of powder than used in studies reported in Table 1; some differences observed.
[b] Based on mean of five determinations.

Is tensile strength a sufficient indicator of the tableting properties? The answer is no! Problems of capping and lamination are not solely identified with bond strength. When present, they are fractures at affected "tensile" strength. These result from brittleness and may occur even when the bond strength is excellent. Perhaps a more usable term than tensile strength would be compactibility,* but this too has been used to indicate an absolute strength [6]. To establish a scale for bonding propensity that is applicable to all materials, one might use a dimensionless number based on the tensile strength obtained per unit of compression pressure, σ_c. Unfortunately, the amount of the applied σ_c that is actually contributing to added σ_T varies over a very large range as the ρ_r changes. At very high values of ρ_r, a significant fraction of σ_c in the usual uniaxial die system is counteracting die wall friction or is present as an ineffective hydrostatic stress component within the die. The ineffective portion of the hydrostatic stress is the intraparticle (microregion) hydrostatic stress, which does not produce plastic deformation of the particle† and, therefore, does not produce a change in

*It is preferred that compressibility not be confused with compactibility. Gases are highly compressible, but do not readily make compacts; compressibility refers to volume change.
†From strength of materials, one learns that stress is divided into hydrostatic and deviatoric stresses. Only the deviatoric component produces plastic deformation.

bond strength at the contact regions. Only an unknown portion of the total σ_c may be considered to be the bonding active compression stress (BACS).

C. The Indentation Hardness

It is believed that the indentation hardness measurement correlates more consistently with the magnitude of the BACS. It offers several advantages over the total compression pressure. Figures 1 through 5 show data obtained with a few different materials (also see Table 1). Two hardness values are shown: one the impact or, essentially, instantaneous value, H_0, and the other a long indenter dwell time value, H_{30}. Within the solid fraction range readily accessible in most pharmaceutical laboratories, the log H versus ρ_r plots are nearly linear. Note that σ_c usually is between H_0 and H_{30}.

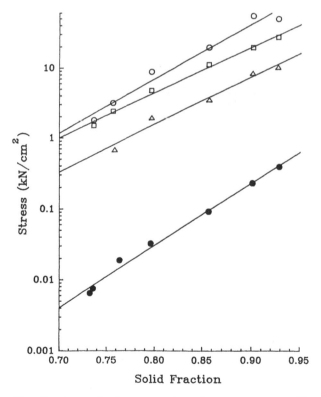

Fig. 1 Strength of sucrose: ○ — impact hardness; □ — compression pressure; △ — hardness, 30-min dwell time; ● — tensile.

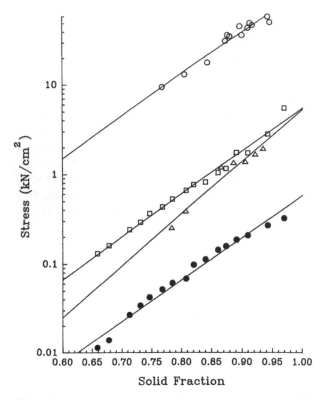

Fig. 2   Strength of sorbitol: ○ — impact hardness; □ — compression pressure; △ — hardness, 30-min dwell time; ● — tensile.

This is assumed to be a result of the difference of dwell time. However, the relative location of the σ_c data and its slope do not fit an understood or predictable pattern. It is assumed that effects from the total confinement of the powder within the die often contribute much to σ_c. While not used in the indices, hardness values for two additional dwell times for the indenter are shown in Table 1.

   For use in a bonding index, the indentation hardness test should produce additional consolidation under the indented surface, within the compact, similar to the compression pressure; i.e., the stress must be sufficient to produce the necessary for bonding, plastic deformation of particles. However, the free surface around the indenter may heave, and, thereby, the excessive hydrostatic stress under the indenter is limited. For most of the materials shown in Figs. 1–5 the maximum compression stress

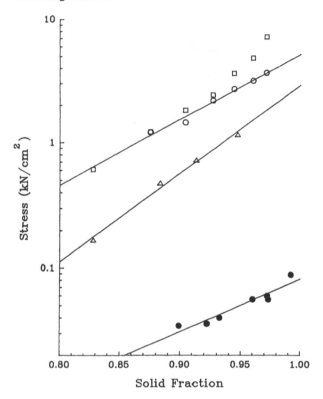

Fig. 3 Strength of aspirin: ○ — impact hardness; □ — compression pressure; △ — hardness, 30-min dwell time; ● — tensile.

applied was not sufficient to introduce extensive departure from a linear log σ_c versus ρ_r relationship. (Earlier experience has shown this departure to occur at high solid fraction for σ_c but not for H_0 [7].) However, with aspirin, log σ_c versus ρ_r clearly is nonlinear even at relatively low solid fractions. The reason for this is not known.

For materials that are difficult to tablet, an extended punch dwell time during the making of the compact may enable one to produce a flawless compact. This freedom to vary the compression pressure in any way needed is important. The use of σ_c in the bonding index would restrict this variation of strain rate used during compaction and could limit the success at making flawless compacts. The indentation hardness measurements do not have these constraints. This is an added reason for using hardnesses instead of σ_c in any bonding index.

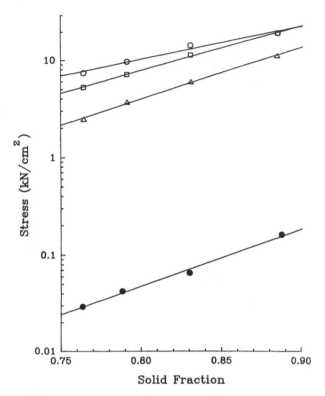

Fig. 4 Strength of calcium sulfate-dihydrate: ○ — impact hardness; □ — compression pressure; △ — hardness, 30-min dwell time; ● — tensile.

IV. THE BONDING INDICES

Based on the above arguments, σ_T/H_0 had been designated [8] as a bonding index; H_0 is the impact hardness. More recently [9], it has been called the worst-case bonding index (BI_w) because $H_0 > H_t$, where t designates a strain rate slower than the impact method of measuring hardness.

A. Hardness and Indenter Dwell Time

H_0 is obtained with a dwell time of the indenter on the compact of less than a millisecond [10, 11]. thus, the viscoelastic decay of the stress during the dwell time is minimal. However, using a different apparatus, the dwell time at maximum penetration for obtaining an indentation hardness can be made any selected length of time. If that dwell time is 30 min, the hardness

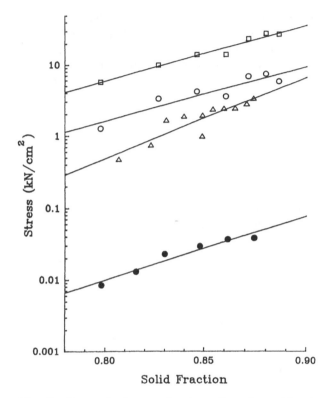

Fig. 5 Strength of acetaminophen: ○ — impact hardness; □ — compression pressure; △ — hardness, 30-min dwell time; ● — tensile.

will be designated as H_{30}. H_{30} would be expected to be less than σ_c. For this case of constant displacement for a selected dwell time, an equivalent strain rate would be hard to estimate, but qualitatively the similarity of long dwell time and a very slow strain rate are obvious.

Since tableting performance of a material is not expected to involve extremely slow strain rates or very long dwell times such as 30 min, a bonding index based on the ratio σ_T/H_{30} would represent a practical [9] best-case bonding index (BI_b). $H_0 > H_{30}$ and $\sigma_T/H_{30} > \sigma_T/H_0$. Real tableting operations probably will correlate closer with the rank order of BI_w than BI_b. However, the theory of tablet bonding indicates that viscoelastic materials produce much better bond than nonviscoelastic materials [2, 3]. Therefore, both the magnitude of BI_w and BI_b should be considered, especially the difference between the two. A large difference is desired, especially when BI_w is on the low side, but it will mean that the tableting performance could be influenced dramatically by the speed of the tableting machine. Table 1 shows data obtained using several dwell times. The choice to use H_{30} for the best case bonding index was arbitrary. Probably an equally useful value could be obtained by comparing shorter dwell time data.

B.   When H_0 Is Inappropriate for Use in the BI

The discussion of the BACS concluded that H is a more suitable parameter for the denominator of the bonding index than is σ_c. When making the test specimens, the dwell time of the punch (elapsed time at maximum displacement) is 1 to $1\frac{1}{2}$ minutes. Seldom is $H_0 < \sigma_c$ in the linear portion of the semilog plots; whenever it is, one has a material that has the potential for tableting problems. If $H_0 << \sigma_c$, the material has very undesirable tableting characteristics. Furthermore, the bonding indices based on H values will not be meaningful.

How can H_0 be less than σ_c when the dwell times are so different? A possible, but unproved, explanation can be given. It is believed that when the particles are very hard and the bonding between them is weak, particle movement, sliding over each other, during the formation of the dent occurs at relatively low stresses. The shear strength of the compact is weak and relieves the stress before the applied stresses becomes sufficient to produce plastic deformation of the particles; therefore, $H_0 < \sigma_c$. Fortunately, the observation that $H_0 << \sigma_c$, alerts the experimentalist that he has a special, unusual case, in which the hardness should not be used in the BI. When this occurs, the bonding index is estimated for the regions where the log H plot is nearly parallel with the log σ_c plot by using the ratio σ_T/σ_c. An extreme example of this is with acetaminophen shown in Fig. 5; the compression pressure is nearly 4.2 times the impact indentation hardness in the regions of higher solid fraction. If the bonding index were based on the hardness, BI_w would be approximately 0.0075. However, if based on the compression pressure, the estimate for the bonding index becomes approximately 0.0018. The ratio H_0/H_{30} for acetaminophen is only ~1.2, a relatively small value; i.e., it is not very viscoelastic. Nevertheless, one must ask is the 0.0018 value a BI_b or a BI_w bonding index. Since σ_c is based on a $1\frac{1}{2}$-min punch dwell time, it may be neither, but probably it is closer to the best case than to the worst case. If it is taken as the best-case bonding index, the worst case should be $0.0018/(H_0/H_{30}) = 0.0015$.

Acetaminophen has very undesirable tableting characteristics. The fact that one must settle for a substitute for the impact indentation hardness is an indication that the material is a difficult one to tablet. Fortunately, this extreme is a very rare case. With phenacetin and maize starch, σ_c was only slightly larger than H_0; the bonding indices were calculated in the usual way. However, one should recognize that whenever $\sigma_c > H_0$, the material may not tablet satisfactorily. Possibly an index of tableting performance could be defined using this difference; but in the absence of a certain explanation for the relative values of σ_c and H_0, such an index has not been defined. However, it clearly is an indicator of problems that accompany low shear strength.

V. A BRITTLE FRACTURE INDEX

A. Strength When a Macroscopic Defect Is Present

Size is a key consideration for the defect. Microscopically, the perimeter of the true contact regions between particles is a region of stress concentration. However, these are "building units" for the compact and are always present. For the stress at the defect to be treated in a continuum sense, the defect must be large compared to the pores within the compact. The compact fails by brittle fracture even when the interparticle bonding may be a ductile process [2,3]. The macroscopic defect only enhances this brittle behavior. Thus the brittle fracture index is a measure of degree of brittleness. When the brittle fracture index is low, the compact is not sufficiently brittle to be a serious problem in processing.

The Griffith crack theory for the fracture of brittle materials is based on the concept that a crack will be initiated when the incremental change of elastic energy during crack growth provides the incremental gain of surface energy for the new surfaces. Furthermore, it is assumed that the origin of the crack is from a defect site where the elastic stress is concentrated. It is higher than the nominal stress, and therefore it would be the first region to reach the stress level needed for the crack to grow. Since the crack tip may continue to be a stress concentrator, the crack often continues to propagate. Therefore, the material fails at a much lower applied stress than expected from the theoretical bond strength and/or theoretical shear strength. However, if a material is able to relieve by plastic deformation some of the stress in the region of concentrated stress, then the stresses may not build rapidly to the required level for the crack to propagate; i.e., the measured strength will approach closer to the nondefect value. For these cases the Griffith theory is modified to include the absorption of energy by the plastic deformation. Thus, the force required to produce fracture is increased over the Griffith case; i.e., the fracture strength is increased. This is the basis on which a brittle fracture index (BFI) is developed [8, 9, 12]. A stress concentrator, a small (1.1-mm diameter) round hole, is put in the center of the tablet and the measured tensile strength is compared with that of a tablet without a hole. The hole is made during the compression by a retractable, spring-loaded pin extending from the center of one punch. The ratio of the two values depends on the ability to "consume" elastic energy around the hole by plastic deformation, i.e., to increase the work of separation. For this measurement, the platen movement rate is kept the same as for the defect-free compact to which its strength is compared.

Elasticity theory for an isotropic material indicates that for the biaxial stress state induced during the transverse compression tensile test, the stress concentration factor is about 3.2 times.* Since compacts are not isotropic homogeneous bodies, the factor would vary because of these density gradients within the compacts. The specific stress concentration factor for each compact is unknown. For simplicity, the factor 3 was adopted as the stress concentration factor at the edge of the hole in a compact. Thus, it is assumed that in the absence of stress relief by plastic deformation around the hole, a small round hole at the center of the compact would result in an apparent strength of one-third that of a compact without the hole. However, if the material is capable of relieving the stress around the hole by plastic deformation before it fractures, then the stress around the "defect" cannot build up to three times the surrounding stress. Among various materials the stress relief will range from zero to nearly total. Thus the tensile strength of the compact with the hole in it will range from one-third to nearly equal that of the compact without the hole. The brittle fracture index is based on this, but the numbers are approximately normalized by using the equation BFI = $(\sigma_T/\sigma_{T_0} - 1)/2$, where σ_{T_0} is the apparent tensile strength when the hole is present.† Experience teaches that when the BFI is less than 0.20, there will not be a problem with fracture during tableting on a rotary press unless the bonding is very weak. The indices values provide the warning of this problem combination. (It is desirable to have the bonding index greater than 0.005.) Since there is some fracture dependence on the bonding index, both numbers should be considered; i.e., one cannot set absolutes for either number.

VI. A STRAIN INDEX

The magnitude of the elastic recovery following the application of enough stress to produce plastic deformation might be stated as the maximum

---

*The compressive stress at the center of the compact is 3.7 times (and normal to) the induced tensile stress [4].

†Some materials have been observed to have a brittle fracture index greater than one. The scale still is valid. It is a very brittle material and for tableting use will require a diluent that reduces brittleness to an acceptable level, less than ~0.2.

elastic strain for that specific material. For two spheres, if one substitutes $H_0 \pi a^2$ for the force in Hertz's equation for elastic deforming spheres, one obtains the equation $Z_e = 3\pi a H_0/4E'$, where Z_e is the elastic recovery distance of the center of the sphere, a is the chordal radius of the contact of the two surfaces before recovery, and $1/E' = \Sigma(1 - \nu^2)/E$ for the two materials, ν is Poisson's ratio, and E is Young's modulus; or $Z_e/a \propto H_0/E'$. While not a true strain, either Z_e/a or H_0/E' provides a relative value for the elastic strain following plastic deformation. If one views the particles as spherical indenters acting on each other, the possible relationship to the magnitude of the elastic recovery following compression of a compact is apparent; hence, it is a valid strain index. Since the strain index (SI), H_0/E', is obtained from the impact hardness experiment [8] without doing any additional experiments, it is available for use; however, the principal use has been to show that the tendency of compacts to fracture does not correlate with the amount of elastic displacement. Instead, fracture propensity correlates with the brittleness of the material as indicated by the BFI.

VII. EXPERIMENTAL PROCEDURES*

A. The Tablet Press, the Test Specimen

Because compacts made from some materials develop fracture lines, the usual uniaxial compression-decompression cycle is not a satisfactory way to make a test specimen compact. With square compacts, it is possible to split the die across a diagonal so that one can use a triaxial decompression process, i.e., reduce the die wall pressure simultaneously with the punch pressure. This reduces the shear stresses (continuum perspective) within the compact during the decompression. Because compacts are not isotropic homogeneous bodies, the internal shear stresses will not be zero at all points even under the best of conditions. However, if the mean die wall and the mean punch pressures are held equal, the internal shear stresses should be below the strength of the compact, in which case no fracture lines are introduced. Satisfactory test specimens, compacts free from macroscopic defects, can be made from nearly any material by using nearly equal die wall and punch pressure conditions during decompression.

The triaxial decompression press shown in Fig. 6 was built specifically to provide the needed control. This system permits the selection of compact thickness, dwell time of compression, the ratio of die wall and punch

---

*Versions of the apparatuses described and used are covered by U.S. Patents 4,880,373, 4,885,933, and 4,957,003 and owned by The Upjohn Company, Kalamazoo, MI. Applications have been filed in other countries. Licensing to build from Upjohn plans is available at reasonable cost.

Fig. 6 Triaxial decompression tablet press.

pressure, and the duration of decompression. The powder is poured into the die; care is taken to distribute it as evenly as possible. The compression force is applied and the maximum force held constant for a fixed time, perhaps 1 or $1\frac{1}{2}$ min. The triaxial decompression rate is controlled by selecting the time interval over which decompression will occur, usually 1 min is chosen. This operation has been automated by using a computer to control the timing and the pressures during unloading.

B. Solid Fraction Determination

The apparent density is calculated from the weight and dimensions of the compact, which is divided by the absolute density of the material to give the solid fraction. The tablet dimensions are measured immediately after the indentation hardness test or immediately before the tensile strength test. The average of five thickness values is used. The time for measurement is chosen to minimize the effect of the unavoidable change of dimensions with time that may occur after removal from the die. The solid fraction at the time of the test is the critical value. When the absolute density is unknown, a gas pycnometer is used to determine it.

C. Indentation Hardness Measurements

For the tableting indices, the indentations are produced by a 2.54-cm-diameter steel sphere. The chordal diameter of the dent is kept to less than 4 mm and is made in the center of the square face of the test specimen. Because a large spherical indenter is used to make a small dent, the normal body stresses are oriented within a few degrees of normal to the original surface. (Friction between the compact and the indenter surface would introduce a limited shear stress.) The large indenter is used to avoid the dislodging of particles on the surface. Also it is intended to provide the needed consolidation below the indenter and to diminish the pile up around the indenter.

Figure 7 shows the dynamic indentation hardness apparatus, and Fig. 8 shows the longer dwell time fixture on the drive assembly common to it and the tensile strength test. H_0 is obtained by bouncing the sphere off the surface of the compact. H_{30} is obtained by pushing the spherical indenter into the compact and then holding its position fixed for 30 min while the force decays. Because the average pressure under the indenter is used, the distribution of stress under the indenter, though believed to be very different in the two cases, is not considered. (The stress distribution under the indenter is unknown but will depend on the viscoelastic properties of the material and will be influenced by the dwell time as well as by the porosity.) In practice, a comparison of the magnitudes of H_0 and H_{30} will indicate important differences among materials. An organic material whose properties are $H_0 \geq 5H_{30}$ would be a strongly bonded material. However, if $H_0 \leq 1.2H_{30}$, the bonding will be weak. Since the bonding of a viscoelastic material will be influenced by the dwell time of the press in which the tablet is made, the magnitudes of the differences between the two are needed to indicate the magnitude of this effect; indeed both values are important.

To avoid fracture during the hardness tests the tablet is placed in a split die and some die wall pressure applied before the indentation is made. Especially important with the dynamic test is a backup support, a backup punch is used to push the tablet to the front of the die after some die wall pressure has been applied. Thus, the square tablet is firmly supported on five surfaces during the indentation. An evaluation was done of whether the die wall pressure would influence the hardness value observed. While the answer is yes, for the cases studied the effect was so small that it was insignificant. Of course when the die wall pressure is needed to avoid fracture, the magnitude of the effect cannot be determined easily because the "window" between the attainable (with the current apparatus) die-wall pressure that prevents fracture and that failing to prevent fracture may be small.

Fig. 7 Impact indentation hardness apparatus.

D. Dent Size, Important Considerations

The relative dent size to compact size must be within certain limits. Arbitrarily, the dent is made to have a chordal diameter between 2 and 4 mm. For this dent size the thickness of the compact should be at least 9.5 mm; otherwise, the supporting punch on the opposite side from the dent will significantly influence the force required to make the dent. The square compacts are 19.05 mm on a side. With this width, the dent must be made in the very center region; otherwise, the die wall could influence the result. Tabor [13] discusses the influence of the elastic deformation of the "hinterland" away from the indenter and indicates that there are effects when Young's modulus for the material being indented is low [13]. With the compacts the discontinuity at the die wall must be considered. The elastic modulus of the steel die wall is much greater than the elastic modulus of the compact. Also, limits on the size of the indentation must be imposed. From experience, the following is considered satisfactory. The distance from the edge of the dent to the specimen edge should not be less than $5a$, the chordal radius, and the thickness should be at least five times the depth of the dent.

Fig. 8 Long dwell time hardness apparatus.

These recommended values are somewhat different from Tabor's recommendations. Possibly because of the porosity changes under the indenter, the requirements are less stringent. Furthermore, reducing the amount of an expensive medicament that is needed for the indices determination is sufficient motivation to accept a slightly smaller than ideal compact for this applied use. This last is important. Formulators need to know the values of the indices for small lots of material used in clinical studies, when the supplies may be very scarce, so these properties can be matched when production scale-up occurs. This avoids clinical studies having to be repeated with a new, different formulation.

E. Dent Size Determination

For both indentation tests the chordal radius of the dent produced must be determined. The simplest procedure for this is to observe the chordal radius, a, using a low-power microscope with oblique angle, surface illumination. Care must be exercised to eliminate differences among operators. The average of five readings, taken by rotating the compact approximately 72° between each reading, is used. If the compact surface is grainy, the microscopic method may not be as accurate as wanted. A more precise value for the chordal radius may be obtained by using a surface roughness analyzer (e.g., Federal Surfanalyzer 150*), an instrument whose probe moves across the dent and yields data to produce a much enlarged profile of the dent (depth is enlarged much more than diameter). Figure 9 shows the arrangement and the probe of the apparatus used with a computer for handling the data. The tip of the probe used with the tablets is a 0.79-mm- (1/32 in.) diameter sphere. (A much smaller probe tip is used for the instrument's original purpose, for surface roughness measurements of metals.) Current practice uses the major portion of the profile to obtain a radius of curvature. A circular segment, least-squares fit is used with the dent depth, measured from the amplified profile, to calculate the chordal radius of the dent. This is done with a menu-driven computer program to handle the data. The design reduces the operator subjectivity of the measurement.

F. Hardness and Strain Index Calculations

H_{30} is the mean pressure under the indenter after a dwell time of one half-hour. Specifically, it is the force observed by the load cell, after the force has decayed for one half-hour, F_{30}, divided by the chordal area of the dent; $H_{30} = F_{30}/\pi a^2$, where a is the chordal radius of the dent.

---

*Federal Products Corporation, 1144 Eddy Street, Providence, RI 02940-9400.

Fig. 9 Apparatus for dent chordal radius measurement.

The equations used to calculate H_0 assume that the work of forming the dent is equal to the hardness, H, times the volume of the dent [10, 11]. Also assumed is Hertz's laws of elasticity during the elastic recovery. Based on experimental evidence [10], Eq. (1) is believed to provide good estimates of the hardness.

$$H_0 = \frac{4mgrh_r}{\pi a^4} \left(\frac{h_i}{h_r} - \frac{3}{8} \right)$$ (1)

where

m = indenter mass
r = indenter radius
h_r = indenter rebound height
h_i = indenter initial height

The strain index may be calculated from the same data [8] using Eq. (2):

$$\frac{H_0}{E'} = \frac{5a/6\pi r}{h_i/h_r - \frac{3}{8}}$$ (2)

A SI for the relaxed case requires additional data, viz., the radius of curvature of the unloaded surface. Because the radius is changing rapidly immediately after unloading, meaningful data are not easy to obtain. Fortunately H/E' for the relaxed cases is not important for the use of indices.

G. Tensile Strength Measurements

The fixture used for the transverse tensile strength is shown in Fig. 10; it is mounted on the common drive mechanism. For this arrangement, the tensile stress is 0.16 times the mean stress at the platen-compact surface [4]. Since the maximum tensile stress exists at the center of the compact, it is assumed that any crack will start in that region. If stress concentrations at the edge of the platens initiate the fracture, the test has not measured the tensile strength. (The stress at the edge of the platen is not known.) Fortunately the shape of the fragments identify these rare cases. Usually this can be overcome by placing pads, blotter or cardboard, between the compact and the platen. When the failure occurs along a central line normal to the platens, the test has been a measure of the tensile strength. The platen force versus time is recorded and used to obtain the time constant of the tensile fracture test.

VIII. NEEDED PRECISION

For commercial application, it is desirable to reduce the need for many duplicate tests. A realistic demand on the needed precision of the data contributes to this. The criterion is to provide a rugged formulation, one that will manufacture satisfactorily even if some variation of the material properties occurs. The experimental precision of the measurements will vary with the material being evaluated; minimum variation occurs with good tableting materials. Fortunately, there is only limited need for precision for clearly problem materials or for very good ones. The greatest precision may be needed to prove that an acceptable elimination of problems has been attained by the excipients in the formulation. Fortunately, this is the range where little variation of the data occurs.

If one wishes to know how the property varies with solid fraction, semilog plots usually are nearly linear, viz. log H versus ρ_r and log σ_T versus ρ_r, over the range of interest. These plots can be used to help spot problems with the data, and when nonlinear curve fitting is used, confidence intervals can be estimated. It is useful to establish the minimum target values for the experiments.

A word of caution is needed relative to sampling. There are standard procedures for obtaining representative samples. The use of mixtures, granu-

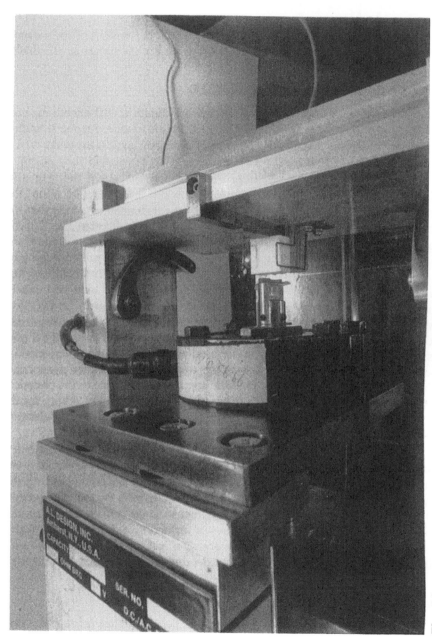

Fig. 10 Apparatus for transverse compression, tensile strength measurement.

lated materials, and direct compaction mixtures requires special care. Often, poor sampling practice introduces much scatter of the data. This becomes obvious when the plots of log H versus ρ_r or log σ_T versus ρ_r are made.

IX. USE OF THE TABLETING INDICES

The tableting indices provide a quantitative evaluation of the tableting performance of a powder or powder mixture. This is obtained without adding "property diluting" excipients. They will detect mechanical property differences among lots that meet all chemical specifications; and they permit a rational approach to the selection of excipients. Perhaps if the user asks himself whether any other system provide as much information about the tableting properties of each individual ingredient, he will conclude that the tableting indices are uniquely useful. Selected data obtained in The Upjohn Company laboratories are shown in Table 3. Hypothetical illustrations of the potential benefits from having a quantitative evaluation of tableting properties have been given [9]. This information could be especially desirable when the dosage is high and the volume of excipient must be kept very small. Without the indices values, it may not be easy to convince the chemical division to produce a product that has better tableting characteristics than the by-chance properties of the most convenient or most economical process. Actually the cost of making the medicament may not be increased by just using carefully controlled operations to produce the needed properties. With the tableting indices, it is possible to provide guidance to the chemical division as they experiment with different crystallization and drying conditions to attain the objective. Also, the indices are useful when evaluating materials available from more than one source.

Values for the BI_w have been observed in the range of 0.001 to 0.04. Currently with only limited studies, BI_b values have been observed in the range 0.002 to 0.14. While tablets can be made from the materials with the lower BI_w values, they may be very soft and excessively friable. The acceptable value is not independent of the brittleness. Usually if the final formulation has a BFI below 0.2, it indicates a formulation that would not give fracture problems. However, if the bonding index value is too low, e.g., 0.001, a slightly lower BFI value would provide a better margin of safety.

X. CONCLUSIONS

The fundamental considerations used to justify the selection of the tableting indices, the strain index, the brittle fracture index, and the bonding indices, have been described. The second bonding index was justified be-

Table 3 Examples of Values Observed for Tableting Indices

Material $\rho_r = 0.9$	BI_w $\times 10^2$	BI_b $\times 10^2$	BFI	SI $\times 10^2$
Avicel[a]	3.4	13.5	0.03	2.3
Sorbitol	0.46	13.7	0.03	0.94
Lactose, Sp. Dr.	0.36	1.1	0.12	1.8
Sucrose	0.40	2.3	0.68	1.5
Ibuprofen[b]	0.76	4.1	0.06	0.6
Aspirin	2.1	5.1	0.19	0.7
Caffeine	1.4	4.5	0.47	1.3
Phenacetin	0.88[c]	1.4	0.43	1.0
Acetaminophen	0.15[d]	0.18[e]	0.03	1.4
$CaSO_4 \cdot 2H_2O$	0.79	1.3	0.08	1.2
Starch (maize)	0.76[c]	0.67	0.80[f]	1.9

[a] PH-101, microcrystalline cellulose.
[b] Lot-to-lot variation regularly observed.
[c] Compression pressure slightly $> H_0$.
[d] $BI_w \times H_{30}/H_0$.
[e] Based on σ_T/σ_c.
[f] At $\rho_r = 0.8$, BFI is 0.44.

cause of the very important role of viscoelasticity in the development of bond strength. Special conditions are described where the indentation hardness should not be used in the bonding index term. Cautious use of the compression pressure is indicated in these cases. The necessary precautions and consideration when using the experimental procedures are described.

REFERENCES

1. H. Leuenberger, *Int. J. Pharm.* 27: 127–138 (1985).
2. E. N. Hiestand, *Int. J. Pharm.* 67: 217–229 (1991).
3. E. N. Hiestand and D. P. Smith, *Int. J. Pharm.* 67: 231–246 (1991).
4. R. Berenbaum and I. Brodie, *Br. J. Appl. Phys. 10*: 281 (1959).
5. E. N. Hiestand and C. B. Peot, *J. Pharm. Sci. 63*: 605–612 (1974).
6. H. Leuenberger, E. Hiestand, and H. Sucker, *Chem.-Ing. Tech. 53*: 45–47 (1981).
7. E. N. Hiestand, Int. Conference on Powder Technology and Pharmacy, Basel, Switzerland, 1978.
8. E. N. Hiestand and D. P. Smith, *Powder Technol. 38*: 145–159 (1984).
9. E. N. Hiestand, *Pharm. Tech. Int. 1*: 22–25 and *Pharm. Tech. 13*(9): 54–66 (1989).

10. E. N. Hiestand, J. M. Bane, Jr., and E. P. Strzelinski, *J. Pharm. Sci. 60*: 758–763 (1971).
11. D. Tabor, *The Hardness of Metals*, Clarendon Press, Oxford, UK, 1951.
12. E. N. Hiestand, J. E. Wells, C. B. Peot, and J. F. Ochs, *J. Pharm. Sci. 66*: 510–519 (1977).
13. D. Tabor, *Rev. Phys. Technol. 1*: 145–179 (1970).

10

Particle Dimensions

Göran Alderborn

Uppsala University, Uppsala, Sweden

I. PHYSICAL STRUCTURE OF PHARMACEUTICAL COMPACTS

A powder can physically be described as a special type of disperse system consisting of discrete, solid particles which are surrounded by or dispersed in air. However, the particles are normally in contact with each other. The interparticulate attractions at the points of contact are relatively weak and the powder is thus characterized by exhibiting a low mechanical strength. The transformation of a powder into a compact is the result of a reduction in the porosity of the powder system, and thus particle surfaces will be brought into close proximity to each other. As a consequence, the number and strength of interparticulate attractions will increase with a subsequent increased coherency of the powder system. The result of the compaction procedure is thus a solid specimen of a certain porosity, normally in the range of 5% to 35% for pharmaceutical compacts.

Although the mechanical strength of a compact is considerably higher than that of a powder, an examination of the literature indicates that it is also adequate to physically describe the compact in terms of or similar to a disperse system, consisting of solid particles surrounded by a gas phase. Several supports for this model can be found in the literature.

1. A comparison between external surface areas of coherent tablets and powders obtained by deaggregation of tablets have indicated a similar

surface area for both these powder systems [1,2]. Hence, although the creation of solid bridges between particles during compression is possible, i.e., the formation of a continuous solid phase between particles, the cross-sectional area of these bridges seems to be small compared to the external surface area of the particles in the compact.

2. The same conclusion has been drawn based on surface area and mechanical strength analysis of compacts after the addition of a lubricant [3]. Furthermore, the removal of magnesium stearate from the compact by dissolution [4] can dramatically increase the compact strength. This indicates that a substantial part of the interparticulate attractions in a compact are formed between solid surfaces which can physically be described as being separated; i.e., separation distances between solid surfaces in the compact are in many cases considerably larger than the separation distances between the molecules and ions which form the solid material.

3. Compaction and the subsequent strength analysis of compacts in liquids [5,6] have also supported that the dominating attraction force between the solid surfaces in the compact is intermolecular attractions forces acting over distances; i.e., the formation of solid bridges between particles is limited.

4. Qualitative inspection of tablets, both upper and fracture surfaces, by electron microscopy has supported [7–10], at least for plastically deforming materials, that the compact consists of discrete particles packed very closely to each other.

5. Finally, direct relationships between the mechanical strength and the pore size characteristics of compacts of nonporous particles [11–14] as well as for compacts of porous, granulated particles [15,16] have been presented. This observation can be explained by the model that the compact consists of small particles cohered to each other and that the size of and separation distance (relative position) between those "compact particles" governs the compact strength.

Thus, a pharmaceutical compact can thereby normally be described as an aggregate of smaller particles which are strongly co- or adhered to each other. The gas phase in the compact can be described as the continuous phase and consists of a three-dimensional network of connected pores. The compact can therefore be physically described by both the characteristics of the interparticulate pore system (e.g., porosity, pore size distribution, and pore surface area) and by the properties of the particles constituting the compact (e.g., surface area, particle size distribution, and the packing characteristics or relative positions of the particles). The distinction between what is normally referred to as a powder or as a compact will in practice be based on the mechanical properties of the assembly of particles or the specimen while the physical structure in a broader sense is similar.

II. THE CONCEPT OF COMPACTIBILITY OF POWDERS

The compactibility of a powder mass has been defined as [17] the ability of the powder to be transformed into a compact of a certain mechanical strength. The compactibility of the powder is thus an essential and fundamental property of a tablet mass and a determining factor for successful tablet production. According to this definition, the term compactibility is related to the ability of the powder to cohere during the actual compression phase (i.e., a process-related parameter). Thus, compactibility is normally assessed by the relationship between compact strength and a process variable, mostly the maximum force or pressure applied on the powder during the compaction. However, other process-related factors, such as the time periods for the different parts of the compaction cycle, can affect the obtained mechanical strength.

The concept of compactibility of a powder is complicated by the fact that the time period which elapses between compaction and strength analysis as well as the storage conditions (primarily the relative humidity of the environment) during this period can dramatically affect the strength of the compact [13,14,18–20]. Thus, the compactibility of a powder can refer to two different powder properties—the ability to cohere during the actual compression phase or the ability to cohere during compaction and subsequent storage of the compact. Therefore, it seems valuable to distinguish those two definitions when the properties of powders are evaluated with respect to their compactibility. Unfortunately, there is only a limited literature on this subject. Thus, in this chapter, the importance of post-compaction strength changes for the relationship between dimensions of particles before compaction and the mechanical strength of the tablets are not considered.

Another complication is that mechanical strength is not a constant property of a compact. A series of reasons for this can be identified, such as the stress direction used during the analysis of compact strength (e.g., tensile strength in radial and axial directions [21,22]) and the type of stress applied to the specimen (e.g., tensile or compressive strength [23]). Whether such aspects of mechanical strength analysis of a specimen are important for the relationship between a particle property and the compact strength is not well known. Thus, it seems practical to treat the concept of the mechanical strength of the compact as a well-defined property of the specimen as long as the experimental procedure used for the evaluation of the compact strength is similar between different studies. In most studies, the diametral compression test has normally been used and presented as either the crushing/breaking strength of the specimen or recalculated to the tensile strength [24]. This seems the most suitable alternative for two reasons: (1) A fracturing of the tablet in a plane parallel to the axis of formation of the tablet will result in a measure which represents the mean strength of the compact [25]. (2) A tensile failure of the compact seems to be the most attractive type of test situation when the strength of a compact is described in terms of the particle-particle interactions in a compact.

III. RELATIONSHIP BETWEEN DIMENSIONS AND VOLUME REDUCTION PROPERTIES OF PARTICLES FOR COMPACT STRENGTH

It has been argued that a suitable physical model of a pharmaceutical compact is an aggregate of smaller particles, which can be described as discrete units strongly cohered to each other. The consequence of this model is that the tensile strength of a given plane within a compact can theoretically be described in terms of the sum of all interparticulate attractions in that plane. When the compact is stressed across that plane, a failure will thus occur when the applied stress exceeds the total attraction strength in the plane, i.e.,

$$\sigma_c = F_b n_b \tag{1}$$

where

σ_c = tensile strength of compact (N/m^2)
F_b = mean bonding force of individual interparticulate bond (N)
n_b = number of interparticulate bonding zones per cross-sectional area (m^{-2})

In practice, the interparticulate bonding forces will vary in strength across the plane and can thus be described as a distribution in interparticulate bonds, i.e., the number of bonds as a function of bonding force.

The validity of Eq. (1) requires probably that the fracture plane be formed instantaneously; i.e., all bonds over the whole cross-sectional area of the fracture plane are separated simultaneously. However, in accordance with fracture mechanics, the failure in practice is often the result of a kinematic process involving the initiation and propagation of the fracture [26]. The existence of a kinematic process during fracturing diametrically has also been demonstrated for pharmaceutical compacts [27]. The practical consequence is that there might be a discrepancy between the theoretical strength of a specimen and the stress needed in practice to fracture a compact, and this discrepancy is probably related to the mechanical properties of the particles constituting the compact. The importance of the kinematic fracturing process for the strength of pharmaceutical compacts in relationship to the particle-particle interactions and the mechanical properties of the particles from which the compact is formed is, however, not yet well understood. In the following discussion, the strength of a compact will for simplicity be described as the sum of the interparticulate attractions in the fracture plane.

The variables F_b and n_b in Eq. (1) have not yet been quantified with respect to pharmaceutical compacts. Fundamental studies on the mechanical strength of compacts have therefore almost entirely been focused on the effect of secondary factors on the compact strength (i.e., material and process factors in relationship to compact strength). The approach here is mainly qualitative, and few attempts to quantitatively relate the properties of the material to the strength of the compact have been presented. However, for loosely packed beds of particles, quantitative expressions which describe the relationship between the strength of the powder bed and the packing characteristics of the particles in the bed have been presented. The theoretical basis of this field has been presented by Rumpf, who developed the following expression [28] for the strength of an aggregate of monodispersed spheres:

$$\sigma = \frac{9(1 - e)kH}{8\pi d^2} \tag{2}$$

where

σ = tensile strength of aggregate (N/m^2)
e = porosity of aggregate (—)
k = coordination number, i.e., number of contact points for one sphere (—)
H = bonding force at a point of contact (N)
d = diameter of sphere (m)

The factors governing the strength of such powder beds are consequently the size of the particles, their relative positions, and the force of cohesion between particles. Deviations from this simple model make a quantitative description more difficult, but it has been stated that both the distribution in particle size as well as the shape characteristics of the particles affect the mechanical strength of the bed [29].

The physical analogy between beds of powders and compacts discussed above leads to the conclusion that for compacts the dimensions of the particles constituting the compact and their relative positions might also be of direct importance for the mechanical strength of the compact. These characteristics of the compact will probably affect both the bonding force and the number of the interparticulate bonds.

The characteristics of the "compact particles" as well as their relative positions are difficult to assess. However, it is reasonable to assume that the particulate characteristics of the particles before compaction are important for the characteristics of the "compact particles" and thus affects the compactibility of the powder. Thus, an understanding of the effect of a certain particle characteristic for the mechanical strength of compacts can form the basis for a fundamental understanding of the compaction process. Furthermore, this knowledge is also of practical interest in the formulation of direct compactible formulations, e.g., with respect to the importance of batch variations of the raw materials.

The assumption that the particulate characteristics of the particles before compaction are related to the characteristics of the "compact particles" seems to form the basis for attempts to quantify relationships between tablet strength and original particle size. For example, the following expression was used by Shotton and Ganderton [30] to relate compact strength to original particle size:

$$F = Kd^{-a} \tag{3}$$

where

F = force needed to break compact (N)
d = diameter of particle (m)
K, a = constants

The compaction properties of compacts of sodium chloride and hexamin could be described by Eq. (3), but the value of the exponent differed between the materials (Fig. 1). They concluded that this difference was related to the properties of the materials. One such important material property is the volume reduction behavior of the material during the compression.

The compaction of particles into a coherent compact normally requires such high compaction pressures and such a dramatic reduction in volume of the powder bed that the dimensions of the particles can be dramatically changed. The incidence of such changes is often described as the volume reduction mechanisms which a material exhibits while compressed and is probably related to the mechanical properties of the particles (i.e., strength and deformability). A discussion of the effect of particle dimensions on compact strength must thus consider at least the volume reduction properties of the material and be described as a relationship between three variables—original particle dimensions, main volume reduction behavior, and compact strength.

For powders consisting of more-or-less nonporous particles, the main volume reduction mechanisms of the particles during the compaction are particle rearrangement (repositioning), deformation, and fragmentation. If

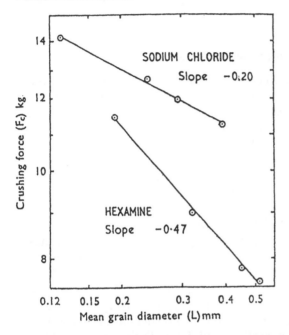

Fig. 1 Log-log relationships between the diametral compression strength of compacts of sodium chloride and hexamin as a function of original particle size. (From Ref. 30 with permission of the copyright holder.)

the first mechanism dominates the volume reduction process (e.g., during the compaction of very fine particulate powders), it is reasonable that the characteristics of the particles before compaction will affect the compact strength in accordance with the general statements which are valid for loosely packed beds of powders. However, a permanent deformation of the whole or parts of a particle as well as particle fragmentation during the compression phase can result in other types of relationships.

The aim of this chapter is to give some examples of experimentally observed relationships between particle dimensions before compaction and the mechanical strength of the resultant compact. It is obvious that the compaction of a powder into a compact and a controlled analysis of compact strength are complex procedures and generalized rules concerning factors of importance for compact strength are therefore difficult to establish. Nevertheless, the aim of this chapter is also to make, based on the model of the physical character of pharmaceutical compacts discussed above, an attempt to present a fundamental explanation for the effect of particle dimensions on compact strength.

IV. PARTICLE SIZE

A. Effect on Volume Reduction Characteristics

It is generally assumed that a change in particle size not only affects the external surface area of particles but also the mechanical properties of the particles. Generally, a decrease in particle size increases its mechanical strength [31]. The underlying reason for the change in particle strength with particle size is generally attributed to a reduced probability of the existence of defects in the crystal structure. These defects can act as sites at which a crack is initiated during a fracture. It has also been reported [32] that when the particle sizes are reduced markedly, a transition from a brittle to a plastic behavior takes place—in other words, a change in the dominating mechanical characteristic of a particle.

The size effects on the mechanical properties of particles indicates that the volume reduction characteristics (i.e., the propensity of the particles to fragment or deform during compression) are also affected by a change in particle size. The effect of particle size on volume reduction behavior of a powder has been assessed primarily by studying the relationship between applied pressure and volume of powder during the actual compression phase by Heckel profiles and force-displacement profiles. Duberg and Nyström [33] found that the linearity of the Heckel profiles (Chapter 3) increased with a decreased particle size during compaction of a series of materials. This can be interpreted as a reduced degree of fragmentation of the particles during compaction.

There are, however, few reports with the object of directly assessing the fragmentation propensity of a material as a function of particle size. By the aid of air permeability measurements of tablets compacted at a series of compaction pressures, Alderborn and Nyström [34] found that the tablet surface area as a function of compaction pressure increased more markedly with a reduction in original particle size (Fig. 2). However, this corresponded to a reduced relative increase in surface area with pressure; i.e., the changes in tablet surface area in relationship to the starting surface area decreased with a decreased particle size. A similar observation was reported by Leuenberger et al. [35] for the compaction of lactose (Fig. 3). It can thus be concluded that the readiness of individual particles to fracture into smaller units is reduced with a reduced particle size, although the absolute change in surface area is larger due to the small dimensions of the involved particles.

As seen in Fig. 2, almost linear relationships between tablet surface area and compaction pressure were obtained, and the slopes of these profiles can thus be used to express the rate of changes in surface area with pressure, dS/dP (S = tablet surface area and P = compaction pressure).

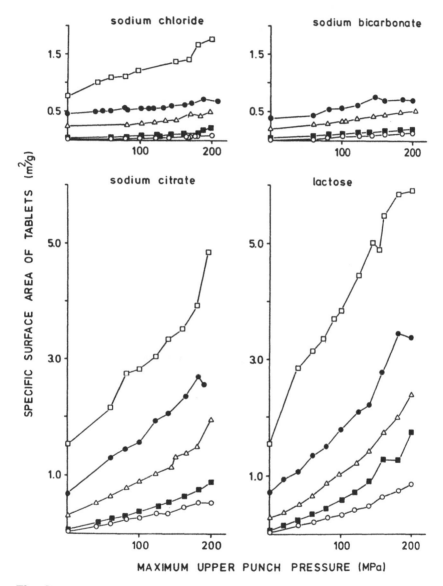

Fig. 2 The effect of compaction pressure on the permeametry surface area of tablets, compacted of different particle size fractions: ○ = 5–10 μm, ■ = 10–20 μm, △ = 20–40 μm, ● = 63–90 μm, □ = 125–180 μm. (From Ref. 34 with permission of the copyright holder.)

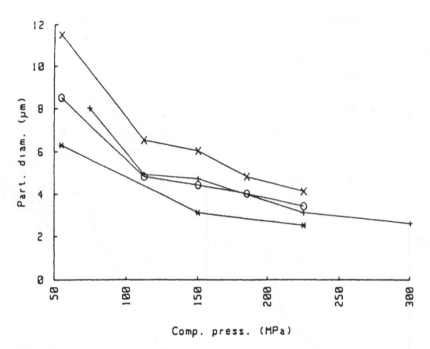

Fig. 3 The effect of compaction pressure on the estimated mean particle diameter in compacts of α-monohydrate lactose, compacted of different particle size fractions: ✳ = 32–45 μm, + = 100–125 μm, ○ = 125–160 μm, × = 315–400 μm. (From Ref. 35 with permission of the copyright holder.)

This rate of change increased with increased original surface area in a nonlinear way, and the relationship between rate of changes and original surface area can be expressed with the differential equation

$$\frac{dS}{dP} = kS^n \tag{4}$$

where

> k = material-related coefficient describing the propensity of the material to fragment
> n = constant

Thus, the compression process is described in terms of a size reduction process, and the pressure required to obtain a given reduction in particle size is related to the starting particle size and the propensity of the material to fragment. In the literature on the comminution of particulate solids, a number of attempts to quantify the relationship between energy consumption and the reduction in particle size during comminution has been presented. The most well-known expressions are the equations given by Kick, Rittinger, and Bond [36]. However, it has been argued [36] that all these expressions are interpretations of a single general equation:

$$\frac{dE}{dD} = cD^n \tag{5}$$

where

$$
\begin{aligned}
E &= \text{energy consumed} \\
D &= \text{particle size} \\
c, n &= \text{constants}
\end{aligned}
$$

Equations (4) and (5) express the same approach to relate size reduction to the actual particle size at hand and to a process characteristic. In the case of compaction, the choice of pressure or stress applied to the powder seems a reasonable approach to characterize the process since this is the most common parameter in studies on the compression/compaction properties of materials. Furthermore, applied stress is commonly used in relationship to strain in the evaluation of mechanical properties of specimens.

Alderborn and Nyström found that a value for n in Eq. (4) of ⅔ gave a linear relation between dS/dP and S, and the following expression was developed:

$$S_p^{1/3} - S_0^{1/3} = \frac{kP}{3} \tag{6}$$

where

$$
\begin{aligned}
S_p &= \text{tablet surface area at pressure } P \text{ (e.g., cm}^{-1}) \\
S_0 &= \text{original surface area, i.e., powder surface area before compaction (e.g., cm}^{-1}) \\
P &= \text{compaction pressure (e.g., MPa)}
\end{aligned}
$$

The ratio ⅓ was denoted the fragmentation propensity coefficient (C_{FP}) and was regarded as a material constant describing the propensity of the material to fragment while compressed independent of original particle size (Fig. 4).

In analogy with the discussion above, it seems reasonable that also the readiness of a particle to deform under stress will be reduced with a reduction in particle size. However, the question regarding the degree of deformation, both plastic and viscoelastic, in relationship to particle size and tablet compaction is not fully understood. However, it has been ob-

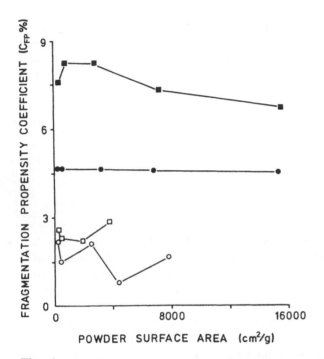

Fig. 4 Relationships between the fragmentation propensity coefficient, calculated according to Eq. (6), and the original powder surface area for sodium chloride (○), sodium bicarbonate (□), sodium citrate (●), and α-monohydrate lactose (■). (From Ref. 34 with permission of the copyright holder.)

served [37] that a reduction in particle size reduces the effect of the rate of compression on the relationship between tablet porosity and applied compression pressure. A reasonable interpretation of the data is a reduced deformability (viscoelastic) of the particles with a reduced particle size.

B. Effect on Tablet Strength

1. Relationships Between Original Particle Size and Compact Strength

It has been claimed that one of the most important variables for the mechanical strength of a compact is the size of the particles before compaction [38], and, as a rule of thumb, it is normally assumed that a decreased original particle size increases the tablet strength. Based on the physical characteristics of a pharmaceutical compact described above, the rule of thumb seems a reasonable statement provided that the compression process only affects the appearance of the particles to a limited extent. The relationship between original particle size and tablet strength has also been extensively studied in the literature. However, an examination of the presented data can give a somewhat confusing picture with respect to this relationship. For some materials (e.g., lactose and sodium chloride [12,30,39–41]) an increased tablet strength with a decreased particle size has been found. However, the reverse, or more complex, relationships have also been reported for compacts of sodium chloride [42,43]. Also for other materials, complex relationships with, for example, a minimum in the compact strength–particle size profile have been presented [42]. Finally, an almost unchanged tablet strength with variations in particle size has been observed for lactose [44], aggregated calcium phosphate [42], and sodium bicarbonate [45].

In the following discussion, some examples of data will be presented in more detail which forms the basis for the mechanistic discussion of the effect of original particle size on compact strength. The procedure used for preparing suitable model materials has normally been the classification of unmilled particles, e.g., by dry sieving.

Alderborn and Nyström [42] studied the effect of particle size on the mechanical strength of tablets of a series of materials consisting of relatively coarse particles (90–1000 μm). The substances were chosen to represent materials with varying volume reduction characteristics during compression, from low- to high-fragmenting materials. The classification of materials with respect to their degree of fragmentation was also later supported by tablet surface area measurements [46].

The most dramatic or complex relationships between original particle size and tablet strength were associated with a change in shape characteristics of the particles with a change in particle size. This was due to the formation of secondary particles with an increased particle size or to a change in the number of primary particles constituting the aggregate. It is obvious that such changes in particle structure can cause a shape-related change in the particle-particle interactions (see below) but also change dramatically the volume reduction behavior of a powder. This change in the appearance of the particles formed the basis for the explanation for the dramatic effect of particle size on the mechanical strength of tablets of dendritic sodium chloride and of Starch 1500.

For the other materials used in the study, the volume reduction properties, as evaluated by the ratio between the tensile strength measured parallel and normal to the compression axis during the tablet formation, did not seem to change dramatically with particle size. Furthermore, the particle shape was similar for all particle sizes used for the respective mate-

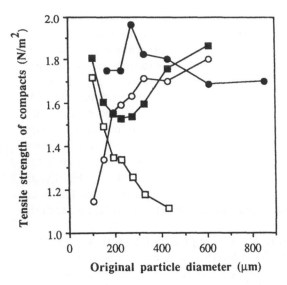

Fig. 5 The effect of original particle size on the tensile strength of compacts of α-monohydrate lactose (□, compaction pressure of 215 MPa), sodium chloride (○, compaction pressure of 160 MPa), saccharose (●, compaction pressure of 265 MPa), and Emcompress (■, compaction pressure of 160 MPa). (Drawn from data given by Alderborn and Nyström [42].)

rial. For these materials, three types of relationships between particle size and tablet strength were found.

The most common relationship found, e.g., for lactose and sodium citrate, was an increase in tablet strength with a decrease in particle size, and the effect was most pronounced at the lower-particle-size range (Fig. 5). A tablet strength more-or-less independent of particle size was observed for compacts of aggregated calcium phosphate dihydrate and saccharose (Fig. 5). Finally, for the plastically deforming or low-fragmenting material cubic sodium chloride, the general tendency was an increased tablet strength with increased particle size with the most pronounced effect at the lower-particle-size range (Fig. 5). In a later study [45], a tablet strength more-or-less independent of particle size was observed for another low-fragmenting material, sodium bicarbonate.

Researchers in Groningen, Holland, have in a series of papers studied the compaction properties, including the relationship between particle size and tablet strength, of different types of lactose. Both for α-monohydrate lactose and α-anhydrous lactose [41], a decreased particle size increased the compact strength (Fig. 6). The tablet strength for compacts of the anhydrous

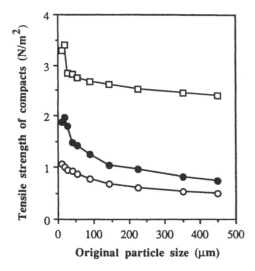

Fig. 6 The effect of original particle size on the tensile strength of compacts of alpha monohydrate lactose (○ = compaction pressure of 151 MPa and, ● = compaction pressure of 226 MPa) and alpha anhydrous lactose (□ = compaction pressure of 151 MPa). (Drawn from data given by Vromans et al., ref. [41].)

lactose was generally higher, which was later attributed [11] to a higher degree of fragmentation during compression. In a later paper [12], the question of the effect of particle size on compact strength was addressed once again during the compaction of α-monohydrate lactose. In this chapter, the surface area of the tablets was measured by mercury intrusion. A more of less linear relationship was found between compact strength and tablet surface area (Fig. 7). Thus, it seems that the compact strength was governed by the surface area of the tablet and the increased tablet strength with decreased original particle size was due to the fact that the finer original particles gave a higher tablet surface area. Thus, a linear relationship between surface area and mechanical strength of compacts has been suggested for compacts of different types of lactoses [11–14]:

$$F_p = kS_p + i \qquad (7)$$

where

F_p = fracture or tensile strength of compact at pressure p (N/m²)
S_p = tablet surface area at pressure p (m⁻¹)
k = constant
i = intercept (normally negative for reported values)

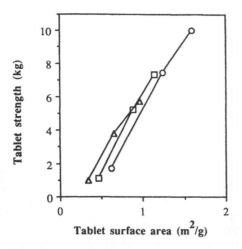

Fig. 7 Relationship between the diametral compression strength and tablet specific surface area, assessed by mercury intrusion, for compacts of alpha monohydrate lactose, compacted of three different particle size fractions: ○ = 32–45 μm, □ = 125–160 μm, △ = 315–400 μm. (Drawn from data given by de Boer et al., ref. [12].)

Equation (7) can be thought of as a relationship between the surface area of the particles constituting the compact and the total area of interparticulate bonds in the compact; i.e., an increased exposed particle surface area increases the bonding area. However, the observed relationship between tablet strength and tablet surface area can alternatively be explained in terms of a relationship between the compact strength and the number of particles exposed in the fracture plane. An increased surface area corresponds to a reduced "compact particle" size and, thus, an increased number of particles (n) exposed within a cross section of the compact, i.e., $n = f(S_p)$. This will, in combination with the porosity of the compact, govern the number of interparticulate bonding zones which must be ruptured when the compact is fractured during strength analysis.

Based on the suggested relationship between tablet strength and tablet surface area (Fig. 7 and Eq. (7)), the effect of type of lactose (Fig. 6) can be explained as follows: an increased degree of fragmentation of the material will stress the relationship between compact strength and original particle size with respect to the absolute difference in strength of compacts of different original particle size.

Results obtained at Uppsala University [47] on the relationship between tablet strength and tablet surface area, as measured by air permeametry, for four materials with varying compression behavior while compacted also generally gave nearly linear relationships. However, the relationship for the respective original particle size did not coincide, as found by Vromans et al. [11–14]. On the contrary, separate relationships dependent on the original particle size were obtained, i.e., an increased original powder surface area gave lines more and more to the right on the abscissa. Such an effect of original powder surface area seems reasonable if one consider the surface area and the strength of the powder as a part of the relationship. The powders represent samples with markedly varying surface areas but, in relative terms, with similar tensile strengths close to zero. The generality of the suggested relationship between tablet strength and tablet surface area, as suggested for the lactose tablets [11–14], can therefore be questioned.

The results discussed show that the relationship between original particle size and tablet strength is complex. However, it seems that the most common effect of a reduced particle size is an increased tablet strength. The results on the compaction of different types of lactose indicate that an increased fragmentation propensity of the material will give a more marked effect of particle size in absolute terms. However, there are reports for materials [42] which are not consistent with these conclusions, e.g., compacts of saccharose, which is a relatively high fragmenting material. In that paper, Alderborn and Nyström suggested that this type of relationship was caused by the high degree of fragmentation of the particles during the compression, which masked differences in particle size before compression. However, alternative explanations for this observation can be found.

Firstly, it is possible that the physical structure, in terms of the size of the compact particles and their relative positions, of the compact is similar, although the original starting particle size varied. A possible explanation for this effect is that the fragmentation propensity was reduced with a decreased particle size, i.e., the concept of fragmentation propensity as a material property cannot be applied to all materials. However, this seems to be a reasonable possibility for a material which consists of aggregates of small primary particles.

Secondly, a more generalized explanation is that the unexpected small influence of original particle size on the compact strength for the fragmenting materials is due to the fact that the original particle size can affect the interparticulate bond structure in a compact in terms of both the number and the bonding force of the interparticulate bonds: i.e., the force needed to rupture the interparticulate bond cannot be treated as a material-specific characteristic but can vary between compacts of different original particle size. This suggestion also forms the basis for the explanation of the observed relationships between tablet strength and original particle size for sodium bicarbonate [45] and sodium chloride [42].

If an increase in original particle size affects the character of the interparticulate bonding zone in such a way that the force needed to rupture the individual interparticulate bond increases, a distribution of bonds in the compact characterized by a reduced number of bonding points but of a higher bonding force will be obtained. The net effect can be a tablet strength independent of particle size (saccharose and sodium bicarbonate) or an increased tablet strength with increased particle size (sodium chloride). In Fig. 8, the effect of particle size on the distribution of bonds in the compact for these two types of idealized situations is schematically illustrated.

Karehill et al. [4] reported ratios between tablet strength and tablet surface area, i.e., a surface-specific tensile strength of tablets, as a measure of the character of the individual interparticulate bonds in compacts of three particle size fractions of sodium chloride. They found (Table 1) an increased ratio with increased particle size, which can be interpreted as support for the suggestion that the force needed to rupture the individual interparticulate bonds in compacts of sodium chloride increases with an increased original particle size. For the other test materials used in the study, the ratios were similar to the observed ratio for the most fine particulate quality of sodium chloride.

An increased particle size will generally reduce the number of contact points in the compact which transmit the applied stress through the powder bed during the compression phase. The result will be an increased stress at the particle-particle contact points which can facilitate local particle defor-

Table 1 Ratio between Tensile Strength and Permeametry Surface Area for Compacts (as a measure of the bonding force of individual bonds between compact particles) of Three Sieve Fractions of Sodium Chloride Compacts at 150 MPa

Sieve fraction (μm)	Ratio (kPa cm^{-2})
<63	0.9
250–355	2.4
425–500	4.2

Source: From Ref. 4

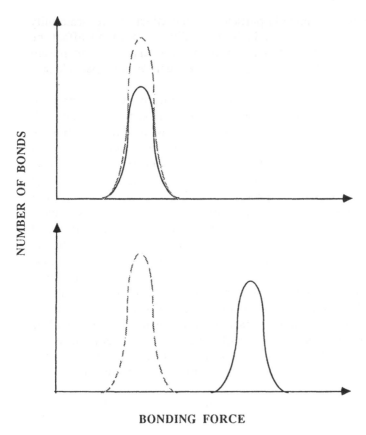

BONDING FORCE

Fig. 8 Hypothetical types of distribution in bonding force illustrating the discussion on the changes in bonding characteristics in tablets compacted of a coarse and a fine particulate fraction of the material. Dashed line = fine particle size, solid line = coarse particle size. Upper panel illustrates a material showing an increased compact strength with decreased original particle size while the lower panel illustrates a material showing a decreased compact strength with decreased original particle size.

mation. The degree of local deformation at the interparticulate junctions can affect the area of bonding for intermolecular forces or the formation of solid bridges between particles, i.e., a bridge constituting of a continuous solid phase of material which bonds particles to each other. The formation of such bridges can be the result of a drastic increase in mobility of the molecules or ions at particle surfaces due to, e.g., increased local temperatures in the compact with a subsequent local surface fusion. It was also suggested above that a change in particle size can affect the deformability of particles. Thus, such a variation in deformability can probably affect the bonding force of the interparticulate bonding zone due to a decreased contact area and thus affect the relationship between original particle size and tablet strength.

In most of the studies discussed above, comparatively coarse particulate materials were used. The preparation of finer particulate materials normally requires a milling procedure before size classification. The bulk of the particles can be affected by inducing dislocations or cracks in the crystal structure, which can affect both the deformability and the fragmentation propensity of the particles. However, more important might be the change in particle surface properties as a result of milling: Firstly, on a molecular level, the surface of the particles will probably show a disordered, amorphous structure [48], and, secondly, on a microscopic level, the surface of the particles will probably show a rough texture with surface protuberances. These two phenomena might lead to a change in the mechanical properties of the surface and markedly increase the surface deformability; i.e., the surface deformability might be higher than the bulk deformability of the particles. Consequently, the bonding force will increase due to an increased area of bonding between particles which is larger than could be predicted from the bulk deformability of the particles. In addition, it has been suggested that a disordering of surfaces, either before or during the compaction, will affect the changes in mechanical strength during storage of the compacts [13,14].

Eriksson and Alderborn [49] studied the effect of particle size on the mechanical strength of sodium chloride tablets compacted of particles of a wide size range. The finest particles were prepared by milling and classification. To avoid the effect of postcompaction strength changes on the interpretation of the data, the compact strength was measured immediately after compaction. An increased tablet strength (Fig. 9) with a decreased original particle size was observed with a more pronounced effect with increased compaction pressure. These results thus contradict the results discussed above for sodium chloride. A possible interpretation is that for this quality of sodium chloride, an increase in the number of interparticulate bonds due to a reduced original particle size governed the compact strength and was more important than a change in the bonding force with a change in original particle size. However, the finest particles were prepared by milling. It could thus be assumed that those particles possess a comparatively rough surface with a disordered solid-state structure. This might lead to a changed local deformability at the particle surface with an improved binding capacity and thus an improved bonding force of the

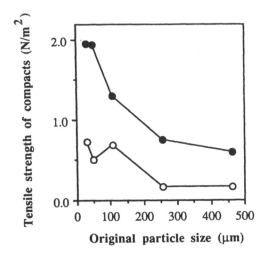

Fig. 9 The effect of original particle size on the tensile strength of compacts of sodium chloride, measured immediately after compaction. ○ = compaction pressure of 75 MPa and ● = compaction pressure of 150 MPa. (Drawn from data given by Eriksson and Alderborn, ref. [49].)

individual bonds. This might thus explain the good compactibility of the fine sodium chloride particles and constitute a general explanation for the commonly accepted statement that smaller particles form stronger tablets.

In the discussion so far, no consideration has been given to elastic deformation of particles. It seems reasonable, however, that if the degree of elastic expansion of particles is limited or does not vary markedly between different particle size fractions, a direct relationship between the interparticulate bonds formed during the actual compression phase and the bonding characteristic in the ejected tablet will be obtained. Thus, the mechanistic discussion on the relationship between volume reduction behavior, original particle size, and compact strength will also be applicable when particles deform elastically to some extent. The results obtained [30] for the mechanical strength of compacts of a capping proned material, Hexamin, support the assumption that in a restricted compaction pressure range the effect of particle size could be predicted also for a material which could be expected to be markedly elastic (Fig. 10). However, after capping, it is not meaningful to relate compact strength to original particle size. It is worth noting, however, that capping occurs at a lower compaction pressure for the finest particulate quality of hexamin.

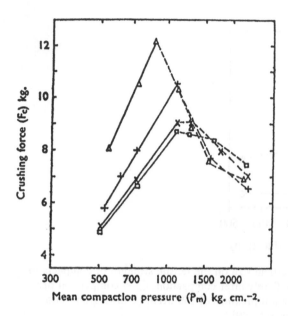

Fig. 10   The effect of compaction pressure on the diametral compression strength of hexamin tablets, compacted of four different particle size fractions: ☐ = 20–30 mesh, ✕ = 30–40 mesh, + = 40–60 mesh, △ = 60–80 mesh. (From ref. [30] with permission of the copyright-holder.)

2.   *Effect of Compaction Pressure*

In some of the papers on the relationship between compact strength and original particle size, the importance of the compaction pressure has also been addressed. It seems generally that an increased compaction pressure stresses the relationship between original particle size and tablet strength in absolute terms, e.g., for lactose (Fig. 11) [41] and hexamin (Fig. 10) [30]. However, for saccharose and sodium chloride, Alderborn and Nyström [42] observed only a limited effect of compaction pressure on the relationship between compact strength and original particle size (Figs. 12 and 13). A possible interpretation of these results is that the stress applied to the powder during the compression affects both the number of bonds per cross-sectional area of the compact and the bonding force of the individual interparticulate bonds. An increased compaction pressure will probably increase the degree of fragmentation which the particles undergo during compression. Thus, the number of interparticulate bonds per cross-sectional area of compact will increase. However, it seems reasonable that an increased compaction pressure will also increase the degree of particle

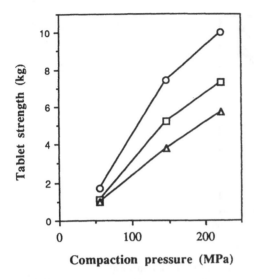

Fig. 11 The effect of compaction pressure on the diametral compression strength of alpha monohydrate lactose tablets, compacted of three different particle size fractions: ○ = 32–45 μm, □ = 125–160 μm, △ = 315–400 μm. (Drawn from data given by de Boer et al., ref. [12].)

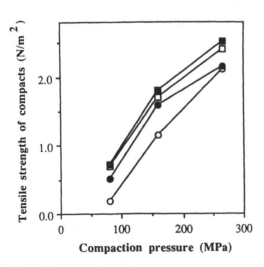

Fig. 12 The effect of compaction pressure on the tensile strength of sodium chloride tablets, compacted of four different particle size fractions: ○ = 90–125 μm, ● = 212–250 μm, □ = 355–500 μm, ■ = 500–710 μm. (Drawn from data given by Alderborn and Nyström, ref. [42].)

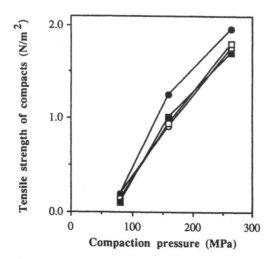

Fig. 13 The effect of compaction pressure on the tensile strength of saccharose tablets, compacted of four different particle size fractions: ○ = 125–212
μm, ● = 250–300 μm, □ = 355–500 μm, ■ = 710–1000 μm. (Drawn from
data given by Alderborn and Nyström, ref. [42].)

deformation with an increased bonding force as the result. Moreover, an
increased difference in bonding force in compacts prepared of particles of
different original size might be obtained. Thus, it is possible that there will
be a relationship between original particle size and compaction pressure
with respect to both the number of bonds and the bonding force.

V. PARTICLE SHAPE

A. Effect of Particle Geometrical Shape on Volume
Reduction Properties and Tablet Strength

The shape of a particle is a complex characteristic, and its importance in
relationship to powder properties is therefore difficult to assess. In practice, it is feasible to describe particles as of the same size if they exhibit the
same particle diameter defined in some way, e.g., projected area diameter.
Two particles defined as the same size can, thus, differ with respect to the
relative magnitude of their main dimensions (i.e., length, breadth, and
thickness) as well as with respect to their surface texture. In this chapter, an
attempt is made to distinguish between two main shape characteristics of a
particle. The term "geometrical shape" refers to the main dimensions of a
particle, and the term "surface geometry" to the texture of the surface. It is

then assumed that the dimension of an asperity is small compared to the dimension of the whole particle.

There is a limited number of studies in the pharmaceutical literature which have specifically discussed the relationship between particle shape and compact strength for a specific material. In these studies, two approaches for the preparation of powders which consist of particles with different shapes have been used. In the first case, different qualities of the same material have been produced by crystallization under different conditions or by using different crystallization batches. The other approach is based on the preparation of two shape fractions from the same batch of a material. This has been achieved by either sorting particles by shape or by producing different shapes by a combination of milling and sieving.

Lazarus and Lachman [50] compared the compactibility of different batches of potassium chloride. The compactibility of the different batches varied and it was concluded that the batches consisting of larger and more irregular particles gave tablets of a higher mechanical strength. They also found that the addition of a lubricant to the powders decreased the diametral compression strength of the compacts, indicating that potassium chloride particles fragment to a limited degree during compression.

Shotton and Obiorah [51] found that dendritic sodium chloride gave compacts of a higher mechanical strength compared to the cubic form, which was also shown later [42]. Shotton and Obiorah attributed the observed difference in compactibility to the difference in particle shape; i.e., the more irregular dendritic sodium chloride gave stronger tablets. The compression characteristics of cubic sodium chloride are well studied in the literature, and it has been shown that this material fragments to a limited degree during compression and thus reduces in volume mainly by particle rearrangement and particle plastic deformation. It has, however, been suggested [42] that the dendritic quality of sodium chloride fragments during compression due to its more complex physical character; i.e., dendritic sodium chloride particles are aggregates of small sodium chloride particles, and the aggregates have a complex, porous structure. Hence, the increased compactibility can alternatively be due to a changed compression behavior characterized by a fragmentation of the original particles followed by plastic deformation of the formed sodium chloride particles.

By using a combination of milling-sieving procedures, Alderborn and Nyström [52] and Alderborn et al. [45] prepared two powder qualities for a series of model materials. The two qualities (A and B) for each material had particles of the same sieve size but with different shape characteristics. The model materials were chosen to represent materials with different degrees of particle fragmentation during compression.

The particulate characterization indicated generally an increased irregularity of shape quality B, as evaluated by the Heywood shape coefficient and surface area analysis for powders of sodium chloride, sucrose, and sodium citrate [52]. The same conclusion was drawn from microscopy studies of sodium bicarbonate powders [45].

Compaction of the powders into tablets showed that for the materials which fragmented to a limited degree during compression, the particle shape affected the compact strength; i.e., a more irregular particle improved the compactibility (Figs. 14 and 15). However, for materials which fragmented markedly during compression, the shape of the particles before compaction did not affect compact strength (Fig. 14). Hence, the effect of particle shape on the compact bonding characteristics and, thus, the compact strength is dependent on the volume reduction properties of the material. For a material which fragments to a large extent, the physical structure of the formed compact is to a limited extent affected by variations in particle shape before compaction, provided that the articles can be described as the same particle size.

For the materials which fragmented to a limited extent during compression (sodium chloride and sodium bicarbonate), the increased strength of compacts prepared of the more irregular particles indicates a change in the distribution of interparticulate bonds with a change in original particle shape. It is possible that the bonding characteristics within the compact can be affected by the original shape of the particles with respect to both the number of interparticulate bonds and the force needed to rupture the individual bonds.

It is reasonable that with an increased particle irregularity, the number of possible interparticulate attraction zones in a compact, and consequently the compact strength, will increase, although the packing properties of the particles are similar. It is also possible that during the compression process, the particles can undergo an increased degree of deformation due to an increased particle irregularity, especially at edges and corners of the particles. In this case, the degree of deformation at such sites could be further enhanced due to the existence of lattice defects, primarily dislocations, created during the milling of the material [53]. The consequence of the increased local particle deformation will be an increased bonding force of the attraction zones between compact particles. For sodium chloride, an alternative explanation is that an increased stress or degree of deformation at the interparticulate contact points during the compression process facilitated the formation of solid bridges between particles, which thus increased the bonding force and compact strength.

Both for sodium chloride and sodium bicarbonate, an increased compaction pressure increased the absolute difference in strength of compacts

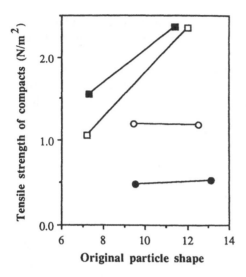

Fig. 14 The effect of original particle shape (expressed as the Heywood surface-volume shape coefficient) on the tensile strength of compacts of sodium chloride I (□), sodium chloride II (■), saccharose (○) and sodium citrate (●). Tablets compacted at 160 MPa. (Drawn from data given by Alderborn and Nyström, ref. [52].)

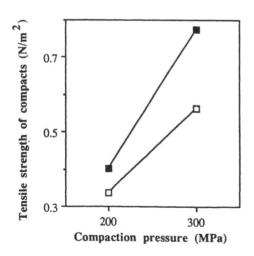

Fig. 15 The effect of compaction pressure on the tensile strength of sodium bicarbonate tablets, compacted of particles of two different shapes: □ = regular particles, ■ = irregular particles. (Drawn from data given by Alderborn and Nyström, ref. [45].)

of different original particle shape (Fig. 15). This can be interpreted as that the difference in the character of the individual bonding zone will be stressed with an increased compaction pressure.

For one of the test materials, sodium bicarbonate, the volume reduction characteristics of the different shape qualities were assessed by porosity-pressure profiles [45], performed as in-die measurements. This evaluation indicated that the volume reduction behavior was more-or-less identical for the two powders. This can be interpreted as support for the explanation that the improved compact strength with increased particle irregularity is related to an effect of particle shape on the number of interparticulate attraction zones in the compact. However, it is possible that a porosity-pressure profile reflects mainly the bulk behavior of the particles; i.e., differences in the degree of local particle deformation can occur without affecting the overall compression profile.

As a result of the milling procedure used during the preparation of the model materials, also the surface characteristics of the particles can change. It has been suggested [48] that the fracturing of a solid, crystalline material can cause a disordering or amorphization of the solid material at the particle surfaces. Such changes in the solid-state structure can cause an improved ability of the material to create interparticulate attractions, e.g., as a result of an increased local deformability or an increased surface energy. Hence, the findings described above might also be explained by a disordering of the material as a result of milling. However, it seems reasonable that during storage of "activated" materials, the disordered regions will crystallize in order to reduce the energy of the particle surfaces. Thus, for the sodium chloride powders, compacts were also made after storage of the powder for one year. The compactibilities of the powders were identical after the storage period. It therefore seems likely the observed effect was related to the changes in particle geometrical shape rather than a disordering of their solid-state structure.

Wong and Pilpel [54] prepared particles with different shape characteristics by the use of a shape sorting table. Also in these studies, materials with different volume reduction characteristics were used. Their observations were consistent with the findings by Alderborn et al; i.e., for materials which fragment markedly during compression, the shape characteristics of the particles did not affect the tablet strength, while the converse applied for materials which showed limited fragmentation (Figs. 16 and 17). In addition to the evaluation of compactibility, the volume reduction characteristics of the test materials were assessed by porosity-pressure profiles and by tablet permeametry surface area–pressure profiles. They observed an increased degree of fragmentation for the fragmenting materials with increased particle irregularity (Table 2). Thus, it seems like more irregular

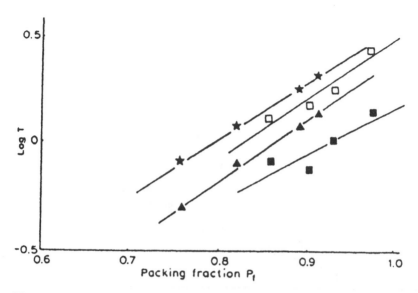

Fig. 16 The effect of the relative compact density on the tensile strength of tablets of sodium chloride (■ = regular shape of original particles, □ = irregular shape of original particles) and Starch 1500 (▲ = regular shape of original particles, ★ = irregular shape of original particles). (From ref. [54] with permission of the copyright-holder.)

particles were slightly more prone to fragment, which might be due to the fact that corners and asperities of the particles correspond to weak parts of the particles which can be fractured or knocked off the rest of the particle relatively easily. However, the observed difference in degree of fragmentation was not marked enough to cause any difference in the mechanical strength of the resultant compacts. The low-fragmenting materials did not show a tendency to an increased degree of fragmentation with a change in particle shape, but the compressibility of the powder increased with increased particle irregularity. They thus concluded that the irregular particles showed a more marked deformation during the compression phase, which especially is due to deformation at particle asperities. This increased deformation can thus produce an increased total bonding strength between particles and increase the tablet strength.

In the papers discussed hitherto, no considerations were taken of postcompaction events in relation to the effect of particle shape on compact strength. As stated, the mechanical strength of compacts can change dramatically during a limited time period after compaction. It is possible that such a phenomenon might be responsible for the observed effects

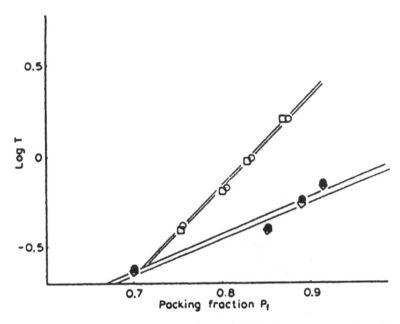

Fig. 17 The effect of the relative compact density on the tensile strength of tablets of lactose ($\Diamond$ = regular shape of original particles, ● = irregular shape of original particles) and Emcompress ($\Box$ = regular shape of original particles, $\bigcirc$ = irregular shape of original particles). (From Ref. 54 with permission of the copyright holder.)

Table 2 Fragmentation Propensities, Estimated from Profiles of Tablet Permeametry Surface Area as a Function of Compaction Pressure, and Yield Pressures, Estimated from Profiles of Tablet Porosity (measured in-die and expressed as the Heckel function) as a Function of Compaction Pressure, for Four Materials with Two Shape Characteristics of the Particles Before Compaction

Material	Fragmentation propensity $(m^{-1} kN^{-1})$	Yield pressure (MPa)
Sodium chloride, regular	0.303	83.0
Sodium chloride, irregular	0.287	72.3
Starch 1500, regular	0.136	42.0
Starch 1500, irregular	0.128	34.0
Lactose, regular	0.507	154
Lactose, irregular	0.680	150
Emcompress, regular	0.603	264
Emcompress, irregular	0.869	261

Source: From Ref. 54.

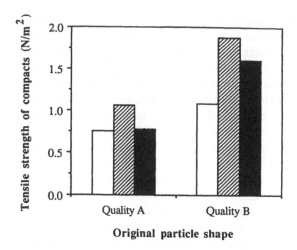

Fig. 18 The effect of original particle shape on the tensile strength of compacts of sodium chloride, compacted at 100 MPa. Open = tablet strength measured immediately after compaction, gray = tablet strength measured after 60-min storage at 0% relative humidity, black = tablet strength measured after 60-min storage at 57% relative humidity. Quality A = regular particles, quality B = irregular particles. (Drawn from data given by Eriksson and Alderborn [49].)

discussed here. Eriksson and Alderborn [49] therefore studied the compactibility of two qualities of sodium chloride powders prepared by the milling-sieving procedure. An increased particle irregularity increased the tablet strength measured directly after compaction (Fig. 18); i.e., the earlier observed differences were not related entirely to postcompaction effects. However, after storage of tablet for one hour at 0% and 57%, the difference in mechanical strength of tablets of the two shape qualities increased further. Hence, the effect of particle shape on tablet strength for sodium chloride seemed to be the combined effect of differences in compaction and postcompaction behavior. It seems also that the relative humidity of the environment can affect the changes in tablet strength which occurs during tablet storage.

B. Effect of Particle Surface Geometry on Tablet Strength

It seems reasonable that also the characteristics of the surface of particles is a factor of importance for the mechanical strength of the compact. The composition and energetics of the surface, affected by, e.g., the presence of contaminants and the solid-state structure of the surface, might affect directly the strength of the interparticulate bonds. Furthermore, as discussed, the geometry of the surface can affect the number of attraction zones between particles in a compact and their volume reduction behavior with respect to their deformability and fragmentation propensity. As also discussed, there is no clear distinction between geometrical shape, related to the main dimensions of particles, and surface geometry with respect to surface roughness. The effect of particle surface geometry on tablet strength is therefore difficult to evaluate due to the difficulty of preparing suitable model materials. Milling, for example, will not change only the shape of the particles, but also the geometry and the energetics of the surface. However, an attempt to assess the importance of the geometry of the surface in relationship to powder compactibility have been presented by Karehill et al. [10] by the use of double-layer compacts. A series of materials, representing both fragmenting and nonfragmenting materials, was compacted into tablets at a series of compaction pressures. An increased compaction pressure resulted in a flatter or less rough upper surface of the compact; i.e., the series of compacts represented material surfaces of different geometry. Without ejecting the tablet from the die, more powder was poured into the die and a double-layer tablet was compacted. This double-layer tablet was ejected and its strength was assessed by axial tensile testing.

For the plastically deforming materials, the double-layer compact failed in the contact zone between the individual tablets while stressed, and the axial tensile strength decreased with increased compaction pressure for the lower tablet (Fig. 19), i.e., a decreased surface roughness. For the fragmenting materials, the double-layer compact failed generally in the lower compact while stressed; i.e., the weakest plane of the whole compact was not in the contact zone between the individual tablets. Nevertheless, an increased compaction pressure reduced the compact strength (Fig. 19), probably due to the creation of weak zones in the first tablet during the second loading.

The results can be interpreted as follows: for materials possessing a limited fragmentation during compression, the possibility of forming interparticulate attractions depends on the geometry of the particle surfaces. An increased roughness of the surface increases the possibility for a particle to find a position at an adjacent surface which promotes the formation of a large number of bonds, i.e., cavities at particle surfaces. If the particle should be able to develop attractions more-or-less independent of geometry of the surface of adjacent particles, a very marked particle deformation is probably necessary. The test materials used by Karehill et al. probably do not show such extensive deformation while loaded. Pharmaceutically rele-

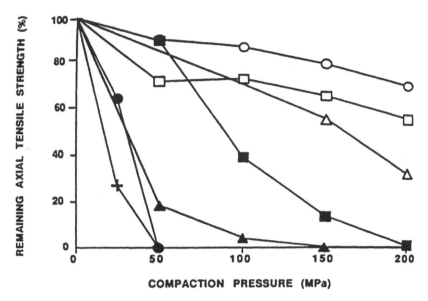

Fig. 19 The axial tensile strength of double-layer compacts, expressed as the relative strength in relationship to the strength of a single-layer compact, as a function of the compaction pressure of the first part of the double-layer compact (○ = Emcompress, □ = Lactose, △ = saccharose, ■ = sodium chloride, ● = Avicel PH 101, + = Avicel PH 101 < 10 μm, ▲ = starch 1500. (From Ref. 10 with permission of the copyright holder.)

vant materials possessing this characteristic might be polymers used as binders in granulated powders.

For fragmenting materials, the results indicate that the fracturing of the particles during compaction creates a situation where the total bonding between particles is more-or-less independent of the geometry of the surface of adjacent particles. Thus, a similar dependence between particle shape and particle surface geometry, on one hand, and the volume reduction characteristic of the material, on the other, in relationship to powder compactibility seems to exist.

VI. CONCLUDING REMARKS

In this chapter, it has been argued that a pharmaceutical compact can be described in terms of a large aggregate of particles which are strongly co- or adhered to each other, i.e., a special case of a disperse system. The gas phase in the compact can thus be described as the continuous phase and constitutes a three-dimensional network of connected pores. This is probably valid at least for relatively porous compacts, which are applicable to "normal" pharmaceutical tablets. The consequence of this model is that bonds within particles will be stronger than bonds between, and the compact will thus fail mainly around, rather than between, particles. At the fracture surface of the compact, a number of closely packed but individual particles will be exposed. The tensile strength of the compact will thus theoretically be governed by the sum of the bonding forces of all individual interparticulate bonds in the failure plane of the compact, i.e., Eq. (1). This distribution in interparticulate bonds in a compact is governed by two main factors: the properties of the material and the characteristics of the compaction process (Fig. 20). Two of the most important material properties for tablet strength are compression behavior of the particles and their dimensions before compaction.

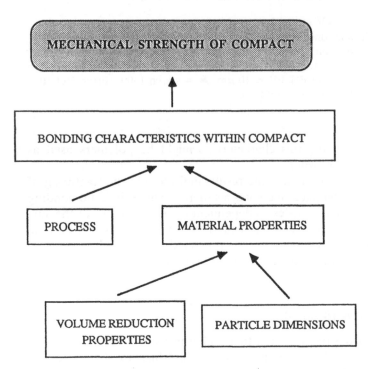

Fig. 20 Proposed qualitative relationships between the mechanical strength of the compact and the dimensions of the particles before compaction.

Concerning the compression behavior of the material it is suggested that both fragmentation and plastic deformation of the particles are bond-forming compression mechanisms. The physical significance of these mechanisms seems to be that fragmentation will affect mainly the number of interparticulate bonds, while deformation will affect mainly the bonding force of the interparticulate bonds. In this chapter it has also been suggested that the dimensions of the particles before compaction can affect both the degree of fragmentation and the degree of deformation which the particles undergo during compression and thus affect the number and the bonding force of the interparticulate bonds.

Concerning the effect of the size of the particles before compaction on the tablet strength, it is normally suggested that the most common type of relationship is that a reduced original particle size will increase the tablet strength, and the magnitude of this effect, in absolute terms, will increase with an increased compaction pressure. Deviations from this pattern are, however, commonly reported both for highly fragmenting and for non-fragmenting materials. This chapter suggests that the relationship between original particle size and tablet strength must be explained in terms of the effect of particle size on the number and bonding force of the interparticulate bonds. A reduced particle size can increase the bonding potential in a cross section of the compact by increasing the number of bonds and by increasing the bonding force of the bonds due to a high surface deformability of rough, disordered particle surfaces (small particles are often prepared by milling). However, a reduced original particle size can also decrease the bonding force due to a decreased local deformation of particles at the interparticulate contact points during the compression. This can be caused be a reduced contact force at the interparticulate contact points during compression and a decreased particle deformability. Examples of mechanistic explanations for relationships between original particle size and compact strength for tablets are given in Fig. 21.

Concerning the effect of the shape of the particles before compaction on the compact strength, it seems that only in the case when the material fragments to a limited degree during the compression phase a change toward more irregular particles increases the compact strength. Furthermore, it is suggested that an increased deformability of the material or an increased propensity to form solid bridges will stress the effect of particle shape on compact strength.

Finally, concerning the geometry of the solid surface, the literature indicates that for fragmenting materials the propensity to establish interparticulate attractions is relatively independent of the geometry of the adjacent surfaces, while the converse applies to nonfragmenting materials. For such materials, an increased irregularity of the adjacent surface im-

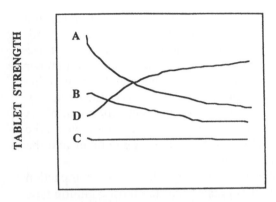

ORIGINAL PARTICLE SIZE

Fig. 21 Examples of explanations for the relationship between original particle size and tablet strength: (A, B) Increased particle surface deformability with a reduced particle size or marked importance of the number of bonds for tablet strength, (C) No particle size effect on particle deformability or particle surface deformability, (D) Increased particle deformability with increased particle size.

Table 3 Proposed Effects on Tablet Strength of Original Particle Shape and Surface Geometry for Different Types of Materials[a] with Respect to Volume Reduction Mechanisms (no capping or lamination tendencies)

Type of material[b]	Effect of increased particle irregularity	Effect of increased surface irregularity
A	No effect	No effect
B	Increase	Marked increase
C	Increase	Limited increase

[a]For materials with a high propensity to form solid bridges while compacted, it seems that an increased particle irregularity facilitates the formation of solid bridges during the compression process (e.g., sodium chloride)
[b]A = intermediate or high fragmenting
B = low fragmenting, limited plastic deformability
C = low fragmenting, high plastic deformability

proves the formation of interparticulate attractions. However, an increased deformability of the material might result in less importance of the surface geometry. A summary of the proposed effects of original particle shape or surface geometry on the compact strength for tablets of materials with different properties is given in Table 3.

REFERENCES

1. G. Alderborn, *Acta Pharm. Suec. 22*:177 (1985).
2. G. Alderborn and M. Glazer, *Acta Pharm. Nord. 1*:11 (1990).
3. C. Nyström and P. G. Karehill, *Powder Technol. 47*:201 (1986).
4. P. G. Karehill, E. Börjesson, M. Glazer, G. Alderborn, and C. Nyström, Studies on direct compression of tablets. XX. Investigation of bonding mechanisms of some directly compressed materials by addition and removal of magnesium stearate, published in thesis by P. G. Karehill, Uppsala University, 1990.
5. M. Luangtana-anan and J. T. Fell, *Int. J. Pharm. 60*:197 (1990).
6. P. G. Karehill and C. Nyström, *Int. J. Pharm. 61*:251 (1990).
7. J. S. Hardman and B. A. Lilley, *Nature 228*:353 (1970).
8. C. Führer, E. Nickel, and F. Thiel, *Acta Pharm. Technol. 21*:149 (1975).
9. G. R. B. Down, *Powder Technol. 35*:167 (1983).
10. P. G. Karehill, M. Glazer, and C. Nyström, *Int. J. Pharm. 64*:35 (1990).
11. H. Vromans, A. H. de Boer, G. K. Bolhuis, C. F. Lerk, K. D. Kussendrager, and H. Bosch, *Pharm. Weekblad Sci. Ed. 7*:186 (1985).
12. A. H. de Boer, H. Vromans, C. F. Lerk, G. K. Bolhuis, K. D. Kussendrager, and H. Bosch, *Pharm. Weekblad Sci. Ed. 8*:145 (1986).
13. C. Ahlneck and G. Alderborn, *Int. J. Pharm. 56*:143 (1989).
14. G. Alderborn and C. Ahlneck, *Int. J. Pharm. 73*:249 (1991).
15. M. Wikberg and G. Alderborn, *Int. J. Pharm. 63*:23 (1990).
16. M. Wikberg and G. Alderborn, *Int. J. Pharm. 69*:239 (1991).
17. H. Leuenberger, *Int. J. Pharm. 12*:41 (1982).
18. J. E. Rees and E. Shotton, *J. Pharm. Pharmacol. Suppl. 22*:17S (1970).
19. P. J. Rue and P. M. R. Barkworth, *Int. J. Pharm. Technol. Prod. Manuf. 1*:2 (1980).
20. P. G. Karehill and C. Nyström, *Int. J. Pharm. 64*:27 (1990).
21. C. Nyström, K. Malmqvist, J. Mazur, W. Alex, and A. W. Hölzer, *Acta Pharm. Suec. 15*:226 (1978).
22. J. M. Newton, G. Alderborn, and C. Nyström, *Powder Technol. 72*:97 (1992).
23. J. M. Newton, G. Alderborn, C. Nyström, and P. Stanley, *Int. J. Pharm.*, to appear.
24. J. T. Fell and J. M. Newton, *J. Pharm. Sci. 59*:688 (1970).
25. G. Alderborn and C. Nyström, *Acta Pharm. Suec. 21*:1 (1984).
26. M. A. Mullier, J. P. K. Seville, and M. J. Adams, *Chem. Eng. Sci. 42*:66 (1987).
27. A. B. Mashadi, Mechanical properties of compacted powders, Ph.D. thesis, University of London, 1988.
28. H. Rumpf, in *Agglomeration* (W. A. Knepper, ed.), Interscience, New York, p. 379, 1962.
29. D. C.-H. Cheng, *Chem. Eng. Sci. 23*:1405 (1968).
30. E. Shotton and D. Ganderton, *J. Pharm. Pharmacol. Suppl. 13*:144T (1961).
31. K. Schönert and H. Rumpf, in *Symposion zerkleinern*, Verlag Chemie, Weinheim, p. 108, 1962.

32. J. J. Benbow, in *Enlargement and Compaction of Particulate Solids* (N.G. Stanley-Wood, ed.), Butterworths, London, p. 161, 1983.
33. M. Duberg and C. Nyström, *Acta Pharm. Suec. 19*:421 (1982).
34. G. Alderborn and C. Nyström, *Powder Technol. 44*:37 (1985).
35. H. Leuenberger, J. D. Bonny, C. F. Lerk, and H. Vromans, *Int. J. Pharm. 52*:91 (1989).
36. G. C. Lowrison, in *Crushing and Grinding*, Butterworths, London, p. 49, 1974.
37. R. J. Roberts and R. C. Rowe, *J. Pharm. Pharmacol. 38*:567 (1986).
38. J. A. Hersey, G. Bayraktar, and E. Shotton, *J. Pharm. Pharmacol. Suppl. 19*:24S (1967).
39. J. T. Fell and J. M. Newton, *J. Pharm. Pharmacol. 20*:657 (1968).
40. A. McKenna and D. F. McCafferty, *J. Pharm. Pharmacol. 34*:347 (1982).
41. H. Vromans, A. H. de Boer, G. K. Bolhuis, C. F. Lerk, and K. D. Kussendrager, *Acta Pharm. Suec. 22*:163 (1986).
42. G. Alderborn and C. Nyström, *Acta Pharm. Suec. 19*:381 (1982).
43. E. Shotton and C. J. Lewis, *J. Pharm. Pharmacol. Suppl. 16*:111T (1964).
44. O. Alpar, J. A. Hersey, and E. Shotton, *J. Pharm. Pharmacol. Suppl. 22*:1S (1970).
45. G. Alderborn, E. Börjesson, M. Glazer, and C. Nyström, *Acta Pharm. Suec. 25*:31 (1988).
46. G. Alderborn, K. Pasanen, and C. Nyström, *Int. J. Pharm. 23*:79 (1985).
47. M. Eriksson and G. Alderborn, *Pharm. Res.*
48. M. J. Kontny, G. P. Grandolfi, and G. Zografi, *Pharm. Res. 4*:104 (1987).
49. M. Eriksson and G. Alderborn, *Int. J. Pharm., 109*: 59 (1994).
50. J. Lazarus and L. Lachman, *J. Pharm. Sci. 55*:1121 (1966).
51. E. Shotton and B. A. Obiorah, *J. Pharm. Pharmacol. Suppl, 25*:37P (1973).
52. G. Alderborn and C. Nyström, *Acta Pharm. Suec. 19*:147 (1982).
53. C. Führer, *Acta Pharm. Technol. Suppl. 6 24*:129 (1970).
54. L. W. Wong and N. Pilpel, *Int. J. Pharm. 59*:145 (1990).

11

Mechanical Properties

Ray C. Rowe and Ron J. Roberts

Zeneca Pharmaceuticals, Macclesfield, Cheshire, United Kingdom

I. INTRODUCTION

The vast majority of drugs when isolated exist as crystalline or amorphous solids. Subsequently they may be admixed with other inactive solids (excipients) and finally compacted to form tablets. The process of compaction involves subjecting the materials to stresses causing them to undergo deformation. The reaction of a material to a deformation stress, σ_d, is dependent on both the mode of deformation and the mechanical properties of the material; e.g.,

(a) for elastic deformation

$$\sigma_d = \epsilon E \tag{1}$$

where E is Young's modulus of elasticity of the material and ϵ is the deformation strain;

(b) for plastic deformation

$$\sigma_d = \sigma_y \tag{2}$$

where σ_y is the yield stress of the material, and

(c) for brittle fracture

$$\sigma_d = \frac{A K_{IC}}{\sqrt{d}} \tag{3}$$

where

K_{IC} = critical stress intensity factor of material (an indication of the stress required to produce catastrophic crack propagation)

d = particle size (diameter)

A = constant depending on geometry and stress application

For compression of rectangular samples with large cracks $A = \sqrt{32/3}$ or 3.27 [1], but for other geometries A varies between 50 and 1 [2].

It is evident from the discussion above that in order to be able to predict the compaction behavior of a material it is essential that methods be derived to measure the following for powdered materials:

(a) Young's modulus of elasticity (E).
(b) Yield stress (σ_y); this is directly related to the indentation hardness, H, since for a plastic material

$$\sigma_y = \frac{H}{3} \tag{4}$$

(c) Critical stress intensity factor (K_{IC}); this is directly related to the fracture toughness, R, since for plane stress

$$K_{IC} = (ER)^{1/2} \tag{5}$$

In this chapter various methods which have been specifically applied to pharmaceutical materials are reviewed.

II. YOUNG'S MODULUS OF ELASTICITY

If an isotropic body is subjected to a simple tensile stress in a specific direction it will elongate in that direction while contracting in the two lateral directions, its relative elongation being directly proportional to the stress. The ratio of the stress to the relative elongation (strain) is termed Young's modulus of elasticity. It is a fundamental property of the material directly related to its interatomic or intermolecular binding energy for inorganic and organic solids respectively and is a measure of its stiffness.

Young's modulus of elasticity of a material can be determined by many techniques several of which have been used in the study of pharmaceutical materials, namely flexure testing using both four- and three-point beam bending, compression testing, and indentation testing on both crystals and compacts.

A. Flexure Testing (Beam Bending)

In flexure testing, a rectangular beam of small thickness and width in comparison to its length is subjected to transverse loads and its central deflection due to bending is measured. The beam may be supported and loaded in one of two ways (Fig. 1). If the beam is supported at two points and is loaded at two points it undergoes what is known as four-point bending, while if it is supported at two points but is loaded at one point it undergoes what is known as three-point bending. Equations for the calculation of Young's modulus from the applied load F and the deflection of the midpoint of the beam ξ can easily be derived:

for four-point bending [3] $$E = \frac{F}{\xi}\frac{6a}{h^3 b}\left(\frac{l^2}{8} + \frac{al}{2} + \frac{a^2}{3}\right) \qquad (6)$$

for three-point beam bending [4] $$E = \frac{Fl^3}{4\xi h^3 b} \qquad (7)$$

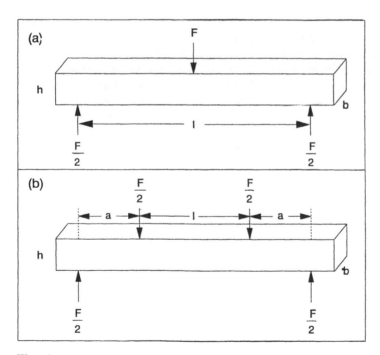

Fig. 1 Geometries for (a) three-point and (b) four-point beam bending: F, applied load; h, beam thickness; b, beam width; l and a, distance between loading points as shown.

where h and b are the height (thickness) and width (breadth), respectively, and l and a are as in Fig. 1. In both cases the value for Young's modulus for the specimen under test can either be calculated from a single point determination for more commonly from the slope of total load (F) versus central deflection (ξ).

The four-point beam bending test was first used for pharmaceutical materials by Church [3] and has since been adapted by Mashadi and Newton [5,6], Bassam et al. [7,8,9], Roberts et al. [10], and Roberts [11]. Generally beams of varying height have been used—100 mm long by 10 mm wide or 60 mm long by 7 mm wide—prepared using specially designed punches and dies. Beams are generally prepared at varying compression pressures to achieve specimens of varying porosity. Although the height of the beam (at constant porosity) is an experimental variable, Bassam et al. [8] have shown that it does not have a significant effect on the measured modulus. Generally low rates of testing are used (≈ 1 mm min^{-1}), but Bassam et al. [8] have shown an independence of loading rate up to rates of 15 mm min^{-1}.

A disadvantage of the four-point beam method is that it invariably requires large specimens and hence large quantities of materials (15–20 g). In addition, high-tonnage presses are needed to prepare the specimens, thus exacerbating problems with cracking and lamination on ejection from the die. It was to overcome these difficulties that Roberts et al. [10] developed a three-point beam testing method that uses beams prepared from 200 mg of material. The beams in this case are 20 mm long by 7 mm wide and are stressed by applying a static load of 0.3 N with an additional dynamic load of $\pm$ 0.25 N (at a frequency of 0.17 Hz) using a thermal mechanical analyzer (Mettler Instruments TMA40). In operation a calibration run is first performed to eliminate distortions in the sensor and other parts of the displacement measuring system and then 20 measurements of specimen displacement are undertaken to an accuracy of $\pm$ 0.005 μm. Young's modulus of elasticity of the specimen is then calculated from the mean displacement corrected for distortions using Eq. (7) where in this case F is the applied dynamic load. Extensive testing by Roberts et al. [10] and Roberts [11] has shown equivalence between this test and the conventional four-point beam test.

A problem associated with the analysis of data for specimens prepared from particulate solids is in the separation of the material property from that of the specimen property, which by definition includes a contribution due to the porosity of the specimen. All workers have found that for all materials there is a decrease in Young's modulus with increasing porosity (Fig. 2). Numerous equations have been published which describe this relationship [13]. Certain equations are based on theoretical considerations

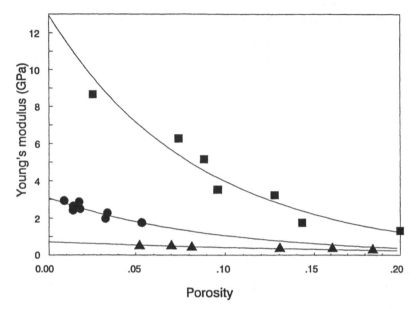

Fig. 2 The effect of porosity on the measured Young's modulus of ▲, PTFE; ●, testosterone propionate; ■, theophylline (anhydrous). (Data generated using three-point beams, taken from Roberts et al. [4].)

[14,15], while others are empirical curve-fitting functions [16,17]. Recently Bassam et al. [8] have reported a comparison of all the equations currently in use for data generated from four-point beam bending on 15 pharmaceutical powders ranging from celluloses and sugars to the inorganic calcium carbonate, and have concluded that the best overall relationship is the modified two-order polynomial [17]

$$E = E_0(1 - f_1P + f_2P^2)$$ (8)

where

E_0 = Young's modulus at zero porosity
E = measured modulus of the beams compacted at porosity P
f_1, f_2 = constants

However, such a conclusion may not be universal since during extensive studies on a wider range of materials including drugs, Roberts [11] has concluded that an exponential relationship [16] is the preferred option for data generated using the three-point beam testing method; i.e.

$$E = E_0 e^{-bP}$$ (9)

Table 1   Young's Modulus at Zero Porosity Measured by Flexure Testing

Material	Particle size (μm)	Young's modulus (GPa)	Reference
Drugs			
Theophylline (anhydrous)	31	12.9	4
Paracetamol DC	120	11.7	11
Caffeine (anhydrous)	38	8.7	4
Sulfadiazine	9	7.7	4
Aspirin	32	7.5	4
Ibuprofen	47	5.0	4
Phenylbutazone	50	3.3	4
Testosterone propionate	85	3.2	4
Sugars			
Sorbitol instant	—	45.0	5
α-lactose monohydrate	20	24.1	4
α-lactose monohydrate	63	3.2	8
Lactose β-anhydrous	149	17.9	8
Lactose β-anhydrous	149	18.5	9
Lactose (spray dried)	—	13.5	12
Lactose (spray dried)	125	11.4	8
Dipac sugar	258	13.4	8
Mannitol	88	12.2	8
Polysaccharides			
Starch 1500	—	6.1	12
Maize starch	16	3.7	8
Celluloses			
Avicel PH-101	50	10.3	6
Avicel PH-101	50	9.7	12
Avicel PH-101	50	9.2	8
Avicel PH-101	50	9.0	7
Avicel PH-101	50	7.8	10
Avicel PH-101	50	7.6	10
Avicel PH-101	50	7.4	10
Avicel PH-102	90	8.7	8
Avicel PH-102	90	8.2	7
Avicel PH-105	20	10.1	7
Avicel PH-105	20	9.4	8
Emcocel	56	9.0	7
Emcocel	56	7.1	8
Emcocel (90M)	90	9.4	7
Emcocel (90M)	90	8.9	8
Unimac (MG100)	38	8.8	7
Unimac (MG100)	38	8.0	8
Unimac (MG200)	103	8.0	7
Unimac (MG200)	103	7.3	8
Elcema (P100)	—	8.6	12

Table 1 Continued

Material	Particle size (μm)	Young's modulus (GPa)	Reference
Inorganics			
Calcium carbonate	8	88.3	8
Calcium phosphate	10	47.8	8
Polymers			
PVC	—	4.4	9
PVC	—	4.1	4
Stearic acid	62	3.8	4
PTFE	—	0.81	10
PTFE	—	0.71	9
PTFE	—	0.71	10

where b is a constant. It is interesting to note that on average the extrapolated values of E_0—Young's modulus at zero porosity—calculated using both equations are only marginally different [7,8].

The analysis of Young's modulus at zero porosity thus provides a means of quantifying and categorizing the elastic properties of powdered materials, Table 1 shows literature data for a variety of pharmaceutical excipients and drugs determined using beam-bending methods. It can be seen that the values vary over two orders of magnitude ranging from hard rigid materials with very high moduli (e.g., the inorganics) to soft elastic materials with low moduli (e.g., the polymeric materials). As a result a rank of increasing rigidity of tabletting excipients can be listed: starch < microcrystalline celluloses < sugars < inorganic fillers with variations in the groups dependent on chemical structure as well as the preparation and pretreatment routes (including particles size).

The effects of particle size are distinguishable within the celluloses and α-lactose monohydrate. In the former there is a small increase with decreasing particle size, while the latter the increase is much greater. Whereas in the former the effect is probably due to an increase in contact area, in the latter the effect is due to specimen defects in that the specimens used by Bassam et al. [8] contained microscopic flaws and cracks. Recent work by Roberts [11] has shown that it is necessary to eliminate all specimens with cracks otherwise the extrapolated modulus values to zero porosity are inconsistent.

For the lactose samples the rank order of increasing rigidity is spray dried < β-anhydrous < α-monohydrate, consistent with the findings of workers describing the compaction properties of the materials using instrumented tabletting machines [18,19]. The variations in the cellulose samples can be attributed to subtle differences in the manufacturing preparative technique. However, it is known that this factor can also affect the equilibrium moisture content of the samples, with Unimac samples attaining a lower equilibrium moisture content than Avicel samples. Since it is known that increasing moisture content can lead to a decrease in Young's modulus for microcrystalline cellulose [8] the differences in modulus between sources listed in Table 1 would be expected to increase if all materials were compared at equivalent moisture contents.

Recently Roberts et al. [4] have investigated the relationship between Young's modulus of a variety of drugs and excipients using three-point beam bending and their molecular structure based on intermolecular interactions using the concept of cohesive energy density (CED). They found a direct relationship of the form

$$E_0 = 0.01699\text{CED} - 2.7465 \tag{10}$$

where CED is expressed in units of MPa and Young's modulus in units of GPa (Fig. 3). The significance of this finding is in the recognition of the validity of both the test method and the data manipulation (i.e., the extrapolation to zero porosity).

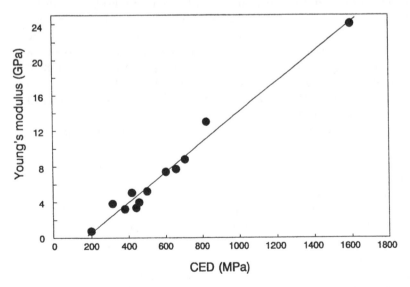

Fig. 3 Young's modulus versus cohesive energy density for various drugs and excipients (Taken from Roberts et al. [4].)

B. Compression Testing

In compression testing a compressive stress is applied to either a crystal or compacted specimen and the corresponding strain measured, the ratio of the stress to strain being a measure of the compressive Young's modulus of elasticity.

The loading system depends on the size of specimen, ranging from a microtensile testing instrument providing a load up to 5 N and a minimum displacement of 5 nm, as used for single crystals by Ridgway et al. [20], to an Instron physical testing instrument providing a load upto 50 kN, as used for flat-faced cylindrical compacts (8 mm diameter) by Kerridge and Newton [21]. Young's modulus in compression can be determined from the equation

$$E = \frac{L}{(X - C)A} \qquad (11)$$

where

L = specimen length
X = slope of strain versus stress
C = machine constant (strain versus stress for loading of the machine without a specimen
A = specimen cross-sectional area

While for crystals a single measurement is all that is necessary, for compacts measurements on specimens prepared at different porosities are required. In the latter test Young's modulus at zero porosity can be determined using Eq. (9).

Values for the compressive modulus of a variety of materials in Table 2 are lower than those measured by other techniques. Furthermore, for the

Table 2 Young's Modulus Measured by Compaction Testing

Material	Specimen	Young's modulus (GPa)	Reference
Aspirin	Crystal	0.1	20
Aspirin	Compact	2.5	21
Sodium chloride	Crystal	1.9	20
Potassium chloride	Compact	9.2	21
Avicel PH-102	Compact	4.7	21
Sucrose	Crystal	2.2	20
Salicylamide	Crystal	1.3	20
Hexamine	Crystal	0.9	20

test involving crystals Ridgway et al. [20] indicated that cracks were important and that this factor caused variation in the results. This factor may also
account for the low values of modulus when compared with those from
flexure testing. In view of this and the fact that the compressive modulus
should always be greater than the tensile modulus, the results from these
tests must be viewed with some scepticism.

C. Indentation Testing

In indentation testing a hard indenter made of either diamond, sapphire, or
steel and machined to a specific geometry (either a square-based pyramid
(Vickers indenter) or a spherical ball (Brinell indenter) is pressed under
load into the surface of a material either in the form of a crystal or compacted specimen. Although well recognized as a method for measuring
hardness (see later), it may also be used to measure Young's modulus,
except that in this case it is the recovered depth after removal of the load
that is important (Fig. 4). The test may be either static [22,23,24] or dynamic involving a pendulum [25].

For crystals a Vickers indenter is generally used, and in this respect
Duncan-Hewitt and Weatherly, [23,24] have used a Leitz-Wetzler Miniload
hardness tester on preselected single crystals. Load was applied over 15 sec

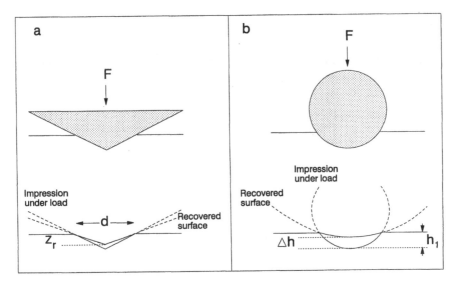

Fig. 4 Loading geometry and recovery of indent for (a) Vickers and (b)
Brinell hardness testers used to measure Young's modulus.

and was maintained for 10 sec. In the case of sucrose crystals, hardness anisotropy of the various faces was demonstrated.

The equation used by Duncan-Hewitt and Weatherly [24] involves measuring the recovered depth of indentation and the length of the Vickers diagonal and substituting in the equation of Breval and Macmillan [26]:

$$\left(\frac{2Z_r}{d} \right)^2 = 0.08168 \left[1 - 8.7 \left(\frac{H_v}{E} \right) + 14.03 \left(\frac{H_v}{E} \right)^2 \right] \tag{12}$$

where

H_v = indentation (see later for equation)
Z_r = recovered depth of the indentation
d = length of Vickers diagonal (Fig. 4).

Data on several materials (Table 3) show values of the same order as those determined by flexure testing. The value of sodium chloride compares favorably with that reported in the literature (37 GPa [27]).

For compacts Ridgway et al. [22] have used a pneumatic microindentation apparatus to measure both the initial depth of penetration (h_1) and recovery (Δh) after application of small loads (4 g) using a sapphire ball indenter 1.5 mm in diameter and substituting in a modified Hertz equation [28]:

$$E(GPa) = 1.034 \frac{F}{\Delta h \sqrt{h_1}} \tag{13}$$

where h_1 and Δh are in micrometers and F, the indentation load, is in grams. The values quoted for aspirin are very small, two orders of magnitude below the figure given in Table 1 and hence must be viewed with some suspicion.

In the dynamic method used by Hiestand et al. [25] a metre long pendulum with a steel sphere of 25.4 mm diameter strikes the face of a compact of cross-sectional area of 14.52 cm^2. The modulus of elasticity can be calculated from a knowledge of the indentation hardness, H_b and the

Table 3 Young's Modulus Measured by Indentation on Crystals

Material	Indenter	Young's modulus (GPa)	Reference
Sodium chloride	Vickers	43.0	24
Sucrose	Vickers	32.3	24
Paracetamol	Vickers	8.4	24
Adipic acid	Vickers	4.1	24
Aspirin	Brinell	0.32	22

strain index, ξ_i, a measure of relative strain during elastic recovery that follows plastic deformation using the equation

$$E = \frac{H_b(1 - \nu^2)}{\xi_i} \qquad (14)$$

where ν is Poisson's ratio. The strain index ξ_i is calculated using values from the indentation experiment:

$$\xi_i = \frac{5a}{6\pi r} \left(\frac{h_i}{h_r} - \frac{3}{8} \right)^{-1} \qquad (15)$$

where

a = chordal radius of the indent
r = radius of spherical indenter
h_i, h_r = initial and rebound heights of indenter

Assuming a Poisson's ratio of 0.3 it is possible, using the calculated values for hardness and strain index [29] to determine Young's modulus for spray-dried lactose and microcrystalline cellulose (Avicel PH-102)—25.5 and 6.2 GPa respectively. These values compare favorably with those obtained from flexure testing (Table 1).

III. INDENTATION HARDNESS AND YIELD STRESS

The hardness and/or yield stress of a material are measures of its resistance to local deformation. For plastic materials the two parameters are directly related, the yield stress being one-third of the hardness (Eq. (4)). While hardness is generally measured by means of indentation, yield stress may be determined by more indirect methods, e.g., data from compaction studies. Both will be discussed in this section.

A. Indentation Hardness Testing

The most common method of measuring the hardness of a material is the indentation method. In this a hard indenter (e.g., diamond, sapphire, quartz, or hardened steel) of specified geometry is pressed into the surface of the material. The hardness is essentially the load divided by the projected area of the indentation to give a measure of the contact pressure. The most commonly used methods for determining the hardness of pharmaceutical materials are the Vickers and Brinell tests. In general the former has been used for measurements on compacts whereas the latter technique has been solely applied to single crystals.

In the standard Brinell test [30] a hard steel ball (usually 2mm in diameter) is pressed normally on to the surface of the material. The load F is applied for a standard period of 30 sec and then removed. The diameter, d, of the indentation is measured and the Brinell hardness, H_b, determined by the relationship

$$H_b = \frac{F}{(\pi D/2)(D - \sqrt{D^2 - d^2})} \tag{16}$$

where D is the diameter of the ball.

In the Vickers test [31] a square-based diamond pyramid is used as an indenter. It is capable of measuring hardnesses over the entire range from the softest to the hardest materials. The Vickers hardness, H_v, is determined from the following equation, where F is applied force and d is the length of the diagonals of the sqare impression:

$$H_v = \frac{2F \sin 68°}{d^2} \tag{17}$$

The specimen thickness should be at least one and a half times the diagonal length with greatest accuracy being obtained with high loads, but loads as small as 0.01 N can be used. However, at lower loads the elastic recovery is of greater importance. For loads lower than 0.5 N the technique is usually described as microhardness testing.

An improved analysis of the Vickers hardness of compacts was performed by Ridgway et al. [32] using a Leitz microhardness tester with loads of between 5 and 2000 g. Accurate diagonal lengths were determined after lightly dusting the indent with graphite powder. Of the materials studied, potassium chloride, hexamine, and urea showed little change in hardness with increasing compaction pressure, while for aspirin and sodium chloride the hardness increased possibly due to work hardening. The maximum hardness values reported were 29, 17, 16, 12, and 8 MPa for sodium chloride, potassium chloride, aspirin, hexamine, urea, respectively. These values are an order of magnitude lower than those generally accepted and hence must be viewed with suspicion.

In many studies on compacts spherical indenters have been used either fitted to commercially available instruments or custom-built equipment. An example of the former is the pneumatic microindentation apparatus described by Ridgway et al. [22]. This instrument can apply loads of between 4 and 8 g using a spherical indenter 1.5 mm in diameter and measure depths of penetration of 1–6 μm. Using this apparatus on aspirin compacts, the authors were able to show that hardness measurements at the center of compacts were higher than those at the periphery, a property for flat-faced compacts confirmed by Aulton and Tebby [33]. However, for compacts prepared using concave punches hardness distribution tends to vary with the degree of curvature [33].

In a modification to the original microindentation apparatus [22], Aulton et al. [34] added a displacement transducer to measure the vertical displacement. They suggested that the elastic quotient index, which is the fraction of the indentation which rebounds elastically, was a measure of the ability of materials to form tablets.

In a further study involving a larger range of materials, Aulton [35] found that hardness measurements were generally higher on the upper face of the compact than on the lower. However, the differences were material dependent and the hardness measurements reported were 62, 54, 51, 36, 19, and 13 MPa for sucrose, Sta-Rx, Emcompress, Avicel PH-101, lactose β-anhydrous, and paracetamol respectively, were low in comparison with other measurements.

Probably the most relevant of all the work carried out on hardness measurement on compacts is that of Hiestand et al. [25], who used a spherical indenter attached to a pendulum, and that of Leuenberger [36] and his co-workers [37,–41], who used a spherical indenter attached to a universal testing instrument. In the dynamic pendulum method of Hiestand et al. [25] a spherical ball of 24.5 mm diameter falls under the influence of gravity and the rebound height and indent dimensions measured. The hardness is calculated from the expression

$$H_b = \frac{4mgrh_r}{\pi a^4}\left(\frac{h_i}{h_r} - \frac{3}{8}\right) \tag{18}$$

where

m = the mass of the indenter
g = gravitational constant
r = radius of the sphere
a = chordal radius of the indent
h_i = initial height of the indenter
h_r = rebound height of the indenter

In the case of the indenter attached to the universal testing instrument [36] loads of 3.92 and 9.81 N were applied to a sphere of 1.76 mm diameter at a velocity of 0.05 cm min^{-1}, indent diameters being determined from scanning photomicrographs.

In both cases indentation hardness was found to be dependent on the compaction pressure and hence relative density (porosity) of the compact in an exponential manner. While Hiestand et al. [25] considered extrapolation to unit relative density to be questionable, Leuenberger [36] has used the relationship between Brinell hardness, H_b, of a compact and its relative density, D, to develop a measure of compactibility and compressibility, namely

$$H_b = H_{b\,max}(1 - e^{-\lambda \sigma_c D}) \tag{19}$$

where

$H_{b\,max}$ = theoretical maximum hardness as the compressive stress σ_c approaches infinity and the relative density of the compact approaches unity

λ = rate at which H_b increases with increasing compressive stress

In the equation Leuenberger [36] has suggested the $H_{b\,max}$ describes the compactibility and λ the compressibility.

Data on indentation hardness using both methods are shown in Table 4. As with the modulus of elasticity values for hardness vary over two orders of magnitude from the very hard materials (e.g., Emcompress) to very soft waxes. The drugs have intermediate hardness values.

It should be noted that the values recorded are also variable due to the

1. Intrinsic variability in the specimens [37]
2. Work hardening as the compaction pressure is increased [40,42]
3. Increase in hardness with increasing indentation load [40]
4. Rate of measurement [39]

The earliest reported study on the hardness of pharmaceutical crystals was carried out by Ridgway et al. [20] using the Leitz microhardness tester with a pyramidal indenter. The crystals were mounted in heat-softened picene wax on a mounting slide to ensure that the surfaces were horizontal. It is interesting to note that the authors observed that aspirin and sucrose showed cracking and regarded this as a problem with the technique. Hardness values from this study are presented in Table 5. In general, softer materials showed the most variation in results due to a decline in the definition and quality of the indent.

The next reported study of crystal hardness using a Vickers hardness tester was carried out by Ichikawa et al. [43]. In this study the majority of the materials (with the exception of sucrose and urea) were recrystallized. Indentation was performed on the crystal face possessing the largest area (i.e., the face that grows the slowest during crystallization) since it was inferred that this face would have the most influence on the compaction properties. Ichikawa et al. [43] considered differences in crystal hardness as a reflection of the mechanism of deformation during compaction. They

Table 4 Indentation Hardness Measured on Compacts

Material	Indentation hardness (MPa)	Reference
Drugs		
Paracetamol DC	265	37
Caffeine (anhydrous)	290	37
Caffeine (granulate)	288	37
Oxprenolol succinate	262	41
Hexamine	232	36
Phenacetin	213	36
Sitosterin	198	36
Metamizol	91	37
Aspirin powder	91	37
Aspirin FC	87	37
Aspirin	60	36
Aspirin	55	36
Ibuprofen (A)	35	36
Ibuprofen (B)	162	36
Sugars		
α-lactose monohydrate	515	36
α-lactose monohydrate	534	37
Lactose β-anhydrous	251	36
Sucrose	1046-1723	36
Sucrose (250–355#)	493	37
Alkali halides		
NaCl	653	36
NaCl	313	37
NaCl (rock salt)	358	37
KCl	99	38
KBr	69	37
Others		
Emcompress	752	37
Avicel PH-102	168	37
Starch 1500	78	41
Sodium stearate	37	40
PEG 4000	36	40
Castor oil (hydrogenated)	32	41
Magnesium stearate	22	40
Sodium lauryl sulphate	10	40

Table 5 Indentation Hardness Measured on Crystals

Material	Indentation hardness (MPa)	Reference
Drugs		
Paracetamol	421	24
Paracetamol	342	43
Sulfaphenazole	289	43
Hexamine	133	20
Hexamine	42	43
Sulfadimethoxine	231	43
Phenacetin	172	43
Salicylamide	151	20
Salicylamide	123	43
Aspirin	87	20
Sugars		
α-lactose monohydrate	523	43
Sucrose	645	24
Sucrose	636	20
Alkali halides		
NaCl	212	20
NaCl	213	24
NaCl	183	43
KCl	177	20
KCl	101	43
Others		
Urea	91	20
Urea	83	43

also showed that the reciprocal of crystal hardness correlated with the slope *K* from the Heckel equation. Data reproduced in Table 5 represent the mean value from three applied loads.

More recently Duncan-Hewitt and Weatherly [23,24] measured the Vickers hardness on single crystals of number of materials using a Leitz-Wetzlar Miniload tester with a load of 147 mN. The surfaces of sucrose were preconditioned, and this may account for their higher hardness values compared with compacts (Table 4), since in this case the materials could be fully work-hardened solids. However, the authors did not indicate whether the other crystals were preconditioned. Furthermore, Duncan-Hewitt and Weatherly [23] showed that the sucrose crystals were anisotropic in that different crystal faces gave different hardness values.

B. Yield Stress from Compaction Studies

Compaction studies, because they mimic the tabletting process, offer an
ideal method for assessing the mechanical properties of powders. In powder compaction a specific method used to evaluate the average stress of a
material during compression relies on the observations of Heckel [44,45],
who found that for materials that plastically deform the relative density of
a material, D, could be related to the compaction pressure, P, by the
equation

$$\ln\left[\frac{1}{1-D}\right] = KP + A \tag{20}$$

where K and A are constants.

Unfortunately, considerable deviations of the experimental data occur at both low and high pressures due to particle rearrangement and
strain hardening, respectively, but at least over the middle pressure
range a straight-line relationship exists between $\ln[1/(1-D)]$ and P
(Fig. 5).

Equation (20) has been reappraised by Hersey and Rees [46], who
suggested that the reciprocal of K can be regarded as numerically equal to
the mean yield stress of the powder. However, as pointed out by Roberts et
al. [47], this is only a specific case, and the reciprocal of K can be regarded
as a mean deformation stress be it a plastic deformation stress (equal to the
yield stress) for materials that deform plastically, a fracture stress for mate-

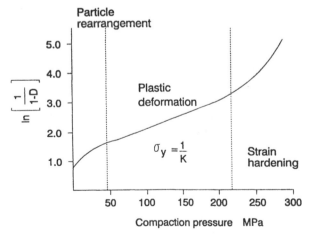

Fig. 5 Schematic diagram of the Heckel plot. (From Refs. 44 and 45.)

rials that undergo fracture, or a combination of the two. This approach implies that provided the experiment is carried out on materials close to or below the brittle-ductile transition (i.e., the material deforms plastically [47]), then the reciprocal of K will be numerically equal to the yield stress of the powder.

Historically two approaches to the analysis of Heckel data have been used and these are generally referred to as "at-pressure" and "zero-pressure" measurements. Pressure/relative density measurements determined during compression are clearly at-pressure measurements, while those from relative density measurements on the compact after ejection are zero-pressure measurements. In the original publication Heckel [44] found that for the metals iron, copper, nickel, and tungsten there was no difference between the two measurements, but for graphite the zero-pressure measurement was higher, attributable to the elastic recovery of the compact causing a lower relative density. Support for this hypothesis can be obtained from studies on the pharmaceutical materials dicalcium phosphate dihydrate (8% increase [49]), lactose (30% increase [18]), microcrystalline cellulose (56% increase [49]), and starch (177% increase [49]) where the magnitude of the percentage increase in yield pressure due to elastic recovery is indirectly related to the modulus of the material, i.e., the larger the increase the lower the modulus.

In the light of the discussion above the findings that other factors such as punch and die dimensions [50,51], state of lubrication [52,53], and speed of compaction [54,55], can have an effect on the measurement, it is not surprising that a great deal of controversy and confusion surrounds the use of data from Heckel plots. However, recent work by Roberts and Rowe [56] has clearly shown that, provided measurements are carried out "at pressure" with lubricated punches and dies on material that is below its brittle/ductile transition and at a very slow speed, the reciprocal of K is identical to the yield stress of the material comparable with that calculated from indentation hardness measurements using Eq. (4).

Although early measurements were generally performed on either instrumented punches and dies in physical testing machines [18] or instrumented single-punch table machines [52] with relatively unsophisticated data capture and analysis, recent measurements have been generally carried out using tablet compression simulators [55,57].

Yield stress data for a number of excipients and drugs are shown in Table 6. The trends mimic those for indentation hardness with the inorganic carbonates and phosphates showing very high values of yield compared to the polymers with low yield values. The variation seen in the drugs may well be due to slight differences in the particle sizes tested.

Table 6 Yield Stresses Measured by Compaction Studies

Material	Experimental details	Yield stress (MPa)	References
Drugs			
Paracetamol DC	Single punch	108	58
Paracetamol DC	Single punch	81	59
Paracetamol DC	Simulator	109	56
Paracetamol	Single punch	79	60
Paracetamol	Single punch	99	61
Paracetamol	Single punch	127	62
Paracetamol	Simulator	102	55
Sulfathiazole	Hydraulic press	109	63
Theophylline (anhydrous)	Single punch	75	61
Aspirin	Single punch	25	60
Aspirin	Single punch	73	62
Tolbutamide	Single punch	24	60
Phenylbutazone	Single punch	24	60
Ibuprofen	Simulator	25	64
Sugars			
α-lactose monohydrate	Hydraulic press	179	65
α-lactose monohydrate	Hydraulic press	183	66
α-lactose monohydrate	Simulator	178	67
Lactose β-anhydrous	Simulator	149	56
Lactose (spray dried)	Simulator	178	57
Lactose (spray dried)	Simulator	147	55
Mannitol	Simulator	90	56
Inorganic			
Calcium phosphate	Simulator	957	56
Calcium carbonate	Simulator	851	56
Calcium carbonate	Hydraulic press	610	68
Magnesium carbonate	Simulator	471	56
Dicalcium phosphate dihydrate	Simulator	431	56
Polymers			
PVC/vinyl acetate	Simulator	70	56
Polyethylene	Simulator	16	56
PTFE	Simulator	12	56
Others			
Sodium chloride	Single punch	89	69
Sodium chloride	Simulator	89	47
Avicel PH-101	Single punch	50	59
Avicel PH-101	Simulator	46	66
Avicel PH-102	Simulator	49	66
Avicel PH-105	Simulator	48	66
Maize starch	Simulator	40	56
Stearic acid	Hydraulic press	4.5	66

Compaction simulators allow the measurement of yield stress over a wide range of punch velocities (Fig. 6). To compare materials, Roberts and Rowe [55] proposed a term—the strain rate sensitivity—describing the percentage decrease in yield stress from a punch velocity of 300 mm sec^{-1} to one of 0.033 mm sec^{-1}. This was later modified [56] to a percentage increase in yield stress over the same punch velocities:

$$\text{SRS} = \frac{\sigma_{y300} - \sigma_{y0.033}}{\sigma_{y0.033}} \tag{21}$$

where σ_{y300} and $\sigma_{y0.033}$ are the yield stresses measured at punch velocities of 300 and 0.033 mm sec^{-1} respectively.

Data on a number of materials are shown in Table 7. Some materials, such as the inorganic carbonates and phosphates, show little rate dependence, while others, such as starch an mannitol, show a large strain rate

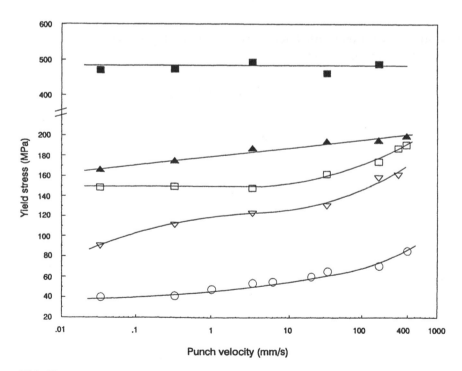

Fig. 6 The effect of punch velocity on yield stress for ■, magnesium carbonate; ▲, α lactose monohydrate; □, lactose β-anhydrous; ▽, mannitol; ○, maize starch.

Table 7 Strain Rate Sensitivities (SRS) of
Some Excipients and Drugs

Material	SRS (%)
Maize starch	97.2
Mannitol	86.5
PVC/vinyl acetate	67.5
Sodium chloride	66.3
Avicel PH-101	63.7
Lactose β-anhydrous	25.5
Lactose (spray dried)	23.8
α-lactose monohydrate (fine grade)	19.4
Paracetamol	11.9
Paracetamol DC	1.8
Heavy magnesium carbonate	0
Calcium carbonate	0
Calcium phosphate	0

dependence consistent with differences in the time-dependent properties of
the material during compaction.

IV. CRITICAL STRESS INTENSITY FACTOR K_{IC}

The critical stress intensity factor, K_{IC}, describes the state of stress around
an unstable crack or flaw in a material and is an indication of the stress
required to produce catastrophic propagation of the crack. It is thus a
measure of the resistance of a material to cracking. Since it is related to the
stress and the square root of crack length it has the dimensions of MPa m$^{1/2}$.

All methods used to measure K_{IC} involve specimens containing induced notches and/or cracks and for pharmaceutical materials include
three- or four-point beam bending (commonly known as the single-edged
notched-beam (SENB) test), double torsion, radial-edge cracked tablet or
disc, and Vickers indentation. The choice of the test and its associated
specimen geometry depends on the rate of testing, the ease of formation of
the specimen and the porosity of the specimen.

A. Single-Edge Notched Beam

In this test a prenotched rectangular beam of small thickness and width in
comparison to its length is subjected to transverse loads and the load at
fracture measured. As with the beams used for the determination of

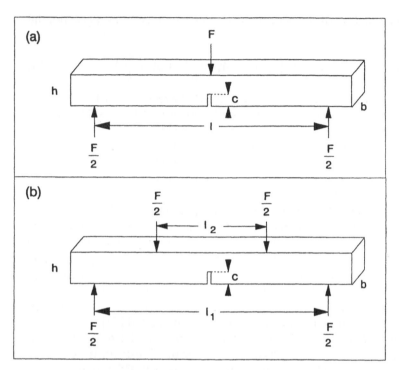

Fig. 7 Geometries for (a) three-point and (b) four-point single-edge notched beam: F, applied load; h, beam thickness; b, beam width; l and a, distances between loading points as shown; c, crack length.

Young's modulus, loading can be by either three- or four-point (Fig. 7). The single-edge notched-beam test has been the subject of much research leading to the specification of standard criteria for test piece geometry dependent on the material to be studied [70,71].

Equations for the calculation of the critical stress intensity factor from the applied load, F, and geometry of the beam can be derived, e.g., for four-point beam bending [5,6]:

$$K_{IC} = \gamma \frac{3Fc^{1/2}(l_1 - l_2)}{2bh^2} \tag{22}$$

and for three-point beam bending [72]:

$$K_{IC} = \gamma \frac{3Fc^{1/2}l}{2bh^2} \tag{23}$$

where γ is a function of the specimen geometry expressed as a polynomial of the parameter c/h:

$$\gamma = A_o + A_1 \left(\frac{c}{h} \right) + A_2 \left(\frac{c}{h} \right)^2 + A_3 \left(\frac{c}{h} \right)^3 + A_4 \left(\frac{c}{h} \right)^4 \qquad (24)$$

where the coefficients A_0, A_1, A_2, A_3, and A_4 have values $+1.99$, -2.47, $+12.97$, -23.17, $+24.8$ for the four-point beam and $+1.93$, -3.07, $+14.53$, -25.11, $+25.8$ for the three-point beam respectively.

The four-point single-edge notched beam was first used for pharmaceutical materials by Mashadi and Newton [5,6] and has since been adopted by York et al. [73]. In all cases large rectangular beams 100 mm long by 10 mm wide of varying height are prepared using the same punches and dies as those used to prepare specimens for the determination of Young's modulus. Notches of varying dimensions and profiles have been introduced by cutting either using a simple glass cutter [5,6] or a cutting tool fitted into a lathe [73]. The latter method has allowed notches of different profiles and dimensions to be accurately cut and investigated. While an arrowhead-type notch did appear to influence the measured value of K_{IC} for beams of microcrystalline cellulose, the effect was much reduced from straight-through notches and hence the latter were recommended [73].

The load required for failure of the specimens under tension is measured using the same testing rig as that described previously for the determination of Young's modulus. Loading rates of between 0.025 mm min^{-1} [5,6] and 100 mm min$^-$ [73] have been used—the latter workers noting a small rise (approx. 10%) in the measured K_{IC} of beams of microcrystalline cellulose for a 100-fold increase in applied loading rate.

As with the measurement of Young's modulus the four-point test requires large beams and consequently large amounts of material. In order to minimize the latter, especially for the measurement of the critical stress intensity factor for drugs under development, Roberts et al. [72] have developed a three-point test using specimen dimensions and testing rig similar to that described earlier for Young's modulus, although in this case a tensometer has been used to stress the specimen. Using beams of dimensions 20 mm long by 7 mm wide of varying height with two types of notches (a V notch induced by a razor blade pressed into the surface and a straight-through notch induced by a small saw blade), the authors were able to show equivalence with data generated by York et al. [73] for the four-point single-edge notched beam.

As can be seen from Fig. 8, specimen porosity has a significant effect on the measured K_{IC}. As porosity decreases, K_{IC} increases, indicating more resistance to crack propagation. It is obvious from the results in Fig. 8 that the relationship between K_{IC} and porosity is not linear as suggested by

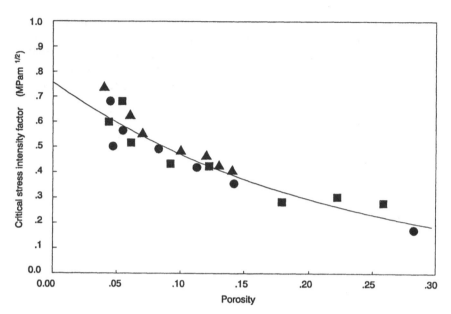

Fig. 8 The effect of porosity on the measured critical stress intensity factor for, ●, V notch; and ■, straight-through notch of Avicel PH-101 under three-point loading. ▲ = data ex York et al. [73] for same material under four-point loading. The line represents Eq. (26) for all points.

Mashadi and Newton [5,6,74]. In this respect York et al. [73] have investigated the application of both the two-term polynomial equation of the type

$$K_{IC} = K_{IC0}(1 - f_1P + f_2P^2) \tag{25}$$

and the exponential equation of the type

$$K_{IC} = K_{IC0}e^{-bP} \tag{26}$$

where

K_{IC0} = critical stress intensity factor at zero porosity
K_{IC} = measured critical stress intensity factor of the specimen at porosity P
b, f_1, f_2 = constants

These equations are analogous to those used previously for Young's modulus. Both relationships give low standard errors and high correlation coefficients for microcrystalline cellulose from various sources. As with Young's modulus the exponential relationship is the preferred option for pharmaceutical materials [72].

A number of pharmaceutical materials have been measured using both three and four-point beam bending (Table 8). Of all the excipients tested, microcrystalline celluloses exhibited the highest values of K_{IC0}, real differences existing between the materials obtained from different sources. In addition there is also a particle size effect in that, for each of the three

Table 8 Critical Stress Intensity Factors Measured Using a Single-Edge Notched Beam

Material	Critical stress intensity factor ($MPam^{1/2}$)	Particle size (μm)	Reference
Drugs			
Ibuprofen	0.10	47	72
Aspirin	0.16	32	72
Paracetamol DC	0.25	120	72
Paracetamol	0.12	15	72
Sugars			
Lactose β-anhydrous	0.76	149	72
α-lactose monohydrate	0.35	20	72
Sucrose	0.22	74	72
Sorbitol instant	0.47	—	5
Celluloses			
Avicel PH-101	1.21	50	6
Avicel PH-102	0.76	90	73
Avicel PH-101	0.87	50	73
Avicel PH-105	1.33	20	73
Emcocel (90M)	0.80	90	73
Emcocel	0.92	56	73
Unimac (MG200)	0.67	103	73
Unimac (MG100)	0.80	38	73
Avicel PH-102	0.91	90	73
Avicel PH-101	0.99	50	73
Avicel PH-105	1.42	20	73
Emcocel (90M)	0.83	90	73
Emcocel	0.80	56	73
Unimac (MG200)	0.76	103	73
Unimac (MG100)	1.05	38	73
Avicel PH-101	0.76	50	72
Others			
Sodium chloride	0.48	20	72
Adipic acid	0.14	176	72

sources of material, the critical stress intensity factor increased with decreasing particle size. The sugars exhibit intermediate values with relatively low values for the drugs. It is interesting that anhydrous β-lactose has a critical stress intensity factor approximately twice that of α-lactose monohydrate (cf. indentation hardness measurements (Table 4)).

B. Double Torsion Testing

In this test the specimen is a rectangular plate (Fig. 9) with a narrow groove extending its full length supported on four hemispheres. The load is applied by two hemispheres attached to the upper platen. Controlled precracks are introduced in the specimen by preloading until a "pop-in" is observed (a pop-in is a momentary decrease in load and is an indication of crack growth). It should be noted that the groove is necessary to help guide the crack and ensure it remains confined within the groove itself. The critical

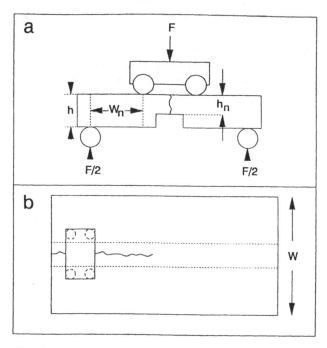

Fig. 9 Geometry for the double-torsion method showing (a) end view and (b) plan view for the measurement of critical stress intensity factor: h, plate thickness; W, plate width; h_n and W_n, distances as shown.

stress intensity factor is then calculated from the load F required to cause catastrophic crack propagation leading to failure by the expression

$$K_{IC} = FW_n \left[\frac{3(1 + \nu)}{Wh^3h_n} \right]^{1/2} \tag{27}$$

where ν is Poisson's ratio and l, h, h_n, W, and W_n are the dimensions of the specimen given in Fig. 9.

The double-torsion method has only been used for microcrystalline cellulose (Avicel PH-101) and Sorbitol "instant" by Mashadi and Newton [74]. As with the single-edge notched-beam specimens, measurements varied with specimen porosity, and using linear extrapolation Mashadi and Newton [74] calculated values of K_{IC0} of 1.81 and 0.69 MPa·m$^{1/2}$ for the two materials respectively. The higher results obtained for theses two materials as compared to the data obtained for single-edge notched beams (1.21 and 0.47 MPa·m$^{1/2}$ respectively) have been explained in terms of the specific geometries and stress uniformity of the two techniques. However, it is known that values of K_{IC} determined from double-torsion techniques are generally greater than those from notched-beam specimens [75].

A specific practical problem of the double-torsion method for pharmaceutical materials is the preparation of the specimen and the very large compaction pressures needed to produce specimens of low enough porosity. For microcrystalline cellulose, Mashadi and Newton [74] were only able to produce specimens of greater than 25% porosity.

C. Radial Edge Cracked Tablets

All the techniques for the determination of the critical stress intensity factor so far described involve the preparation of compressed rectangular beams or plates requiring special punches and dies and some cases high-tonnage presses. The ideal specimen shape for pharmaceutical materials is the right-angled cylinder or a flat-faced tablet. Such a shape has recently been investigated by Roberts and Rowe [76] using microcrystalline cellulose as a model material. Two test methods were investigated, (a) edge opening (Fig. 10a), (b) diametral compression (Fig. 10b). Both the tests involve the introduction of a precrack into the edge of the disc. In edge opening the critical stress intensity factor is given by [77]

$$K_{IC} = \frac{F}{t} \left[\frac{c}{0.3557 \, (d - c)^{3/2}} + \frac{2}{0.9665(d - c)^{1/2}} \right] \left[\frac{c}{2d} \right]^{-1/2} \tag{28}$$

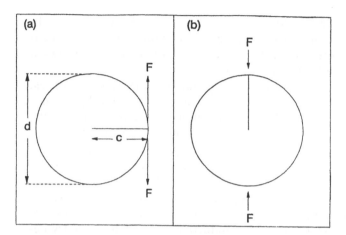

Fig. 10 Geometries for (a) edge opening and (b) diametral compression method for measuring critical stress intensity factors for radially edge cracked tablets: F, applied load; d, tablet diameter; c, crack length.

where F is the peak load for cracking as in Fig. 10a, while for diametral compression the critical stress intensity factor is given by [77]

$$K_{IC} = \frac{F}{dt}\left[\frac{\pi}{2c}\right]^{-1/2}\frac{1.586}{[1-(c/d)]^{3/2}} \tag{29}$$

where F is the compressive force for cracking (Fig. 10b). In both cases c is the crack length, t is the tablet thickness, and d is the tablet diameter.

In testing microcrystalline cellulose (Avicel PH-101) Roberts and Rowe [76] first prepared tablets of varying porosity using 15mm flat-faced punches on an instrumented tablet press. After precracking using a scalpel blade the cracked tablets were stressed using either edge opening, in which tablets were gripped using the air jaws of a tensometer and pulled apart or diametrally compressed with the crack vertical between the platens of a tensometer. During extensive testing it was concluded that, of the two techniques, edge opening was the preferred option since it gave the most stable crack propagation and the effects of crack length were minimal. However, the diametral compression test was found to have some value provided crack lengths were limited to between c/d values of 0.34–0.6.

Extrapolation of the measured values of critical stress intensity factors of specimens over the porosity range 7–37% using linear, exponential (Eq. (26)), and two-term polynomial (Eq. (25)) gave values for edge opening of 1.91, 2.24, and 2.31 MPam$^{1/2}$, respectively, and for diametral compression 2.11, 2.98, and 2.35 MPam$^{1/2}$, respectively. In all cases values obtained from diametral compression were higher than those from edge opening. However, all values are significantly higher than those determined from measurements on beams and plates.

The reasons for this variation in the results from these different test procedures have been discussed by York et al. [73] specifically for microcrystalline cellulose. Basically the problem lies in the difficulties in the introduction of a two-dimensional sharp crack into a specimen and accurately measuring its length and velocity on the application of load. Ideally K_{IC} should be independent of crack length and for those materials which exhibit flat crack growth resistance curve all methods of measurement should produce equivalent data for K_{IC0}. However, many materials, especially ceramics [78], have been shown to exhibit rising crack resistance curves and hence the method of crack induction and notch geometry become critical. For these measurements of specimens with sawn or machined notch will always produce lower values of K_{IC} than those where the crack is introduced by a controlled flaw. This is the case for the double-torsion method and radially edge-cracked tablet since in these methods the total amount of crack extension at maximum load is always higher. This is direct evidence that microcrystalline cellulose has a rising crack growth resistance curve [76].

D. Vickers Indentation Fracture Test

The most distinctive feature of the indentation of brittle materials by the Vickers or pyramidal indenter is the appearance of cracks emanating from the corners of the indent (Fig. 11), and it is from the measurement of the lengths of these cracks that it is possible to determine the critical stress intensity factor for indentation cracking or K_c.

Before describing the technique it is important to realize the assessment of K_c is strictly semiempirical because, firstly, the analytical solutions of the stress field around indentations have not been solved and only approximate solutions have been derived and secondly, the deformation field is not homogeneous and anisotropy and fracture complicate the problem.

Only two papers have examined the indentation fracture test as a means of determining the critical stress intensity factor of pharmaceuticals [23,24]. In the first of these Duncan-Hewitt and Weatherly [23] evaluated the indentation test using sucrose crystals. Microindentation was performed using a Leitz-Wetzlar Miniload hardness tester (Vickers pyramidal diamond indenter) applying loads of 147 mN, with the indentations and cracks measured using a light microscope (Leitz). Large crystals were prepared (1–4 mm diameter) by slow evaporation of saturated aqueous solution at 23°C

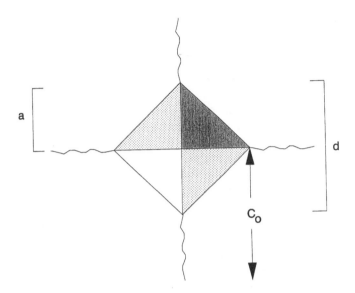

Fig. 11 Schematic diagram showing cracking around a Vickers indent: C_0, crack length; d, length of a diagonal of indent; a, length of half-diagonal indent.

over three to six months. The prismatic crystals were washed in ethanol to remove traces of crystallization solution and were stored under controlled conditions before testing. Furthermore some specific crystal faces [(100), (010), and (001)] were prepared by either abrading with decreasing grades of emery paper or by cleavage. Crystals were mounted in plasticine prior to testing. The authors found that fracture appeared anisotropic and that if crystals were tested immediately after polishing, fracture was either suppressed or significantly decreased. Duncan-Hewitt and Weatherly [23] used two equations to calculate the critical stress intensity factor:

$$K_c = \Phi \left(\frac{E}{H_v} \right)^{1/2} \frac{F}{C_0^{3/2}} \tag{30}$$

as derived by Antis et al. [79] and

$$K_c = \Phi \left(\frac{E}{H_v} \right)^{2/3} \left(\frac{F}{C_0^{3/2}} \right) \left(\frac{\alpha}{l} \right)^{1/2} \tag{31}$$

as derived by Laugier [80], where

Φ = calibration constant equivalent to 0.016 and 0.0143 in Eqs. (30) and (31) respectively

C_0 = crack length (as in Fig. 11)
l = $C_0 - a$
F = applied load
E = Young's modulus
H_v = Vickers hardness

The authors found that the values of K_c using the two equations gave similar values (the means for the various faces were 0.078 MPa m$^{1/2}$ and 0.089 MPa m$^{1/2}$ for Eqs. (30) and (31) respectively). Furthermore they reported that Eq. (31) [80] appeared to emphasize the apparent fracture anisotropy. The fracture plane with the lowest value was the (100) in agreement with the easiest to cleave plane (although the (101) has the lowest K_c value) and the plane with the greatest K_c was the (001) plane. The (100) plane has the hardest surface, whereas the (001) plane is the softest [23].

In their most recent paper Duncan-Hewitt and Weatherly [24] published further indentation critical stress intensity factors calculated from the Antis et al. [79] relationship (Eq. (30)). These are shown in Table 9 with the corresponding data from single-edge notched-beam (three-point) beam testing [72]. Although the rank order is the same, the differences in magnitude of the values are large (with the exception of sodium chloride). A possible explanation for these differences is that in the indentation test the theory is not exact and the equations are derived from calibration with ceramics, i.e., materials with considerably different plastoelastic properties than pharmaceutical materials.

Despite its shortcomings the indentation technique has certain advantages over other testing methods since it can be used on small samples, e.g., single crystals; specimen preparation is relatively simple, and the indentation hardness and Young's modulus can be determined simultaneously.

Table 9 Comparison of Critical Stress Intensity Factors Measured Using Indentation (K_c[24]) and Single-Edge Notched Beams (K_{IC0} [72])

Material	K_c(MPa m$^{1/2}$)	K_{IC0}(MPa m$^{1/2}$)
Sodium chloride	0.50	0.48
Sucrose	0.08	0.22
Paracetamol	0.05	0.12
Adipic acid	0.02	0.14

V. EFFECT OF PARTICLE SIZE—BRITTLE/DUCTILE TRANSITIONS

It has been known for many years that if Heckel plots were constructed for a material of varying particle size the reciprocal of the gradient over the central linear portion of the graph (now defined as the deformation stress, σ_d) varied, either remaining constant or increased as particle size decreased [46,66]. Extensive study [67,48,47] has shown that the effect of particle size is even more complex depending on the material under test (Fig. 12). For a material known to undergo plastic deformation (e.g., microcrystalline cellulose) no effect of particle size could be seen; for a material known to undergo brittle fracture (e.g., dolomite), the deformation stress increased with decreasing particle size, and for materials known to undergo a combi-

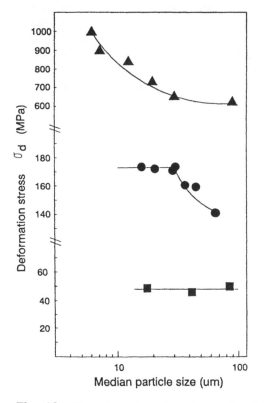

Fig. 12 The effect of particle size on the deformation stress as measured using Heckel plots for ▲, dolomite; ●, α-lactose monohydrate; ■, microcrystalline cellulose. (Adapted from Roberts and Rowe [48].)

nation of brittle fracture and plastic deformation (e.g., α-lactose monohydrate) the deformation stress increased with decreasing particle size to a plateau value.

This is as would be expected from theory (eqs. (2) and (3)) as shown in Fig. 13. The transition from brittle to ductile behavior will occur at a critical size d_{crit} given by the equation [47,48]

$$d_{\text{crit}} = \left(\frac{A K_{\text{IC}}}{\sigma_y} \right)^2 \tag{32}$$

where $A = 3.27$, for compression of rectangular samples with large cracks [1]. Experiments on the compression of sodium chloride has confirmed the applicability of this equation in tabletting, in that particles above 33 μm in diameter fractured under compression, while particles below this size did not show any fragmentation but deformed plastically under compression [47].

Calculated critical sizes (Eq. (32)) for a variety of excipients and drugs using yield stress data and critical stress intensity factors given in previous tables are shown in Table 10. It can be seen that critical particle sizes can vary over several orders of magnitude. The values appear reasonable in the light of experimental findings for microcrystalline cellulose and

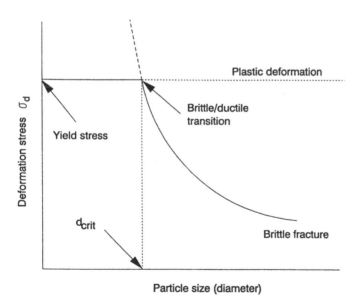

Fig. 13 Schematic diagram showing the effect of particle size on the deformation stress for materials that undergo brittle fracture and/or plastic deformation.

Table 10 Values of Indentation Hardness (H), Critical Stress Intensity Factors (K_{IC0}, from Single-Edge Notched Beams [72]) and Critical Particle Size (d_{crit}) for Some Excipients and Drugs

Material	H (MPa)	K_{IC0} (MPa·m$^{1/2}$)	d_{crit} (μm)
Microcrystalline cellulose	168 [37]	0.7569	1949
Lactose β-anhydrous	251 [36]	0.7597	873
Ibuprofen	35 [36]	0.1044	854
Aspirin	87 [20]	0.1561	309
Paracetamol DC	265 [37]	0.2463	83
α-Lactose monohydrate	515 [36]	0.3540	45
Sucrose	645 [24]	0.2239	12
Paracetamol	421 [24]	0.1153	7

α-lactose monohydrate in Fig. 12. The value for microcrystalline cellulose is similar to that determined for other polymeric materials [1]. The differences in the two values for the two lactoses are consistent with the findings of workers describing their compaction properties [19]. Of the three drugs paracetamol has the lowest critical particle size consistent with it being very brittle. However, the addition of a polymeric binder to paracetamol (i.e., Paracetamol DC) increases its critical particle size by an order of magnitude, causing it to become plastic in nature.

VI. EFFECT OF MOISTURE CONTENT

The moisture content of a material can affect its mechanical properties generally by acting as an internal "lubricant" facilitating slippage and plastic flow [81]. Obviously this will lead to a decrease in both yield stress and Young's modulus of elasticity as adequately demonstrated for microcrystalline cellulose [82,8]. In the case of yield stress, increasing the moisture content caused a linear decrease independent of the source and batch of the material under test (Fig. 14). Similar results have been recorded for paracetamol [83].

The effect of moisture content on the critical stress intensity factor is more complex. Bassam et al. [8] have shown that for microcrystalline cellulose there is a decrease with increasing moisture content; i.e., the material became less resistant to crack propagation. Similar results have been found for glass by Wiederhorn [84], who suggested that the effect was caused by water interacting at the crack tip.

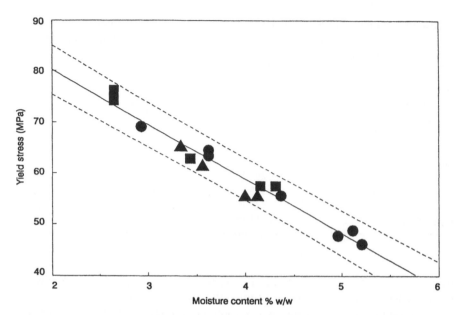

Fig. 14 The effect of moisture content on the yield stress of microcrystalline cellulose for ●, Avicel; ■, Unimac; ▲, Emocel. (Adapted from Roberts and Rowe [82]). The dotted line represents the 95% confidence limits for the fit of the data points.

VII. PREDICTION OF COMPACTION BEHAVIOR

At present a fundamental knowledge and theory describing the compaction of particulate solids, specifically pharmaceuticals, is lacking. While it is generally recognized that the mechanical properties of such materials are critical in directing compaction behavior, a predictive capability is absent. In a recent paper Roberts and Rowe [56] have presented a pragmatic approach to predicting the consolidation mechanism of pharmaceutical materials based on a knowledge of Young's modulus of elasticity, yield stress, hardness, and strain rate sensitivity (Fig. 15). This approach combined with further measurements on critical stress intensity factor enabling the prediction of critical particle sizes, d_{crit} (to account for particle size effects), will enable prediction of the consolidation mechanism of any material of known particle size. This knowledge will allow tablet formulators to take a more scientific approach to formulation. Indeed a combination of a predictive approach based on mechanical property measurements combined with heuristics should enable the development of quality expert systems for tablet formulation.

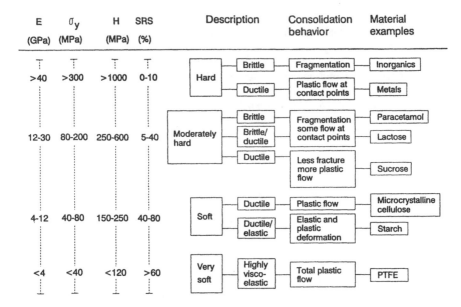

Fig. 15 Relationship between material properties and compaction behavior. (Adapted from Roberts and Rowe [56].)

REFERENCES

1. K. Kendall, *Nature 272*: 710 (1978).
2. K. E. Puttick, *J. Phys. D: Appl. Phys. 12*: 2249 (1980).
3. M. S. Church, Mechanical characterisation of pharmaceutical powder compacts, Ph.D. thesis, Nottingham, 1984.
4. R. J. Roberts, R. C. Rowe, and P. York, *Powder Technol. 65*: 139 (1991).
5. A. B. Mashadi and J. M. Newton, *J. Pharm. Pharmacol. 39*: Suppl., 67P (1987).
6. A. B. Mashadi, and J. M. Newton, *J. Pharm. Pharmacol. 39*: 961 (1987).
7. F. Bassam, P. York, R. C. Rowe, and R. J. Roberts, *J. Pharm. Pharmacol. 40*: Suppl., 68P (1988).
8. F. Bassam, P. York., R. C. Rowe, and R. J. Roberts, *Int. J. Pharm. 64*: 55 (1990).
9. F. Bassam, P. York., R. C. Rowe, and R. J. Roberts, *Powder Technol. 65*: 103 (1991).
10. R. J. Roberts, R. C. Rowe, and P. York, *J. Pharm. Pharmacol. 41*: Suppl., 30P (1989).
11. R. J. Roberts, The elasticity, ductility and fracture toughness of pharmaceutical powders, Ph.D. thesis, Bradford, 1991.
12. R. J. Roberts, and R. C. Rowe, *Int. J. Pharm. 37*: 15 (1987).
13. E. A. Dean, and J. A. Lafez, *J. Amer. Ceram. Soc. 66*: 366 (1983).

14. J. C. Wang, *J. Mater. Sci. 19*: 801 (1984).
15. K. Kendall, M. McN. Alford, and J. D. Birchall, *Proc. R. Soc. Lond. A 412*: 269 (1987).
16. J. M. Spriggs, *J. Am. Ceram. Soc. 44*: 628 (1961).
17. S. Spinner, F. P. Knudsen, and L. Stone, *J. Res. Nat. Bur. Stand. (Eng. Instr.) 67C*: 39 (1963).
18. J. T. Fell, and J. M. Newton, *J. Pharm. Sci. 60*: 1866 (1971).
19. H. Vromans, A. H. De Boer, G. K. Bolhuis, and C. F. Lerk, *Drug Dev. Ind. Pharm. 12* 1715 (1986).
20. K. Ridgway, E. Shotton, and J. Glasby, *J. Pharm. Pharmacol. 21*: Suppl., 19S (1969).
21. J. C. Kerridge, and J. M. Newton, *J. Pharm. Pharmacol. 38:* Suppl., 78P (1986).
22. K. Ridgway, M. E. Aulton, and P. H. Rosser, *J. Pharm. Pharmacol. 22*: 70S (1970).
23. W. C. Duncan-Hewitt, and G. C. Weatherly, *Pharm. Res. 6*: 373 (1989).
24. W. C. Duncan-Hewitt, and G. C. Weatherly, *J. Mat. Sci. Lett. 3*: 1350 (1989).
25. E. N. Hiestand, J. M. Bane, Jnr., and E. P. Strzelinski., *J. Pharm. Sci. 60*: 758 (1971).
26. E. Breval, and N. H. Macmillan, *J. Mater. Sci. Lett. 4*: 741 (1985).
27. G. Simmons, and H. Wang, *Single Crystal Elastic Constants and Calculated Aggregate Properties: A Handbook*. M.I.T., Cambridge, 1971.
28. M. E. Aulton, *Manufact. Chem. Aerosol News 48*: 28 (1977).
29. E. N. Hiestand, *J. Pharm. Sci. 74*: 768 (1985).
30. D. Tabor, *The Hardness of Metals*, Clarendon Press, Oxford, p. 6, 1951.
31. R. Smith, and G. Sandland, *Iron Steel Inst. 1*: 285 (1925).
32. K. Ridgway, J. Glasby, and P. H. Rosser, *J. Pharm. Pharmacol. 21*: 24S (1969b).
33. M. E. Aulton, and H. G. Tebby, *J. Pharm. Pharmacol. 27*: Suppl., 4P (1975).
34. M. E. Aulton, H. G. Tebby, and P. J. P. White, *J. Pharm. Pharmacol. 26*: Suppl., 59P (1974).
35. M. E. Aulton, *Pharm. Acta Hel. 56*: 133 (1981).
36. H. Leuenberger, *Int. J. Pharm. 12*: 41 (1982).
37. W. Jetzer, H. Leuenberger, and H. Sucker, *Pharm. Tech. 7*: April, 33 (1983).
38. W. Jetzer, H. Leuenberger, and H. Sucker, *Pharm. Tech. 7*: Nov., 33 (1983).
39. W. E. Jetzer, W. B. Johnson, E. N. Hiestand, *Int. J. Pharm. 26*: 329 (1985).
40. H. Leuenberger, and B. D. Rohera, *Pharm. Acta Helv. 60*: 279 (1985).
41. B. Galli, and H. Leuenberger, *VDI Ber. 583*: 173 (1986).
42. M. E. Aulton, and I. S. Marok, *Int. J. Pharm. Tech. Prod. Mfr. 2*: 1 (1981).
43. J. Ichikawa, K. Imagawa, and N. Kaneniwa, *Chem. Pharm. Bull. 36*: 2699 (1988).
44. R. W. Heckel, *Trans. Metall. Soc. A.I. M. E. 221*: 671 (1961).
45. R. W. Heckel, *Trans. Metall. Soc. A.I.M.E. 221*: 1001 (1961).
46. J. A. Hersey, and J. E. Rees, *Particle Size Analysis* (M. J. Groves and J. L. Wyatt-Sargent, eds.), The Society for Analytical Chemistry, London, p. 33, 1972.

47. R. J. Roberts, R. C. Rowe, and K. Kendall, *Chem. Eng. Sci. 44*: 1647 (1989).
48. R. J. Roberts, and R. C. Rowe, *Int. J. Pharm. 36*: 205 (1987).
49. P. Paronen, *Pharmaceutical Technology: Tabletting Technology*, Vol. 1 (M. H. Rubinstein, ed.), Ellis Horwood, Chichester, p. 139, 1987.
50. P. York, *J. Pharm. Pharmacol. 31*: 244 (1979).
51. K. Danjo, C. Ertell, and J. T. Cartensen, *Drug Dev. Ind. Pharm. 15*: (1989).
52. H. DeBoer, G. K. Bolhuis, and C. F. Lerk, *Powder Technol. 20*: 75 (1978).
53. G. Ragnarsson, and J. Sjogren, *Acta. Pharm. Suec. 21*: 141 (1984).
54. J. E. Rees, and P. J. Rue, *J. Pharm. Pharmacol. 30*: 601 (1978).
55. R. J. Roberts, and R. C. Rowe, *J. Pharm. Pharmacol. 37*: 377 (1985).
56. R. J. Roberts, and R. C. Rowe, *Chem. Eng. Sci. 42*: 903 (1987).
57. S. D. Bateman, M. H. Rubinstein, R. C. Rowe, R. J. Roberts, P. Drew, and A. Y. K. Ho, *Int. J. Pharm. 49*: 209 (1989).
58. M. S. H. Hussain, P. York, and P. Timmins, *Int. J. Pharm. 70*: 103 (1991).
59. P. Humbert-Droz, D. Mordier, and E. Doelker, *Pharm. Acta Helv. 57*: 136 (1982).
60. P. Humbert-Droz, R. Gurny, D. Mordier, and E. Doekler, *Int. J. Pharm. Tech. Prod. Mfr. 4*: 29 (1983).
61. F. von Podczeck, and U. Wenzel, *Pharm. Ind. 51*: 542 (1989).
62. M. Duberg, and C. Nystrom, *Powder Technol. 46*: 67 (1986).
63. R. Ramberger, and A. Burger, *Powder Technol. 43*: 1 (1985).
64. S. D. Bateman, M. H. Rubinstein, and P. Wright, *J. Pharm. Pharmacol. 39*: Suppl. 66P (1987).
65. H. Vromans, and C. F. Lerk, *Int. J. Pharm. 46*: 183 (1988).
66. P. York, *J. Pharm. Pharmacol. 30*: 6 (1978).
67. R. J. Roberts, and R. C. Rowe, *J. Pharm. Pharmacol. 36*: 567 (1986).
68. O. Ejiofor, S. Esezobo, and N. Pilpel, *J. Pharm. Pharmacol. 38*: 1 (1986).
69. G. Ragnarsson, and J. Sjogren, *J. Pharm. Pharmacol. 37*: 145 (1985).
70. W. F. Brown, and J. E. Srawley, *ASTM Special Tech. Publ.* No. 410 (1966).
71. British Standards Institution, *Plane Strain Fracture Toughness (K_{IC}) of Metallic Materials*, BS 5447, BSI, London, 1977.
72. R. J. Roberts, R. C. Rowe, P. York, *Int. J. Pharm. 91*: 173 (1993).
73. P. York, F. Bassam, R. C. Rowe, and R. J. Roberts, *Int. J. Pharm. 66*: 143 (1990).
74. A. B. Mashadi, and J. M. Newton, *J. Pharm. Pharmacol. 40*: 597 (1988).
75. A. G. Evans, *Fracture Mechanics of Ceramics, Vol. 1: Concepts, Flaws and Fractography* (R. C. Bradt, D. P. H. Hasselman, and F. F. Lange, eds.), Plenum Press, New York, p. 17, 1974.
76. R. J. Roberts, and R. C. Rowe, *Int. J. Pharm. 52*: 213 (1989).
77. K. Kendall, and R. D. Gregory, *J. Mater. Sci. 22*: 4514 (1987).
78. D. Munz, in *Fracture Mechanics of Ceramics, Vol. 6: Measurements, Transformations, and High-Temperature Fracture* (R. C. Bradt, A. G. Evans, D. P. H. Hasselman, and F. F. Lange, eds.), Plenum Press, New York, p. 1, 1983.
79. G. R. Antis, P. Chantikul, B. R. Lawn, and D. B. Marshall, *J. Am. Ceram. Soc. 64*: 533 (1981).

80. M. T. Laugier, *J. Mater. Sci. Lett.* *6*: 355 (1987).
81. K. A. Khan, P. Musikabhumma, and J. P. Warr, *Drug Dev. Ind. Pharm.* *7*: 525 (1981).
82. R. J. Roberts, and R. C. Rowe, *J. Pharm. Pharmacol.* *39:* Suppl., 70P (1987).
83. J. S. M. Garr, and M. H. Rubinstein, *Int. J. Pharm.* *81*: 187 (1992).
84. S. M. Wiederhorn, *J. Am. Ceram. Soc.* *50*: 407 (1967).

12

Granule Properties

Göran Alderborn

Uppsala University, Uppsala, Sweden

Martin Wikberg

Kabi Pharmacia Therapeutics Uppsala, Uppsala, Sweden

I. COMPACTIBILITY OF GRANULATED POWDERS

The traditional way of producing tablets normally involves two size enlargement processes in sequence, i.e., a granulation of the fine particulate drug, often mixed with a filler, followed by the compaction of the granulated powder. The rationale for granulating the powdered drug before tabletting is normally technological. The relationship between bulk density and flowability of a powder mass and the size of the particles constituting the powder is well known and thus, by the aid of a granulation procedure, these characteristics can be improved. However, a granulated powder often exhibits suitable compaction characteristics and powder granulation is thus performed also to ascertain a good compactibility of the tablet mass, i.e., the ability of the powder to be compacted into a compact of a certain mechanical strength [1].

The mechanical strength of a compact is a function of the properties of the material and the treatment of the material during the compaction process, i.e., applied pressure and the time events involved in the tablet formation process. When powders consisting of primary particles are compacted, the mechanical properties of the particles, such as their deformation properties and fragmentation propensity, will be of decisive importance for the mechanical strength of the specimen produced (see Chapters 10 and 11). Also when granules are compacted, the mechanical characteristics of the pri-

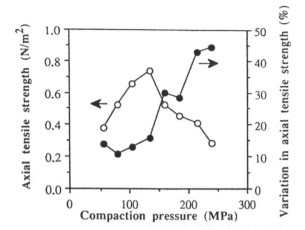

Fig. 1   The axial tensile strength and variability in axial tensile strength of compacts of granules of paracetamol/gelatin (98/2% wt.) as a function of compaction pressure. The reduced strength and increased strength variability of the compacts indicate lamination of the compact (Drawn from data given by Alderborn and Nyström, ref. [2]).

mary particles will probably affect the compactibility of the mass. For example, it is a common experience that capping prone material will show capping tendencies also in the granulated form of the powder (Fig. 1) [2]. However, it has been demonstrated that the compactibility of a granulated powder can be affected by other factors related to the granulation procedure, such as choice of granulation method (Fig. 2) [3,4] and process conditions during the granulation (Fig. 3) [5,6]. For a granulation consisting of a given type of substrate material and a binder, all these factors can be classified into two ways for controlling the compactibility of the granulated powder:

1. Changing the type or amount of binder (normally an amorphous polymer) in the granulation
2. Changing the physical properties of the granules by variations in the process conditions during the granulation procedure or by changing the dimensions of the substrate particles

In practice, formulations are often optimized with respect to their compactibility by the first approach. However, the second approach of optimizing the compaction behaviour is interesting because it allows the design of tablet masses with more limited amounts of excipients and, thus, the production of high-quality tablets at lower cost. Consequently, the pharmaceutical scientist has the possibility to optimize the compaction char-

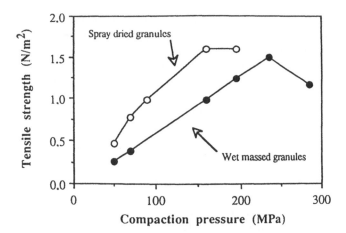

Fig. 2 The tensile strength of compacts of granules of paracetamol and a binder (93/7 % wt.) as a function of compaction pressure. The granules were produced by two different granulation methods. (Drawn from data given by Rue et al., ref. [3].)

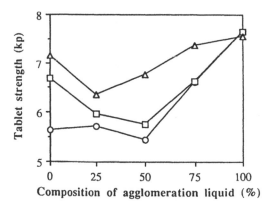

Fig. 3 The effect of the composition of the agglomeration liquid (percentage amount of ethanol in ethanol/water mixtures) on the diametral compression strength of compacts of granules of acetylsalicylic acid/polyvinylpyrrolidone. Compaction pressures: circles = 150 MPa; squares = 225 MPa; triangles = 300 MPa. (Drawn from data given by Wells and Walker, ref. [5].)

acteristics of the powder by manipulating the physical characteristics of the drug particles (in this case the granules). This is normally not the case when a direct compactible formulation is developed. However, this requires firstly, a good understanding of the granulation procedure and secondly, a fundamental knowledge of the relationship between the physical properties of granules and their performance while being compacted.

Although the extensive use of granulated powders in tablet production and, as a consequence of this, the large number of publications on the compactibility of granulated materials, there is still a lack of such fundamental knowledge. The primary reason for this is that there seems to have been few attempts to systematically investigate and understand the relationship between formulation and physical granule properties, on the one hand, and the compactibility of the granulation, on the other.

A survey of a substantial part of the literature on the compaction characteristics of granulated powders has recently been presented by Wikberg [7], and such an overview will not be given here. The aim in this article is to focus on the importance of the physical granule properties for the compactibility of a granulation. This discussion will be based on a selected fraction of papers from the literature.

II. MECHANICAL PROPERTIES OF PHARMACEUTICAL GRANULES

Granules are formed by an aggregation of primary particles into secondary particles. The physical properties of such granules are related to the principle of particle size enlargement and the actual process condition used during the granulation. Today, a broad range of particle size enlargement methods exists and a number of these have been utilized in the production of pharmaceutical granulations. However, the most frequently used granulation methods is based on the formation of secondary particles by the aid of a liquid phase during agitation of particles by convection or by fluidization. The stresses applied to the particles during such agglomeration procedures are probably not high enough to cause deformation or fracturing of the particles. Thus, pharmaceutical granules can be described as a cluster of discrete particles (Fig. 4) and the size and shape characteristics of these primary particles are similar to the original particle properties and therefore definable. An exception is when granules are formed by slugging when marked deformation and fragmentation of the primary particles probably occur during the preparation of the granules.

As a consequence of the technological requirements of a granulation, the most critical physical characteristics of granules for their performance

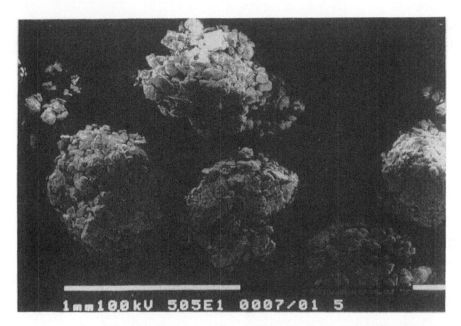

Fig. 4 S.E.M. photomicrograph of a granule of lactose/polyvinylpyrrolidone (95/5 % wt.) produced by wet agglomeration with ethanol as agglomeration liquid. Longer white bar corresponds to a length of 1000 μm.

when processed are their dimensions, porosity, and mechanical strength. These characteristics affect the flowability of the powder, the bulk density, and the ability of the granules to withstand stresses (which can cause fracturing or attrition of the granules) during processing before the actual compaction process (e.g., mixing and transportation of the powders). However, these physical characteristics can also be of importance for the ability of the granules to cohere into compacts.

The porosity of the granules is a reflection of the packing of the primary particles constituting the granules. The porosity is in practice affected by the size of these primary particles as well as the conditions used during the wet agglomeration [8]—i.e., the combined effect of the stress applied to the granules during their formation and the interactions between the primary particles. Generally, a decreased primary particle size tends to increase the granule porosity and an increased degree of liquid saturation and agitation intensity tend to decrease the granule porosity.

The theoretical basis for the prediction of the mechanical strength of an aggregate consisting of defined particles has been presented by Rumpf, who developed the following expression [9] for the tensile strength of an aggregate of monodispersed spheres:

$$\sigma = \frac{9\,(1 - e)kH}{8\pi d^2} \tag{1}$$

where

σ = tensile strength of aggregate (N/m^2)

e = porosity of aggregate (—)

k = coordination number, i.e., number of contact points for one sphere (—)

H = cohesion force at a point of contact (N)

d = diameter of sphere (m)

By assuming that the coordination number is expressed by the ratio π/e [9], the equation is simplified to

$$\sigma \approx \left(\frac{1 - e}{e}\right)\frac{H}{d^2} \tag{2}$$

Although the expression is based on an idealized situation, it can be concluded that the main factors contributing to the strength of an aggregate are the granule porosity and the size of the primary particles. According to the physical character of a pharmaceutical granule described above, it seems reasonable to assume that for a granule which does not include a binder, granule porosity and primary particle size will govern the tensile strength.

According to our experience, few studies in the literature have specifically focused on strength-porosity profiles of pharmaceutical granules irrespective of whether the granules include a binder. In the following discussion, two examples of such a relationship will be discussed.

In Fig. 5, the fracture force-porosity profile for a series of granules of microcrystalline cellulose is shown [10]. Mainly for experimental reasons, the mechanical strength was assessed by diametral compression and the force needed to fracture the granule was used as a measure of the granule strength. The granules did not include a binder and were produced by extrusion-spheronization and generally showed a nearly spherical shape. A reduced porosity increased the fracture force of the granules in an almost linear relationship. Thus, the relationship is generally consistent with Eq. (2), although the physical structure of the granules, with regard to the granule pore structure, varies in a more complex and unpredictable way with the porosity for real granules, compared to idealized granules.

The dominating mechanisms of particle-particle bonds in granules without binders are probably intermolecular attraction forces and solid bridges.

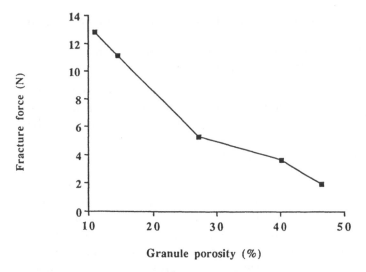

Fig. 5 The effect of granule porosity on the fracture force of spherically shaped microcrystalline cellulose granules produced by extrusion and spheronization. (Drawn from data given by Johansson et al., ref. [10])

However, a binder is normally included in pharmaceutical granules for two reasons: Firstly, in order to secure a sufficient mechanical strength of the granules during processing before the actual compaction procedure, and, secondly, in order to increase the binding capacity of the granules during compaction. The binder is normally mixed with the substrate particles as a relatively viscous liquid, i.e., dissolved in a solvent, although it can also be added as a dry powder. Both the type and the relative amount of binder as well as the procedure used for adding the binder to the substrate particles will affect the mechanical strength of the granules and several reports on these issues exist in the literature ([7],[11]). However, there is still a lack of fundamental understanding of how different binders affect granule strength, although factors such as the interaction between binder and substrate and the mechanical properties of the binder have been addressed.

In an idealized situation, the binder forms a film layer around each substrate particle within the granule. The binder layers can at points be in contact with each other and thus form a binder bridge. However, interactions between binder layers can probably also occur over a certain distance and contribute to the strength of the interparticulate attractions in the granules. When such a granule is stressed, a failure, i.e., a shearing of primary particles or a separation of particles from each other, can be localized to three different places within the granule:

1. At the interface between the substrate and the binder (i.e., an adhesive failure)
2. At the binder-binder interface (i.e., a cohesive failure)
3. Across a binder bridge (i.e., a cohesive failure)

Based on these types of failure, the mechanisms of attraction between particles in a granule can be described as either intermolecular attraction forces (failure 1 and 2) or solid bridges (failure 3). Consequently, the relationship between the strength and the porosity of substrate-binder granules might be dependent on the localization of the failure within the granule and is thus difficult to predict.

The question of the localization of the failure has been the object of examination in the literature. Cutt et al. [12] studied the failure characteristics of granules by SEM examination of fractured granules of glass with different surface properties (hydrophilic and hydrophobic) granulated with various binders. They observed fractures both across binder bridges (primarily for hydrophilic glass beads) as well as at the interface between the binder and the substrate particle (primarily for hydrophobic glass beads). They attributed the localization of the failure to the relative strength of the cohesive and adhesive forces acting within the granules. This in turn is probably related to the dimensions of the binder bridge formed and the relative surface energetics of the binder and the substrate. They also stated that the hydrophilic glass beads gave granules of higher strength, which might be due to the difference in localization of the failure (Table 1). However, the porosity of the granules in relation to strength was not discussed.

Mullier et al. [13] also studied the localization of the failure of granules subjected to stresses. For granules of glass beads and polyvinylpyrrolidone, they observed that a failure at the binder-substrate interface predominated when the granules fractured. However, for granules of sand and PVP, the cohesive bridge failure seemed to predominate. They attributed this difference in relationship to the glass particles to the difference in

Table 1 Friability and Fracture Strength of Granules of Glass Beads and Polyvinylpyrrolidon (2.72% w/w)[a]

Glass surface	Granule friability (%)	Granule strength (g)
Hydrophilic	5.18	202
Hydrophobic	13.8	115

[a]The glass beads possessed different surface characteristics.
Source: From Ref. 12.

surface geometry of the substrate particles; i.e., an irregular particle surface promoted the strength of the adhesive binder-substrate interaction.

Wikberg and Alderborn [14] studied the mechanical strength, assessed by diametral compression, of a series of granules consisting of a given combination of a substrate and a binder. Two substrate materials were used in the study (lactose and dipentum), characterized by a difference in substrate particle size and compaction behavior. For both substrate materials used, the granules was considered to respond to the applied force as brittle specimens; i.e., no ideal plastic flow region nor a marked deformation of the granules before fracture was observed. For the lactose granules, a reduced porosity generally increased the granule strength (Fig. 6). The relationship tended to be exponential; i.e., at the upper part of the studied porosity range the effect of porosity was small, while at the lower a marked increase in strength was observed.

For the dipentum granules (Fig. 6), only a slight tendency to an increased granule strength with reduced porosity was observed. A possible explanation for this is that for these granules, the porosity range obtained corresponds to the part of the strength-porosity profile where porosity variations only affect the strength to a limited degree. One can also observe that the dipentum granules generally were of higher strength, probably due

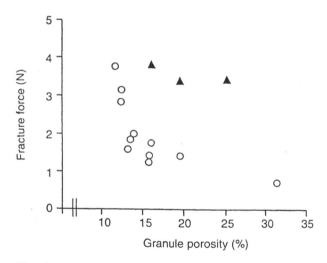

Fig. 6   The effect of granule porosity on the fracture force of granules of lactose/polyvinylpyrrolidone, 95/5 % wt. (cirles) and dipentum/polyvinylpyrrolidone, 94/6 % wt. (triangles). The granules were produced by wet agglomeration. (From ref. [14] with permission of the copyright holder.)

to the lower size of the primary particles of this substrate. This is consistent with the theoretical Eq. (2) as well as with other reports [15] on the strength of lactose granules.

Based on this limited number of publications, the literature also indicates that for binder-substrate granules a reduced porosity and substrate particle size increase the fracture strength of granules. Both these changes in the characteristic of a granule will probably increase the number of bonding points between the primary particles. Thus, a reasonable explanation of this observation is that the granules used in the studies failed by a rupturing of the cohesive binder-binder interactions. The function of the binder will thus be to form a layer around the substrate particles and increase the binding capacity between the substrate particles. However, it is also possible that for substrate-binder granules, failure predominates at the adhesive substrate-binder interface. If this interface constitutes the weakest part of the granule, it seems reasonable that a granule strength more-or-less independent of the granule porosity can be obtained.

As stated above, the compression strength analysis of a series of pharmaceutical granules [14] indicated that these granules could be described as brittle specimens. However, the nature of granules (i.e., a specimen consisting of particles surrounded by a network of pores) indicates that granules may be prone to deform when stressed. It seems reasonable that the primary particles constituing the granules can reposition when subjcted to an external stress which can affect the shape characteristics of the granule (deformation) as well as the granule porosity. In an attempt to assess this propensity of the granules to deform, Wikberg and Alderborn [14] quantified the nearly linear force-displacement curves obtained from the diametral compression measurements of the dipentum granules by calculating the slopes of the profiles (Table 2) (compare also [11]). An increased porosity increased the displacement of the cross-head per applied

Table 2   Deformability, as Measured by the Slope of the Force-Displacement Profile (interquartile range within brackets) During Diametral Compression, of Granules of Different Porosities

Granule porosity (%)	Granule deformability (kN/m)
25.1	69.8 (9.60)
19.4	87.3 (15.2)
15.9	109 (16.1)

Source: From Ref. 14.

force unit which indicates that the porosity of the granules affects their propensity to deform. It seems reasonable that a reduced granule porosity will reduce the void space between the primary particles and, hence, increase the bonding force of the interparticulate attractions. Both the change in interparticulate void space and in the bonding force will probably affect the propensity for densification and deformation of the granule and the stress needed to cause such effects.

The resistance of the granules toward attrition is also of importance during handling and processing of the granules. For the series of lactose granules used in the study in the relationship between granule fracture strength and porosity [14], Wikberg and Alderborn also assessed their resistance to attrition. The friability did not correlate with the porosity or the fracture force of the granules (Fig. 7); i.e., the fracture force and attrition strength of granules are not necessarily related to each other. A probable explanation is that the attrition strength is a mechanical characteristic dependent to a larger extent, compared to the fracture strength, on the shape properties of the granules, primarily the surface geometry. It seems reasonable that particles or small clusters of particles can more easily be abraded from the surface of a granule when the irregularity of the surface increases. It has also earlier been shown [11] that spherical granules possessed a lower friability, compared to more irregular granules produced by traditional wet granulation. Consequently, granules of high porosity can

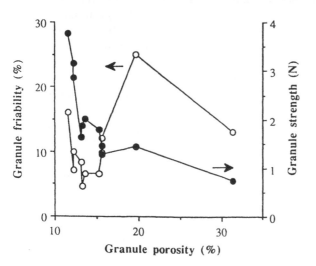

Fig. 7 The effect of granule porosity on the fracture force and the friability of granules of lactose/polyvinylpyrrolidone, 95/5 % wt. (Drawn from data given by Wikberg and Alderborn, ref. [14 and 24].)

probably possess high attritional strength if they are characterized by a regular geometric shape and a smooth surface texture.

The discussion above consequently indicates that the mean separation distance between particles in the granule, and thus the bonding force of the interparticulate attractions, will be of decisive importance for the fracture strength of the granule. However, concerning the attrition strength of granules, the shape characteristics can dominate over the interparticulate bonding force. Some literature reports indicate that a change in the strength of these attractions can also affect the granule friability. For example, it has been reported that an increased amount of binder in the granules can increase the attritional strength of granules [12,16] as well as can a change of agglomeration liquid which increases the solubility of the substrate in the agglomeration liquid [5,17]. Finally, Cutt et al. [12] observed an increased granule friability when the character of the substrate surface was changed from hydrophilic to hydrophobic (Table 1).

III. PHYSICAL STRUCTURE OF COMPACTS OF GRANULES

Pharmaceutical granules are normally relatively porous particles with a porosity in the range of 5–30%. When a bed of such particles is formed by spontaneous packing, the pores within the bed will be distributed between and within the granules (i.e., inter- and intragranular pores). Hence, the pore structure can be described as dualistic. The size distribution in pores is probably in most cases bimodal with a number of large intergranular pores and a number of small intragranular pores. Thus, when the bed of granules is stressed, a failure plane will be obtained between the granules and thus the mechanical properties of the bed of granules are a function of the dimensions and the packing characteristics of the granules.

When granules are compressed by tabletting, the forced reduction in bed porosity will change the characteristics of the pore system of the powder bed. This change will be a reflection of the changes in the physical characteristics and packing of the granules constituting the bed (i.e., the volume reduction behavior of the granules).

It is reasonable that during the initial volume reduction phase, the granules can rearrange in the die and cause a slight reduction in powder bed porosity although the appearance of the granules does not change. However, pharmaceutical granules are normally relatively coarse particles and will probably spontaneously pack to such a low voidage that granule repositioning will contribute only to a limited extent to the total porosity reduction during compression of a bed of granules. As a consequence, irreversible changes in the physical characteristics of the granules will constitute the major part of the volume reduction process.

Van der Zwan and Siskens [18] studied the volume reduction characteristics of granules of a ceramic and a mineral. Their analysis was based mainly on qualitative inspection by SEM of the upper and fracture surfaces of the compacts. For compacts produced at low pressures, individual granules could be clearly distinguished in the compacts but they seemed to be locally deformed at the intergranular contact points. They argued also that the intergranular pore space had already reached a low value at low compaction pressures. Hence, further compression of the mass was associated with a reduction in the porosity of the granules (i.e., granule densification). With increased pressure, it became difficult to distinguish individual granules in the compact.

The authors summarized their findings in the following suggested list of volume reduction mechanisms for granules when compressed:

1. Filling of holes between the granules (i.e., granule rearrangement)
2. Fragmentation and plastic deformation of granules
3. Filling of holes between the primary particles
4. Fragmentation and plastic deformation of primary particles

Mechanism 3 can also be described in terms of a reduction in porosity of the granules—i.e., granule densification. The authors stressed in their discussion that all mechanisms can occur simultaneously in a powder bed during compaction and will not necessarily occur in sequence.

The study by Van der Zwan and Siskens indicated that a compact of granules can be described physically as an aggregate of smaller, cohered granules; i.e., it seems that the granules tend to keep their integrity when compacted. However, with increased compaction pressure, the integrity of the granules will be progressively lost when granules are deformed and fragmented and the intergranular separation distance approaches the distance between the primary particles. In their study, nonpharmaceutical materials were used. However, results have been presented in the literature which support the view that this model of a compact prepared from granules can also be applied to pharmaceutical compacts.

Selkirk and Ganderton [19] compared the pore size distributions, assessed by mercury intrusion, for compacts of a fine powder and the granulated form of the powder. Two materials, sucrose and lactose, were used and the granulations were produced by wet-massing with water. For both materials, similar observations were obtained. The granulation resulted in a considerable widening of the tablet pore structure; i.e., the proportion of coarse pores increased markedly (Fig. 8), compared with compacts formed from the ungranulated particles. Thus, it seems that the granules kept their integrity to some extent when compressed, resulting in intergranular pores which are comparatively coarse. However, the authors

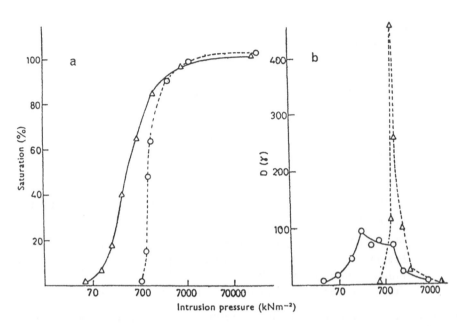

Fig. 8 Cumulative (left graph) and frequency (right graph) pore size distributions for compacts prepared from sucrose granules (circles) and ungranulated sucrose particles (triangles). (From ref. [19] with permission of the copyright holder.)

also pointed out that an increased compaction pressure will probably affect primarily the coarse, intergranular pores and result in a compact of a more even pore size distribution. They also demonstrated, by the aid of tablet air permeability measurements, that the conditions during the granulation procedure, such as amount of agglomeration liquid [20] and slugging pressure [21], affected the pore structure obtained.

Carstensen and Hou [22] studied the pore size distribution of compacts of granules of tricalcium phosphate prepared at a series of compaction pressures. They suggested that the reduction in total tablet porosity, with the subsequent increase in compact strength, was mainly due to the deformation of the granules; i.e., the compact consists of small cohered granules. They argued further that the intragranular pore space seemed to be more-or-less unaffected by the compression procedure.

Wikberg and Alderborn [23] assessed the pore size distribution of compacts of lactose granules by mercury intrusion. The granules were produced by agglomeration with a binder and were of different porosity. The granules of the lower porosity gave a wider pore size distribution with a larger number of coarse pores. The distribution tended to be bimodal

(Fig. 9), which can be interpreted as a reflection of a duality in the pore system of the compacts. For the granules of the higher porosity, a more closed, unimodal distribution was obtained. The observed differences in pore size distributions resulted also in differences in air permeability of the compacts and, thus, the calculated mean pore diameters.

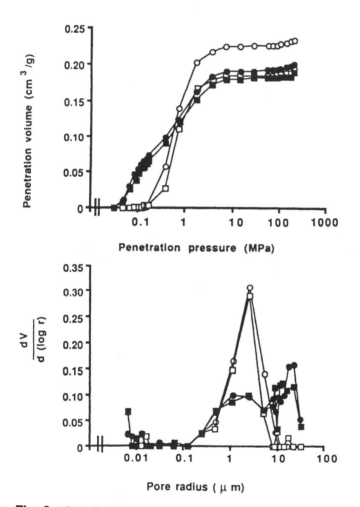

Fig. 9 Cumulative (upper graph) and frequency (lower graph) pore size distributions for compacts prepared from lactose/polyvinylpyrrolidone granules of a porosity of 12.3% (closed symbols) and of 31.3% (open symbols). Compacts were prepared with a compaction pressure of 50 MPa. (From ref. [23] with permission of the copyright holder.)

In another paper [24], Wikberg and Alderborn studied the upper surfaces of compacts of granules of lactose and a high-dosage drug. At relatively low compaction pressures, individual granules could clearly be distinguished at the tablet surface, supporting the model described above. At higher compaction pressures, the authors reported that there was still a tendency that individual granules could be distinguished but the borders between the granules were almost lost. A similar observation was reported by Millili and Schwartz [25] for compacts of pelletized granules of microcrystalline cellulose. This finding was later supported [10] by Johansson et al. by examination of both the upper and the fracture surfaces of compacts of pelletized microcrystalline cellulose granules (Figs. 10 and 11).

In a series of papers [3,26,27], Rue et al. studied the physical properties of a series of granules and of the compacts formed from these granules. The physical structure of the compacts was studied by either a procedure including a removal by dissolution of all compounds in the compact except the binder, followed by an examination of the compact by SEM, or a direct examination of the compact by SEM. They concluded that the compacts of all types of granules used in the study consisted of granules bonded to-

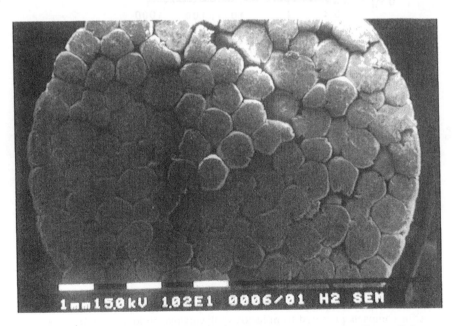

Fig. 10 Upper surface of a compact prepared from microcrystalline cellulose granules of a porosity of 14.4% with a compaction pressure of 100 MPa. White bar corresponds to a length of 1000 μm. (From Johansson et al., ref. [10].)

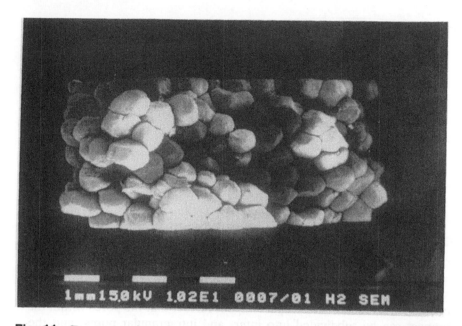

Fig. 11 Fracture surface of a compact prepared from microcrystalline cellulose granules of a porosity of 14.4% with a compaction pressure of 100 MPa. White bar corresponds to a length of 1000 μm. (From Johansson et al., ref. [10].)

gether by intergranular bonds. They also argued that the bonding force of these intergranular bonds will govern the strength of the compact.

Recently, Riepma et al. [28] presented pore size distributions, obtained by mercury intrusion, of tablets of granules formed by slugging. Depending on the applied force during the slugging procedure, uni- and bimodal pore size distributions in the tablets were obtained.

To summarize, an examination of the literature on the physical structure of a compact formed from pharmaceutical granules under normal tabletting conditions indicates that the compact can be described physically as a large aggregate of small, strongly cohered granules. Thus, there is a physical analogy between a powder bed of granules and a compact formed from granules, the main difference being that the mean separation distance between the granules is much smaller and the area of "contact" between two adjacent granules is much larger in the compact. The consequence will be a marked increased coherency of the assembly of granules. The compact can thus be described in terms of the properties of the granules constituting the compact (i.e., size, shape, and porosity) and their relative positions (i.e., intergranular separation distance and number and areas of contact

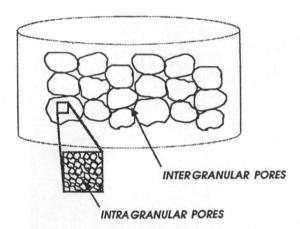

INTERGRANULAR PORES

INTRAGRANULAR PORES

Fig. 12 A simplified schematic illustration of the physical structure from a compact prepared of granules.

zones). This model of the compact implies further that the pores within the compact can be subdivided into inter- and intragranular pores, and the compact will possess a dualistic pore structure. The size distribution of these pores will be bimodal when the intergranular separation distance is large compared to the separation distance between the primary particles constituting the "compact granules." This is probably valid for loosely packed beds of granules and for compacts of granules produced at low compaction pressures. However, for compacts produced at higher compaction pressures, the discrepancy between the two types of pores will probably in many cases be lost, due to the similarity in separation distance between granules and between intragranular particles. A schematic of the suggested physical structure of a compact of granules, illustrating the duality of the pore structure, is given in Fig. 12. A logical consequence of this physical model of a compact of granules is that the dimensions of the granules *before* compaction and the changes in the characteristics of these granules *during* the compression phase, namely their volume reduction behavior, will be of decisive importance for the physical structure of the compact formed.

IV. ASSESSMENTS OF THE VOLUME REDUCTION PROCESS

Other than inspections by SEM of compacts formed from granules (which is discussed above), there seems to be mainly three methods used in the literature with the aim of characterizing the volume reduction behavior of granules.

A. Pore Structure of Compacts

A common procedure for evaluating the volume reduction behavior of granules was based on the characterization of the pore system of the formed compact, e.g., by mercury intrusion [19,23], liquid penetration [20–21], and air permeability [17,20]. As discussed above, a series of arguments can be found in the literature which support a model of a tablet compacted of granules whereby the compact consists of a number of strongly interacting granules. A reasonable consequence of this model is, at least for compacts produced at relatively low compaction pressures, that the largest pores in the compact are found in the intergranular space. With increased compaction pressure, the size of these pores will approach the size of the intragranular pores and the duality of the pore system will be lost progressively. Thus, the characteristics of the intergranular pore space and how it changes with applied pressure will be a reflection of the changes in the properties of granules and their relative position when compressed. Fast and simple methods for the assessment of the pore size characteristics of a compact are air permeability techniques. This approach was used by Ganderton and Selkirk [20] and later by researchers at Uppsala University for the characterization of the volume reduction behavior of granules.

Wikberg and Alderborn [17] studied the volume reduction process of granules by following the changes in tablet permeability and tablet surface area with compaction pressure over a range of relatively low compaction pressures. The authors used both tablet permeability data and the deduced tablet surface areas, calculated by the Kozeny-Carman equation corrected for slip flow, for the assessment of the volume reduction behavior. The reason for using both these measures was that the calculation of tablet surface area involves an error due to the difficulty of defining the effective porosity of the pore system which take part in the airflow. The background is that, for compacts formed at low pressures, the air will probably flow mainly in the intergranular pores. Thus, the specific surface area of these pores could theoretically be calculated if the porosity of this part of the pore system could be inserted in the permeability equation. However, this effective porosity could not be easily defined. The authors therefore used the apparent particle density of the granules in the calculations but pointed out that this will give an overestimated tablet surface area. They therefore also used the tablet permeability as a complement to the tablet surface area data. The limitation is that a change in air permeability could be governed by either a change in the size characteristics of the pores or a change in the porosity of these pores. It will thus be difficult to relate an observed change in tablet permeability to a specific volume reduction mechanism—i.e., fragmentation or deformation of the granules.

In Fig. 13, the tablet air permeability and the deduced tablet surface area is presented as a function of the compaction pressure for compacts of two granules of different porosity. The tablet permeability is reduced in an exponential way with compaction pressure, while the tablet surface area

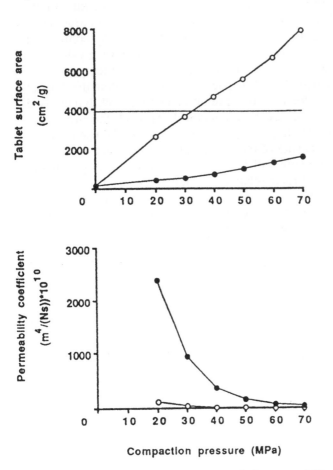

Fig. 13 The changes in surface area and air permeability of compacts of lactose/polyvinylpyrrolidone (95/5 % wt.) granules of a porosity of 12.3% (closed circles) and of 31.3% (open circles) as a function of maximum applied pressure during compaction. The horizontal line in the upper graph represents the surface area of the ungranulated lactose substrate particles. (From ref. [7] with permission of the copyright holder.)

relates almost linearly to the pressure of formation. The reduction in tablet permeability and increase in surface area reflects the formation of a more closed pore structure of the compact. A quantification of these profiles can thus be used as a measure of the volume reduction behavior of the granules. The authors used the area under the curve and the slope of the respective profiles as such measures [17]. The former value can also be seen as a mean tablet permeability over the pressure range studied and can be used to compare different granulations provided that the size and shape characteristics of the original granules are similar; i.e., the original bed of granules will possess a similar air permeability. The profiles in Fig. 13 indicate that the two types of granules respond differently to the applied pressure and it is suggested that the granules which compact into tablets of a more closed pore structure undergo a higher degree of deformation and/ or fragmentation during the volume reduction process.

In the next section the relationship between physical properties of granules, especially size, porosity, and strength of granules, and their volume reduction behavior will be discussed. Most of the interpretations of the degree of fragmentation and deformation which the granules undergo during the compression phase are based on measurements of the air permeability of compacts. Despite the limitation concerning the possibilities to calculate tablet surface areas from permeability results from compacts of granules [17], a permeability method represents a simple technique with high sensitivity for the assessment of the volume reduction characteristics of granules.

B. Tablet Volume—Applied Pressure Relationships

The second approach for assessing the volume reduction behavior of granules is based on the study of the relationship between applied force and punch displacement or tablet volume during the actual compression phase. The techniques used are upper punch force–upper punch displacement profiles [4], upper punch force–die wall force profiles [29], and tablet porosity–applied upper punch pressure profiles [30]. This type of approach can give important information on the ability of the granules to reduce in volume when compressed.

In studies on the compression characteristics of nonporous particles, the tablet porosity–applied pressure relationship according to the Heckel function (see Chapter 3) has been used with the aim of determining the incidence of fragmentation and deformation of particles during the compression process. In this relationship, the pores are considered as reactants as a function of applied pressure. It has been suggested [31,32] that the curvature of the profile over a restricted range of applied pressures can be used as a measure of the degree of fragmentation of the particles and the reciprocal of the slope of the profile as a measure of the degree of deformation of the particles. This way of treating the Heckel profiles has been evaluated for powders of nonporous particles and the whole tablet porosity has been considered to represent the porosity of the reacting pores.

The problem of applying this interpretation also to Heckel profiles from compression of porous particles (i.e., granulated materials) concerns the definition of the reactant pore system. Due to the duality of the pore system, as defined above, a fraction of the porosity represents intragranular pores and the other fraction the intergranular pores. Carstensen and Hou [22] suggested that the pore space of interest in relationship to the Heckel equation is the intergranular pore space and they found that the use of effective particle density (i.e., granule density) values, instead of the apparent particle density, linearized the initial curved part of the Heckel profiles for granules of tricalcium phosphate.

In a study on the compression characteristics of granules of a high-dosage drug, Alderborn et al. [6] observed that quantitative interpretations of Heckel profiles, based on the use of total tablet porosity measurements, gave similar conclusions regarding the volume reduction mechanisms of the granules, as did some other measures. However, in a later paper on the compression characteristics of lactose granules, Wikberg and Alderborn [17] observed the opposite. A possible explanation is that the application of tablet porosity data for the evaluation of volume reduction mechanisms is difficult, due to the complexity of the pore system of the compact. Wikberg and Alderborn also suggested, based on tablet porosity and permeability measurements, that the granules densified markedly in the early phase of the compression process. Thus, it seems that granule densification is an important volume reduction mechanism for granules which is consistent with the observation by Van der Zwan and Siskens [18] (compare also Riepma et al. [28]). The consequence of this is that the effective particle density of the granules, measured before compression, cannot be used to calculate the fraction of the tablet porosity which corresponds to the intergranular pores. It seems therefore reasonable to conclude that the use and interpretations of Heckel data, with respect to the mechanisms involved in volume reduction, should be done with care when granulated materials are studied.

C. Lubricant and Dry Binder Additions

A third approach for studying the volume reduction behavior of particles when compacted is to compare the strength of tablets prepared with and without a second component. Lubricants especially have been used as this second component (e.g., [31–33]) but fine particulate dry binders have been used also for this purpose [34]. Alderborn et al. [6] used both these types of materials to assess the volume reduction characteristics of some granules and in a later study, Wikberg and Alderborn [24] studied the effect of lubricant additions on compact strength. An example is given in Table 3, where the effect of dry binder additions on the relative tablet strength increase for three granulations of a high-dosage drug is summarized. The differences in response to the addition of the excipients indicate that the granules are characterized by differences in their volume reduction behavior, in this case their propensity to expose new surfaces during compression. The same conclusion was drawn from the studies on the effect of lubricant additions [24].

Thus, the approach of assessing the effect of excipients on the compatibility of granules can give valuable information on the incidence of deformation and fragmentation during compression. However, it should be stated that the strength increase and the strength decrease due to the additions of dry binders and lubricants respectively are also related to the distribution of the excipient on the surface of the granules. Vromans et al. [35] pointed out that the degree of strength reduction due to the addition of magnesium stearate was related to the degree of surface coverage of the substrate particles by the lubricant. The surface coverage is a function of the properties of the substrate particles (i.e., their size, shape, and surface roughness [36]) the conditions during the mixing procedure [37] and perhaps also the interaction between substrate and lubricant [38]. The conse-

Table 3 Ratios Between Tensile Strength of Tablets, Compacted from Tablet Masses with and Without Addition of a Dry Binder (denoted a, b, and c) at a Concentration of 8.6% wt[d]

Granulation	Tensile strength ratio (—)		
	a	b	c
I	1.31	1.65	1.49
II	1.27	1.42	1.26
III	1.12	1.28	1.13

a = Avicel PH-101
b = Avicel PH-105
c = Crosslinked polyvinylpyrrolidone
[d]Tablet masses were based on three different granules of a high-dosage drug and a polymeric binder, with the same composition but with different volume reduction behavior.
Source: From Ref. 6.

quence will be that for granules, which might show a large batch variation with respect to the granule surface roughness, the interpretations of strength increase and strength reduction data must be done with care but constitutes an interesting approach for evaluating the volume reduction behavior of granules.

V. PHYSICAL GRANULE PROPERTIES OF IMPORTANCE FOR THE VOLUME REDUCTION PROCESS

A. Granule Porosity and Strength

Ganderton and Selkirk [19–21] studied the pore structure of tablets compacted of granules of lactose and sucrose by air permeability, mercury intrusion, and liquid penetration. Granules with different physical characteristics were prepared by using different methods of granulation (slugging or wet granulation) or by varying the processing conditions during the granulation (slugging pressure or amount of water during agglomeration). The granules consisted generally of one component; i.e., a binder was not included in the granules.

For the granules prepared by wet granulation, an increased amount of agglomeration liquid gave granules which compacted into tablets of a more open and permeable pore structure, i.e., an increased median pore size (Figs. 14 and 15). The authors related this difference in volume reduction behavior of the granules to a decreased porosity and an increased mechanical strength of the granules. However, measures of these granule characteristics were not presented. They suggested that a decreased granule porosity and an increased granule strength reduced both the degree of deformation and fragmentation of the granules when compacted and that these changes in volume reduction behavior of the granules affected the pore structure of the compact formed.

For the granules prepared by slugging [21], an increased slugging pressure gave granules which compacted into more permeable tablets with larger pores. An increased slugging pressure corresponded to a decreased granule porosity and the same explanation of the relationship between porosity and strength of granules, on one hand, and the volume reduction behavior, on the other, was given.

Based on the analysis of liquid penetration into the compacts [20], the authors also concluded that the characteristics of the pores at the surface of the compact were not atypical of the pores in the compact generally. This suggestion supports the validity of the conclusions discussed above concerning the physical structure of compacts of granules based on the observation of the upper surfaces of the compacts.

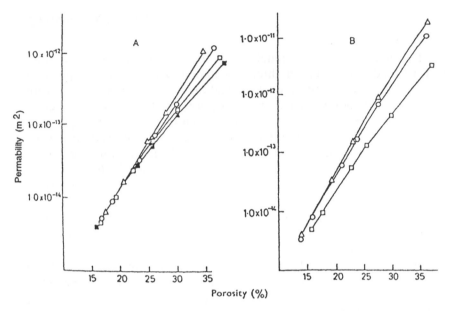

Fig. 14 The air permeability of compacts of lactose granules (left graph) and sucrose granules (right graph) as a function of tablet porosity. The granules were prepared with different amounts of water during the wet agglomeration, i.e. for lactose granules; closed squares = 13%; open squares = 17%; circles = 21%; triangles = 25%, and for sucrose granules: squares = 5%; circles = 7%; triangles = 9%. (From ref. [20] with permission of the copyright holder).

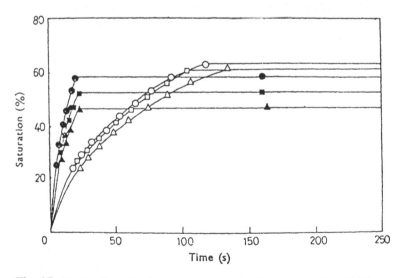

Fig 15 Saturation due to water penetration into compacts of high porosity (closed symbols) and low porosity (open symbols) as a function of penetration time. Compacts were compacted from granules which were prepared with different amounts of water during the wet agglomeration, i.e.: triangles = 5%; squares = 7%; circles = 9%. (From ref. [20] with permission of the copyright holder).

By varying the process conditions during the wet agglomeration of a substrate-binder combination, three granulations of a constant composition were produced by Alderborn et al. [6] and their tabletting properties were subsequently analyzed. The changes in pore structure with compaction pressure was assessed by air permeametry and the surface area–pressure profiles obtained depended on the type of binder solvent used during the agglomeration. The air permeability results correlated well with the magnitude of changes in tablet strength due to the additions of a lubricant and some dry binders (Table 3). Thus, it seems that the changes in pore structure of compacts with applied pressure are associated with the formation of new extragranular surfaces. This formation of new surfaces can be due to either a fragmentation of granules into smaller aggregates or to a marked deformation of the granules.

Alderborn et al. did not address the question of the relationship between the volume reduction behavior of the granules and some other physical characteristics of the granules, such as the porosity. However, Wikberg and Alderborn [17,24] studied the relationship between the porosity of the granules before compaction and their volume reduction behavior, as evaluated by the relationship between air permeability or permeametry surface area and compaction pressure. In these papers, granules of constant proportions of a substrate (lactose) and a binder (polyvinylpyrrolidone) were produced by wet agglomeration in convective mixers. By varying the process conditions during the agglomeration, the porosity of the granules could be varied. The relationship between volume reduction behavior of the granules and the granule porosity was also studied for the dipentum granules used earlier by Alderborn et al. [6].

In both papers, a correlation between granule porosity and the volume reduction behavior of the lactose granules was obtained. In Fig. 16, data from both papers are included in the same graph, and all results fit a unique relationship well. It seems thus that an increased granule porosity before compaction promotes the formation of a pore structure characterized by smaller intergranular pores or by a lower intergranular pore space, i.e., a more closed pore structure. The same type of relationship between granule porosity and tablet pore structure was obtained also for granules of dipentum.

Since a consistent granule size was used in the studies, the observed difference in tablet pore structure due to the variation in granule porosity is caused by variations in the behavior of the granules when compressed. It seems that both differences in degree of fragmentation and deformation between the granules can be responsible for the observed effect. By the use of lubricated granules, Wikberg and Alderborn suggested [24] that the volume reduction process also involved the formation of new extragranular

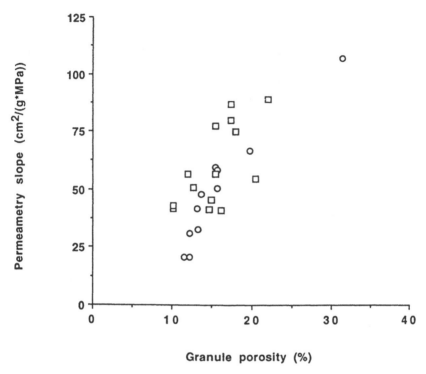

Fig. 16 The slope from table surface area–compaction pressure profiles as a function of the porosity of lactose/polyvinylpyrrolidon (95/5 % wt.) granules produced by wet agglomeration in a planetary mix (squares, from ref. [17]) and in a high shear mixer (circles, from ref. [24]). (Drawn from data given by Wikberg and Alderborn, ref. [17 and 24].)

surfaces. This phenomenon can also be caused by both fragmentation and deformation of the granules during the volume reduction process. In the latter case, a very marked deformation is probably necessary to explain the observed effects, i.e., a deformation which is not comparable with the deformation of nonporous pharmaceutical particles. Marked deformation seems, however, to be a reasonable possibility for granules, since they contain spaces between the substrate particles which allow the primary particles to rearrange, with a subsequent effect on both porosity and shape characteristics of the granules.

Based on the evaluation of the volume reduction behavior of the series of lactose and dipentum granules, Wikberg and Alderborn concluded that both granule fragmentation and deformation occurred during the volume reduction phase. Furthermore, both phenomena probably occurred at low compaction pressures and determined the changes in tablet pore structure with compaction pressure. In an attempt to establish whether the incidence of fragmentation differed between granules of dipentum and lactose, the authors continued to study [14] the mechanical properties of the individual granules, assessed by diametral compression, and the relationship between granule fracture force and tablet permeability.

For the lactose granules, an increased fracture force of the individual granules corresponded to a change in their compression behavior toward a reduced ability to form compacts of a relatively closed pore structure (Fig. 17). The fracture force of a granule can be seen as a measure of the propensity of the granules to fragment into smaller aggregates when loaded. Thus, the good correlation between tablet pore structure and granule strength can be used to support the assumption that the observed

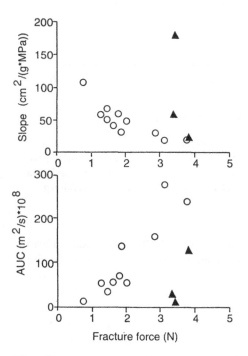

Fig. 17 The slope from tablet surface area–compaction pressure profiles (upper graph) and the area under the curve from tablet permeability-compaction pressure profiles (lower graph) as a function of the fracture force of individual granules of lactose/polyvinylpyrrolidone 95/5 % wt. (circles) and of dipentum/polyvinylpyrrolidone, 94/6 % wt. (triangles). (From ref. [14] with permission of the copyright holder).

differences in tablet pore structure are caused by differences in the degree of fragmentation of the granules during compression; i.e., this mechanism is the significant volume reduction mechanism for the compression process in these granules.

For the dipentum granules, a tendency to a change in tablet pore structure with variation in granule fracture force, was also observed (Fig. 17) but the effect was less pronounced. The results can be interpreted as that the incidence of fragmentation is less pronounced for the dipentum granules compared to the lactose granules and, thus, that densification and deformation of granules is the dominating mechanism, responsible for the changes in pore structure of the compact with applied pressure, for dipentum granules. However, this conclusion concerning the differences in the main volume reduction mechanism was difficult to support by the aid of SEM examinations of the upper tablet surfaces.

In a later paper [10], Johansson et al. studied the volume reduction properties of almost spherical granules of microcrystalline cellulose. Granules of a series of porosities were prepared which, compared to the lactose and dipentum granules, showed a high fracture strength (compare Figs. 5 and 6). The cellulose granules also showed a less distinct failure when compressed diametrically as individual granules, compared to the lactose and dipentum granules; i.e., they could be described as less brittle. This might be related to differences in the mechanical properties of the primary particles within the granules.

Inspection of both upper and fracture surfaces of compacts of the granules indicated that the cellulose granules reduced in volume by deformation (Figs. 10 and 11). This was supported by the fact that the granules also showed a marked sensitivity to addition of a lubricant with respect to compactibility. Also for the cellulose granules, compacts of different pore structure were obtained dependent on the porosity of the granules before compaction (Fig. 18). This finding is thus consistent with the results for lactose and dipentum granules; i.e., an increased granule porosity promotes the formation of a more closed pore structure in the compacts. For the cellulose granules, it is thus suggested that the degree of deformation which the granules undergo during compression to a given applied pressure is governed by the original granule porosity.

Tablets formed at a given pressure tended to be more permeable for compacts prepared from cellulose granules, compared to tablets from lactose and dipentum granules. The differences in size and shape characteristics of the granules before compaction possibly contributes to the difference. However, it might also be related to a difference in the volume reduction behavior of the granules. The granules of microcrystalline cellulose seemed to reduce in volume more or less entirely by densification and

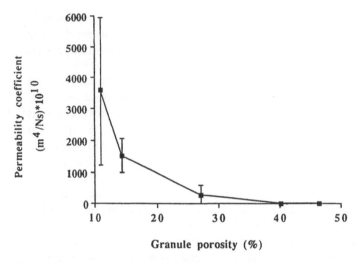

Fig. 18 The permeability coefficient of compacts, prepared at 100 MPa, of microcrystalline cellulose granules of varying porosity. (Drawn from data given by Johansson et al., ref. [10]).

deformation, while for granules of the other materials, fracturing and attrition of granules are probably involved in the volume reduction process to some extent.

The incidence of fragmentation can be affected by both the particulate and the mechanical characteristics of the granules. A change to more irregular granules of a more irregular surface geometry might promote the incidence of fragmentation and attrition during the compression process. A reduction in the granule mechanical strength might also cause a reduction in the propensity of the granules to fracture when compacted. Finally, a change in the mechanical characteristics of the granules (i.e., their brittleness) might also affect their propensity to fail by fracture under load.

The results indicate further that granule porosity will govern the volume reduction phase irrespective of whether the granules show some fragmentation during compression. This might be interpreted that the porosity affects both the degree of granule fragmentation, through a relationship with the granule fracture strength, and the degree of granule deformation. However, it is possible that also for the more brittle granules (lactose and dipentum), the effect of granule porosity on the difference in tablet pore structure could be explained in terms of a relationship between porosity and degree of deformation of granules. However, fragmentation and/or attrition of the granules might still occur to a significant extent during compression, perhaps in the second part of the compression phase. The low sensitivity of the lactose granules of the addition of lubricants can be affected also by difficulties to uniformly distribute the lubricant on the surface of the granules.

B. Intragranular Binder Distribution

It has been suggested [3] that the binder distribution within the individual granules is one physical characteristic of granules which is of decisive importance for their compactibility—i.e., their ability to cohere into compacts. The mechanism behind this effect will be discussed below. However, a localization of a binder at the surface of granules might not only affect the ability of granules to cohere into compacts but also their volume reduction behavior.

Wikberg and Alderborn [39] studied the effect of the intragranular binder distribution on the volume reduction characteristics of granules. Two sets of granules were produced, consisting of the same proportion of substrate and binder (95/5% wt) but with different distribution of the binder within the granules; i.e., in the first set (denoted 5% granules), the binder was relatively homogeneously distributed within the granules, while in the second (denoted 1 + 4% granules), the binder was localized mainly on the granule surface. In addition, a set of granules with a lower amount of binder (1% wt) uniformly distributed in the granules was also studied. These granules were used as "cores" in the preparation of the 1 + 4% granules.

The volume reduction behavior of the granules was evaluated by air permeability measurements. A comparison between the 5% and the 1 + 4% granules indicate that a localization of a substantial amount of binder at the surface of the granules did not affect the assessed changes in pore structure with compaction pressure (Fig. 19). Furthermore, if the 1% and 5% granulations were compared with respect to the relationship between tablet permeability and original granule porosity, the compression characteristics of the granules seemed independent of the amount of binder in the granules. The compressive strength of the granules tended to increase with an increased amount of binder, probably due to an effect of binder content on the bonding force of the interactions between the primary particles within the granules. However, the fracture strength of the individual granules was not markedly affected by the variation in binder content or intragranular binder distribution.

It seems thus that the pore structure of the compacts, prepared from these sets of granules, was governed by the porosity (or the strength) of the granules before compaction, although the physical characteristics of the granules vary with respect to the size of the primary particles, the binder

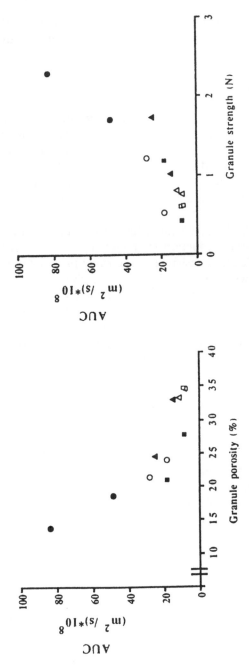

Fig. 19 The area under the curve from tablet permeability–compaction pressure profiles as a function of granule porosity (left graph) and fracture force of individual granules (right graph) for compacts of a series of lactose/polyvinyl-pyrrolidone granules, prepared with different procedures:

closed circles = 200 mesh substrate particle size, 5% binder content
open circles = 200 mesh substrate particle size, 1+4% binder content
closed squares = 200 mesh substrate particle size, 1% binder content
closed triangles = 450 mesh substrate particle size, 5% binder content
open triangles = 450 mesh substrate particle size, 1+4% binder content
open squares = 450 mesh substrate particle size, 1% binder content
(Drawn from data given by Wikberg and Alderborn, ref. [39]).

content, and the intragranular binder distribution. The significance of this observation is demonstrated further by comparing the relationship between granule porosity and volume reduction behavior for all lactose granules used by Wikberg and Alderborn [24,39] (Fig. 20). All granules tend to fit a general relationship between the changes in pore structure of compact with applied pressure and granule porosity. However, it must be pointed out that there was a small variation in the original granule size for these granules. The profile (Fig. 20) indicates that the porosity of the granules before compaction is one of the most critical physical characteristics of granules for their volume reduction behavior, and, thus, the pore structure of the compacts formed.

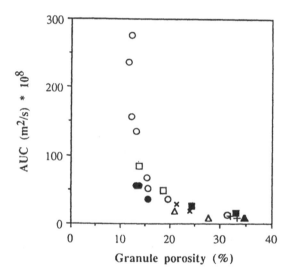

Fig. 20 The area under the curve from tablet permeability–compaction pressure profiles as a function of granule porosity for compacts of a series of lactose/polyvinylpyrrolidone granules, prepared by different procedures:
Open circles = 200 mesh substrate particle size, 5% binder content (agglomeration liquid: ethanol, ref. [24])
Closed circles = 200 mesh substrate particle size, 5% binder content (agglomeration liquid: water, ref. [24])
Open squares = 200 mesh substrate particle size, 5% binder content
Crosses = 200 mesh substrate particle size, 1+4% binder content
Open triangles = 200 mesh substrate particle size, 1% binder content
Closed squares = 450 mesh substrate particle size, 5% binder content
Plusses = 450 mesh substrate particle size, 1+4% binder content
Closed triangles = 450 mesh substrate particle size, 1% binder content
(Drawn from data given by Wikberg and Alderborn, ref. [24 and 39]).

VI. MECHANISMS OF VOLUME REDUCTION OF GRANULES

Based on the discussion above, it is suggested that a series of physical processes is involved in the compaction of pharmaceutical granules, which will affect the physical structure and mechanical strength of the compact formed. A summary of these, classified into two main groups according to the granules and the substrate particles of which the granules consist, is given in Table 4.

During the initial loading phase, some repositioning of granules will probably occur over a range of very low compaction pressures. Thereafter, a further reduction in compact porosity is associated with irreversible changes in the characteristics of the granules. Densification will result from repositioning of the primary particles constituting the granule. A parallel effect to the densification of granules will probably be that the relative dimensions of the granules will change (i.e., a deformation of granules). Thus, these two mechanisms will probably occur simultaneously as a result of the same process: a change in the relative positions of the primary particles within the granules. The loading might also cause a fracturing or fragmentation of granules into smaller aggregates. It is also suggested that primary particles or small aggregates of the primary particles can be abrased from the surface of the granules. This process can be described as an attrition of granules and can be treated as a separate volume reduction mechanism, although it is mechanistically similar to granule fragmentation. Finally, as a consequence of a reduced porosity and size of the granules, the stress exerted on the primary particles will probably increase. The result might be an increased incidence of deformation of the primary particles.

Densification, deformation, and fracturing of granules all seem to be of relevance in the volume reduction process and can occur at low compaction pressures. All these mechanisms require either the formation of a

Table 4 Suggested Physical Mechanisms Involved in the Volume Reduction Process of a Powder Mass Consisting of Porous Granules

---

A. *Volume reduction mechanisms for the granules*
   1. Rearrangement
   2. Densification and deformation
   3. Fragmentation
   4. Attrition
B. *Volume reduction mechanisms for the substrate particles*
   1. Deformation (elastic/plastic)
   2. Fragmentation

---

shear plane where primary particles slide against each other, or the formation of a fracture plane causing a more-or-less instantaneous separation of primary particles from each other. However, the dominating mechanism within the compression cycle, including the sequence of the mechanisms involved, has not been established.

The compression induced changes in the physical characteristics of the granules, in terms of their porosity and dimensions, is associated with the following two changes in physical structure of the compact with applied compression stress:

1. A reduction in intergranular separation distance. The most extreme reduction is a situation where the integrity of the granules is lost and the intergranular separation distance is consequently similar to the separation distance between the primary particles; i.e., the duality of the pore structure is lost.

2. A formation of new extragranular surfaces. These surfaces originate from the interior of the granules and will be exposed during the volume reduction process.

These changes in the physical character of the compact will probably affect the formation of intergranular bonds in the compact, with respect to both the bonding force and the number of bonds, and, thus, the compact strength. It is also suggested that the degree of deformation and fragmentation which the granules undergo during compression will be of special importance for the changes in physical structure of the compact which occur during the compression. The degree of deformation and fragmentation seems in many cases to be related to the porosity of the granules before compaction, i.e., the porosity might affect the degree of granule deformation as well as the degree of granule fragmentation, eventually through a relationship with the strength of the granules. However, the question of the importance of the porosity and the strength of granules for their compression behavior is complex and not yet satisfactorily elucidated and at least two factors might affect this relationship. Firstly, it is possible that the strength of the granules can form thresholds below and above which variations in granule porosity and strength do not affect the compression behavior of the granules. At low granule strengths, the granules might collapse entirely when compressed, and below this strength threshold a variation in granule strength and porosity will thus not affect the compression behavior. Furthermore, granules of a very high strength might be very resistant to both deformation and fragmentation, and above this threshold the granules will show similar compression behavior independent of the porosity and strength. Most of the granules discussed in this chapter will consequently belong to a domain between those strength thresholds (which might be the most common situation for granules prepared by wet agglomeration). Secondly, the earlier discussion on the localization of the failure during loading of individual substrate-binder granules is probably applicable also to the process of compaction; i.e., the surface geometry of the substrate particles and the relative surface energies of the binder and the particles in relation to the strength of the binder-binder attractions will determine the localization of the failure plane. It is possible that a localization of the failure at the interface between binder and substrate will have the consequence that the granules compress in a similar way, although the separation distance between the primary particles, and thus the granule porosity, will vary. An increased knowledge of these issues can contribute to an understanding of the apparently conflicting results reported in the literature on the effect of granule strength and granule porosity on the compactibility of granules.

The model of a tablet used in the discussion in this chapter indicates that dimensions of the "compact granules" within the tablet are of relevance for the tablet pore structure. Consequently, not only the compression induced changes in the dimensions of granules but also the original granule dimensions before compaction, primarily the granule size, will affect the pore structure of a compact [28]. Thus, these two factors will in combination govern the changes in physical structure of the compact while being compressed. The discussion on the relationship between volume reduction behavior of granules and the physical structure of the compact formed is summarized qualitatively in Fig. 21.

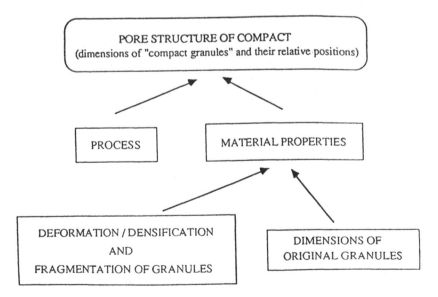

Fig. 21   Proposed main factors contributing to the pore structure of compacts prepared from granules.

VII. PHYSICAL GRANULE PROPERTIES OF IMPORTANCE FOR COMPACT STRENGTH

A. Volume Reduction Behavior

The mechanical strength of a tablet can be described as a function of the bonding force of the bonds between the particles constituting the tablet and the total number of the bonds (see Chapter 10). Both the bonding force and the number of bonding zones will be affected by the dimensions of and the separation distance between the "compact particles."

To achieve small separation distances, it is important that the particles can be brought into close contact without a subsequent elastic recovery; i.e., the volume reduction behavior of the particles will be critical for the relative positions and the area of contact between particles in the compact. A pronounced plasticity of the particles is generally believed to improve the strength of the tablet especially when it occurs at points of interparticle contact [40]. By studying physical characteristics of compacts of different types of lactose [41], Vromans et al. suggested that the degree of fragmentation of particles will be of decisive importance for compact strength.

The degree of deformation or fragmentation which the granules undergo during compression will affect the dimensions of the granules constituting the compact, as well as their relative positions. Consequently, it is reasonable to assume that the volume reduction behavior of the granules will affect the tensile strength of the compact produced.

In Fig. 22, the tensile strength of compacts prepared from nearly spherical granules of microcrystalline cellulose [10], produced by extrusion-spheronization, is shown as a function of the total tablet porosity. The granules showed a range of porosities (Fig. 5) before compaction and the relationship between original granule porosity and volume reduction behavior have been discussed above (Fig. 18). Although the total tablet porosity was similar between the compacts, the tensile strength varied considerably. Thus, it is not possible in this case to explain the differences in compactibility between the granules by differences in the total tablet porosity, i.e., the total degree of bulk volume reduction obtained. The compact strength must be related in this case to the quality of the pore structure.

In Fig. 23, the mechanical strength of tablets produced at 100 MPa from the microcrystalline cellulose granules is shown as a function of the air permeability of the compacts. A direct relationship between the measure of the pore structure and the tensile strength of the compacts was obtained. Similar observations have also been obtained for compacts of a series of lactose-PVP granules [24] of varying porosity (Fig. 24). In this case, the